FOURTH EDITION

KRASNER'S
MICROBIAL
CHALLENGE

A PUBLIC HEALTH PERSPECTIVE

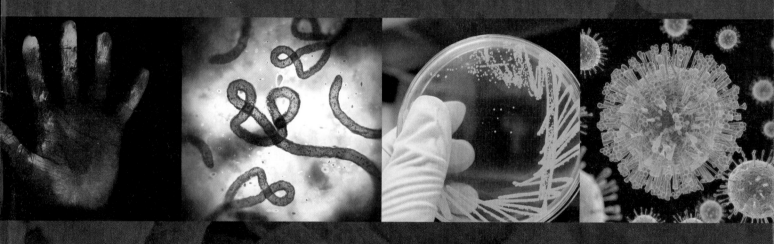

TERI SHORS, PhD

University of Wisconsin Oshkosh

JONES & BARTLETT
LEARNING

World Headquarters
Jones & Bartlett Learning
5 Wall Street
Burlington, MA 01803
978-443-5000
info@jblearning.com
www.jblearning.com

Jones & Bartlett Learning books and products are available through most bookstores and online booksellers. To contact Jones & Bartlett Learning directly, call 800-832-0034, fax 978-443-8000, or visit our website, www.jblearning.com.

Substantial discounts on bulk quantities of Jones & Bartlett Learning publications are available to corporations, professional associations, and other qualified organizations. For details and specific discount information, contact the special sales department at Jones & Bartlett Learning via the above contact information or send an email to specialsales@jblearning.com.

15927-1

Production Credits
VP, Product Management A&P: Amanda Martin
Director of Product Management: Laura Pagluica
Product Specialist: Audrey Schwinn
Product Assistant: Loren-Marie Durr
Senior Production Editor: Nancy Hitchcock
Marketing Manager: Lindsay White
Production Services Manager: Colleen Lamy
Manufacturing and Inventory Control Supervisor: Amy Bacus
Composition: Exela Technologies
Cover Design: Kristin E. Parker
Text Design: Michael O'Donnell
Director of Rights & Media: Joanna Gallant
Rights & Media Specialist: John Rusk
Media Development Editor: Troy Liston
Cover Image: 1_Colored SEM: © Science Photo Library - STEVE GSCHMEISSNER/Brand X Pictures/Getty Images; Cover Image 2_UV light of hand: © Science Photo Library/ Getty Images; Cover Image 3_Ebola virus under microscope: © Henrik5000/iStock/Getty Images; Cover Image 4_Yeast cultivation on agar plate: © Trinset/Shutterstock; Cover Image 5_Illustration of Zika Virus: © AuntSpray/Shutterstock
Printing and Binding: LSC Communications
Cover Printing: LSC Communications

Library of Congress Cataloging-in-Publication Data
Names: Krasner, Robert I., author. | Shors, Teri, author.
Title: Krasner's microbial challenge : a public health perspective / Robert I. Krasner, Teri Shors.
Other titles: Microbial challenge
Description: Fourth edition. | Burlington, Massachusetts : Jones & Bartlett Learning, [2020] | Preceded by The microbial challenge / Robert I. Krasner, Teri Shors. 3rd ed. c2014. | Includes bibliographical references and index.
Identifiers: LCCN 2018040880 | ISBN 9781284139181
Subjects: | MESH: Communicable Diseases, Emerging–prevention & control | Communicable Diseases, Emerging--microbiology | Microbiological Phenomena
Classification: LCC QR46 | NLM WA 110 | DDC 616.9/041—dc23
LC record available at https://lccn.loc.gov/2018040880

6048

Printed in the United States of America
22 21 20 19 18   10 9 8 7 6 5 4 3 2 1

DEDICATIONS

This edition of *The Microbial Challenge* is dedicated to Robert I. Krasner. The publishing team at Jones & Bartlett Learning has chosen to honor his memory and passion for lifelong learning by updating the title of this edition to *Krasner's Microbial Challenge*. It is our hope that students and instructors using this book will continue to be inspired by Dr. Krasner's enthusiasm for new discoveries, connecting with the world around us, and learning both inside and outside the classroom.

From Teri Shors

To the late Elaine (Motschke) Gross, my mother, *Ich vermisse dich jeden Tag.*

To John Cronn, my undergraduate microbiology mentor, colleague, and friend who opened my eyes to the invisible world of microbes and viruses.

To the hundreds of students I have taught, past and present.

"We know nothing of what will happen in the future, but by the analogy of experience."
—*Abraham Lincoln*

BRIEF TABLE OF CONTENTS

TABLE OF CONTENTS

13   Viral and Prion Diseases   390

16   Control of Microbial Diseases   540

PART 5   CURRENT MICROBIAL CHALLENGES   563

17   Harnessing the Power of Microbes: Peril and Promise   564

PREFACE

Birth and Development of *The Microbial Challenge*

After 50 years in the classroom at Providence College in Rhode Island teaching microbiology to biology majors, Dr. Robert Krasner decided to develop a microbiology course for nonbiology majors. Outbreaks of disease were in the news frequently, and, judging by the questions students in a nonmajors, general biology course asked, it was apparent that they, too, needed to know more about microbes and human–microbe interactions. Actually, he had been thinking about teaching a nonmajors course for a number of years, but to his surprise there was no text available. Hence, he used handouts, online resources, magazine and newspaper articles, and videos. He even called in a few speakers to supplement his lectures. This strategy worked, but it was cumbersome, required too many handouts, and resulted in confusion; so he decided to write his own text.

At about that time, Dr. Krasner decided to study public health microbiology and was accepted into the Harvard School of Public Health for the 1999–2000 academic year, 41 years after he had completed his PhD. He had wanted to do this for many years, but raising a family and educating his children were his top priority. Now it was his turn! As far as he was able to determine, Dr. Krasner was the oldest student on a full-time basis to earn the Master of Public Health (MPH) degree at Harvard. His classmates were primarily young medical students who had postponed their fourth year of medical school to earn a MPH before completing medical school. His Harvard studies included 6 weeks in a tropical disease laboratory in Brazil culminating in a 2-week field trip to Manaus in the Amazon.

While at Harvard's School of Public Health, every day Dr. Krasner passed by an inscription that reads, in several languages, "*The highest attainable standard of health is one of the fundamental rights of every human being.*" This inscription, his studies at Harvard, and his travel experiences were major factors in the birth and in the public health perspective of his text, *The Microbial Challenge*.

During the development of the third edition, Dr. Teri Shors, Professor in the Department of Biology and Microbiology at the University of Wisconsin Oshkosh, signed on as a coauthor. Her creativity, judgment, and knowledge resulted in a text Dr. Krasner continued to be proud of. Her specialty is virology, and together the two of them brought many years of teaching experience to this text. Dr. Krasner passed away in 2014 at the age of 84. Dr. Shors has taken up the torch to carry on Dr. Krasner's vision to provide nonmajors with a better understanding of the microbial world we live in. For this Fourth Edition she enlisted the help of two contributors: Dr. Terri Hamrick of Campbell University to assist with updating the chapters on immunity and microbial control and Dr. Nancy Boury of Iowa State University to assist with the chapter on biological weapons, innovations, and technology.

Text Overview

Microbes are as much a part of our biological world as are the more familiar plants, animals, and insects. They are an extremely diverse group, consisting of thousands of species, including viruses, which are not even considered to be "alive." Most microbes are not pathogens that cause infections. The majority are necessary for the maintenance of all life. A number of species have been exploited in the food industry, in genetic engineering, in the research and development of pharmaceuticals to treat infections and vaccines to prevent them, environmental applications, and in many areas of research. This book focuses on the relative "handful" of microbes that cause infectious diseases in humans.

Annihilation of microbes is not a possibility, a goal, or a desirable outcome, but learning to live in harmony with microbes is realistic and necessary. All students, not only biology or microbiology majors, will benefit from understanding microbes and those factors that lead to collisions between microbes and humans. As a potential parent, you will deal with immunizations and the rashes, fevers, ear infections, and sore throats that your child will develop. Further, as history has shown, epidemics and pandemics are a constant threat, and prevention utilizes knowledge-based preparation. Your generation has not known a world without HIV/AIDS, avian influenza, Zika virus disease, Ebola virus disease, Middle East respiratory syndrome, Nipah encephalitis, multidrug-resistant tuberculosis, healthcare-associated infections, Powassan encephalitis, and many other infectious diseases that remain a source of great concern throughout the world.

The following are just a few microbe-related news stories that were making headlines at the time of this writing (2018):

- An epidemic of Ebola virus disease in the Democratic Republic of Congo (DRC) was continuing to slow but had not stopped.

- More than 2,173 laboratory-confirmed cases of cyclosporiasis tied to contaminated romaine lettuce, basil, and cilantro present in store-bought premade salads and an unknown source associated with trays of vegetables in 33 states were reported from May through August 2018.

○ The United Kingdom reported its first human case of monkeypox in September 2018. The patient was infected by monkeypox virus while in Nigeria before traveling to the United Kingdom.

○ Centers for Disease Control and Prevention (CDC) experts expressed concerns over the prevalence of the increasing number of infections caused by the multi-drug-resistant fungus *Candida auris* in New York City hospitals. *C. auris* cases continue to spread, causing large outbreaks in Europe.

○ Rare *Capnocytophaga canimorsus* infections transmitted through a dog lick caused flesh-eating disease in three Wisconsinites (including one death) in September 2018.

○ During fall, 2018, the CDC was investigating cases involving nearly 200 young children located in 24 states across the U.S. suffering from acute flaccid paralysis (AFP). The likely culprit in a non-poliovirus strain of enterovirus that commonly circulates in summer and fall. It is a rare disease that has devastating effects on families.

At the time of this writing, ecological disturbances could potentially facilitate infectious disease outbreaks. In mid-September 2018, Hurricane Florence ravaged North and South Carolina. Flooding caused more than 100 pig farm lagoons to overflow, contaminating shallow groundwater sources and crops with pig manure. More than 900,000 households get their drinking water from a private well supplied by the contaminated groundwater. Approximately two dozen drinking water systems were forced to halt operations, and an additional two dozen facilities were given "boil water" advisories. Any crops exposed to floodwaters were considered contaminated and could not be sold or enter the food supply. More than 5,500 pigs and 3.4 million chickens or turkeys were killed by flooding. North Carolina was labeled a cesspool of pig feces and urine; displaced native species, such as poisonous snakes; chicken waste; animal corpses; toxic chemicals, such as coal ash containing mercury and arsenic; and fecal pathogens, such as *E. coli, Salmonella, Campylobacter, Vibrio*, noroviruses, and adenoviruses. CDC experts warned hurricane survivors to avoid walking in flooded areas. As the floodwaters recede, North and South Carolina residents will face germinating fungal spores and deadly pathogens in the mud. Researchers have detected high levels of *E. coli* from samples of sediments collected in streets and homes. The post-Florence cleanup will be challenging.

Text Format

The chapters are arranged into five logical and sequential parts:

Part 1: Discovery of Microbes and the History of Public Health (Chapters 1–3)

Part 2: The Microbial Challenge (Chapters 4–9)

Part 3: Microbial Disease (Chapters 10–14)

Part 4: Meeting the Microbial Challenge (Chapters 15–16)

Part 5: Current Microbial Challenges (Chapters 17–18)

Taken as a whole, the dynamics of the interactions between microbes and humans unfold. Per recommendations by reviewers of this text, chapters containing the history of microbes in health and disease were added. Part 1 (Chapters 1–3) addresses the discovery of microbes and healthcare practices before and after the medical community accepted the germ theory of disease. It dissects the reasoning and situations in history that resulted in the development of strategies used in microbial disease control based on sanitation and clean water.

Part 2 (Chapters 4–9) considers the appearance of new (or emerging) and reemerging infectious diseases and factors that contribute to their presence, including world population growth, technological advances, human behavior, and ecological disturbances. The array of microbes that constitute the microbial world and their distinctive properties are introduced. Lest students think that all microbes are "bad guys and out to get us," the emphasis here is on "the other side of the coin"—the beneficial aspects of microbial life in health and biogeochemical cycles. The concept of the human microbiome and the biology of bacteria, viruses, and prions are described, as well as basic aspects of microbial genetics, with an emphasis on mechanisms of genetic exchange in bacteria that contribute to pathogenicity or antibiotic resistance of bacterial pathogens. It is stressed that microbes do not "seek us out," but that the association between microbes and their hosts is accidental—a chance collision that may result in harm to the host. Further, the mechanisms of virulence and the stages of disease are discussed.

Part 3 (Chapters 10–14) focuses on the concepts of microbial disease; epidemiology; the cycle of microbial disease; and healthcare-associated infections (HAIs), an increasing worldwide problem. Chapters 12, 13, and 14 present a sample of "the challengers"—namely, bacteria (Chapter 12), viruses and prions (Chapter 13), and protozoans, helminths, and fungi (Chapter 14)—and the infectious diseases they cause. These chapters focus on modes of transmission; each chapter is divided into foodborne and waterborne, airborne, sexually transmitted, contact, soilborne, and arthropodborne infectious diseases.

Part 4 (Chapters 15 and 16) embellishes how microbial challenges are met. Chapter 15 considers the strategies of microbial disease control based on technologies used to prevent contamination, including disinfection methods, the importance of handwashing, and treatment of microbial infections with antibiotics and antiviral agents. The development of antibiotic resistance is also discussed. The immune system is the topic of Chapter 16; the chapter describes the anatomy and physiology of the body's defenses, including the mechanisms by which molecules embedded in microbes or released (toxins) by microbes, seen by the host immune system as "foreign," are targeted for elimination.

Immunization and examples of microbes or treatments that impair the body's defenses are also fodder for discussion.

Part 5 (Chapters 17 and 18) portrays the current challenges faced in public health. Chapter 17 describes the power and peril of microbes, the promising use of microbial processes in genetic engineering, and the ethical dilemmas scientists and policymakers must consider as new discoveries bring both new opportunities and new challenges. Chapter 18 recognizes the burden of disease and the future of public health. Epidemics and pandemics are most effectively prevented and controlled by partnerships at all levels, ranging from the local level to the national level, to the international level, and to the private sector.

New to the Fourth Edition

Care was taken to preserve Robert I. Krasner's voice, tone, and level of this text. The inside cover of *Krasner's Microbial Challenge, Fourth Edition* contains a listing of World Health Days to raise global awareness of specific health themes related to microbiology that are of concern to the World Health Organization (WHO). The number of parts in this edition have been updated to five and are sequentially presented to assist the student in following the logic of the narrative. All of the examples that illustrate key principles in the chapters have been updated, and the art, photos, and overall design have been improved throughout the text.

New to this text is the opening of each chapter with a case study that includes questions and activities and a set of learning objectives that spans the entire chapter. Nearly all case studies are based on case reports in the primary literature. References are included with each case study. Sometimes fictitious names have been used to improve the reading comprehension of students. Topics of case studies focus on contemporary examples that will pique the interest of both nonmajor biology students and students preparing for healthcare-related careers. Examples of case studies are:

- Wounded Civil War Soldiers That Glowed in the Dark
- Saved by a Syringe Full of Dodge Pond Bacteriophages
- Take Two Fecal Pills and Call Me in the Morning
- Don't Touch That Armadillo!
- From Sea to Sepsis
- Surviving Ebola
- Killer Bagpipes
- A Pain in the Back
- Fighting the World's Deadliest Animal

The text was revised to address feedback from instructors and students. Accordingly, Part 1 is new to this edition, per the request of reviewers. In contrast to other microbiology texts, this text did not contain a stand-alone chapter pertaining to the history of microbiology pioneers and their discoveries. Historical perspectives were intertwined throughout various chapters as microbiology concepts and microbial diseases were discussed. Toward this end, many historical accounts scattered throughout the text have been removed and combined into Part 1 of this revised edition. Emphasis has been placed on the discovery of "germs," or microbes, and their role as causative agents in disease; how battlefield medicine led to innovations in medicine and public health practices; the birth of nursing; the construction of hospitals in the 18th and 19th centuries; the role of the scientific method in proving disease causation; the modern era of microbiology; improved sanitation, water quality, and food safety; and infectious disease surveillance.

Part 2, "The Microbial Challenge" (Chapters 4–9), is consistent with the prior edition of this text. It discusses the beneficial aspects of microbes as well as the worldwide challenge posed by the different types of microbes, including viruses. Quorum sensing, fecal microbiota transplants, probiotics, prebiotics, metagenomics, engineering probiotic bacteria, and dysbiosis have been introduced in this edition, with detailed examples provided in Chapters 4, 7, and 9. A number of new boxes have also been incorporated into Part 2, such as:

- The Mystery of the *Elizabethkingia* Outbreak in Wisconsin
- Talking Starter Cultures and Sourdough Bread
- Microbes Clean Up Lead, South Dakota
- Claimed Medical Benefits of Good Gut Microbiota
- Killing Cancer with Oncolytic Viruses
- Promiscuous Bubonic Plague Bacteria
- Engineering Live Microbes as Therapies

Part 3, "Microbial Disease," was updated to reflect more information on the human microbiome and healthcare-associated infections. Box 10.1, "You Need Guts to Survive," was expanded upon to discuss bacterial diversity and the influences of the composition of gut microbiota on health and disease. Chapters 11 and 12 describe the hospital environment as a source for bacterial pathogens causing healthcare-associated infections. The One Health concept, introduced in Chapter 11, is also new to this edition of the text.

The chapters in Part 3 underwent several rearrangements. In prior editions, tuberculosis, anthrax, smallpox, HIV/AIDS, and influenza were discussed as "Current Challenges." These infectious diseases have been moved into their associated microbial disease chapters in Part 3 of this edition. For example, Chapter 12, "Bacterial Diseases," discusses bacterial diseases (tuberculosis and anthrax were moved into Chapter 12). The rationale behind this rearrangement was to aid instructors who preferred to teach infectious diseases by microbial groups (e.g., bacterial diseases or viral diseases and prions). In prior editions, instructors using this strategy would be required to use sections of two or three different chapters to cover bacterial diseases or viral diseases.

New information in Chapters 12, 13, and 14 addresses the impact of climate change on pathogenic microbes and viruses and their distribution in the environment, in addition to emerging and reemerging infectious diseases. Examples include new topics such as healthcare-associated infections caused by bacteria; noroviruses; cytomegalovirus infections; Ebola virus disease; Middle East respiratory syndrome; Chikungunya virus disease; Zika virus disease; measles virus outbreaks; Powassan encephalitis; monkeypox; increases in sexually transmitted infections; hypersensitivity pneumonitis; *Candida auris* infections; and prevention of HIV infections in light of recent complacency regarding the disease. Box 12.3, "1979: The Year of the Biological Chernobyl!"; Box 13.1, "The Coming Flu Pandemic?"; and updated Box 14.2, "Sushi Eaters Beware!" represent new content in the microbial disease chapters.

Part 4, "Meeting the Challenge" (Chapters 15 and 16), was strengthened in content by inviting Dr. Terri Hamrick to revise and contribute within her expertise and teaching experiences. Chapter 15, which covers the immune system, includes new content pertaining to the body's recognition of foreignness (e.g., invading microbes) at a molecular level, the variations of immune status within an individual and between individuals, the role of normal microbiota as protection from microbial infections, and vaccinations. Chapter 15 includes a new box, "Send in the Monoclonal Search Team." Topics were removed from Chapter 15 (Third Edition) to create Chapter 3, which is now focused on the achievements made toward sanitation, clean water, hand hygiene, food safety, and infectious disease surveillance.

Chapter 16 was redesigned to expand upon topics in the previous edition and to provide new content. It is focused on disinfection and disease control, antibiotics and the development of antibiotic resistance, antibiotic-resistance mechanisms, the clinical challenges of biofilms and antibiotic treatment, possible solutions to antibiotic resistance, and antivirals. Chapter 16 contains two new boxes, "MRSA, VRE, CRE, and Others: A Very Dangerous Alphabet Soup" and "Drug Development."

Part 5, "Current Microbial Challenges," consists of the final two chapters (Chapters 17 and 18). Dr. Nancy Boury revised and updated Chapter 17, "Harnessing the Power of Microbes: Peril and Promise." This chapter introduces new topics such as past and future influenza threats and topics that expose the controversies around genetically modified microbes, including CRISPR/Cas technology and bioethics catching up with technology. It includes new Box 17.3, "Sulfanilamide and the Birth of the Modern FDA."

Chapter 18 is a recombination of Chapters 14 and 17 from the Third Edition of this text, now titled "Partnerships in the Control of Infectious Diseases: The Future of Public Health." New topics include the WHO's Blueprint Priority Disease list, the Global Virome Project, zoonotic diseases, the species barrier and spillover, infectious disease hotspots, syndrome/symptom surveillance, real-time surveillance, mobile health, updated information about the CDC's Vessel Sanitation Program, mosquito control, FEMA responses to catastrophic events, and the One Health initiative.

The text underwent a modest facelift with a new interior design and updated and new tables, illustrations, photos, and self-evaluation questions at the end of each chapter.

The Student Experience

Learning is a difficult, time-consuming, and often tedious task that justifies the inclusion of strategies to help the student; therefore, we include a variety of "assists" in this Fourth Edition. Each chapter begins with a **content outline**; a **case study** followed by references, questions, and activities; **learning objectives**; and a **chapter preview**, allowing students to look ahead and to stay focused on the material; and each is concluded with a broad summary and a sampling of questions for self-evaluation. Website URLs are provided throughout the text to support content. The list of World Health Days located in the inside cover of the text may be used as a way to involve students in global awareness activities.

Key terms are highlighted in bold within the chapters and defined in a glossary at the back of the book. The design of the book has also been changed to make the combination of art and text more user-friendly. Numerous feature boxes with human-interest items, **author notes**, and boxes containing microbiology-related information are scattered throughout the text to pique student interest, and we use humor to break the monotony of study. A number of unique photographs were taken by Robert I. Krasner over the years to depict the microbial diseases described in the text, in particular, many photographs in Chapter 14 ("Protozoan, Helminthic, and Fungal Diseases") were taken by Dr. Krasner. He was provided support in 1999 to study neglected tropical diseases in developing countries. Some of the microbial disease photographs are unpleasant to look at, but they are included to let readers know that Robert I. Krasner had "been there." The Author's Notebook replaced the Author's Note which provides annectodes written by all of the authors of the text as another means to reinforce that they have "been there."

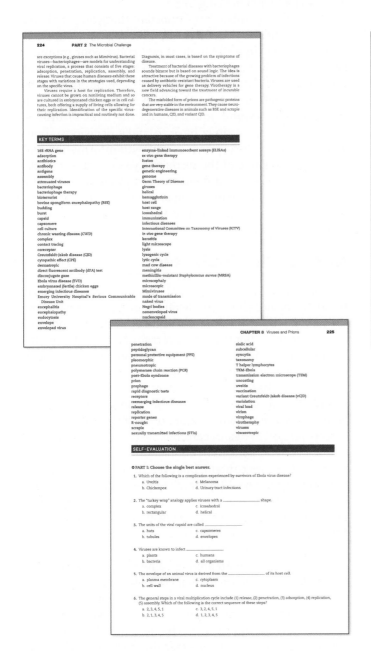

Krasner's Notebook

In writing this text the original author was initially in a dilemma. Should the discussion of the microbes and their virulence mechanisms precede explanation of the body's immune defense, or the other way around? It is the "what came first, the chicken or the egg?" puzzle. Because a strong focus in this text is on disease prevention, it seemed to him more logical to first present what it is that the immune system is combating. It makes sense to me, and I hope you agree!

Krasner's Notebook

The photo in Figure 14.10b is of a row of traditional huts in Botswana, Africa, which has a high incidence of Chagas disease and kissing bugs. In a similar village, more than 50 bugs were isolated from a wooden bed frame, the only piece of "furniture" in the one-room hut. The bugs were examined for the presence of trypanosomes, and most of the insects were positive.

Krasner's Notebook

Frequently, to make a point, the authors have asked in their respective classes for a show of hands as to "Who feels tired and not so great?" Just about every hand goes up (including the author's). Is this the beginning of an infectious disease or conditions related to academic life, including preparing papers, studying (or lack of it), boredom, lack of sleep, and . . . (you can fill in the rest)?

Online Resources

A student companion website has been developed exclusively for this text and is accessible at go.jblearning.com/MicrobialChallengeCWS. The site includes robust practice quizzes, web links, and flashcards.

Note to the Student

During our many years of teaching (70+ years of experience among the authors), we have witnessed that the lack of frequent study, combined with the lack of organizational and time management skills, are the primary cause of academic disappointment. College students face many distractions (e.g., social media, jobs) and more types of technology (e.g., PCs, smartphones, tablets, and smartwatches).

Cramming a few days before an exam will not earn you the best grade you can achieve. Perhaps an analogy will help illustrate what we believe to be the best strategy of action for success. Maybe in your younger years you took music lessons. If so, you would have learned quickly that the time spent practicing between lessons was at least as important as the lessons themselves; the key to improvement and accomplishment was the frequent repetition of the musical exercises assigned. And so it is in handling your college coursework. Studies demonstrate that those students who review course material within 24 hours of a class lecture have higher test scores.

Regular lecture attendance is imperative, but equally so is the effort spent between lectures. More and more students today are working while going to school. Jobs take up study time and time away from social interactions with other students. We encourage you to form study groups as you prepare for exams. Quizzing each other will help to identify which material may need more attention and effort. Some students may understand certain concepts better than others. Studying in groups allows students to help each other learn the material. Lasting friendships will also form and may continue past graduation and many years into the future. After college, networking will play an important role as you seek employment. Studying in groups is how networking begins.

First and foremost, it is important that you get plenty of rest, eat healthy foods (fruits and vegetables!), exercise, and take time to laugh (it can boost your immune system). Handwashing is one of the best ways to prevent the spread of infectious diseases. Developing health-conscious habits (many of which are emphasized in this text) can extend your life and the lives of others. Beyond college, you may have to make decisions about your health. This text provides you with information and points you to some of the resources you can use in making certain decisions and encourages you to be a lifelong learner. Always remember, most microbes are good. You need them to remain healthy. Very few microbes cause disease. Best wishes to you!

Note to the Instructor

The organization of *Krasner's Microbial Challenge: A Public Health Perspective* allows for flexibility in course design. This text, unlike many on the market today, is not intended to be encyclopedic, but to allow coverage of most of the material in a one-semester nonmajors biology or microbiology course. Because this text provides contemporary examples of how microbes impact our daily lives, it has been well received by students who are not biology majors. It can be used for pre-nursing majors or students interested in careers related to health care, but overall it serves as a lab-science general education course. Chapters 4 through 11 make up the core content, but even here there is room for flexibility. Chapters 12, 13, and 14 present approximately 60 diseases: some instructors assign them all, while others pick and choose representative diseases for each mode of transmission or choose those diseases in which students express the most interest.

Instructors can streamline content to fit the needs of the student population in their courses. For example, if the text is used for nonbiology majors who are not healthcare oriented, many topics in Part 1 of the text can be omitted from the course and the core chapters would be the focus for this group. For instructors teaching courses in which the majority of the students are pre-nursing majors, Chapters 6, 12, and 17 may be covered more quickly, with more time being spent on Parts 3 and 4 of the text (microbial diseases and meeting the challenge of microbial diseases). Instructors teaching an honors-level general biology course may spend more time on content related to bacterial genetics, the mechanisms of antibiotics, and the power and peril of microbes in applications and bioethics (e.g., development of vaccines, therapeutics, or food microbiology applications).

Depending on the class size and backgrounds of students in the course, the case studies can be used in group exercises or assigned separately. Instructors can pick and choose which questions students should answer. References to primary literature and URLs are provided for instructors but will be useful to students interested in more details about the subject matter.

The text uses a global approach with examples of disease outbreaks that occur in other parts of the world besides the United States. Those infectious diseases that might be endemic to the area where your students live would be of particular interest. Semester time restraints may dictate that a few chapters, or parts of chapters, be eliminated or assigned as self-study, depending on your own course design. This text can be easily adapted to a two-semester course by the addition of scientific papers, class discussion and debates, digital resources, demonstrations, and "hands-on" exercises that can be performed in the classroom or in the laboratory. Case studies in the primary literature, daily news broadcasts, and Internet resources, including relevant podcasts and YouTube content, can serve as excellent and timely supplements to the text.

Teaching Tools

We are pleased to offer a number of Teaching Tools to instructors using this book to help them prepare for their courses. These were updated for the Fourth Edition by Dr. Kathleen Seiler of Champlain College. All are available for digital download by contacting your Jones & Bartlett Learning Account Manager at go.jblearning.com/findmyrep.

- **Lecture Outlines in PowerPoint format** provides outline summaries of each chapter. The slide set can be customized to meet your classroom needs.
- The **Image Bank in PowerPoint format** provides all the illustrations and photos (to which Jones & Bartlett Learning owns the copyright or has permission to

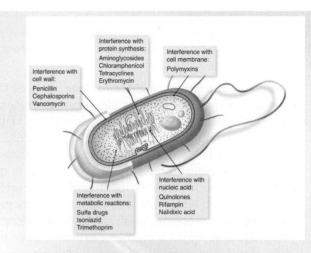

Figure 13.F16: Mechanisms of antimicrobial activity.

reprint digitally) inserted into PowerPoint slides. With the Microsoft® PowerPoint program you can quickly and easily copy individual image slides into your existing lecture slides.

- The **Test Bank** contains more than 650 test items. A typical chapter file contains 15 multiple-choice objective questions, 15 short-answer questions, and 5 essay questions.
- A complete **Instructor's Manual** includes the Learning Objectives from the text for easy reference, a detailed Chapter Outline, Key Terms list, and a number of suggested In-Class Discussions and Activities for every chapter.
- A **Transition Guide** has been prepared to assist instructors who have used previous editions of the text with conversion to this new edition.
- Hand-selected lists of **Web Links** for each chapter will direct students and instructors to relevant Internet resources
- An **Answer Key to the Self-Evaluation Questions** has been prepared by the author and contributors.

Laboratory Component

Most likely, this text will be used in courses that do not have a scheduled laboratory session. Nevertheless, you may be able to squeeze in some exercises that can be done as a demonstration or as a "hands-on" exercise in the classroom or in the laboratory. For example, you could use antibiotic disks to demonstrate antibiotic activity; you can even use prestreaked plates. The presence of bacteria in the environment and on and in the body can be shown by swabbing the floor, desk, a doorknob, and body parts (skin, throat, ear) and streaking the swabs on agar plates. Exercises using market-purchased yeasts are safe and inexpensive and can be used to demonstrate fermentation or disease transmission and other principles.

Translation: mRNA to Protein
- mRNA, in the language of nucleic acid
  (i.e., the 4 bases, A, G, C, and U), is **translated** by ribosomes into the 20 amino acid language of protein
- **Amino acid** structure:
  - a central carbon atom with one of **20 side chains**
  - **amino group** (NH_2)
  - **carboxyl group** (COOH)

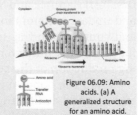

Figure 06.09: Amino acids. (a) A generalized structure for an amino acid.

If your course does include a lab component or you are interested in more hands-on activities, Jones & Bartlett Learning has both print and digital resources to help you and your students succeed:

○ *Laboratory Fundamentals of Microbiology* has been the trusted resource for providing undergraduate students a solid foundation of microbiology laboratory skills. Now, the completely modernized Eleventh Edition represents a lab manual revolution built for today's learners, focusing on the student's experience in the lab. Access to more than 100 minutes of 34 instructor-chosen, high-quality videos of actual students performing the most common lab skills, procedures, and techniques provides a seamless experience for the user. Within the manual, Sections and Exercises open with a list of relevant videos, and icons identify where students should refer to them to best prepare for each exercise. This encourages students to read, see, do, and connect with the material.

In addition to the integration of videos, other significant updates to the Eleventh Edition include the new, full-color, easy-to-navigate interior design, with images from the videos found throughout the manual. Labs have been expanded and reorganized into new sections, such as "Laboratory Safety," "Population Growth," and "Immunology." The all-new Laboratory Safety section emphasizes a "culture of safety" approach to the microbiology lab. *Laboratory Fundamentals of Microbiology, Eleventh Edition* is the perfect companion to any modern microbiology course.

○ If you don't need a printed lab manual for your course but still want to expose students to key laboratory skills, our *Fundamentals of Microbiology Laboratory Video Series* is available as a stand-alone product. In addition to 110 minutes of high-quality videos showing real students performing the most common lab skills, procedures, and techniques, Skills Checklists are available to record progress and assignability.

Acknowledgments

Writing and revising a textbook requires much focus and energy. Without the support of people in our daily lives, it would be nearly impossible to remain focused during the toughest of times. A few kind words of encouragement are what authors draw upon when their reserve energies are low. I especially acknowledge Sami Saydjari for his support of this writing project, enthusiasm for the impact of microbes and viruses on life (especially humans), and his willingness to critically review case studies and boxes from another textbook author's perspective outside of the field of microbiology despite his busy schedule. The support of this text and communications pertaining to microbiology-related topics and life in general with my former undergraduate mentor, colleague, and friend John Cronn were precious. It kept me grounded.

Developing a textbook is a daunting task requiring a harmonious partnership between authors and a publisher. Each has particular ideas, and there is no one right way to go about producing the best text possible. Ultimately, in a spirit of compromise, a book comes together of which all involved can be proud and that will enhance the college experience of the students who read it. In the preparation and publication of this text, Drs. Terri Hamrick and Nancy Boury were invited to revise and update chapters in the text. They breathed new life and content into this text, and I especially thank them for all of their hard work and efforts that extended during production of the book.

I had the opportunity to work with a group of very talented and dedicated people at Jones & Bartlett Learning. Director of Product Management Matt Kane was instrumental in seeing this book through to its finished reality. Product Specialist Audrey Schwinn prepared the revised manuscript for production. I extend a very heart-felt thanks to Production Manager Nancy Hitchcock, the nicest and most capable and wonderful person with whom to work. Nancy went beyond the call of duty to ensure accuracy up to the very last minutes of produc-tion. Her work ethic was tireless. She was so patient and good at delegating tasks as they were needed, taking care not to overwhelm with too many tasks at once. Nancy's appreciation for microbiology set the tone for a great partnership as the book went through the various stages of production. Thank you for the "flowers and chocolates" that arrived from the Jones & Bartlett Learning team as an emotional boost when I was dragging my feet with one chapter left to revise! Special thanks to Jenny Corriveau, Director of Project Management, who came on board to facilitate wrapping up production of this text.

We thank other members of the talented team: copy-editor Jennifer Coker, who is an angel for also volunteering to revise the glossary. Her grasp of the content was a huge plus in the copyediting process. We also thank proofreader Kim Driscoll, photo researcher John Rusk, and the com-positor, Exela Technologies. Special thanks to Troy Liston in managing the art and persistence in making sure it was satisfactory from an author's perspective.

A number of instructors and students who used the First, Second, and Third Editions of *The Microbial Challenge* provided feedback and valuable suggestions as to how the text could be improved in future editions. Thank you for your input.

Cindy Gustafson-Brown, University of California, San Diego

Cheryl Ingram-Smith, Clemson University

Jeffrey A. Hughes, Millikin University

Mark Kainz, Ripon College

Kim LeBard-Rankila, University of Wisconsin–Superior

Michael R. Leonardo, Coe College

Barry Margulies, Towson University

Stacey Massulik, SUNY Onondaga Community College

Melanie Beth Meyer, Community College of Vermont

Karen Grandel Nakaoka, Weber State University

Gary B. Ogden, St. Mary's University

Kathleen Page, Southern Oregon University

Karen Palin, Bates College

Ofra Peled, National Louis University

Jeffrey C. Pommerville, Glendale Community College

Rebecca Rowoth, Ohio Dominican University

Carsten Sanders, University of Pennsylvania

Joyce A. Shaw, Endicott College

Jeffrey J. Sich, Maryville University

Crystal Sims, University of Arkansas–Cossatot

Mike Seong Son, Plymouth State University

Burton E. Tropp, Queens College, City University of New York

Robert L. Wallace, Ripon College

Stephanie A. Yarwood, Oregon State University

Brenda Zink, Northeastern Junior College

I also express thanks to Roger and Sylvia Gasser, Fran Widmer, and Amanda Prigan for their support (including some photographs used as figures). Additional thanks to colleagues at University of Wisconsin Oshkosh, includ-ing David Dilkes, Todd Kostman, Sheldon Cooper, Morgan Churchill, and Jess Lucas who I undoubtedly bored to death about this text during its preparation. Lastly, spe-cial thanks to the thousands of students I have taught, many of whom are employed in health care or preparing for careers in health care. Former students Laura Jaeger, Dustin Winnekens, and Aaron Smith reviewed some con-tent or provided resources for new content. Other former students who have been an inspiration in the preparation of this text are Jaime Antonio Castillo, Jaime Hernandez, Boda Zhao, Yujian Weng, Kaitlin Galow, Dao Vang, Pedro Gonzales, Elias Flor Martinez, Nathan Books, Ashley Utech, and Alyssa Liebenow.

Teri Shors, MS, PhD
Professor
Department of Biology
University of Wisconsin–Oshkosh

ABOUT THE AUTHORS

Robert I. Krasner

Professor Emeritus Robert I. Krasner, a member of the Department of Biology at Providence College (PC) in Rhode Island, retired after 50 years of teaching and research starting in 1958. During his tenure, he developed new courses and mentored many students in research, many of whom went on to graduate and medical schools. Dr. Krasner's courses were popular, demanding, and embellished with humor. He was recognized on several occasions for excellence in teaching. The Robert I. Krasner Teaching Award was established upon his retirement at PC to recognize outstanding graduating seniors. He is the author of *20th Century Microbe Hunters*, many scientific papers, and a contributor to other scholarly works.

Dr. Krasner's love for travel was sparked by his service as a young army medical officer in Japan. He has spent sabbaticals and leaves of absence from PC at numerous domestic and foreign institutions, including Fort Detrick Army Biological Laboratories, Georgetown University School of Medicine, and those in Israel, Paris, Brazil, and London. At 69 years of age, he was accepted into the Harvard School of Public Health and earned a Master of Public Health (MPH) degree and is the oldest full-time student to have accomplished this.

Dr. Krasner founded and directed the Summer Science Program for high school students at PC from 1975 to 2006; the program hosted approximately 1,000 students in its 31 years of operation. During this time he also developed and directed several grant-funded microbiology and biotechnology workshops for high school teachers.

Over the years, Dr. Krasner presented over 60 research papers in the United States and abroad, including at numerous annual meetings of the American Society for Microbiology (ASM). Teaching remained his major interest, and when asked by colleagues "what (research) he was working on," his favorite reply was "students." His initiative in 1980 led to the establishment of the Division for Microbiology Educators within the ASM. After retirement he continued to lecture occasionally and enjoyed gardening, pet therapy, studio art, and playing the harmonica. Dr. Krasner passed away in 2014.

Teri Shors

Teri Shors has been a member of the Department of Biology and Microbiology at the University of Wisconsin–Oshkosh since 1997; she was promoted to the rank of Professor in 2010. Dr. Shors is a devoted teacher and researcher at the primarily undergraduate level and has been a recipient of university awards, including a distinguished teaching award and two endowed professorships. She has taught a variety of courses and laboratories and has made a strong contribution to the development of new courses in microbiology and molecular biology.

Dr. Shors' graduate and postgraduate education is virology based and is reflected in her research. Before teaching at University of Wisconsin–Oshkosh, she was a postdoctoral fellow in the Laboratory of Viral Diseases under the direction of Dr. Bernard Moss in the National Institute of Allergies and Infectious Diseases (NIAID) at the National Institutes of Health (NIH). While her expertise centers upon the expression of vaccinia virus genes, she was involved with research investigating cranberries and other fruits for the presence of antiviral compounds. This antiviral research was funded by a variety of granting agencies, including a prestigious Merck/AAAS award. She has mentored many students engaged in independent research and related readings projects.

Dr. Shors was the major contributor to the Fourth Edition of *Krasner's The Microbial Challenge*. She is also the author of *Understanding Viruses* (2013), now in its third edition, and *Encounters in Virology* (2012). She was a coauthor of *AIDS: The Biological Basis* (2015). Dr. Shors has contributed to and authored a variety of other texts and scientific papers.

Initiative, creativity, humor, networking, using current events and the latest technology in her courses, and leading collaborative, cross-disciplinary studies are hallmarks of Dr. Shors' talents and makes her popular among students in the classroom. She has recently developed and taught an online virology course for undergraduates.

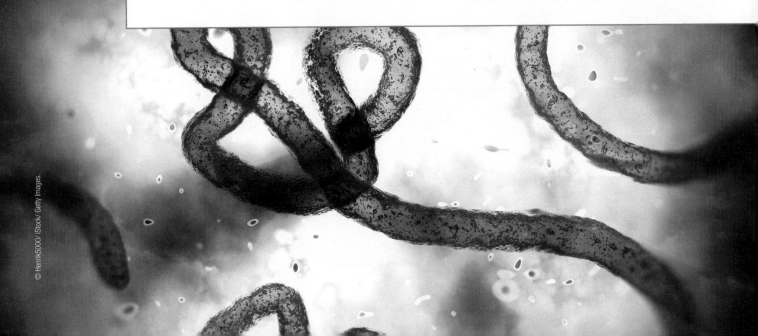

PART 1

DISCOVERY OF MICROBES AND THE HISTORY OF PUBLIC HEALTH

PRE-GERM THEORY, MICROBIOLOGY, AND MEDICINE

> "*The rapidly evolving outbreak of Zika warns us that an old disease that slumbered for 6 decades in Africa and Asia can suddenly wake up on a new continent to cause a global health emergency.*"
>
> —World Health Organization (WHO)
>
> *Director-General Dr. Margaret Chan, May 23, 2016*

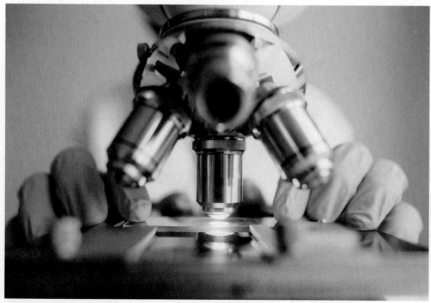

© Kkoloso/iStock/Getty Images Plus/Getty.

Visualizing stained bacteria through a light microscope remains a common practice in clinical microbiology laboratories today.

LEARNING OBJECTIVES

1. Explain why it is important to know that microbes can cause infectious diseases.

2. Identify why it was necessary to visualize microbes with a microscope.

3. Summarize changes that occurred in hygiene practices during the Crimean War and the American Civil War.

4. Describe the barriers that female nurses encountered in the field of health care during the 19th century.

5. List at least three pioneers in medical microbiology and describe their contributions toward creating safer hospitals, obstetrics and neonatal care, and surgery practices before the germ theory of disease was established.

6. Summarize the history of hospitals and nursing in the United States and England in the 19th century.

7. Compare and contrast the organization of hospitals in the 19th century with that of hospitals today.

8. Describe the significance of the introduction of handwashing and antisepsis for wound treatment, surgical procedures, and management of childbirth.

Case Study: Wounded Civil War Soldiers Who Glowed in the Dark

The American Civil War (1861–1865) was the bloodiest war in U.S. history. Twice as many soldiers died of **infectious diseases** (diseases caused by **microorganisms** that can be transmitted to others) during the war than of battlefield injuries. Approximately 752,000 soldiers died, and countless others were left disabled. After the war ended, the state of Mississippi spent 20% of its annual budget on artificial limbs for its veteran soldiers.

The war was fought in over 10,000 locations. The Battle of Shiloh (also called the Battle of Pittsburg Landing) was fought April 6th and 7th, 1862, in Hardin County, Tennessee. Hardin County is located in the southwestern part of the state. The Confederate Army was defeated even though the Union Army had more casualties. Both sides were shocked at the carnage; the number of deaths was four times higher than it was at the Battle of Bull Run, just 10 months prior.

According to Confederate military records, at the Battle of Shiloh 1,728 were killed, 8,012 wounded, and 959 missing or captured, for a combined total of 10,699 Confederate casualties. Tragically, the cries of wounded and dying soldiers lying on the battlefield could be heard for days (**Figure 1**). Ulysses S. Grant wrote in his memoirs that:

I saw an open field, in our possession on the second day, over which the Confederates had made repeated charges the day before, so covered with dead that it would have been possible to walk across the clearing, in any direction, stepping on dead bodies, without a foot touching the ground.

An unusual phenomenon was observed among the wounded soldiers that was not witnessed among those injured in other battles. It was referred to as the **Angel's Glow**. As soldiers lied on the battlefield waiting for treatment, their wounds glowed faintly blue at night. Interestingly, the wounds that glowed in the dark healed more quickly than those wounds that did not glow. Hence, the term *Angel's Glow* to describe the wounds. The cause of these glow-in-the-dark wounds was a mystery until 17-year-old Bowie High School student Bill Martin and his family visited the Tennessee Civil War battlefield in 2001. As soon as he heard the tales of the enigmatic glowing wounds, he immediately thought about his mother's research on a glowing, or **bioluminescent**, bacterium, *Photorhabdus luminescens*. Bill's mother was a microbiologist at the Agricultural Research Service (ARS) Plant Science Institute in Beltsville, Maryland. He asked his mother if *Photorhabdus* could have caused the soldiers wounds to glow (because *Photorhabdus* emits light) and heal more quickly. His mother suggested that Bill, along with his friend Jonathon Curtis, collect soil from

(continues)

Case Study: Wounded Civil War Soldiers Who Glowed in the Dark (continued)

Library of Congress Civil War Collection.

Figure 1 Casualties at Shiloh. Watercolor by artist Adolph Metzner, dated April 7, 1862, illustrating the bloodshed of this battle.

the battlefield and try to isolate bioluminescent bacteria in the research laboratory.

Bill and Jonathon isolated three different strains of *P. luminescens* from the swampy Tennessee battlefield soil. However, because *Photorhabdus* is a soil **bacterium**, the two high school seniors wondered if the bacteria could multiply in the wounds of soldiers at body temperature (37°C or 98.6°F). Experiments in which they incubated the bacterial cultures at 37°C (98.6°F) resulted in no colonies on bacteriological media. While this seemed to put a damper on their hypothesis that *Photorhabdus* was multiplying in the wounds of the soldiers, they decided to further research the weather and temperature conditions during and after the Battle of Shiloh. They learned that there was a thunderstorm on the eve of the first day of battle. Wounded soldiers were stuck in the cold muck for a day or longer as they waited for medics to carry them off the battlefield and attend their wounds. They developed **hypothermia**. Hypothermia occurs when the body rapidly loses heat, causing the body temperature to deteriorate below 35°C (95°F). Therefore, the idea emerged that *Photorhabdus* could possibly multiply in the wounds of cold, hypothermic soldiers.

As they continued their research, Bill was curious if and how *Photorhabdus* could be involved in the healing process of battle wounds. First, they gathered research to learn more about the traits of *P. luminescens* and what is known about its role in the soil ecosystem. Typically, bioluminescent bacteria such as *Vibrio*, *Shewanella*, and *Photobacterium* live in the guts of marine animals, whereas *P. luminescens* is a terrestrial (soil) bacterium.

P. luminescens lives in the midgut of a species of **nematodes**, *Heterorhabditis bacteriophora* (**Figure 2**). The two organisms maintain a mutualistic relationship. A **mutualism** is a form of **symbiosis** in which both organisms benefit by living together. The bacteria infect and colonize the midgut of nematodes present in the soil. The nematode serves as a **host** for the symbiont bacteria to multiply when the appropriate nutrients are available. The nematodes infect insect larvae present in the soil and subsequently regurgitate the *P. luminescens* bacteria. The bacteria release insecticidal **crystal toxins** into the hemolymph (similar to the bloodstream in humans) of the larvae, killing the larvae but not harming their nematode host (therefore, the nematodes benefit from hosting the bacteria). At the same time, the bacteria produce antibiotics that kill or inhibit scavengers and competing bacteria, preventing them from colonizing the nematode host. The regurgitated *Photorhabdus* bacteria secrete **exoenzymes** (**proteases** and **lipases**) that convert the carcass of the larvae into nutrients that both the nematode and bacteria can utilize, allowing both to grow and multiply. The nematodes reproduce inside of the dead larvae, and as adults search for more prey. The life cycle illustrating the symbiotic phase and insect pathogenic phase is shown in **Figure 3**.

Bill and Jonathon were excited to learn that *P. luminescens* produces antibiotics in order to deter other scavenger and insect microbes from colonizing the nematodes or stealing nutrients from the dead insect carcass. One of the main antibiotics produced by *P. luminescens* are **carbapenems**, **broad-spectrum antibiotics** that kill both **Gram (+)** and **Gram (–) bacteria**.

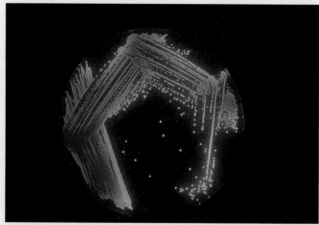

© KPWangkanont / Shutterstock.

Figure 2 These colonies of bioluminescent bacteria growing on the solid media in the petri plate glow light blue, similar to the bioluminescence emanating from the wounds of soldiers at the Battle of Shiloh.

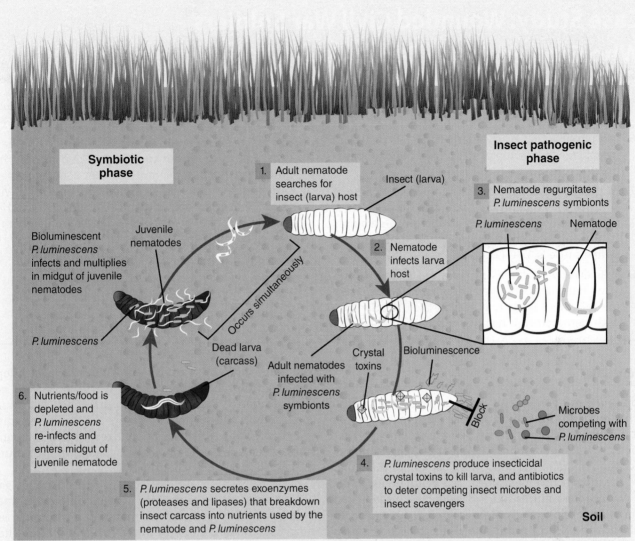

Information from M. E. Hoinville & A. C. Wollenberg. (2018). Changes in *Caenorhabditis elegans* gene expression following exposure to *Photorhabdus luminescens* strain TT01. *Developmental and Comparative Immunology, 82,* 165–176.

Figure 3 Illustration of the life cycle between *Heterorhabditis bacteriophora* nematodes and *Photorhabdus luminescens* that depicts their mutualistic relationship. Bioluminescence is most intense when *P. luminescens* is free-living (due to being regurgitated from the nematode inside of a larval host) outside of its host during the insect-pathogenic phase. The insecticidal insoluble crystal toxin complexes produced by *P. luminescens* are similar to toxins produced by *Yersinia pestis,* the bacterium that caused the **plague,** or **Black Death,** that swept through Asia and Europe during the mid-1300s, killing 25 million people.

Bill and Jonathon believed it was highly likely that the antibiotics produced by *P. luminescens* could kill bacteria causing infections in the wounds of soldiers. They performed experiments to demonstrate that their isolated strains of *P. luminescens* also produced antibiotics. The mystery of the Angel's Glow had been solved. Bill Martin and Jonathon Curtis entered their research into the 2001 Intel International Science and Engineering Fair, an annual science fair competition for innovative high school students around the world. They won first prize!

Questions and Activities

1. Define *symbiosis.* Describe the symbiotic relationship between *P. luminescens* and *H. bacteriophora.*

2. What happens if *H. bacteriophora* nematodes are not infected with *P. luminescens*? Can the nematodes survive without the bacteria? What challenges do the nematodes face if they can survive without it? Explain.

3. *P. luminescens* cannot be directly isolated from soil. How were the students able to isolate and identify three different strains of this bacterium from the Civil War battlefield in Tennessee? (Hint: think about symbiosis.)

4. What is bioluminescence? List four bacterial genera that produce bioluminescence and identify their hosts.

(continues)

Case Study: Wounded Civil War Soldiers Who Glowed in the Dark (continued)

5. Why did the glowing wounds of soldiers heal faster than the wounds that did not glow?

6. Is it possible that *P. luminescens* could be used as an insecticide to kill crop pests in agricultural applications? Explain.

7. In the laboratory, *P. luminescens* multiplies best at 30°C (86°F). Explain why there was the possibility that these bacteria could multiply in the wounds of the soldiers given the fact that normal body temperature is 37°C (98.6°F).

8. *P. luminescens* colonies produce red pigments and emit blue light (bioluminescence). Hypothesize the role these traits may play in their relationship with nematodes. What role might they play with competing microbes in the environment? How could you test for this in the laboratory?

9. Some bacteria communicate with each other through **quorum sensing (QS)**. When present in high density, or a *quorum*, quorum-sensing bacteria perform group behavior(s) or metabolic activities. What group activities are *P. luminescens* bacteria involved in after being regurgitated from their nematode hosts? (See Figure 3.)

Information based on F. L. Inman III, S. Singh, & L. D. Holmes. (2012). Mass production of the beneficial nematode *Heterorhabditis bacteriophora* and its bacterial symbiont *Photorhabdus luminescens. Indian Journal of Microbiology, 52,* 316 -324; Glowing wounds, Science NetLinks. Retrieved from http://sciencenetlinks.com/science-news/science-updates/glowing-wounds/

Preview

At one time or another, everyone has been reminded to wash their hands in order to prevent the spread of germs that can cause disease. It may be hard to imagine that it took centuries to debunk the theory of **spontaneous generation** or dispel the belief that "bad air" or **miasmas** were the main source of sickness and death in hospitals.

This chapter retraces some of the earliest insights into the prevention of infectious diseases before the **germ theory of disease** and **Koch's postulates** were established within the medical community. The notion that microbes can cause infectious diseases did not come with the invention of the microscope, but instead was based on critical observations of hospital infection-control practices used in surgery and delivering babies. The attempts to solve the problem of infections were unrelenting. Medicine remains an experimental science that takes dedication, persistence, and the willingness to make changes in patient care and hygiene practices in order to prevent infections and their spread.

The Origin of Life and Spontaneous Generation

The origin of life was a subject of debate for early thinkers. Two trains of thought predominated: (1) all life was derived from preexisting life and (2) living things emerge from nonliving things, also known as *spontaneous generation*. According to spontaneous generation, plants, insects, and animals emerged from nonliving objects. This was the generally accepted view for centuries. People based beliefs on their interpretations of untested observations of the world around them. They did not use the **scientific method**. For example, it was believed that frogs and salamanders spontaneously arose from mud. This is because people observed that in the springtime rivers would flood certain locations. When the floodwaters receded, large numbers of frogs and salamanders were observed in the mud. Scholars such as Aristotle rationalized that the mud must have given rise to the frogs and salamanders. Of course, today, we know this is not the case.

Other beliefs were that sewage and dirty rags gave rise to rats, fireflies came from morning dew, rotting meat was a source of flies, oysters materialized from the sea, and fish and eels came from riverbeds and sand. These beliefs became fodder for recipes to create life. In the 17th century, Belgian physician and chemist Johannes (also spelled Jean or Johann) Baptiste van Helmont (1579–1644) devised recipes for producing mice, bees, and scorpions (**FIGURE 1.1A–C**) and set up a 5-year willow tree experiment in which he justified that water created the wood, bark, and leaves of the willow tree (**FIGURE 1.1D**). His experiments at the time seemed to reinforce support for the belief in spontaneous generation. The view that spontaneous generation explained the origin of life was generally believed for more than 2,000 years, from 340 B.C. to approximately 1870 A.D.

Visualizing Germs

Even though there were observations and reports of microorganisms by individuals in the 1600s, the accounts were ignored for about 200 years. Robert Hooke and Anton van Leeuwenhoek are credited with developing the first instruments for biological investigations in which one could observe and document microorganisms that were too small to be seen by the naked eye. However, neither Hooke nor van Leeuwenhoek made the

(a) Spontaneous generation of bees recipe

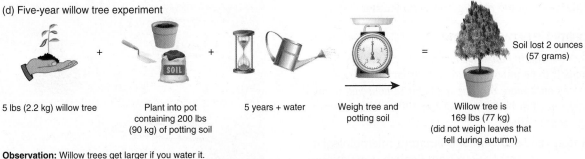

Kill bull with a stone to its head

+

Bury the bull in standing position with horns sticking up and out of the ground

1 month

Saw off the tops of the horns

=

Swarms of bees fly out of the bull carcass through its horns

Observation: Bees were observed burrowed in the eye sockets and rotted flesh of dead animals.
Rationalization: The bees came from the rotting carcass.

(b) Spontaneous generation of mice recipe

Wheat germ

+

Sweaty/soiled clothing

Place into barrel

21 days

=

Mice

Observation: Farmers stored grain in barrels in barns where there are lots of mice around.
Rationalization: The mice came from the grain/wheat germ.

C. Spontaneous generation of scorpions recipe

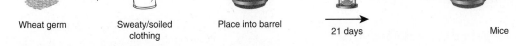

Chisel a groove into a brick

+

Fill groove with crushed basil

+

Cover with another brick

Place into sun for a few days

=

Scorpions

Observation: Scorpions are found hiding under rocks in desert regions.
Rationalization: Fumes from the basil will act as a leavening agent (similar to how bread is made), transforming the crushed basil into scorpions.

(d) Five-year willow tree experiment

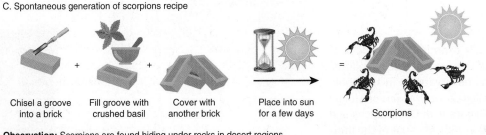

5 lbs (2.2 kg) willow tree

+

Plant into pot containing 200 lbs (90 kg) of potting soil

+

5 years + water

Weigh tree and potting soil

=

Willow tree is 169 lbs (77 kg) (did not weigh leaves that fell during autumn)

Soil lost 2 ounces (57 grams)

Observation: Willow trees get larger if you water it.
Rationalization: 164 lbs of wood, bark, roots etc. arose from water.

FIGURE 1.1 Van Helmont's spontaneous generation recipes and experiments during the 1600s. His conclusions were based on superficial observations. **(a)** Recipe for bees. **(b)** Recipe for mice. **(c)** Recipe for scorpions. **(d)** Five-year willow tree experiment.

connection that microorganisms play a role in causing infectious diseases.

Robert Hooke

Robert Hooke (1635–1703) wore many different hats. He was a curator and fellow of the Royal Society in Lon-don, a natural philosopher (today the term used would be scientist), an architect, an astronomer, surveyor of the City of London, and an inventor. He is credited with coin-ing the term *cell*, which is still used in biology today. Hooke was one of the inventors of the first **compound microscope**, which he used to magnify and resolve living things

© The Royal Society.

FIGURE 1.2 Copperplate engraving of an illustration based on the microscopic observation of blue mold growing on leather in Micrographica as Figure Schem XII. It is the first published image of a microorganism. The mold is likely of the genus Mucor. Sporangia, reproductive structures, are identified by letters A–D.

35 times their normal size. He examined thin slices of cork from the bark of an oak tree, bones, insects, plants, and mold. He published his observations in his book, *Micrographica: or Some Physiological Descriptions of Minute Bodies Made by Magnifying Glasses with Observations and Inquiries Thereupon*, published in 1665. Because of the limitations of magnification at 35×, it was difficult for Hooke to learn much about the internal structures and organization of cells. **FIGURE 1.2** is Hooke's illustration of the microscopic view of blue mold growing on leather.

Anton van Leeuwenhoek

"I'm well aware that my writings will not be accepted by some, as they judge it to be impossible to make such discoveries. . . . [but] I will say once more that 'tis my habit to hold fast to my notions only until I'm better informed or till my observations make me go over to others."

—Anton van Leeuwenhoek, amateur microbiologist (excerpt taken from English translations given by Dobell, 1932)

Anton van Leeuwenhoek (1632–1702; also spelled Antoni, Antonie, and Antony) was born into a large family in Delft, Holland. Like Hooke, van Leeuwenhoek also wore many hats. At the age of 15 or 16, he was sent to school in Amsterdam to learn how to master the linen-draper trade (wholesaler of linen goods and related items). After completing school, he set up his own drapery business in Delft, married twice, had children, and remained there his entire life. He also became the town surveyor, an honorary sheriff, the executor of an artist's estate, and the official wine gauger.

The wine gauger tasted and measured all wine and spirits entering Delft. Van Leeuwenhoek lived to be 91 years old. Is it possible that his position as the wine gauger contributed to his longevity?

At the age of 36, during a trip to London, van Leeuwenhoek was given a copy of *Micrographia*, which he read with the help of English friends who acted as translators. He was literally "hooked"! van Leeuwenhoek began his journey as an amateur microbiologist. He ground and polished lenses made from glass, sand, and rock crystal and built **simple microscopes**. In his lifetime, he made about 500 simple microscopes and lenses that could magnify specimens from 32× to 266×. The microscopes were tiny handheld devices like the one illustrated in **FIGURE 1.3**. A lens was mounted between two brass plates and was held close to the eye. He also held it up toward the bright light while observing samples.

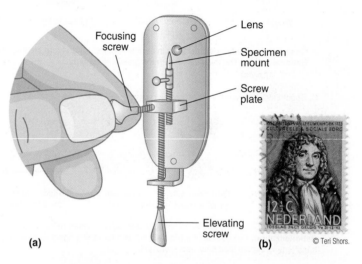

(a)

(b)

© Teri Shors.

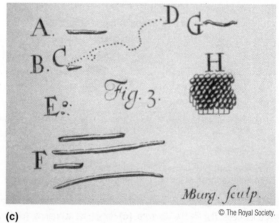

(c)

© The Royal Society.

FIGURE 1.3 **(a)** Illustration of a replica of van Leeuwenhoek's simple microscope being held by a human hand. Observations of specimens were made by applying the eye at a distance of 1 centimeter (0.4 inches) or less. Note that it is so small (the entire instrument was less than 50 millimeters, or 2.2 inches, in length) that the brass body of the instrument was the size of a postage stamp! **(b)** 1937 Anton van Leeuwenhoek Nederland Commemorative Postage Stamp. **(c)** Anton van Leeuwenhoek sketched bacteria from his mouth, which are shown here. The dotted lines from the letter C to D in the illustration represent movement.

van Leeuwenhoek discovered the invisible world. Even though he had no medical or anatomy training, his record keeping was exceptional. He kept detailed sketches and notes describing everything he observed microscopically (Figure 1.3C). He is credited with reporting the first known descriptions of microorganisms. Many of his discoveries were communicated by letter to the Royal Society of London, of which he became a non-British member. The letters were published in the most prestigious scientific publication at the time, *Philosophical Transactions of the Royal Society*. It was the first journal in the world dedicated solely to science.

For over 60 years, as an amateur microbiologist, van Leeuwenhoek described freshwater samples that had hundreds of tiny moving particles, which he thought were small, living animals, which he called **animalcules**. He also observed samples that included teeth scrapings, saliva, soil, rainwater, well water, seawater, spices, mold, algae, wine, blood, rabbit feces, plant matter, food, millet seeds, bone, teeth, vinegar eels (nematodes), lice, frog eggs, and even the sperm of rabbits, lambs, and codfish. His passion for microscopy was indefatigable. He was able to observe a wide variety of microorganisms, including bacteria. At the time of his discoveries, van Leeuwenhoek did not make the connection that the animalcules, or **microbes**, also caused disease. For this reason, even though his scientific contributions to society were impressive at the time of his work, very little was known about them until 200 years after his death.

Battlefield Medicine

The Crimean War of 1853–1856 and the American Civil War of 1861–1865 were two of the bloodiest wars in human history. It is estimated that more than 1 million soldiers died during the Crimean War and about 752,000 during the American Civil War. Countless soldiers were left disabled after the wars ended. Not enough doctors were available to handle battlefield injuries and disease. The doctors themselves were woefully unprepared. Medical training was not standardized. Surgeons were inexperienced. Doctors were practicing before the germ theory of disease (the recognition that microbes caused infectious diseases that could spread from person to person) was established. Few medicines were available. Knowledge of sterile technique or the use of antiseptics was nonexistent. Poor sanitation and overcrowding in camps contributed to deaths from wound infections and non-combat diseases.

Despite this, many medical innovations occurred as a result of dedicated medical staff. *Both wars gave birth to the modern nursing profession.* Capable nurses were needed to care for large numbers of sick and wounded soldiers. Before this time, nurses in the United States cared for the sick within the home, which usually occurred when epidemics swept through towns and cities.

The Crimean War and Florence Nightingale, the Sanitarian

In 1853, the Russian military invaded Turkey. Out of concern over Russia's growing power, Britain and France deployed troops to support the Turkish military. Most of the war was fought on the Crimean Peninsula. The most famous engagement of the Crimean War, known as the Charge of the Light Brigade, took place on October 25, 1854, at the Battle of Balaclava (also spelled Balaklava). The British, led by Lord Cardigan, along with allied soldiers from Turkey and France, fought against the Russians.

Approximately 670 British soldiers on horses were armed with sabers and ordered (through a miscommunication blunder) to proceed down the valley to capture Russian guns in the surrounding hills. Alfred Lord Tennyson wrote the poem titled "*The Charge of the Light Brigade*" to memorialize the event. He wrote: "*Theirs not to reason why, theirs but to do and die, into the valley of Death rode the six hundred.*" The British were forced to retreat. Hundreds of soldiers and horses were wounded and killed by Russian cannon and rifle fire or captured as they charged to bottom of the valley in a mere 7.5 minutes of battle.

The Turks offered the British the use of their army barrack hospital located in Scutari (now Uskudar, Turkey) to care for the wounded and sick soldiers of the Light Brigade. Florence Nightingale (1820–1910) was asked by the head of Britain's War Office, Sydney Herbert, to be the superintendent of all female nurses in the British military hospitals in Turkey. She was to train female nurses to care for the soldiers in Scutari. Prior to this, she was the superintendent at the Institute for the Care of Sick Gentlewomen on Upper Harley Street, London, and a volunteer nurse to cholera patients at the Middlesex Hospital during the epidemic of 1854. Each volunteer nurse was offered a uniform, free meals, weekly wages, and a bed for her work. Nightingale was also provided with food and supplies, and she brought with her a sizeable amount of funding collected through her personal efforts.

Florence Nightingale arrived at Scutari 5 days after the Battle of Balaclava with a group of 38 voluntary nurses from England. They went directly to the Selimiye Barrack Hospital. Prejudiced surgeons who believed female nurses were risqué and drank too much alcohol did not welcome them and prevented them from caring for the wounded and sick soldiers. (There were British male nurses at the time, but their numbers were few and their work was limited to British insane asylums because of their superior strength to restrain violent patients.)

Nightingale immediately tasked the nurses with preparing meals using portable stoves. She delegated the nurses to sew large bags together and fill them with straw to use as beds, and to wash and mend the tattered bed linens. The men were lacking basic needs such as cutlery for eating. Nightingale purchased needed supplies. She did not allow the nurses to enter the sick wards until the doctors asked for help. When hundreds more wounded soldiers arrived from the Battle of Inkerman, the surgeons changed their minds (because Nightingale had financial support and supplies, they saw her as a savior) and allowed their aid.

Nightingale and the voluntary nurses walked into unsanitary conditions. The Selimiye Barrack Hospital was partially burnt down and falling apart. There were rotting and decaying corpses. At any one time, there were more than 1,000 patients in the hospital. Wounded soldiers were brought in to die in pain without any medical attention. The wounded soldiers were still wearing their blood-stained uniforms. They lay crowded together on the bare floors or on pallets of straw saturated with blood oozing from their wounds. Many had broken arms or legs. Jagged splinters of bone could be seen protruding through human flesh. Soldiers had maggot-infested sores. Supplies such as bandages, soap, and clean clothing were not available, and there were shortages of beds, blankets, and food. Inadequate nutrition caused scurvy (vitamin C deficiency). The dilapidated barracks were filthy and rat infested. The air was filled with mosquitoes and smelled of gangrene and the stench of overflowing unusable latrines. Diseases such as typhus, typhoid, cholera, and dysentery were rampant among the wounded, resulting in deaths. The mortality rate was as high as 40%, but that was reduced to 2% after Nightingale's persistent leadership and efforts.

The challenging war conditions were said to be responsible for the birth of modern nursing practices. Nightingale was a brilliant organizer and statistician. She recognized that many of the wounded soldiers died of bacterial infections instead of battle wounds. She believed improving the quality of **sanitation** could prevent infections. The female nurses faced continuous derogatory treatment. To dispel the doctors' negative beliefs about female nurses, she did not allow the other nurses to go into the wards at night. Instead, Nightingale made rounds at night to comfort the men. She carried a paper Turkish candle lantern called a *fanoos* (**FIGURE 1.4**). Henry Wadsworth Longfellow's poem "Santa Felomena" describes her tireless energy and compassion as she carried out her duties caring for the men at Scutari. Longfellow wrote: "*A lady with a Lamp will stand in the great history of the land, a noble type of good heroic womanhood.*"

Nightingale was a **sanitarian**. She emphasized the need for adequate ventilation and cleanliness to promote healing of patients. She insisted that windows and fireplaces be left open in order to allow fresh air to freely circulate in the wards. She believed that many factors determined the health outcomes of patients, including proper nutrition, pure air, clean water, efficient drainage (e.g., latrines), general cleanliness (hospital rooms; patient dressings be changed, etc.; handwashing), a quiet and peaceful atmosphere (speaking negatively about a patient's condition within hearing range was condemned), and adequate light. Dark rooms were considered unhealthy and the cause of **rickets** (a vitamin D deficiency in humans). She stressed that the rooms of patients be bright and that patients should be able to look out of a window. Nightingale wrote extensive reports during the war on proper nursing techniques, stressing cleanliness. After the war was over, she continued writing and authored the book *Notes on Nursing:*

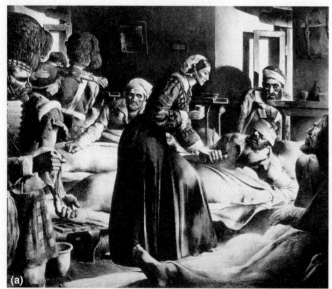

(a)

Courtesy of the National Library of Medicine.

(b)

© Alamy stock photo.

FIGURE 1.4 **(a)** Lithograph of Florence Nightingale caring for wounded soldiers belonging to the British Light Brigade being treated at Scutari. **(b)** Turkish candle lantern (*fanoos*) on display at the Florence Nightingale Museum in London, England. Florence Nightingale used a similar lantern as she made rounds at night in Scutari.

What It Is and What It Is Not, published in 1860. Within 2 months, 15,000 copies of the book had been sold. The book highlighted the core elements of sanitation that are still considered key concepts in public health today, but modified for the needs of patient care in the 21st century.

Another heroic female nurse of the Crimean War who is rarely mentioned is Mary Jane Seacole, or Mother Seacole (1805–1881). Seacole was born in Kingston, Jamaica, the daughter of a Scottish father and Creole mother. She was one of three black female nurses who applied to serve under Florence Nightingale in the Crimean War, but Nightingale's committee rejected her. She was determined to serve. She raised funds, secured a business partner, and set up a hotel, store, and restaurant behind the lines of the Crimean war zone. Seacole nursed and fed wounded soldiers in the hotel. She acquired knowledge about herbal remedies in the Caribbean and applied that knowledge to treat soldiers who suffered from diarrhea, dysentery, and other camp illnesses (**FIGURE 1.5**). A British officer remembered her well and was quoted as saying, "*She was a wonderful woman. All the men swore by her, and in case of any malady would seek her advice and use her herbal medicines, in preference to reporting themselves to their own doctors.*" Mother Seacole returned to England bankrupt after the war. She wrote her autobiography, *Mrs. Seacole's Wonderful Adventures in Many Lands* (published in 1857). It became a bestseller, and she was able to recoup her financial losses from the Crimean War.

Medical Innovations During the American Civil War

Even though the American Civil War (1861–1865) occurred after the Crimean War had ended, little attention was paid to the lessons learned in medical care during the Crimean War. It would take medical staff 4 bloody years of war for them to understand and apply what Florence Nightingale had learned at Scutari. Before the war began, the United States had 16,000 soldiers. The army had 113 doctors. At the start of the war, 24 Union Army doctors went south, and 3 were dismissed for disloyalty. At the end of the war, the Union Army had more than 12,000 doctors and the Confederate Army had 3,000. Before the war, the largest military hospital in the north was a two-story, six-room smallpox hospital located in Washington, D.C. The largest hospital in the south was a 40-bed military hospital at Fort Leavenworth, Kansas.

The first battle of the Civil War was the Battle of Bull Run. It took place at Manassas, Virginia, on July 21, 1861. Neither side was prepared for the 481 killed and 1,011 wounded Union soldiers and the 387 killed and 1,582 wounded Confederate soldiers. Wounded soldiers remained on the battlefield for days (the first 2 days in the rain) because battlefield medical care was not organized. Civilians drove ambulances but they fled after the first shots in battle were fired. Very quickly, changes were essential, the first being the development and implementation of a military ambulance system for the evacuation of wounded to temporary battlefield tent hospitals (**FIGURE 1.6**). Organization of medical treatment was needed. Tent/field hospitals were set up within a mile of the front lines that focused on trauma injuries (e.g., amputations). Later in the war, patients were transported to large general hospitals located in big cities by train or ship. Between 1887 and 1915, the number of hospitals in the United States grew from fewer than 200 to 5,000.

The germ theory of disease was not established until 1870 by Louis Pasteur; Koch's postulates, developed

FIGURE 1.5 Mary Jane Seacole, heroic nurse of the Crimean War, wrote one of the earliest autobiographies of a mixed-race woman.

FIGURE 1.6 Ambulance corps demonstrating removal of wounded soldiers during the Civil War.

Courtesy of the National Archives and Records Administration (79-T-2265).

FIGURE 1.7 Surgical tent at the headquarters of the U.S. Sanitary Commission at Camp Letterman in Gettysburg, Pennsylvania, in 1863. A Civil War surgeon with a Liston knife prepares for an amputation procedure as a soldier is being held down on a table. Liston knives were renowned for their sharp steel. The knives were named after pioneering Scottish surgeon Robert Liston (1794–1847), who was described as "the fastest knife in the West End." He could amputate a leg in less than 2 minutes. He would begin an operation by exclaiming, "time me, gentlemen, time me."

by Robert Koch to prove disease causation by bacterial pathogens, in 1882. Antibiotics had yet to be discovered. Had it been common knowledge that microbes caused infectious diseases, many infections and deaths could have been prevented. Medical care at the start of the Civil War was primitive. Three out of every four surgeries performed were amputations (**FIGURE 1.7**).

All doctors were called surgeons in the Civil War. The surgeons operated in blood- and pus-stained aprons. Everything about the operation was **septic** (contaminated with microbes). The surgeon would hold his knife in his mouth when he helped a patient on or off a table. If he dropped it, he picked it up and rinsed it off with bloody water and continued to work. After loose pieces of bone and tissues were removed, the wound would be packed with unsterile raw cotton and bandaged with wet, unsterilized bandages. A few days after the operation, the incision would fill with **laudable pus**. It was thick and creamy and considered "good pus," in contrast to a thin or watery and bloody surgical site wound.

The general observation by surgeons in the battlefield was that soldiers with laudable pus that formed in gunshot wounds and surgical incisions had a higher probability of surviving than the soldiers with the watery pus. The laudable pus was thought to be involved in the healing process. Nurses and doctors transferred laudable pus into postoperative wounds of soldiers that did not have it. Bandages (which were often rags at the start of the war) were reused between patients. Unfortunately, in all likelihood, the laudable pus was a **Staphylococcus aureus** bacterial infection. Every wound was infected by bacteria. The watery pus was probably a *Streptococcus* infection. Joseph Lister's paper on the use of **antiseptics** was not in the medical literature until 1867. According to D. Vaughan, who wrote, *The Everything Civil War Book* published in 2000, "*In the eyes of many, those who died in battle were luckier than those who were wounded.*"

As the war proceeded, doctors made changes based on observations and experience. The surgeons noticed that mortalities decreased if they applied certain chemicals to wounds. For example, the surgeons had success in combating gangrene (caused by *Clostridium perfringens*, a bacterium commonly found in soil) by packing gunshot wounds with bromine-soaked dressings (bromine is a chemical that is related to chlorine that is still used in disinfectants today), and they isolated patients with gangrene in separate tents with a separate bandage supply. Nurses dressed the wounds and would wash their hands with chlorinated soda between patients. Silk sutures were in short supply so Confederate surgeons would boil horse hair to make it more pliable for suturing wounds and blood vessels, which may have reduced infections. They also boiled sponges to remove dried blood before they reused the sponges.

Prior to the Civil War, nursing was practiced by women caring for the sick in their homes. In order to volunteer for the army, one had to be male and at least 18 years old. Women were not allowed to join as soldiers; however, it has been suggested that as many as 400 women disguised themselves as men and became soldiers. The Civil War gave birth to the modern concept of nursing in the United States. Approximately 2,000 women volunteered to be nurses (**FIGURE 1.8**). **TABLE 1.1** lists some of the most well-known nurses in the Civil War.

The Union's victory and the ending of slavery overshadowed the death of approximately 752,000 soldiers between 1861 and 1865. More soldiers died from camp diseases than from battle wounds. Losses from disease usually exceeded half of the state mortality totals (**FIGURE 1.9**). Many soldiers from rural farming communities were not immune to diseases such as measles, chickenpox, mumps, rubella, and smallpox because they had never been exposed to the viruses that caused them. Therefore, the rural infantrymen were vulnerable to infections by these viruses and were associated with their spread in the camps. Army drills were suspended and entire regiments were temporarily disbanded during measles outbreaks. Military records of Union soldiers documented 21,676 cases of measles, including 551 deaths during the first year of the conflict. Infectious diseases contracted by Union and Confederate soldiers during the Civil War are listed in **TABLE 1.2**. A number

Courtesy of the Library of Congress. Courtesy of the Library of Congress.

FIGURE 1.8 Portraits of prominent Civil War nurses. **(a)** Susie King Taylor was a well-known nurse and teacher. While caring for soldiers, she taught many how to read and write. She authored and self-published *Reminiscences of My Life in the Camp with the 33d United States Colored Troops Late First S. C. Volunteers* (Boston, 1902). (The book can be downloaded and read at: http://lcweb2.loc.gov/service/gdc/scd0001/2008/20081001004re/20081001004re.pdf). **(b)** Civil War nurse Clara Barton, also known for being the founder of the American Red Cross. This portrait was created during the Civil War and is how she wished to be remembered.

TABLE 1.1 Esteemed Nurses of the American Civil War

Nurse	Year of Birth–Year of Death	Comments/Contributions
Clara Barton	1821–1912	Union nurse who tended to sick and injured soldiers on the front lines. Appointed Superintendent of the Union Nurses in 1864. Founded the American Red Cross.
Mary Ann Bickerdyke	1817–1901	Union nurse on the front lines; performed surgeries.
Helen Gilson	1836–1868	Union nurse who set up a tent hospital to care for black Civil War soldiers even though it was controversial at the time.
Cornelia Hancock	1839–1926	Was turned away as a volunteer nurse at Gettysburg because she was only 23 years old but made her way to the battlefield anyway and gained praise and admiration for her dedication to and care for Union wounded soldiers.
Mary Jewett Telford	1839–1906	Was granted special permission to be a Union nurse even though she was too young (age 14) when the war broke out. She was the only female nurse in Hospital No. 8 in Nashville, Tennessee.
Susan King Taylor	1848–1912	First black teacher for freed African-American students on St. Simon's Island in Georgia. She met and married an army sergeant. She served as a Union nurse in the same army regiment as her husband's brother. She taught many of the troops to read and write. She wrote and self-published a book about her experiences caring for soldiers in the 1st South Carolina Colored Troop, later called the 33d United States Colored Infantry.

(continues)

Nurse	Year of Birth–Year of Death	Comments/Contributions
TABLE 1.1 Esteemed Nurses of the American Civil War (*continued*)		
Annie Etheridge	1839–1913	Union nurse nicknamed "Gentle Annie" and commended for her bravery under fire, she was an excellent equestrian and would ride fearlessly to the front lines with medical supplies. Her horse was killed by gunfire from underneath her twice.
Phoebe Pember	1823–1913	Widowed at an early age, she accepted an invitation to work at Richmond's Chimborazo Hospital as a Confederate nurse in 1862 and was in charge of one of the hospital divisions. Wrote her memoirs, *A Southern Women's Story: Life in Confederate Richmond*.
Abigail Hopper Gibbons	1801–1893	Union nurse who joined the U.S. Sanitary Commission training base at David's Island Hospital in New York. Distributed medical supplies out of the Washington, D.C. office hospital.
Dorthea Dix	1802–1887	Union nurse nicknamed "Dragon Dix," she became Union Superintendent of the Female Nurses in 1861.

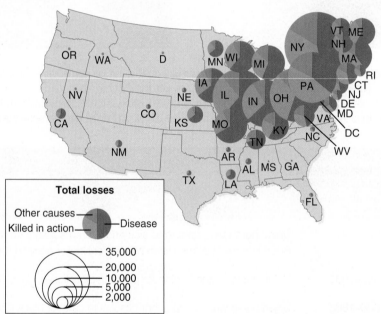

Information from Smallman-Raynor, M.R. and A.D. Cliff, 2004. "Impact of infectious diseases on war." Infectious Disease Clinics of North America 18:341-368.

FIGURE 1.9 Geographic distribution of mortalities (killed in action/other causes/disease) plotted by the state of origin of troops during the American Civil War (1861–1865). Losses from disease commonly exceeded 50% of the state totals.

of factors contributed to the high rates of infectious diseases among the soldiers:

- Unsanitary conditions of the military camps, battlefields, and abandoned plantations
- Spoiled food and inadequate hygiene
- Fly-infested latrines located next to cooking tents
- Coinfections among the soldiers
- Dietary deficiencies (malnutrition), which caused immune suppression
- Polluted waterways (latrines were dug too close to streams, contaminating water supplies for drinking and hygiene practices)
- Overcrowding
- Migrations of troops
- Unburied bodies of humans and horses
- Limited medical treatment (no antibiotics)
- No knowledge of aseptic technique

TABLE 1.2 Infectious Diseases in Union Army Camps During the American Civil War*		
Infectious Disease	Pathogen Type	Morbidity (Cases) and Mortality (Deaths) of Union Soldiers
Influenza and bronchitis	Virus or bacterium	1,765,000 cases and 45,000 deaths
Malaria	Protozoan	1,316,000 cases and 10,000 deaths
Severe diarrhea/dysentery	Bacterium, virus, and/or parasite	360,000 cases and 21,000 deaths
Sexually transmitted infections (STIs), such as syphilis and gonorrhea	Bacterium	182,482 cases and 158 deaths
Typhoid	Bacterium	149,000 cases and 35,000 deaths
Measles	Virus	76,318 cases and 5,177 deaths
Epidemic jaundice (hepatitis)	Virus	71,691 cases with a 0.5% mortality rate
Tuberculosis	Bacterium	29,510 cases and 6,946 deaths
Smallpox	Virus	18,952 cases and 7,058 deaths
Diphtheria	Bacterium	8,053 cases and 777 deaths
Meningitis	Virus or bacterium	3,999 cases and 2,660 deaths
Yellow fever	Virus	1,371 cases and 436 deaths
Scarlet fever	Bacterium	696 cases and 72 deaths
Staphylococcus aureus infections (e.g., boils)	Bacterium	Prevalent but rarely caused deaths

*Values represent infections reported among Union forces. Similar data for mortalities for Confederates are lacking, but cases and mortalities are likely proportional.
Information from J. S. Sattin. (2007). Civil War medicine. The toll of bullet and bacteria. *Gundersen Lutheran Medical Journal*, *4*, 79–83; A. J. Bollett. (2004). The major infectious epidemic diseases of the Civil War soldiers. *Infectious Disease Clinics of North America*, *18*, 291–309.

With so many soldiers lying dead on the battlefield, the handling of corpses became an urgent public health issue (**BOX 1.1**).

Contagion in Hospitals

The idea that germs, or *microbes*, were the cause of infectious diseases did not begin with the invention of the microscope that van Leeuwenhoek and others used to observe microbes. Instead, surgeons and hospital staff in the 1800s believed that "bad air," or miasmas, were the source of **contagion** and cause of death in hospitals. The **miasma theory** was the dominant belief, going back hundreds of years, that infection was the result of noxious or toxic vapors emanating from sewage or filth and trash in towns. The term *contagion* originated in the 1620s. It means "infectious contact." Surgeons considered causes beyond themselves to account for deaths or infections related to surgery, such as overcrowding in hospitals or poor ventilation. During this time, some hospitals in Paris had wards in which three to four postoperative patients shared one large bed! Sir James Young Simpson (1811–1870), professor and physician of midwifery (obstetrics and gynecology) at

the University of Edinburgh, coined the term **hospitalism**. He observed that hospitals had a negative effect on patient outcomes, referring to the phenomenon as *hospitalism*. At the time, based on the data he collected, a patient had a greater chance of dying on the operating table in a hospital than a patient being operated on in a private or country practice. He reported his findings as an article in the *Edinburgh Medicinal Journal* in 1869.

Dr. Simpson suggested that hospitals should be restructured from wards into rooms for each patient in order to provide sufficient space, air ventilation, and isolation from other sick patients. As you learned in the previous section of this chapter, the building of the first hospitals in the United States coincided with the American Civil War. The hospitals had large, one-room wards like the one depicted in **FIGURE 1.10A**. Airborne diseases that are transmitted to individuals inhaling air droplets containing infectious agents that cause infectious diseases such as tuberculosis, influenza, and measles could spread easily among patients in the large open wards. Today, hospitals are no longer organized as large wards unless there is a medical emergency, such as the Ebola epidemic in West Africa in

BOX 1.1 Burial on the Battlefield During the American Civil War

Disposing of the bodies of the soldiers and animals that died in action presented a challenge on the battlefield. Cumulative estimates are that about 752,000 soldiers died in action or due to infections as a result of battle wounds or disease in the camps. Mules and horses were utilized for officers and cavalry in battle, to drive ambulances on the battlefield, to transport cannon artillery, and to carry supplies during movements (**Figure 1**). Historians estimate that 1.5 million horses and mules were wounded or killed during the war. Care of the wounded and dead soldiers were of obvious priority. The winning side (the side that held the ground at the end of the battle) buried soldiers with as much dignity as possible, which meant in their uniforms and in pine boxes, if they were available. The losing side of a battle had to retreat, leaving their dead to be laid to rest by the enemy. Enemy soldiers were buried in mass graves or common graves on the battlefield. Disposing of animal carcasses, the silent casualties of the war, was a real problem. After the battle of Gettysburg, 5 million pounds (2.2 million kilograms) of horse flesh were removed from the battlefield and burned.

If you have ever observed tombstones of Civil War soldiers in a local cemetery, you may not realize that the remains of the soldiers buried there were likely soldiers that survived the war (**Figure 2a**). There is a tradition of leaving coins on gravestones. A coin left is a message to the family of the deceased that someone has visited the grave to pay respect (**Figure 2b**). Local cemeteries often contain memorials for those comrades who died in action or of disease during the war (**Figure 2c**).

Toward the end of the war, embalming was used to prepare the bodies of dead soldiers for transport home or to nearby cemeteries (**Figure 3**). Graveyards next to military hospitals became a more organized way and new form of sacred space in which to bury soldiers who died of war wounds and/or disease in the hospitals. These cemeteries, along with cemeteries located near or on major battlefields, are registered today as National Cemeteries to honor those who served in the Civil War.

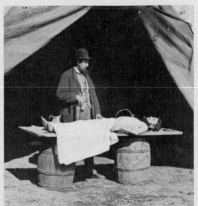

Courtesy of the Library of Congress.

Courtesy of the Library of Congress.

Figure 1 Benson's Horse Battery near Fair Oaks, Virginia, June 1862.

Figure 3 Dr. Richard Burr, an embalming surgeon with the Union Army, demonstrating the procedure on a deceased Civil War soldier. Later in the war, embalming was introduced on the battlefield and in military camps, which made it possible for the preserved remains of soldiers to be transported home to relatives for interment.

(a)

© Teri Shors.

(b)

© Teri Shors.

(c)

© Teri Shors.

Figure 2 Riverside Cemetery, Oshkosh, Wisconsin. **(a)** Soldier's burial lot for remains of indigent Civil War veterans of the Grand Army of the Republic (G.A.R.). The photo was taken a few days before Memorial Day. For Memorial Day, a U.S. flag is placed at every grave in which a war veteran has been laid to rest. **(b)** Lower right of the photograph shows a penny was placed on the gravestone of a Civil War veteran. Leaving a coin is a way to remember the dead and commemorate their life. **(c)** Photograph of a Civil War memorial from Nurse Clara Barton, Tent No. 3, Daughters of Union Veterans of the Civil War. In the background, on the right-hand side, is another memorial for Union soldiers overlooking the Fox River.

There are approximately 130 sites or registered Civil War cemeteries, such as Forest Hill Cemetery Soldiers Lot located in Madison, Wisconsin. The remains of 240 Union soldiers who died in the military hospital or Madison's general hospital are interred there in a section referred to as the Union Soldier's Lot (Figure 4a). Even though no battles were fought in Madison, Camp Randall was a Union Army training facility and a prisoner of war camp. Approximately 140 Confederate troops or prisoners of war were buried in this cemetery in a section known as Confederate Rest (Figure 4b).

© Teri Shors.

Figure 4 Remains of Civil War soldiers and prisoners of war (Confederate soldiers) buried in Forest Hill Cemetery, Madison, Wisconsin. **(a)** Union Soldiers' Lot. **(b)** Confederate Rest.

2015. Emergency hospitals called **emergency treatment units (ETUs)** or isolation wards were set up to specifically care for patients suffering from Ebola virus disease (EVD). **FIGURE 1.10B** is a photograph of facilities that were upgraded for use as an isolation ward to treat patients with EVD in 2014 in Lagos, Nigeria.

Surgery wards were also located near the postmortem rooms in which autopsies were performed on patients who had died from infections. Keep in mind, none of the hospital personnel practiced **barrier technique**, including wearing gloves and a mask. They did not practice handwashing. The doctors and hospital staff were spreading diseases caused by microbes from patient to patient. Medical students and teachers often performed autopsies on patients who had died from infections. It was a relatively dangerous practice, as was cadaver dissection, due to the potential for disease transmission through contact with **pathogens** present on or inside of a cadaver or corpse.

The idea that microbes caused infectious diseases was unproven. However, some pioneering figures in medicine recognized that cleanliness could prevent contagion and

Courtesy of the Library of Congress.

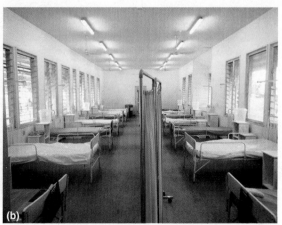

Courtesy of CDC/Bryan Christensen.

FIGURE 1.10 (a) Ward K of the Army Square Hospital located in Washington, D.C. (photographed in 1865). It was a 1,000-bed complex and one of the largest Civil War hospitals in America. The hospital was located where the Smithsonian National Air and Space Museum stands today on the Mall in Washington, D.C. **(b)** Upgraded facilities containing a one-room Ebola virus disease isolation ward in Lagos, Nigeria, in 2014. The Ebola outbreak spread across the West African countries of Guinea, northern Liberia, Sierra Leone, and Nigeria.

mortalities. A few of those medical geniuses, including Joseph Lister, John Snow, Ignaz Semmelweis, Florence Nightingale, Carl Siegmund Franz Crede, and Jean-Antoine Villemin are introduced in this chapter.

Semmelweis and Handwashing

Ignaz Philipp Semmelweis (**FIGURE 1.11**) (1818–1865) graduated from the Vienna University Medical School in 1844 at the age of 26. He started his medical practice in Austria, specializing in obstetrics at a large maternity hospital. He was not an easy doctor to work with. Dr. Semmelweis was cantankerous, arrogant, and abrasive. What he lacked in personality, he made up for in intelligence and his astute observation skills.

The death rate of healthy women from **childbed fever** was so high that women begged to deliver their babies at home rather than risk a hospital delivery. Semmelweis placed the blame, often with insults, on the doctors' and medical students' unsanitary procedures. He was ridiculed by his colleagues, his hospital privileges were limited, and his academic rank was reduced. "Bad blood" and miasmas were thought to be the cause of childbed fever and other diseases; these conditions certainly were not caused by the doctors. Upwards of 30% of women were dying of childbed fever after delivering babies in the hospital, whereas women who delivered their babies at home with the aid of a midwife rarely died of childbed fever.

Childbed fever is sometimes called **puerperal fever**. Symptoms of the disease usually began on the second or third day after delivery. The new mothers experienced a violent shivering fit or fever followed by pain in the uterus radiating toward the abdomen that was tender to the touch. The pulse was rapid and the pain became excruciating. Patients complained that the pain was greater than that they suffered during labor. They stopped lactating. As the infection progressed, patients produced cloudy, putrid urine, when they could urinate. They produced a foul smelling vaginal discharge. Some vomited and had diarrhea. Their tongues became white, and the patients became thirsty. As the disease progressed, the pain and agony were unbearable. As the suffering continued, they became confused and delirious. Some doctors ordered frequent bleeding or purging procedures and opium for the pain. Today it is known that childbed fever is an infection of the endometrium following childbirth or an abortion caused by group A hemolytic streptococci.

After a woman died of childbed fever, the body was moved to a nearby dissection room—the "death house." Surgeons and medical students would perform autopsies. They smelled of the death houses. They did not wear gloves. Their hands were covered with pus containing bacteria from corpses. The bloodier and dirtier their laboratory coats, the prouder they became. The smell and filthy coats represented evidence of their superior skills as physicians and interns.

Dr. Semmelweis observed that the obstetricians and medical students performing autopsies on the wombs of postpartum women who died of childbed fever went directly to the delivery rooms to perform routine vaginal examinations. He believed they carried "death particles" with them to the birthing rooms. Because of his difficult personality, he used his authority to order all obstetricians and medical students to wash their hands with chloride of lime before entering the maternity ward. The results were dramatic! The doctors and interns no longer smelled of death. The morbidity rate plummeted to less than 2%!

Dr. Semmelweis was ignored and ridiculed by many in the medical community. He believed that cleanliness was critical in the hospitals; however, he could not handle criticism. He was difficult and dogmatic, retaliating by writing angry letters. His term of appointment at the hospital expired in 1849, and he was dismissed. Thirteen years later, he published his treatise, *The Etiology, Concept, and Prophylaxis of Childbed Fever*, which is dated 1861

© Robert I. Krasner.

FIGURE 1.11 Ignaz Semmelweis, sometimes called the "savior of mothers." Before the establishment of microbes as causative agents of disease, Semmelweis realized that childbed fever was transferred from physicians to mothers during delivery.

but was actually published in 1860. The treatise of over 500 pages contains passages of great clarity interspersed with lengthy, muddled, repetitive, and bellicose passages in which he attacks his critics.

In 1865, Semmelweis was committed to an insane asylum. He had become an uncontrollable psychotic—possibly due to tertiary syphilis or Alzheimer's disease. He died at the age of 47, ironically, of a wound infection that may have been caused by the same bacterium that causes childbed fever. A tragic ending for a medical doctor ahead of his time. Irrespective of his difficult nature, he is credited with the practice of **handwashing**, a simple, standard **aseptic technique**. In his honor, Semmelweis University, a medical school located in Budapest, Hungary, is named after him.

Handwashing is the single most important practice in preventing **healthcare-associated infections (HAIs)**, as well as infectious diseases in daily living. **Global Handwashing Day**, dedicated to increasing the awareness and understanding the importance of handwashing with soap as an easy, effective, and affordable way to prevent diseases and save lives, was established in 2008. It is celebrated annually on October 15. Over 120 million children in more than 70 countries washed their hands with soap and water for 30 seconds in 2008 in order to spread the word about handwashing. The theme for Global Handwashing Day 2017 was "Our hands. Our future!" The **World Health Organization (WHO)** called upon health facilities to adopt a resolution on improving the prevention, diagnosis, and treatment of **sepsis** at the 70th World Health Assembly held May 5, 2018 (**FIGURE 1.12**). The theme was: "Save lives: clean your hands!"

Lister

Joseph Lister (1827–1912) was an English surgeon. Like Semmelweis, Lister worried about the 46% mortality rates of patients when he was performing surgeries at the Glasgow Royal Infirmary from 1864 to 1866. Many surgeons observed that fatal sepsis was more prevalent in hospitals than in private homes and occurred more often in the city than in smaller towns.

After researching and familiarizing himself with Louis Pasteur's experiments that disproved spontaneous generation, Lister speculated that particles in the air of the operating room were contaminating the surgical incision sites of patients, causing fatal complications of surgery such

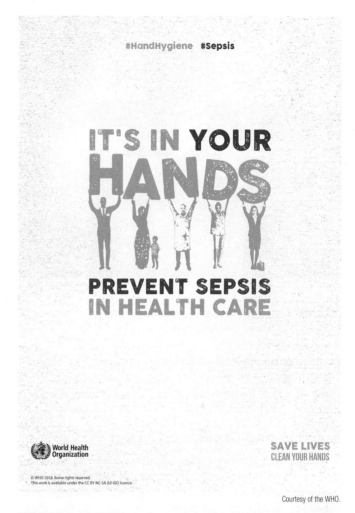

FIGURE 1.12 WHO poster to promote handwashing to prevent sepsis in healthcare settings.

as **gangrene**, **lockjaw**, or some other septic scourge. Lister reasoned that if he could apply dressings with antiseptics that destroyed the floating air particles contaminating surgical incision sites, postoperative sepsis would be reduced.

In 1865, he applied the first **carbolic acid (phenol)** dressing to a patient's compound fracture (**FIGURE 1.13**). The wound was left undisturbed for 4 days. Upon removal of the dressing there was no sign of infection. The incision site healed and the patient recovered. After treating 11 more patients with dressings presoaked in carbolic acid, the surgical sites of 9 out of 11 patients healed and the patients recovered.

Lister presented his results at the Annual Meeting of the British Medical Association in Dublin on August 9, 1867; his findings were also reported in a medical journal titled *The Lancet*. In contrast to Semmelweis, Lister received accolades throughout his career, and his principles of antisepsis were well received by the medical community. He soon transformed surgery and the management of childbirth into safe procedures through the use of chemically treated dressings and carbolic acid sprays as a preoperative measure to cleanse the hands of the

Krasner's Notebook

The story of Ignaz Semmelweis is one of my favorites. His life was a struggle spent trying to convince his colleagues to wash their hands before delivering babies. I visited the Semmelweis Museum in Budapest during the summer of 2006. A small garden in the courtyard of the museum displays a statue of a mother holding her baby in her arms and raising her eyes in gratitude toward Semmelweis.

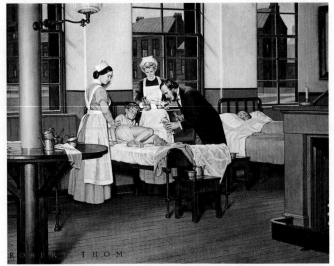

Robert A. Thom, artist from *A History of Medicine in Pictures;* Parke, David & Company. Courtesy of the U.S. National Library of Medicine.

FIGURE 1.13 Lister introduces antisepsis on a patient with a compound fracture at the Glasgow Royal Infirmary in 1865.

surgeons and surgical assistants and to sanitize surgical instruments. Antiseptic products, including **Listerine®** (named after Lister), have been marketed since 1889 (**FIGURE 1.14**).

Villemin and Crede

The 19th century was a time of upheavals in social order and battles among medical doctors over different beliefs in the causes of diseases and mortalities, especially with regard to childbirth and surgeries. In the end, the scientific method prevailed in overcoming ideas that failed to be supported through research. Two individuals who are not typically mentioned in the history of microbiology in textbooks who were pioneers in recognizing that diseases are caused by germs (and therefore can be prevented from spreading) before the germ theory of disease was established are Jean-Antoine Villemin and Carl Siegmund Franz Crede.

Jean-Antoine Villemin (1827–1892), a French military doctor, was able to produce **tuberculosis** in laboratory rabbits 17 years before Robert Koch's famous presentation, *Die Aetiologie der Tuberculose*, to the Berlin Physiological Society on March 24, 1882. Villemin inoculated rabbits with tuberculous matter from a human corpse. By reproducing tuberculosis symptoms and pathology in rabbits, it implied that tuberculosis was contagious. Tuberculosis was also referred to as **consumption**.

Villemin presented his research findings to the Academy of Medicine (the most prestigious medical experts at the time). The Academy members had a lively debate over the report, but in the end were not inclined to accept such a radical idea and his study was met with fierce criticism. One of the harshest critics was Academy member Dr. Hermann Pidoux. Pidoux later became notorious for opposing the germ theory of disease. Ultimately, science prevailed, the germ theory of disease emerged and was

© Teri Shors.

FIGURE 1.14 Many products today, especially mouthwashes, contain a small percentage of phenol. The next time you see Listerine® on store shelves, remember to acknowledge Joseph Lister.

triumphant. Pidoux's resistance would be forgotten. Jean-Antoine Villemin received the Leconte Prize posthumously. Louis Pasteur described his research as the most epoch-making of their time.

Carl Crede (1819–1892) was born in Berlin. His father was a French immigrant who held a high position in the Ministry of Health and Education in Berlin. Carl Crede studied medicine at the University of Berlin and went on to become a distinguished and respected doctor in obstetrics throughout Europe in the 19th century. He was known as a likable, well-rounded individual who was dedicated to medical science. His claim to fame was developing a prophylactic treatment to prevent **gonococcal ophthalmia neonatorum**, a form of blindness in newborns that was later proven to be caused by *Neisseria gonorrhoeae*, the bacterium that causes gonorrhea in two-thirds of opthalmia neonatorum cases (**FIGURE 1.15**). **Gonorrhea** was common in 19th-century Europe. Crede knew that women who married men with a history of gonorrhea became infected. About 35% of pregnant women in European public hospitals had gonorrhea. Crede hypothesized that if ophthalmia was transmitted by the infected mother during delivery to the newborn child, it would be possible to prevent it.

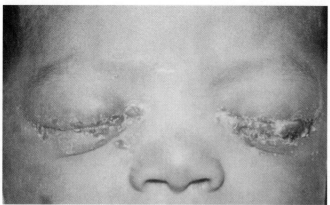

FIGURE 1.15 Newborn with gonococcal ophthalmia neonatorum caused by maternally transmitted *Neisseria gonorrhoeae* infection.

Crede tried several different experiments and collected statistics on the rates of ophthalmia in newborns. He tried cleansing the vagina of the pregnant mother through vaginal douches of a dilute borax solution, but it did not eliminate cases of ophthalmia neonatorum. In December 1879, he switched to the application of a single drop of 2% silver nitrate solution into each eye of the newborn. In the 3 years after he began this prophylactic treatment in newborns, there were 1,160 live births from women under his care, and only two cases of ophthalmia neonatorum! The concentration of silver nitrate was later reduced, but due to its success rate it became a part of standard practice in obstetrics in most parts of the world and remained so for many years.

Contagion and Public Health

John Snow and Cholera

Public health is the science of protecting and improving the health of people and their communities. English physician John Snow (1813–1858) was a leader in promoting the use of anesthesia and hand hygiene in performing surgeries and delivering babies. He laid the groundwork for modern methodologies of epidemiology. **Epidemiology** is the study of epidemics and how they effect a community. In 1849, a major epidemic of cholera occurred in the Soho district of London. It was a major public health threat, causing about 500 deaths in the span of only 10 days. **Cholera** is a bacterial disease manifested by diarrhea so pronounced that life-threatening amounts of water are lost from the body in a short time, causing death from dehydration. Snow's epidemiological detective work showed that most of the cholera victims lived in the Broad Street area and drew their water from the Broad Street pump (**FIGURE 1.16**). Further investigation

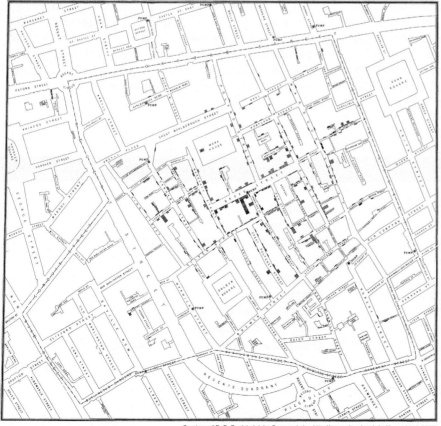

(a)

(b)

FIGURE 1.16 (a) A portion of the map created by Dr. John Snow of the Broad Street area. The lines plot how many people died at each address. **(b)** The Broad Street Pump memorial.

Krasner's Notebook

In 1990, I visited the site of the old Broad Street pump. At the site is now a pub called the John Snow Pub in commemoration of the pump. I had a few beers and reminisced about history.

revealed that the pump was contaminated with raw sewage, and when the pump handle was removed, the cholera epidemic was halted.

A second outbreak of cholera occurred in London in 1854. Snow's sleuthing revealed that most of the cholera victims purchased their drinking water from the Southwark and Vauxhall Company; the company's source of water was the Thames River downstream from the site where raw sewage was discharged into the river. The Lambeth Company, another water supplier, obtained its water further upstream; the incidence of cholera in the population using Lambeth water was much lower. The cause of the disease was not known and led to conjecture and imagination (**FIGURE 1.17**). The actual bacterial contaminant in the water in both outbreaks proved to be *Vibrio cholerae*.

Summary

This chapter focuses on the history of microbiology and medicine before the germ theory of disease and Koch's postulates were embraced by the medical community. The medical community believed in the theory of spontaneous generation. Surgeons thought that sepsis was caused by poor ventilation and bad air, or miasmas. The Crimean War and the American Civil War were fundamental events that resulted in the building of hospitals and the birth of nursing. Other medical

© Wellcome Library, London.

FIGURE 1.17 "**Monster Soup**," commonly called Thames Water. A satirical cartoon created by William Heath, c. 1928.

innovations also occurred during these wars. Thousands of soldiers were killed or wounded on the battlefield. Military surgeons and volunteer nurses learned quickly through experience. Florence Nightingale worked tirelessly to improve sanitation at Scutari in an effort to prevent disease.

Even though van Leeuwenhoek visualized bacteria in the 1600s, it took 200 years for his discoveries to be acknowledged. The work of Nightingale, Mother Seacole, Semmelweis, Lister, Crede, Villemin, and Snow are examples of medical pioneers who worked diligently to reduce infection rates before it was proven that specific pathogens cause infections. These 19th-century trailblazers prevailed in their efforts through persistence and criticism. In the end, the scientific method outweighed the myths of spontaneous generation and miasmas.

KEY TERMS

Angel's Glow
animalcules
antiseptics
aseptic technique
bacteria
barrier technique
bioluminescent
Black Death
broad-spectrum antibiotics
carbapenems
carbolic acid
childbed fever
cholera
compound microscope
consumption

contagion
crystal toxins
emergency treatment units (ETUs)
epidemiology
exoenzymes
gangrene
germ theory of disease
Global Handwashing Day
gonococcal ophthalmia neonatorum
gonorrhea
Gram (+) bacteria
Gram (–) bacteria
handwashing
healthcare-associated infections (HAIs)
hospitalism

host
hypothermia
infectious disease
Koch's postulates
laudable pus
lipases
Listerine®
lockjaw
miasma
miasma theory
microbe
microorganism
Monster Soup
mutualism
nematodes
pathogen
phenol

plague
proteases
public health
puerperal fever
quorum sensing (QS)
rickets
sanitarian
sanitation
scientific method
septic
sepsis
simple microscope
spontaneous generation
Staphylococcus aureus
symbiosis
tuberculosis
World Health Organization (WHO)

SELF-EVALUATION

O PART I: Choose the single best answer.

1. Infectious diseases are caused by _____.
   - a. chemicals
   - b. miasmas
   - c. microbes
   - d. temperature changes

2. Who was known as "The Lady with the Lamp" in the Crimean War?
   - a. Dorothea Dix
   - b. Clara Barton
   - c. Florence Nightingale
   - d. Phoebe Pember

3. _____ was likely the bacterium present in laudable pus in the wounds of Civil War soldiers.
   - a. *Streptococcus*
   - b. *Clostridium*
   - c. *Staphylococcus*
   - d. *Bacillus*

4. Hooke coined the term _____.
   - a. bacterium
   - b. microbe
   - c. germ
   - d. cell

5. Which of the following diseases caused the most cases and deaths among the soldiers in the American Civil War military camps?
   - a. Influenza and bronchitis
   - b. Measles
   - c. Typhoid
   - d. Smallpox

6. What term did van Leeuwenhoek use to describe bacteria that he visualized with a simple microscope?
   - a. Germs
   - b. Animalcules
   - c. Bugs
   - d. Cells

7. Union soldiers were not exposed to which of the following infectious diseases in the camps?
   a. malaria
   b. measles
   c. *Staphylococcus aureus* infections of wounds
   d. West Nile encephalitis

8. Dressings soaked in _____ prevented gangrene in the gunshot wounds of some soldiers during the Civil War.
   a. carbolic acid (phenol)
   b. chlorine
   c. triclosan
   d. bromine

9. Semmelweis recommended what practice in hospitals in order to prevent childbed fever?
   a. Handwashing
   b. Antibiotic therapy
   c. Soaking bedsheets in olive oil
   d. Giving the mother a laxative before birthing a baby

10. What public health crisis did John Snow investigate in the 19th century in London?
    a. Syphilis crisis
    b. Climate change
    c. Tuberculosis epidemic
    d. Cholera epidemic

○ PART II: Fill in the blank.

1. Crede put drops of silver nitrate into the eyes of newborns to prevent _____.

2. _____ is the best way to prevent sepsis in healthcare settings.

3. _____ is the science of protecting and improving the health of people and their communities.

4. _____ is the bacterium that caused the "Angel's Glow" in the wounds of soldiers who fought in the Battle of Shiloh in the American Civil War.

5. Before the Civil War, female nurses cared for patients located in the _____.

6. Crede's research suggested that _____ was a contagious disease.

7. _____ was the Superintendent of Female Nurses at Scutari during the Crimean War.

8. _____ in the wounds of Civil War soldiers was called "good pus."

9. Lister soaked dressings and made sprays containing _____ to prevent sepsis.

10. _____ is the study of epidemics and how they affect a community.

○ **PART III: Answer the following.**

1. Explain what caused the Angel's Glow in the wounds of soldiers who fought at the Battle of Shiloh.

2. Discuss the impact of the Crimean War and the American Civil War on medical practices, the building of hospitals, and the field of nursing.

3. Explain the theory of spontaneous generation and summarize why people believed the theory to be true for such a long time period in human history.

4. Before the germ theory of disease was established, what practices by 19th-century surgeons caused sepsis in postoperative patients and mothers during childbirth?

5. Explain why handwashing is important in health care and in daily living.

6. Explain how Lister's use of antiseptics revolutionized medicine.

7. Explain how a hospital ward with 100 beds in one large room differs from a ward that contains separate rooms for individual patients in terms of spreading infectious diseases.

8. Even though 19th-century surgeons did not see microbes, some of them believed that the microbes were in the air and caused disease. What kinds of observations led them to believe in microbes even though the microbes were unseen?

9. List five individuals and their contributions to the field of microbiology or medicine.

10. Discuss why sepsis has not been totally eliminated today in health care.

CHAPTER 2

POST-GERM THEORY, MICROBIOLOGY, AND MEDICINE

OUTLINE

"*Not a single year passes without [which] . . . we can tell the world: here is a new disease!*"

—Rudolf Virchow
(1821–1902)

German physician, 1867

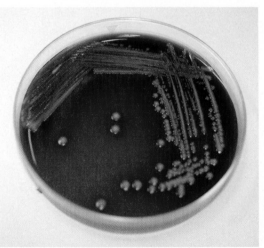

Courtesy of the CDC/Megan Mathias and J. Todd Parker.

Colonies of *Yersinia pestis* on sheep's blood agar medium. *Y. pestis* is the bacterium that causes the bubonic plague, or the Black Death, which swept through Europe in the 14th century, killing millions of people. Alexandre Yersin (1863–1943) isolated this bacterium, grew it in pure culture, and identified it with the microscope.

ShutterStock, Inc. / happykanppy.

1. Define *phage therapy*, and explain why there is renewed interest in phage therapy applications.

2. Describe Pasteur's experiments and how they were used to disprove spontaneous generation.

3. List Koch's postulates and summarize their relevance.

4. Explain why the germ theory of disease led to new discoveries in medical microbiology.

5. Summarize five important discoveries or inventions during the golden age of microbiology, and state their relevance in history and their continuing impact on the field of microbiology.

6. Compare how medicine has changed from the middle of the 18th century to the 21st century.

7. Justify why the elucidation of the structure of DNA was so important for the development of molecular biology tools.

8. Discuss how metagenomics has impacted our knowledge of microbial diversity and the field of medical microbiology.

9. Describe the human microbiome and its association with diseases caused by microbes and changes in the immune system.

Case Study: Saved by a Syringe Full of Dodge Pond Bacteriophages

A 76-year-old Connecticut man underwent surgery to replace an aortic arch containing a bulging aneurysm with a synthetic graft to strengthen the aortic arch and increase blood flow through the aorta and blood vessels of his heart (**Figure 1**). The surgery went well, and the patient returned home. However, 2 days later, he developed a raging fever and had to be rushed back to the hospital. His chest cavity was full of blood and pus. The patient suffered from a postoperative **hospital-associated infection (HAI)** of the aortic graft caused by the bacterium *Pseudomonas aeruginosa*.

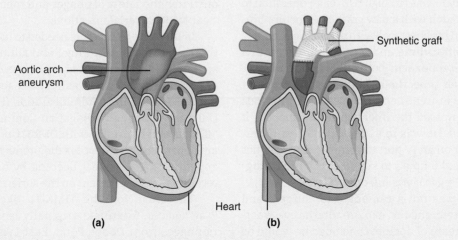

Figure 1 **(a)** The patient's weakened aortic arch contained a bulging aneurysm that could burst at any time. **(b)** Surgeons repaired the aortic arch with a synthetic graft.

(continues)

Case Study: Saved by a Syringe Full of Dodge Pond Bacteriophages (continued)

The bacteria were growing as a green **biofilm** on the graft. *P. aeruginosa* bacteria produce a green pigment, so the biofilm looked like a patch of moss on his aorta. A biofilm is a community of microbes that adhere to surfaces such as river rocks in nature. They also form on implanted medical devices in humans, such as mechanical heart valves and stents, as well as on catheters, ventilators, sutures, and grafts. The biofilms that form may cause infections. The microbes within the biofilm may consist of a single type of microbial species or many different types.

The outside of the bacteria is surrounded by a layer of extracellular material, called a **capsule**, that acts as a protective coating. This makes it difficult to treat biofilms, because **antibiotics** cannot penetrate the protective coating to kill the microbes effectively. *P. aeruginosa* thrives in natural environments, such as lakes and rivers, as well as in hospitals and household environments (e.g., sink drains). Individuals with surgical wounds are particularly at risk for developing *P. aeruginosa* infections that are acquired in the hospital.

The patient had endured a surgery that had saved his life, but the infection lingered. Three months later, he was allowed to return home with an IV port in his chest that would be used to administer antibiotics three times a day into his heart. Eventually, surgeons were able to close the patient's chest wall, but he had to be operated on multiple times in order to **debride** and wash the remaining infected tissues (**Figure 2**).

Over the next year, the patient was hospitalized three more times for **bacteremia** (bacteria in the bloodstream) caused by the same bacterial culprit, *P. aeruginosa*. After consultation with multiple experts, the decision was made that the patient was too high risk for an operation to replace the aortic arch with a new graft. The doctors conservatively treated his chest wall with a multiweek course of intravenous **ceftazidime** and sent him home with a prescription for **ciprofloxacin** (both drugs are antibiotics). Unfortunately, over time, the bacteria became resistant to the ciprofloxacin treatment. Courses of intravenous ceftazidime suppressed the infection for a time, but it always rebounded. He was in a medical state of uncertainty, dependent on an IV port to administer antibiotics into his chest every 8 hours to suppress the **multidrug-resistant (MDR)** *Pseudomonas* infection (**Figure 3**).

The patient was not a candidate for surgery and expressed the preference to explore alternative treatment options. A team of doctors concluded he would be an excellent candidate for **phage therapy**. Phage therapy

Courtesy of Dr. Deepak Narayan, Cardiac Surgery, Yale School of Medicine.

Figure 2 The open chest wall with the infected aortic graft. An arrow points to a region infected with *P. aeruginosa*. This photo was taken during an operation to debride and wash the infected tissues.

is the use of **bacteriophages** (**phages** for short) to kill bacteria. Phages are **viruses** that infect and kill bacteria. In this case, the doctors sought to use phages to kill the bacteria causing the *P. aeruginosa* infection. Western medicine has shown renewed interest in bacteriophage therapy to treat MDR bacterial infections, especially given that clinical trials using phage therapy have demonstrated the safety of phages and their effectiveness to treat MDR bacterial infections.

In this case, the doctors needed to use bacteriophages that would specifically target and kill the MDR *P. aeruginosa*. The patient consented to the use of phage therapy, and the medical team was granted special permission by the Food and Drug Administration (FDA) and the Yale University Human Investigation Committee to treat the MDR bacterial infection with **OMKO1 phages**. All viruses are dependent on a **host** for the production of new progeny viruses. They infect bacteria by binding to a host **receptor** molecule present on the surface of bacterial cells.

The host for the OMKO1 bacteriophages is *P. aeruginosa*. Scientists originally isolated the bacteriophages from Dodge Pond, East Lyme, Connecticut. Subsequently, the researchers purposely further

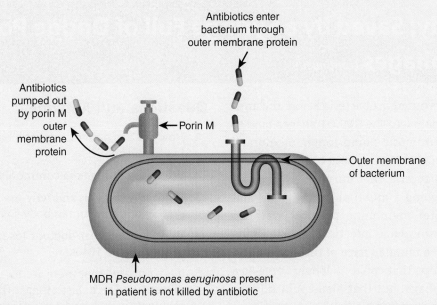

Figure 3 The mechanism of antibiotic resistance developed by the bacterium causing the infection in this patient. Antibiotics enter the bacterium, but they are pumped out of the cell by an outer membrane protein called porin M. Hence, the antibiotics are not effective in killing the bacterium.

screened the phages in order to isolate one that binds to a surface protein known as the **outer membrane porin M protein**. They named the phage OMKO1, a shorthand for *outer membrane porin M knockout dependent phage 1*. Once the medical team received the phages, they were able to confirm in the clinical laboratory that the OMKO1 phages could lyse (kill)

the *P. aeruginosa* strain isolated from the patient's chest cavity.

The OMKO1 phage targets and kills its *P. aeruginosa* host by binding to the bacterial outer membrane porin M proteins (**Figure 4**). *The outer membrane porin M proteins are responsible for the antibiotic-resistant properties of the bacterium. The proteins act like a pump,*

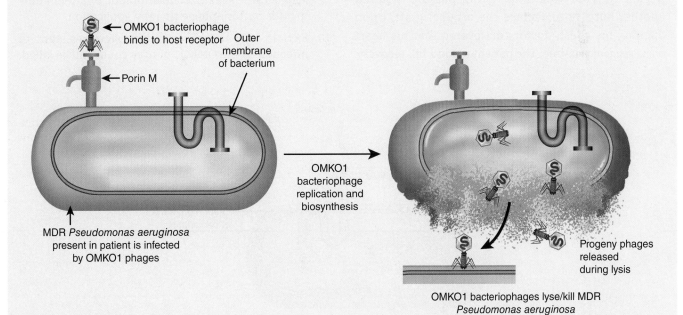

Figure 4 The OMKO1 bacteriophages infect the MDR *Pseudomonas aeruginosa* bacterium by adhering to porin M, which acts as a receptor for the entry of the phage's genome into the cell. During the course of replication and biosynthesis of phages, the fully assembled phages break out of the cell, causing cell lysis/death.

(continues)

Case Study: Saved by a Syringe Full of Dodge Pond Bacteriophages (continued)

flushing antibiotics out of the bacteria before the antibiotics can cause damage. The OMKO1 phages bind to the bacterium's antibiotic pump (outer membrane porin M), killing its MDR P. aeruginosa host.

Pseudomonas bacteria that survived this treatment after undergoing a mutation in the gene that codes for the outer membrane porin M were no longer susceptible to phage infection. The surviving bacteria possessed a mutated form of the outer membrane porin M pumps that could no longer serve as a receptor for the phage, but that also could no longer flush out antibiotics, rendering them sensitive to the effects of antibiotics! Essentially the antibiotic-resistant Pseudomonas bacteria developed sensitivity to the effects of antibiotics in order to prevent infection and their death by OMKO1 phage attack, but in doing so, the antibiotics would kill the remaining mutated Pseudomonas bacteria (**Figures 5** and **6**). What an ingenious plan!

Doctors injected a mixture of 100,000,000 phages and ceftazidime into the **fistula** of the patient's heart (Figure 5). Four weeks later, the patient required another surgery to replace the original graft. Upon opening his chest, doctors discovered that his heart was Pseudomonas-free! The patient made a full recovery.

Questions and Activities

1. How did this patient acquire a bacterial infection in his chest cavity?

2. Where is P. aeruginosa commonly found?

3. What is a biofilm, and why are ones adhered to a graft inside the human body so difficult to treat?

4. Why did Western doctors lose interest in phage therapy in the 1940s?

5. It is estimated that there are 1×10^{31} bacteriophages present on Earth. List at least three different environments in which you could isolate bacteriophages.

6. What role does the outer membrane porin M protein of P. aeruginosa play in antibiotic resistance?

7. What protein do the OMKO1 bacteriophages bind to in order to infect P. aeruginosa?

8. Why did researchers specifically isolate phages from Dodge Pond that would bind to the outer membrane porin M protein present on the surface of P. aeruginosa?

9. Explain the mechanism by which the MDR P. aeruginosa became sensitized to antibiotic treatment during this bacteriophage therapy application.

10. Explain why the bacteria that were resistant to infection by the OMKO1 phages could still be killed,

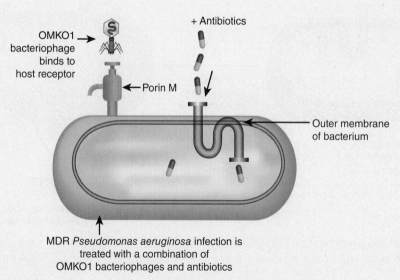

Figure 5 The patient's infection was treated with a combination of OMKO1 bacteriophages and antibiotics.

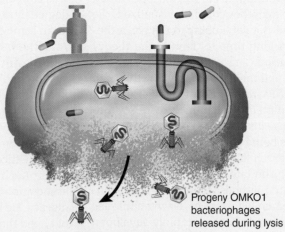

Progeny OMKO1 bacteriophages released during lysis

OMKO1 bacteriophages infect bacterium through porin M and lyse/kill MDR *Pseudomonas aeruginosa*

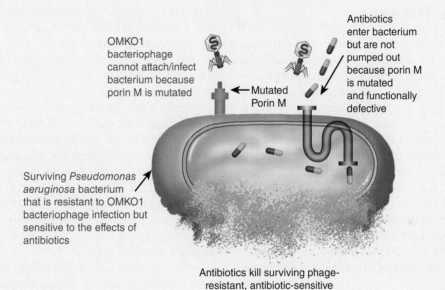

OMKO1 bacteriophage cannot attach/infect bacterium because porin M is mutated

Mutated Porin M

Antibiotics enter bacterium but are not pumped out because porin M is mutated and functionally defective

Surviving *Pseudomonas aeruginosa* bacterium that is resistant to OMKO1 bacteriophage infection but sensitive to the effects of antibiotics

Antibiotics kill surviving phage-resistant, antibiotic-sensitive *Pseudomonos aeruginosa*

Figure 6 The OMKO1 bacteriophages were designed by researchers to work with antibiotics. The OMKO1 bacteriophages kill MDR *Pseudomonas aeruginosa*, and any surviving bacteria are antibiotic sensitive and susceptible to the effects of antibiotics (in this case, ceftazidime will kill the survivors).

resulting in the elimination of the bacterial infection in this case study.

11. Perform an Internet search. List other examples of successful phage therapy cases in North America. What type of bacterial infection was being treated? How did the infection occur? What phages were used to treat the patient?

12. Explain why phage therapy is less toxic to the body than the use of most antibiotics.

Based on B. K. Chan, P. E. Turner, S. Kim, H. R. Mojibian, J. A. Elefteriades, & D. Narayan. (2018). Phage treatment of an aortic graft infected with *Pseudomonas aeruginosa*. *Evolution, Medicine, and Public Health, 1*(1), 60–66, doi:10.1093/emph /eoy005; B. K. Chan. (2016). Phage selection restores antibiotic sensitivity in MDR *Pseudomonas aeruginosa*. *Scientific Reports*, 6, 26717, doi:101038/srep26717.

Preview

Before the germ theory of disease and Koch's postulates were established, people considered disease and death a part of life and nature's way. The cause was often thought to be **miasmas**, a term meaning "bad air" or "swamp air."

By the 19th century a link had been made between filth and the lack of sanitation and what is now referred to as **infectious diseases**. This observation led to the recognition that unhygienic conditions foster the prevalence of disease, and that **vectors**, such as rodents, fleas, ticks, lice, and mosquitoes, carry infectious agents.

Even though the invention of the microscope meant that bacteria and other microbes could be visualized, the medical community had not yet embraced the germ theory of disease. It took a long time to debunk the **theory of spontaneous generation**. This chapter focuses on the history of microbiology and medicine that was influenced by the establishment of Pasteur's germ theory of disease (1870) and Koch's postulates (1882), which were used to prove that **microbes** cause disease. Koch's postulates were dependent on scientists being able to cultivate bacteria in **pure culture** in the research laboratory. Many of the procedures devised during the late 19th century and early 20th century are standard methods still used today in clinical diagnostic and research laboratories. New techniques to study bacterial and human genomes rapidly developed after the structure of DNA was elucidated in 1953.

Scientific Rigor Proves Germs Cause Disease

Between 1865 and 1900, the idea that germs caused disease was fiercely debated. One can page through and read the British medical journals and minutes of meetings of medical societies of the period and see that the idea of germs was in the collective consciousness of those practicing medicine. Surgeons worried about infections caused after accidental injuries or surgeries. Doctors struggled with how to treat **tuberculosis** (also known as **consumption**) in the absence of antibiotics. Debates focused on the ideas that chemical poisons, miasmas, **ferments**, elusive disease agents, or degraded cells were the cause of human disease and infection. Measures to prevent disease, including sanitary practices, were used before scientists identified and isolated the germs that caused the main killers of the era, such as tuberculosis, **cholera**, **typhoid**, **anthrax**, and the **bubonic plague**.

Louis Pasteur and Robert Koch

Toward the end of the 19th century, Robert Koch (1843–1910), Louis Pasteur (1822–1895), and other early microbe hunters established the role of bacteria as causative agents of specific diseases. Pasteur and Koch would become scientific rivals while trying to solve the causative agent of anthrax, a disease affecting sheep and other hoofed animals such as cows, goats, and even deer, but which can also cause disease in humans. They came at the problem from two different trains of thought. Pasteur was highly interested in immunity and disease prevention through **vaccination**; Koch favored the use of public health measures to control infectious diseases. Some of the rivalry between the two men was related to political antagonism between France and Germany. Even though the men were rivals, both devoted their intellectual prowess and hearts in service to humanity.

Neither understood the other's native language. Pasteur, a French chemist and microbiologist, was appointed professor of chemistry and dean of the science faculty at the University of Lille in 1854. He had access to a well-equipped research laboratory and began by studying wine and beer fermentations. Koch, a German physician,

was 20 years younger than Pasteur. He earned his medical degree from the University of Gottingen in 1866 and served as an army doctor in the Franco-Prussian war (1870–1871). After the war, at the age of 29, he was appointed the district medial officer of Wollstein while maintaining an extensive private medical practice in Wollstein (now Wolsztyn in modern-day Poland). Koch created a modest home laboratory equipped with a microscope, an incubator to cultivate bacteria, and a microtome to slice tissue samples for observation. In 1878, he secured a position in the Imperial Health Office in Berlin where he set up a bacteriology laboratory. Koch methodically conducted experiments on the association between a particular microbe and a specific disease. He began by microscopically observing the blood of sheep that had died from anthrax.

Pasteur challenged the belief in spontaneous generation by designing an elegant series of experiments in 1861. He was a keen observer and took a methodical approach to solving problems. Pasteur hypothesized that microbes that appeared in broth that had been sterilized (he called this an **infusion**) were from the "air."

He created two different types of flasks. The first was a **swan-necked flask** that had a curve in it, which would trap microbes from the air, preventing them from entering the sterile broth. He snapped the neck off the second flask, creating a straight neck so that microbes from the air could enter the broth. He put broth in both flasks, boiled the broths to kill any microbes present, and then observed both flasks over time. Microbes only appeared in the broth present in the flask with the snapped off neck. The microbes caused the broth to appear cloudy. This experiment proved that microbes did not spontaneously appear in the broth, but instead entered from the air and began growing in the infusion (**FIGURE 2.1**).

Pasteur spent a good part of his life investigating problems with fermentation of beer and wine and diseases of silk moths, all of which were practical applications that were important to the economy of France. He learned that heating wine or beer to 55°C (131°F) for 30 minutes would not alter the taste of the alcoholic fermentations and would also prevent them from becoming contaminated by microbes that would spoil them. This heating procedure became known as **pasteurization**. The process was subsequently applied to milk in the late 1880s and is still used today for many foodstuffs (**FIGURE 2.2**). Raw milk was found to harbor the bacteria that caused tuberculosis, also known as the **white plague**.

Later in his career Pasteur reasoned that if microbes could cause these maladies, perhaps they were responsible for certain diseases of higher animals and humans. In short order, by 1870, he proved that microbes were responsible for cholera in fowl and anthrax in sheep. Pasteur's scientific approach resulted in the development of his **germ theory of disease**. His crowning achievement was immunization against the dreaded disease **hydrophobia**, now known as **rabies**.

It was Koch, a German physician, who experimentally proved that specific microbes were the causative agents

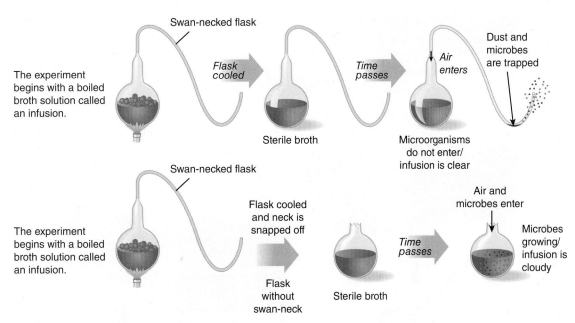

FIGURE 2.1 Pasteur's experiment that disproved spontaneous generation.

for specific diseases. Koch's work initially centered on tuberculosis, a disease that was rampant in his time. Koch established that this disease was caused by a bacterium, later named *Mycobacterium tuberculosis*, in 1882. This was the first proven causal relationship between a microbe and a particular disease. Koch's work extended well beyond tuberculosis and led to the development of a series of four postulates, known as **Koch's postulates**, that are still used today to establish that a particular organism is the cause of a particular disease (**FIGURE 2.3**):

1. *Association:* The causative agent must be present in every case of the specific disease.

Photographed by Jack Delano. Courtesy of the Library of Congress.

FIGURE 2.2 Pasteurization was discovered because of Pasteur's interest in keeping beer from going bad, but the process is best known for its application to milk. This photograph depicts raw milk being pasteurized at the United Farmers' Cooperative Creamery in Sheldon Springs, Vermont, September 1941. Many small towns in rural America had a creamery that pasteurized raw milk before it could be distributed and sold for consumption.

2. *Isolation:* The causative agent must be isolated in every case of the disease and grown in pure culture.

3. *Causation:* The causative agent in the pure culture must cause the disease when inoculated into a healthy and susceptible laboratory animal.

4. *Reisolation:* Microbes identical to those identified in postulate 2 are isolated from the dead animal.

These postulates have been invaluable to medical microbiologists and physicians throughout the 20th century and into the 21st century and continue to play an important role in the identification of the causes of new and reemerging infections. They are not perfect, and sometimes have to be taken with a grain of salt. For example, Dr. Gerhard Henrik Armauer Hansen (1841–1912), a Norwegian scientist, reported in 1873 that bacilli or rod-shaped bodies (known today as *Mycobacterium leprae*) were present in the fluids from lepromatous nodules collected from lepers. He concluded that these bacilli caused leprosy. *Mycobacterium leprae* has one of the slowest **generation times** of any bacterial pathogen. It has a doubling time of 13 days. Hansen's attempts to grow the bacterium on bacteriological media in the laboratory failed (and is still not cultivatable *in vitro* today). His attempts to inoculate laboratory animals such as rabbits or rats and reproduce the disease in the animals also failed (the animals were naturally resistant to *M. leprae* infection), making it difficult for him to prove *M. leprae* caused leprosy. The medical world laughed at Hansen because he was unable to satisfy Koch's postulates.

The lesions of lepers were mostly present on cooler areas of the body such as the extremities (e.g., feet and hands). It wasn't until the 1960s that Dr. Charles C. Sheppard picked up on this observation and experimented with other ways to grow the bacteria. He inoculated the *cooler* footpads of mice with fluid containing the leprosy bacilli and subsequently the mice developed lesions. He had reproduced

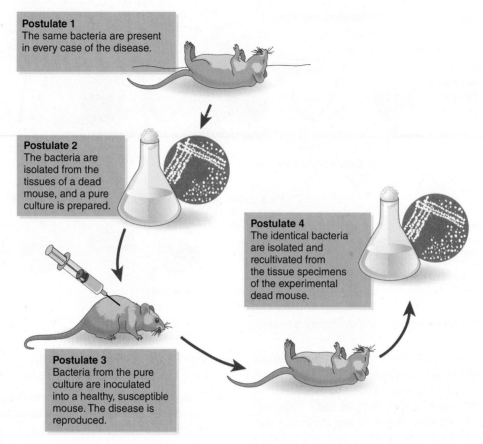

Postulate 1
The same bacteria are present in every case of the disease.

Postulate 2
The bacteria are isolated from the tissues of a dead mouse, and a pure culture is prepared.

Postulate 4
The identical bacteria are isolated and recultivated from the tissue specimens of the experimental dead mouse.

Postulate 3
Bacteria from the pure culture are inoculated into a healthy, susceptible mouse. The disease is reproduced.

FIGURE 2.3 Koch's postulates.

leprosy in those mice! Leprosy researchers wanted an animal model to study how the bacteria affected the body. Now that the medical research community knew and accepted that leprosy could be reproduced in mice foot pads, they looked for more suitable animal models that had a lower body temperature. Oddly enough, armadillos had been used in medical research since the mid-1800s. Armadillos, which have a body temperature of about 32–35°C (89.6–95°F), attracted the attention of leprosy researchers and it became the ideal animal model to study the replication and pathogenesis of *M. leprae*. Who would have thought of that creature as a candidate? Syphilis has long been known to be caused by a corkscrew-shaped (spirochete) bacterium, which still has not been routinely grown in culture medium.

What about viruses? Viruses cannot be grown on lifeless culture media; nevertheless, there are countless examples of viruses associated with specific diseases. Viruses were not described until the 1920s and 1930s. With the advent of the techniques of molecular biology, it is now possible to identify microbes through their DNA fingerprints without growth in culture, and, "classical" Koch's postulates may not always be fulfilled.

Yersin the Underdog and the Bubonic Plague

Rivalry between two microbiology giants was also involved in the race during the late 1800s to determine what bacterium caused the bubonic plague. Alexandre Yersin (1863–1943) was a lesser-known French physician who had worked with

Emile Roux (1853–1933) to characterize diphtheria toxin in Pasteur's research laboratory. Roux and Pasteur developed the first rabies vaccine. In contrast, the Japanese bacteriologist Shibasaburo Kitasato (1853–1931) was more senior and famous for his discovery of the bacterial cause of tetanus while working in Koch's laboratories in Berlin.

In 1894, a raging bubonic plague epidemic resulting in 60,000 deaths in the city of Canton (modern-day Guangzhou) had spread along an important trade route to Hong Kong (which was under British rule). Kitasato was given orders by Japanese government officials to investigate the cause of bubonic plague in Hong Kong. Yersin, a young and naturally curious doctor, wanted to conduct field investigations in new locations. If he were alive today, he would most certainly be involved with **Doctors Without Borders** (*Médecins Sans Frontières*), a group that does humanitarian work to help victims of major disasters, disease outbreaks, and wars. When the plague epidemic spread to Hong Kong, Yersin asked permission to investigate the epidemic in the field. Pasteur had sent a letter of support to the French Ministry of Foreign Affairs through the English Ambassador in Paris in support of his desire to travel in order to investigate the cause of diseases, which resulted in grant support, but Yersin had difficulty obtaining permission to travel to Hong Kong from government officials in British Indochina.

Kitasato arrived in Hong Kong with a team of six assistants, 3 days before Yersin. Yersin, the underdog, was alone with a servant (a language translator), carrying a microscope, sterilizer, and culture supplies in his baggage. Kitasato and Yersin shared German as a common language but exchanged little information. As soon as Kitasato arrived, he had access to autopsies of plague victims. Immediately, he observed bacteria microscopically in the blood of plague victims and began to inject animals with blood samples. Yersin, a short and shy man, spoke no English and thus was unable to communicate with the British authorities. He was denied access to autopsied victims for several days.

It is said that during an autopsy while the two were together, Yersin was surprised that Kitasato was examining blood samples rather than buboes. **Buboes** are swellings that occur in the lymph nodes, especially in the groin or armpit of someone suffering from the bubonic plague (**FIGURE 2.4**). Upon Yersin's suggestion through one of Kitasato's assistants, Kitasato also examined the buboes. Yersin took accurate notes, describing the disease symptoms of plague victims. He drew blood and fluid from the buboes but focused on the bacteria present in the buboes of patients and bodies during autopsy. He correctly identified the presence of a Gram (–), rod-shaped bacteria in the buboes; Kitasato insisted the bacteria in the blood were diplococcal shaped and Gram (+). Given that the bacteria in the buboes were rod-shaped, Kitasato's results were confusing.

As the months and years proceeded, Kitasato stressed that his rod-shaped bacterium from the buboes was not the same as Yersin's. Eventually, Yersin was credited for discovering the bacterium that causes the bubonic plague. The name of the bacterium underwent several name

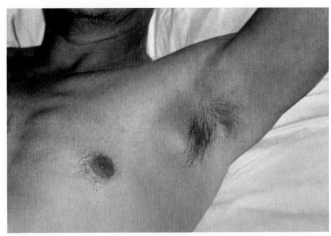

Courtesy of the CDC/Margaret Parsons and Dr. Karl F. Meyer.

FIGURE 2.4 Bubo and **edema** in the armpit of a bubonic plague patient. Buboes contain *Yersinia pestis*.

changes. Initially named *Bacterium pestis*, the name was changed to *Bacillus pestis* in 1900. In 1923, it was renamed *Pasteurella pestis*; in 1970, it was designated with its current name, *Yersinia pestis*, in honor of Alexandre Yersin, posthumously. During this golden age of microbiology, an explosion of investigations led to the identification of many bacterial pathogens that cause disease in humans (**TABLE 2.1**). During this time period, "Typhoid Mary" made U.S. history. She was a healthy carrier of *Salmonella typhi*, the cause of typhoid fever (see **BOX 2.1**).

TABLE 2.1 Bacterial Diseases Identified During the Golden Age of Microbiology

Year	Disease	Investigator, Country
1873	Leprosy (Hansen's disease)	Gerhard Henrik Armauer Hansen, Norway
1877	Anthrax	Robert Koch, Germany
1879	Gonorrhea	Albert Neisser, Germany
1880	Typhoid fever	Karl Eberth, Germany
1882	Tuberculosis	Robert Koch, Germany
1883	Diphtheria	Edwin Klebs, Germany
1884	Tetanus	Arthur Nicolaier, Germany
1885	Infant diarrhea	Theodore Escherich, Germany
1886	Pneumonia	Albert Frankel, Germany
1887	Undulant fever	David Bruce, Australia
1887	Cerebrospinal meningitis	Anton Weichselbaum, Austria

(continues)

TABLE 2.1 Bacterial Diseases Identified During the Golden Age of Microbiology (*continued*)

Year	Disease	Investigator, Country
1888	Food poisoning/salmonellosis	A. A. Gartner, Germany
1892	Gas gangrene	William Welch and George Nuttall, United States
1894	Bubonic plague	Alexandre Yersin, France, and Shibasaburo Kitasato, Japan (independent discoveries)
1896	Botulism	Emile van Ermengem, Belgium
1898	Bacterial dysentery	Kiyoshi Shiga, Japan
1903	Syphilis	Fritz Schaudinn and Erich Hoffman, Germany
1906	Whooping cough/pertussis	Jules Bordet and Octave Gengou, France
1911	Tularemia	George McCoy and Charles Chapin, United States

BOX 2.1 Typhoid Mary: A Public Health Dilemma

Had you accepted an invitation to dinner in the summer of 1906, at the rented vacation home of Charles Henry Warren, a wealthy New York banker, in Oyster Bay, New York, you might have become seriously ill or died from typhoid fever. The culprit would have been the bacterium *Salmonella typhi*, harbored unknowingly (at least initially) in the digestive tract of **Mary Mallon** (1869–1938), the Warrens' cook.

Over the next several years, Mary became stigmatized as **Typhoid Mary**. During that time, as a **healthy (asymptomatic) carrier** of *Salmonella*, she was responsible for approximately 50 cases of typhoid fever, at least 2 of which ended in death. Mary had vowed never to stop working as a cook and refused a gallbladder operation that would have rid her of the bacteria. Consequently, she was forced to spend 26 years of her life in relative isolation, shunned by her fellow humans, in a tiny cottage on North Brother Island, a small island in the East River in New York City. Mary, not *Salmonella*, became the culprit. This was hardly the life she envisioned when, as a teenager, she emigrated from Ireland. In a legal sense she was never tried for her "crime."

Mary Mallon was born in 1869 in Ireland and immigrated to the United States in 1883. The first suspicion that she was a healthy typhoid carrier was during her brief stint as a cook for the Warrens, during which, in a 1-week period, 6 of 11 people in the household (Mrs. Warren, 2 daughters, 2 maids, and the gardener) developed typhoid fever. Typhoid and other diseases associated with "filth" were simply "not nice" for the elite of Oyster Bay where the Warrens summered; hence, George Soper, a sanitary engineer, was hired to determine the cause of the outbreak. Soper quickly ruled out soft clams as the source and found a pattern of typhoid in families where Mary had worked as a cook, dating back to 1900 and totaling 26 cases. Soper was unsuccessful in convincing Mary to be tested to determine if she was a *Salmonella* carrier. She proclaimed that her health was excellent. She had no history of typhoid, and hence saw no reason for compliance.

The case was referred to Herman Biggs, medical officer of the New York City Health Department. In 1907, Biggs authorized police to take Mary against her will to a contagious disease unit in a New York hospital. Examination of her feces revealed a high number of *Salmonella* bacteria, and she was removed to North Brother Island. Three years later she was released by the health commissioner on her promise that she would not cook again, but after about 2 years she used the name "Mary Brown" and was hired to cook, her only employment skill, at a hospital.

In 1915, she was sent back to the island for the rest of her life. During her 4 years of freedom she was responsible for another 25 cases of typhoid fever, including 2 deaths. She became the butt of jokes and cartoons (**Figure 1**). "Typhoid Mary" appeared in medical dictionaries, as a disease carrier. She died on November 11, 1938, not as Mary Mallon, but as Typhoid Mary, a public menace blamed for a total of 53 cases of typhoid fever, including 3 deaths, and with the dubious distinction of being the first healthy *Salmonella* carrier in the United States.

The account of Mary Mallon exposes the dilemma of the public's health and safety versus the rights of the individual. The circumstances of this case and its outcome (Mary Mallon's confinement) can be viewed as an application of *objective utilitarianism*, a consequence-based moral philosophical position succinctly stated as "the greatest good for the greatest number." Proponents of utilitarianism are clear that all individuals count

Courtesy of New York Public Library Digital Collections.

Figure 1 Typhoid Mary depicted in a cartoon published in the *New York American*, Sunday, June 30, 1907.

and count equally, but, nevertheless, the sacrifice of some (Mary, in this case) is justified for the sake of the many (the public). At the time, those charged with the public's health exhibited sex and social class prejudice in considering a broad spectrum of factors and in arbitrarily deciding that liberty was an impossible privilege to allow Mary Mallon. In the case of other disease carriers, particularly males, other options were explored.

Do you believe Mary was justly treated? This case serves as a paradigm for today's "Typhoid Mary"—"AIDS Sam" or "Resistant TB Joe." Health officials have not only the right but the obligation to develop public health policies, but, in so doing, they incur the responsibility to provide realistic and appropriate alternatives, including health care, housing, and reemployment training, to those individuals in society who are afflicted. **Public health** is the science of protecting and improving the health of an entire population or community through education, policy-making, and practices (e.g., vaccination) through organized efforts. AIDS, in some respects, parallels the Typhoid Mary dilemma.

This poem appeared in *Punch*, a British magazine, in 1909:

> In U.S.A. (across the brook)
> There lives, unless the papers err,
> A very curious Irish cook
> In whom the strangest things occur;
> Beneath her outside's healthy glaze
> Masses of microbes seethe and wallow
> And everywhere that MARY goes
> Infernal epidemics follow.

The Golden Age of Microbiology

The golden age of microbiology was a time of major advances in the 18th and early 19th centuries. During this period, scientists addressed four main microbiological questions:

1. Can life emerge from nonlife (spontaneous generation)?
2. Do microbes cause infectious diseases?
3. How diverse is the microbial world?
4. Do microbes in the soil and water carry out beneficial activities?

Most of the focus of this chapter is in response to the first two questions. However, microbes are involved in all aspects of life. Less than 1% of microbes on Earth cause disease. What helped to drive the identification of bacterial pathogens forward was the development of basic methodology, such as pure culture, **aseptic technique**, and diagnostic tools such as microscopy and Gram staining. Even though the vast majority of discoveries were associated with bacterial pathogens, Louis Pasteur developed the first vaccine to prevent rabies virus infection (**BOX 2.2**).

BOX 2.2 Pasteur and the Development of the Rabies Vaccine

When one thinks of the famous names associated with the history of science, the name Louis Pasteur ranks among the greatest. Pasteur's phenomenal success in diverse fields of research is without parallel, but his development of the **rabies vaccine** was his crowning glory. Nobel laureate Selman A. Waksman, discoverer of streptomycin, the first antibiotic effective against tuberculosis, wrote in *The New York Times* on February 5, 1950, "Pasteur was not only the great scientist who was largely responsible for the creation of the science of microbiology, he was its high priest, preaching and fighting for the recognition of its importance in health and in human welfare."

In the early 1870s, at about the age of 50, Pasteur and his research assistant Emile Roux began to conduct research related to human infection. By this time, he was well renowned and respected by the scientific community in France and beyond for his work on crystallography, fermentation, diseases affecting the production of wine and beer, spontaneous generation, silkworm disease, cholera, and anthrax. It is not possible to cite with certainty the event or events that resulted in Pasteur's entry into rabies research. One story describes how, as a young boy, Pasteur had witnessed and heard screams of a victim bitten by a rabid wolf; the victim was undergoing cauterization of the wounds with a red hot iron.

Before the mid-1880s, Pasteur succeeded in cultivating the virus of rabies in the brainstem of rabbits. He then **attenuated** the rabies virus (weakened its virulence) by suspending a portion

(continues)

BOX 2.2 Pasteur and the Development of the Rabies Vaccine (continued)

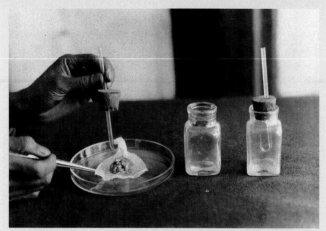

Figure 1 The first step in the preparation of the rabies vaccine. Weighed rabbit brainstem (infected by rabies virus) was placed on mosquito netting. It was subsequently hung on the glass hook in the sterile glass vial and then suspended in ether. Photo circa 1910 at the Pasteur Institute, India.

of the rabbit's virus-infected brainstem on a curved glass rod in a sterilized vial (**Figure 1**). As the brainstem preparation dried, the virulence of the virus gradually decreased, and after 14 days the crude preparation was totally without virulence. Dogs that were immunized with this crude vaccine and experimentally bitten by rabid dogs or had the active rabies virus applied directly onto their brains did not develop rabies. Pressure was immediately brought to bear on Pasteur to use the vaccine on humans, but Pasteur resisted the temptation, realizing that the vaccine was not yet ready for a trial on human subjects.

Monday, July 6, 1885, was a momentous day for Pasteur and for the history of medicine. It was on that day that Joseph Meister, a 9-year-old boy, was brought to Pasteur's laboratory by his mother. On his way to school, 2 days earlier, Joseph had been severely attacked by a rabid dog and suffered 14 wounds on his hands, legs, and thighs. The child's wounds were treated with carbolic acid 12 hours after the incident by a doctor who advised that Joseph be brought to Paris to be seen by Pasteur. The boy's mother pleaded to Pasteur for help for her doomed son. Pasteur's reputation was on the spot. Here was this mother pleading with him to save her child, but he worried that his vaccine, which had been tried only on dogs, would not be effective. Pasteur relented and administered the rabies vaccine to his first human subject. Young Meister received 13 inoculations of Pasteur's rabies vaccine over the next several days and survived his ordeal.

Jean-Baptiste Jupille, age 15, was the second patient treated by Pasteur, under less favorable circumstances. Jupille was in a group of six shepherd boys who were attacked by a rabid dog. One of the many biographers of Pasteur describes the event:

The children ran away shrieking, but the eldest of them, Jupille, bravely turned back in order to protect the flight of his comrades. Armed with his whip, he confronted the

969 — PARIS. — Institut Pasteur.
Statue du berger Jupille.

Figure 2 Statue of shepherd boy Jean-Baptiste Jupille at the Pasteur Institute in Paris. Jupille was attacked while attempting to protect younger shepherd boys from a rabid dog. He was the second person to receive the vaccine.

infuriated animal which flew at him and seized his left arm. Jupille wrestled the dog to the ground and succeeded in kneeling on him, forcing his jaws open in order that he might disengage his left hand, and in so doing, his right hand was seriously bitten in its turn; finally, having been able to get hold of the animal by the neck, Jupille called his little brother to pick up his whip which had fallen in the struggle and secured the animal's jaws with the whip. He then took his wooden shoe, with which he battered the dog's head.*

Jupille was brought to Pasteur for treatment 6 days after being bitten, in contrast to the 2-day interval in the Meister case. Pasteur was particularly reluctant to treat Jupille because of the extended time that had elapsed since the attack. Nevertheless, Pasteur again put his reputation on the line and injected Jupille with the vaccine. Jupille did not develop rabies, and the boy's act of bravery was commemorated with an impressive statue depicting the struggle between a boy and a rabid dog, which stands today on the grounds of the Pasteur Institute (**Figure 2**).

*Rene Vallery-Radot. (1911). La vie de Pasteur. Paris: Librairie Hachette Et Cie.

Establishing Pure Cultures of Bacteria in the Laboratory

Cultivating bacteria as pure cultures in the laboratory was the key to identifying a bacterium that was the cause of a specific infectious disease. Pure cultures contain a homogeneous population of one bacterial species or clones of the same bacterium. Pasteur's experiments involved the use of **nutrient broths**, which he called *infusions*. The infusions were not pure cultures. They contained mixed cultures of fermenting bacteria. He observed liquid suspensions of microbes microscopically and performed other experiments (e.g., pasteurization; the use of swan-necked flasks to disprove spontaneous generation). The cultures he was working with were not pure.

Fanny Hesse's Cooking Tip Revolutionized Bacteriology

To confirm his postulates, Koch and researchers in his laboratory developed methods to obtain pure cultures of suspected disease-causing bacteria or microbes (**pathogens**). In order to isolate pure cultures, a solid medium was needed. Research assistants streaked bacteria on solid media containing beef stock, with gelatin as the solidifying agent. Visible colonies would form on the smooth, solid medium. Different types of colonies could then be separated from each other. These were removed from the medium and then grown in isolation to grow colonies that were all produced from the same type of bacterium. Unfortunately, sometimes the bacteria would produce **enzymes** that would break down the gelatin and the medium would become a liquefied mess. On hot summer days the higher temperatures also caused the gelatin to melt or liquefy. Experiments were ruined, slowing the progress of their research.

One hot day in Germany, Walther Hesse (1846–1911), an assistant in Koch's lab, vented his frustrations to his wife, Fanny. Fanny Angelina Hesse (1850–1934) was an unpaid assistant in the laboratory, working alongside Walther, assisting him in preparing media, cleaning equipment, and creating illustrations for scientific publications. Fanny's culinary knowledge became advantageous in the bacteriological media kitchen. She suggested the use of a Japanese seaweed known as **agar-agar** (referred to as **agar** in the microbiology lab) instead of gelatin to solidify the medium.

Fanny explained that agar-agar was a heat-stable gelling agent used in puddings and jellies that remained solid at higher temperatures. It could be mixed with beef broth, melted at 100°C (212°F), and when allowed to cool it would resolidify at room temperature. Agar-agar was used in Southeast Asia and India as a gelling agent in desserts. Fanny learned about agar-agar when she spent time in New York making puddings with a Dutch neighbor who had emigrated from Java (Indonesia).

Agar is a gelatinous substance derived from red seaweed. It is inert. In other words, microbial enzymes cannot break it down. Walther tried it, and everyone in the lab substituted agar for gelatin to prepare solid bacteriological media. In 1882, the use of agar to prepare media was mentioned in a short communication about the isolation of the bacterium that causes tuberculosis. Agar is still used today as the solidifying agent to prepare solid bacteriological media (**FIGURE 2.5**). It is purchased in powder form.

The Invention of the Petri Dish

Julius Richard Petri (1852–1921) was a military doctor conducting research in Koch's research laboratory in Germany during the 1880s. At the time, bacteriological media (nutrients that bacteria need to grow) containing agar (a solidifying agent) were heated (so that the agar melted) and then poured into round, shallow, glass dishes and placed under a bell jar and allowed to solidify. The plates were then inoculated with bacteria using a sterilized loop to streak the bacteria across the solid medium in order to obtain isolated colonies of bacteria. After inoculation, the open plates were placed under a glass bell jar (**FIGURE 2.6**). After bacterial colonies grew on the plates, the microbiologists would remove the plates for viewing, exposing them to air. Plates were readily contaminated with other microbes from the air, causing frustration when using the bell jar method. Prior to this, all bacteria were grown in broths or slants in test tubes (**FIGURE 2.7**).

Petri had the idea to create a slightly larger lid over the top of the shallow round glass dishes. This method proved to prevent microbial contamination from the air and be an overall better way to contain bacterial cultures. Hence, the **petri dish** (also called a *petri plate*) was born! Petri's invention lives on in that petri dishes are still used today. The disposable polystyrene plates are mass-produced in many different sizes. They are sterilized by either gamma irradiation or in an ethylene oxide gas sterilizer (because autoclaving melts the polystyrene dishes) (**FIGURE 2.8**).

Gram's Timely Stain

Danish physician Hans Christian Gram (1853–1938) arrived in Berlin, Germany, in 1883, to conduct research in the laboratory of the eminent German pathologist and microbiologist Karl Friedlander (1847–1887). Friedlander was interested in isolating the bacteria that caused pneumonia (ironically, he died prematurely at the age of 40 from a respiratory infection likely caused by the bacterium he discovered). Gram was interested in developing a **differential staining** procedure that would detect bacteria present in diseased tissues of the lungs and kidneys. Up to this time, the staining methods used stained the tissues and bacteria equally; there was no visual color distinction between the two. Gram wasted no time. Immediately upon his arrival, he began staining lung tissues from patients who had recently died of pneumonia and lung tissues from experimental animals in attempts to distinguish cocci-shaped bacteria from the lung tissues.

© Teri Shors.

FIGURE 2.5 **(a)** Agar powder. **(b)** Agar powder (close-up). **(c)** Warm bacteriological media cooling on a countertop in a college microbiology media prep kitchen. It was prepared with agar as a solidifying agent. The media is poured into sterile plastic petri dishes and stacked on a countertop. Stacking the plates helps to prevent condensation inside the lids of the plates. **(d)** In honor of Fanny Hesse, consider making dragon fruit agar-agar. This photograph shows dragon fruit agar-agar. However, agar-agar can be combined with any fruit or flavor. Just boil and chill!

© Teri Shors.

FIGURE 2.6 **(a)** The bell jar method used before the petri dish was invented. The bell jar contains a brass stopper and is covering a plate of bacteria.
(b) Left: Close-up of bell jar covering plate of bacteria. Right: Glass petri dish containing solid media.

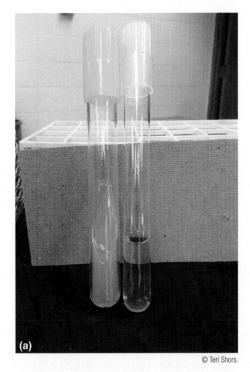

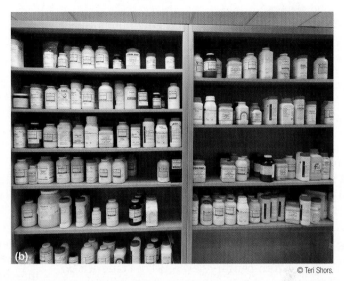

© Teri Shors.   © Teri Shors.

FIGURE 2.7 Bacteriological media. **(a)** Left: Test tube with solid media referred to as a "slant." The media is poured into the test tube and allowed to cool at an angle instead of being upright. In doing so, the media lasts longer and is used to maintain stock cultures of bacteria. It is not conducive to creating pure cultures of bacteria. Right: Test tube containing nutrient broth. **(b)** Many different types of media can be used to cultivate bacteria in pure cultures. The medium is purchased in powder form. It is weighed and then water is added. The medium is then sterilized in an autoclave. This photograph shows a typical college microbiology lab prep kitchen. Many types of commercial media are on the shelving.

It is said that his now famous invention of the **Gram stain** occurred as the result of an accident and his keen observations. Some historians speculate that after Gram added gentian, or **crystal violet**, to the diseased lung tissues, followed by rinsing the unbound stain off with

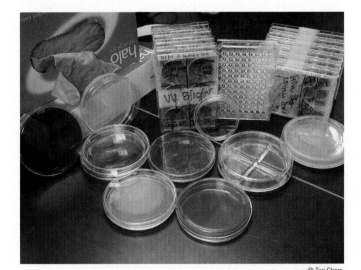

© Teri Shors.

FIGURE 2.8 The invention of the petri dish revolutionized the microbiology laboratory. Dishes of different sizes can be used to prepare solid bacteriological media and also be used to do experiments with viruses (mammalian viruses or bacteriophages) and diagnostic testing such as immunological assays.

water, he accidently spilled Lugol's solution (potassium triiodide solution) onto the diseased lung tissues, which acted as a fixative. In attempts to remove the Lugol's solution with alcohol, he keenly observed that some bacterial cells were resistant to destaining (**decolorization**), which later were described as being **Gram positive**, or **Gram (+)**, and others were not, and thus were considered **Gram negative**, or **Gram (−)**.

Gram applied this method to pure cultures of bacteria. He prepared a liquid smear of rod-shaped bacteria on a slide and dried it over a Bunsen burner (a procedure referred to as **heat fixation** today) before adding the crystal violet solution. The bacteria that he isolated and stained from lung tissues is known today as *Klebsiella pneumoniae*, a Gram (−) rod-shaped bacterium. Around the same time, another scientist isolated Gram (+) cocci-shaped bacteria from pneumonia patients that were thought to be the cause of a **secondary bacterial infection** and that are known today as *Streptococcus pneumoniae*. Gram reported his results in 1884. He did not use a counterstain at the time. Pathologist Carl Weigert (1845–1904) from Frankfurt, Germany, improved on Gram's method a few years later by counterstaining with **safranin** as the final step of the procedure. The Gram stain continues to be a standard procedure used to identify bacteria in medical technology laboratories today (**FIGURE 2.9**).

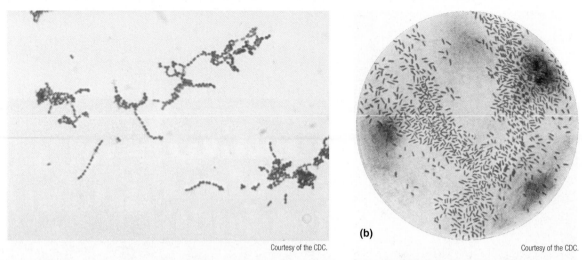

Courtesy of the CDC.

(b)

Courtesy of the CDC.

FIGURE 2.9 **(a)** Gram stain of *Streptococcus pneumoniae* viewed by light microscopy (1,000× magnification). The bacteria are Gram (+) cocci arranged in chains. **(b)** Gram stain of a Gram (−) rod-shaped bacterium, *Haemophilus influenza*, viewed by light microscopy (1,000× magnification).

Phage Biology

During the early 20th century (1915–1917), Frederick W. Twort (1877–1950) and Felix d'Herelle (1873–1949) used **Chamberland porcelain ultrafilters** to independently isolate "ultrafilterable" agents (**FIGURE 2.10**) that could "eat" bacteria. They observed small clear areas, called **plaques** (**FIGURES 2.11** and **2.12**), that appeared on the surface of **bacterial lawns** growing on plates containing solid media when infected with bacteriophages. D'Herelle is credited with coining the term *bacteriophage*, frequently

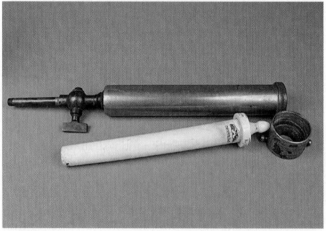

Courtesy of the National Museum of American History.

FIGURE 2.10 Chamberland porcelain ultrafilters were used to retain or trap bacteria on the ultrafilter to purify drinking water. The bacteria were stuck on the ultrafilter and could not pass through it into the drinking water. Pasteur and others, including Twort and d'Herelle, discovered that unfilterable germs or bacteria were retained on the ultrafilter but infectious agents, such as bacteriophages (i.e., bacterial viruses), could pass through the porcelain ultrafilter to the bottom of a flask containing the filtrate (e.g., drinking water). The viruses could not be grown like bacteria in the lab on solid media or in broth cultures, but they could infect living organisms and potentially cause disease. All viruses require a host for replication, such as a bacterium, cell culture, plant, or animal.

shortened to *phage*. He realized the potential therapeutic use of phages as a natural process to control bacterial infections. In the 1930s, the government ordered wells in India to be treated with phages to prevent cholera.

Phage therapy has an almost 100-year history. It might be the answer to the crisis resulting from the misuse of antibiotics in human medicine and on the farm to promote the growth of farm animals. Antibiotic resistance has been cited as a significant factor in emergence and reemergence of microbial diseases. *Each year, an estimated 2.5 million people acquire antibiotic-resistant infections that claim the lives of at least 50,000 people in Europe and the United States alone.* Meanwhile, biologists are constantly on the search for new antibiotics and new therapies, including a renewed interest in phage therapy because the drug pipeline to combat these emerging threats is scarce.

Perhaps had antibiotics not come into widespread use in the 1940s, the early promise of phage therapy might have further developed. However, in Eastern Europe, including Russia, phage research continued over the years, particularly in research institutes in Poland and in Tbilisi, Georgia. Phage therapy was used to successfully treat Russian soldiers suffering from dysentery both during and after World War II. In fact, an **Intestiphage®** cocktail is routinely used today in Georgia to prevent gastrointestinal infections in pediatric hospitals.

Successes in the early years of phage therapy were reported, but so were disappointments and failures. In retrospect, these negative results may well have been due to flawed procedures resulting from the relatively meager knowledge available at the time regarding phage–bacteria interactions. The use of phages has several inherent advantages that warrant continued research into their potential to combat bacterial infections:

o Phages can be targeted to specific bacteria, minimizing disturbance to the body's normal **microbiota** (e.g., patients on antibiotics may develop oral or vaginal

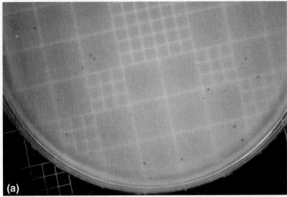

© David B. Fankhauser, PhD, University of Cincinnati, Clermont College.

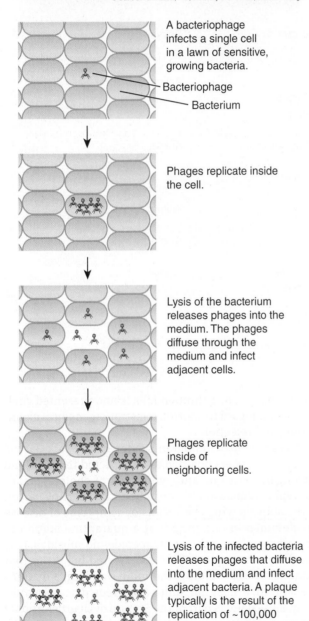

A bacteriophage infects a single cell in a lawn of sensitive, growing bacteria.

— Bacteriophage

— Bacterium

Phages replicate inside the cell.

Lysis of the bacterium releases phages into the medium. The phages diffuse through the medium and infect adjacent cells.

Phages replicate inside of neighboring cells.

Lysis of the infected bacteria releases phages that diffuse into the medium and infect adjacent bacteria. A plaque typically is the result of the replication of ~100,000 phages. The region of visible lysis is called a plaque.

(b)

FIGURE 2.11 (a) A bacterial lawn with plaques (clearings), regions where bacteriophages infected the bacteria and lysed the cells. **(b)** The formation of plaques within a lawn of bacteria present on solid media.

© David B. Fankhauser, PhD, University of Cincinnati, Clermont College.

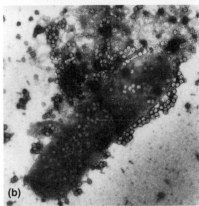

© Biozentrum, University of Basel/Science Source.

FIGURE 2.12 (a) Bacterial cells growing in nutrient broth (top) being lysed (killed) by bacteriophages (bottom). The broth changes from being cloudy or turbid with bacteria to clear (the bacteria are lysed/blown apart resulting in bacterial death). The bacteria burst due to the phage infection. **(b)** Antibiotic effect on bacterium. Colored transmission electron micrograph of the effect of an antibiotic drug on the bacteria Salmonella typhimurium. The drug polymyxin B caused the outer membrane (blue) to form long surface projections. This is the early stage of lysis, the bursting of the cell wall and release of the cell contents, killing the bacterium. Polymyxin B is a broad spectrum antibiotic against Gram-negative bacteria. Salmonella typhimurium is a major cause of food poisoning in humans. The bacteria are transmitted in contaminated food, causing fever and diarrhea.

thrush, a painful condition caused by yeasts that are part of the normal microbiota of the vagina or mouth because of their overgrowth due to antibiotic-induced suppression of the normal microbiota).

- Phages are easily grown and purified and are relatively inexpensive.
- Small doses can be used because their numbers increase exponentially as they spread from bacterium to bacterium.
- Toxicity to humans is not an issue because phages invade bacterial cells, not human cells.
- The replication and activity of phages introduced into the body are self-limited. Once their bacterial targets are destroyed, they gradually disappear from the body.

In recent years, researchers have reported on promising new phage-based approaches to combat bacterial infections as indicated in the Case Study for this chapter and in **TABLE 2.2**.

TABLE 2.2 Recent Preclinical (Animal) and Human Clinical Phage Therapy Studies

Disease Symptoms	Causative Bacterial Pathogen	Type of Study (Preclinical/Animal or Clinical/Human)
Diarrhea	*Escherichia coli*	Human
	Shigella sonnei	Animal
Otitis (ear infection)	*Pseudomonas aeruginosa*	Animal and human
Venous leg ulcers	*Pseudomonas aeruginosa*	Human
	Staphylococcus aureus	
	Escherichia coli	
Endocarditis (inflammation of the inner lining of the heart and heart valves)	*Pseudomonas aeruginosa*	Animal
Peritonitis (inflammation of the inner wall of the abdomen that covers the abdominal organs)	*Escherichia coli*[a]	Animal
	Staphylococcus aureus[a]	
	Pseudomonas aeruginosa	
Respiratory infections	*Pseudomonas aeruginosa*	Animal
	Esherchia coli	
	Klebsiella pneumoniae	
Bacterial meningitis	*Esherichia coli*	Animal
Osteomyelitis	*Staphylococcus aureus*	Animal
Skin infections	*Staphylococcus aureus*[b]	Animal
	Klebsiella pneumoniae	

[a] Use of genetically engineered or synthetic bacteriophages.
[b] Use of genetically engineered and natural bacteriophages.
Information from D. R. Roach & L. Debarbieux. (2017). Phage therapy: Awakening a sleeping giant. *Emerging Topics, 1,* 93–103.

Ellis Island Federal Immigration Station and the 6-Second Medical Examination

The United States is often called "a nation of immigrants." An estimated 100 million Americans can trace their family origins to Ellis Island. **Ellis Island**, situated in New York Harbor, opened its doors as a **federal immigration station** in 1892. As the doors opened, the field of medicine also was changing. In this post–Civil War era, medical training for doctors was becoming standardized and measures to control infectious diseases were improving. The germ theory of disease was just starting to attain credibility in the United States.

The main building on Ellis Island burned down in 1897 and was reconstructed with bricks. The new, fireproof, main building opened in 1900. Shipload after shipload of people, totaling approximately 12 million immigrants came through Ellis Island, mostly Europeans. The highest number of immigrants processed on a single day was 11,747 on April 17, 1907. The sheer number of immigrants passing through Ellis Island presented challenges with regard to control, registration, medical examination, and rejection.

As each ship of immigrants entered New York Harbor, quarantine officers boarded and inspected the crew and passengers for signs and symptoms of cholera, smallpox, typhus, yellow fever, and bubonic plague. Any person on the ship showing overt signs or symptoms of disease was detained (**quarantined**) at a quarantine station in the harbor until the threat of an epidemic subsided. Ellis Island was a complex of interconnected buildings that included a ferry building, the main building, a quarantine station, a hospital, and contagious disease wards. Today, Ellis Island is part of the Stature of Liberty National Monument and can be toured (https://www.nps.gov/elis/index.htm).

Passengers who were not detained at the quarantine station were ferried to the main building on Ellis Island (**FIGURE 2.13**). Immigrants, along with their luggage, children, and extra layers of clothing, were steered into the

Courtesy of the Library of Congress.

Courtesy of The New York Public Library.

FIGURE 2.13 (a) Immigrants cleared as being free of disease on the boat are ferried to Ellis Island. Notice the crowded ferry, c. 1915–1920. **(b)** Ferry of immigrants docked at the U.S. Immigration Station for registration and examination, c. 1902–1913.

high-ceilinged main building (**FIGURE 2.14**). They left their luggage on the first floor and were guided to the second floor to the Great Hall or Registry Room, where they underwent a **6-second medical examination** (**FIGURE 2.15**). The medical line inspectors would look at their feet, legs, body, hands, arms, face, eyes, and head for deformities. If the immigrant was healthy, he or she was steered into the line to the waiting room. If the immigrant was "chalk-marked" by a line doctor, a more rigorous examination was performed. The following are examples of some of the chalk marks that were used:

- "X" on the front side of the right shoulder for a mental defect
- "X" further down the right shoulder for disease or deformity
- "X" within a circle signified definite disease

- "E" for eye defects or **trachoma** (a bacterial chlamydia infection that can cause blindness; although not fatal, at the time, it was considered "dangerous contagion")
- "H" for heart defects
- "B" for back problems
- "G" for struma (thyroid problem or goiter)
- "F" for favus (ringworm) on the scalp
- "L" for lice on the scalp
- "Pg" for pregnancy

About 15% to 25% of immigrants were chalk-marked for further examination. Of those immigrants who were detained, 80% had trachoma. X-rays and stethoscopes were not used in the line inspections but were routinely used during examinations of chalk-marked immigrants after

Courtesy of the Library of Congress.

Courtesy of The New York Public Library.

FIGURE 2.14 (a) Immigrants on a steamboat were transported to Ellis Island on a barge (instead of a ferry). They are walking on the boardwalk toward the high-ceilinged main building on Ellis Island. The large building in the background is the hospital. **(b)** Immigrants seated on long benches on the first floor of the main building, U.S. Immigration Station, c. 1902–1913.

Courtesy of the Library of Congress.   Courtesy of the Library of Congress.

FIGURE 2.15 **(a)** Immigrants waiting on the second floor of the Main Hall for examination, c. 1907–1921. **(b)** Physical examination of the eyes of a female immigrant at Ellis Island in 1911.

1910. Chalk-marked immigrants went through more rigorous examinations in small rooms on the second floor or the Great Hall.

Most immigrants passed the initial inspection and were ushered to the registration table. They answered questions about their origin, destination, and the amount of money in their pocket. They were given a landing card to American soil (**FIGURE 2.16**). Those immigrants who were detained in dormitories, including pregnant women, were treated at local hospitals or at the Ellis Island general hospital. From the 1890s through 1914, less than 1% of detained immigrants required hospitalization. The medical examinations were set up to yield results rather quickly. In reality, the practice of "keeping the line moving" was not the best screening method. Trachoma, a nonfatal disease, caused more deportations than infectious diseases such as syphilis or tuberculosis, which in reality posed a greater public health threat.

Courtesy of the Library of Congress.

FIGURE 2.16 Immigrants walking across the pier from the bridge, leaving the Great Hall.

Discovery of Antibiotics

The first "wonder drugs" were the **sulfonamide (sulfa) drugs**. These drugs, although they are antimicrobials, are not antibiotics because they are synthetic compounds, not products of microbes. Sulfa drugs saved millions of lives in World War II; when sulfa drugs were not available, medics were taught to sprinkle sulfur on wounds sustained on the battlefields.

The story of the discovery of **penicillin** centers on the observations of Alexander Fleming (1881–1955). Others had previously described the antibacterial properties of the *Penicillium* mold, but it was Fleming who followed through on a serendipitous (chance) observation. (Serendipity has been a significant factor in several important scientific discoveries. Pasteur wrote, "*In the field of observation, chance favors only the prepared mind.*" In other words, it is not just sheer luck but rather the ability to recognize the significance of the unexpected.)

Fleming was studying staphylococci and left a petri dish streaked with this organism on his lab bench while he went away on a 2-week vacation. On returning, he noted that the plate was contaminated with a common (*Penicillium*) mold and that the staphylococci failed to grow only in the vicinity of the mold; the mold had produced an inhibitory substance—penicillin (**FIGURE 2.17**). Fleming was a humble person, later stating, "*Nature created penicillin. I only found it.*"

Penicillin's therapeutic potential was not fully investigated until several years after its discovery, when Ernst Chain, Howard Florey, Edward Abraham, and Norman Heatley produced purified penicillin and successfully cured eight mice that had been injected with fatal doses of bacteria. The antibiotic era was ushered in with the first use of penicillin in humans. On February 12, 1941, Police Constable Robert Alexander of Oxford, England, was the first person in

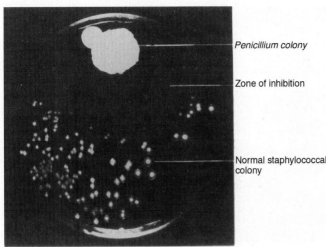

Penicillium colony

Zone of inhibition

Normal staphylococcal colony

© National Library of Medicine.

FIGURE 2.17 Alexander Flemings' petri dish showing the inhibition of staphylococcal colonies in the immediate vicinity of a mold contaminant known as *Penicillium*. The *Penicillium* produced penicillin, which killed the staphylococci.

the world to receive penicillin (**FIGURE 2.18**). Alexander was seriously ill with a staphylococcal infection that started with a small sore at the corner of his mouth. Despite treatment with sulfonamides, the staphylococci spread uncontrollably into his bloodstream, resulting in **sepsis** and numerous abscesses over his body and the spread of infection to the

Courtesy of National Library of Medicine.

FIGURE 2.18 Early equipment and materials used to cultivate *Penicillium* and prepare penicillin in the laboratory.

rest of his face, eyes, and scalp, necessitating removal of his left eye. Death seemed imminent.

Miraculously, 24 hours after receiving penicillin he was much improved—his lesions showed signs of healing, his elevated body temperature dropped toward normal, and within several days his right eye was almost normal. Unfortunately, the small amount of penicillin that was available was insufficient for continued treatment. In a heroic effort to save the patient's life, doctors extracted penicillin from his urine and injected it back into his bloodstream. But the microbes gained the upper hand, and Alexander's condition deteriorated. He died on March 15, 1941.

During the next few weeks, more penicillin was purified and used to treat five more patients, four of whom were children with streptococcal or staphylococcal infections that had not responded to sulfa drugs or surgery. All five patients had a grim prognosis. Only two out of five died, and no toxicity was reported. More human trials were initiated and proved to be very successful. World War II triggered the large-scale production of penicillin, saving thousands of soldiers' lives. By 1944, supplies of penicillin were abundant and were released for the civilian population. In 1945, Fleming was awarded the Nobel Prize in Physiology or Medicine, along with Chain and Florey, who helped develop penicillin into a widely available medicinal product.

The Modern Era of Microbiology

The field of microbiology developed quickly in the 20th century. The scientists who delineated the structure of DNA and proved that it is the hereditary material of living cells used bacteria and bacteriophages in their experiments. Many new powerful laboratory techniques also became available, such as the use of cell cultures to cultivate viruses that cause disease in humans, tools enabling the manipulation of DNA, and the sequencing and mapping of microbial and viral genomes.

Cell Culture Paved the Way for the Development of Poliovirus Vaccines

The field of biology has numerous examples of where basic research has led to major advances in biology and medicine. The research of John Enders, Thomas Weller, and Frederick Robbins is one such case. Their 1959 discovery that poliovirus could be grown in cell cultures of a variety of nonnerve tissues opened the door for research on virus vaccines, including one for poliovirus (**FIGURE 2.19**).

Previously, poliovirus could be grown only in cultures of human embryonic brain or nerve tissue, which was difficult to obtain and to culture, imposing severe limitations on scientists' research. Cell cultures became the host system used to cultivate polioviruses for the development and preparation of poliovirus vaccines by Jonas Salk and Albert Bruce Sabin. Poliovirus is close to being eradicated from the face of Earth. In the United States and across

Sven Hoppe/ShutterStock, Inc.

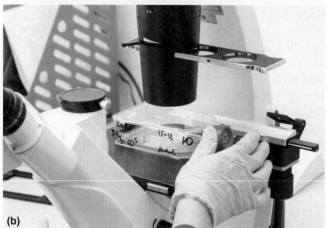

sruilk/ShutterStock, Inc.

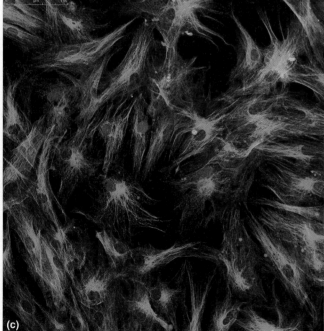

Vshivkova/ShutterStock, Inc.

FIGURE 2.19 **(a)** Cultivation of cell cultures in the laboratory, which are used in research on viral diseases and development of vaccines. **(b)** Scientist using an inverted microscope to observe tissue culture cells. **(c)** Fibroblasts labeled with fluorescent dyes and viewed with a confocal microscope.

Courtesy of the CDC/Stafford Smith.

FIGURE 2.20 Two workers are standing next to a cart loaded with boxes of poliovirus vaccines. During the early 1950s, more than 20,000 cases of polio were reported in the United States each year. Once vaccination against poliovirus began in 1955, the number of cases dropped dramatically, falling to 3,000 cases in 1960 and only 10 cases in 1979.

the world, summers before 1960 were seasons of dread because of the fear that polio might strike (**FIGURE 2.20**). The following is excerpted from the presentation speech by S. Gard, member of the Staff of Professors of the Royal Caroline Institute:

> Your Majesties, Your Royal Highnesses, Ladies and Gentlemen.
>
> The principles of cultivation of bacteria were laid down in the late 1870s by Robert Koch. Since that time the bacteriologists could study systematically the diseases caused by bacteria, isolate the causative agents in pure culture, and make themselves familiar with their nature. With the aid of the culture technique they . . . could produce therapeutic sera and prophylactic vaccines. . . .
>
> Turning to the virus diseases we meet an entirely different picture . . . many virus diseases are on the increase, a tendency particularly evident in poliomyelitis. . . . Poliomyelitis in this country is now responsible for almost one fifth of all deaths from acute infections. . . .
>
> It is not difficult to find the reason why the virologists have failed where the bacteriologists were so successful. . . . Unlike bacteria and other microorganisms, virus is incapable of multiplying in artificial lifeless culture media.
>
> Then, in 1949 there appeared from a Boston research team a paper, modest in size and wording but with a sensational content. John Enders . . . and his associates Thomas Weller and Frederick Robbins reported the successful cultivation of the poliomyelitis virus in test-tube cultures of human tissues. A new epoch in the history of virus research had started. . . .
>
> . . . Other scientists had previously attacked the problem with very moderate success. It was generally held that the final word had already been said by Sabin and Olitsky who in 1936 tried to grow the virus in Maitland cultures

of various tissues from chick embryos, mice, monkeys, and human embryos. Their results remained completely negative except in cultures of human embryonic brain tissue in which the virus at least seemed to maintain its activity. These findings were taken as a definitive confirmation of the accepted concept of the virus as a strictly neurotropic agent, i.e., capable of multiplying in nerve cells exclusively. Accordingly, the hopes of a practicable method for the cultivation of the poliomyelitis virus were temporarily shelved. Of all tissues, nerve tissue is the most specialized, the most exacting, and consequently the most difficult to cultivate. As, at that, there seemed to be no alternative to the use of human brain tissue, the general resignation is easily understood.

. . . Enders, Weller, and Robbins decided to repeat Sabin and Olitsky's experiment with an improved technique. In their first experiments they used human embryonic tissue. To the great surprise of everybody . . . they registered a hit in their first attempt. The virus grew not only in brain tissue but equally well in cells derived from skin, muscle, and intestines. Furthermore, in connection with the multiplication of the virus, typical changes appeared in the cellular structure, finally leading to complete destruction, easily recognizable under the microscope. This observation furnished a convenient method of reading the results. . . . Enders, Weller, and Robbins found that . . . all tissues except bone and cartilage seemed to be equally suitable. Finally they tried to isolate the virus from various specimens directly in tissue cultures. This was likewise achieved. In the latter observation probably the greatest practical importance of their discoveries is to be found. The virologists finally had a tool in the same class as the culture technique of the bacteriologists.

Dr. John Enders, Dr. Frederick Robbins, Dr. Thomas Weller. Karolinska Institute has decided to award you jointly the Nobel Prize for your discovery of the capacity of the poliomyelitis virus to grow in test tube cultures of various tissues. Your observations have found immediate practical application on vitally important medical problems, and it has made accessible new fields in the realm of theoretical virus research.[1]

Discovery of the Structure of DNA and the Proof of DNA as the Genetic Material

The race to discover the structure of **deoxyribonucleic acid (DNA)** is a story of collaboration and competition that began in 1951 at King's College, London (**FIGURE 2.21**). Rosalind Franklin (1920–1958), a physical chemist trained in the technique of **x-ray crystallography**, began doing research in Maurice Wilkin's (1916–2004) laboratory. She was highly skilled, very thorough, and reluctant to

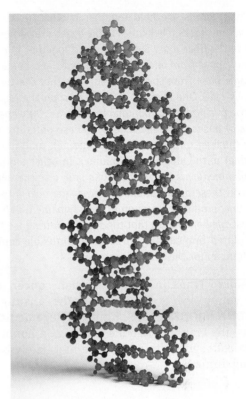

© Mopic / ShutterStock, Inc.

FIGURE 2.21 Highly detailed model of the structure of DNA.

share data until she was 100% confident in her results. Franklin's research was focused on determining the structure of DNA fibers.

Rosalind's competition was James Watson (1928–), along with Francis Crick (1916–2004). Watson was in a hurry to earn fame and greatness, which he knew would happen if he were the first to unravel the structure of DNA. Watson referred to Rosalind Franklin as bad-tempered "Rosy." Wilkins was frustrated with Franklin's secretive nature of data collection. Feeling left out, without Franklin's permission, he provided Watson and Crick with the x-ray photograph (#51) that contained the precise measurements in determining the helical structure of DNA. Watson and Crick's intuition quickly discerned that DNA was a **double helix** and that the two strands of DNA were **antiparallel**. In April 1953, the peer-reviewed scientific journal *Nature* published Watson and Crick's article, "A Structure for Deoxyribose Nucleic Acid." The article took up slightly more than one page of the journal, yet it became the benchmark of molecular biology. What follows are excerpts from their article:

We wish to suggest a structure for the salt of deoxyribose nucleic acid (DNA). . . . The structure has two helical chains each coiled round the same axis. . . . Both chains follow right-handed helices, but the two chains run in opposite directions. . . . The bases are on the inside of the helix and the phosphates on the outside. . . . The novel feature of the structure is the manner in which the two chains are held together by the purine and pyrimidine bases. The planes of the bases are perpendicular to the fiber axis. They are joined together in pairs, a single base from one chain being hydrogen-bonded to a single base from the other chain, so that the two lie side by side. One of the pair must be a purine and the other a pyrimidine for bonding to occur. . . . Only specific pairs of bases can bond together. These pairs are adenine (purine) with thymine (pyrimidine), and guanine (purine) with cytosine (pyrimidine). In other words, if an adenine forms one member of a pair, on either chain, then on these assumptions the other member must be thymine; similarly for guanine and cytosine. . . . It has not escaped our notice that the specific pairing we have postulated immediately suggests a plausible copying mechanism for the genetic material. . . .[2]

Rosalind Franklin died of ovarian cancer in 1958 at the young age of 37. She was the unsung heroine of DNA (**FIGURE 2.22**). King's College dedicated a building, the Franklin–Wilkins building, in honor of her and the colleague with whom she had barely been on speaking terms.

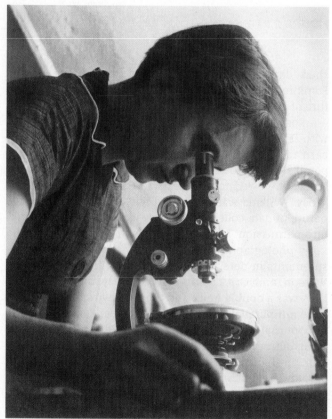

© Science Source.

FIGURE 2.22 Rosalind Franklin.

---

[2]From J. D. Watson and F. H. C. Crick. (1953). A structure for deoxyribose nucleic acid. *Nature, 171,* 737–738.

The proof that DNA is the genetic material was based on research using bacteria and bacteriophages. Ask students who discovered DNA and its significance and the likely answer will be James Watson and Francis Crick. DNA, however, was first isolated in 1869, by Friedrich Meischer from fish sperm and from puss in wounds. At the time there was no hint that "nuclein" (DNA) would prove to be the repository of genetic information; biologists had begun to speculate on the nature of the genetic material, but all bets were on protein.

Eighty-one years after Meischer, three classical experiments, all using microbes, established DNA as the genetic material, the "blueprint" of life. In 1928, Frederick Griffith (1879–1941), an English bacteriologist, was investigating the role of streptococci in pneumonia. Griffith observed two strains of *Streptococcus pneumoniae*, each with distinctive characteristics. Cells of one strain, the **S strain of S. pneumoniae**, are encapsulated, produce "smooth" (S) and raised colonies, and are **virulent** (cause pneumonia and death) in mice. In contrast, cells of the other strain, the **R strain of S. pneumoniae**, exhibit "rough" (R) and flat colonies on solid media because the colonies are not encapsulated, and are **avirulent** (do not cause disease) in mice (**FIGURE 2.23**).

These characteristics serve as markers of differentiation. Griffith observed, as expected, that heat-killed S bacteria (incapable of growth) did not cause pneumonia and death in mice. Surprisingly, however, when a mixture of heat-killed S bacteria and viable avirulent R streptococci was injected into mice, the mice developed pneumonia and died. *By some mechanism the R cells were transformed into virulent S-like cells.* Although Griffith was not able to explain the process by which the R cells picked up an unidentified factor from the S cells (this factor was named the **transforming principle**), *it was the first demonstration of the transfer of genetic information.*

Sixteen years later, in 1944, Oswald Avery, Colin MacLeod, and Maclyn McCarty identified DNA as the "active" factor in Griffith's transforming principle (**FIGURE 2.24**). They divided the **carbohydrate**, **protein**, **nucleic acid** (DNA + RNA), and other components of the transforming principle from the S cells into separate fractions. To each fraction, R (avirulent) cells were added and injected into mice. The result was that only the nucleic acid fraction with intact DNA mixed with the avirulent R cells established virulence. Confirmation experiments using the enzyme **DNase** to degrade DNA negated the virulence of the R cell–nucleic acid mixture (RNA has no effect on virulence), establishing DNA as the genetic material. However, some scientists still held to the idea that protein, not DNA, was the genetic material. Their logic was that proteins consist of sequences of 20 different amino acids, allowing for greater diversity than that possible with a four-nucleotide alphabet.

Finally, in 1952, Alfred Hershey (1908–1997) and Martha Chase (1927–2003) confirmed DNA as the genetic material in an experiment using bacteriophages and

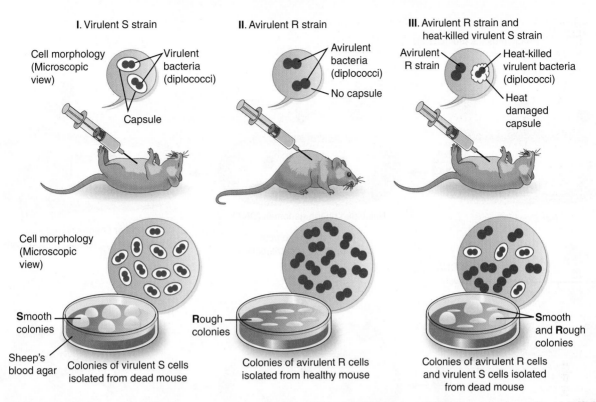

FIGURE 2.23 Characteristics of S and R strains of *Streptococcus pneumoniae* used in the Griffith experiment (1928). *S. pneumoniae* are Gram (+) **diplococci** (cocci arranged in pairs). A mixture of heat-killed S bacteria and viable avirulent R streptococci was injected into mice; the mice developed pneumonia and died. By some mechanism the R cells were transformed into virulent S-like cells. The mechanism was determined by Avery, MacLeod, and McCarty's experiments in 1944.

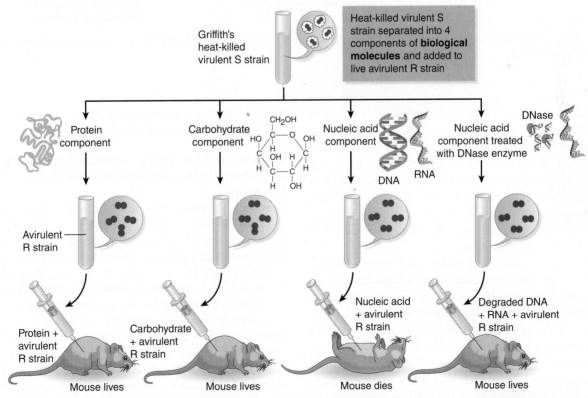

FIGURE 2.24 The heat-killed S strain of *Streptococcus pneumoniae* was fractionated into different components of biological molecules: protein, **carbohydrates**, and nucleic acids (RNA + DNA). Another fraction containing nucleic acids (DNA + RNA) was treated with DNase, an enzyme that digests DNA. The fractions mixed with the avirulent R strain and the untreated nucleic acid component were injected into mice and the mice died. Intact DNA from the heat-killed S strain transformed the avirulent R strain of *S. pneumoniae* into a virulent S strain, causing pneumonia in mice and their death. In this 1944 experiment, Avery, MacLeod, and McCarty discovered the transforming principle that turned an avirulent bacterial strain into a virulent one was DNA.

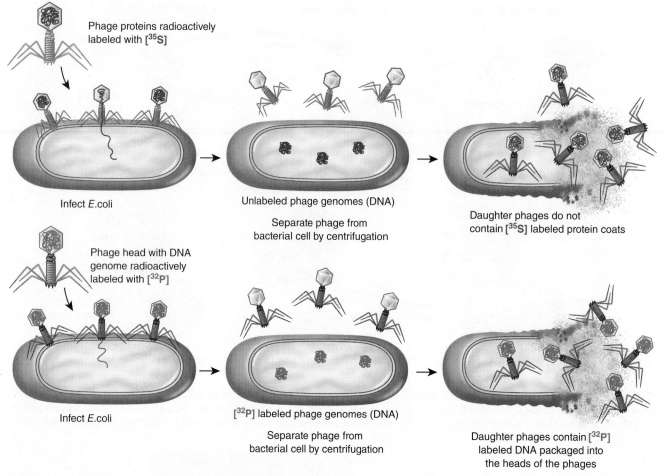

Phage proteins radioactively labeled with [^{35}S]

Infect *E.coli*

Unlabeled phage genomes (DNA)

Separate phage from bacterial cell by centrifugation

Daughter phages do not contain [^{35}S] labeled protein coats

Phage head with DNA genome radioactively labeled with [^{32}P]

Infect *E.coli*

[^{32}P] labeled phage genomes (DNA)

Separate phage from bacterial cell by centrifugation

Daughter phages contain [^{32}P] labeled DNA packaged into the heads of the phages

FIGURE 2.25 Hershey and Chase (1952) confirmed DNA as the genetic material. Progeny phages contained [^{32}P] radioactively labeled DNA genomes packaged in the phage heads.

Escherichia coli (**FIGURE 2.25**). The nucleic acid is enclosed within the protein coat within the phage. The phage then replicates within E. coli, resulting in new phage units identical to the original infecting bacteriophage. In the Hershey–Chase experiment the phage DNA was labeled with radioactive phosphorous and the phage proteins with radioactive sulfur to determine whether it was the proteins or the nucleic acid within the protein coat that directed the phage replication.

Because radioactive DNA and not radioactive protein was found in the daughter phage, Hershey and Chase conclusively demonstrated that the DNA carried the genetic information, resulting in the synthesis of new and identical copies of the phage. Their investigations established DNA as the genetic material. Concurrently (in 1952), Joshua Lederberg (1925–2008) coined the term **plasmid** as an independent DNA molecule that passes on traits (the transforming principle in Griffith's experiments). It took 20 years for scientists to recognize the importance of plasmids. Over time, plasmids have impacted many areas of biological research and the development of molecular biology methods. **Molecular biology** is the study of DNA,

RNA, proteins, and other organic molecules involved in the functioning of cells (and viruses, too!).

Restriction Enzymes, Genetic Engineering, DNA Sequencing, and Polymerase Chain Reactions

By the time Watson and Crick worked out the structure of the DNA molecule in 1953, the significance of DNA as the genetic material had been established. The field of bacterial genetics began to flourish. Many technical developments in the laboratory were made, including advances in **transmission electron microscopy (TEM)** and **scanning electron microscopy (SEM)**, to study the structures of cells and viruses. TEM transmits a narrow beam of electrons through a prepared ultra-thin specimen, creating detailed images of viruses, cells, and parts of cells. TEM has a resolving power of about 0.2 nanometer (nm), which is about 1,000 times greater than the resolving power of a light microscope, which resolves in the 0.2-micrometer (μm) range. It was first used in botany, but by the 1960s it was being used in clinical microbiology to identify human viruses such as poxviruses (e.g., variola virus, which causes smallpox) and visualizing the ultrastructure of

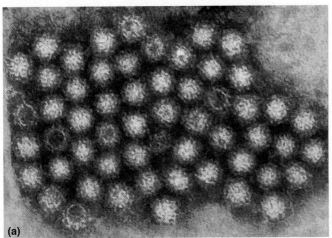

Courtesy of the CDC/Charles D. Humphrey.

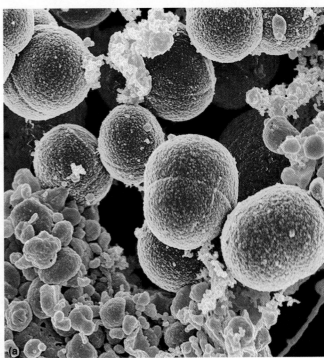

(a)

Courtesy of the CDC/Frank DeLeo, National Institutes of Allergy and Infectious Diseases.

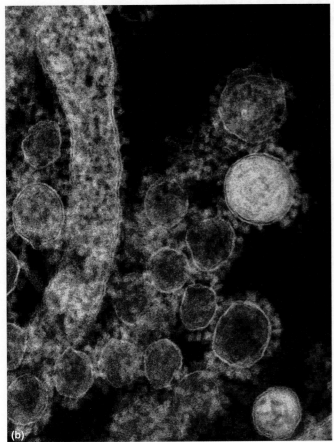

(b)

Courtesy of the CDC/The National Institutes of Allergy and Infectious Diseases.

FIGURE 2.26 Transmission electron micrographs. **(a)** Digitally colorized TEM of noroviruses magnified ~100,000×. **(b)** Digitally colorized TEM of Middle East respiratory syndrome-coronaviruses (MERS) magnified ~100,000×.

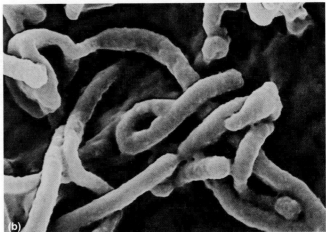

(b)

Courtesy of the CDC/The National Institutes of Allergy and Infectious Diseases.

FIGURE 2.27 Scanning electron micrographs (SEMs). **(a)** Digitally colorized SEM of mustard-colored, coccus-shaped *Staphylococcus aureus* bacteria attempting to escape destruction by blue-colored white blood cells. Magnification ~ 20,000×. **(b)** Digitally colorized SEM of Ebola viruses budding from the surface of VERO cells (an African green monkey kidney epithelial cell line). Magnification ~25,000×.

different types of human tumor cells. It is especially useful for the visualization of viruses that cannot be cultivated in the laboratory, such as noroviruses (**FIGURE 2.26**). SEM is a form of electron microscopy in which the beam of electrons scans the surface of specimen, resulting in a three-dimensional image of a specimen (**FIGURE 2.27**).

The discovery of **restriction enzymes** made possible genetic engineering and the development of DNA sequencing. Restriction enzymes act as molecular scissors to recognize and cut specific short stretches of nucleotides in DNA. The sequences recognized by the restriction enzymes are called **palindromes** because the nucleotide bases have the same sequence on both DNA strands when read forward or backward. Restriction enzymes are used to create **recombinant DNA molecules**, DNA molecules containing "foreign DNA fragments or **foreign genes**"

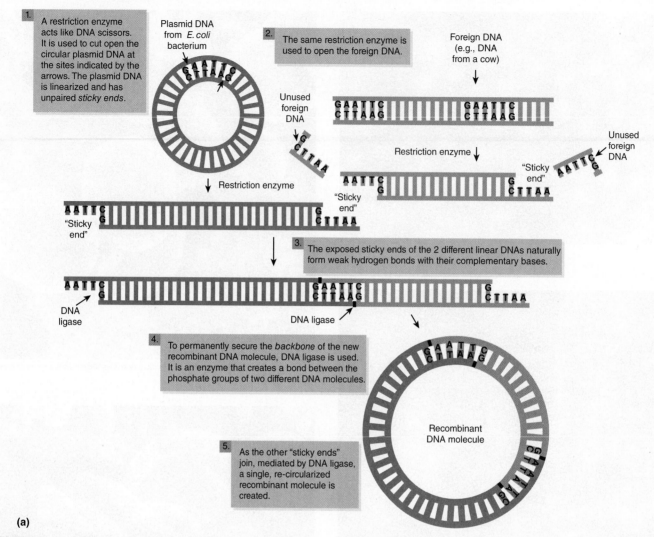

1. A restriction enzyme acts like DNA scissors. It is used to cut open the circular plasmid DNA at the sites indicated by the arrows. The plasmid DNA is linearized and has unpaired *sticky ends*.

Plasmid DNA from *E. coli* bacterium

2. The same restriction enzyme is used to open the foreign DNA.

Foreign DNA (e.g., DNA from a cow)

Unused foreign DNA

Restriction enzyme

Unused foreign DNA

Restriction enzyme

"Sticky end"

"Sticky end"

"Sticky end"

3. The exposed sticky ends of the 2 different linear DNAs naturally form weak hydrogen bonds with their complementary bases.

DNA ligase

DNA ligase

4. To permanently secure the *backbone* of the new recombinant DNA molecule, DNA ligase is used. It is an enzyme that creates a bond between the phosphate groups of two different DNA molecules.

Recombinant DNA molecule

5. As the other "sticky ends" join, mediated by DNA ligase, a single, re-circularized recombinant molecule is created.

(a)

FIGURE 2.28 (a) Construction of recombinant DNA molecules.

from two or more organisms (**FIGURE 2.28A**). Recombinant DNA molecules are used in genetic engineering to create commercial products such as human insulin from bacteria. The bacteria produce human insulin, which is then purified and used by patients suffering from diabetes (**FIGURE 2.28B**).

Fifteen years after the elucidation of the structure of DNA, the first methods were developed to determine the sequence of DNA molecules (**DNA sequencing**). These discoveries have led to DNA sequencing becoming an automated, rapid, and routine laboratory procedure. DNA sequencing has made it possible to determine the mechanisms by which some genes function.

Kary B. Mullis invented **polymerase chain reaction (PCR)** during the 1980s. PCR transformed research in the biological sciences and diagnostic microbiology. PCR is a technique that allows for the rapid synthesis of DNA molecules in a test tube that contains **template DNA**, **primers** (short fragments of single-stranded DNA), free nucleotides, and the enzyme **DNA polymerase**. PCR radically transformed research and development in the biological sciences, enabling endeavors such as the Human Genome

Project, the development of newer-generation vaccines, and diagnostic microbiology. Today, bacterial or viral infections can be identified through PCR from a very small clinical sample containing nucleic acids. PCR can also be used to screen for genetic disorders and in forensic applications such as analyzing traces of blood, semen, or hair from a crime scene in order to identify victims or crime suspects (**FIGURE 2.29**).

Medical Microbiology and the Future of Medicine

Medical microbiology got its start when bacteriological staining methods such as Gram staining became available and progressed to incorporating robotic instrumentation systems that were originally introduced to detect life on Mars (e.g., **Vitek systems** used to identify bacterial pathogens and screen for their susceptibility to antibiotics used in clinical laboratories).

Today, medical microbiology is at a crossroads. Medicine is approaching a radical change in diagnostics

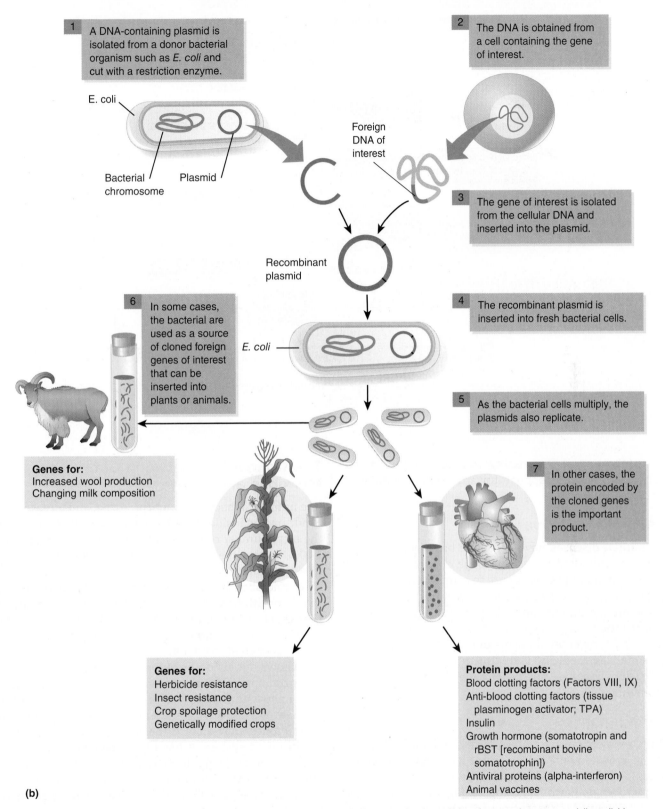

1 A DNA-containing plasmid is isolated from a donor bacterial organism such as *E. coli* and cut with a restriction enzyme.

2 The DNA is obtained from a cell containing the gene of interest.

E. coli

Bacterial chromosome

Plasmid

Foreign DNA of interest

3 The gene of interest is isolated from the cellular DNA and inserted into the plasmid.

Recombinant plasmid

6 In some cases, the bacterial are used as a source of cloned foreign genes of interest that can be inserted into plants or animals.

E. coli

4 The recombinant plasmid is inserted into fresh bacterial cells.

5 As the bacterial cells multiply, the plasmids also replicate.

Genes for:
Increased wool production
Changing milk composition

7 In other cases, the protein encoded by the cloned genes is the important product.

Genes for:
Herbicide resistance
Insect resistance
Crop spoilage protection
Genetically modified crops

Protein products:
Blood clotting factors (Factors VIII, IX)
Anti-blood clotting factors (tissue plasminogen activator; TPA)
Insulin
Growth hormone (somatotropin and rBST [recombinant bovine somatotrophin])
Antiviral proteins (alpha-interferon)
Animal vaccines

(b)

FIGURE 2.28 **(b)** Basic overview of genetic engineering, a method for inserting foreign genes into bacteria in order to produce commercially available products.

and therapy. *Identifying the infectious agent causing a disease is only half of the problem.* We can no longer ignore the response of a patient's immune system toward a particular pathogen and how the pathogen changes the patient's immune system. Further complicating the

nature of infectious diseases are **emerging** (**new**) and **reemerging pathogens**, along with an increase in the number of patients who are **immunocompromised** and susceptible to hospital-associated infections. The future of medical microbiology will involve identifying

The Basics of Polymerase Chain Reaction (PCR): Synthesizing DNA in a Test Tube

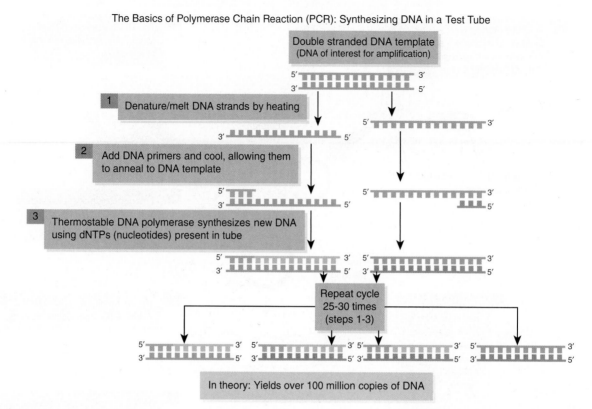

FIGURE 2.29 The PCR process.

the pathogens and assessing changes related to the patient's immune response toward and/or modulated by the pathogen.

Genomic Medicine

All human diseases have a genetic basis, whether the genes are inherited or occur through changes caused by the environment (e.g., exposure to carcinogens that alter one's genes, increasing a person's cancer risk). Sequencing of the human genome and all major pathogens is beginning to impact diagnosis, treatment, and prevention of diseases. We have entered the era of personalized medicine. **Personalized medicine** is treatment uniquely developed for each patient based on genetic, diet, lifestyle, and environmental factors.

Metagenomics

Only a very small fraction of microbes found in nature can be grown in pure culture in the laboratory. Therefore, our knowledge and understanding of the genetic diversity of microbes on Earth is lacking. **Metagenomics** is a new approach toward the assessment of **microbial diversity**. It involves isolating DNA from environmental samples and sequencing it, without any attempt to culture the microbes in the sample. Metagenomics has allowed researchers to begin to catalog the microbial diversity present in soil, aquatic environments, and human body (e.g., gut, skin, and oral microbiota) and has resulted in the discovery of new and reemerging pathogens.

The Connection Between Gut Microbes and Health

The number of bacteria, fungi, protozoa, and viruses that reside on and inside of the human body vastly outnumber the total number of human cells in the body. More than 1,000 different species of microbes live inside (e.g., gastrointestinal tract) and on the surface of the body. The majority of microbes are **commensal** and **symbiotic microbes** that live with us without causing harm. Indeed, many of these bacteria benefit human health. The microbes have coevolved with humans and their **adaptive immune system**, lifestyles, and diet.

Knowledge about the human **microbiome** is strengthening our appreciation that microbes make major contributions toward human health and well-being. From the moment we are born, we are inhabited and surrounded by microbes that dictate the healthy development of the immune system. A few fun facts about the human microbiome are listed in **FIGURE 2.30**. Low diversity of gut microbes is associated with a plethora of diseases, including obesity, inflammatory diseases (e.g., arthritis, irritable bowel disease, and asthma), diabetes, hypertension (high blood pressure), and even neuropsychiatric disorders such as Parkinson's disease and anxiety disorders.

Fecal microbiota transplantation, the transfer of feces from a healthy donor to a patient suffering from an altered gut microbiome, has been successful in restoring health in patients with recurrent *Clostridium difficile (C. diff.)* **infections**. Who would have thought that swallowing a poo

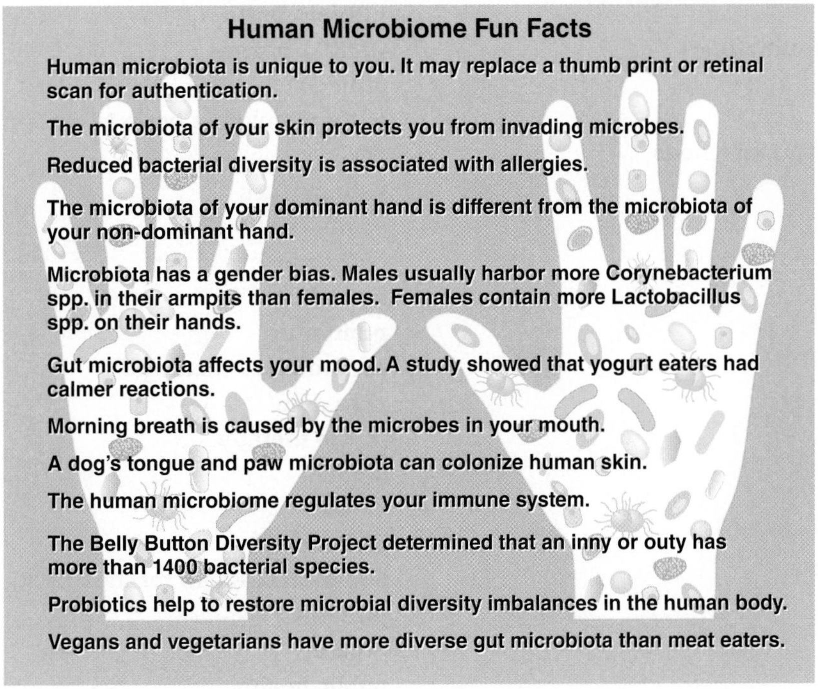

Human Microbiome Fun Facts

Human microbiota is unique to you. It may replace a thumb print or retinal scan for authentication.

The microbiota of your skin protects you from invading microbes.

Reduced bacterial diversity is associated with allergies.

The microbiota of your dominant hand is different from the microbiota of your non-dominant hand.

Microbiota has a gender bias. Males usually harbor more Corynebacterium spp. in their armpits than females. Females contain more Lactobacillus spp. on their hands.

Gut microbiota affects your mood. A study showed that yogurt eaters had calmer reactions.

Morning breath is caused by the microbes in your mouth.

A dog's tongue and paw microbiota can colonize human skin.

The human microbiome regulates your immune system.

The Belly Button Diversity Project determined that an inny or outy has more than 1400 bacterial species.

Probiotics help to restore microbial diversity imbalances in the human body.

Vegans and vegetarians have more diverse gut microbiota than meat eaters.

FIGURE 2.30 Human microbiome fun facts.

capsule or drinking a specific cocktail of microbes tailored to treat you specifically may become the norm in medicine in the not too distant future? Keep in mind the words of molecular biologist Dr. Bonnie Bassler (1962–): "*We mostly don't get sick. Most often, bacteria are keeping us well.*"

Summary

This chapter reflects upon the changes in medical microbiology and healthcare practices after Koch's postulates and the germ theory of disease were embraced by scientists and physicians during the latter part of the 19th century. The acceptance that microbes caused disease spearheaded changes in hygienic practices in health care and research on and development of microbiology techniques during the golden age of microbiology (late 1800s and the beginning decades of the 1900s) that are still used in laboratories today (e.g., pure culture technique, use of bacteriological media, and Gram staining). Phage therapy and the discovery of antibiotics were instrumental in reducing infections and increasing the average human life span.

The modern era of microbiology was driven by the study of the biological molecules that comprise cells. After the elucidation of the structure of DNA, the genomes of microbes and even humans could be studied and analyzed because of the development of molecular biology tools for the analysis of DNA and how genes function. Today, medicine is at a crossroads in which diagnostic testing and therapies may be based on the assessment of a patient's microbiome and immune status.

The popular terms *bugs* and *germs* are used in a collective sense, but there is no basis for lumping these diverse microbes together. Further, these terms have a negative connotation, because they are usually used to describe infectious diseases, but it is important to remember that only a handful of microbes cause disease. Microbial diversity is the key to healthy living.

KEY TERMS

adaptive immune system
agar
agar-agar
anthrax
antibiotic
antiparallel
aseptic technique
asymptomatic carrier
attenuated

avirulent
bacteremia
bacterial lawn
bacteriophages
biofilm
buboes
bubonic plague
capsule
carbohydrates

ceftazidime
Chamberland porcelain ultrafilters
cholera
ciprofloxacin
Clostridium difficile (C. diff.) infections
commensal microbes
consumption
crystal violet
debride
decolorization
deoxyribonucleic acid (DNA)
differential staining
diplococci
DNA polymerase
DNA sequencing
DNase
Doctors Without Borders
double helix
edema
Ellis Island
emerging pathogen
endocarditis
enzyme
fecal microbiota transplantation
Federal Immigration Station
ferments
fistula
foreign gene
generation time
germ theory of disease
Gram negative or Gram (–)
Gram positive or Gram (+)
Gram stain
healthy carrier
heat fixation
hospital-associated infection (HAI)
host
hydrophobia
immunocompromised
infectious diseases
infusion
Intestiphage®
Koch's postulates
Mary Mallon
metagenomics
miasmas
microbes
microbial diversity
microbiome
microbiota
molecular biology
multidrug-resistant (MDR)
new pathogen
nucleic acid

nutrient broth
OMKO1 phages
otitis
outer membrane porin M protein
[^{32}P]
palindromes
pasteurization
pathogen
penicillin
peritonitis
personalized medicine
petri dish
phage therapy
phages
plaques
plasmid
polymerase chain reaction (PCR)
primers
protein
public health
pure culture
quarantine
R strain of *S. pneumoniae*
rabies
rabies vaccine
receptor
recombinant DNA molecules
reemerging pathogen
restriction enzymes
S strain of *S. pneumoniae*
safranin
scanning electron microscopy (SEM)
secondary bacterial infection
sepsis
6-second medical examination
sulfonamide (sulfa) drugs
swan-necked flask
symbiotic microbes
template DNA
theory of spontaneous generation
trachoma
transforming principle
transmission electron microscopy (TEM)
tuberculosis
typhoid
Typhoid Mary
vaccination
vectors
virulent
viruses
Vitek systems
white plague
x-ray crystallography

SELF-EVALUATION

O PART I: Choose the single best answer.

1. _____ is the hereditary material of a cell.
   - a. Protein
   - b. Carbohydrate
   - c. Lipid
   - d. DNA

2. Scientists have renewed interest in the use of _____ to treat multidrug-resistant bacterial infections.
   - a. antibiotic therapy
   - b. fecal transplants
   - c. phage therapy
   - d. radiation therapy

3. _____ is a differential staining method used to distinguish between different types of medically important bacteria.
   - a. Simple staining
   - b. Gram staining
   - c. Fingernail staining
   - d. Heat fixation

4. _____ is a solidifying agent used in solid media today.
   - a. Gelatin
   - b. Rubber cement
   - c. Diatomaceous earth
   - d. Agar

5. Penicillin is a(n) _____ produced by a fungus.
   - a. pigment
   - b. dye
   - c. carbohydrate
   - d. antibiotic

6. The _____ was used to disprove the theory of spontaneous generation.
   - a. swan-necked flask
   - b. Gram stain
   - c. Chamberland ultrafilter
   - d. microscope

7. _____ is a new approach toward the assessment of microbial diversity.
   - a. Vaccine development
   - b. Metagenomics
   - c. Histological staining
   - d. Pure culture technique

8. Less than _____ of microbes cause disease.
   - a. 1%
   - b. 10%
   - c. 20%
   - d. 50%

9. _____ is used to study the ultrastructure of microbes.
   - a. Light microscopy
   - b. PCR
   - c. DNA sequencing
   - d. Electron microscopy

10. _____ is a technique that allows for the rapid synthesis of DNA molecules in a test tube.
   - a. Gram staining
   - b. Electron microscopy
   - c. PCR
   - d. DNA sequencing

○ **PART II: Fill in the blank.**

1. The rabies vaccine is credited to _____.

2. The human _____ comprises the bacteria, protozoa, fungi, and viruses that live on the skin and inside of the human body.

3. _____ act as molecular scissors to recognize and cut specific short stretches of nucleotides in DNA.

4. _____ are DNA molecules containing DNA fragments from two or more organisms.

5. A(n) _____ is an independent DNA molecule that passes on traits (the transforming principle in Griffith's experiments).

6. _____ were used to label the proteins or DNA of bacteriophages in order to determine that DNA was the hereditary material.

7. The x-ray crystallography results of _____ contributed to the elucidation of the structure of DNA.

8. _____ is treatment uniquely developed for each patient based on genetic, diet, lifestyle, and environmental factors.

9. _____ was a process applied to milk in order to prevent tuberculosis.

10. In the mice in Griffith's experiments, *Streptococcus pneumoniae* had to possess a(n) _____ in order to cause disease.

○ **PART III: Answer the following.**

1. Explain why medical screening of immigrants entering America through Ellis Island Federal Immigration Station during the late 1800s and early 1900s was inadequate for protecting public health.

2. List four individuals and summarize their contributions to medical microbiology.

3. Why is microbial diversity important?

4. Explain how the development of different DNA sequencing approaches revolutionized bacterial and human genetics.

5. Summarize why the use of mammalian cell cultures was essential to the development of a vaccine to prevent poliomyelitis.

6. Competition drives progress. Discuss competition in the race to determine the structure of DNA and how it affected future developments in diagnostic microbiology. What scientists were involved, and what new knowledge did they contribute?

7. Explain how DNA is synthesized during a PCR reaction.

8. Why are restriction enzymes, considered the workhorses of molecular biology, used in genetic engineering and the development of DNA sequencing? Describe their function in your answer.

9. What is pure culture technique, and why is it important in identifying bacterial pathogens?

10. Explain how a person's microbiota is unique. What factors impact the diversity of the human microbiome?

CHAPTER 3

CONTROLLING THE SPREAD OF INFECTIOUS DISEASES

"*Infectious diseases have become less prominent as causes of disease and disability in regions of improved sanitation and adequate supplies of antibiotics.*"

—Rosalyn Sussman Yalow (1921–2011)

American scientist

Courtesy of the CDC/Dawn Arlotta.

It is important to vigorously wash fresh, uncooked produce before consumption in order to remove soil, microbes (some may be potential pathogens), and environmental pollutants such as pesticides. The woman pictured here is thoroughly cleaning carrots and radishes.

ShutterStock, Inc. / happykanppy.

1. Summarize the changing roles of health departments over the 20th century.

2. Explain why human waste should not be deposited near water sources.

3. Explain why proper sanitation can prevent potential diarrheal disease outbreaks.

4. List the two leading causes of death worldwide in children younger than 5 years and describe the measures that are needed to prevent these deaths.

5. Identify at least five pathogens that can cause infectious diseases associated with drinking contaminated water.

6. Summarize how changes in food-processing technologies, personal eating habits, and food distribution have contributed to the presence of new emergent pathogens.

7. Discuss the message of the CDC hand hygiene campaign, "Clean hands save lives!"

8. List and distinguish the three types of infectious disease surveillance strategies.

Case Study: The Romans and Toilet Phobia

Amanda Pasteur and Tony Cassetti spent a summer on the west coast of Italy with relatives after they graduated from college. Both would be attending medical school in the fall. They felt fortunate to be able to tour the ancient Roman ruins. One day, they spent time in Ostia Antica, a large archeological site close to modern Ostia. Two thousand years ago, Ostia Antica was a thriving port town of 60,000 Romans. Unfortunately, Rome fell, the river changed course, and the port filled up with mud, becoming a mosquito-infested swamp. However, the ruins were well preserved.

While walking through the ruins, Amanda and Tony stumbled upon the public latrine (*forica*), which was located between two shops near the Forum Baths (**Figure 1**). This ancient city at one point had more than 15 public baths. The Roman baths were places of entertainment, politics, and socializing. As they toured the ruins they ran into another tourist and managed to strike up a conversation. They discovered that the fellow tourist taught microbiology at a college in the United States. The professor was full of knowledge about the public toilets.

The professor explained that the public toilets were large rooms with marble bench seats. A person would sit over the top of the plate-sized hole located on the marble toilet, which was connected to the city's main sewer system. Another hole provided access to a **tersorium**, or "butt brush," which was used for wiping. The tersorium consisted of a sea sponge from the Mediterranean Sea that was attached to the end of a stick. After each use, it was rinsed in an open gutter that had a small continuous stream of water running through it. The stone gutter ran parallel to

© Benedictus/ShutterStock, Inc.

Figure 1 Public latrine located in the Roman ruins of Ostia Antica.

the seats along the ground. After the water rinse, the tersorium was rinsed in a bucket filled with salt water or vinegar (as a disinfectant) and placed back into the hole. Yes, you read that right, the tersoria were shared butt cleaners!

While the plumbing appears to be quite sophisticated, *Romans feared the public toilets.* The sewer systems could get blocked with silt, requiring regular cleaning. The system did not have traps to keep flies out, so the flies had easy access to human waste. The flies could transfer fecal matter and pathogens to humans. Besides insects, rodents could lurk in the sewers and toilets, invading private homes with toilets connected to the main sewer. Blocked drainage sewers resulted in the accumulation of odors and toxic bacterial end products such as explosive

(continues)

Case Study: The Romans and Toilet Phobia (continued)

methane gas. It is possible that Romans saw flames coming out of a hole when entering a public loo.

The public latrines have very little graffiti. Is it because no one wanted to spend much time there? Many of the Roman latrines have small shrines to the goddess Fortuna, who was believed to protect the latrines from danger, such as disease-causing demons. Researchers analyzed latrine samples collected from a sewage channel under a Roman bath complex. They identified evidence of roundworms and whipworms (which cause malnutrition) and *Giardia duodenalis*, a protozoan that causes dysentery or diarrhea in humans and some mammals. Therefore, it is possible that people using the latrines at the baths spread intestinal parasites among each other through the use of tersoria, an **oral–fecal mode of transmission**. *Giardia* cysts could survive the disinfection methods. Amanda and Tony finished their tour with a greater appreciation for toilet paper and today's sanitation systems.

Questions and Activities

1. What are tersoria, and why are they no longer used today?

2. Define *oral-fecal mode of transmission*. Provide at least three examples of infectious diseases that can be transmitted this way.

3. Why did the Romans fear public toilets?

4. How are the plumbing/toilets in your residence more hygienic than private ancient Roman toilets?

5. What were the inherent problems associated with the sewer systems in ancient Rome?

6. When was toilet paper invented? Who invented it? (Hint: You will need to do some research for this answer.)

7. How are the portable toilets used today at outdoor entertainment events more hygienic than the public toilets in ancient Rome?

Information from A. O. Koloski-Ostrow. (2018). *The archaeology of sanitation in Roman Italy: Toilets, sewers, and water systems.* Reprint ed. Studies in the History of Greece and Rome. Chapel Hill: University of North Carolina Press; C. Wald. (2016). *The secret history of ancient toilets: By scouring the remains of early loos and sewers, archaeologists are finding clues to what life was like in the Roman world and in other civilizations. Nature, 533*, 456–458.

Preview

Advances in public health made during the 20th century have decreased the burden of infectious diseases on a worldwide basis, particularly in the United States and in other industrialized nations. Sanitation, clean water, hand hygiene, food safety, and infectious disease surveillance are major factors in the control of infectious diseases. Each of these factors is discussed in this chapter.

Sanitation

Part of the daily routine each morning as you prepare for the day is attending to matters of **sanitation** and personal **hygiene**, including showering, brushing your teeth, and using a clean flush toilet. These activities are taken for granted, but they are luxuries. On the other side of the globe is Bangladesh that welcomed a sudden influx of more than 900,000 Rohingya refugees fleeing ethnic cleansing (mass killings and other abuses of an unwanted ethnic or religious group in a society) in Myanmar in 2017. They have continued to let another 11,000 refugees in at the beginning of 2018 (the time of this writing). The refugees are in the Cox's Bazar area located in the southern tip of Bangladesh.

The mega camp is severely overcrowded. The hurried and unorganized construction of the Kutupalong–Balukhali refugee camp has inadequate numbers and positions of latrines (toilets), which are too close to drinking water sources, increasing the risk for outbreaks of diarrhea

and other diseases. Over half of the drinking water samples tested were contaminated with microbes that can cause disease. According to an interview of refugee Noor Haba by members of the Human Rights Watch (https://www.hrw.org/):

Nearly every refugee interviewed by Human Rights Watch put the lack of safe drinking water at the top of their list of living condition problems. There is not enough water, people get sick after drinking it, and they have to walk long distances and stand in long queues to get it. "To get drinking water we have to go to the other side of the main road," said Noor Haba, a 26-year-old mother of four. She claimed three or four people were struck and killed by cars while crossing the road to get water, but Human Rights Watch could not confirm the information.

The Bangladesh government and humanitarian partnerships have worked hard to try to vaccinate refugees to prevent diseases, but despite their efforts, 8,000 cases of diphtheria were reported as of July 4th, 2018. Another Human Rights Watch interview of a refugee, referred to as Tasmin stated:

"The children have gotten diarrhea and other sicknesses from impure water. Our health conditions are deteriorating. I feel weak. We have gone to clinics when we get sick and received medicine. When I take the medicine, I feel better, but then I get sick again."

According to the **World Health Organization (WHO)** and the **United Nation's Children Fund (UNICEF)** Progress on Drinking Water, Sanitation, and Hygiene 2017 Update and SDG Baselines Report, in 2015, 2.3 billion of the world's citizens (30% of the world's population) lacked basic hygiene and sanitation facilities and more than 844 million people were without basic drinking water service. Diarrheal diseases are the second leading cause of death in children younger than 5 years. These diseases are both preventable and treatable. **Diarrhea** is the leading cause of malnutrition in children younger than 5 years. Every year, about 525,000 children under the age of 5 die of diarrheal diseases because they lack safe drinking water and access to the most basic sanitation. *In most developing countries, cell phone technology is expanding at a faster rate than sanitation!*

World leaders agree that hygienic means of sanitation and a safe supply of drinking water are basic human needs. Kofi Annan, a past UN Secretary General, said it all in his words: "*We shall not finally defeat AIDS, tuberculosis, malaria, or any of the other infectious diseases that plague the developing world until we have also won the battle for safe drinking water, sanitation, and basic health care.*" Clean water, improved sanitation, and hygiene have the potential to prevent at least 11% of world diarrheal diseases and 7% of all deaths.

Development of Sanitation

Urbanization is not a new phenomenon but one that dates back millennia to when hunter-gatherers first saw a benefit in pooling their meager resources by living together in villages. These villages evolved into today's cities. Population growth and the resulting **urbanization** that began in the mid-19th century and continues today are major factors contributing to the challenge of controlling **infectious diseases**. In the mid-19th century, the **germ theory of disease** and the idea of **contagion** had yet to be developed, and the rapidly growing urban centers struggled to establish infrastructure to keep pace with the burgeoning masses. Little regard was paid to **public health** measures. Industrial development and immigration led to an influx into the cities, which, in turn, fueled poor housing, overcrowding, lack of clean water, and limited means for the disposal of human waste. John Cairns, a British biologist, termed cities the "*graveyards of mankind.*" The 19th-century outbreaks of **cholera** in London illustrate the consequences of inadequate sanitation and hygiene. Arno Karlen, in his book, *Man and Microbes* (Simon & Schuster, 1996), wrote the following:

> *The city's seven sewer systems were uncoordinated and relied on defective pipes. They received tons of human and animal feces, dead animals, waste from abattoirs (slaughter houses), effluvia from hospitals and tanneries, the occasional human corpse, and contaminated ground water from cemeteries.*

All of London's human waste ended up in the Thames River, which provided most of the city's drinking water. Cholera was the result. The city had eight separate water companies and just one experimental filtration system. Water not taken from the Thames River came from wells, many as badly polluted as the river. The residents of the city drank, cooked, and washed in their own filth. With the crowding and dirt, once a waterborne disease was established, further person-to-person transmission was virtually assured.

London was not the only city to be so afflicted. Filth and squalor prevailed in Europe and around the world. Cities in the United States were hardly models of cleanliness and sanitation. Sewage disposal systems were few, and outhouses, overflowing cesspools, and garbage-littered streets flourished, as did **tuberculosis**, **diphtheria**, **scarlet fever**, and **typhoid fever**. Somewhere about the middle of the 19th century a sanitary reform movement gradually arose from the ashes of human corpses, debris, and human and animal wastes, perhaps fueled by the third cholera epidemic to hit London (on the heels of the second epidemic). The combination of disease, filth, and lack of shelter (**FIGURE 3.1**) led to the enactment of laws relating to

© Vishal Shah/Shutterstock, Inc.

© National Library of Medicine.

FIGURE 3.1 Factors in the spread and emergence of infectious disease. **(a)** People living in squalor without adequate shelter and basic sanitation are at increased risk for infectious disease. **(b)** Despite major advances over the past century, substandard levels of living such as those shown in this old drawing persist in underdeveloped countries and in pockets of developed countries.

© National Library of Medicine.

FIGURE 3.2 At the beginning of the 20th century, the age of sanitation began to emerge.

sanitation, including sewage and water treatment, garbage collection, and other public health measures. In the 1880s, Pasteur and Koch triumphed with their discoveries. The germ theory of disease was established by Louis Pasteur, and **Koch's postulates**, developed by Robert Koch, proved that microbes were linked to sanitation and microbial or infectious diseases.

The germ theory of disease was embraced in Europe and in the United States. Sanitation was "in," and sanitary engineers and bacteriologists (a term that preceded "microbiologists") flourished (**FIGURE 3.2**). In 1887, the Marine Hospital Service was established and charged

with monitoring cholera in immigrants on ships coming into New York. This facility was the forerunner of today's **National Institutes of Health (NIH)**. Other public health laboratories and organizations were established in major cities around the world, and bacteriologists and sanitary engineers worked in concert. Public health statutes promoting sanitation and good hygiene were passed and implemented, and by 1900, 40 states had health departments.

Over the 20th and 21st centuries the focus of health departments changed from meeting the urgent and immediate needs of basic sanitation to delivering health services (**TABLE 3.1**). The 1920s through the 1950s witnessed great strides in public health strategies to control infectious diseases (**FIGURE 3.3**). Great attention was paid to the construction of water and sewage treatment facilities, chlorination, better housing, control of tuberculosis and venereal diseases (now called **sexually transmitted infections**, or **STIs**), food production and distribution, animal and pest control measures, and garbage disposal (**FIGURE 3.4**). The public was bombarded with information regarding the evil of "germs" and their transmission from the sick to the healthy by various modes of transmission (**FIGURE 3.5**). The "*gospel of germs*" was accepted, and rub-a-dub-dub, scrub, dust, and clean-clean-clean were heralded. Somewhere along the line the expression "*Cleanliness is next only to godliness*" became a household dictate; some say this expression is attributed to Mahatma Gandhi of India as his "battle cry" during his efforts in the 1920s and 1930s to clean up the villages. These new efforts paid off. **Malaria**, **plague**, tuberculosis, and other infectious diseases were markedly reduced; the last major outbreak of plague in the United States occurred during 1924 and 1925, in Los Angeles.

The 20th century could be termed the "golden age of public health" based on the gain of more than 60% life expectancy, which is directly correlated with public health

TABLE 3.1 Changing Role of Health Departments	
Period	Services
Health department services in 1900 (driven by urgent needs)	Sewer construction
	Water supply inspection
	Sewage disposal
	Nuisance and pest control
	Privy inspection or removal
	Milk supply inspection
	Infectious disease control (tuberculosis, diphtheria and croup, scarlet fever, smallpox, and typhoid fever)

Period	Services
Health department services in 1999 (developing an organized approach)	Monitoring community health status to identify potential hazards
	Investigating disease outbreaks and safety hazards in the community
	Mobilizing community partnerships to solve health problems
	Developing policies and plans that support individual and community health efforts
	Enforcing laws and regulations that protect health and ensure safety
	Linking populations with needed personal health services and ensuring the provision of health care when otherwise unavailable
	Ensuring a competent public health and personal healthcare workforce
	Evaluating effectiveness, accessibility, and quality of personal and population-based health services
	Researching new ideas and innovative solutions to health problems
U.S. state public health agencies top programs and functions since 2005	Preparedness and disease monitoring
	Data collection: vital statistics
	Maintaining public health laboratories
	Tobacco prevention and control
	Environmental health
	Food safety
	Health facilities, drinking water, and environmental regulations

Information from CDC and L. M. Beitsch, R. G. Brooks, N. Menachemi, & P. M. Libbey. (2006). Public health at center stage: New roles, old props. *Health Affairs, 25*(4), 911–922.

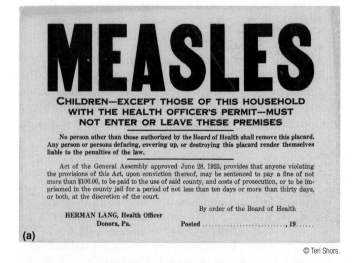

© Teri Shors.

FIGURE 3.3 Public health statutes were passed and enforced by health officers. **Quarantine signs** were required by law to be placed on homes in which an individual was suffering from a "contagious" disease. **(a)** Measles sign used in 1920s. **(b)** Chickenpox sign used in the 1940s.

© Robert I. Krasner.

FIGURE 3.4 Garbage disposal is an important public health measure. Garbage accumulation attracts rats and other rodents and animals that serve as **reservoirs** and **vectors** of infectious disease.

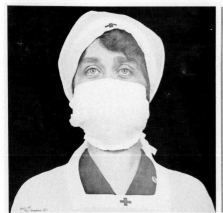

1918 New Haven, Connecticut, *Illustrated News*. Courtesy of the U.S. National Library of Medicine.

To Prevent
Influenza!

Do not take any person's breath.
Keep the mouth and teeth clean.
Avoid those that cough and sneeze.
Don't visit poorly ventilated places.
Keep warm, get fresh air and sunshine.
Don't use common drinking cups, towels, etc.
Cover your mouth when you cough and sneeze.
Avoid Worry, Fear and Fatigue.
Stay at home if you have a cold.
Walk to your work or office.
In sick rooms wear a gauze mask like in illustration.

FIGURE 3.5 In the early and mid-twentieth century, departments of health fostered in the population an appreciation of good personal hygiene in an effort to control the dissemination of infectious disease such as influenza. A Red Cross nurse is pictured with a gauze mask over her nose and mouth. The text on the right offers tips in how to prevent the spread of influenza.

knowledge of infectious diseases and vaccinations, along with improved sanitation, food and water safety practices, and regulations. Today, the breadth and scope of public health programs are vast, ranging from disease surveillance and data collection to environmental regulation and medical or mental health and education programs (TABLE 3.1).

Human Waste Disposal

Your first impulse may be to laugh at learning there is actually a World Toilet Organization (http://worldtoilet .org), founded in 2001 to promote sanitation. Almost 2.3 billion people around the world lacked access to appropriate toilet facilities in 2015. Further, 200 million tons of human waste is not collected and treated around the world because of the lack of toilets (**FIGURE 3.6**). **World Toilet Day** has been celebrated every November 19th since the inception of the World Toilet Organization. Some of you may enjoy camping in the woods. In rustic settings, you should take precautions when using modest sanitation facilities (**BOX 3.1**).

© Robert I. Krasner.

© Robert I. Krasner.

FIGURE 3.6 Human waste disposal. **(a)** Contamination of food and water by fecal material is a major cause of many infectious diseases. **(b)** Primitive and simple toilets are relatively inexpensive and efficient if properly constructed. Outer barrel is for privacy; inner barrel is for urination only.

The safe disposal of human excreta is central to sanitation, and its significance cannot be overemphasized. The General Assembly of the United Nations (UN) determined that the 2015 goal to halve the proportion of people living without sanitation was running 150 years behind schedule.

Krasner's Notebook

In New Delhi, India, the **Sulabh International Museum of Toilets** features a collection of artifacts, pictures, and objects illustrating the historical development of toilets since the year 2500 B.C. On display at the museum is a replica of the throne of King Louis XIV with its built-in commode he used to defecate while conducting court sessions. For a virtual tour of the museum, check out this website: http://www.sulabhtoiletmuseum.org.

BOX 3.1 When Nature Calls in the Woods

Sanitation facilities may not be available if you are camping, hiking, or rock-climbing in the great outdoors. Public toilets, also known as *porta-potties*, *Port-a-John's*, or *portaloos* (in the United Kingdom), are plastic portable restroom enclosures that contain a chemical toilet. They are used as temporary toilets for large gatherings such as festivals and concerts, at trailheads, or for construction sites (**Figure 1a**). An outhouse, also known as an *Earth closet*, a *privy*, a *bog* (in the United Kingdom), or a *dunny* (in Australia), is a small structure located outdoors as a pit latrine or dry toilet (**Figure 1b**). American outhouses usually have a crescent moon-shaped hole in the door that provides lighting and ventilation (**Figure 2**).

(a)

© Teri Shors.

(a)

© Teri Shors.

(b)

© Teri Shors.

Figure 1 Examples of modest public sanitation facilities in the United States. **(a)** Porta-potties at a summer event, Sawdust Days, Oshkosh, Wisconsin. **(b)** Public restrooms can be found in county parks, such as this one in Manitowoc County, near Cato Falls in Wisconsin.

(b)

Author's photos with permission by Roger and Sylvia Gasser. © Teri Shors.

Figure 2 Even though hand sanitizer will not inhibit all viruses and microbes, it is the best option for shared sanitation facilities such as this outhouse that does not have a sink with running water for handwashing. **(a)** Outhouse located in a private, rural area near Omro, Wisconsin. **(b)** Inside of the outhouse shown in (a).

If the smell inside of a porta-potty or outhouse does not kill you, the **pathogens** inside might. Noroviruses, enteroviruses, hepatitis A and E viruses, adenoviruses, herpesviruses, papilloma viruses, and influenza viruses can lurk in portable toilets. If you do not pick up a virus, the bacteria that cause salmonellosis, shigellosis, traveler's diarrhea, and cholera (rarely) also lurk

(continues)

BOX 3.1 When Nature Calls in the Woods *(continued)*

in public facilities. Follow these tips when using a porta-potty or outhouse:

- Assume that the door handle is contaminated. No one washes their hands before entering the bathroom!

- If possible, use a paper towel to turn off the sink (sinks are one of the germiest places in a bathroom or public restroom).

- Don't put bags on the floor. Floors harbor microbes, including pathogens.

- Assume that the soles of your shoes will become contaminated from the floor. Avoid touching the soles of your shoes.

- Use hand sanitizer or wash your hands with soap and water for 20 seconds after using the restroom. Soap and water is always best for removing pathogens from your hands.

- Sanitize your hands before you prepare food and/or eat!

But what do you do if you are an adventurer, backpacking in the natural wilderness for days or weeks? Or, there may be a time one day when you are walking in the woods and you *really* need to go to the bathroom. If you didn't bring a portable toilet with you, then the following suggestions may come in handy:

- If you brought a backpacker's trowel, dig a hole about 6 to 8 inches (15 to 20 centimeters) deep. Make sure you are 200 feet (60 meters) away from a water source (e.g., river, stream, or lake). Cover your poop in the hole with soil, leaves, and twigs.

- If you don't have anything to wipe with, a smooth rock is your best alternative (don't use leaves—too many people have regretted the use of poison ivy leaves by mistake and ended up with a rash in a sensitive body location).

- If you are camping, be sure that you do your business at least 300 yards (900 meters) from your campsite to keep wildlife away from you and your food supply.

- What if you are scaling a rocky cliff and you can't dig a hole? Toss your poo down the mountain for disposal or wipe the remains on a rock so that the sun's rays can sanitize it.

These tips may help you prepare for an adventure in the wild.

Fifteen percent of the world population, or 1 billion people, particularly in developing countries and in poverty pockets in developed countries, still practice open defecation because they lack proper sanitation facilities. They have no choice but to squat and defecate in the open directly onto the ground. According to the eighth Secretary General of the United Nations, Ban Ki Moon: *"Every dollar invested in water and sanitation leads to $4 in economic returns."*

In India, more than half of the population still defecates in the open. According to a 2011 census, 400,000 schools in India lack basic functional toilets and 23% of girls drop out of school because of lack of toilets. Further, more than 238 million gallons (900 million liters) of urine and 298 million pounds (135 million kilograms) of fecal material need to be disposed of each day. About half a million children die every year in India because of dehydration resulting from diarrheal diseases that are frequently traceable to open defecation. The subject of toilets and defecation is hardly dinnertime conversation, but it is a fact of life and an integral part of the history of human hygiene.

Organizations such as the United Nations and the **World Bank** are working on the development of sewage disposal systems in developing countries. Bathrooms and flush toilets, such as our society is accustomed to, are not necessary goals; certainly, less luxurious and primitive facilities, be they outdoor or indoor, are affordable and effective (Figure 3.6). Programs to improve poor sanitation are in progress in slums and squatter settlements of the world's poorest countries. The **Kampung Improvement Program** in Indonesia, a highly successful program, has focused on covering open sanitation drains and on bringing reasonably clean water to families. The **Orangi Pilot Project** has reached thousands of people in a poor neighborhood in Karachi, Pakistan. In 2006, 50% of Karachi's 16 million people lived below poverty line (which was 10% of Pakistan's total population). According to a public health official, *"People don't need a flush toilet in every home or a faucet in every room. But with a standpipe for every three units, with adequate pit latrines, and other forms of [waste] treatment, the services can be there and the health of the children maintained."*

A 2017 update about the progress on sanitation published by the WHO and UNICEF includes the following statistics:

- 2 out of 5 people (1.2 billion) using safely managed sanitation (i.e., human waste was disposed of properly) lived in rural areas.

- Worldwide, 892 million people (12% of global population) still defecate in the open (**FIGURE 3.7A**).

- 39% of the world population (2.9 billion people) used a safely managed sanitation service.

- 68% of the world population (5 billion people) used at least a basic sanitation service.

The improved sanitation facilities worldwide are depicted in the map in **FIGURE 3.7B**. Virtually the entire population in developed regions uses improved facilities. This is not true for developing regions, in which only around half the population use improved sanitation.

The average human produces roughly 132 gallons (500 liters) of urine and 13.2 gallons (50 liters) of feces every year. The United States has 14,748 wastewater treatment plants that convert human waste into **sludge** (also referred to as **biosolids**). Fifty years ago, thousands of cities dumped their raw sewage directly into the nation's rivers, lakes, and bays. Since the 1990s, sludge has been sold or given away free to farmers for use as fertilizer to improve soils and stimulate plant growth. The **Environmental Protection Agency (EPA)** has promoted the use of sludge as an environmentally friendly way to recycle sewage. Local governments decide whether to recycle the biosolids as fertilizer or to incinerate or bury it in a landfill. About 50% of all sludge is recycled to land and takes place in all 50 states. The biosolids are used on less than 1% of the nation's agricultural land.

Given the fact that human waste contains microbes, including a small percentage that may be pathogenic (**TABLE 3.2**), rigorous testing and stringent guidelines were created to address the risk of infectious disease and its consequence to public health. The National Academy of Sciences reviewed the current practices, public health concerns, and regulator standards and have concluded that the use of sludge in the production of crops for human consumption, in conformance with federal and state regulations, presents negligible risk to the consumer, to crop production, or to the environment. There are two classes of biosolids. Class A biosolids are treated or sanitized and contain no detectable levels of pathogens. Class B biosolids are treated but still contain detectable levels of pathogens. Class B biosolids are restricted regarding public access and crop harvesting.

Clean Water

"*Water, water everywhere, nor any drop to drink.*" This famous line from the classic 1798 poem, "The Rime of the Ancient Mariner," by Samuel Taylor Coleridge, depicts the desperate plight of an old mariner surrounded by a sea of undrinkable water. It can also serve to depict the desperation of one-fourth of the world's population who have only limited access to water that may or may not be safe (**FIGURE 3.8A**). Less than 2.5% of the Earth's water can be used and reused as freshwater. This is in sharp contrast to the United States and in other developed countries where water is plentiful and wasted by the gallon. Row after row of jugs and bottles of water are displayed in markets (**FIGURES 3.8B** and **3.8C**) despite the availability of clean water in just about every household. In general, there is no shortage of clean water for bathing, showering, cleaning the dishes, and washing the dog.

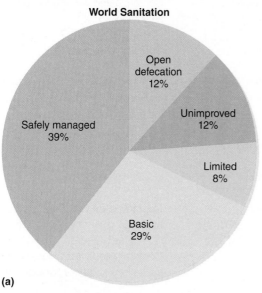

(a)

Data from World Health Organization and United Nations Children's Fund. (2017). *Progress on Drinking Water, Sanitation and Hygiene: 2017 Update and SDG Baselines.* Geneva: World Health Organization and United Nations Children's Fund.

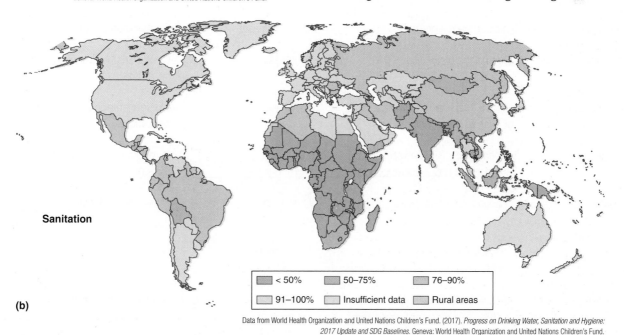

(b)

Data from World Health Organization and United Nations Children's Fund. (2017). *Progress on Drinking Water, Sanitation and Hygiene: 2017 Update and SDG Baselines.* Geneva: World Health Organization and United Nations Children's Fund.

FIGURE 3.7 (a) World sanitation coverage. Two out of five people (39% of the global population) used safely managed sanitation services in 2015. **(b)** World map showing the use of basic sanitation facilities in 2015. By 2015, 154 countries had achieved over 75% coverage with basic sanitation services.

TABLE 3.2 Water-, Sanitation-, and Hygiene-Related Infectious Diseases in Humans*

Type of Pathogen	Pathogen	Disease	Association
Virus	Norovirus	Gastroenteritis	Contaminated water
Virus	Enterovirus	Gastroenteritis	Contaminated water and poor sanitation
Bacterium	*Campylobacter jejuni*	Campylobacteriosis	Contaminated water
Bacterium	*Chlamydia trachomatis*	Trachoma	Poor sanitation and hygiene
Bacterium	*Escherichia coli*	Gastroenteritis	Contaminated water and poor sanitation
Bacterium	*Leptospira* spp.	Leptospirosis	Contaminated water
Bacterium	*Mycobacterium ulcerans*	Buruli ulcer	Contaminated water related to environmental changes (e.g., increased water temperatures)
Bacterium	*Salmonella* spp.	Salmonellosis	Contaminated water
Bacterium	*Salmonella typhi*	Typhoid fever	Contaminated water
Bacterium	*Shigella sonnei*	Shigellosis	Contaminated water and poor hygiene
Bacterium	*Vibrio cholerae*	Cholera	Contaminated water, inadequate sanitation, and poor hygiene
Protozoan	*Cryptosporidium* spp.	Cryptosporidiosis	Contaminated water, inadequate sanitation, and poor hygiene
Protozoan	*Cyclospora cayetanensis*	Cyclosporiasis	Contaminated water
Protozoan	*Entamoeba histolytica*	Amebiasis	Contaminated water and poor sanitation
Protozoan	*Giardia intestinalis, Giardia lamblia,* or *Giardia duodenalis*	Giardiasis	Contaminated water and poor sanitation
Helminth	*Ascaris* spp., *Trichuris* spp., *Anclostoma* spp., *Necator* spp.	Soil-transmitted helminthiasis	Inadequate sanitation and poor hygiene
Helminth	*Dracunculus medinensis*	Dracunculiasis (guinea-worm disease)	Contaminated water
Helminth	*Fasciola hepatica* (common liver fluke)	Fascioliasis	Contaminated water
Helminth	*Schistosoma mansoni, S. haematobium,* or *S. japonicum*	Schistosomiasis	Contaminated water
Helminth	*Wuchereria bancrofti*	Lymphatic filariasis	Inadequate sanitation and poor hygiene

Type of Pathogen	Pathogen	Disease	Association
Fungi	*Tinea* spp.	Ringworm	Inadequate sanitation and poor hygiene
Arthropod	*Pediculus humanus capitis* (head louse); *Pediculus humanus corporis* (body louse, clothes louse); *Pthirus pubis* ("crab" louse, pubic louse)	Lice	Inadequate sanitation and poor hygiene
Arthropod	*Sarcoptes scabiei* var. *hominis* (mites)	Scabies	Inadequate sanitation and poor hygiene

*List does not include all arthropod-associated diseases associated with water.
Information from CDC Global Water, Sanitation & Hygiene (WASH) https://www.cdc.gov/healthywater/global/

The world map in **FIGURE 3.9** shows that Africa faces the greatest challenge in increasing the use of improved drinking water. As of 2015, 181 countries had achieved over 75% coverage with at least basic drinking water services.

The WHO estimates that about 29% of the world's population lacked access to clean water in 2015; in some villages, people, primarily women and children, spent more than 30 minutes of their day collecting buckets of clean water. **Safely managed drinking water** is defined as the use of an improved drinking water source that is located on the premises, available when needed and free of fecal and chemical contamination, in contrast to **basic drinking water service**, which means it takes no more than 30 minutes per round trip from an improved water source. A 2017 update about the progress on drinking water published by the WHO and UNICEF includes the following statistics for 2015:

- 7 out of 10 people used safely managed drinking water services (**FIGURE 3.10**).
- 1 in 3 people (1.9 billion people) who used safely managed drinking water services lived in rural areas.
- 844 million people still lacked a basic drinking water service.
- 3 out of 4 people (5.4 billion people) used improved water sources *free from contamination*.
- 263 million people spent more than 30 minutes per round trip to collect water from improved resources.

The WHO estimates that at any given time, perhaps one-half of all people in the developing world are suffering from one or more of the six main diseases associated with drinking contaminated water (Table 3.2). The poorest people in the world are paying many times more than their richer compatriots for the water they need to live and are getting more than their share of deadly diseases because supplies are dangerously contaminated. Control of waterborne diseases is a particularly difficult problem when human feces and urine are disposed of in a way that allows them to gain access to water sources and food supplies. Water may also be contaminated with chemicals such as mercury and arsenic. In Bangladesh, people have

© Marcus Brown/ShutterStock, Inc.

© Robert I. Krasner.

© Robert I. Krasner.

FIGURE 3.8 **(a)** Clean water is a luxury. For much of the world's population, water is not piped into their homes and must be transported in containers. Bodies of water may serve for clothes washing, bathing, water for animals, and drinking water. **(b)** Rows of water jugs. **(c)** Rows of regular and carbonated drinking water at a superstore.

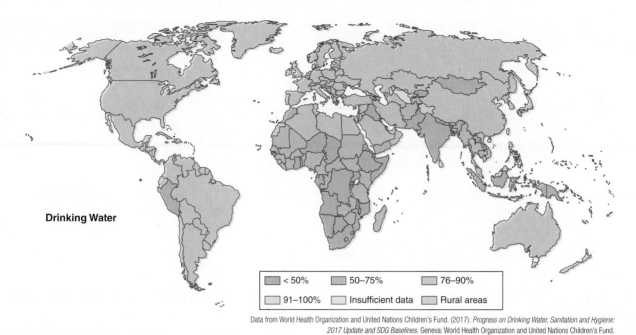

Drinking Water

☐ < 50%	☐ 50–75%	☐ 76–90%
☐ 91–100%	☐ Insufficient data	☐ Rural areas

Data from World Health Organization and United Nations Children's Fund. (2017). *Progress on Drinking Water, Sanitation and Hygiene: 2017 Update and SDG Baselines.* Geneva: World Health Organization and United Nations Children's Fund.

FIGURE 3.9 World map showing the improved drinking-water sources (2015). Sub-Saharan Africa continues to face the greatest challenges.

developed serious symptoms from drinking arsenic-contaminated water (**BOX 3.2**).

Improvement of water quality was, and continues to be, a major public health priority aimed at the decline of waterborne and water-associated diseases. Anywhere from a 20% to 80% decline in morbidity and mortality is possible with improved water sanitation. In some

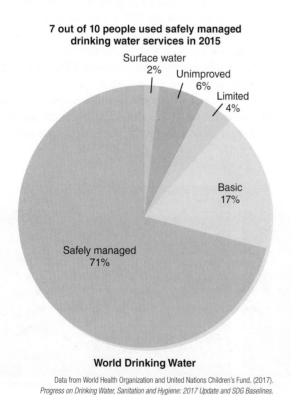

7 out of 10 people used safely managed drinking water services in 2015

Surface water 2%
Unimproved 6%
Limited 4%
Basic 17%
Safely managed 71%

World Drinking Water

Data from World Health Organization and United Nations Children's Fund. (2017). *Progress on Drinking Water, Sanitation and Hygiene: 2017 Update and SDG Baselines.* Geneva: World Health Organization and United Nations Children's Fund.

FIGURE 3.10 World drinking water coverage. Seven out of 10 people used safely managed drinking water services in 2015.

countries, people are infected with guinea worms, which bite their way through their victim's flesh, or with cholera so devastating that in several hours their life is threatened by severe dehydration resulting from massive loss of water through diarrhea. In 1986, The **Carter Center** led an international campaign to eradicate guinea worm disease by partnering with the **Centers for Disease Control and Prevention (CDC)**, the WHO, UNICEF, and many others. **Guinea worm disease**, also known as **dracunculiasis**, is on track to become the first parasitic disease eradicated and the first disease eradicated without the use of a vaccine or medicine. In 1986, an estimated 3.5 million people in 21 countries in Africa and Asia were infected with the Guinea worm each year. Today, the incidence has been reduced by 99.99%—to 30 cases in 2017!

However, even in developed countries vigilance needs to be maintained. A safe water supply can never be taken for granted, as the citizens of Milwaukee, Wisconsin, painfully discovered in 1993, when *Cryptosporidium parvum* caused the largest outbreak of waterborne disease in U.S. history. When developed countries experience fires, flood, hurricanes, wars, and other catastrophic events, they suffer the same misery of land and water becoming fouled with human and animal wastes and carcasses as the developing countries. In 2017, the devastating hurricanes Harvey, Irma, Maria, and Nate hit the United States and its associated territories, causing deaths, months of power outages, and billions of dollars in damages (**TABLE 3.3**, **FIGURE 3.11**). Some hurricane evacuees and rescue workers contracted **necrotizing fasciitis** and **leptospirosis**, which are associated with infected cuts or scratches, while walking in contaminated flood waters.

The CDC and the U.S. Environmental Protection Agency (EPA) have maintained a collaborative surveillance system since 1971, and continue to report on the occurrence of

BOX 3.2 Arsenic in the Well and in the Woods

Sometimes, in an effort to alleviate a problem, well-meaning public health officials initiate interventions that backfire. Consider the following example.

Until about 40 years ago, millions of people in Bangladesh and West Bengal, poor and densely populated regions, drank surface water contaminated with pathogens from shallow hand-dug wells, streams, and ponds, resulting in a high burden of disease. To combat the high incidence of death and disease resulting from contaminated drinking surface water, international aid agencies such as UNICEF and local health officials installed tube wells to tap groundwater.

A **tube well** is a simple device constructed of steel pipes sunk deep into the ground and fitted with a pump handle (**Figure 1**). You might find one at a roadside picnic area where piped water is not available. The pump is sealed topside to prevent water from leaking back down the pipe. Microbes are filtered out as groundwater trickles through the aquifer, resulting in microbiologically safe water.

An estimated 3.5 million wells gave millions of people in the area access to the groundwater and was expected to be the answer to the epidemics associated with the use of contaminated unsafe surface water. Although the number of water-borne diseases was markedly reduced, the price was too high; the groundwater was contaminated with naturally occurring **arsenic** in concentrations well above the accepted levels. As it turns out, farmers were fertilizing their soil two to three times per year with a fertilizer that contained 20 milligrams of arsenic per kilogram, resulting in contaminated groundwater in the tube wells. By the mid-1990s, thousands of people had been diagnosed with arsenic poisoning, and a new crisis existed.

Chemically, arsenic is categorized as a heavy metal, as is mercury and lead. It is usually excreted from the body, but if excess amounts are ingested it accumulates. Arsenic is very toxic and interferes with essential enzyme systems, resulting in death due to multiple-organ failure. Historically, the use of arsenic as a poison for political assassinations dates back several centuries. Some historians believe Napoleon was killed by food and beverage tainted with arsenic. (*Arsenic and Old Lace* was a hilarious and highly popular play that opened on Broadway in 1941. It was a comedy about two elderly sisters

© Kennerth Kullman/Shutterstock.

Figure 1 Manual water pump that draws water from below ground. The water has not been treated in any form and can become contaminated with chemical pollutants and microbes in the environment.

who poisoned lonely old men with elderberry tea containing arsenic, strychnine, and a "touch" of cyanide). Interestingly, **salvarsan**, developed about 1909, one of the first drugs to treat syphilis, was an arsenic-containing compound.

It seems that the people of Bangladesh and West Bengal have unwittingly gone "from the frying pan into the fire." The choice may be between drinking arsenic-free, microbiologically contaminated water and drinking arsenic-contaminated, microbiologically safe water. This problem continues but there are immediate alternatives such as using arsenic-free safe water traps at shallow depths and rainwater harvesting. Since the discovery of arsenic-contaminated groundwater in Bangladesh, recent studies have found arsenic contamination to be a worldwide occurrence. Arsenic can be found in the groundwater in the United States, contaminating aquifers and wells. For example, arsenic in groundwater is a known public health issue in Maine.

waterborne diseases with the goal of characterizing and identifying the causative agents. Over the past century, legislation has been implemented in an effort to regulate the nation's water supply. In 1972, the Clean Water Act came into effect as a response to the pollution of water in the United States with industrial and human wastes; bodies of water were becoming the nation's dumping ground. The goal was to reduce waterborne diseases and other adverse outcomes. Two years later, in December 1974, the Safe Drinking Water Act became effective and established

measures to ensure the safety of drinking water at the tap. The act was updated in 1986 and in 1996.

Krasner's Notebook

Over the last several years we have watched students trudge into class with water bottles of all sizes, shapes, colors, and labels to get through the upcoming class of less than an hour. It almost appears like an army of soldiers readying for an extended march through Sinai or some other desert area.

TABLE 3.3 Effects of the Brutal Hurricane Season of 2017

Hurricane (Dates)	Location	Strength	Human Deaths	Damage
Harvey (August 17–September 1)	Texas, Louisiana	Category 4	68 (all in Texas)	$180 billion
Irma (August 30–September 12)	Florida, Georgia, South Carolina, Caribbean Islands	Category 5	47	$150 to $200 billion
Maria (September 16–30)	Puerto Rico, U.S. Virgin Islands	Category 5	Officially, 64; unofficially, ~4,600 (many from delayed or interrupted medical care). At least 74 residents fell ill with leptospirosis.	$90 billion Power outages as long as 7 months; second largest blackout in world history. Knocked out one-third of the sewage treatment plants; 4% without clean water 9 months later
Nate (October 4–8)	Louisiana, Mississippi, Alabama	Category 1	0	$6.5 billion

Information from N. Kishore, D. Marques, A. Mahmud, M. V. Kiang, I. Rodriguez, . . . C. O. Buckee. (2018). Mortality in Puerto Rico after Hurricane Maria. *New England Journal of Medicine*, May 29.

Sara Armos/ShutterStock, Inc.

michelmond/ShutterStock, Inc.

michelmond/ShutterStock, Inc.

michelmond/ShutterStock, Inc.

FIGURE 3.11 The 2017 hurricane season took its toll on the United States and its associated territories, including the loss of lives, diseases associated with flooding, indirect deaths related to power outages, and other damage. **(a)** Telephone poles snapped like toothpicks in Puerto Rico during the winds of Hurricane Maria. Power outages lasted for months. Some individuals were dependent on power for medical support and died as a result of power losses that cut off life-sustaining equipment. **(b)** Emergency service vehicles crossing flooded streets in Houston, Texas, during the heavy rains caused by Hurricane Harvey. **(c)** First responders walking through flooded streets caused by heavy rains of Hurricane Harvey to inspect flooded houses in Sienna Plantation, Missouri City, Texas. **(d)** Another shelter opens at the NRG Center in Houston, Texas, during Hurricane Harvey. Refugees are pictured seeking safety. New evacuees are waiting to check in.

Hand Hygiene

From 2006 to 2016, the number of worldwide deaths attributed to unsafe water, poor sanitation, and lack of **hand hygiene** (handwashing with soap and water) decreased by 25%. In developing countries, lack of handwashing is still the third largest contributor to the spread of infectious disease. In 2015, most countries in Africa had less than 50% coverage with basic handwashing facilities (**FIGURE 3.12**). It is estimated that inadequate hand hygiene results in 300,000 deaths each year, with the majority of deaths being children younger than 5 years.

Handwashing with soap can reduce the prevalence of pneumonia and diarrhea, the two leading causes of child **mortalities** (deaths) for those younger than 5 years. Interventions promoting handwashing with soap and hygiene education reduce the incidence of diarrhea, respiratory infections, and in turn, increase school attendance in developing countries. In health care, poor handwashing compliance is one of the major contributors to **healthcare-associated infections (HAIs)**.

Global Handwashing Day occurs October 15th of every year (it conveniently occurs during influenza season). This campaign was initiated in 2008 to reduce child mortality rates due to diarrheal and respiratory diseases. "Clean hands save lives!" (**FIGURE 3.13**). When should you wash your hands? The CDC Handwashing Page (https://www.cdc.gov/handwashing/) lists the following situations in which you should wash your hands:

- Before eating food
- Before, during, and after preparing food
- Before and after caring for someone who is sick
- Before and after treating a wound or cut

Courtesy of the CDC/Kimberly Smith.

FIGURE 3.13 Although people around the world clean their hands with water, soap is not always available for handwashing. Using soap removes microbes and viruses much more effectively. To prevent the spread of infections, wash your hands with soap and water for at least 20 seconds. If you don't have a clock or timer, sing the *Happy Birthday* song twice. Wet, lather, scrub, rinse, and dry your hands.

- After using the toilet
- After changing diapers or cleaning up a child who has used a toilet
- After blowing your nose, coughing, or sneezing
- After touching an animal, animal feed, or animal waste
- After handling pet food or pet treats
- After touching your garbage

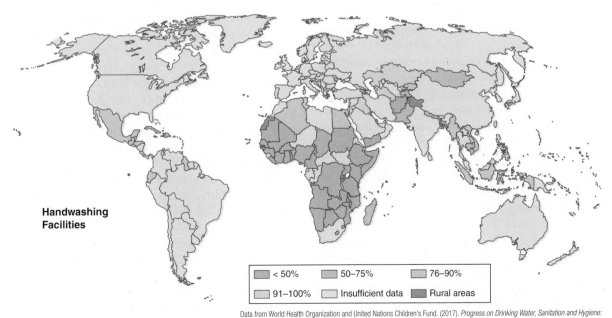

Handwashing Facilities

| < 50% | 50–75% | 76–90% |
| 91–100% | Insufficient data | Rural areas |

Data from World Health Organization and United Nations Children's Fund. (2017). *Progress on Drinking Water, Sanitation and Hygiene: 2017 Update and SDG Baselines*. Geneva: World Health Organization and United Nations Children's Fund.

FIGURE 3.12 Proportion of the world population with handwashing facilities, including soap and water at home, in 2015.

Handwashing education can reduce the number of people getting sick with diarrhea by 25% to 30%. It can reduce diarrheal illness in people with HIV by 50%. It can also reduce respiratory illnesses in the general population by 20%.

Food Safety

Leaders of the sanitation movement, initiated in the early 1900s, recognized that typhoid fever, tuberculosis, scarlet fever, botulism, and other diseases were transmitted by food (including milk) and water and advocated safer food-handling procedures, including **pasteurization**, refrigeration, hand hygiene, safer food processing, and pesticide application. *The Jungle*, the powerful 1906 novel by Upton Sinclair, portrayed the unsanitary practices in the Chicago meatpacking industry. An excerpt from the book follows:

> *There were some [cattle] . . . that had died, from what cause no one could say; and they were all to be disposed of here in darkness and silence. "Downers," the men called them; and the packing houses had a special elevator upon which they were raised to the killing beds, where the gang proceeded to handle them, with an air of business-like nonchalance which said plainer than any words that it was a matter of everyday routine. . . . and in the end Jurgis saw them go into the chilling rooms with the rest of the meat, being carefully scattered here and there so that they could not be identified.*

The public reacted so strongly that within a year after publication of Sinclair's book the Pure Food and Drug Act was passed. Even today, alarming stories are featured by news media about food processing. Despite the advances in the safety of the food supply, foodborne diseases remain a major cause of morbidity and mortality. Well over 30 microbes, including bacteria, viruses, protozoans, **prions**, and worms, are infectious agents that cause foodborne diseases. Many of these are tracked by public health systems that monitor diseases and outbreaks.

Further, changes in food-processing technologies, personal eating habits, and food distribution have contributed to the presence of new emergent pathogens. In 2011, the CDC estimated that 1 in 6 Americans (or 48 million people) suffered from a foodborne illness in the United States each year, with 128,000 hospitalizations and 3,000 deaths. Safer foods are considered one of the 10 great public health achievements of the United States during the 20th century. President Obama signed the FDA Food Safety Modernization Act (FSMA) on January 11, 2011. It was the most extensive reform of food safety in more than 70 years. The law will be fully implemented over time. It calls for rules and regulations related to the design, production, labeling, promotion, manufacturing, and testing of regulated food products. The WHO has expanded its food safety initiatives in response to new challenges, including health implications of genetically engineered foods. In the United States, the Food Safety Inspection Service of the **U.S. Department of Agriculture (USDA)** is charged with the safety, labeling, and packaging of the nation's commercial supply of meat, poultry, and eggs. Nevertheless, outbreaks of foodborne disease continue to occur.

How does food become contaminated along the pathway from farm to plate? What are the sources of contamination? In 2010, the American Society for Microbiology issued a report called *Global Food Safety: Keeping Food Safe from Farm to Table* that cited urgently needed technologies to improve food safety at the preprocessing stage; these technologies are listed in **TABLE 3.4**.

Ultimately, foods are prepared and consumed in the home—the last stage in the journey from farm to plate. Cleanliness in the kitchen has received considerable attention in recent years. A biologist who calls himself the "Sultan of Slime" claims that *"our bathrooms may be cleaner than our kitchens"*; perhaps a course in "kitchen microbiology" is not a bad idea. Environmental microbiologist Charles Gerba, also known as "Dr. Germ," became famous for his study on toilet splatter and continues to test offices, households, and public places for bacterial pathogens. Dish towels, sponges, counter surfaces, and hands are all culprits involved in the spread of foodborne pathogens. Using common sense can minimize the problem.

New measures to increase food safety, some of which are "gimmicky," are in vogue. The (uninspiring) debate regarding the use of wooden versus plastic cutting boards still exists. One entrepreneurial company now advertises "cut and toss" disposable cutting boards. Toxin Alert, a Canadian company, has developed an "intelligent" food wrap, covered with antibody sensors, that changes color when it is contaminated with *Salmonella*, *Campylobacter*, *E. coli*, and other bacteria. Pasteurized eggs are available and marketed under the brand name Davidson's Pasteurized Eggs®; they look and taste like unpasteurized eggs. This new technique, allowing in-shell pasteurization, may "egg on" other producers to make pasteurized eggs more available.

The most common causes of foodborne outbreaks that occurred in the United States from 2009 to 2016 are shown in **FIGURE 3.14**. According to the CDC's National Outbreak Reporting System (NORS), from 1998 to 2016, there were 19,991 foodborne outbreaks in the United States, resulting in the reporting of 388,238 illnesses, 27,909 hospitalizations, and 1,290 deaths.

One can take some comfort in knowing that city and state health departments, as well as agencies at the national and international levels, including the CDC's NORS, the USDA, and the WHO, are all working to develop more efficient strategies of **surveillance** and of standardization of sampling procedures. In the United States, the largest meat and poultry processing plants have been required by the USDA since 1998 to implement the Hazard Analysis and Critical Control Point program, which uses a systematic approach to identify, evaluate, and control food safety hazards.

TABLE 3.4 Cost-Effective Technologies Used to Significantly Prevent Food Contamination

Point of Use	Technology
Prevention at initial source	Effective vaccines for livestock and inoculants to prevent pathogen colonization
	Pathogen-resistant varieties of plants and animals
	Better means for managing potential **mycotoxin** contamination in the field
Harvesting and processing	Improved produce-harvesting equipment to reduce risk of contamination that can be easily sanitized on a regular basis
	Effective decontamination techniques for dried products and fresh produce to prevent uptake of pathogens by produce during post-harvest processing
Storage	Better storage structures in developing countries to prevent contamination by pests
	Better means for managing potential mycotoxin contamination during storage
Water contamination	Means to detect water quality breaches
	Practical water treatment devices

Information from the American Society for Microbiology. (2010). *Global food safety: Keeping food safe from farm to table*. Washington, DC: American Academy of Microbiology. (The report is based on a colloquium sponsored by ASM that convened April 24–26, 2009, in San Francisco, California.)

The first significant foodborne pathogen, *E. coli* O157:H7, was recognized in 1982. In 1996, **PulseNet** (https://www.cdc.gov/pulsenet/index.html), a network designed to track foodborne illnesses at a national level, was established, allowing comparison of the DNA fingerprint of bacteria from the patient that makes them sick to determine if the DNA fingerprint matches DNA fingerprints or clusters from anywhere in the United States in order to identify the contaminated food source. About a year later, PulseNet revealed that isolates of *Listeria* responsible for outbreaks of infection involving an estimated 100 cases and 22 deaths were linked to contaminated hot dogs and deli meats from a single processing plant.

In 2001, PulseNet became a nationwide system: all 50 state public health laboratories were trained and certified in methods to detect foodborne pathogens. In 2016, the *American Journal of Preventive Medicine* published an evaluation that suggested that PulseNet prevents at least 270,000 illnesses from *Salmonella*, *E. coli*, and *Listeria* each year, ultimately saving $500,000 each year in reduced medical costs. The CDC uses three types of data to track illnesses to contaminated foods in order to solve outbreaks:

- Epidemiologic
- Traceback
- Food and environmental testing

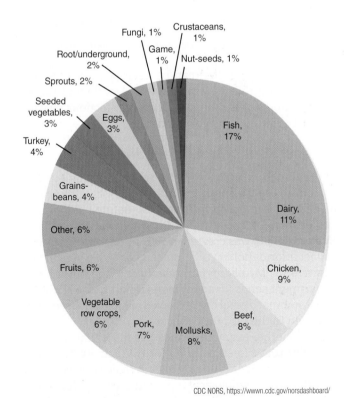

CDC NORS, https://wwwn.cdc.gov/norsdashboard/

FIGURE 3.14 CDC reports foodborne illnesses in the United States, 2009 to 2016.

FIGURE 3.15 explains the steps of an outbreak investigation and data collection. Health officials will evaluate all three types of data collected to try to identify the source of the outbreak. Further actions will be used to protect the public when there is convincing evidence that links a foodborne illness to a contaminated food product. Not every case is solved, because sometimes outbreaks end before enough information is gathered to investigate and solve outbreaks faster.

Food Safety News is a web-based publication dedicated to reporting illnesses associated with food safety. The CDC also has a Food Safety web page with food safety tips and information on laws and regulations, prevention, education, current safety challenges, and outbreaks (https://www.cdc .gov/foodsafety/). It is convenient to rely on the food industry and governmental agencies as guardians to protect consumers from the perils of foodborne diseases. However, the specter of a disease of cattle, **mad cow disease** (also known as bovine spongiform encephalopathy, or BSE) and its form in humans, variant Creutzfeldt-Jacob Disease (vCJD), are prion-caused diseases, and continued outbreaks of foodborne disease sound a word of caution the world over. Consumers need to share in the responsibility of minimizing the consumption of contaminated foods by exercising common sense without becoming food safety fanatics.

Infectious Disease Surveillance

Improving sanitation and hand hygiene, providing access to clean water, and ensuring food safety are basic measures needed in order to improve human health and prevent the spread of infectious diseases. *According to the WHO, infectious diseases are the second leading cause of deaths in humans worldwide and continue to remain a major public health problem.* HIV/AIDS, tuberculosis, and malaria have been nicknamed the "big three" because of their large impact on global human health.

Surveillance systems must be in place to collect data to monitor the presence of disease outbreaks and identify pathogens causing disease outbreaks. The three surveillance strategies are **disease-specific surveillance**, **syndrome/symptom-based surveillance**, and **event surveillance**. The following types of data are collected for surveillance:

- Laboratory data (e.g., diagnostic testing)
- Demographic data (e.g., vital statistics, cases, deaths, absenteeism from work and schools)
- Data from physicians (e.g., signs and symptoms reported on medical records)
- Hospital and clinical laboratory data (e.g., hospitalizations and discharge data, disease reports)
- Pharmacy data (e.g., type and quantity of prescriptions sold)
- Other types of data (e.g., wild and domestic animal health data, Internet queries, health hotlines, etc.)

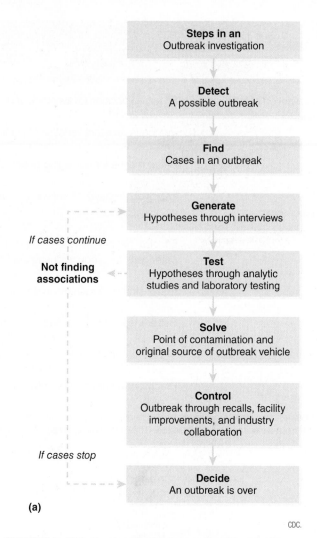

CDC.

FIGURE 3.15A The steps of a foodborne outbreak investigation.

Most data collected can be used by more than one surveillance strategy (**FIGURE 3.16**). Developing surveillance systems to combat and prevent infectious diseases is a fast-growing and evolving field.

Disease-Specific Surveillance

Surveillance is achieved through numerous tracking systems. Disease-specific surveillance targets a specific disease or set of symptoms in a defined population. For example, the **National Tuberculosis Surveillance System (NTSS)** was created in 1953 in the United States to collect information on tuberculosis cases. If a person tests positive for *Mycobacterium tuberculosis*, the state health department sends an anonymous report to the NTSS. The NTSS summarizes all data received, which is subsequently provided to the CDC for publication.

As another example, France has more than 4,300 laboratories that screen patients for HIV activity. Their data are shared through a weekly *Institut National de Veille Sanitaire* (NVS) epidemiological report. In 2004, the **Gonococcal Antimicrobial Surveillance Programme (Euro-GASP)** was implemented to provide data on the antibiotic susceptibility of *Neisseria gonorrhoeae* strains circulating in Europe.

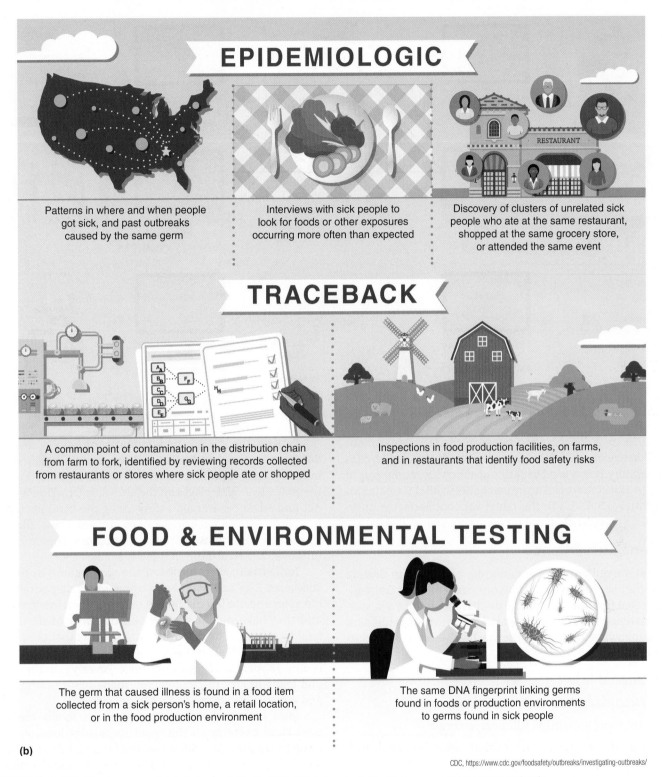

(b)

CDC, https://www.cdc.gov/foodsafety/outbreaks/investigating-outbreaks/

FIGURE 3.15B The three types of data that are collected by CDC officials in order to identify the source of a foodborne outbreak.

Disease-specific surveillance can be done with a wide range of pathogens and is useful for following disease trends.

Syndrome/Symptom-Based Surveillance

These surveillance systems are based on **real-time data collection**. The real-time (or near real-time) tracking of data on the occurrence of syndromes or symptoms to identify potential public health threats. The data collected are nonspecific indicators of the health status of a population, such as school or work absenteeism rates. These reporting or tracking systems can be automatically generated and used to assist public health leaders in making decisions about the implementation of programs and policies for the prevention and control of infectious diseases. The **Electronic Surveillance System for Early Notification of Community-Based Epidemics (ESSENCE I)** was initiated

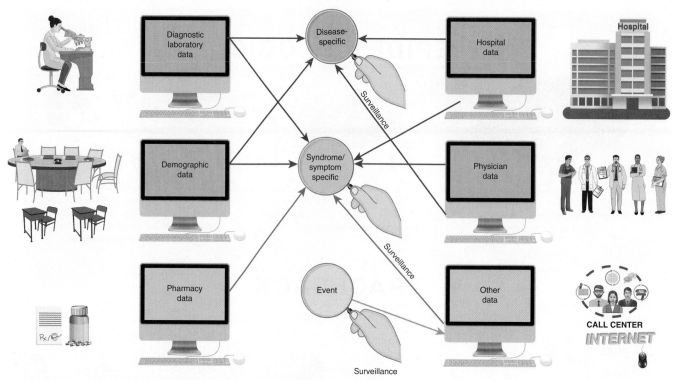

Information from C. Abat, H. Chaudet, J.-M. Rolain, P. Colson, & D. Raoult. (2016). Traditional and syndromic surveillance diseases and pathogens. *International Journal of Infectious Diseases, 48,* 22–28.

FIGURE 3.16 Summarizes the main types of data available for each surveillance strategy.

in 1999, as a collaboration between the U.S. Department of Defense and John's Hopkin's University Applied Physics Laboratory. It was used to perform worldwide monitoring of the health status of all army personnel in all U.S. treatment facilities. ESSENCE II, the latest version, performs automated surveillance of military and civilian medical data.

Event Surveillance

Event surveillance is the gathering of data from diverse Internet sources in real time or near-real time. Data are collected from news or online discussion platforms in various languages to detect potential or confirmed epidemics. The following are some of the most well-known event surveillance systems:

- **ProMED-mail**: https://www.promedmail.org
- **HealthMap**: http://www.healthmap.org/en/
- **Flu Near You**: https://flunearyou.org/#!/
- **Emergency and Disaster Information Service**: http://hisz.rsoe.hu/alertmap/index2.php

Event surveillance systems also can be used to monitor developing countries and gather data in multiple languages in order to detect true global epidemics.

Summary

The 20th century witnessed an increase in life expectancy in many nations of the world. The increase is attributable to public health achievements in sanitation, clean water, hand hygiene, food safety, and infectious disease surveillance systems—the topics of this chapter.

During the 1900s, departments of public health designed to promote health and longevity flourished in industrialized countries of the world. Implementation of the 1972 Clean Water Act and the 1974 Safe Drinking Water Act and advances in food safety along the complex path from farm to table were major steps forward. The availability of penicillin, followed by other antibiotics, dramatically improved the treatment of microbial diseases.

Industrialized countries of the world share in these successes. The sad part is the tremendous disparities in life span and quality of life that exist between the "haves" and the "have-nots"—a highly significant global public health problem. The disparities are so great that the average life span in Macau is 84 years, whereas in Swaziland it is only 32 years, a 52-year difference. Poverty is at the root and leads to lack of clean water, inadequate sanitation, and a host of other public health deficiencies.

More than half of Americans perceive foodborne illness from bacteria as the most important food safety issue today. President Obama signed the FDA Food Safety Modernization Act on January 11, 2011. It was the most significant reform of food safety in more than 70 years. The law will be fully implemented over time. It calls for rules and regulations related to the design, production, labeling, promotion, manufacturing, and testing of regulated food products.

Infectious disease surveillance systems are rapidly evolving in order to combat the challenges of emerging diseases and world population growth. Computer networks and real-time sharing of data worldwide allow for the detection of new disease outbreaks, which will allow health officials to respond with appropriate and timely control and prevention measures.

KEY TERMS

arsenic

basic drinking water service

biosolids

Carter Center

Centers for Disease Control and Prevention (CDC)

cholera

contagion

diarrhea

diphtheria

disease-specific surveillance

dracunculiasis

Electronic Surveillance System for Early Notification of Community-Based Epidemics (ESSENCE I)

Emergency and Disaster Information Service

Environmental Protection Agency (EPA)

event surveillance

Flu Near You

Food Safety News

germ theory of disease

Global Handwashing Day

Gonococcal Antimicrobial Surveillance Programme (Euro-GASP)

Guinea worm disease

hand hygiene

handwashing

HealthMap

healthcare-associated infections (HAIs)

hygiene

infectious diseases

Kampung Improvement Program

Koch's postulates

leptospirosis

mad cow disease

malaria

mortalities

mycotoxin

National Institutes of Health (NIH)

National Tuberculosis Surveillance System (NTSS)

necrotizing fasciitis

oral–fecal mode of transmission

Orangi Pilot Project

pasteurization

pathogens

plague

prions

ProMED-mail

public health

PulseNet

quarantine signs

real-time data collection

reservoirs

safely managed drinking water

salvarsan

sanitation

scarlet fever

sexually transmitted infections (STIs)

sludge

Sulabh International Museum of Toilets

surveillance

surveillance systems

syndrome/symptom-based surveillance

tersorium

tube well

tuberculosis

typhoid fever

U.S. Department of Agriculture (USDA)

United Nation Children's Fund (UNICEF)

urbanization

vectors

World Bank

World Health Organization (WHO)

World Toilet Day

SELF-EVALUATION

O PART I: Choose the single best answer.

1. _____ were responsible for the greatest percentage of foodborne outbreaks in the United States from 2009 to 2016.

   a. Beef

   b. Fruits

   c. Fish

   d. Dairy

2. The Kampung Improvement Program in Indonesia focused on _____.

   a. immunization programs

   b. providing antibiotics

   c. improving sanitation

   d. food safety

3. The _____ century in America could be called the "golden age of public health."
   a. 18th   c. 20th
   b. 19th   d. 21st

4. Less than _____ of Earth's water can be used and reused as freshwater.
   a. 2.5%   c. 25%
   b. 10%   d. 50%

5. On a typical day, more than half of the hospital beds in sub-Saharan Africa are occupied by patients suffering from _____ diseases.
   a. respiratory   c. liver
   b. fecal-related   d. kidney

6. _____ is an example of a disease-specific surveillance system.
   a. ProMED mail   c. HealthMap
   b. Euro-GASP   d. UNICEF

7. _____ is *not* a disease associated with contaminated water.
   a. Influenza   c. Dysentery
   b. Cholera   d. Giardiasis

8. _____ is a disease associated with improper sanitation.
   a. Gastroenteritis/diarrhea   c. Necrotizing fasciitis
   b. Mad cow disease   d. West Nile encephalitis

9. _____ is a method used to reduce the number of pathogenic microbes in milk.
   a. Autoclaving   c. Freeze-drying
   b. Pasteurization   d. Fermentation

10. _____ prevents the spread of infectious diseases.
   a. Blood-letting   c. Handwashing
   b. Consuming apples   d. Weight-lifting

○ PART II: Fill in the blank.

1. The sharing of _____ by the Romans in public toilets, likely contributed to the spread of diseases.

2. _____ outbreaks have been associated with hurricanes and flooding.

3. Epidemologic, traceback, and _____ types of data are gathered to link illnesses to foodborne disease outbreaks.

4. HealthMap reports disease outbreaks in _____ time or _____ time.

5. Improving _____ is an example of a basic measure needed in order to improve human health and prevent the spread of infectious diseases.

6. Pathogens that cause _____ may be present in outhouses or porta-potties.

7. _____ has become a serious contaminant of well water in Bangladesh and ground-water in the United States.

8. One should _____ for 20 seconds before preparing food for consumption.

9. The first significant foodborne pathogen was _____.

10. Climate change may be associated with an increase in _____ diseases.

O PART III: Answer the following.

1. Sanitation was "in" during the 20th century. Provide examples that support this statement.

2. Describe the work of local health departments as partners in the control of infectious diseases.

3. Explain why washing your hands with soap and water is more effective in combating pathogens than using hand sanitizer.

4. Why are hygienic means of sanitation and a safe supply of drinking water basic human needs?

5. Explain why life expectancy increased in many nations over the 20th century.

6. Approximately 892 million people still defecate in the open worldwide. Discuss why this is happening. What steps would you take to solve this problem?

7. List problems that compromise human health in large refugee camps.

8. How can "clean hands save lives"?

9. List and compare the top 10 countries that do not meet proper sanitation, the availability of clean water, and hand hygiene facilities. Discuss the challenges these countries face in order to improve these basic human needs for healthy living.

10. Explain why disease surveillance systems are rapidly growing and evolving.

PART 2

THE MICROBIAL CHALLENGE

IDENTIFYING THE CHALLENGE

We need to address the growing problem of drug-resistant infections as the global medicine cabinet is becoming increasingly bare.

—**Professor Dame Sally Davies**

Chief Medical Officer for England (2017)

Courtesy of the CDC.

This photograph was taken during an investigation of a monkeypox outbreak that occurred over 1996–1997 in the Democratic Republic of Congo (formerly Zaire). Veterinarian R. Joel Williams is pictured wearing a suit of medical scrubs, a face mask, and latex gloves as he handles two Gambian rats collected for study.

ShutterStock, Inc. / happykanppy.

1. Explain why infectious diseases are a global challenge.

2. State the significance of emerging and reemerging infectious diseases, and list three examples of each.

3. Detail the factors responsible for emerging infectious disease outbreaks.

4. Explain why the human population explosion is the "hub" of the problem of increased disease outbreaks.

5. Summarize how urbanization and poverty result in an increase of infectious diseases.

6. Describe the effects of climate change on the emergence of infectious diseases.

Case Study: Don't Forget to *Safely* Wash Your Hands!

In 2016, the **U.S. Food and Drug Administration (FDA)** issued a final ruling requiring manufacturers to remove **triclosan** and 18 other antimicrobial chemicals from liquid and bar soaps (**Figure 1, Table 1**). Though late in coming, the ruling was made for two reasons: (1) *there was no scientific evidence to prove that antibacterial soaps were more effective at preventing infections caused by human pathogens than ordinary soap and water, and (2) manufacturers did not demonstrate that the ingredients were safe for long-term daily use.* Companies had 1 year to remove the ingredients from their soap products; many sought to reformulate them with alternative ingredients such as benzethonium chloride. The FDA gave manufacturers 1 year to provide more data on the safety of benzalkonium chloride, benzethonium chloride, and chloroxyenol, the most likely alternatives to the banned chemicals.

©Teri Shors.

Figure 1 Common personal care and first aid products that contain antimicrobial compounds. Some of the items contain chemicals recently banned by the FDA. Studies are required to determine the need and speed for regulating triclosan and triclocarban.

In 1906, the medical pioneer Paul Ehrlich (1854–1915) was aware that both chlorine and phenol had antiseptic properties and suggested the use of chlorinated phenols as chemotherapy agents. However, it was not until the late 1930s and early 1940s that triclosan and a similar chemical compound, **triclocarban**, were synthesized by chemists in the laboratory by substituting hydrogen atoms present on the aromatic rings of phenol with chlorine atoms to yield a novel group of chlorinated phenols known as **organohalides** (**Figure 2**).

The organohalides were first developed as **biocides**. The term *biocide* was used in 1945 to describe any chemical that destroyed life, such as a pesticide, herbicide, or fungicide. In 1952, researchers demonstrated the germicidal activity of synthetic organohalides. Triclocarban and triclosan were patented in 1957 and 1964, respectively, as antimicrobials available for commercial applications. However, besides potent biocidal activity, most of the other organohalides synthesized in the 1940s that contained six substituted chlorines, such as hexachlorophene, had adverse effects, such as human toxicity. They accumulated in the environment, resulting in **ecotoxicity**, and thus many were banned by the 1970s.

In the 1970s, triclosan was first introduced into hospital soap scrubs and was able to effectively kill resistant staphylococci bacteria. Relaxed regulation, aggressive and widespread advertising by manufacturers, and media reports of serious and sometimes lethal infections acquired in everyday life resulted in consumer demand. Soon triclosan made its way into a plethora of products, such as laundry detergents, toothpastes, mouthwashes, skincare products, deodorants, shampoos, wipes, first aid medicated sprays, medical devices (e.g., surgical sutures, catheters, stents), plush toys, children's building blocks, children's pacifiers, clothing (e.g., socks, undergarments, surgical scrubs), kitchenware

(continues)

Case Study: Don't Forget to *Safely* Wash Your Hands!
(*continued*)

TABLE 1 Antimicrobial Compounds Banned from Bar and Liquid Soaps by the FDA*

Category of Antimicrobial Compound	Banned Antimicrobial Compound
Organohalides	Triclosan Triclocarban Hexachlorophene
Iodophors (iodine-containing compounds)	Iodine complex (ammonium ether sulfate and polyoxyethylene sorbitan monolaurate) Iodine complex (phosphate ester of alkylaryloxy polyethylene glycol) Nonylphenoxypoly (ethyleneoxy) ethanoliodine Poloxamer-iodine complex Povidone-iodine 5–10% Undecoylium chloride iodine complex
Phenolic compounds	Phenol (greater than 1.5%) Phenol (less than 1.5%)
Other	Cloflucarban Fluorosalan Hexylresorcinol Tribromsalan Methylbenzethonium chloride Secondary amyltricresols Sodium oxychlorosene Triple dye

*This ruling does not affect hand sanitizers or wipes or antimicrobial products used in healthcare settings.
Information from FDA. Final Ruling retrieved from: https://www.federalregister.gov/documents/2016/09/06/2016-21337/safety-and-effectiveness-of-consumer-antiseptics-topical-antimicrobial-drug-products-for

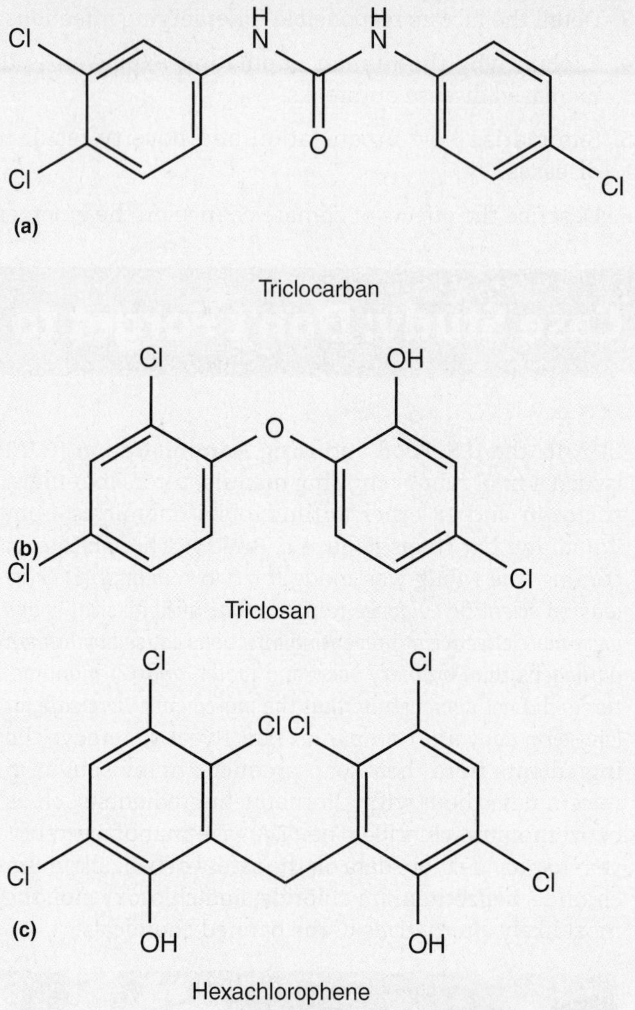

Figure 2 Structure of organohalides: **(a)** triclocarban, **(b)** triclosan, and **(c)** hexachlorophene. All of them possess antimicrobial properties. Due to toxic effects and bioaccumulation, hexachlorophene was banned from use in the United States in the 1970s. In 2016, the FDA banned triclosan and triclocarban in household liquid and bar soaps.

(e.g., cutting boards), sports gear (e.g., hockey helmets, fitness mats, earplugs), food packaging plastic, office and school products (e.g., binders, markers, scissors, calculators), carpets, humidifiers, and even paint. Triclosan is marketed under **Microban®** when incorporated into the manufacture of plastics and clothing and as **Biofresh™** when added to acrylic fibers.

By 2001, more than 700 antimicrobial products containing triclosan were available on U.S. supermarket shelves. In 2002, a U.S. Geological Survey listed triclosan among the seven most frequently detected chemicals in streams across the United States. In 2014, more than 2,000 products containing triclosan or triclocarbon were available for sale in the United States. Over 70% of liquid hand and dish soaps on supermarket shelves contained triclosan. The antimicrobial chemicals were prevalent pollutants in the environment, contaminating soil, water, and wildlife.

By 2001, researchers recognized that triclosan and triclocarbon caused **cross-resistance** toward medically important antibiotics used to treat infections. Today, an important area of research is the human **microbiome**, the microbes (and viruses) that live inside or on the human body. Researchers are attempting to discover associations between microbes and human diseases and to determine whether the presence or absences of microbes cause or contribute to human diseases. Because triclosan can effectively kill bacteria, it raises the possibility that its daily or long-term use could alter the bacterial species present on the skin and/or colonizing the gut. Disruption of human gut bacteria has been linked to a wide variety of diseases and metabolic disorders, including childhood obesity, irritable bowel syndrome (IBS), liver disease, rheumatoid arthritis, asthma, food allergies, Parkinson's disease, and Alzheimer's disease (**Figure 3**).

Before birth the skin is devoid of microbes. Bacterial colonization begins the instant a baby enters the world. If the baby is born from a cesarean section, the baby's microbiome will differ from that of a baby born through the mother's birth canal. The microbiomes of breastfed versus formula-fed infants also differ. The microbiome continues to change with alterations in diet/nutrition; exposures to people and the environment; antibiotic use; and major events such as puberty, pregnancy, and menopause. The microbiome of the skin is complex. It is unclear if prenatal and postnatal triclosan exposure results in an adapted microbiome, raising the possibility that the child will be more vulnerable to the aforementioned diseases and harder to treat skin infections.

Triclosan has been detected in human blood. U.S. studies in 2007–2008 detected triclosan in 75% of urine and 97% of breastmilk samples collected from human participants. It is absorbed through skin or ingested (e.g., toothpaste or mouthwash). Animal studies provided evidence that triclosan is an **endocrine disruptor**. Endocrine disruptors interfere with thyroid function and alter estrogen and testosterone regulation. Human and environmental exposure to triclosan may lead to developmental and/or reproductive problems in mammals, including humans. Some studies have suggested that triclosan negatively affects fetal neurodevelopment, decreases muscle strength, promotes early puberty, reduces sperm production, decreases infertility, and plays a role in the development of breast and liver cancers.

Triclosan exposure is ubiquitous, starting as early as prenatal exposure. More than 98% of infants in the United States are delivered in hospitals where triclosan and disinfectants containing other antimicrobials are used more frequently and in higher concentrations than in the home. Continued research is needed to determine the impact of triclosan on the human microbiome and the endocrine system and to what extent human health and well-being are adversely affected. How much triclosan and triclocarban is acceptable on planet Earth? The demand for antimicrobial agents continues to increase

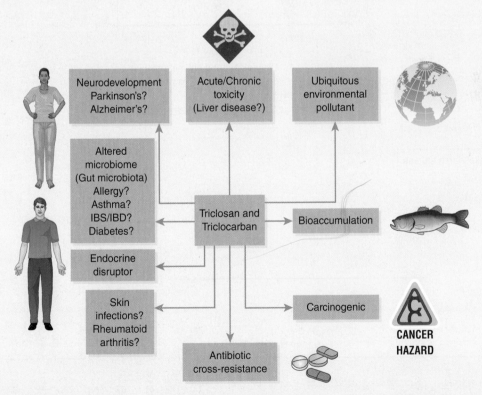

Figure 3 Triclosan and triclocarban accumulate and impact the environment, potentially causing a variety of adverse effects in humans and aquatic and terrestrial organisms. The future may hold greener, more sustainable antimicrobials that degrade rapidly in the environment, do not foster antimicrobial resistance, and are highly effective but low in toxicity to life. This approach promises to yield important benefits to humans and the planet.

(continued)

Case Study: Don't Forget to *Safely* Wash Your Hands!
(continued)

(**Figure 4a**). The number of studies to determine the health risks of triclosan also is increasing (**Figure 4b**). More than a decade into the accelerated use of triclosan and triclocarban, the Centers for Disease Control and Prevention (CDC), the U.S. Environmental Protection Agency (EPA), the FDA, and other agencies in the United States and Canada, and indeed worldwide, have increased their scrutiny because of their widespread contamination of the environment, wildlife, and human populations.

Remember, you only need to remove the microbes from your hands and wash them down the drain of the sink; *killing* them is not necessary. Procedures used in

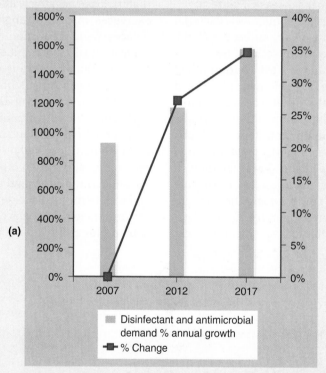

(a)

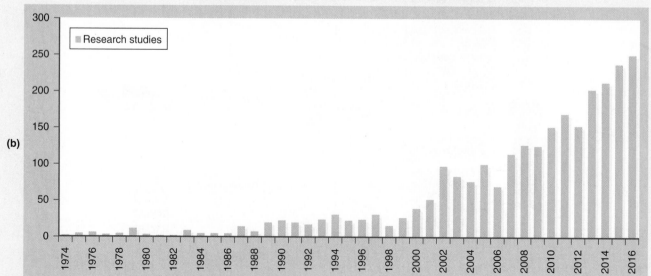

(b)

Figure 4 **(a)** The demand for disinfectants and antimicrobial chemicals continues to increase dramatically. The market for these products was forecasted in 2013 to rise 6.1% annually, becoming a $1.6 billion market. **(b)** A PubMED search for "triclosan" yields 2 reports in 1974 compared with 250 in 2016. A PubMED search for the effects of triclosan on human health (search string: "triclosan toxicity human") resulted in 122 reports through 2016. The first report occurred in 1989. Of the 122 reports, only 24 reports occurred between 1989 and 2010 (data not shown).

wastewater treatment plants will kill the removed microbes, rendering them harmless. **Global Handwashing Day** (http://globalhandwashing.org) is October 15th. It is a global advocacy day aimed to increase awareness and understanding about the importance of handwashing with soap as an effective and affordable way to prevent the spread of infectious diseases and save lives.

Questions and Activities

1. Can you find products in your residence that contain any of the chemicals listed in Table 1? If so, create a list of the products and the antimicrobial compounds they contain (remember that the antimicrobial compounds are present in many other products besides soaps).

2. Explain why antimicrobial compounds are both a boon and threat to human health.

3. Why did the FDA rule to ban 19 antimicrobials from household soaps but not from healthcare settings? Do you agree with this ruling? Why or why not?

4. Why are there concerns about the fate and effects of triclosan in the environment?

5. Create a list of greener and safer next-generation antimicrobial personal care alternatives.

6. Plan an event to celebrate Global Handwashing Day or share handwashing messages through social media.

Information from Halden, R. U. et al., 2017. "The Florence Statement on Triclosan and Triclocarban." *Environmental Health Perspectives 125*:064501.

Preview

This text is about a challenge—a worldwide challenge—posed by microbes, invisible marauders that inhabit Earth, many of which cause illness and death. Classically, there are five distinct kinds of microbes: bacteria, viruses, protozoans, fungi, and unicellular algae. More recently, prions have been added to the list, bringing the total of "infectious agents" to six. It should be emphasized at the outset that relatively *few members of each of these groups pose the potential for infection*. Most microbes are beneficial, and many are essential to the cycles of nature, without which higher life forms could not exist. In many cases, microbes have been harnessed for the benefit of humankind.

But this text, by intent, has a bias because its theme relates to those few microbes that cause disease. In the language of medical microbiology, they are referred to as **pathogens** or *virulent microbes*. Why some of these microbial diseases now represent an increased challenge is the subject matter of this chapter and sets the stage for the remaining chapters.

The Challenge

The need for this text continues. New or **emerging infectious diseases** are diseases that have appeared in humans in recent years; **reemerging infectious diseases** are those diseases that were present decades ago, subsequently disappeared, and then have recently reappeared in the population. It is possible that some emerging diseases have occurred throughout human history but were only recently recognized as specific diseases caused by a microbe, such as peptic ulcers caused by the bacterium *Helicobacter pylori*. More than 20 new **infectious diseases** (diseases caused by microbes) have emerged in the past decade, including Ebola virus disease (EVD), invasive candidiasis caused by

Candida auris, healthcare-associated *Elizabethkingia anophelis* infections, Zika virus disease (ZVD), Middle East respiratory syndrome (MERS), and Lyme disease associated with a new species of bacteria, *Borrelia mayonii*. Other infectious diseases have reemerged in recent decades, such as yellow fever, measles, whooping cough, multidrug-resistant (MDR) and extensively drug-resistant (XDR) tuberculosis, and multidrug-resistant gonorrhea, to name only a few (**FIGURE 4.1**).

Movies, books, and articles about microbial diseases intrigue large numbers of viewers and readers. Popular news magazine programs, including BBC and PBS Frontline documentaries, CBS 60 Minutes, ABC News Nightline, and CNN Anderson Cooper 360° frequently air segments relating to dangerous microbes; newspaper articles and news broadcasts appear almost daily and further alert the public to threats posed by microbes. The movie *Contagion*, based on a deadly viral epidemic, debuted in September 2011 and thrilled theatergoers about a pandemic caused by an airborne chimeric (hybrid) influenza-Nipah-like virus. In 2016, the TV series *Containment*, which follows an epidemic that begins in Atlanta, Georgia, aired in the United States. At the time of this writing, *The Walking Dead* TV series is in its eighth season. A virus is spread among humans, causing them to turn into "zombies" or "walkers."

Today, concerns have increased with regard to pathogens transmitted through the bite of infected black-legged, or "deer," ticks. The CDC has issued warnings because in the past 13 years, the diagnosis of tick-borne diseases have more than doubled in U.S. history, which may be due to warmer winters and an abundance of acorns fueling the population of white-footed mice, which are a preferred host of deer ticks. Deer ticks can be infected with a variety of microbial pathogens, including the bacteria that cause Lyme disease, ehrlichiosis, and anaplasmosis; babesiosis caused by parasites; and viral diseases caused by Powassan, Bourbon, Heartland, and

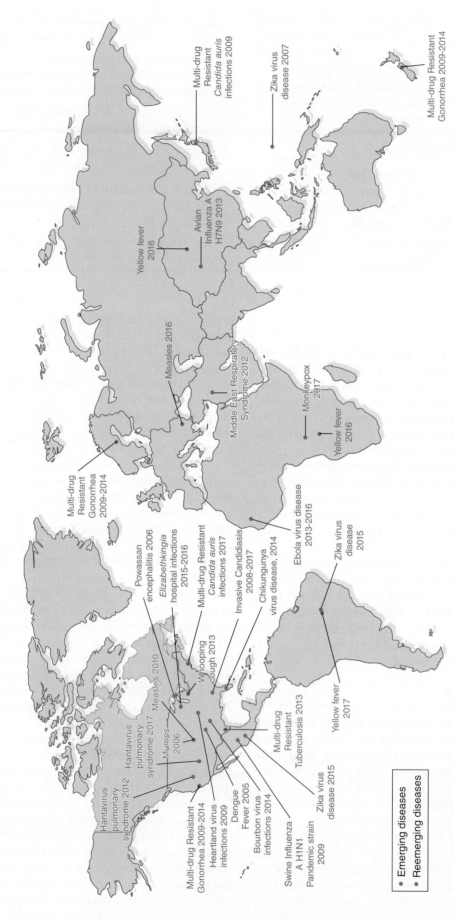

FIGURE 4.1 Recent emerging and reemerging diseases. For more recent outbreaks, see HealthMap (http://healthmap.org/en/).

Colorado tick fever viruses. Infected deer ticks spread disease when they bite and feed on animals in the wild. Most people become infected through tick bites during the spring and summer months. Tickborne diseases are becoming a serious problem in the United States because more and more people are building homes in previously uninhabited wilderness areas where ticks and their animal hosts live.

Global climate change and increased air travel also are influencing the emergence or reemergence of infectious diseases transmitted through the bite of infected mosquitoes. The mosquitoes are adapting to rising temperatures that influence the range of mosquito breeding and distribution of infectious diseases. Some diseases transmitted through mosquito bites are caused by parasites, such as *Plasmodium* spp., the cause of malaria, with 214 million new cases being identified in 2015. Chikungunya virus disease and dengue fever are mosquitoborne infectious diseases that are reemerging in regions around the world where they have been absent for years. Chikungunya virus, transmitted by *Aedes* spp. mosquitoes, spread to the Americas for the first time in 2013. Zika virus disease surfaced in South America in 2015. Zika virus is transmitted through the bites of infected *Aedes* spp. mosquitoes. The size of the outbreaks and their association with neurological damage such as **microcephaly** in newborns of infected pregnant women led the World Health Organization (WHO) to announce a Public Health Emergency of International Concern on February 1, 2016. Since December 2016, more than 50 countries in the Americas reported active local transmission of Zika viruses.

Measles, which had been eliminated in the United States in 2000, is now on the rise in the United States and Europe. Measles is the most contagious infectious disease; if you have never had the measles and are in contact with someone with the measles, you have a 98% chance of becoming infected with measles virus! Even though an effective vaccine is available to prevent measles, cases continue to emerge because of parental opposition to vaccination (**FIGURE 4.2**). Most cases in the United States are the result of international travel. At the time of this writing, more than 14,000 cases of measles had been reported in Europe since January 2016.

As noted by the American author and journalist David Quammen, "*Ebola doesn't disappear. It just goes into hiding.*" That being said, in 2013, an unprecedented

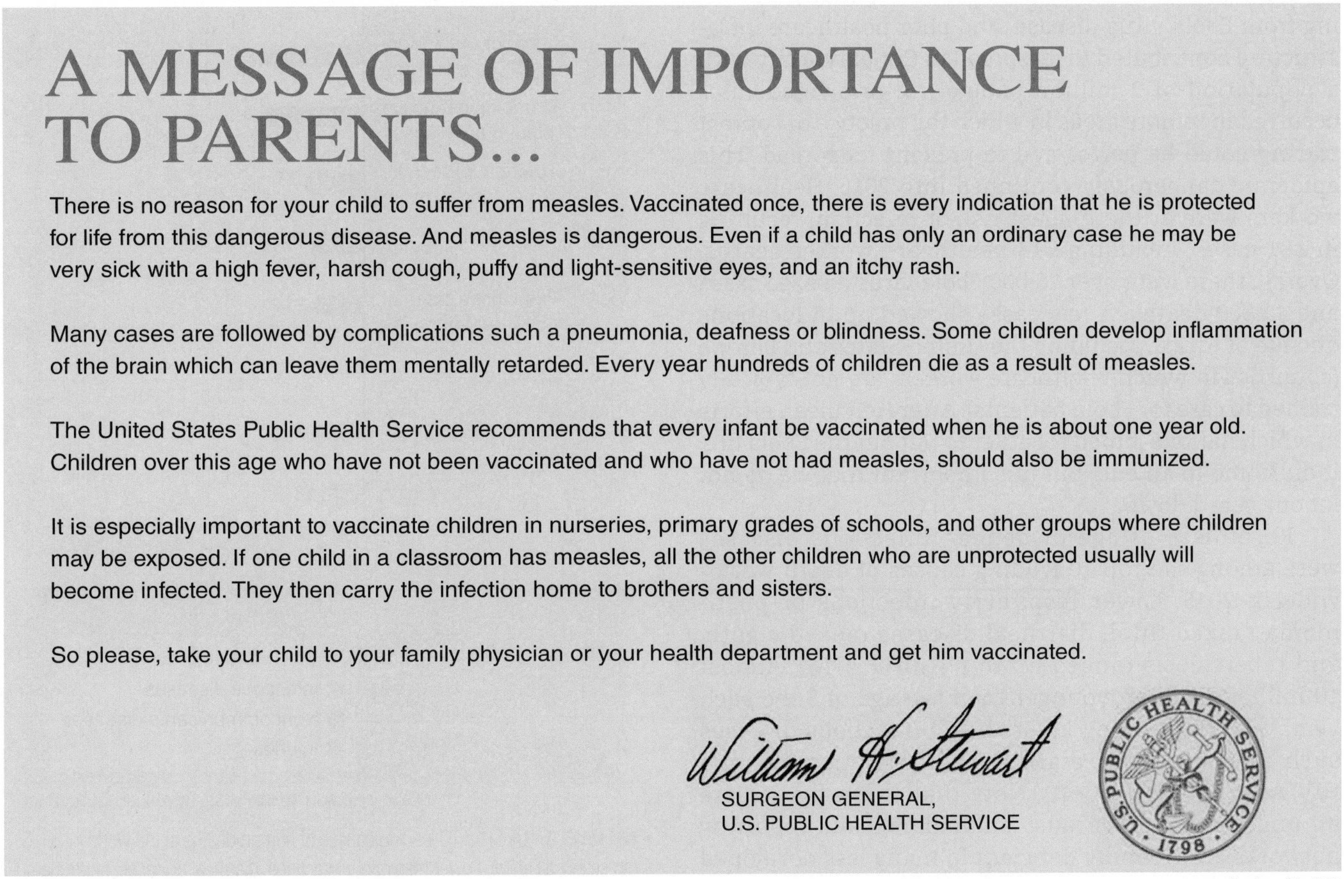

A MESSAGE OF IMPORTANCE TO PARENTS...

There is no reason for your child to suffer from measles. Vaccinated once, there is every indication that he is protected for life from this dangerous disease. And measles is dangerous. Even if a child has only an ordinary case he may be very sick with a high fever, harsh cough, puffy and light-sensitive eyes, and an itchy rash.

Many cases are followed by complications such a pneumonia, deafness or blindness. Some children develop inflammation of the brain which can leave them mentally retarded. Every year hundreds of children die as a result of measles.

The United States Public Health Service recommends that every infant be vaccinated when he is about one year old. Children over this age who have not been vaccinated and who have not had measles, should also be immunized.

It is especially important to vaccinate children in nurseries, primary grades of schools, and other groups where children may be exposed. If one child in a classroom has measles, all the other children who are unprotected usually will become infected. They then carry the infection home to brothers and sisters.

So please, take your child to your family physician or your health department and get him vaccinated.

SURGEON GENERAL,
U.S. PUBLIC HEALTH SERVICE

Courtesy of the CDC.

FIGURE 4.2 This U.S. Public Health Service announcement from the Office of the Surgeon General, Dr. William H. Stewart, was used during the 1960s to educate the public about the complications associated with measles and the importance of measles vaccination.

(a)

Courtesy of CDC/Umid Sharapov, M.D., M.Sc.

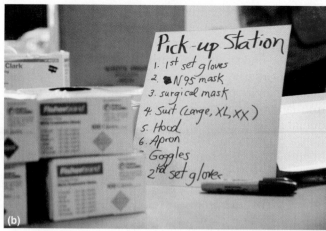

(b)

Courtesy of the CDC/Cleopatra Adedeji, RRT, BSRT.

FIGURE 4.3 (a) A burial team in Liberia on their way to perform a safe and dignified burial of Ebola victims during the unprecedented Ebola outbreak in West Africa. Despite the scorching heat, all four members of the team were dressed in full **personal protective equipment (PPE)**. **(b)** Station of PPE paraphernalia and a list of items that trainees were required to pick up to complete their PPE. The training station was located inside of a mock Ebola virus disease treatment unit (ETU) used in a CDC Domestic Training Course for healthcare workers.

Ebola epidemic started in Guinea, West Africa. It was the first time in human history in which the epicenter of an Ebola epidemic took place in West Africa. Medical teams struggled to curb its spread (**FIGURE 4.3**). Superstition, misinformation about how to care for patients suffering from Ebola virus disease, and poor healthcare infrastructure contributed to its spread to Conakry, a city with a population of 2 million people! All prior epidemics occurred in remote areas in which the practice of **contact tracing** could be performed to prevent its spread. This epidemic dangerously continued into 2016. Healthcare workers were at the highest risk of infection, resulting in 881 cases, including 513 healthcare worker deaths. Overall, there were over 28,000 Ebola virus disease cases and 11,300 deaths. A few cases showed up in locations outside of Africa, including the United States and Europe (countries in which healthcare workers are not typically trained to care for Ebola patients). After Herculean efforts in which nations joined together to put out the Ebola fire, it did come to an end, but it is important that we do not let our guard down.

Regardless of age or gender, infectious diseases were among the top 10 leading causes of death worldwide in 2015. Lower respiratory infections or pneumonia ranked third, diarrheal diseases ranked eighth, and tuberculosis ranked ninth (**FIGURE 4.4A**). Almost 10 million children younger than the age of 5 die each year, with the leading killers being infectious diseases such as pneumonia, diarrhea, malaria, measles, and HIV/AIDS (**FIGURE 4.4B**). Note that these figures are an underestimate, because surveillance and reporting networks are woefully deficient in many less developed countries. The leading infectious killers in the world, according to the WHO, include bacterial, viral, protozoan, and worm diseases. Initially, one would attribute these

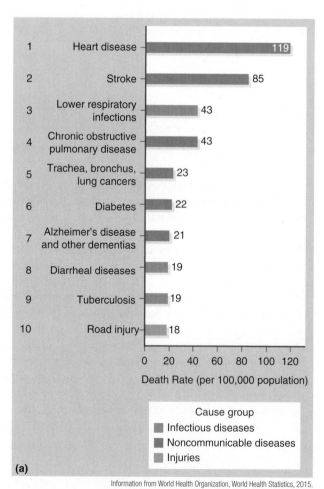

(a)

Information from World Health Organization, World Health Statistics, 2015.

FIGURE 4.4A The 10 leading causes of human deaths in the world in 2015 (both sexes, all ages). Lower respiratory infections, diarrheal diseases, and tuberculosis are infectious diseases ranked in the top 10 causes of death. Heart disease, stroke, chronic obstructive pulmonary disease, trachea, bronchus, lung cancers, diabetes, Alzheimer's disease and other dementias are not communicable diseases.

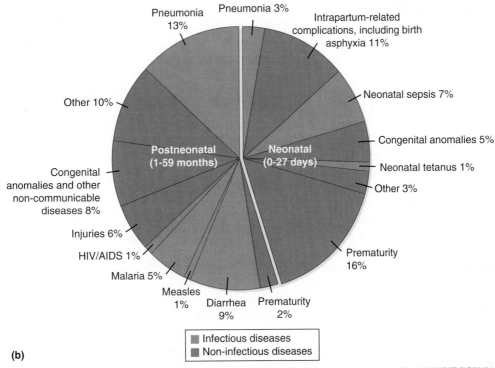

(b)

Information from WHO: MCEE Methods and Data Sources for Child Causes of Death 2000–2015. Global Health Estimates Technical Paper WHO/HIS/IER/GHE/2016.1.

FIGURE 4.4B Top causes of death worldwide in children younger than 5 years of age in 2015.

devastating statistics to the poverty associated with developing nations, but, as surprising as it may seem, despite the tremendous strides in infectious disease control over the past century, 2015 data from the U.S. National Vital Statistics Systems indicate that microbial disease also remains in the top 10 causes of death in the United States (**FIGURE 4.5**). No wonder that the statement of the director-general of the WHO in a 1996 report remains true to this day, "We stand on the brink of a global crisis in infectious diseases. No country is safe from them. No country can any longer afford to ignore this threat."

So, what's the bottom line? What grade would the world now be awarded in terms of its success in coping with microbial diseases? Certainly, under the leadership of the United Nations, WHO, the CDC, and other organizations, the burden of infectious diseases around the world can be lessened and a higher grade achieved.

Why are new diseases emerging and older ones reemerging with a vengeance, as it sometimes appears? The 1992 Institute of Medicine's report, *Emerging Infections: Microbial Threats to Health in the United States*, warned that microbes were winning the battle and that our previous complacency and optimism had weakened our ability to counterattack. Essentially, it appeared that the choreography of adaptation between microbes and humans was beginning to come apart at the seams

because of a variety of linked and overlapping factors: world population growth, urbanization, ecological disturbances, technological advances, microbial evolution and adaptation, and human behavior.

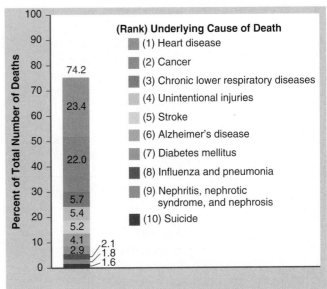

National Center for Health Statistics. Health. United States, 2016: With Chartbook on Long Term Trends in Health. Hyattsville, MD, 2017.

FIGURE 4.5 Leading causes of death in the United States in 2015. Average age of life expectancy was 78.8 years, and there was an average of 5.9 infant deaths per 100,000 live births.

Factors Responsible for Emerging Infections

Infections are a part of civilization and actually predate civilization. Microbes and humans coexist and share the same ecosystem. Infectious diseases have threatened the human population since ancient times, as evidenced by numerous references in the Old Testament and other sources of antiquity to "pestilence" and "plague." The **disease triangle model** addresses the interactions among the host (e.g., humans), infectious agent, and environment that produce infectious disease. Major causal factors that are involved in the emergence and reemergence of infectious diseases are discussed in this section (**TABLE 4.1**). All it requires is a change in one of the three elements of the disease triangle model (host, pathogen, or environment) to result in an infection and subsequent epidemic (**FIGURE 4.6**).

World Population Growth

Planet Earth is rapidly approaching a population of 7 billion people, with an annual growth rate of 1.10%. The United Nations estimates that the world population will reach 9.8 billion by 2050, and the most recent long-range projection is that it will reach 11.2 billion by 2100 (**FIGURE 4.7**).

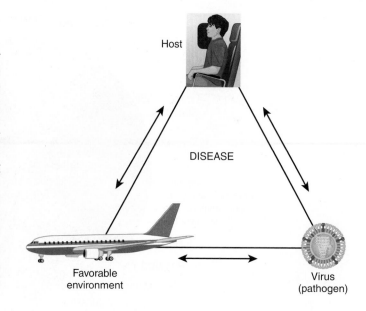

FIGURE 4.6 Disease triangle model of disease causation. Whenever there is a change in the host, pathogen, or environment, an infectious disease will occur. In this example, air travel can contribute to the global spread of respiratory infections such as influenza. Changes or causal factors that favor emerging infectious diseases are listed in Table 4.1.

Element of Disease Triangle Model	Causal Factor
Human host	Age (very young and elderly) Immune suppression Coinfection Genetic factors Behavior and attitudes (e.g., complacency and unprotected sex and sexually transmitted infections)
Microbe/pathogen	Microbial evolution and adaptation Resistance to antimicrobial therapies Evasive of host immune system

Environment	World population growth
	 Urbanization
	 Ecological disturbances
	   Deforestation
	   Climate change
	   Natural disasters (drought, floods)
	 Technological advances
	   Air travel
	   Transfusion of unsafe blood

TABLE 4.1 Factors Responsible for Emerging and Reemerging Infectious Diseases

To add to the problem of the burgeoning population, 80% of the population is living in less developed countries (of which 60% are tropical and subtropical areas) with a diminished capacity to cope with population increase. Several factors have been cited for the current crisis of new or emerging infectious diseases, but the human population explosion is central to the issue (**FIGURE 4.8**).

But there is another negative consequence of overpopulation: transmission of infectious diseases. The total human population of a country or a region, in itself, is not as crucial as its population density—the number of people per square kilometer (or per 0.39 square mile) in a defined area. Infectious diseases can be transmitted by person-to-person contact, by **biological vectors**, including mosquitoes, ticks, lice, and flies, and by animal-to-human contact (**zoonotic diseases** or **zoonoses**); whatever the mode of transmission, population density is a significant factor. Consider, for example, a classroom with fixed dimensions and assume that one person in the class has a cold, but only 10 other students are randomly spaced throughout the room, then consider the same classroom with 60 students. Clearly, the chain of transmission is fostered in the larger population, simply because respiratory droplets are able to traverse the shorter distance from contact to contact when population density is higher.

The age distribution of the population is also of considerable significance in terms of risk factors. For example, people older than 65 years comprised 38.6% of the U.S. population in 2010. Elderly populations are more susceptible to infectious diseases, presumably because of declines in the strength of the immune system with advancing age (see **BOX 4.1**). They serve as an increasing source of

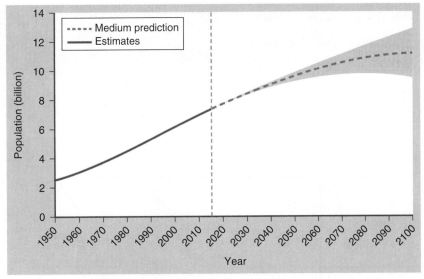

Information from United Nations, Department of Economic and Social Affairs, Population Division. (2017). World Population Prospects: The 2017 Revision. New York: United Nations.

FIGURE 4.7 World population estimates for 1950–2015 and median projections for 2015–2100. Projections are based on a growth rate of 1.10% per year, resulting in an additional 83 million people annually. The world's population was 7.53 billion in 2017 and is projected to increase further to 9.8 billion in 2050 and 11.2 billion people by 2100.

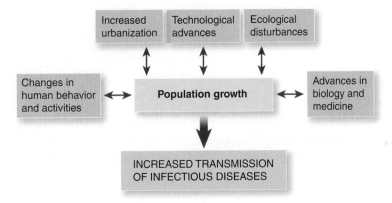

FIGURE 4.8 Population explosion: the "hub" of the problem.

BOX 4.1 The Mystery of the *Elizabethkingia* Outbreak in Wisconsin

In January 2016, epidemiologists from the CDC, the Wisconsin Division of Public Health, and the Wisconsin State Laboratory of Hygiene began a joint investigation on a healthcare-associated outbreak of bloodstream infections (**sepsis**) caused by the bacterium *Elizabethkingia anophelis*. Most patients were older than age 65 with a history of health issues such as cancer, diabetes, kidney dialysis, or chronic obstructive pulmonary disease (COPD) and had been treated at a healthcare facility in 1 of 12 different counties in Wisconsin (**Figure 1**). A total of 66 laboratory confirmed cases and 19 deaths occurred between November 1, 2015, and mid-May, 2016, in Wisconsin (63 cases), Illinois (2 cases), and Michigan (1 case). The patients suffered from invasive bloodstream infections and

presented with symptoms of fever, chills, shortness of breath, or cellulitis.

Infections are rarely caused by *Elizabethkingia* spp. because it is not a very virulent bacterium. It causes **opportunistic infections** in newborns, the elderly, or others with preexisting conditions, including immune suppression. Each year, 5 to 10 cases of *Elizabethkingia* spp. infections are confirmed in Wisconsin. An outbreak of this size caused by the same species and strain of *Elizabethkingia* was unusual. *Elizabethkingia* spp. infections are usually resistant to antibiotic therapy. Fortunately, laboratory testing was able to determine that the strain in this outbreak was susceptible to fluoroquinolones, rifampin, and trimethoprim.

(continues)

BOX 4.1 The Mystery of the *Elizabethkingia* Outbreak in Wisconsin
(continued)

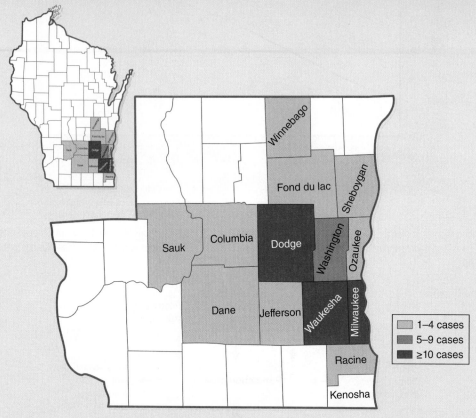

Figure 1 Map of Wisconsin counties affected by the *Elizabethkingia anophelis* outbreak (2015–2016). Confirmed cases (58 of the 63) were in Sauk (1), Columbia (1), Dane (1), Winnebago (1), Fond du lac (2), Sheboygan (2), Racine (2), Ozaukee (2), Washington (5), Dodge (9), Milwaukee (14), and Waukesha (18) counties.

The name of this bacterium is the combined first and last names of the CDC medical bacteriologist Elizabeth O. King (1912–1966; **Figure 2**), who determined that *E. meningoseptica* was associated with newborn meningitis cases in 1959. The three medically important species are *E. anophelis*, *E. meningoseptica*, and *E. miricola*. *Elizabethkingia* spp. infections are known to cause meningitis in newborns and meningitis, bloodstream infections, and respiratory infections in immune-compromised adults. *Elizabethkingia* spp. are ubiquitous in the environment. The bacteria are commonly found in fresh water, soil, and plants. A relative, *E. endophytica*, was isolated from sweet corn! Medically important *Elizabethkingia* spp. also have been found in hospital environments, with bacteria being isolated from sinks, contaminated medical devices such as catheters, flushing solutions used to clean mechanical ventilators, and infant pacifiers and comforters. *E. miricola* was also isolated from condensation water from inside the Russian space station MIR in 1997.

E. anophelis was isolated from the gut of *Anopheles gambiae* mosquitoes native to McCarthy Island, The Gambia,

West Africa, in 2011. Even though the bacterium was isolated from the midgut of mosquitoes, speculation that *E. anophelis* infections were transmitted by mosquitoes was ruled out in a 2012 neonatal meningitis case. In that case, a 33-year-old woman was suffering from **chorioamnionitis**, a complication of pregnancy caused by a bacterial infection of the fetal amnion and chorion membranes. The bacteria infected her newborn child through transplacental spread. Genomic DNA of the *E. enophelis* bacterial isolates collected from the mother and child clinical specimens contained the exact same DNA sequences. Therefore, maternal infection, not mosquitoes, was the most likely source of the neonatal *E. anophelis* infection.

The first documented case of an infection in humans caused by *E. anophelis* was reported in 2011, when an infant died from meningitis in the Central African Republic. In 2012, the first *E. anophelis* **healthcare-associated infection** occurred in a hospital in Singapore, involving five patients. In this cluster of cases, researchers isolated 14 different strains of *Elizabethkingia* spp. from contaminated

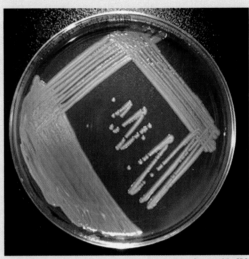

Courtesy of the CDC.   Courtesy of the CDC.

Figure 2 **(a)** The genus of the bacterium responsible for the Wisconsin epidemic was named after medical microbiologist Elizabeth O. King. She was the first researcher to identify the bacterium associated with neonatal meningitis cases in 1959. King was a researcher at the CDC from 1948 until her death in 1966. She earned a reputation as an expert on unusual bacteria that were difficult to identify in the laboratory. **(b)** A bacterial culture of *Elizabethkingia anophelis* growing on blood agar.

aerators on handwashing faucets in the intensive care units and the postsurgery wards to which the patients had been transferred.

The 2015–2016 Wisconsin epidemic, which was caused by a single strain of *E. anophelis*, had the largest number of cases and deaths to date. Investigators assessed potential point sources such as healthcare products, personal care products, food, tap water, and person-to-person transmission. Nose and throat swabs were collected from patients, family members, other household contacts, and healthcare workers. To date, all environmental, nose, and throat specimens have tested negative for *E. anophelis*. Investigators formed a "social network" to examine any commonalities shared between patients, healthcare facilities, and shared locations or activities in the community. To date, many unanswered questions remain about this common bacterium that seldom infects humans. No potential sources or routes of transmission have been identified despite an extensive CDC investigation.

infection for family and community members. The point is that those seeking to predict the emergence and spread of infectious disease need to take into account not only the total population and population density but also the age demographics in that population.

The increase in the world population will have a number of consequences that will foster an increase in infectious diseases:

- Increased person-to-person contact

- Accelerated climate change, which could expand habitats of infectious disease vectors

- Larger numbers of travelers

- More frequent wars and accompanying infection-fostering conditions

- Greater numbers of refugees and displaced people bringing infectious diseases to new areas

- Overcrowding in major cities and other urban areas that would increase person-to-person contact in areas without clean water and sufficient sanitation

- Increased construction of large dams, irrigation projects, and deforestation will expand the range of infectious disease vectors (e.g., mosquitoes) and result in a loss of habitat for species that carry zoonotic pathogens (e.g., bats carrying rabies or Ebola viruses), increasing their proximity to human populations

- More people living in poverty without proper sanitation or access to health care

- Increased hunger and malnutrition, resulting in weakened immune systems

- Inadequate potable water supply, facilitating the spread of waterborne diseases

The last three of these could potentially be prevented or minimized by new technologies. For example, Dhaka, the capital city of Bangladesh, has many slum areas. Bangladesh is one of the world's most populous countries, with approximately 163 million people in 2016. Population control is Bangladesh's most urgent problem, along with the low personal-attendant care (skilled nurses, doctors) per capita income. As would be expected, high levels of malnutrition exist, with much of the population getting less than one-third of the normal food intake because agricultural production has not been able to keep pace with population growth. The consequences in terms of infectious diseases, particularly diarrhea and pneumonia, due to poverty and poverty-related conditions have been dramatic.

Urbanization

At an international conference, Gerard Piel (1915–2004), an authoritative scientific journalist and the publisher of *Scientific American* magazine, founded in 1948, stated that "*the world's poor once huddled largely in rural areas. In the modern world they have gravitated to the cities.*" The following story is indicative of the depth of despair suffered by many in the world:

> *Zaynab Begum lives in Bangladesh in a Dhaka slum, along with her husband and three children in a primitive hut, less than 6 square meters in size, constructed of bamboo and makeshift materials. There is no electricity, running water, or toilet, and an open sewer runs outside the hut. Zaynab's husband is a rickshaw puller; her fate is shared by millions of others who have left their villages and migrated to the cities in search of work and a better life. It's a cruel realization that many who migrate to cities in search of a better life now live in urban poverty characterized by crowded and substandard housing lacking safe drinking water, inadequate toilets, and few tenant's rights.*

The trend toward urbanization dates back to early in the 20th century. In 1900, just 13% of the world's population lived in urban areas (towns or cities). According to the United Nations' *The World's Cities in 2016: Data Booklet*, the percentage increased to 54.5% in 2016 and is expected to climb to 60% by 2030 (**FIGURE 4.9**). In 2016, 23% of the world's population lived in a city with at least 1 million people. The world's 10 largest megacities are listed in **TABLE 4.2**. As of 2016, of the world's 31 megacities (cities with 10 million people or more), 24 are located in less developed regions or the "Global South." China alone was home to six megacities in 2016, while India had five. The magnitude of the effect of urbanization on communicable diseases varies dramatically in developed and developing countries, as a function of the economy and the public health infrastructure necessary to cope with the stress of increasing population density. The challenge of maintaining acceptable standards of sanitation and hygiene is far more difficult in developing countries, but it is also important to keep in mind that pockets of poverty and despair also exist in the United States and other developed nations (**FIGURE 4.10**).

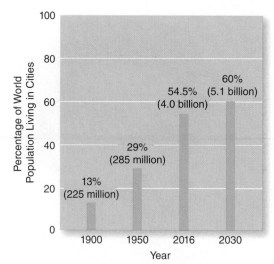

Information from United Nations: Department of Economic and Social Affairs, Population Division (2016) The World's Cities in 2016—Data Booklet (ST/ESA/SER.A/392).

FIGURE 4.9 Progressive urbanization of our planet.

TABLE 4.2 World's 10 Megacities, 2016	
1	Tokyo, Japan
2	Delhi, India
3	Shanghai, China
4	Mumbai (Bombay), India
5	São Paulo, Brazil
6	Beijing, China
7	Ciudad de Mexico (Mexico City), Mexico
8	Kinki M. M. A. (Osaka), Japan
9	Al-Qahirah (Cairo), Egypt
10	New York-Newark, USA

Information from United Nations: Department of Economic and Social Affairs, Population Division (2016). The World's Cities in 2016—Data Booklet (ST/ESA/SER.A/392)

Urbanization frequently leads to poverty and together set up a cycle of infectious disease (**FIGURE 4.11**). The drain on natural resources, including safe drinking water, is excessive, whereas at the same time problems of pollution, including human waste disposal and sanitation, are magnified. Untreated human wastes are dumped by the tons into the rivers, streams, and oceans. Ultimately, slums and shantytowns develop. The United Nations Human Settlements Programme estimates more than half of the urban population in sub-Saharan Africa

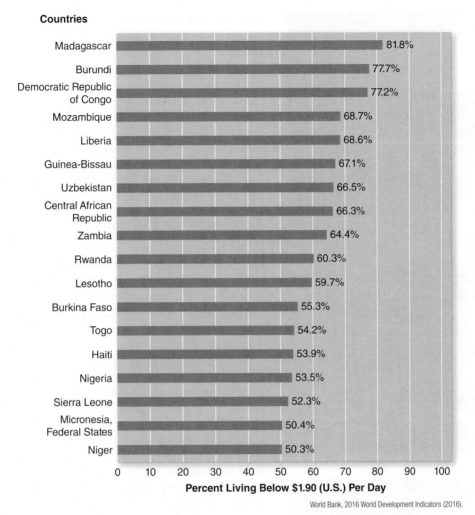

Countries

Madagascar — 81.8%
Burundi — 77.7%
Democratic Republic of Congo — 77.2%
Mozambique — 68.7%
Liberia — 68.6%
Guinea-Bissau — 67.1%
Uzbekistan — 66.5%
Central African Republic — 66.3%
Zambia — 64.4%
Rwanda — 60.3%
Lesotho — 59.7%
Burkina Faso — 55.3%
Togo — 54.2%
Haiti — 53.9%
Nigeria — 53.5%
Sierra Leone — 52.3%
Micronesia, Federal States — 50.4%
Niger — 50.3%

Percent Living Below $1.90 (U.S.) Per Day

World Bank, 2016 World Development Indicators (2016).

FIGURE 4.10 World's poorest countries in which 50% of a country's population earns less than $1.90 per day. Poverty is especially serious where rapid population growth occurs.

lives in slum conditions. Countries that have faced civil war report the highest rates: in the Central African Republic, South Sudan, and Sudan, more than 90% of the urban population live in slums. People live in filth

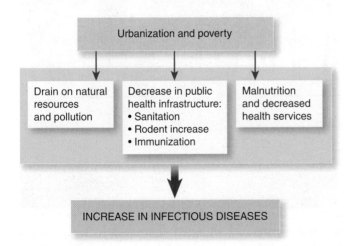

FIGURE 4.11 Relationships among poverty, urbanization, and infectious disease.

and squalor (**FIGURE 4.12**). Rodent populations increase as sanitation decreases, and the cycle of disease is perpetuated. Rodents may harbor fleas, which transmit pathogens that cause a variety of diseases, including the Black Death of 14th-century Europe, now known simply as "the Plague."

Numerous studies have concluded that city dwellers get sick more often than their rural counterparts and that people living in poverty are sick more often. As the world urbanizes, declining air quality in fast-growing regions is placing a growing burden on people's health. According to the 2013 Global Burden of Disease study, exposure to outdoor air pollution (as measured by levels of particles less than 2.5 microns in diameter) is responsible for 2.9 million deaths per year, or about 1 every 10 seconds! Outdoor air pollution levels are highest in East Asia and Pacific and South Asia, which have more than four times the particulate levels than the WHO's recommended guideline. Average pollution levels are estimated by comparing satellite of airborne particles observations covering both cities land rural areas with models of atmospheric chemistry. Prevention of communicable diseases is a

©Robert I. Krasner.

© Reuters/Athar Hussain/Landov.

© meunierd/ShutterStock, Inc.

FIGURE 4.12 Slums and shantytowns. Poverty is associated with a lack of sanitary facilities, an increase in rodent populations, a lack of safe drinking water, and other circumstances that contribute to infectious diseases. **(a)** A shack in rural Panama. **(b)** A shantytown in Karachi, Pakistan. **(c)** Individuals walking in the mud of a Nairobi slum, Kenya (October 2011).

major component of public health and is more problematic in cities than in rural areas and more open spaces.

However, it does not necessarily follow that infectious diseases run rampant in the megacities. Consider, for example, Tokyo, ranked as the world's largest urban area (TABLE 4.2), which is hardly poverty stricken or disease ridden. In fact, population statistics indicate that the Japanese people enjoy the longest life expectancy

(estimated to be 83.84 years in 2015). The country's economy and public health infrastructure make it possible for the country to cope with urbanization. By contrast, in most developing countries the crush of humanity and the tide of urbanization are overwhelming and beyond the financial resources necessary to construct sewage systems and to develop and maintain a public health infrastructure. Pulitzer Prize–winning science journalist Laurie Garrett in *The Coming Plague*, originally published in 1994, refers to cities as "microbe magnets" and "microbe heavens." "Graveyards of mankind" is a term used by British physician and molecular biologist John Cairns (1922–).

Ecological Disturbances

○ Deforestation

Almost half of Earth's forests either no longer exist or have been damaged, possibly to the point of no return, as a result of agriculture, settlement, logging, and mining over the past 8,000 years. The driving force, ultimately, is deliberate and financial—to make money—although wildfires and other natural phenomenon play a significant role.

Deforestation is a major factor in the eruption of emerging and reemerging infectious diseases. Wilderness habitats serve as **reservoirs** for a large variety of insects and other animals that harbor infectious agents. When the village or town becomes too crowded, whether it is in a poor and developing country or in a developed country, expansion occurs into the surrounding areas for tracts of land on which to build. Generally, the first event to give notice that construction is about to take place is the whine of the chainsaws signaling deforestation, followed by bulldozers moving in to uproot the tree stumps (**FIGURE 4.13**). Every time a tree is felled or a bulldozer digs up the soil to create another shopping or housing development, microbes and other soil and plant organisms are displaced. Fungi and bacteria and their spores are released into the environment and may alight and colonize on a human or animal and possibly give rise to a new or reemerging

©Robert I. Krasner.

FIGURE 4.13 Deforestation. As people move into areas that were formerly forests, they have increased contact with animals, including insects that harbor infectious microbes. Further, the displaced animals return to neighborhoods that were once their lands in search of food.

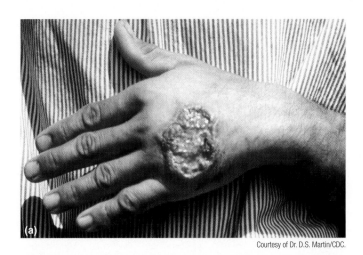

Courtesy of Dr. D.S. Martin/CDC.

©Robert I. Krasner.

FIGURE 4.14 Leishmaniasis is a protozoan infection transmitted by infected sand flies. **(a)** A leishmaniasis skin ulcer on the hand of a Central American villager. **(b)** Primitive living conditions in a village in Central America. It occupies an area that was formerly a forest and is encircled by a perimeter of trees. Sand flies are poor fliers but can traverse the short distance from their forest habitat.

disease. Examples of disease outbreaks of certain fungal diseases, such as coccidiodomycosis, also known as Valley fever, have been reported in construction workers, particularly in the southwestern United States. Perhaps this is what Louis Pasteur was referring to 150 years ago when he advised, "*the microbe is nothing; the terrain is everything.*"

Deforestation favors human intrusion into the environment and fosters contact with wildlife and insects and plays a major role in the migration of these displaced species into villages, communities, and backyards in search of food. An example is the rise of rabies in the eastern part of the United States as a result of raccoons infected by rabies virus foraging for food in the garbage cans of suburban and rural communities. Chagas disease is a protozoan disease carried by beetles, commonly called "kissing bugs," because they bite on the face and lips where the skin is thin. They are particularly prevalent in Brazil and other areas of South America. In the early 1900s, construction of the Central Railroad in Brazil was undertaken through the heavily forested tropical wilderness, a project that necessitated large-scale deforestation. You can guess the outcome—the indigenous mammals were displaced, as were the beetles that fed on them for their **blood meal**. Humans and their domesticated animals took up the slack and became infected, as did rodents. Experts later associated the Chagas disease cases with a species of beetles that inhabits housing in urban populations.

In 1998 and 1999, a new and deadly virus named "Nipah," killed more than 100 people in Malaysia after first showing up on a pig farm. It is speculated that the pigs ate dropped dates from trees contaminated with Nipah virus, which was then spread to farmworkers. Fruit bats, also known as "flying foxes," are the world's largest bats and have been identified as the natural reservoir for Nipah virus. In the years preceding the outbreak, massive deforestation took place and scores of date trees were destroyed causing the bats to forage elsewhere, including that remote pig farm surrounded by date trees. The bats foraged on the dates, leaving their saliva behind that was laden with Nipah viruses. The pigs ate the fallen and partially eaten dates that were contaminated with fruit bat saliva.

Leishmaniasis, a disease spread by sand flies infected with a protozoan, is a striking example of the consequences of deforestation and the emergence of urban infectious diseases. Even though leishmaniasis was once limited to mammals of the forest, the disease is now urban, primarily as a result of deforestation. The circumstances are similar to those described for Chagas disease (**FIGURE 4.14**).

Although it can be an attempt to relieve suffering and death, the intrusion of humans into ecosystems can backfire. Inadequate assessment of the public health impact can inadvertently increase infectious diseases. This was the case in the construction of Egypt's billion-dollar Aswan High Dam, a 10-year project completed in 1970. The dam harnessed the uncontrolled Nile River by creating Lake Nasser. The lake was named after Gamal Abdel Nasser, who was the Egyptian president from 1956 to 1970.

Krasner's Notebook

In the spring of 1999, I spent 6 weeks in Salvador, Brazil, at the Institute for Tropical Medicine in completion of the requirements of the M.P.H. degree at the Harvard School of Public Health to study tropical diseases in the natural context of their host. The last week of my stay was in the Amazon and included a visit to a small village with a high incidence of leishmaniasis. It was readily apparent why this was the case; the village bordered a heavy forest (FIGURE 4.14). Trees had been felled to allow for the construction of primitive dwellings, and yet, less than a mile away, leishmaniasis was not prevalent. The reason—sand flies are poor fliers and could not fly far from the forest. As a result of the deforestation, the village inhabitants and their dogs provided a blood meal for the sand flies. The old expression, "You can't see the forest because of the trees" has its virtues.

Conservationist Wangari Maathai founded the Green Belt Movement in Kenya to counteract deforestation. The movement resulted in the planting of 45 million trees by 900,000 poor women who received a few shillings for the work. Maathai won the Nobel Peace Prize in 2004 for her work on environmental strategies (and other accomplishments). She died on September 25, 2011, at the age of 71.

© Jan van der Hoeven/ShutterStock, Inc.

FIGURE 4.15 The interspecies leap. AIDS, which originated in Africa, is presumed to have jumped the species barrier from infected chimpanzees or sooty mangabeys to humans. Other infectious diseases of humans have made a leap from animals to humans.

TABLE 4.3 Infectious Diseases Linked to Climate Changes

Infectious Disease	Pathogen	Transmission
Lyme disease	Bacterium	Ticks
Anaplasmosis	Bacterium	Ticks
Powassan encephalitis	Virus	Ticks
Hantavirus pulmonary syndrome	Virus	Mice
Cholera	Bacterium	Waterborne
Gastroenteritis caused by *E. coli* infection	Bacterium	Waterborne
Cryptosporidiosis	Protozoan	Waterborne
Viral hepatitis	Virus	Waterborne
Leptospirosis	Bacterium	Waterborne
Chikungunya virus disease	Virus	Mosquitoes
Dengue (breakbone) fever	Virus	Mosquitoes
West Nile encephalitis	Virus	Mosquitoes
Yellow fever	Virus	Mosquitoes
Zika virus disease	Virus	Mosquitoes
Malaria	Protozoan	Mosquitoes

Unfortunately, the walls of the dam served as a new and convenient habitat for snails. The population of snails boomed, resulting in an increase in the incidence of schistosomiasis, a disease caused by a parasitic worm with a complicated life cycle requiring snails for its completion.

Rift Valley fever, a viral hemorrhagic disease transmitted by mosquitoes carrying Rift Valley fever virus, is another example of the downside of the Aswan High Dam project. An epidemic of this viral disease occurred close to the dam area, resulting in illness in 200,000 people and more than 500 deaths. The epidemic was the result of a thriving mosquito population in the flood lands created by the dam.

The most contemporary example of the potential consequences of humans' intrusion into the forest is that of AIDS. Most scientists agree that the origin of the human immunodeficiency virus (HIV) is the result of the simian immunodeficiency virus (SIV) that crossed the **species barrier** from infected chimpanzees and sooty mangabeys to humans (**FIGURE 4.15**).

◔ Climate Changes

What about the effect of **climate changes** on the emergence of infectious diseases? Ample evidence suggests that climatic changes cause ecological disturbances that affect the incidence and distribution of infectious diseases (**TABLE 4.3**). The 20th and 21st centuries have witnessed an increase in average global temperature attributed largely to the burning of fuels and forests, resulting in an increase in carbon dioxide and other heat-trapping greenhouse gases. The effects are seen not only in human health but also in the disruption of ecosystems and the resulting interference with food productivity. A meeting of world leaders was held in Kyoto, Japan, in 1997, to develop countermeasures against the impending threat of greenhouse gas emissions caused by human activity; these talks resulted in the **Kyoto Protocol**.

The following year, 1998, was the warmest year worldwide since 1880, when fairly accurate recordings began. The first 2 months were dominated by a record-breaking El Niño–influenced weather pattern, with wetter than normal conditions across much of the southern third of the United States and warmer than normal conditions across much of the northern two-thirds of the country. The increasingly high temperatures exacerbated the extreme regional weather and climate anomalies associated with

El Niño. That year brought into focus what scientists had long hypothesized, namely, that climate change could favor outbreaks of a variety of infectious diseases.

Temperatures measured on land and at sea for more than a century show that the Earth's average surface temperature is rising. Since 1970, the average global Earth surface temperature has risen at about the rate of 0.17°C (0.3°F) per decade, which is more than twice as fast as the average 0.07°C (0.1°F) rise per decade observed from 1880 to 2015. The year 2016 marked the fifth time in the 21st century that a new annual record high temperature was set (along with 2005, 2010, 2014, and 2015). The five warmest years have all occurred since 2010!

The United Nations created 17 sustainable development goals to transform Earth. Implementation of the Paris Agreement is needed to fulfill these goals. The Paris Agreement (http://www.un.org/sustainabledevelopment /climatechange/) was adopted in 2016 by over 100 countries to address global climate change.[1] The major goal is to limit the global temperature rise well below 2°C (3.5°F), striving for 1.5°C (2.6°F). **Earth Day** is an annual worldwide event that is celebrated on April 22nd to support environmental protection. Many communities celebrate **Earth Week** with this annual event. The **March for Science** occurred on Earth Day, April 22, 2017, and was followed by the **People's Climate Mobilization Day** on April 29, 2017.

In the case of vectorborne diseases, the vector or the microbe, or both, may be influenced by the temperature. **TABLE 4.4** summarizes data on the world population

TABLE 4.4 Status of Major Vectorborne Diseases

Infectious Disease	Insect Vector	World Population at Risk or Burden (Infected)
Malaria	Mosquito (*Anopheles* spp.)	3.4 billion people at risk
Dengue fever	Mosquito (*Aedes aegypti*)	2.5 billion people at risk
Chikungunya virus disease	Mosquito (*Aedes* spp., *Culex annulirostris*, *Mansonia uniformis*, and *Anopheles* spp.)	2.5 billion people at risk
Lymphatic filariasis	Mosquito (*Aedes, Culex,* and *Anopheles* spp.)	More than 120 million people infected
Yellow fever	Mosquito (*Haemagogus* and *Aedes* spp.)	200,000 people infected every year; 30,000 deaths every year
Zika virus disease[a]	Mosquito (*Aedes aegypti* and *Aedes albopictus*)	2.5 billion people at risk
Schistosomiasis (bilharzia)	Four types of freshwater snails (*Biomphalaria, Bulinus, Oncomelania,* and *Neotricula* spp.)	700 million people at risk
Onchocerciasis	Black fly (*Simulium* spp.)	37 million people infected
Leishmaniasis	Sand fly (phlebotomine species)	1.3 million new cases per year; 20,000 to 30,000 deaths every year
Chagas disease[b] (American trypanosomiasis)	*Triatominae* bugs (kissing bugs)	10 million people infected (mostly in Latin America)
Tickborne encephalitis[c]	Ticks (*Ixodes* spp.)	Europe, Russia, Siberia, and Asia ~5,000 to 13,000 people *reported* infected each year
Lyme disease[b]	Deer ticks (*Ixodes* spp.)	7.9 cases per 100,000 people infected in the United States per year; also occurs in rural areas of Asia and northwestern, central, and Eastern Europe

[a]Region-specific estimates.
[b]Estimates assumed to be similar to Chikungunya and dengue.
[c]Estimates from the 2018 *CDC Yellow Book*, Chapter 3: Infectious Diseases and Travel. All other information is from: *A Global Brief on Vector-Borne Diseases*, WHO, 2015; Document number: WHO/DCO/WHD/2014.1.

---

[1]President Donald Trump announced on June 1, 2017 that the United States was pulling out of the Paris Agreement, a vastly unpopular decision.

risk or burden of vectorborne diseases, all of which are likely sensitive to climate changes. Malaria, a protozoan disease transmitted by mosquitoes, is at the top of the list; an increase in both temperature and rainfall extends habitats favorable to mosquitoes. Conversely, increased temperature and decreased rainfall favor the distribution of sand flies, the vectors responsible for transmission of leishmaniasis, a protozoan disease. Estimates are that an increase in mean ambient temperature in central Africa by 2°C (3.5°F) would extend the range of the vectors of sleeping sickness, filariasis, and leishmaniasis, allowing for these diseases of the tropics to invade marginal temperate zones. Further, higher temperatures may push malaria transmission to higher altitudes, causing epidemics, as has occurred in the highlands of Ethiopia and Madagascar. In Rwanda in late 1987, malaria incidence increased by 337% over the previous 3-year period as a result of increases in temperature and rainfall. The reported cases of malaria (the form of malaria with the highest fatality rate) in the North-West Frontier Province of Pakistan rose from a few hundred in 1983 to more than 25,000 in 1990. This dramatic rise is attributed to unusually high temperatures at the end of the normal malaria season that extended the season. Malaria, tickborne encephalitis, and leishmaniasis (carried by sand flies) are also on the upswing in Italy as a result of climate change.

Hantavirus, the cause of a potentially fatal disease, emerged in the Four Corners area of the United States (where New Mexico, Utah, Colorado, and Arizona meet). Hantavirus infection occurs through airborne or droplet transmission of aerosolized urine or feces of deer mice, which contain hantaviruses. Deer mice are the natural reservoir and host of the hantaviruses. The mice feed on pine kernels. Higher than normal humidity favored an abundant crop of the pine kernels, which, in turn, led to a 10-fold increase in the deer mouse population between 1992 and 1993. This is an excellent example of a climate change triggering a chain of events resulting in the emergence of an infectious agent. *It can happen and will happen again.*

Diseases in which infectious agents cycle through insects to complete their development are particularly sensitive to subtle climate variations compared with diseases spread from human to human. Hence, it is imperative that consideration be given to and appropriate measures be enacted regarding the influence of global climate change on microbes and their vectors.

O Natural Disasters

Floods, hurricanes, earthquakes, drought, landslides, tsunamis, and volcanoes are environmental disturbances that place populations at risk of an increased burden of infectious diseases. In March and April of 2000, severe floods put the people of Mozambique and other southern African countries at risk for several diseases, particularly malaria and cholera. Up to 250,000 people in Mozambique alone were endangered by these two diseases. According to a WHO press release:

> The threat of a malaria epidemic in the country is increasing and will be at its most dangerous in around three to six weeks' time as flood waters gradually subside, the rain stops, and warm temperatures return—ideal breeding conditions for mosquitoes. . . . Before the floods, there were between 6 and 10 cholera cases a week; since the floods it has increased to 120 cases per week.

Myanmar (formerly Burma) was hit by tropical cyclone Nargis on May 2, 2008, resulting in over 100,000 deaths because of heavy rains and 12-foot water surges unleashed by the storm. Mudslides were triggered, contaminating wells that were a source of drinking water and blocking latrines, raising public health concerns as a result of a breakdown in sanitation.

The news and photographs from Somalia are truly devastating, particularly those featuring the terribly malnourished and dehydrated. There is little doubt that most, perhaps all, of the population harbors at least one species of worms. For over 30 years, Somalia has experienced poverty, armed violence, political insecurity, and natural disasters, leading to underdevelopment and the failure of the government to meet the humanitarian needs of its people. In 2015, the average life expectancy of Somalians was 55 years. Every year, natural disasters continue to occur in Somalia, affecting hundreds of thousands of people. A severe drought in November 2016 displaced 740,000 Somalians. These are in addition to the more than 1.1 million Somalians who have been displaced by civil unrest. On May 11, 2017, the WHO called for immediate action to save lives in Somalia in response to the ongoing drought that has plunged the country further toward famine, infectious disease, and health insecurity.

On December 6, 2004, the world was shocked to witness mountains of water cascading on the northwest coast of the island of Sumatra, Indonesia, triggered by an earthquake measuring 9.2 (out of 10) on the Richter scale. Natives and tourists ran to high ground for their lives, but an estimated 230,000 didn't make it and died. Half a million people were displaced from their homes, and thousands remain unaccounted for. The rapid response and level of relief measures from the international community was unprecedented. Almost immediately, groups worked to prevent infectious disease epidemics by providing "clean" water and bed nets (to prevent diseases transmitted by mosquitoes), initiating a measles vaccination program, and working to prevent and treat soil-transmitted worm infections. Although gaps in the public health infrastructure of the area and in the management of catastrophic events were uncovered, no large-scale outbreaks of infectious disease occurred, and mortality from disease was lower than anticipated.

Another natural disaster occurred when Hurricanes Katrina and Rita inundated large sections of New Orleans

(a)

(b)

Courtesy of Jocelyn Augustino/FEMA.

Courtesy of Andrea Booher/FEMA.

FIGURE 4.16 **(a)** A flooded neighborhood in New Orleans as a result of Hurricane Katrina. **(b)** A New Orleans resident inspects mold damage.

and surrounding parishes within a month of each other, August 29 and September 24, 2005, respectively (**FIGURE 4.16**). The pictures on television and in the newspapers and magazines were horrific; thousands of desperate people crowded into the New Orleans Convention Center; others clung to trees and rooftops hoping for rescue from the swirling waters. Surprisingly, no major outbreaks of infectious disease occurred, although there were cases of wound and gastrointestinal infection primarily due to exposure to contaminated floodwaters. The major microbial culprit was mold. As the waters receded, a black mold, *Stachybotrys*, thrived and grew in the high humidity and excess moisture. Anyone exposed risked respiratory infections. Local doctors dubbed patients' lingering respiratory symptoms, including cough, runny nose, and sinus problems as "Katrina cough." Further, the CDC reported the occurrence of 18 cases of wound-associated

illness caused by two bacterial species of *Vibrio*, 5 of which resulted in deaths (**FIGURE 4.17**). These infections generally result when open wounds are exposed to warm seawater containing specific *Vibrio* spp.; those with weakened immune systems and the elderly are particularly at risk. More recently, in 2011, wildfires in Texas and Arizona devastated the lives of scores of people and increased the potential for microbial diseases, as did Hurricane Irene. Joplin, Missouri, was hit by a tornado in 2011, which ripped through the city. In the aftermath, an unusual fungal skin infection caused by *Apophysomyces trapeziformis* broke out. Don't even try to pronounce the name. Also, be assured, the fungus is not related to trapeze artists!

Meteorologists agree that climate change makes hurricanes worse. Hurricanes thrive over warm ocean water and strengthen in intensity. The oceans have warmed 1°F to 3°F (-17°C to -16°C) over the past 100 years, and sea

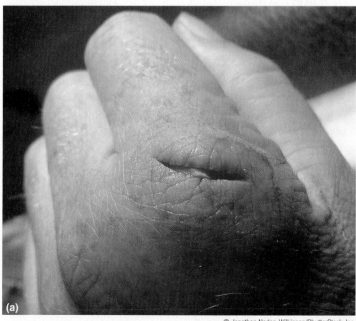

(a)

© Jonathan Noden-Wilkinson/ShutterStock, Inc.

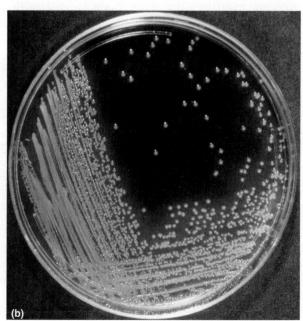

(b)

Courtesy of CDC.

FIGURE 4.17 **(a)** An open wound on a hand. **(b)** A diagnostic culture of *Vibrio cholerae*, the cause of cholera.

Courtesy of FEMA News Photo.

Courtesy of FEMA News Photo.

FIGURE 4.18 **(a)** Members of FEMA's Urban Search and Rescue Nebraska task force comb through neighborhoods impacted by Hurricane Harvey flooding. **(b)** Submerged vehicle on a road outside of Beaumont, Texas, after flooding caused by Hurricane Harvey.

level has risen about 18 centimeters (7 inches) in this period of time. Hurricane Harvey ravaged the greater Houston, Texas, area, followed by Hurricane Irma's storm surge and heavy rains that stretched across the entire state of Florida to Charleston, South Carolina, in 2017. Harvey dumped a record-breaking 129 centimeters (51 inches) of rainfall as it sat over Texas and Louisiana (**FIGURE 4.18**). Hurricane Irma was the longest-lasting Category 5 hurricane on record, hitting 297 kilometer per hour (185 mile per hour) winds or higher for a total of 37 hours (**FIGURE 4.19**). More than 4 million residents of Florida were without power for over a week. Eight nursing home residents died of heat exhaustion due to nursing homes lacking power for air conditioning, and at least six people died of carbon monoxide poisoning from generators.

After the hurricanes were gone and the flooding receded, even healthy individuals faced health hazards and infectious disease risks. In addition to heat exhaustion and carbon monoxide poisoning from generators, additional health hazards were related to contaminated floodwaters, polluted air, mold, mosquitoes, and infected wounds. People in the shelters were at risk of human-to-human spread of intestinal and respiratory diseases. Just days after Hurricane Harvey, a first responder faced a life-threatening infection caused by *Streptococcus pyogenes*, the "flesh-eating bacteria." The hurricane hero recovered after a 10-day hospital stay and multiple surgeries.

Technological Advances

○ Air Travel

Human activities lead to technological advances that may pose public health risks; jet travel is an example. It has been well documented that air travel plays a significant role in the transmittance of infectious diseases from continent to continent. **TABLE 4.5** lists approximate flying

Courtesy of FEMA News Photo.

Courtesy of FEMA/Robert Kaufmann.

FIGURE 4.19 **(a)** Members of FEMA Urban Search and Rescue Team Arizona Task Force One (AZ-TF1) take a brief break from the sweltering heat while performing door-to-door searches in a Fort Myers, Florida, neighborhood impacted by Hurricane Irma. **(b)** A neighborhood in Fort Myers, Florida, is scattered with debris following Hurricane Irma.

TABLE 4.5 Jet Travel: Microbes Without Passports	
Approximate Flying Time from New York City	Incubation Period for Selected Infectious Diseases
Sydney, Australia: 22 hours (1 stop)	Whooping cough: 7–10 days
Tokyo, Japan: 14 hours (nonstop)	Gonorrhea: 2–6 days
Tel Aviv, Israel: 10 hours (nonstop)	*Salmonella* food poisoning: 8–48 hours
Nairobi, Kenya: 16 hours (1 stop)	Ebola virus disease: 2–21 days (usually 8–10 days)
Karachi, Pakistan: 23 hours (nonstop)	Measles: 10–14 days
Delhi, India: 22 hours (1 stop)	Chickenpox: 2–3 weeks (commonly 14–16 days)
Moscow, Russia: 10 hours (nonstop)	Influenza: 24–48 hours
Monrovia, Liberia: 9 hours (nonstop)	Tuberculosis: weeks to years
Tokyo, Japan: 13 hours (nonstop)	Norovirus: 4 hours to 3 days
Brasilia, South America: 8.5 hours (nonstop)	Zika virus disease: 3–12 days

times from New York City to distant places. The farthest destination is Sydney, Australia, taking 22 hours, less than the incubation time for many infectious diseases. This means that an infectious traveler could board a commercial jet and arrive at any world destination in less than the time it takes for that passenger to show symptoms (**FIGURE 4.20**). Such an incident occurred in the spring of 2000: A tourist left Tel Aviv bound for Newark International Airport, an approximately 10-hour nonstop flight, and died of bacterial meningitis approximately 2 hours after landing. Fortunately, there were no reports of other passengers acquiring the disease. Bacterial meningitis has an incubation period of only a few hours to about 2 days.

Not so lucky were 13 travelers infected by a passenger with tuberculosis, often referred to as "TB," on a flight from Russia to New York. In May 2007, an international TB scare occurred when an Atlanta man previously diagnosed with an extremely antibiotic-resistant strain of TB (XDRTB) traveled abroad. The good news is that none of his fellow passengers on the aircraft became infected. In these cases the best that can be done is to notify other passengers to seek medical advice.

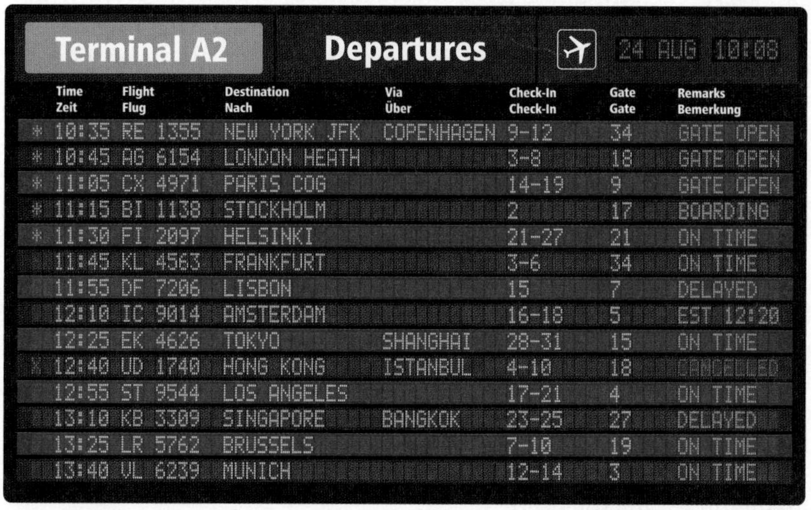

Besides other respiratory infections such as influenza and the common cold that can spread by airborne transmission among humans inside of a commercial jet, today the majority of infectious diseases related to travel in the United States are measles and Zika virus disease cases. Travelers have been infected while in locations such as Brazil where natural cases of Zika virus disease have occurred. They then present with symptoms upon their return to the United States. Typically, unvaccinated children contract measles when their families are vacationing in other countries and then show symptoms after they return. In 2015, a multistate outbreak occurred of 189 measles cases in 24 states and Washington, D.C. The epidemic was linked to exposure to a visitor infected with measles virus who went to Disneyland or Disney California Adventure Park in Anaheim, California.

© Scanrail / iStock / Getty Images Plus/Getty.

FIGURE 4.20 A flight departure board at an international airport. Jet aircraft, a major technological advance of the 20th century, serve as vectors for microbes around the world.

Probably one of the most concerning travel-related cases in history was the spread of Ebola virus disease through international travel in 2014. As the world was scrambling to put out the largest Ebola fire in human history, "sparks," or **imported cases**, began to hit countries outside of West Africa through international travel. Thomas Eric Duncan flew from Monrovia, Liberia, to Texas after being infected by Ebola virus. Duncan did not have any symptoms of Ebola virus disease and passed three separate temperature airport screenings and answered "no" on surveys in reference to having contact with anyone who had Ebola while in Liberia before he boarded a commercial plane and landed in Texas on September 20, 2014. Dallas, Texas, was ground zero for a potential Ebola outbreak! Duncan died of Ebola virus disease at Texas Health Presbyterian Hospital. Two nurses who cared for him were also infected with Ebola virus. Fortunately, both survived, and no healthcare workers were infected during their care. The United States had dodged the Ebola bullet.

Now consider the implications of the Airbus A380, nicknamed "Superjumbo"—introduced into service on October 25, 2007, with flight number SQ380 between Singapore and Sydney. It is the world's largest commercial airliner—a double-decker, four-engine craft with a wingspan almost as big as a football field. It has a cruising speed of 560 miles per hour (900 kilometers per hour). In a three-class seating configuration it can carry 525 passengers, but in a one-class economy seating configuration it can accommodate 853 passengers. From an epidemiological point of view, the aircraft is a nightmare—it serves as a huge potential *mechanical vector* capable of bringing infected people to any part of the world. Infected people can carry infectious agents to many different, final destinations, and in this way an epidemic can be triggered. In June 2014, more than 65 million passengers had flown on the A380. Three airlines are the primary users of the aircraft. As of August 31, 2017, Airbus had received 317 firm orders for the plane and 215 A380s were built and delivered.

Microbes can be harbored and transported across borders not only in their human hosts but also in their baggage and personal items. Further, **insect vectors** harboring infectious agents can also travel; fleas can be carried in rugs transported by jet cargo from the Middle East and Asia. Public health officials inspect many items being transported from country to country and are authorized to impose **quarantine** in an effort to minimize the risks.

○ Medical Advances

The use of whole blood and blood products is a life-giving and lifesaving practice. Unfortunately, in some countries blood and blood products may be hazardous to your health. According to the WHO, most of the countries with an unsafe blood supply are developing nations, in which the chances of acquiring infectious diseases are highest. As population pressure increases, so does the demand for blood. In some countries blood is not screened and may harbor causative agents of AIDS, hepatitis, syphilis, malaria, West Nile encephalitis, and trypanosomiasis.

To some extent advances in medical technology that make organ transplantation possible contribute to the burden of infectious diseases. Recipients are at increased risk for infection because they are on a regimen of immunosuppressive drugs to minimize organ rejection. It appears that face transplants (both partial and full) are on the map. The first face transplant was performed in France on a woman who was severely mauled by her Labrador dog. The first full facial transplant in the United States was at Brigham and Women's Hospital in March 2011. Other conditions leading to immunosuppression include AIDS, certain inherited diseases, and malnutrition.

Prostate cancer is the second leading cause of cancer-related deaths in the United States. Transrectal ultrasound-guided biopsies of the prostate gland are common diagnostic procedures. According to the CDC, approximately 624,000 of these procedures are performed annually. On July 21, 2006, the CDC reported on four cases of infection caused by *Pseudomonas aeruginosa* after transrectal ultrasound procedures. The infections were caused by contamination of the biopsy equipment that had not been properly sterilized. The bacterial strains recovered from patients matched the strains recovered from the lumen of the biopsy needle. This is an excellent example of a healthcare-associated infection.

Microbial Evolution and Adaptation

The 1940s ushered in the dawn of **antibiotics**—agents that were rightfully called "wonder drugs." Penicillin was the first antibiotic, and numerous others quickly followed; some were tailored to be effective against a broad spectrum of bacteria, whereas others were more specific. It should be emphasized that antibiotics are not effective against viruses and hence should not be prescribed for viral infections. The number of lives saved worldwide over the past 70 years because of antibiotic therapy is beyond estimation. An individual today who is infected with a variety of life-threatening bacteria has a fighting chance, assuming antibiotics are administered promptly, whereas an individual infected 70 or 80 years ago had little chance of recovery.

In the United States, each year 2 million people are infected with bacteria that are resistant to antibiotic treatment, and at least 23,000 people die each year as a direct result of these infections. The development of antibiotics was a major factor leading to the optimism of the 1970s. Many dreaded diseases, so it seemed, were about to become vanquished. But it turns out that the tables are turning—antibiotics are losing their punch, and increasing numbers of microbes are resistant. The expression "I'm resistant to such and such an antibiotic" has no meaning; people do not become resistant to antibiotics—their microbes do.

Emblazoned on the cover of the September 12, 1994, issue of *Time* magazine was the headline "Revenge of the Killer Microbes" and the question "Are we losing the war

TABLE 4.6 Antibiotic-Resistant Bacteria: Biggest Threats in the United States	
Antibiotic-Resistant Bacterium	CDC Ranking by Level of Concern
Clostridium difficile (CDIFF) Carbapenem-resistant *Enterobacteraceae* (CRE) *Neisseria gonorrhoeae*	Urgent

Multidrug-resistant *Acinetobacter*
Drug-resistant *Campylobacter*
Extended-spectrum beta-lactamase
  producing (ESBL) Enterobacteriaceae
Vancomycin-resistant *Enterococcus* (VRE)
Multidrug-resistant *Pseudomonas*
  aeruginosa
Drug-resistant nontyphoidal *Salmonella*
Drug-resistant *Shigella*
Methicillin-resistant *Staphylococcus*
  aureus (MRSA)
Drug-resistant *Streptococcus*
  pneumoniae
Drug-resistant *Mycobacterium tuberculosis*
	Serious

Vancomycin-resistant *Staphylococcus*
  aureus
Erythromycin-resistant Group A
  Streptococcus
Clindamycin-resistant Group B
  Streptococcus
	Concerning

Information from CDC Antibiotic/Antimicrobial Resistance.

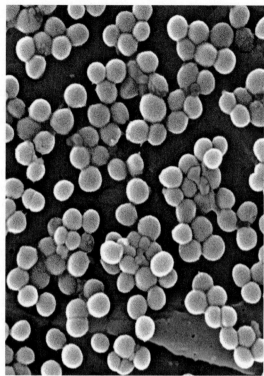

Courtesy of Janice Haney Carr/Jeff Hageman, M.H.S./CDC.

FIGURE 4.21 A scanning electron micrograph of methicillin-resistant *Staphylococcus aureus*, commonly referred to as MRSA.

against infectious diseases?" The answer to this question might be an uncomfortable "yes" (**TABLE 4.6**). Resistance to antimicrobial and antiviral agents is at a crisis level worldwide. **Vancomycin**, an antibiotic considered by many to be the last stronghold in certain situations, is no longer effective against many bacterial strains that responded 10 years ago. Some refer to antibiotic-resistant bacteria as "**superbugs**" (**FIGURE 4.21**). What's happening? In a nutshell, the forces of natural selection are in play. Antibiotics have been grossly overprescribed, which has promoted the emergence of antibiotic-resistant organisms in a Darwinian fashion. The antibiotic-resistant strains are the result of chance mutations, and their survival is favored by the presence of antibiotics. Antibiotics are the "selecting," not the "causing," agent.

The battle against the natural process of microbial adaptation and change, whether exhibited by resistance against antibiotics or by evasive strategies, is an ever-present and ongoing struggle for survival. Failure to meet the challenge affords microbes the upper hand.

Insect vectors are also able to adapt to a changing environment, bringing up the issue of climate change as previously discussed. Malaria, a mosquitoborne protozoan disease, was thought to be a disease of the past, thanks to the application of the insecticide dichlorodiphenyltrichloroethane (DDT). Little did scientists realize that the forces of natural selection would again interfere as a result of the misuse of the insecticide. DDT-resistant mosquitoes emerged with a vengeance, and other vectorborne diseases shared their triumph over DDT. West Nile virus, a mosquitoborne agent, now threatens all of the contiguous United States, prompting ground and aerial spraying. Could insecticide-resistant mosquitoes carrying West Nile virus emerge?

Human Behavior and Attitudes

○ Complacency

How easy it is to cut corners on health-related matters when it appears that progress and improvement have taken place, leading to the false assumption that prevention and control are no longer necessary? Complacency is the belief that "it can't happen to me." The failure of people to complete their full dose of antibiotics because they are feeling better is a prime example.

Advances in food technology make it much easier to eat "on the run" by buying prepared foods in markets and frequenting fast-food restaurants. Obesity is recognized as a major health problem in the United States and elsewhere around the globe. Obesity favors the development of numerous health problems, including susceptibility to microbial diseases. Lung cancer is the leading cause of cancer death and the second most common cancer among men and women in the United States. Many adults have switched to the use of electronic cigarettes (e-cigarettes), commonly referred to as "vaping," as a "safer" nicotine alternative to smoking tobacco/cigarettes. Vaping among middle and high school students tripled from 2013 to 2014 in the United States. Research has shown that e-cigarettes can cause adverse health effects and are an emerging public health concern.

A dramatic example of complacency is evident in the threatened resurgence of AIDS, particularly among young gay men, because of a return to risky sexual behavior fueled by glowing reports of new drug therapies for the management of AIDS. At least 5 million Americans have sex and/or drug habits that put them at high risk for HIV infection. The number of cases in the United States has fallen dramatically since the peak of the 1980s; the decrease is primarily attributed to safer sex habits and avoidance of dirty needles by drug abusers, but public health officials worry that the decrease in cases could cause complacency and result in an increase in the number of cases as people return to unsafe sex practices.

Epidemiologists did not predict that the worst HIV outbreak in the United States since the 1980s would occur in a small rural town in Indiana in 2015. Many of those infected lacked HIV awareness. The Austin, Indiana, outbreak was associated with injection drug use. Generations of families were injecting the prescribed opiate painkiller known as oxymorphone (Opana®), and some were also injecting heroin or methamphetamine. To prepare for injection, the 40-milligram tablets of Opana® were crushed, dissolved in water, and injected. Because it is illegal to purchase needles in Indiana unless one is a diabetic or requires a prescribed drug that must be administered by injection, it was not possible to obtain sterile needles. Users were sharing nonsterile needles to inject drugs. At least 170 individuals tested positive for HIV infection and 80% of them were coinfected with hepatitis C virus. Governor Mike Pence reluctantly declared the outbreak a public health emergency and announced a 1-year legal syringe-exchange program to stop the epidemic from spreading (**FIGURE 4.22**).

People have become complacent about receiving immunizations or keeping their immunization boosters up-to-date. According to the CDC, only 80% of 2-year-olds in the United States have been given the full sequence of currently recommended immunizations, primarily because of parents' complacency and concerns about vaccine safety. Past history reveals that a 10% decline in measles vaccination between 1989 and 1991 resulted in an outbreak of 55,000 cases, several thousand hospitalizations, and 120 deaths, indicating the power of immunization. Measles and mumps outbreaks continue to occur in the United States. The CDC tracks vaccinations and provides resources on vaccination requirements, statistics, and state laws about school requirements for vaccination

(a)

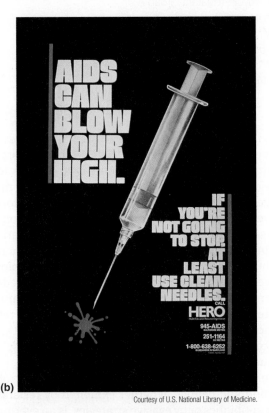

(b)

Courtesy of U.S. National Library of Medicine.

FIGURE 4.22 **(a)** Indiana county map showing the location of Scott County, where the Austin HIV/HCV epidemic occurred. **(b)** 1987 HIV/AIDS and injection drug use awareness poster used during the early years of the HIV epidemic in the U.S. Education plays an important role in preventing the spread of HIV.

(https://www.cdc.gov/vaccines/imz-managers/coverage/schoolvaxview/index.html).

Individuals traveling to foreign countries need to be aware that particular immunizations may be necessary against diseases prevalent in that area. For example, in 1996, tourists traveling to yellow fever areas neglected to be immunized against the disease and were responsible for infecting others with the disease upon their return to the United States and Switzerland.

Human Migration

Human migration is a major factor in the emergence and reemergence of many communicable diseases. The Population Reference Bureau estimates that in the mid-2000s about 191 million people lived outside their native countries. Populations on the move contribute to the emergence of disease beyond that resulting from voluntary urbanization fueled by a search for a better life. Population movement is frequently not a matter of choice but rather a forced movement because of wars and conflicts resulting from political upheavals. The United Nations defines **internally displaced persons (IDPs)** as:

> Persons or groups of persons who have been forced or obliged to flee or to leave their homes or places of habitual residence, in particular as a result of or in order to avoid the effects of armed conflict, situations of generalized violence, violations of human rights or natural or human-made disasters, and who have not crossed an internationally recognized State border.

The term *refugee* is reserved for those who are forced under the same circumstances to cross an international border. The Office of the UN High Commissioner for Refugees estimated the number of forcibly displaced persons at nearly 21,020,000 people in 2007 worldwide. In 2015, the number of refugees more than doubled to an unprecedented number of approximately 59 million people! The dramatic increase was mainly attributed to the Syrian civil war and ongoing deterioration in countries such as Afghanistan, Burundi, Democratic Republic of Congo, Mali, Somalia, South Sudan, and the Ukraine. These displaced persons carry with them their microbes and disease vectors, resulting in an exchange with intermingling populations.

Malaria is an excellent example; refugees migrating through regions where malaria is endemic can acquire the infection and disseminate the disease to other areas. Malaria is a common cause of death among refugees in numerous countries, including Thailand, Somalia, Rwanda, the Democratic Republic of the Congo, and Tanzania.

Masses of refugees are forced to settle in uninhabitable environments without adequate shelter, food, clean drinking water, and latrines. Personal hygiene and sanitation may be virtually nonexistent, and what few facilities are available become quickly overwhelmed. People live in filth and squalor. These camps are hotbeds for epidemics, and their potential spreads as refugees continue to flee from one area to another (**FIGURE 4.23**).

© Northfoto/ShutterStock, Inc.

FIGURE 4.23 A refugee camp. Refugee camps are hotbeds of infection. Crowding and lack of hygiene and sanitation favor the incidence and transmission of disease.

The Darfur conflict in western Sudan was a human catastrophe and a worst-case scenario. The United Nations has estimated that 200,000 to 400,000 have died from violence and disease and another 2.5 million have been displaced to refugee camps. The 2011 drought in Somalia, the worst in 70 years, forced more than 1,000 people per day to migrate to refugee camps in Kenya. The camps were designed to hold 90,000 refugees, but more than 430,000 refugees migrated into the camps. Try to imagine the almost overwhelming public health problems created by the surge of people, many of whom suffer from cholera and other infectious diseases. Maintenance of sanitation and provision of clean water are of the first order.

Wars and civil unrest, in addition to creating refugees and IDPs, disrupt the public health infrastructure and favor the spread of infectious diseases. The decline in water management programs and a lack of treatment facilities contribute to the spread of disease. The destruction of housing leads to increased human-to-human and human-to-vector contact.

Societal Factors

In many societies, particularly in developed countries, family life and structure have changed as a result of economic growth and increased opportunities for women. In most American families both parents work, leading to an increase in child care centers. Millions of children attend day care centers that put them at risk for a variety of **communicable diseases** such as diarrhea, strep throat, conjunctivitis, viral infections that sometimes trigger middle ear infections, and respiratory infections. For example, outbreaks of shigellosis, winter vomiting disease (caused by noroviruses), and other diarrheal diseases have caused problems in many day care centers around the country. (To a large extent, the simple act of hand washing by the staff after they change a diaper is an effective control measure.) Children convey the microbes to their family members, many of whom, in turn, bring their microbes to the workplace.

© PeterSVETphoto/ShutterStock, Inc.

FIGURE 4.24 Tattooing and body piercing. Tattooing and body piercing are a risky part of popular culture. The skin is invaded, potentially resulting in serious infections because of the use of unclean instruments. Tattoos.

As longevity increases, so does the number of elderly citizens requiring nursing homes, adult day care centers, and assisted living environments. Like child day care centers, these facilities are potential hotbeds for the emergence and spread of communicable diseases within the resident population and the staff, visitors, and their contacts.

Food production and dietary habits also affect the spread of infectious diseases. Globalization of the food supply, centralized processing, fast-food restaurants, dining out, and take-out food are all vulnerable to food-related outbreaks. Foodborne diseases are a major public health problem in the United States. During the spring and continuing into the summer of 2008, more than a thousand cases of salmonellosis occurred in the United States, presumably due to certain varieties of contaminated tomatoes. In 2018, there was a multistate outbreak of E. coli O157:H7 associated with contaminated romaine lettuce and 107 countries received frozen vegetables contaminated with *Listeria*.

In a better economy and a family structure in which both parents work, many people rely more on prepared foods to reduce household chores. Fast-food restaurants and take-out restaurants are part of our social structure. Food has increasingly become a source of recreation. Consider, for example, a typical conversation: "So, what'll we do tonight?" Answer: "Let's eat out!" This is all well and good, assuming that personal hygiene and sanitary control measures practiced by food handlers are not compromised. Television news shows have aired segments featuring high-end restaurants that are enough to make you sick!

Tattooing and body piercing are ancient art forms that have continued through the centuries. In developed countries these practices have long been popular with sailors and bikers. Young people in the 1990s and the new millennium have brought the trend into the mainstream. Tattoo and body piercing parlors are found in many countries, including the United States; for a price you can get just about any part of your body tattooed or pierced (**FIGURE 4.24**). The risk of infection with a variety of microbes, particularly staphylococci, is a real possibility, and patrons are often at risk because of nonsterile instruments and poorly trained personnel. The CDC reported 44 cases of methicillin-resistant *Staphylococcus aureus* (MRSA) skin infections in Ohio, Kentucky, and Vermont in 2004–2005 as a result of 13 unlicensed tattooists, presumably because of the use of nonsterile equipment. *Mycobacterium haemophilum* has also been identified as a cause of infection following tattooing. Even if the establishment is certified by a local health authority, let the buyer beware!

Summary

This chapter makes the case that despite the optimism of 50 or 60 years ago, infectious diseases have not been eliminated (with the single exception of smallpox) but continue to flourish as a major cause of mortality and morbidity around the world. The reasons for this are based on world population growth, urbanization, ecological disturbances, technological advances, microbial evolution and adaptation, and human behavior. A quotation from Donald A. Henderson during his tenure as associate director of the U.S. Office of Science and Technology Policy serves as an excellent way to close this chapter:

> *The recent emergence of AIDS and dengue hemorrhagic infections, among others, [is] serving usefully to disturb our ill-founded complacency about infectious diseases. Such complacency has prevailed in this country [USA] throughout much of my career. . . . It is evident now, as it should have been then that mutation and change are facts of nature, that the world is increasingly interdependent, and that human health and survival will be challenged, ad infinitum, by new mutant microbes, with unpredictable pathophysiological manifestations.*

KEY TERMS

antibiotics

biocides

Biofresh™

biological vectors

blood meal

chorioamnionitis

climate change

communicable disease

contact tracing

cross-resistance

deforestation

disease triangle model

Earth Day

Earth Week

ecotoxicity

emerging infectious disease

endocrine disruptor

Food and Drug Administration (FDA)

Global Handwashing Day

healthcare-associated infection

imported case

infectious disease

insect vectors

internally displaced persons (IDPs)

Kyoto protocol

March for Science

Microban®

microbiome

microcephaly

opportunistic infections

organohalides

pathogen

People's Climate Mobilization Day

personal protective equipment (PPE)

quarantine

reemerging infectious disease

reservoirs

sepsis

species barrier

superbug

triclocarban

triclosan

vancomycin

zoonoses

zoonotic diseases

SELF-EVALUATION

○ PART I: Choose the single best answer.

1. Which bacterium recently caused an outbreak of bloodstream infections in a healthcare environment in Wisconsin?

   a. *Escherichia*

   b. *Enterobacter*

   c. *Salmonella*

   d. *Elizabethkingia*

2. Which of the following insects have the potential to carry infectious agents and transmit disease to humans?

   a. Ladybugs

   b. Fire ants

   c. Mosquitoes

   d. Wasps

3. Which of the following antibacterial or chemical compounds were recently banned by the FDA from bar and liquid soaps?

   a. Penicillin

   b. Triclosan

   c. Glycerol

   d. Lavender

4. The construction of the Central Railroad in Brazil led to an increase in _____.

   a. leishmaniasis

   b. malaria

   c. tuberculosis

   d. Chagas disease

5. Which of the following is *not* an emerging or reemerging disease?
   a. Measles
   b. Zika virus disease
   c. Mononucleosis
   d. Whooping cough

6. The _____ model addresses the interactions among the host (e.g., humans), infectious agent, and environment that produce infectious disease.
   a. herd immunity
   b. epidemiology
   c. disease triangle
   d. chain of transmission

7. Flooding is associated with which of the following infectious diseases?
   a. Dengue fever
   b. Coccidioidomycosis
   c. Measles
   d. Mononucleosis

8. _____ and diarrheal diseases are the leading cause of infectious disease mortalities worldwide in children 5 years and younger.
   a. Skin infections
   b. Meningitis
   c. Pneumonia
   d. Leprosy

9. Which of the following practices is used to prevent malaria?
   a. Use of bed nets
   b. Antibiotics
   c. Vaccination
   d. Blood transfusion

10. What is the biological vector that carries the pathogen that causes Zika virus disease?
   a. Ticks
   b. Mosquitoes
   c. Fleas
   d. Sand flies

○ PART II: Fill in the Blank/True or False.

1. Most microbes are beneficial and many are essential to the cycles of nature without which higher forms of life could not exist. *True or false?*

2. In the United States infectious diseases are in the top 10 leading cause of death. *True or false?*

3. There is evidence that _____ are associated with the emergence of infectious diseases.

4. Poverty is especially serious where rapid population growth occurs. *True or false?*

5. According to the most recent United Nations Data Booklet, the world's largest megacity is _____.

6. Air travel plays a significant role in the spread of _____ from continent to continent.

7. _____ has led to a resurgence in HIV infections.

8. Deforestation can play a role in the eruption of emerging and reemerging diseases. *True or false?*

9. Diseases transmitted from animals to humans are called _____ diseases.

10. _____ can result in an increase in sexually transmitted infections.

○ **PART III: Answer the following.**

1. List five reasons why infectious diseases are emerging and increasing.

2. Choose two of the reasons you gave in the previous question and discuss them.

3. Why did the FDA ban certain antibacterial compounds from bar and liquid soaps?

4. Cairns described cities as "graveyards of mankind." Explain.

5. A number of quotations are cited in this chapter. Develop your own quotation that targets the problem of emerging and reemerging infections.

6. Do you believe that a grade of A– should be given to the world for its efforts in coping with infectious diseases? What grade would you award, and why?

7. List changes in the environment that are causal factors in disease outbreaks.

8. Complacency is listed as a major factor responsible for the continued threat of infectious diseases. Describe at least three specific examples, including an example for which you and family members may be "guilty."

9. Create a list containing at least five infectious diseases that are resistant to drug therapy and discuss reasons for their development.

10. Explain how technological advances may pose public health risks.

THE MICROBIAL WORLD

The Microbe is so very small, You cannot make him out at all, But many sanguine people hope, To see him down a microscope. Oh! Let us never, never doubt, What nobody is sure about!

—Hilaire Belloc

More Beasts for Worse Children (1896)

A laboratory technician in a new laboratory in India is setting up diagnostic tests to identify bacteria causing infections. The Centers for Disease Control and Prevention (CDC) is working internationally to build new diagnostic laboratories so that infectious diseases can be diagnosed locally.

Courtesy of the CDC.

LEARNING OBJECTIVES

1. List the characteristics of microbial life.
2. Explain why microbial diversity is important on Earth.
3. Compare and contrast procaryotic and eucaryotic microbes.
4. Summarize how microbes and subcellular agents are measured in size relative to more familiar objects.
5. Describe the two types of subcellular agents.

Case Study: Microbes in the New York City Subway System

Have you ever wondered what microbes are present in public transportation systems? Are you afraid to touch the railings on an escalator going down into a subway station? Young children sit on the floors of trains, putting toys and other objects into their mouths that have touched surfaces in public areas. Do dangerous pathogens lurk in public environments? Every day millions of people (along with the microbes present on them) commute on trains in U.S. cities such as Washington D.C., Boston, San Francisco, Chicago, and New York City.

A collaborative research project led by researchers at Weill Cornell Medical College in New York surveyed the surfaces in the highly trafficked subway system, the Gowanus Canal, and the public parks of New York City. The researchers swabbed surfaces of turnstiles, emergency exits, Metro card kiosks, stairwell handrails, trash cans, doorknobs, poles, and seats in 466 subway stations

for all 24 subway lines of the NYC Metropolitan Transit Authority (MTA), the Staten Island Railway (SIR), 12 sites in the Gowanus Canal, 4 public parks, and 1 closed subway station that was submerged during the 2012 Hurricane Sandy (Superstorm Sandy). The swabs were placed in transport medium and taken back to the laboratory (**Figure 1**). The microbes present were identified through **metagenomics**. Metagenomics is the study of genetic material recovered from environmental samples, in this case the complex community of microbes within the NYC subway system and similar high-traffic areas. Researchers discovered the following:

- 48% of the DNA did not match any known microbes.
- 47% of the DNA matched known bacteria (**Table 1**).
- 0.8% of the DNA matched eucaryotic organisms.

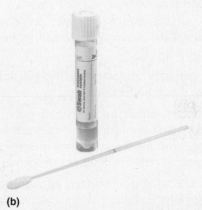

Figure 1 **(a)** View of MetroCard turnstiles leading into the NYC subway system from downtown and Brooklyn. **(b)** Transport media used to swab surfaces in the NYC subway system and public areas for this study.

(continues)

Case Study: Microbes in the New York City Subway System (continued)

TABLE 1 NYC Subway System Study: Identified Bacteria and Bacteriophages

Number of Samples (N = 1,457)	Bacterium	Number of Samples (N = 1,457)	Bacteriophage
1,224	*Pseudomonas stutzeri*	74	*Enterobacteria* phage (φX174)
1,007	*Stenotrophomonas maltophilia*	28	ε15likevirus
939	*Enterobacter cloacae*	13	*Erwinia* phage
728	*Acinetobacter radioresistans*	12	*Enterobacteria* phage (ENT90)
675	*Acinetobacter nosocomialis*	10	*Stenotrophomonas* phage (φSMA7)
555	*Lysinbacillus sphaericus*	9	*Staphylococcus* phage
544	*Enterococcus casseliflavus*	7	*Enterobacteria* phage (mEp235)
460	*Brevundimonas diminuta*	6	*Lactococcus* phage
428	*Acinetobacter lwoffii*	6	*Stenotrophomonas* phage (φSMA9)
427	*Bacillus cereus*	4	*Enterococcus* phage

Information from E. Afshinnekoo, C. Meydan, S. Chowdhury, D. Jaroudi, C. Bover, . . . C. E. Mason. (2015). Geospatial resolution of human and bacterial diversity with city-scale metagenomics. *Cell Systems,* 1, 1–15.

- 0.032% of the DNA matched bacteriophages (viruses that infect bacteria; TABLE 1).

- Only an average of 0.2% of the DNA mapped to human genomes.

- The most dominant bacteria were *Pseudomonas stutzeri, Enterobacter cloacae,* and *Stenotrophomonas maltophilia.*

- The majority of bacteria (57%) are not known to cause human infections.

- About 31% of the bacteria were potential **opportunistic pathogens** that might cause infections in people who are immunocompromised, injured, or susceptible because of other preexisting conditions.

- About 12% of the detected known bacteria were pathogens, including *Yersinia pestis* (cause of the Black Death/bubonic plague), *Bacillus anthracis* (cause of anthrax), and MRSA (methicillin-resistant *Staphylococcus aureus*).

The researchers also determined if the microbes collected on swabs were viable and could be cultured in the laboratory. Indeed, the bacteria were viable, and 28% of the bacteria were resistant to standard antibiotics used to treat infections. The dominant bacteria present on subway surfaces were associated with human skin, followed by the gastrointestinal tract, urogenital tract, airways, oral cavity, blood, and conjunctiva. Even though *Yersinia pestis* DNA and *Bacillus anthracis* were detected and the bacteria could be cultured, no human cases of bubonic plague or anthrax were recorded during or after the NYC subway study.

The researchers established a baseline of microbes present in the NYC subway system. Even though there was some evidence of bacterial pathogens, the lack of reported cases during the study suggested the pathogens represented a normal, urban **microbiome**.

Questions and Activities

1. A New York City subway turnstile is teeming with microbial life. Why can someone remain healthy after touching the turnstile?

2. Explain why the researchers discovered diverse microbes in the subway system environment (i.e., what is the source of the diverse microbes?).

3. *Pseudomonas stutzeri* was the most numerous bacterium detected in this study. Perform an Internet search to determine if *P. stutzeri* is a human pathogen. If so, list what type of infections it can cause and what populations are at risk of infection. List the online resources you used in your answer.

4. Explain why not all of the microbes identified could be cultured in the laboratory.

5. Can the data that the researchers collected help with long-term infectious-disease surveillance, bioterrorism threat mitigation, and health management in an urban environment? Explain.

Information from E. Afshinnekoo, C. Meydan, S. Chowdhury, D. Jaroudi, C. Bover, . . . C. E. Mason. (2015). Geospatial resolution of human and bacterial diversity with city-scale metagenomics. *Cell Systems, 1*, 1–15.

Preview

The term **microbe** has been used extensively because this text is about microbes, particularly those relatively few that are pathogens. Although the definition of the term *microbe* is somewhat imprecise, six groups of microbes are generally identified and discussed in microbiology texts: prions, viruses, bacteria, protozoans, unicellular algae, and fungi. (Worms, biologically known as helminths, also are frequently included in microbiology texts even though they are not microbes because a number of species cause infections resembling microbial infections.)

Some Basic Biological Principles

Cell Theory

To further understand microbes, whether pathogenic or not, it is necessary to review a few very basic concepts of biology, because all microbes are biological packages with certain unique characteristics. **Cells** are considered the basic unit of life, based on the observations of Robert Hooke in 1665. Hooke used the word *cella* in his examination of cork, which revealed tiny compartments that reminded him of the cells in which monks lived. His studies ultimately gave rise to the cell theory, a fundamental concept in biology, as postulated by Matthias Schleiden and Theodor Schwann (1838) and Rudolf Virchow (1858). The major points of the **cell theory** are as follows:

1. All **organisms** are composed of fundamental units called cells.

2. All organisms are unicellular (single cells) or multicellular (more than one cell).

3. All cells are fundamentally alike with regard to their structure and their metabolism.

4. Cells arise only from previously existing cells ("life begets life").

"Life begets life" is a refutation of the theory of **spontaneous generation**, a concept that was disproved by the end of the 19th century. An understanding of the cell theory is the basis for an understanding of life, including microbial life. The cell theory does not apply to viruses and prions; they are described as *acellular*, *subcellular*, or as *biological agents*, terms that are used somewhat interchangeably. Nevertheless, as a matter of convenience, license is sometimes taken, and they are described as microbes or microorganisms. **Viruses** are not considered by scientists as being "alive," but they come close; they are in that gray area between living and nonliving. **Prions** are infectious proteins, making them even less biologically complex than viruses.

Metabolic Diversity

The term *life* is elusive and cannot be given an exact definition; at best, it can only be described. Nevertheless, several attributes are associated with living systems that, collectively, establish life. By one strategy or another all organisms exhibit these characteristics, summarized in **TABLE 5.1**. A major property of life is the ability to constantly satisfy the requirement for energy. It takes energy for every cell to stay alive, whether it is a single cell or a component of a multicellular organism; in the latter case each cell contributes to the total energy requirement of the organism. Your body constantly expends energy.

It takes energy to breathe even during sleep and for the heart to constantly push blood through an interconnected and tortuous maze of blood vessels. Because you do not fill up at the gas station, it is obvious that your energy is derived from the foods you eat. Through a complex series of biochemical reactions, the body metabolizes the **organic compounds** (proteins, fats, and carbohydrates) of your diet and releases the energy stored in their chemical bonds into a biologically available high-energy compound known as **adenosine triphosphate (ATP)**; you live directly off of this and constantly replace it as you take in nutrients. Most organisms, including most microbes, are **heterotrophs**, meaning that they require organic compounds as an energy source; humans are heterotrophs.

TABLE 5.1 Characteristics of Life

Characteristic	Description
Cellular organization	The cell is the basic unit of life; organisms are unicellular or multicellular.
Energy production	Organisms require energy and a biochemical strategy to meet their energy requirement.
Reproduction	Organisms have the capacity to reproduce by asexual or sexual methods and in doing so pass on genetic material (DNA) to their progeny.
Irritability	Organisms respond to internal and external stimuli.
Growth and development	Organisms grow and develop in each new generation; specialization and differentiation occur in multicellular organisms.

Other microorganisms and plant life are **autotrophs** and do not require organic compounds, but they do require energy. Some are able to directly use the energy of the sun (**photosynthetic autotrophs**); others derive energy from the metabolism of **inorganic compounds** (**chemosynthetic autotrophs**). In so doing, autotrophs produce organic compounds and oxygen (O_2). Hence, heterotrophs are dependent on autotrophs for energy (**FIGURE 5.1**)

Requirement for Oxygen

In addition to metabolic diversity, organisms exhibit diversity in their O_2 requirements. The "higher" organisms that are more familiar to you are **aerobes**, meaning that they require O_2 for their metabolic activities. Some bacteria are **strict anaerobes** and do not require oxygen and are actually killed by O_2. **Facultative anaerobes** are bacteria that grow better in the presence of O_2 but can shift their metabolism, allowing them to grow in the absence of O_2. Knowledge of the oxygen requirements of pathogens is important in clinical microbiology. For example, specimens from infections caused by bacteria suspected of being anaerobes must be transported and cultured under anaerobic conditions (**FIGURE 5.2**).

Krasner's Notebook

Even when you doze in class, it takes energy to keep from slithering out of your chair and onto the floor. I have seen this happen only once in all my years at the lecture podium.

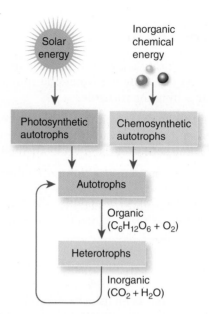

FIGURE 5.1 A pathway map showing heterotroph dependency on autotrophs and the autotrophs' energy sources.

Genetic Information

The genetic information for the structure and functioning of all cells is stored in molecules of **deoxyribonucleic acid (DNA)**, a large and complex organic molecule. **Genes** are segments of the DNA molecule. Since the establishment of DNA as the hereditary material, the expression "life begets life" can be expanded to explain the mechanism by which a particular life form gives rise to the same life form; that is, tomatoes produce tomatoes, humans produce humans, and *Escherichia coli* produces *Escherichia coli*. Each of these groups has its characteristics embedded in DNA that confer its identity. The DNA is transferred, by a variety of reproductive strategies, from parent to offspring.

© Teri Shors.

FIGURE 5.2 Culturing anaerobic bacteria. Some bacteria cannot grow in the presence of oxygen. The GasPak jar is a means of culturing anaerobes. The jar contains a sachet of chemicals that creates an anaerobic environment with greater than 15% CO_2. It includes an oxygen indicator.

What Makes a Microbe?

With these basic biological principles in mind, the term *microbe* (or *microorganism*) can now be better described. The question to be considered is what makes a microbe a microbe? As will become apparent, this question is not easily answered. Your first response may be "they are all too small to be seen by the human eye without the aid of a **light microscope**" or are **microscopic**. Wrong. At first thought this would appear to be true, but what about the algae and the fungi? Are fungi microscopic? No doubt you have seen molds (**FIGURE 5.3**), which are classified as fungi, growing on food left too long in the refrigerator or perhaps on a pair of old sneakers that you forgot about in the dank basement or hidden away in a dormitory closet. They are **macroscopic**; that is, they can be seen with the human naked eye. Hence, "microscopic" is not a distinguishing microbial characteristic. To describe all microbes as being unicellular is also not correct because the fungi and many of the algal forms are macroscopic and clearly multicellular. These organisms must be multicellular; if they were unicellular, that one cell would be enormous—a ridiculous idea! As pointed out later, some fungi, namely yeasts, are unicellular.

There are exceptions to the rule that all bacteria are unicellular and microscopic. This might seem like an amazing fish story, but in 1985, a large cigar-shaped microorganism was found in the guts of the Red Sea brown surgeonfish. This organism was subsequently identified as

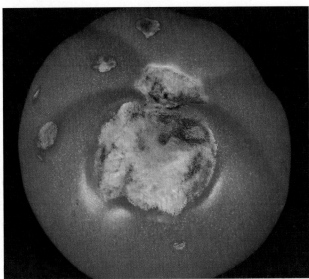

© Jones & Bartlett Learning.

FIGURE 5.3 Mold growing on a tomato.

a bacterium, approximately a million times larger in volume than *E. coli*, and was christened *Epulopiscium fishelsoni*. Twelve years later, in 1997, an even more monstrous bacterium was discovered in sediment samples residing off the coast of Namibia (**BOX 5.1**); the organism has the tongue-twisting name *Thiomargarita namibiensis*, and to date is one for the *Guinness Book of World Records*. They are visible to the naked eye.

BOX 5.1 "Monster" Bacteria

If asked to describe bacteria, just about everyone would reply that they are too small to be seen without a microscope. However, in 1985, *Epulopiscium fishelsoni*, a giant bacterium that can be seen without a microscope, was discovered in the guts of surgeonfish in the warm waters of the Red Sea and off the coast of Australia. The organism can grow to about 500 micrometers (μm), or about the size of the period at the end of this sentence. To give you some idea of size, one scientist projected that "if ordinary bacteria were mouse sized, *E. fishelsoni* would be equivalent to a lion." This organism is referred to as "epulo" for short and was originally thought to be protozoan-like. However, analysis of their DNA revealed that they are, in fact, bacteria.

In 1997, *Thiomargarita namibiensis* stole the prize for size from *Epulopiscium*. This "monster" bacterium, approximately the size of a fruit fly's head, was discovered in samples of sediment in the greenish ooze off the coast of Namibia in Africa. These spherical cells range from 100 to 750 μm in size. Dispersed throughout their cytoplasm are globules of sulfur. The bacteria tend to organize into strands of cells that glisten white from light reflected off their sulfur globules, which explains the name. *Thiomargarita namibiensis* means "Namibian sulfur pearl."

Both epulo and the sulfur pearl are anomalies in the bacterial world. The sizes of cells of all kinds, not only bacterial cells, are limited by the surface area of the cell membrane, because nutrients and waste are transported in and out of the cells by diffusion across the cell membrane. As cells increase in size, both volume and surface area increase, but surface area increases to a lesser degree than does volume. At some point the surface area becomes too limited to allow for sufficient diffusion between the cell and its environment.

So how did *E. fishelsoni* and *T. namibiensis* manage to become so big? What are the physiological adaptations? In the case of epulo, microscopic examination reveals that the cell membrane, rather than being stretched smoothly around the cell, is convoluted (wrinkled), resulting in "hills and valleys," a phenomenon that greatly increases cell surface. This adaptation is not unique to bacterial cells; the surface of the human brain is highly convoluted, resulting in a greater surface area, a factor that correlates with species intelligence. The large size of *T. namibiensis* is attributed to the presence of a large fluid-filled sac occupying over 90% of the cell's interior. The sac is packed with nitrate that the cell uses in its metabolism to produce energy, making it less dependent on constant diffusion across the membrane to transport nutrients and waste.

To give you some idea of size relationships, if an ordinary bacterium were the size of a baby mouse, *E. fishelsoni* would be equivalent to a lion and *T. namibiensis* would be the size of a blue whale, the world's largest animal. The blue whale measures up to 90 feet (29 meters) and weighs about 120 tons. How many cells might make up such an enormous creature? The number would be in the trillions. Each of these cells exhibits the same fundamental life characteristics as the single microbe.

Microbes are sometimes described as "simple" because many consist of only a single cell or are less than a cell (viruses and prions). Consider, however, that this single cell must fulfill all the functions of life. On the other hand, in a multicellular organism (like the whale), although each cell fulfills all the criteria for life, there is a "sharing" of function because of specialization into a variety of cell types (e.g., muscle cells, nerve cells, and blood cells). Perhaps that makes life easier. Hence, single-celled organisms, and even those multicellular organisms consisting of only a small number of cells without evidence of true specialization, are simple only in the sense of numbers and not in a physiological (functional) sense.

So, if microbes are not necessarily microscopic and/or unicellular, then what is a microbe? There really is no unifying principle or precise definition. The term *microbe*, as well as *microorganism*, is a term of convenience used to describe biological agents, in a collective sense, that in general are too small to be seen without the aid of a microscope. The term is also used for microbes that are cultured and identified using similar techniques. Based on what has been presented here, it is clear that these descriptions are not always true. Some biologists consider microbes to be organisms that are at less than the tissue level of organization. This statement requires some explanation and is based on what is referred to as "biological hierarchy," or levels of biological organization.

Recall that a cell is the fundamental unit of biological organization and that groups of cells establish multicellularity. Consider the human, or any other multicellular animal or plant, and it is obvious that in addition to an increase in cell number, the process of differentiation and specialization has taken place. For example, over 200 cell types make up the human, including red blood cells, five categories of white blood cells, epithelial cells, connective tissue cells, nerve cells, and muscle cells. All these cells, as stated in the cell theory, share common fundamental characteristics, but superimposed on their basic structure and function is a specialization of structure and function.

Cells of the same type constitute the **tissue** level of organization, as exemplified by nerve tissue, blood tissue, and connective tissue. Tissues, in turn, constitute **organs**, structures composed of more than one tissue type; the heart, brain, stomach, and kidney are examples. Organs, in turn, constitute **organ systems**, a collection of organs that contribute to an overall function or functions. The digestive system, nervous system, respiratory system, excretory system, and reproductive system are examples familiar to you. This hierarchy is summarized as follows and is further illustrated in **FIGURE 5.4**: cells → tissues → organs → organ systems.

All microbes are devoid of tissues; that is, they are all at the subcellular or cellular level of organization, although fungi and some algae hint at specialization and approach the tissue level of organization. Prions and viruses can be properly placed at the acellular or subcellular level, which, simply put, means that they are less than cells and are at the threshold of life (Figure 5.4).

Procaryotic and Eucaryotic Cells

Biologists recognize the existence of two very distinct types of cells, referred to as **procaryotic** and **eucaryotic cells** (Greek, *pro*, "before," + *karyon*, "nut" or "kernel," + *eu*, "true"). Procaryotic cells have a simpler morphology than eucaryotic cells and are primarily distinguished by the fact that there is no membrane around the nucleus. The cell has a nuclear area rich in DNA that serves as the carrier of genetic information, as in all cells, but that DNA is not enclosed within a nuclear membrane. This DNA-rich area is referred to as a **nucleoid** rather than as a true nucleus. Further, procaryotic cells do not have membrane-bound cellular structures (**organelles**), in contrast to the cellular anatomy of the eucaryotic cells.

Procaryotic and eucaryotic cells are compared in **TABLE 5.2** and in **FIGURE 5.5**. Bacteria are procaryotic microorganisms; protozoans, unicellular algae, fungi, and all other forms of life (except viruses and prions) are composed of eucaryotic cells.

Microbial Evolution and Diversity

Procaryotes date back 3.5 billion years and were the only life forms for 2.5 billion years; they are the ancestors of eucaryotes. Aristotle pondered the relationships among organisms, as do scientists today. In the 18th century, the botanist Carolus Linnaeus (1707–1778) classified all life forms as belonging to either the plant or the animal kingdom. Microbes were largely ignored because little was known about them, but, because they had to be placed somewhere, they were considered plants, probably because those that had been observed possessed cell walls. Various schemes of classification have been proposed over the last

---

Krasner's Notebook

Some years ago, I attended the annual meeting of the American Society for Microbiology in Miami, Florida, and overheard two airport baggage handlers commenting that about 12,000 microbiologists were expected to attend. One asked the other, "What's a microbiologist, anyway?" to which the other replied, "Beats me! I suppose it's a small biologist." Several miles from the airport was a huge billboard with the words "Orkin Pest Control welcomes microbiologists." It was a memorable meeting.

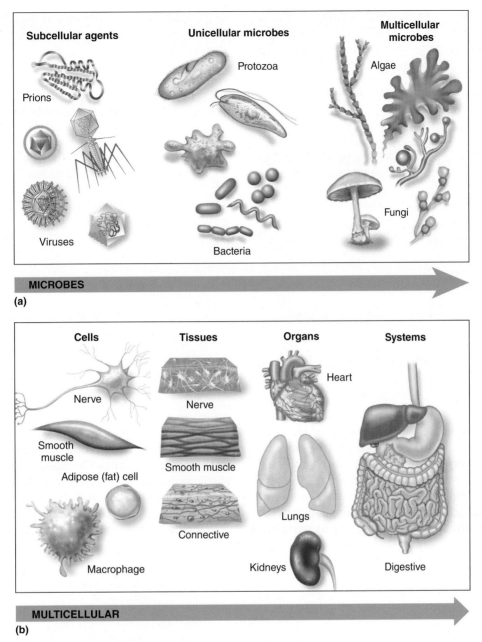

FIGURE 5.4 Levels of biological organization. **(a)** Microbes. **(b)** Multicellular organisms.

few centuries, and **taxonomy**, the science of classification, became more and more complex.

In 1866, Ernst Haeckel proposed a three-kingdom system—animals, plants, and a new kingdom, Protista, a collection to accommodate microbes. In the light of modern biology, it became apparent that even three kingdoms were not enough. In 1969, a five-kingdom system was proposed by Robert Whittaker and initially accepted by most biologists. This classification—**Whittaker's five-kingdom system**—describes organisms as belonging to the kingdoms Monera, Protista, Fungi, Animalia, and Plantae (**FIGURE 5.6**). Recall that microbes consist of six groups accommodated in one of Whittaker's five kingdoms as follows: bacteria are classified as Monera, protozoans and unicellular algae are

classified as Protista, and fungi are classified as Fungi. Note that viruses and prions are not considered in this scheme of classification, because they are neither procaryotic nor eucaryotic cells but are subcellular.

The 1950s ushered in the tide of molecular biology, and its wake introduced new taxonomic techniques. Biologist Carl Woese and his colleagues at the University of Illinois focused in on a ribosomal ribonucleic acid gene (**16s rRNA gene**) unique to bacterial species as a "fingerprint" to identify shared characteristics of bacteria and thus gain insight into their relatedness, which, in turn, would point to their evolutionary history.

In 1990, Woese, along with Otto Kandler and Mark L. Wheelis, proposed a novel scheme of classification

TABLE 5.2 Comparison of Procaryotic and Eucaryotic Cells

Characteristic	Procaryotes	Eucaryotes
Life form	Bacteria, Archaea	Eucarya All microbial cells (with the exception of bacteria, viruses, and prions) and all other cells
Nucleus	Do not contain a nucleus. DNA chromosome not enveloped by membrane	DNA chromosome present inside of a membrane bound nucleus
Cell size	About 1–10 μm	Greater than 10 μm
Chromosomes	Single circular DNA (two chromosomes in a few)	Multiple paired chromosomes present in nucleus
Cell division	Asexual binary fission; no "true" sexual reproduction	Cell division by mitosis; sexual reproduction by meiosis
Internal compartmentalization	No membrane-bound organelles	Organelles bound by membrane
Ribosomes	Present in the cytoplasm and smaller than eucaryotic cells	Larger than procaryotic ribosomes; free ribosomes in the cytoplasm and ribosomes present on the endoplasmic reticulum

based on Woese's analysis. **Woese's three-domain system** assigns all organisms to one of three domains, or "superkingdoms"—the **Bacteria**, **Archaea**, and **Eucarya** (**FIGURE 5.7**), all of which arose from a single ancestral line. (All of Whittaker's five traditional kingdoms can be reassigned among the three domains.) The Bacteria and

the Eucarya first diverged from an ancestral stock, followed by the divergence of the Archaea from the Eucarya line. The domains differ remarkably from one another in their chemical composition and in other characteristics, as summarized in **TABLE 5.3**. The term *bacteria* as commonly used includes both the Bacteria and the Archaea domains.

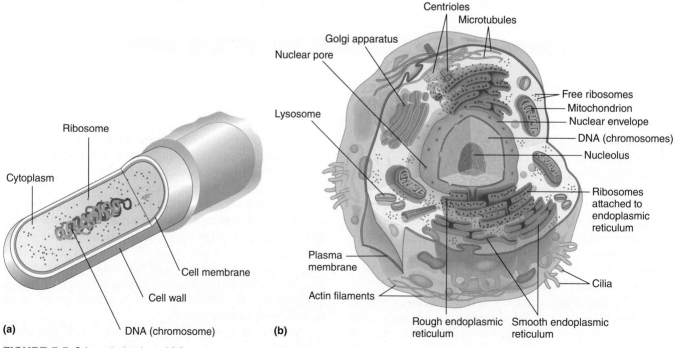

(a)

Ribosome

Cytoplasm

Cell membrane

Cell wall

DNA (chromosome)

(b)

Centrioles

Microtubules

Golgi apparatus

Nuclear pore

Lysosome

Free ribosomes

Mitochondrion

Nuclear envelope

DNA (chromosomes)

Nucleolus

Ribosomes attached to endoplasmic reticulum

Plasma membrane

Actin filaments

Cilia

Rough endoplasmic reticulum

Smooth endoplasmic reticulum

FIGURE 5.5 Schematic drawings of **(a)** a eucaryotic cell and **(b)** a procaryotic cell.

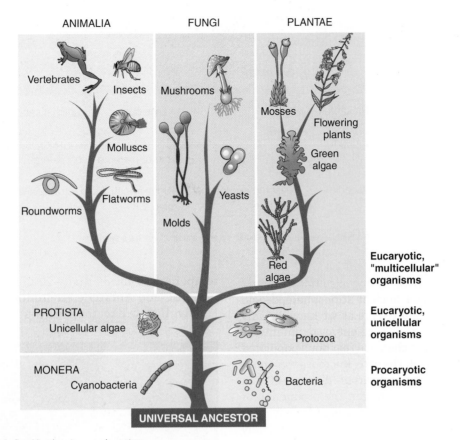

FIGURE 5.6 Whittaker's five-kingdom taxonomic system.

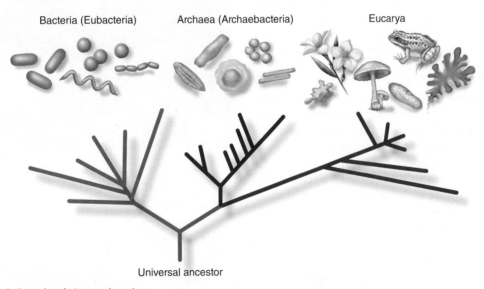

FIGURE 5.7 Woese's three-domain taxonomic system.

It should be apparent that classification, particularly at the level of microorganisms, is not cast in concrete but is constantly under revision as new information becomes available. It is a credit to the scientific process that reevaluation is the name of the game. Admittedly, it is confusing, but to quote William Shakespeare (who probably never even took a course in biology), "What's in a name? That which we call a rose by any other name would smell as sweet." So you need not sweat it too much! No matter what the classification, bacteria were the first forms of life on Earth. Fossilized bacteria have been discovered in **stromatolites**, stratified rocks dating back 3.5 billion to 3.8 billion years, a long time ago in the history of the estimated 4.6 billion-year-old planet Earth.

When life arose, the Earth's ancient atmosphere contained little or no free oxygen but consisted principally of carbon dioxide and nitrogen with smaller amounts of

TABLE 5.3 Comparisons of Bacteria, Archaea, and Eucarya*

Domain	Membrane-Bound Nucleus	Cell Wall	Antibiotic Susceptibility	Characteristics
Bacteria	No	Present	Yes	Large number of bacterial species
Archaea	No	Present	No	"Extreme" bacteria growing in high-salt environments and at extreme temperatures
Eucarya	Yes	Variable	No (some exceptions in fungi)	Algae (most), fungi, protozoans, "higher" animals and plants

*Major differences are present in the biochemistry of cell walls, cell membranes, genetic material, and structures in the cytoplasm.

gases, including hydrogen (H_2), hydrogen sulfide (H_2S), and carbon monoxide (CO). This ancient atmosphere, devoid of O_2, would not have supported life as we know it. Only microbes that were able to meet their energy requirements with non-oxygen-requiring chemical reactions populated the primordial environment. The early microbes were photosynthetic and used water and carbon dioxide (CO_2) in photosynthetic reactions, resulting in the production of O_2 and **carbohydrates**. This process was responsible for the generation of O_2 in Earth's atmosphere approximately 2 billion years ago.

Since their origin on Earth billions of years ago, bacteria have exhibited remarkable diversity and have filled every known ecological niche. Yet, according to some estimates, fewer than 2% of microbes have been identified, and even fewer have been cultured and studied. Perhaps you can recall the old Mother Goose nursery rhyme "Peas Porridge Hot," which states "some like it hot and some like it cold." This is similar to the diversity in bacteria. Bacteria belonging to the domain Archaea continue to be found in environments once considered too extreme or too harsh for life at any level. In most cases these organisms, **extremophiles**, cannot be grown by existing culture techniques; evidence of their presence has been obtained by molecular biology techniques that allow scientists to examine minute amounts of their deposited ribonucleic acid (RNA). Some like it hot and are called **hyperthermophiles** ("heat lovers"). Some hyperthermophiles have been identified in the hot springs in Yellowstone Park where the temperatures exceed 70°C (158°F). Some microbes do best at temperatures even higher, above 100°C (212°F). *Thermus aquaticus* was isolated from hot springs in the 1960s. It produces an enzyme (**Taq polymerase**), which is essential to a very important technique (**polymerase chain reaction**, or **PCR**) in molecular biology that allows for rapid DNA synthesis and sequencing. *Pyrococcus furiosus* lives in boiling water bubbling from undersea hot vents and freezes to death in temperatures below 70°C (158°F) (**FIGURE 5.8a**). Some extremophiles, the **psychrophiles**, like it cold. Psychrophiles have growth

temperatures lower than –20°C (–4°F) and are happy in Arctic and Antarctic environments (**FIGURE 5.8b**). Finally, if you like tongue-twisters, try your tongue

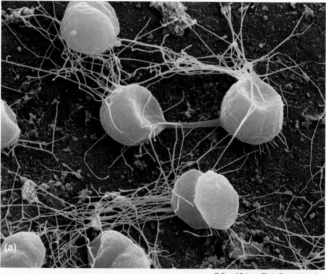
(a)
© Eye of Science/Photo Researchers, Inc.

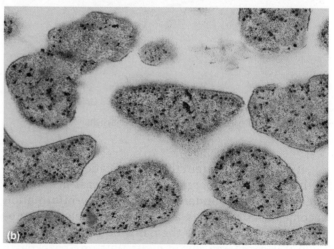
(b)
© Dr. M.Rohde, GBF/Photo Researchers, Inc.

FIGURE 5.8 (a) *Pyrococcus furiosus*, a highly heat-resistant bacterium. **(b)** Psychrophilic *Methanococcoides burtonii* discovered in 1992 in Ace Lake, Antarctica, can survive in temperatures as low as –2.5°C.

around *Psychromonas ingrahamii* and *Colwellia psychrerythraea*, both at home and living comfortably in frigid waters. Some bacteria are extreme **halophiles** ("salt lovers") (**FIGURE 5.9**), and some produce methane gas in their metabolism. These bizarre examples indicate that many of the Archaea live at the extremes of life zones (**BOX 5.2**). Archaea have not been implicated as disease producers and are not further considered in this text.

In 2010, NASA scientists reported the discovery of bacteria that could substitute arsenic for phosphorous, a necessary component of DNA. If correct, the finding would open up a new form of life on Earth. Other scientists were skeptical about the study, so the role of arsenic-loving bacteria remains open.

You may not like to hear this, but the human mouth is considered one of the most diverse ecosystems and rivals the biological diversity of tropical rainforests. Within the past few years, scientists at Stanford University have discovered 37 new organisms in the mouth, pushing the total to more than 500. These new microbes were found in the scum (plaque) in the deep gum pockets between teeth. (Your dentist would love this tidbit!) Their presence remained unknown simply because traditional culture methods do not allow their growth. Enterprising microbiologists (sometimes known as "plaque pickers") extracted DNA from plaque and mapped out DNA sequences, revealing bacteria that had not been previously known to inhabit the mouth. In fact, some new bacterial species were identified, supporting the statement that less than 2% of microbes has been identified.

A comprehensive global microbial survey to identify microbes that make up the **biosphere** is underway thanks to the cooperative effort of the National Science Foundation and the American Society for Microbiology. These two organizations are establishing a network of biodiversity research sites, or "microbial observatories." Other international efforts are in the works to develop a worldwide microbial inventory of genetic sequences.

© Robert I. Krasner.

FIGURE 5.9 The Dead Sea. This sea has a salt concentration well above that found in the Great Salt Lake in Utah; it lies farther below sea level than any other terrestrial spot on Earth. You can lie on your back and float without any effort. Amazingly, this extreme environment is home for a variety of halophilic bacteria.

The origin of life on Earth continues to be a fascinating and mind-boggling question to which the explanation is purely speculative. The general consensus among scientists is that the "primordial soup" hypothesis is the most likely explanation. In this hypothesis, organic compounds formed from a specific combination of atmospheric gases collected in water that was then sparked by an energy source. How exactly this happened is still debated.

Another intriguing possibility is the hypothesis that Earth was seeded by life forms from Mars, the Red Planet. Photographs taken from the Global Surveyor spacecraft orbiting Mars indicated the possibility of water just below the surface of the planet. If, in fact, Mars has water, it is possible that the planet entertains, or entertained, life.

Box 5.2 Some Bizarre Bacteria

Some microbes exhibit an unusual lifestyle and remarkable characteristics that provide fascinating stories and illustrate the tremendous diversity of the microbial world. Consider *Deinococcus radiodurans*, a bacterium further described in **BOX 5.3**, which can survive a dose of radiation greater than 3,000 times the dose that can kill a human. The scientists who study it have dubbed it as "Conan the Bacterium." The Dead Sea, characterized by its extreme salinity, is erroneously named; it is not dead at all but teems with salt-loving (halophilic) bacteria. You will be surprised to learn that microbes can grow in your car's battery acid or that some bacteria thrive on arsenic. How about **magnetotactic bacteria**? They manufacture minute, iron-containing magnetic particles used

as compasses by which the organisms align themselves to Earth's geomagnetic field. These curious microbes prefer life in the deeper parts of their aquatic environment where there is less oxygen. Their magnetic compass points the way.

And then there are the as yet unnamed bacteria living in symbiotic partnership with giant tube worms, as long as 2 meters (6.5 feet), living in the hydrothermal vents of the ocean floor. As these worms mature, their entire digestive tract disappears, including their mouth and anal openings. Now that presents a problem, and it's bacteria to the rescue! The tissues of the worm are loaded with bacteria that obtain energy from the surrounding chemical environment sufficient for their own needs and for those of their worm hosts. In turn, the worms provide a safe harbor for

(continues)

BOX 5.2 Some Bizarre Bacteria (continued)

the bacteria, ensure an adequate environment for energy production, and provide nitrogen-rich waste materials, allowing for synthesis of microbial cellular components—a great mutualistic arrangement.

Here is a strange story about *Serratia marcescens*, a bacterium whose colonies produce a deep red pigment when grown in moist environments. In 1263, in the Italian town of Bolsena, a priest was celebrating Mass. When he broke the communion wafers he found what he thought was blood on them and assumed it to be the blood of Christ. Given the lack of scientific knowledge during the Dark Ages, it is understandable the event was regarded as a miracle. It was not a miracle at all. The red pigment–producing *S. marcescens* had contaminated the wafers during their storage in the dampness of the ancient church (**Figure 1**). Nevertheless, Raphael's painting *The Miracle of Bolsena*, depicting this event, hangs on a wall in the Vatican. Here's another story of an unusual bacterium, *Shewanella*, reported in June 2011. These microbes use metal ions in place of oxygen in their metabolism, and in so doing, can minimize some toxic metals from migrating into soil and groundwater—a nice example of bacteria used for clean-up.

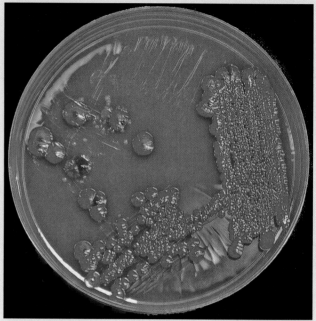

Courtesy of Jeffrey Pommerville.

Figure 1 A culture of *Serratia marcescens*.

According to the Laboratory for Atmospheric and Space Physics at the University of Colorado at Boulder, "Mars meets all the requirements for life." The possibility that life originated on Mars and was subsequently carried to Earth is plausible. Meteors and meteorites are constantly bombarding Earth and some, originating from Mars's surface, could have transported ancestral procaryotic cells. Bacteria have been cultured out of Siberian and Antarctic permafrosts that have been in the deep freeze for millions of years. The National Aeronautics and Space Administration (NASA) is now planning the Mars Sample Return Mission, which will bring Martian rocks back to Earth, and this will help to resolve the question of the beginnings of life. A famous and historic press conference was held by NASA in Washington, D.C. on August 7, 1996, announcing that scientists had found evidence of ancient microbial life in a Mars meteorite known as ALH84001. Bear in mind, however, that the evidence was viewed by some authorities as weak and remains highly refuted.

BOX 5.3 Conan the Bacterium

Conan the Barbarian, a 1982 movie starring Arnold Schwarzenegger, was the first Conan movie. In this fantasy story, from the mythical age of sword and sorcery, Arnie portrays Conan as only Arnie can do!

Deinococcus radiodurans has been nicknamed "Conan the bacterium"; it is one of nature's "toughest cookies." It can survive the rigors of being completely dried out, have its chromosomes disrupted, and be exposed to 1.5 million rads of radiation, a dose 3,000 times greater than that which would kill a human. Further, it can transform toxic mercury into a less toxic form, a feature especially useful at nuclear waste sites. According to Owen White of the Institute for Genomic Research in Rockville, Maryland, "The Department of Energy is very jazzed

about *D. radiodurans*, because the agency has a pretty big toxic cleanup problem at its waste development sites." Genes from bacteria that can digest toxic waste but cannot survive radiation have been genetically engineered into *D. radiodurans*, resulting in bacteria that can transform toxic mercury into a nontoxic form and unstable uranium into a stable form. These genetically engineered bacteria are powerful tools in cleaning up the approximately 3,000 waste sites containing millions of cubic yards of contaminated soil and contaminated groundwater estimated to be in the trillions of gallons. The ability of *D. radiodurans* to repair its own DNA is of interest to biologists because the process provides an insight into the mechanisms of aging and into the biology of cancer.

Introducing the Microbes

Although there is no clear definition of microbes, it is time to introduce those biological agents that fall under the **microbial umbrella** (**FIGURE 5.10** and **TABLE 5.4**).

Algae are not discussed in detail in this text beyond this section, although they are highly significant in terms of food chains and other beneficial aspects. Further, some fungi are human pathogens, and many contribute to the death toll of patients with AIDS. With the exception of

FIGURE 5.10 The microbial umbrella.

Characteristic	Archaea	Bacteria	Protozoans	Fungi	Unicellular Algae
TABLE 5.4 Comparison of Microbial Groups[a]					
Cell type	Procaryotic	Procaryotic	Eucaryotic	Eucaryotic	Eucaryotic
Size	Microscopic	Microscopic[b]	Microscopic	Macroscopic	Microscopic
Cell wall	Present	Present	Absent	Present	Present
Reproduction	Mostly asexual (binary fission)	Mostly asexual (**binary fission**)	Sexual and asexual	Sexual and asexual	Asexual
Energy process	Variable	Mostly heterotrophic	Heterotrophic	Heterotrophic	Autotrophic

[a]Viruses and prions are not cells and therefore are not included.
[b]There are a few exceptions.

viruses and prions, all microbes have both DNA and RNA, as do all cells.

Microbes are measured in very small units of the metric system called **micrometers** (equal to one millionth of a meter), abbreviated as μm, and **nanometers** (equal to one billionth of a meter), abbreviated as nm. A meter is equivalent to about 39 inches, so a micrometer is equal to one millionth of 39 inches. These numbers are probably not very meaningful to you because they do not allow you to appreciate the size of microbes relative to more familiar objects, but there are some points that may help you think in this scale. A spoonful of fertile soil contains trillions of microbes, and *the number of microbes that can be accommodated on the period at the end of this sentence is in the millions.* **FIGURE 5.11** indicates the relative size of microbes. The monstrous bacteria described earlier are exceptions. Bacteria are many times smaller than eucaryotic cells but are about 50 times (or more) larger than most viruses.

Smallness has its advantages. Smallness provides a large surface area per unit volume, allowing for rapid uptake of nutrients from the environment. *E. coli*, for example, has a surface-to-volume ratio about 20 times greater than that of human cells. And now for the introduction—from least to the most complicated.

Prions

Prions are the most recent addition to the microbial list; some texts continue to place them with viruses for lack of a better place. But the awarding of the 1997 Nobel Prize to Stanley Prusiner, who discovered these agents, legitimized them as separate entities. The word *prion* is an abbreviation for "proteinaceous infectious particles." **Prions** are infectious protein molecules and are devoid of both DNA and RNA; their lack of nucleic acid is their major (and most puzzling) biological property. Prions exist normally, primarily in the brain, as harmless proteins. Abnormal prions convert normal proteins into infectious, disease-producing proteins responsible for mad cow disease and dementia-type diseases in humans and in other animals. Questions remain unanswered regarding their biology.

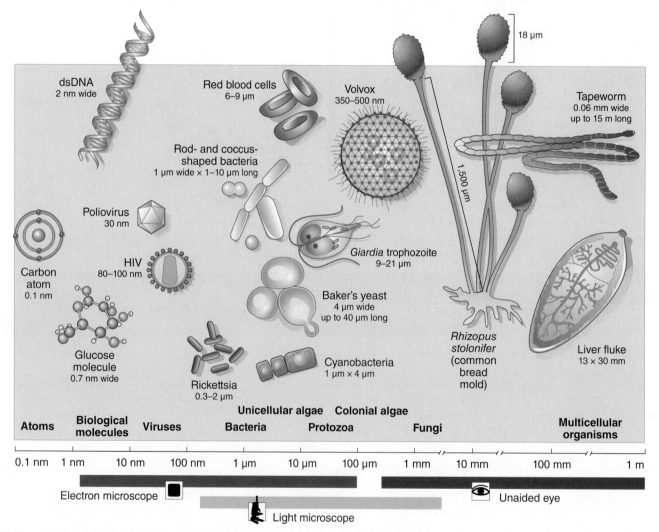

FIGURE 5.11 Comparison of the sizes of atoms, molecules, viruses, cells such as bacteria, yeast, fungi, and multicellular organisms. Bacteria are usually within a 1–10 μm range, which allows us to view them with a light microscope. To see anything smaller than 500 nm in size, an **electron microscope** is needed.

Viruses

As frequently noted, viruses are not organisms; Figure 5.4 indicates their subcellular position. Two major distinguishing characteristics of viruses are that, in contrast to cells, they contain either RNA or DNA and, further, they are submicroscopic particles and can be seen only with an electron microscope (**FIGURE 5.12**). Some have an additional protein coat or envelope encompassing them. Viruses are described as obligate intracellular parasites, meaning that they require a host cell. They must be (obligate), inside of living cells (intracellular) to replicate; they are not capable of autonomous replication. They take over the metabolic machinery and reap the benefits of energy production, without any expenditure of energy, by the **host cell**. Perhaps this is the ultimate in parasitism.

Bacteria

Bacteria are the best known of the microbes. They are microscopic, unicellular, procaryotic, and have cell walls (with the exception of a single subgroup, the *Mycoplasma*). They reproduce asexually by binary fission. In terms of size, they can be seen with a regular (light) microscope (**FIGURE 5.13**). Many bacteria are heterotrophs and use organic compounds as a source of energy. Others are autotrophs and use the energy of the sun, whereas some derive energy from the use of inorganic substances. Although a number of bacteria are pathogens and are the major subject of this text, the vast majority of bacteria are nonpathogenic and play essential roles in the environment without which life would not be possible.

Protozoans

Protozoans are unicellular and eucaryotic and are classified according to their means of locomotion (**FIGURE 5.14**). Their energy generation requires the utilization of organic compounds. Many diseases, including malaria, sleeping sickness, and amoebic dysentery, are caused by protozoans.

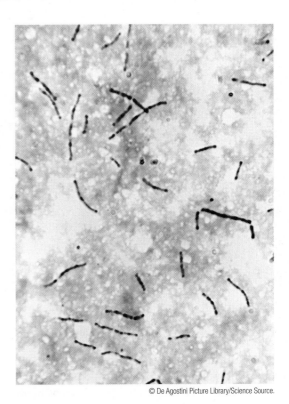

© De Agostini Picture Library/Science Source.

FIGURE 5.13 Stained *Lactobacillus* bacteria visualized with a light microscope at 1000× magnification.

Algae

Algae are photosynthetic eucaryotes and in the photosynthetic process produce oxygen and carbohydrates used by forms requiring organic compounds. Hence, they are highly significant in the balance of nature. **Dinoflagellates** and **diatoms** are examples of unicellular algae and fall under the umbrella of microbes (**FIGURE 5.15**). Dinoflagellates (plankton) are the primary source of food in

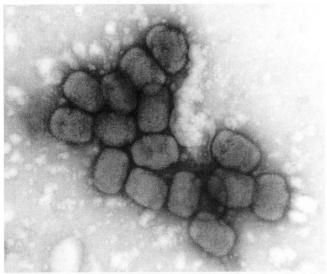

Courtesy of Dr. Fred Murphy/CDC.

FIGURE 5.12 A transmission electron micrograph of smallpox (Variola) viruses.

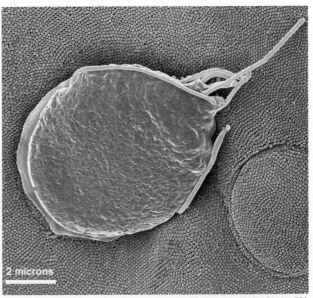

2 microns

Courtesy of Dr. Stan Erlandsen/CDC.

FIGURE 5.14 A colorized scanning electron micrograph of a flagellated *Giardia* protozoan adhering to an intestinal epithelial cell.

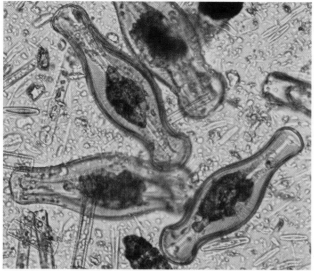

Courtesy of Robert W. Pillsbury.

FIGURE 5.15 Freshwater diatoms.

the oceans of the world. Some algae are pathogenic for humans indirectly. For example, the toxin produced by the dinoflagellate that causes red tide, *Gymnodinium breve*, causes neurological disturbances and death in humans as a result of our consumption of fish and shellfish that have fed on the dinoflagellates. Another species of dinoflagellate, *Pfiesteria piscicida*, also referred to as the "cell from hell," threatened the fishing industry in the eastern United States in 1997. A bloom of these algae resulted in the release of large amounts of neurotoxin, causing neurological symptoms in fishermen and fear among consumers.

Fungi

Fungi are eucaryotes. Morphologically, they can be divided into two groups: the yeasts and the molds. The yeasts are unicellular and are larger than bacteria; many reproduce by budding. Molds are the most typical fungi and are multicellular, consisting of long, branched, and intertwined filaments called **hyphae**. In early schemes of classification fungi were considered plants, primarily because they have cell walls. However, their cell wall composition is quite different from that of plants and from the cell walls of bacteria. Fungi are highly significant in terms of food chains and have certain beneficial aspects. Some are pathogenic

and cause diseases that are difficult to treat; others play a highly significant role as opportunistic pathogens (organisms that are not usually considered to be pathogens), because, as in HIV/AIDS, when the body's immune system is depressed they cause disease. Molds played a major role in the aftermath of Hurricanes Katrina, Rita, Sandy, Harvey, Irma, and Maria rendering houses uninhabitable. Patches of mold threaten the prehistoric paintings of animals in the Lascaux Cave in Dordogne region of southwest France, and museum curators constantly need to guard against mold intrusions. Mushrooms are a well-known group of fungi. Their diversity is unusual, as demonstrated by a wide range of size, colors, and patterns on their caps (**FIGURE 5.16**). Many species are edible.

The terms *bugs* and *germs* are part of our popular speech but have no scientific meaning. It should be clear from the above descriptions that each group of microbes is distinct from the others. When your physician diagnoses you as having a "bug," you might ask what kind.

Summary

The microbial world is remarkable for its extreme diversity, as is evident in the distinct characteristics of the six microbial groups—prions, viruses, bacteria, protozoans, unicellular algae, and fungi. Further, within each group there is considerable diversity. Not all microbes are unicellular and microscopic; some are multicellular and macroscopic, and others are subcellular and microscopic. In recent times, "monster" bacteria have been found that are unique in being unicellular and macroscopic, a rare combination. Viruses and prions are subcellular and are not considered life forms. There is no clear definition of what makes a microbe a microbe, but it is clear that they are all at less than the tissue level of biological organization.

All bacteria are procaryotic; all other microbes are eucaryotic. (Viruses and prions are not cells and are neither procaryotic nor eucaryotic.) Taxonomy evolved from a two-kingdom system (in which bacteria were considered plants) to a five-kingdom system, with various other schemes along the way; the trend has been toward recognizing the uniqueness of microbes. Woese proposed a classification system based on 16s rRNA analysis and assigned bacteria to one of three domains that reflected their evolutionary history.

© Teri Shors.

© Teri Shors.

© Teri Shors.

FIGURE 5.16 Mushrooms growing on tree stumps in woods or in leaf debris.

Since their origin on Earth microbes have adapted to extreme ecological diversity and can be isolated from all environments. Some live at the extremes—from hot springs to permafrost. All organisms must meet a basic requirement for energy, and microbial evolution has fostered a diversity of strategies. Some microbes obtain energy from organic compounds, whereas others use the energy of the sun or derive their energy from the metabolism of inorganic compounds.

The major characteristics of each of the six microbial groups show that each category is distinctive. The popular terms *bugs* and *germs* are used in a collective sense, but there is no basis for lumping these diverse microbial agents together. Further, these terms have a negative connotation, because they are usually used to describe microbial diseases, but it is important to remember that only a handful of microbes are disease producers.

KEY TERMS

16s rRNA gene
adenosine triphosphate (ATP)
aerobes
Archaea
autotroph
Bacteria
binary fission
biosphere
carbohydrates
cell
cell theory
chemosynthetic autotrophs
deoxyribonucleic acid (DNA)
diatoms
dinoflagellates
electron microscope
Eucarya
eucaryotic cells
extremophiles
facultative anaerobes
genes
halophiles
heterotrophs
host cell
hyperthermophiles
hyphae
inorganic compounds
light microscope
macroscopic
magnetotactic bacteria

metagenomics
microbe
microbial umbrella
microbiome
micrometer (μm)
microscopic
nanometer (nm)
nucleoid
opportunistic pathogens
organ systems
organelles
organic compounds
organisms
organs
photosynthetic autotrophs
polymerase chain reaction (PCR)
prions
procaryotic cells
protozoans
psychrophiles
spontaneous generation
strict anaerobes
stromatolites
Taq polymerase
taxonomy
tissue
virus
Whittaker's five-kingdom system
Woese's three-domain system

SELF-EVALUATION

○ PART I: Choose the single best answer.

1. A major distinction between procaryotic and eucaryotic cells is based on the presence of _____.
   a. a cell wall
   b. DNA
   c. a nuclear membrane
   d. a cell membrane

2. Most bacteria are considered to be _____.
   a. harmful   c. autotrophs
   b. anaerobes   d. heterotrophs

3. Which of the following is the smallest unit of measurement?
   a. Millimeter   c. Micrometer
   b. Nanometer   d. Centimeter

4. Which of the following are subcellular agents that require a host cell to replicate?
   a. Bacteria   c. Unicellular algae
   b. Viruses   d. Diatoms

5. Which of the following does not have nucleic acid in its structure?
   a. Viruses   c. Bread mold
   b. Diatoms   d. Prions

6. Bacterial species are identified through _____ gene analysis.
   a. tox   c. lacZ
   b. 16s rRNA   d. mutT

7. According to Woese _____.
   a. Eucarya arose from Archaea   c. Bacteria, Archaea, and Eucarya all arose independently
   b. Archaea arose from Eucarya   d. None of the above is correct

8. Which of the following bacteria produces an enzyme that catalyzes DNA synthesis?
   a. E. coli   c. *Serratia marcescens*
   b. *Thermus aquaticus*   d. *Dienococcus radiodurans*

9. Which of the following microbes are not disease producers?
   a. Fungi   c. Archaea
   b. Bacteria   d. Protozoa

10. _____ is used to identify microbes present in environmental samples through the analysis of DNA.
    a. Microeconomics   c. Genetic engineering
    b. Metagenomics   d. Gene therapy

○ PART II: Fill in the blank.

1. Bacteria, viruses, fungi, and protozoans are microbes. Another group that falls under the microbial umbrella are the _____.

2. The cell theory is credited to _____.

3. Compounds containing carbon are called _____ compounds.

4. Organisms that do not require organic compounds in their metabolism are called _____.

5. The "energy compound" is _____.

6. Strict anaerobes are killed by _____.

7. The term _____ is used to describe organisms too small to be seen without a microscope.

8. _____ is the belief that "life" is derived from "nonlife."

9. All microbes contain _____ as their genetic or hereditary material.

10. _____ are infectious protein molecules.

○ PART III: Answer the following.

1. Criticize the terms *bugs* and *germs* as used in a collective sense to describe microbes. List the categories of microbes, and write a one-sentence description of each.

2. What makes a microbe a microbe?

3. What is the relevance to microbiology of Shakespeare's "What's in a name? That which we call a rose by any other name would smell as sweet"?

4. Explain why heterotrophs are dependent on autotrophs.

5. The Archaea can survive extreme environments. How are they able to do this?

6. Fungi play a highly significant role as opportunistic pathogens. Define *opportunistic pathogens* and identify those who are most susceptible to these types of infections.

7. Explain why viruses and prions are not cells and are neither procaryotic nor eucaryotic.

8. Explain why some microbes can survive in the absence of oxygen.

9. Compare and contrast eucaryotic and procaryotic cells.

10. What do bacteria gain in forming symbiotic partnerships with other organisms?

BENEFICIAL ASPECTS OF MICROBES: THE OTHER SIDE OF THE COIN

CHAPTER 6

There is no field of human endeavor, whether it be in industry or in agriculture, whether it be in the preparation of foodstuff or in connection with problems of shelter and clothing, whether it be in the conservation of human and animal health and the combating of disease, where the microbe does not play an important and often a dominant part.

—Selman A. Waksman, 1943

Nobel laureate, microbiologist

© Teri Shors.

Brewing beer is a traditional food fermentation method in which the yeast *Saccharomyces* produces ethanol as a fermentation end product. The six-packs of beer shown for sale are craft or specialty beers that are manufactured in smaller breweries and are made with unusual ingredients to give the beer an enhanced or distinctive aroma and flavor.

LEARNING OBJECTIVES

1. Discuss the take-home message of this chapter that only a few microbes cause disease and many are beneficial in our daily lives.

2. Explain the roles of microbes as decomposers.

3. Describe the biogeochemical cycles and how microbes are involved in these cyclic processes.

4. Discuss fermentation and explain how microbes are used for the production of fermented foods.

5. List at least five fermented foods and the microbes that are used to produce each.

6. Explain the association between probiotics and prebiotics.

7. Detail the importance of bioremediation to accelerate the degradation of pollutants in the environment.

8. Compare and contrast gene therapy and virotherapy.

Case Study: Take Two Fecal Pills and Call Me in the Morning

Dana McDermott was a 56-year-old professor who had recently finished a course of antibiotics to treat a urinary tract and bladder infection. She developed **colitis**. She had diarrhea 10 to 15 times a day and was diagnosed with an infection caused by *Clostridium difficile*, also referred to as *C. diff*. *C. difficile* releases **exotoxins** that attack the lining of the intestines, triggering colitis. The commensal **microbiome** of her gut was disrupted by the cephalosporins used to treat her earlier bacterial infections. Antibiotics kill both "good" and pathogenic bacteria. **Table 1** is a list of antibiotics that may induce *C. difficile* diarrhea and colitis.

TABLE 1 Antibiotics Associated with Colitis and *Clostridium difficile* Infections		
Rarely Associated	Sometimes Associated	Often Associated
Aminoglycosides	Macrolides	Cephalosporins
Chloramphenicol	Trimethoprim	Clindamycin
Metronidazole	Sulfonamides	Fluoroquinolones
Tetracycline		Penicillin
Vancomycin		

Dana had lost her appetite and was losing weight. Her physician recommended **fecal microbiota transplantation** to restore the "normal" population of bacteria in her gut/colonic environment. At first Dana was repulsed. She instinctively rejected the idea of human feces being introduced into her body through an enema, colonoscopy, or nasogastric (oral) procedure because she couldn't get past the "yuck factor." As her condition worsened, she began doing research and learned that clinical trials in which patients were ingesting oral capsules containing freeze-dried or frozen stool cured more than 15,000 cases of *C. difficile*. Two companies in the United States had stool banks. She was curious about the requirements of a stool donor and the safety of the stool. She did not want to get another infection through somebody else's "poo."

Stool donors may be genetically related, typically a spouse or child, or unrelated to the recipient. Stool from a stool bank is usually shipped frozen or on dry ice overnight. The screening criteria used to assess the donors and the safety of their stool for fecal microbiota transplantation are listed in **Table 2**. The process of preparing donor stool and fecal microbiota transplantation is shown in **Figure 1**.

Dana participated in a fecal microbiota transplantation clinical trial in which she ingested capsules of frozen donor stool to treat colitis/*C. difficile* infection. Her body responded favorably, and Dana's gut microbiota was healthy after treatment using the "poopsickles" approach.

(continues)

Case Study: Take Two Fecal Pills and Call Me in the Morning (continued)

TABLE 2 Criteria for Screening Donors for Fecal Microbiota Transplantation

Patient History Criteria/Considerations	Screening for Infectious Agents
Does the donor have any active infections?	Stool screening for bacteria: *Clostridium difficile*, *Campylobacter*, *Helicobacter pylori*, *Salmonella*, Shiga toxin–producing *E. coli*, *Shigella*
Has the donor been on antibiotics in the past 3 months?	Donor blood testing for bacteria: *Treponema pallidum*
Has the donor been exposed to epidemic diarrheal disease through travel?	Other bacterial screening considered: *Aeromonas*, *Plesiomonas* (formerly known as *Aeromonas shigelloides*), *Listeria monocytogenes*, *Yersinia*, *Vibrio cholerae*, and *Vibrio parahaemolyticus*
Does the donor have gastrointestinal conditions such as inflammatory bowel disease, irritable bowel syndrome, or chronic diarrhea/constipation?	Stool screening for viruses: rotavirus, norovirus
Does the donor have any autoimmune disorders or history of significant allergies?	Donor blood testing for viruses: hepatitis A, B, and C viruses; HIV
Does the donor have other risk factors (e.g., diabetes, metabolic syndrome, exposure to medications that may alter the gut microbiota)?	Stool screening for parasites: *Cryptosporidium*, *Cyclospora*, *Giardia*, *Isospora*
Does the donor have any other risk factors (e.g., body piercing or tattoo in the prior months, high-risk sexual behaviors)?	

Questions and Activities

1. Research the different approaches used in fecal microbiota transplantation (oral/nasogastric, colonoscopy, enema, and ingestion of frozen or freeze-dried capsules) and list the advantages and disadvantages of each.

2. Define *commensalism* as it pertains to the human gut microbiota.

3. List at least five different bacteria that are normal inhabitants of the gut microbiota.

4. Explain why *C. difficile* infections may be associated with chronic antibiotic therapy.

Information from Jiang, Z. D. et al., 2017. "Randomized Clinical Trial: Faecal Microbiota Transplantation for Recurrent Clostridium difficile Infection—Fresh, or Frozen, or Lyophilized Microbiota from a Small Pool of Healthy Donors Delivered by Colonoscopy." *Aliment Pharmacol Ther 45*:899–908.

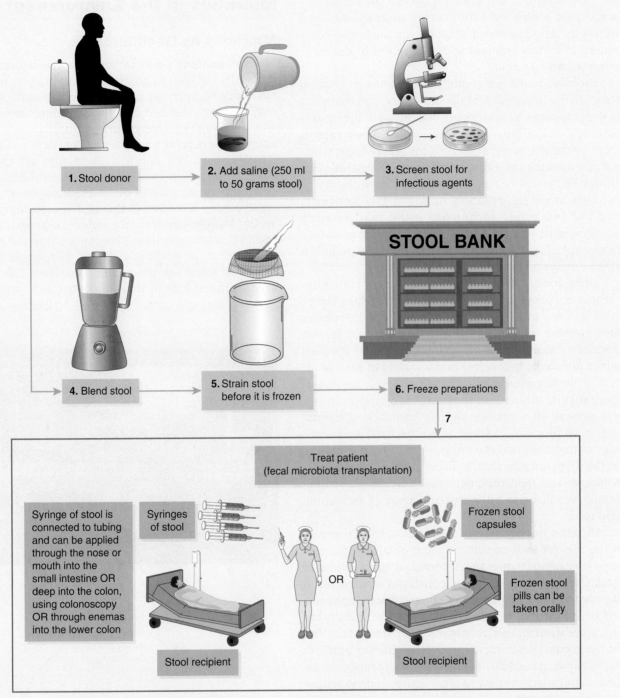

Figure 1 Procedure used to prepare donor stool (unrelated or relative of a patient). The donor stool samples are screened as being "safe from infectious agents" and are subsequently prepared at stool banks for distribution. Fecal microbiota transplantations are performed at hospitals and clinics or sometimes can be done at home (e.g., enema procedure).

Preview

It would be understandable if you are biased against microbes; you may have more reason to hate and fear them than to love them. In this chapter, however, the take-home message is that only a few microbes are disease producers and many are beneficial in our daily lives. The goal of microbiologists is not to annihilate all microbes but to eradicate pathogens or at least to minimize their impact and the burden of microbial disease by avoiding circumstances leading to a collision course. After all, microbes were the first inhabitants of our planet; they are the senior citizens from which evolution to eucaryotic cells and multicellularity proceeded. Microbes are the largest component of Earth's biomass and are present in the most extreme habitats of life.

Microbes make the planet's ecosystems go around. They are the foundation of the biosphere, and many act as **decomposers** or scavengers. Bacteria are the underpinnings of the biogeochemical cycles. Their role in these cycles is unseen and taken for granted, but without microbes the cycles could not be completed, and life, ultimately, would cease. The cycles occur without our initiative or our intervention. In fact, the danger exists that our increasing technology could inadvertently interfere with and shut down certain cycles of nature as a result of nonbiodegradable products and pollution of the environment.

Since antiquity, societies the world over learned to harness microbes for their beneficial aspects long before there was any awareness of a microbial world. Societies were content with the empirical evidence that certain practices simply "worked." The production of distilled spirits (alcoholic beverages) and a variety of food products, including breads, yogurt, and cheeses, are examples. Yogurt, which contains live bacterial cultures, was prescribed centuries ago for "stomach ailments" and, in some cases, seemed to do the trick. As knowledge of microbes and the enzymes they produce in the metabolism increases, the manufacture of alcoholic beverages and foodstuffs becomes increasingly sophisticated, resulting in an increasing array of fermented food products.

Microbes are powerful biological research tools because of the relative ease of culturing and obtaining them in large populations in a short period of time. Evidence establishing DNA as the genetic material was the result of experiments using bacterial viruses (**bacteriophages**) and their **hosts** (bacterial cells). Industry, particularly the pharmaceutical industry, has learned to harness microbes for the production of many products, including antibiotics, vaccines, genetically engineered therapeutics, insecticides, and a large variety of other compounds to control or eliminate microbes.

Bioremediation, the use of microorganisms to clean up polluted environments, is on the increase; microbes played a role in reducing the impact of the *Exxon Valdez* oil spill that occurred off the coast of Alaska on March 24, 1989; they were also employed as remedial agents in the Gulf Coast oil spill in April 2010 and in 1989 to degrade cyanide by treating mining tailings left over from extracting gold from ore at the Homestake Mining Company, Lead, South Dakota. After reading this chapter, perhaps you will rethink your love–hate relationship with the microbial world.

Microbes in the Environment

Microbes as Decomposers

Perhaps at some point in your life you had an aquarium with goldfish or tropical fish. If so, you recall the necessity of properly maintaining the aquarium. You may not have realized it at the time, but the aquarium was a simulated **ecosystem**—a population of organisms in a particular physical and chemical environment (**FIGURE 6.1**). The fish and the plants are the added **biotic** components, whereas the chemical and physical environment constitutes the **abiotic** component. You will recall paying attention to the light source, temperature, acidity, and cleanliness of the water to maintain a healthy and balanced ecosystem. You can assume the presence of microbes as additional biotic components. In the fish tank, the green plants are considered the **primary producers** because of their photosynthetic capabilities

(a)   Courtesy of Dr. Morgan Churchill, UW Oshkosh.

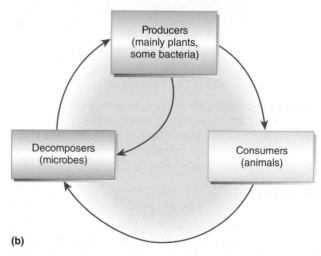

(b)

FIGURE 6.1 **(a)** A fish tank is an artificial ecosystem. **(b)** The cycle of life in an ecosystem: producers, consumers, and decomposers.

that result in the production of organic compounds and oxygen; the fish are the **consumers** and remove oxygen from the water and exhale carbon dioxide and use organic compounds as their nutrient source. The bacteria and fungi are the decomposers and are the link between the producers and the consumers. The microbial population decomposes waste materials of the fish and dead leaves of the plant and, in so doing, functions as recyclers. And so it is in nature—witness plant debris, animal wastes, and the bodies of dead animals.

The greatest recyclers of all time are microbes, without which life would be a dead end and, ultimately, would cease. Imagine a huge garbage dump into which materials are deposited daily and continue to accumulate year after year and generation after generation. Under these circumstances, and without recycling, the resources of the planet would soon run out. If you walk through swampy areas, you may detect the unmistakable odor of methane that smells like sulfur—marsh gas—resulting from the bacterial action on decomposing materials. The microbes involved as scavengers are nonpathogenic and free-living. The resources on Earth are limited and are recycled through food webs with microbes as the decomposers (**FIGURE 6.2**). Nature has always practiced this, but it has only been in the past 30 or 40 years that society has realized the inextricable link between populations, soil, water, air, and energy, all of which are interdependent and dependent on microbes.

Microbes and the Biogeochemical Cycles

As previously stated, bacteria are the basis for the **biogeochemical cycles**, the processes involved in the

recycling of carbon, nitrogen, sulfur, iron, and phosphorus, resulting in the return of these elements to nature for reuse. These cycles are discussed separately but are linked. The carbon and nitrogen cycles are presented as examples.

⊙ Carbon Cycle

Carbon atoms are key elements in living systems and are found in proteins, carbohydrates, fats, and DNA. Most of the carbon used by organisms is present in association with carbon dioxide, a simple compound consisting of one carbon atom attached to two oxygen atoms. Photosynthetic organisms capture the sun's energy and use it for the conversion of atmospheric carbon dioxide and, along with water, produce glucose and other energy-rich carbohydrates. Hydrogen and water are necessary reactants in photosynthesis. An important spin-off of photosynthesis is the release of oxygen from the carbon dioxide back into the atmosphere.

Plants are associated with photosynthesis, but some microbes are also photosynthetic and are the primary producers in the ocean. *Chlorella* is a photosynthetic alga that is found on the surface of ocean water, and **cyanobacteria** are photosynthetic bacteria.

Carbon, captured as carbon dioxide, is ultimately recycled back to its elemental form through food chains. Cellulose, a polymer (chain) of glucose (sugar) molecules, is an energy-rich carbohydrate product of photosynthesis. Bacteria produce enzymes that are able to break down cellulose into single molecules of glucose. **Herbivores** (grazers) feed on plants but lack the necessary digestive enzymes to break down the cellulose. Bacteria come to the rescue! They are a part of the normal **microbiota** residing in the intestinal tract of grazers, and their enzymes digest cellulose, allowing these animals to use cellulose as an energy source. Bacteria in the intestinal tract of termites, as another example, allow termites to lunch on your house. Ultimately, **predators** feed on the grazers, including humans, but one way or another they and their waste material enter the food web. From there, microbial decomposition takes over, and the **carbon cycle** is completed (**FIGURE 6.3**).

⊙ Nitrogen Cycle

Nitrogen is a constituent of amino acids, the building blocks of proteins and of the nucleic acids of microbes, plants, and animals. It is the most common gas in the atmosphere (about 80%), but atmospheric nitrogen cannot be tapped by animals or by most plants. Here again microbes come to the rescue; only bacteria can convert, or fix, nitrogen into a usable form and, ultimately, recycle it back to the atmosphere. The **nitrogen cycle** is illustrated in **FIGURE 6.4**.

The process begins with the fixation of atmospheric nitrogen and its conversion to ammonia by **leguminous**

© Teri Shors.

FIGURE 6.2 Microbes are the ultimate decomposers.

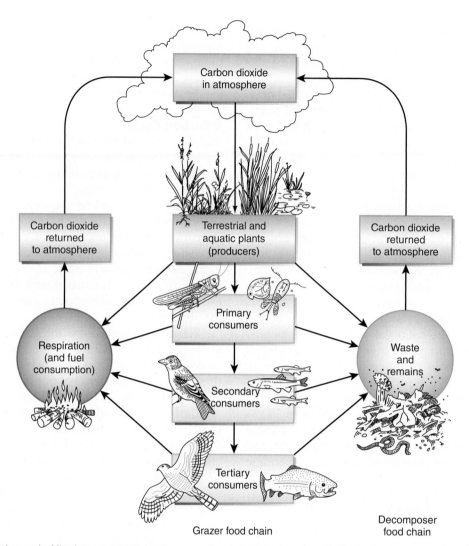

FIGURE 6.3 The carbon cycle. Microbes are essential in the conversion of atmospheric carbon dioxide to organic compounds and back to the atmosphere.

plants. These are plants that have swellings or nodules along their root systems containing *Rhizobium* and other nitrogen-fixing bacteria (**FIGURE 6.5**). Peas, soybeans, alfalfa sprouts, peanuts, and beans are examples of leguminous plants. The association of nitrogen-fixing bacteria and leguminous plants is an example of **symbiosis**.

The next phase of the nitrogen cycle is called **nitrification**; in this process ammonia is converted into nitrates, the form of nitrogen most used by plants. Members of the bacterial genera *Nitrobacter* and *Nitrosomonas* carry out these processes. Other bacteria, as well as fungi, decompose plants and animals and their waste products and in the process convert nitrogen into ammonium, giving meaning to the expression "death yields life."

Urine is a waste product particularly rich in nitrogen. Finally, **denitrifying** bacteria are responsible for the return of nitrogen to the atmosphere as nitrogen gas. Horticulturists and agriculturists have long realized the importance of nitrogen in growing flowers and food crops and use a variety of fertilizers containing nitrogenous compounds.

Krasner's Notebook

You might be interested in seeing these resident bacteria. All you need to do is dig up peas, a patch of clover, or some other leguminous plant, being sure to take some of the root system. Wash away the soil and crush a nodule onto a clean slide. Add a drop of water and a dye, such as methylene blue, and spread the preparation with a toothpick or a matchstick onto the slide to establish a thin film. Allow the preparation to dry and examine it under a microscope. You will observe rod-shaped bacteria, probably members of the genus *Rhizobium*. If you perform a **Gram stain**, the bacterial population will be dominated by Gram-negative (red) rods.

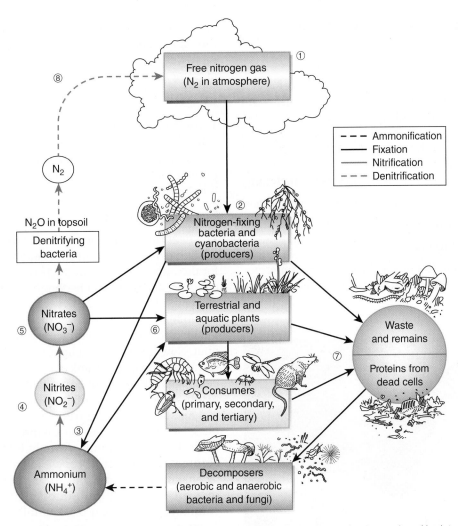

FIGURE 6.4 The nitrogen cycle. Microbes are essential in the conversion of atmospheric nitrogen to organic compounds and back to the atmosphere.

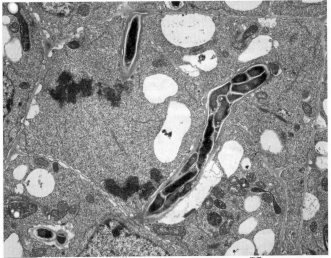

Courtesy of Louisa Howard, Dartmouth College, Electron Microscope Facility.

FIGURE 6.5 Transmission electron micrograph of a cross section through a soybean (*Glycine max*) root nodule. The bacteria infect the roots and establish a nitrogen-fixing symbiosis.

○ Other Cycles

In addition to the carbon and nitrogen cycles, the movement of other elements, including sulfur, phosphorus, and iron, through ecosystems in a cyclical manner depends on microbial communities. *It really is microbes that "make the world go around."* Their role in these biogeochemical cycles, which take place in all imaginable ecosystems and at all extremes of temperature, demonstrates the necessity of microorganisms for sustaining life on Earth.

Krasner's Notebook

Look at a variety of fertilizers and you will see three numbers, for example, 22-3-12. The first number pertains to the nitrogen content, the second pertains to the phosphorus content, and the third pertains to the potassium content. Some farmers might not add fertilizer to their soil but instead may include leguminous crops, which are plowed under during the off-season as a way of enriching the nitrogen content of the soil.

Microbes in Food Production

Foods

Mushrooms have long been recognized as tasteful and nutritionally rich foods whether raised in mushroom farms or picked "in the wild" (**FIGURE 6.6**). But let the mushroom pickers beware: some species of mushrooms are extremely toxic to the point of death. A number of dangerous myths are out there on the distinction between edible and toxic (nonedible) mushroom species, including:

- Poisonous mushrooms cause a silver spoon to turn black.
- Poisonous mushrooms have a pointed cap.
- Poisonous mushrooms taste bad.

The medical benefits of mushrooms are numerous. Studies have shown them to be beneficial as antimicrobial agents, anticholesterol agents, cognitive stimulants, a vitamin source, and in lowering blood sugar levels. Interestingly, before synthetic dyes, mushrooms were the source of numerous textile dyes.

Algae

These microbes are food sources in some societies. In China, people consume more than 70 species of algae, including fat choy, which when dried resembles long, black human hair and has a soft texture. (Sounds good! Order it next time!)

Food Production

The next time you shop at the market, look around at the shelves of foods for those that depend on microorganisms for their production or products that contain live bacterial cultures. Examples can be found in just about all categories of foodstuffs (**TABLE 6.1**). **FIGURE 6.7** presents items dependent upon microbes for their production. Use your imagination and come up with your own microbial banquet. Now it is time to delight in the fanciful images that come to mind as you think about the wonderful foods and beverages whose stimulating tastes and aromas are the result of microbial activities.

As stated in the introduction of this chapter, people have used microbes for the production of fermented foods, knowingly or unknowingly, for thousands of years. **Fermentation** is a series of chemical reactions mediated by enzymes of a variety of strains of bacteria and yeasts that break down sugars into smaller molecules (most commonly lactic acid or ethanol and carbon dioxide). A characteristic of all organisms is the ability to meet their energy requirements to stay alive. Most procaryotic microbes (and most organisms) use energy-rich organic foodstuffs and oxygen and through a complex cyclical series of biochemical reactions "extract" the energy inherent in the bonds of that food and convert it into **adenosine triphosphate (ATP)**—a readily available form of energy. Fermentation is a metabolic path by which some microbes, primarily yeasts, are able to shift their metabolism in the absence of oxygen to produce small but sufficient amounts of ATP. Although it is true that fermentation, an **anaerobic process**, is a far less efficient manner for the conservation of energy, it is an evolutionary advantage in that it allows for survival under anoxic conditions.

The end products of fermentation, such as lactic acid, carbon dioxide, and ethanol, are of considerable value in commercial food and alcoholic beverage industries, as further described in this chapter. In particular, lactic acid and ethanol may act to inhibit the growth of unwanted microbes, thus acting as food preservatives (extending the useful life of these foods). The choice of microorganism to carry out fermentation determines the taste and aroma of the product. The specific microbial strain referred to as the **starter culture** and the process used for many products are carefully guarded secrets;

© Gaby Smith.

FIGURE 6.6 Picking morels during the spring in Wisconsin is very popular. Morels are easy to identify and often found popping up through leaves near a dying elm or by apple trees. After the morels are cleaned, cooks may batter-dip and fry them in butter. (Everyone should experience the glory of the fried morel! They are yummy!)

TABLE 6.1 Foods Produced by Using Microbes

Milk Products	Meats	Breads	Miscellaneous Products	Alcoholic Beverages
Cheese	Bologna	Sourdough bread	Sauerkraut	Beer
Yogurt	Salami	Numerous other breads and rolls	Pickles	Wine
Buttermilk	Country cured ham		Olives	Sake (Japanese rice wine)
Kefir	Sausage		Vinegar	Distilled spirits (e.g., brandy, whiskey, rum, vodka, gin)
Acidophilus milk			Tofu	
Sour cream			Soy sauce	
			Kimchi	

© Teri Shors.

FIGURE 6.7 Thanks to the microbes! Bon appétit!

starter cultures are handed down in families from generation to generation (**BOX 6.1**). Yeasts are unicellular fungi and are efficient fermenters of sugar into alcohol and carbon dioxide, a property exploited in the production of breads and alcoholic beverages. *Saccharomyces cerevisiae* is one commonly used yeast.

Bread Products

Fermentation by yeasts produces the gas carbon dioxide (and ethanol that evaporates during baking), which causes bread dough to rise and increase in size, or leaven, before baking. If you want to make your own bread, you can buy packages of live yeasts inexpensively

at the supermarket (**FIGURE 6.8**). Bread has long been a staple in primitive societies. Bread samples dating back to 2100 B.C. are on display at the British Museum in London.

Unleavened flatbread is central to the 3,000-year-old story of Passover, a Jewish celebration commemorating the time when the pharaoh of Egypt freed the Israelites from bondage. The people left in haste, without time to bake bread for their journey. They took the raw dough and baked it on rocks under the hot sun. With no yeast to raise the dough, it produced flat crackers called "matzo," an unleavened flatbread (**FIGURE 6.9**).

Dairy Products

Cheese production, too, dates back thousands of years, as do fermented milk beverages that have been promulgated for centuries to treat a variety of intestinal tract disorders, from constipation to flatulence (gas) (**BOX 6.2**). A variety of fermented milk products exist, many with centuries old origins in the Middle East (Table 6.1; Box 6.2). They vary in their texture, taste, and aroma, depending on the type of milk, incubation period, and, most significantly, the microbial culture used to carry out the fermentation process. Different species of *Lactobacillus* and a few species of *Streptococcus* are commonly used. Cultured buttermilk is generally made by adding *Streptococcus cremoris* and *Leuconostoc citrovorum* to pasteurized milk. Other microbes produce buttermilk with different flavors. Certain species of lactobacilli or streptococci added to cream results in sour cream. *Lactobacillus bulgaricus* and *Streptococcus thermophilus* are commonly used to produce yogurt and yogurt drinks. Kefir is a cultured milk product made from the milk of

BOX 6.1 Talking Starter Cultures and Sourdough Bread

"The smell of good bread baking, like the sound of lightly flowing water, is indescribable in its evocation of innocence and delight."
—M. F. K. Fisher (1908–1992), American food writer

Who would have thought that bacteria can "talk to each other"? They do so through chemical "words" known as **quorum sensing (QS)** signal molecules. The term *quorum* is Latin; it is used in politics and committee meetings to indicate when enough members are present in order to pass a motion at a committee meeting. As bacteria grow, they release QS signal molecules into the environment. Once a group of bacteria reach a certain size, bacterium-to-bacterium chemical communication occurs as the QS molecules bind to the outside of bacteria that are in close proximity to each other, resulting in a group or social task to be carried out as though the community of bacteria was a multicellular organism. This communication and cooperation in the procaryotic world was first observed when the marine bacterium *Vibrio fischeri* produced bioluminescence (i.e., light) when the bacteria reached high densities.

Today, it is known that bacterial QS group behaviors regulate **biofilm** production (adherence of bacteria to a solid surface such as a catheter or to rocks in a river or stream), **pigmentation** (the secretion of colorful pigments involved in cellular activities such as photosynthesis, protection, and survival), infection/disease processes, and even the quality of fermented foods.

The making of sourdough bread can be traced back to its origins in ancient Egypt. By 5,000 years ago, sourdough bread was a regular part of the European diet. In the United States, sourdough bread became a staple in San Francisco during the California Gold Rush of 1849. Miners brought carefully maintained dough mixtures of fermenting lactic acid bacteria and wild yeast that they used to make sourdough bread. Bakeries opened in San Francisco that used the miner's dough mixtures (i.e., starter cultures). Each bakery maintained its own starter cultures that were referred to as "mother's sponge."

Sourdough is a mixture of wheat or rye flour and water that is fermented by microbiota composed of lactic acid bacteria, mostly of the genus *Lactobacillus*, and yeasts (*Saccharomyces*, *Candida*, and *Kazachstania* spp.). Typically, the ratio of lactic acid bacteria to yeast is 100:1. The *Lactobacillus* spp. used to develop the dough come from natural contaminants of the flour or from a starter culture containing one or more known species of *Lactobacillus* (**Figure 1**). During fermentation, the bacteria produce lactic acid and acetic acid in the dough mixture that results in a sour taste. The fermentation temperature is an important factor in making sourdough bread. The temperature used to make traditional sourdough is typically at ambient temperatures of 20°C to 30°C (68°F to 86°F). Yeasts will not grow

© Teri Shors.

Figure 1 Typical ingredients, including starter cultures, used to make traditional sourdough bread.

in temperatures outside of this range. Yeasts are needed to *leaven*, or ferment and raise the bread. Defined starter cultures can be used to improve the flavor, nutritional quality, and shelf life of the final product (**Figure 2**).

The knowledge of QS systems during fermentation of foods is rather preliminary. However, research indicates that QS systems play an important role in the changes of microbiota during the fermentation of sourdough. It is known that certain starter *Lactobacillus* spp. secrete QS signal molecules

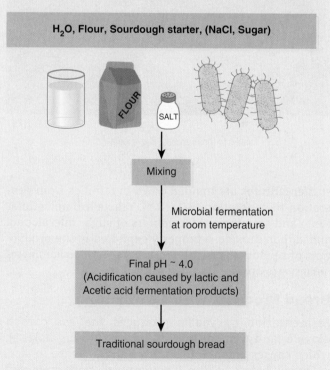

Figure 2 The procedure used to make traditional sourdough bread.

that activate the expression of genes that code for bacteriocins when the bacteria reach a certain density. **Bacteriocins** are bacterial proteins that contain antibacterial properties. They function by inhibiting closely related strains or species of nonstarter *Lactobacillus* spp. that lower the quality of sourdough flavor, dough structure, nutrition, and shelf life (causing spoilage) (**Figure 3**). So, it turns out that it takes a quorum of lactic acid bacteria to produce sourdough breads.

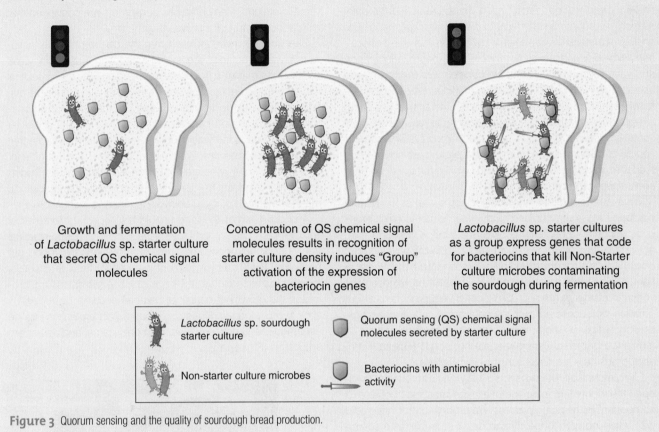

Growth and fermentation of *Lactobacillus* sp. starter culture that secret QS chemical signal molecules

Concentration of QS chemical signal molecules results in recognition of starter culture density induces "Group" activation of the expression of bacteriocin genes

Lactobacillus sp. starter cultures as a group express genes that code for bacteriocins that kill Non-Starter culture microbes contaminating the sourdough during fermentation

Lactobacillus sp. sourdough starter culture

Quorum sensing (QS) chemical signal molecules secreted by starter culture

Non-starter culture microbes

Bacteriocins with antimicrobial activity

Figure 3 Quorum sensing and the quality of sourdough bread production.

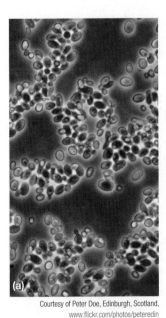

Fleischmann's®
ActiveDry
YEAST®

Fleischmann's®
ActiveDry
YEAST®

All Natural

(b) www.breadworld.com

FIGURE 6.8 Yeast. **(a)** Yeast is used in baking to get the dough to "rise" as a result of carbon dioxide production. **(b)** Baker's yeast (*S. cerevisiae*) is available in various forms at supermarkets.

FIGURE 6.9 Most breads are made with yeast, which causes the bread to rise. Matzo uses no yeast, so the bread stays flat. Matzo forms an integral component of Jewish cuisine eaten during the Passover festival. It is cracker-like with popped bubbles throughout.

BOX 6.2 Claimed Medical Benefits of Good Gut Microbiota

Yogurt is a centuries-old food from Eastern Europe and a particularly popular food in the United States. A look in the dairy section of a supermarket attests to the popularity of yogurt. The shelves are stacked with nonfat; low-fat; "fruit on the bottom"; and flavored yogurts, including mocha latte, peach, and apricot; and yogurt with "mix-ins" that are bundled with the yogurt and added when you eat it. "Do-it-yourself" yogurt-making kits are readily available. Why the yogurt craze? The answer is simple: it is good for you. Many people attempting to lose weight consume a container of yogurt as a meal. Yogurt is an excellent source of calcium, some vitamins, and protein, as indicated on the label. It can be low in fat, low in calories, or low in both fat and calories. The choice is yours!

What is in yogurt that, according to some, makes it a beneficial food? Most yogurts, and many other fermented milk products, contain live bacteria. Look at the label; it will state "live, active cultures," or words to that effect, depending on the brand. *Lactobacillus acidophilus* and other lactobacilli are the predominant live cultures; other bacteria are present as listed on the yogurt container. A gram (0.03 ounce) of yogurt contains about 1 million lactobacilli; a 6-ounce container, intended as a single serving, weighs 170 grams and, therefore, contains about 170 million live bacteria. Generally speaking, about 1 billion live *L. acidophilus* cells are necessary for effectiveness.

Elie Metchnikoff, best known for his work in immunity, developed the theory that toxic bacteria in the intestinal tract were a component in the aging process. He proposed that lactic acid was a key factor in longevity and drank sour milk as a source of lactic acid on a daily basis. Ultimately, interest in his theory led to the production of a variety of fermented foods, including yogurt and probiotics.

Foods supplemented with live microbes are called **probiotics**, defined by the U.S. Food and Drug Administration (FDA) as "live microorganisms or live active cultures which, when administered in sufficient quantities, may improve health." They are primarily in dairy products, but probiotics are also available as juices, tonics, sparkling drinks (**Figure 1a**), and as tablets or capsules that can be purchased in pharmacies, health food stores, markets, and other retail outlets. We tend to think of the presence of microbes in the intestinal tract, other than the normal microbiota, as detrimental to our health. The proposed beneficial effects of probiotics are based on the assumption that consumption of "good" live lactobacilli and certain other bacteria complements, or in some cases partially replaces, the normal microbiota, which helps to keep your digestive system healthy by controlling the growth of harmful bacteria. Advocates of probiotics claim that benefits include reduction in blood pressure, regression of tumors, reduction in allergy symptoms, a healthy digestive system, shortened duration of diarrhea, and decreased gas production. Individuals on antibiotic therapy sometimes suffer from yeast infections, most commonly manifested in the mouth and in the vagina. Some evidence suggests that consumption of yogurt and other probiotics during antibiotic therapy may be of value in preventing and treating yeast infections. Many probiotic consumers are convinced that regular consumption improves their health.

Skeptics, however, maintain that the claimed benefits associated with probiotics are exaggerated and based on weak science, primarily a poor understanding of the intestinal microbiota. Some skeptics say, "they [probiotics] go in at one end of the digestive tract and come out the other, and hopefully something good happens along the way." Take a look at the bags of dog food when you are in the market and note that many contain probiotics. Probiotic supplements are available for cats as well. Those who are not convinced of the value of yogurt and similar products call for research conducted in a scientific manner. Meanwhile, if you enjoy your yogurt, continue to eat it!

And don't forget that one should also consume *prebiotics*! To maintain healthy probiotic levels in the gastrointestinal tract, one should consume prebiotics to "feed" the "good gut bacteria." **Prebiotic**, or fermentable dietary fiber, is essential food for gut microbiota in the large intestine, which, in turn, promotes the growth of beneficial bacteria such as *Lactobacillus* spp. and *Bifidobacterium* spp. Foods rich in prebiotics include asparagus, legumes, bananas, oatmeal, jicama, and Jerusalem artichokes (**Figure 1b**).

(a)

© Teri Shors.

(b)

© Teri Shors.

Figure 1 **(a)** A variety of foods and beverages contain probiotics. **(b)** Food sources of prebiotics.

FIGURE 6.10 A variety of cheeses. Their texture, aroma, and taste are the result of the strain of microbe and the fermentation process used.

© Teri Shors.

© Robert I. Krasner.

FIGURE 6.11 Cheeses of the world. An array of cheeses from many countries is displayed in a supermarket.

cows, sheep, goats, or buffalo and kefir grains (gelatinous clumps of bacteria and yeasts). Kefir dates back many centuries to the shepherds of the Caucasus Mountains, who discovered that fresh milk carried in goatskin bags sometimes fermented into an effervescent beverage. (The Caucasus Mountains are between the Black and Caspian Seas and range through Georgia, Armenia, Azerbaijan, and the southwest region of Russia.)

It is not mice but bubbles of carbon dioxide produced by the fermentative activity of bacteria that put the holes in Swiss cheese. Cheeses are classified as soft, semisoft, hard, and very hard (**FIGURES 6.10** and **6.11**). More than a thousand varieties of cheeses exist in countries around the world, and cheeses from The Netherlands (**FIGURE 6.12**), Switzerland, Italy, and France are particularly popular.

The texture, aroma, and taste of cheese depend primarily on the production process and the microorganisms used. Some varieties of cheese have a wonderful aroma, whereas others really stink! Cheese is made by adding lactic acid–producing bacteria and the enzyme rennin (or bacterial enzymes). The lactic acid sours the milk, and the enzymes coagulate casein, a protein in milk, to the solid curd portion and a watery portion known as whey. The terms "curd" and "whey" are used in the old nursery rhyme about Little Miss Muffett: "Eating her curds and whey." The curd is pressed to remove the whey. Cottage cheese and cream cheese are packaged and sold without further ripening. Other cheeses can be ripened without the addition of other microorganisms, whereas for some cheeses additional microbes are added during the ripening process. Spores of the mold *Penicillium roqueforti* are added during the production of blue cheese and Roquefort cheese. In producing Swiss cheese, bacteria known as propionibacteria are added for the desired taste. The length of time allowed for ripening and the microbes involved in the ripening process determine the consistency of the cheese.

© Robert I. Krasner.

© Robert I. Krasner.

FIGURE 6.12 A cheese factory in a village near Amsterdam, the Netherlands.

Wine, Beer, and Other Alcoholic Beverages

The next time you drink a cold and refreshing beer, enjoy the fragrance and flavor of a fine wine on your taste buds, celebrate an important event by sipping on champagne, or lie on a beach drinking a frozen daiquiri, remember that none of these alcoholic beverages would have been possible without the fermentation of a variety of sugars and grains carried out by strains of S. cerevisiae and other yeasts. Wines and other alcoholic beverages have been imbibed as far back as 6000 B.C. and have been used in religious ceremonies dating back many centuries.

Enology is the science of wine making. Most wines are derived from the sugary juice extracted from grapes, but other fruits can be used; even dandelions are used to make wine. The extracted juice is usually treated with sulfur dioxide to kill naturally occurring yeasts that would produce uncontrolled and undesirable fermentation products. The yeast strain is added, and fermentation is allowed to proceed for a few days at a temperature between 20°C and 25°C (68°F and 77°F), followed by the aging process, which is carried out in wooden casks and takes weeks, months, or even years. During the aging process the flavor, aroma, and bouquet of the wine develop due to the production of a variety of compounds resulting from the metabolism of the yeast. A number of factors are involved in the quality of the wine, including characteristics of the grapes, the strain of yeast, the casks in which the wine is aged, and the duration of aging. Wine connoisseurs pride themselves on knowing a particular year for a fine vintage wine and are prepared to pay hundreds of dollars for this treasure (**BOX 6.3**).

A variety of wines to please every taste is available (**FIGURE 6.13**). All grapes have white juices; red wines are made from red grapes, and the color is due to the pigments of the grape skins. Sweet wines are those in which fermentation is stopped while a significant amount of sugar is still present; in dry wines little sugar remains. Sparkling wines result from continued fermentation that takes place in bottles.

Krasner's Notebook

Believe it or not there is now a new standard allowing grade A Swiss cheese to have smaller holes, or "eyes," to keep the cheese from getting tangled in high-speed slicing machines. The older standard required that the eyes had to be 11/16 to 13/16 inch in diameter, but new regulations reduce minimum eye size to 3/8 (6/16) inch.

BOX 6.3 Wine Tasting

Some people are true connoisseurs of wines (or put on a good act). Next time you have occasion to dine in a fancy restaurant, watch the antics of patrons who appear to be sophisticated in choosing, tasting, and approving a wine once it is brought to the table. Their expressions, as they ceremoniously sniff the cork of the opened bottle and roll the first taste of the wine around in their mouths, are almost comical. Note, also, the manner of presentation of the bottle of wine to the diner; it is held in a way to prominently display the label on which is clearly stated the year. Although the actions of the tasters may seem to the uninitiated somewhat snobbish and frivolous, the rating of wines is a very serious business. A fine bottle of wine in a restaurant could cost several hundred dollars. Wine tasting clubs for the amateur and for the professional are popular.

When tasting wine, connoisseurs consider sweetness; wines are characterized as "sweet" when their taste is dominated by sugars and "dry" when other flavors mask the sugar. Acidity is another attribute; the words "tart," "crisp," and "fresh" are part of the jargon. "Astringency" refers to what is known as the "bitterness" of the wine. Some wine tasters break the process down to the five basic components of color, swirl, nose, taste, and finish:

Color: Color is a reflection of the type of grape used as the source, the age of the wine, and the aging process. White wines increase in color with aging, whereas the color of red wine decreases.

Swirl: Gently swirl the glass of wine to oxygenate the wine. This releases those beautiful aromas characteristic of a good wine and complements the taste. By swirling, the wine is allowed to "breathe"; this can also be accomplished by uncorking the bottle and letting it sit open for a while before drinking.

Nose: Swirling the wine releases the aroma, or "bouquet." Here some subjective and quite imaginative terms such as "bountiful," "cherry," "heady," and "nutty" are used. Some people sniff the cork, but the smell can be just as well detected from the wine in the glass. The main point is that wine may have some unpleasant odors.

Taste: Take a small sip of wine from your glass, taking care not to swallow it. Let it bathe the taste buds on your tongue.

Your taste buds are sending signals to your brain. Are the sensations evoked pleasant ones?

Finish: The finish is the summation of the previous steps. Is the wine satisfying, mellow, and laced with a pleasant taste and no unpleasant aftertaste? If so, give the waiter a pleasant nod and be prepared to pay the price, and be generous in your tip.

Now you can talk like a wine connoisseur and (if you are of legal drinking age) on the next occasion impress your companions as you delicately inspect the color, gently swirl, fashionably "nose" the wine, and then settle back a few seconds for the "finish." You may then say to your waiter, "Yes, the color is perfect, the bouquet is superb, and it has a perfect crispness."

© Teri Shors.

FIGURE 6.13 A drink for all occasions. Alcoholic beverages, including wines, beers, and distilled spirits, are available for all tastes and occasions. Their production depends on the fermentation process carried out by yeasts. Note, from left to right, the wine color is a reflection of the different types of grapes used. The far left glass of dinner wine was made with red grapes, the glass with a pinkish red wine was made from pink moscato grapes, the white wine from white moscato grapes, and the sweet red wine on the far right was made from red grapes.

In the 1860s, the French wine industry was in a state of chaos and collapse due to poor quality wines. Emperor Napoleon III called on Louis Pasteur to seek a solution, and within only about 3 years he determined that the problem was due to contamination of the wines. His solution was simple—heat the wine to 50°C to 60°C (122°F to 140°F), a process later applied to milk and other food products that is now referred to as *pasteurization*. Pasteur's manuscript *Études sur le Vin* (*Studies on Wine*) was published in 1866, and his experience with "sick wines" played a role in his shift to studying disease in humans.

Whereas wines are produced primarily from grape juices, beers are products of the fermentation of cereal grains, including barley (the most common), wheat, and rice. The grains are "malted" by being moistened and kept warm until partially germinated, which begins the enzymatic breaking down of the starch to simpler carbohydrates. The malt is then oven dried; the length of the drying process contributes to the flavor and determines the final color of the beer. The dried, cracked grains are steeped in hot water (mashed) to extract the sugars, starches, and other flavor compounds. The resulting liquid, called **wort**, is boiled to sterilize it, stop enzyme activity, and establish flavor. Hops (dried flowers of the female *Humulus lupulus* vine or their extract) are added at various times for flavor, aroma, preservative qualities, and retention of the head (the foam at the top of a glass of beer). After filtration and cooling, yeast (usually a strain of *S. cerevisiae*) is added to ferment the wort, resulting in production of ethyl alcohol, carbon dioxide, and distinctive aroma and flavor compounds.

Brandy, whiskey, rum, vodka, and gin are referred to as **distilled spirits**. Their production resembles that of wine fermentation. A raw product is used as the starting point; it is fermented by yeast species and then aged in casks. After fermentation, distillation is carried out, yielding a product with a higher alcohol content than beer or wine. The alcoholic content of beer is usually 4% to 6%, that of wine is about 12% to 13%, and that of distilled spirits ranges from 40% to 50%. Scotch whiskey results from the fermentation of barley and rye; brandy results from the fermentation of wine or fruit juice; vodka results from the fermentation of potatoes or grains such as rye or barley; and rum results from the fermentation of molasses.

Harnessing Microbes as Research Tools

Microbes (except viruses and prions) offer biologists packets of life complete with enzymes, energy-generating mechanisms, nucleic acids, structure, and reproductive ability. Viruses, although not cellular, have some of these properties and are equally important as biological tools. Biologists have capitalized on the fact that microorganisms are easy and inexpensive to grow and reproduce rapidly. As knowledge of the microbial world and techniques to manipulate microbes became available over the twentieth century, experimentation with microbes increased, resulting in many of the major advances in biology. Virtually all fields in biology, and many aspects of physics and chemistry, have been enhanced by exploration with microbes. Genetics and molecular biology, in particular, are beneficiaries of the use of microbes in the laboratory. **Genetic engineering**, also known as **recombinant DNA technology**, is a product of these studies.

One of the greatest achievements of the 20th century was the success of the **Human Genome Project**—the mapping of the approximately 25,000 genes in the 23 pairs of human chromosomes. The announcement of the completion of this genetic human blueprint amazed the world and captured the headlines. Earlier efforts at mapping microbes played a major role in this triumph.

The Human Genome Project was initiated in 1990 with the mission of mapping and sequencing the entire human genome—a genetic human blueprint with enormous potential impact on humankind in the coming years. Without microbes, none of this would have happened. The Microbial Genome Program was initiated in 1994 with the goal of sequencing the genomes of medically, environmentally, and industrially significant microbes; this program has led to further success in harnessing these microbes for the benefit of humans. Genetics has come a long way since Mendel's 19th-century observation on the inheritance of color and other characteristics in plants.

A concept that has emerged in recombinant DNA technology is gene therapy. **Gene therapy** is the insertion of modified DNA (genetically engineered into a virus) into a patient's cells to treat disease. A malfunctioning gene may be replaced with a correctly functioning one. A major setback to gene therapy, however, occurred in 1999 when an 18-year-old man died 4 days after the initiation of gene therapy as a result of organ failure, presumably due to a severe immune response toward the viral vector.

Nevertheless, with certain restrictions imposed by government agencies, gene therapy has continued with some very positive results. The National Cancer Institute, a component of the National Institutes of Health, in 2006 successfully engineered lymphocytes, a category of white blood cells, to target and attack melanoma cancer cells; in that same year an international group using gene therapy succeeded in treating two patients with a disorder affecting a particular type of white blood cell. Experimentally, in 2005, researchers at the University of Michigan cured laboratory-induced deafness in guinea pigs by injecting them with a genetically engineered virus carrying a gene to stimulate growth of hair cells in the cochlea; possibly, the procedure will be effective in humans.

Another spinoff of gene therapy is to use **oncolytic viruses** to destroy cancer cells. Oncolytic viruses are manipulated using recombinant DNA techniques so that the viruses cannot cause disease but are still able to infect, replicate inside of, and kill cancer cells. The viral infection also stimulates the body's immune system to destroy the infected cancer cells, potentially shrinking malignant tumors. This experimental cancer therapy is called **virotherapy** (**BOX 6.4**).

Box 6.4 Killing Cancer with Oncolytic Viruses

Most cancers are treated with a combination of chemotherapy and radiation. **Chemotherapy** *is administered intravenously. It is essentially a "poison" that interferes with DNA synthesis in both dividing tumor and healthy cells. Hence, chemotherapy is toxic. Doses of radiation are targeted to damage and destroy tumors, but there will likely be some damage to surrounding healthy tissues. Even today, with the most advanced cancers, remission is brief due to toxicity of treatments and the emergence of resistant cancer cells.*

Even before viruses were discovered as subcellular infectious agents, medical doctors reported some observations of cancerous tumors regressing during a viral infection. For example, in 1904, Dr. George Dock reported the remission of a 4-year-old boy suffering from leukemia after the onset of chickenpox (caused by varicella zoster virus). With new technologies, it was discovered that oncolytic viruses could infect, replicate inside of, and kill cancer cells but not harm healthy tissues and cells. Today, clinical trials are in progress that use virotherapy to treat cancers in advanced stages that are not responsive to chemotherapy and radiation. Virotherapy is an experimental cancer therapy in which oncolytic viruses are used to target and destroy cancer cells.

One of the most notable clinical trials in progress today is the use of a genetically engineered poliovirus to treat individuals who have been diagnosed with aggressive **glioblastoma** (brain cancer). The virus cannot cause polio but it can replicate inside of brain cancer cells and kill them. To date, 3 of the 23 patients in a Duke University study lived more than 3 years after the poliovirus treatment. It is generally believed that the brain tumors regressed because the poliovirus injected into the brain tumor targeted only the cancer cells (not nearby healthy cells) and triggered a longer-lasting immune response that inhibits the growth of the tumor. Nursing student Stephanie Lipscomb was the first patient in this trial. She has been in remission for over 4 years (**Figure 1**).

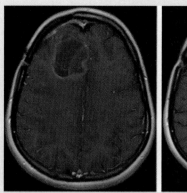
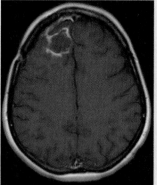
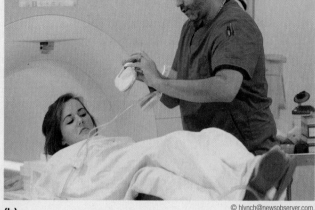

Patient treated on PVS-RIPO
2 months after treatment

Same patient treated on PVS-RIPO
9 months after treatment

(a)   © Duke Medicine.   **(b)**   © hlynch@newsobserver.com.

Figure 1 (a) MRI showing the shrinkage of a brain tumor 2 and 9 months after treatment with the modified poliovirus. **(b)** MRI technician at Duke Cancer Center adjusts Stephanie's headphones on July 15, 2013. She was being prepared to undergo a series of MRI scans of her brain during a follow-up appointment after the virotherapy treatment for brain cancer.

Harnessing Microbes in Industry

In industry, including the pharmaceutical industry, the challenge is to harness microbes as factories and extract their metabolic products. The growth medium depends on the particular microbe and the desired products. The list of products is impressive (**TABLE 6.2**). Antibiotics and other medicinals, food additives, a variety of chemicals, cleaning products, enzymes, proteins, carbohydrates, nucleic acids, and yeasts are examples. Certain characteristics of microbes promote their use as microbial factories:

- The high ratio of surface area to volume leads to rapid replication; the product yield depends on the number of microbes maintained under optimum conditions.

- Microbes are versatile and can be grown in vats on a large scale and under a variety of growth conditions.

- Some products can be produced only by microbes.

- Microbes produce a large variety of enzymes that can be harvested to obtain desired products.

- Microbes can be genetically engineered to produce biological products that are used in the prevention and treatment of cardiac disease and other medical problems.

- Microbes can be genetically engineered to increase their productivity.

- Microbial factories are cost-effective.

In the early years of antibiotic production, only about 5 milligrams of penicillin could be recovered per liter of culture, whereas new strains of *Penicillium* have increased the yield to over 50,000 milligrams per liter. Industrial microbiologists are always on the hunt for microorganisms that synthesize new products or synthesize known products at a greater yield, such as microbial mutants that are not able to control synthesis of a particular product and thereby produce "overruns." The search for antibiotic-producing microorganisms in soil continues, particularly in light of the antibiotic-resistance problem.

A field trip to an industrial plant is a worthwhile experience. You cannot help being amazed at the sheer magnitude—the fermentation vats are two or more stories high—and complexity of industrial microbiology and the skills of the bioengineers. Many companies, including breweries, offer tours of their facilities (and free samples!).

Composting takes advantage of the natural decomposition process to turn organic matter into rich soil (**FIGURE 6.14**). Aerobic bacteria are the primary decomposers aided by a variety of insects and worms. Industrial composting is a big (and wormy) business and is a valuable strategy in reducing the volume of wastes dumped in overcrowded landfills. Home composting has been on the rise for over the past 20 years, and a variety of composting drums and bins are on the market.

The genetic engineering of microbes has enabled the growth of biotechnology and pharmaceutical companies that produce medicinals, including vaccines, antibiotics, hormones, and immune regulatory factors. **Human insulin** and **human growth hormone** are remarkable examples of genetic engineering products (**FIGURE 6.15**),

TABLE 6.2 Products of Genetic Engineering

Product and/or Microbe	Function
Products used in human medicine	
Alpha interferon (*E. coli*)	Treatment for some viral infections
Insulin (*E. coli*)	Treatment for diabetes
Antivenoms (*Bacillus subtilis*)	Treatment of poisonous snake bites
Human growth hormone (*E. coli*)	Treatment for pituitary dwarfism
Interleukin-2 (*E. coli*)	Stimulation of immune system
Tumor necrosis factor (*E. coli*)	Treatment of certain cancers
Epidermal growth factor (*E. coli*)	Treatment of skin wounds and burns
Hepatitis B vaccine (*S. cerevisiae*)	Vaccine used to prevent hepatitis B virus infection
Products used in animal husbandry	
Bovine growth hormone (*E. coli*)	Increases weight gain and milk production
Porcine (swine) growth hormone (*E. coli*)	Increases weight gain
Genetically engineered microbes	
Pseudomonas fluorescens	Carries genes from *Bacillus thuringiensis* that produce an insecticidal toxin (Bt) used in agriculture
Pseudomonas syringae (ice-minus bacterium)	Engineered to remove protein that initiates ice formation on plants and affords protection from frost; also used to make snow for ski slopes.

© jeff gynane/ShutterStock, Inc.

FIGURE 6.14 A large compost heap in a garden.

as are some of the newer recombinant DNA vaccines. Human insulin is now produced in *E. coli* by cloning the human "insulin gene" into its genetic material. Prior to genetic engineering, this insulin was produced from the pancreas of slaughtered cows and pigs; it was not as effective as human insulin and more expensive to produce. Before genetic engineering, human growth hormone for the treatment of dwarfism was obtained from the pituitary glands in human cadavers. Each patient received hormones from two or three batches per year. Each batch was derived from a pool of approximately 16,000 cadaver pituitary glands and posed a severe limitation of supply.

Gene therapy using viral vectors is the wave of the future and has already taken place on a trial basis. Viruses are obligate intracellular parasites, meaning their ecological niche is within a host cell accounting for their usefulness as vectors to deliver replacement genes into defective cells.

The agricultural industry uses **bioinsecticides**—a preparation containing microbes or toxins produced by them. Strains of *Bacillus thuringiensis* are used in a bioinsecticide, commonly called Bt, to control caterpillars and other leaf-eating insects. Each strain produces a particular protein that is toxic for one or a few related species of insect pests; the protein binds to the larval gut, causing the insect to starve.

Bioinsecticides are available that will kill moth, mosquito, fly, and other larvae. Silver leaf and greenhouse white flies are a problem for horticulturists, but it is a pesticide from the fungus *Paecilomomyces fumosoroseus* (another tongue-twister) that comes to the rescue! Some bioinsecticides are virus derived as, for example, baculovirus, which acts exclusively against tomato fruitworm. Bioinsecticides are available to control mosquitoes, Japanese beetles, crickets, and grasshoppers. These products, unlike many chemicals, are highly selective, of low toxicity for humans, and relatively safe for the environment.

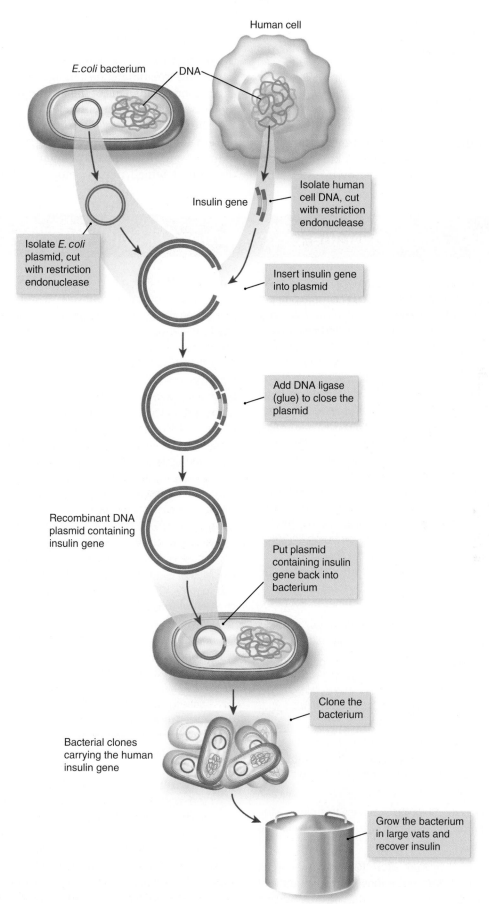

FIGURE 6.15 Production of insulin by genetic engineering. The human insulin gene is cloned and expressed in *E. coli.*

Harnessing Microbes for Bioremediation

Visit Prince William Sound off the coast of Alaska after a storm and you will see beaches tarnished with an oil slick. On March 24, 1989, the oil tanker *Exxon Valdez* ran aground (**BOX 6.5**). Immediately, massive efforts took place to clean up the oil and rescue the thousands of seabirds, sea otters, seals, bald eagles, killer whales, and salmon and herring eggs affected by the spill. The cleanup was mostly effective in removing the oil on the surface of the beaches, but it had little effect on the oil that had sunk beneath the sand and rocks. It is impossible to estimate the damage caused to microbial life and to food chains. One of the techniques used to clean up the mess was bioremediation (**FIGURE 6.16**).

The Environmental Protection Agency (EPA) defines bioremediation as the act of adding materials to the environment, such as fertilizers or microorganisms, to increase the rate at which natural biodegradation occurs. It is

Box 6.5 An Account of the *Exxon Valdez* Oil Spill

The Exxon Valdez departed from the Trans-Alaska Pipeline terminal at 9:12 p.m. on March 23, 1989. William Murphy, an expert ship's pilot hired to maneuver the 300-meter (986-foot) vessel through the Valdez Narrows, was in control of the wheelhouse. At his side was the captain of the vessel, Joe Hazelwood. Helmsman Harry Claar was steering. After passing through Valdez Narrows, Pilot Murphy left the vessel and Captain Hazelwood took over the wheelhouse. The Exxon Valdez encountered icebergs in the shipping lanes, and Captain Hazelwood ordered Claar to take the Exxon Valdez out of the shipping lanes to go around the icebergs. He then handed over control of the wheelhouse to Third Mate Gregory Cousins with precise instructions to turn back into the shipping lanes when the tanker reached a certain point. At that time, Claar was replaced by Helmsman Robert Kagan. For reasons that remain unclear, Cousins and Kagan failed to make the turn back to the shipping lanes, and the ship ran aground on Bligh Reef at 12:04 a.m. on March 24, 1989. Captain Hazelwood was in his quarters at the time.

The National Transportation Safety Board investigated the accident and determined that the top three probable causes of the grounding were the following:

1. Failure of the third mate to maneuver the vessel properly, possibly due to fatigue and excessive workload.

2. Failure of the captain to provide a proper navigation watch, possibly due to impairment from alcohol. The captain was seen in a local bar and admitted to having some alcoholic drinks; a blood test showed alcohol in his blood even several hours after the accident. The captain has always insisted that he was not impaired by alcohol. The state charged him with operating a vessel while under the influence of alcohol. A jury in Alaska, however, found him not guilty of that charge.

3. Failure of the Exxon Shipping Company to supervise the captain and provide a rested and sufficient crew for the *Exxon Valdez*.

The amount of oil spilled was 10.8 million gallons (257,000 barrels, or 38,800 metric tons), roughly equivalent to 125 Olympic-sized swimming pools. The ship was carrying 53,094,510 gallons (1,264,155 barrels or 8,015,825 metric tons) of oil. The *Exxon Valdez* spill was the largest ever in the United States but ranks 34th largest worldwide. It is widely considered the number-one spill worldwide in terms of damage to the environment, however. The timing of the spill, the remote and spectacular location, the thousands of miles of rugged and wild shoreline, and the abundance of wildlife in the region combined to make it an environmental disaster well beyond the scope of other spills.

The spill stretched from Bligh Reef to the tiny village of Chignik, 460 miles (740 kilometers) away on the Alaska Peninsula. The oil affected approximately 1,300 miles (2,100 kilometers) of shoreline. Two hundred miles (321 kilometers) were heavily or moderately oiled (meaning the impact was obvious); 1,100 miles (1,760 kilometers) were lightly or very lightly oiled (meaning light sheen or occasional tar balls). By comparison, the spill region has more than 9,000 miles (17,200 kilometers) of shoreline.

The cleanup efforts spanned more than four summers before it was called off. Some beaches remain oiled today. At its peak, the cleanup effort included 10,000 workers, about 1,000 boats, and roughly 100 airplanes and helicopters, known as Exxon's army, navy, and air force.

To clean the rocky beach shoreline, dozens of people used fire hoses to spray high-pressure cold water and hot water onto the rocks and sand (**Figure 1a** and **1b**). The water, with floating oil, would trickle down to the shore. The oil would be trapped within several layers of boom and either be scooped up, sucked up, or absorbed using special oil-absorbent materials. The hot water treatment was popular until it was determined that the treatment could be causing more damage than the oil. Small organisms were being cooked by the hot water.

Mechanical cleanup was attempted on some beaches. Backhoes and other heavy equipment tilled the beaches to expose oil underneath so that it could be washed out. Many beaches were fertilized to promote growth of microscopic bacteria that eat the hydrocarbons. Known as *bioremediation*, this method was successful on several beaches where the oil was not too thick. Exxon says it spent about $2.1 billion on the cleanup effort. It is widely believed, however, that wave action from winter storms did more to clean the beaches than all the human effort involved.

Photo courtesy of the *Exxon Valdez* Oil Spill Trustee Council.

Courtesy of the Exxon Valdez Oil Spill Trustee Council/NOAA.

Photo courtesy of the *Exxon Valdez* Oil Spill Trustee Council.

Figure 1 **(a)** Using high-pressure water jets to clean oil from seashore rocks. **(b)** Clean up after the Alaskan oil spill. **(c)** Many seabirds were killed by spilled oil.

No one knows how many animals died outright from the oil spill. The carcasses of more than 35,000 birds and 1,000 sea otters were found after the spill, but because most carcasses sink this is considered to be a small fraction of the actual death toll. The best estimates are 250,000 seabirds, 2,800 sea otters, 300 harbor seals, 250 bald eagles, up to 22 killer whales, and billions of salmon and herring eggs.

Oil kills wildlife in three ways:

1. The oil gets on the fur and feathers and destroys the insulation value. Birds and mammals then die of hypothermia (they get too cold).

2. Animals eat the oil, either while trying to clean the oil off their fur and feathers or while scavenging on dead animals. The oil is a poison that causes death.

3. The oil affects the animals in ways that do not lead to a quick death, such as damaging the liver or causing blindness. An impaired animal cannot compete for food and avoid predators.

A professional team and dozens of volunteers, including veterinarians, set up a cleaning and recovery facility (**Figure 1c**). Dawn dishwashing detergent was the cleaning agent of choice. At least 23 species were listed as injured by the spill. And what became of the *Exxon Valdez*? The ship was repaired and renamed the *Sea River Mediterranean*; it is used to haul oil across the Atlantic Ocean. The tanker is banned by law from ever returning to Prince William Sound.

Information from *Frequently Asked Questions about the Oil Spill*, *Exxon Valdez* Oil Spill Trustee Council, 2001.

important to realize that natural biodegradation would occur for many products but at a much slower rate. Biodegradation can also be enhanced by the spraying of nutrients on beaches or on other problem sites to foster the growth of the indigenous microbes to accelerate degradation of the pollutant; this process is called **bioaugmentation**.

Environmental disaster struck again in the form of an oil spill in the Gulf of Mexico on April 20, 2010; this

© Science VU/Visuals Unlimited, Inc.

FIGURE 6.16 Rocks cleaned by bioremediation (right) compared with uncleaned rocks (left) covered with oil from the *Exxon Valdez* spill.

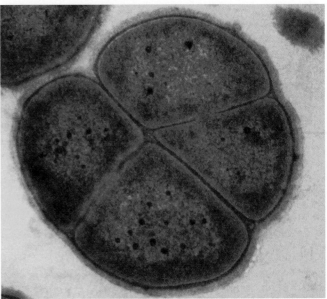

Courtesy of Dr. Michael Daly, Department of Pathology at Uniformed Services, University of the Health Sciences.

FIGURE 6.17 A transmission electron micrograph of *Deinococcus radiodurans.*

event was about 20 times more devastating than the *Exxon Valdez* incident. The spill was the result of an oil gusher stemming from the explosion of the Deep Water Horizon oil drilling rig in the Gulf about 40 miles from Louisiana. Eleven workers were killed and 16 injured. The oil continued to flow for 3 months, releasing almost 5 million barrels. On July 15, 2010, the gushing wellhead was finally capped. The method involved the use of chemical dispersants at the wellhead to facilitate microbial digestion of the oil in order to keep the oil beneath the surface. The logic had its critics, including the National Oceanic and Atmospheric Administration (NOAA).

Microbes exhibit diversity in many ways, as previously noted, underscoring their seemingly unlimited potential in terms of bioremediation. This has spawned the growth of companies in the private sector specializing in remedial services; for example, a company in Florida uses a soil cleaner containing emulsifiers to cause oil, gas, diesel, and other contaminants to be more easily broken down by soil microorganisms into carbon dioxide and water. Consider the use of a bioremediation system using oil- and soap-eating bacteria in a car wash in California, allowing recycling of water instead of discharging it into a septic system. Genetic engineering allows for the creation of microbes specifically designed for bioremediation. *Deinococcus radiodurans* is used in highly radioactive nuclear wastes to digest toluene and toxic mercury (**FIGURE 6.17**). Bioremediation offers an efficient and relatively low-cost alternative approach as compared with the strategy of excavation followed by treatment.

It is difficult to believe that microbes can bring about the decomposition of so many materials, but this is another illustration of microbial diversity. The use of microbes as recyclers to clean up the environment offers several advantages, including cost-effectiveness, self-destruction once the conditions are improved, and minimal disruption of the environment. Cyanide has been used widely in many different industries, including the production of synthetic textiles such as nylon and silk,

plastics, photography, paints, agriculture, food, medicine, and mining. High concentrations of cyanide can disrupt entire ecosystems. A variety of bacteria can be cultured in the laboratory that are able to degrade cyanide compounds (**TABLE 6.3**). Bioremediation was used to remove contaminants discharged as a result of gold mining and milling operations at Lead, South Dakota (**BOX 6.6**). This type of bioremediation application further advanced and played an important role in coping with ecological disturbances.

The possibilities of bioremediation are exciting and seemingly unlimited as microbes from diverse habitats are identified and as new microbes are genetically engineered with the enzymatic capability of breaking down environmental contaminants, including toxic products, petroleum products, landfill wastes, soils, polychlorinated biphenyls, plastics, paper, concrete, and even disposable diapers.

Some evidence shows that **climate change**, an atmospheric disturbance with worrisome consequences, may be at least partially slowed by *Synechococcus*, a cyanobacterium that decreases carbon dioxide released in industrial processes, thus counteracting the greenhouse effect. *Methylosinus trichosporium* is another naturally occurring microbe that conceivably could be harnessed to reduce the process of climate change; the organism breaks down the chlorofluorocarbon (CFC) gases that are products of refrigerants, air conditioners, foam packaging, and spray can propellants. CFCs are greenhouse gases that contribute to climate change and deplete Earth's protective ozone layer.

Sewage and Wastewater Treatment

Until the 1900s and the realization that microbes were the causative agents of a variety of (waterborne) diseases, communities discharged their raw, or untreated,

TABLE 6.3 Cyanide-Degrading Microbes

Bacteria	Fungi	Algae
Pseudomonas spp.	*Trametes versicolor*	*Scenedesmus obliquus*
Bacillus pumilus C1 strain	*Fusarium* spp.	*Micractinium* spp.
Burkholderia spp.	*Acremonium strictum*	
Klebsiella spp.	*Aspergillus awamori*	
Thiobacillus spp.	*Trichoderma* ssp.*	
Azotobacter vinelandii	*Cryptococcus* spp.	
Halomonas spp.		
Rhodococcus spp.		
Halothiobacillus halophilus/ hydrothermalis		
Citrobacter spp.		
Ralstonia spp.		
Acinetobacter johnsonii		
Methylobacterium thiocyanatum		
Paracoccus thiocyanatus		
Thiohalobacter thiocyanaticus		
Thialkalivibrio thiocyanodenitrificans		

*Note: spp. refers to more than one species whereas ssp. refers to subspecies.

sewage into nearby rivers, streams, and marine waters. (The source of cholera epidemics in London in the 1800s was untreated wastewater in the Thames.) However, microbes also play an essential role in the treatment of raw sewage and wastewater, the goal of which is to prevent fecal pathogens from contaminating clean water.

Marine waters, too, need to be protected from fecal contamination. It is not uncommon, particularly in the warmer months of the year, for beaches to have high bacteria counts in the water. This may be primarily the result of flooding and overflow from catch basins and tidal barriers and from careless hygienic habits of persons using the waters for recreation. Because so many **infectious diseases** of humans are waterborne, municipalities are required to establish and maintain facilities to treat their sewage for the protection of the community.

Microbes play an essential role in treating the sewage before it is finally discharged into receiving waters. The process can be divided into three stages: primary, secondary, and tertiary (**FIGURE 6.18**). Initially, raw sewage is filtered through screens to remove sticks, plastics, and other large pieces of debris; the wastewater then flows into a primary treatment settling tank in which heavy particulate matter, referred to as the *primary sludge*, settles to the bottom of the tank. The primary stage is a physical one to accomplish separation of the liquid portion from the solid and particulate matter (which constitutes less than 1% of the wastewater). The microbes are put to use in the secondary treatment to carry out digestion. In this process the liquid portion from the primary tank is passed into the secondary tank; the wastewater is then aerated by a trickling filter. The wastewater passes over and trickles through a bed of fist-sized rocks during which time a biofilm composed of bacteria, protozoa, algae, and viruses carry out aerobic and anaerobic fermentative degradation. An alternative trickling filter system is the activated sludge process; the effluent from the primary treatment tank is aerated by bubbling air through it in the secondary tank. Aerobic microbes, primarily the bacterial species *Zoogloea ramigera*, grow in aggregates called *flocs* and degrade the matter. As in the trickling process, the fluid wastewater from the activated sludge unit passes into the secondary clarifier unit and is chlorinated to kill many of the remaining pathogens as a further safety feature.

Secondary treatment is successful in removing organic matter from wastes but can be a problem because it does not remove the inorganic byproducts of the microbial degradation, which may act as nutrients and stimulate algal blooms. Tertiary wastewater processes can effectively reduce these inorganic materials, but this treatment is expensive and not always used.

A mall of factory outlets in Wrentham, Massachusetts, is a pioneer in "gray" water recycling. It recycles 90% of its 35,000 gallons (132,489 liters) of water a day into clean water that can be reused for toilet flushing. Gray water is nonindustrial wastewater from domestic sources such as sinks, tubs, showers, and bathtubs. It contains lower levels of organic matter and nutrients than does wastewater because urine, fecal material, and toilet paper are not present. About 50% to 80% of wastewater is gray water that can be safely recycled to flush toilets, irrigate landscapes, and in certain industrial processes as a cost-effective alternative to "clean water."

BOX 6.6 Microbes Clean Up Lead, South Dakota

During the 1870s, prospectors discovered gold deposits in fractured rock (ore) in the Black Hills of Dakota Territory (now South Dakota). Very quickly, "Lead City" became a community for generations of miners. Homestake Gold Mine in Lead, South Dakota, opened in 1877. It started out as a 10-acre operation that grew to 2,000 acres of a surface cut and underground mining operation that employed more than 2,000 miners by 1900. As the miners tunneled underground, they dug, hammered, and crushed the rocks with picks, working by candlelight. Carts were filled with ore and pulled by mules or horses out of the mine (Figure 1). In subsequent generations, the ore was drilled and broken apart by explosives. Compressed air and electric locomotives replaced the mules and horses.

The gold ore was low grade. Less than 1 ounce (28 grams) of gold was present in 6 tons (5 metric tons) of ore. Early mining involved crushing the ore into fine rock that was sifted to separate the gold from the ore. The remainder of the fine rock after the sifting process was dumped into large piles called *tailings*. During the early 1900s, "Cyanide Charlie" introduced cyanidation to extract gold from the tailings. Cyanide Charlie's method The remaining cyanide solution was dumped into the nearby

(a)

Courtesy of the Library of Congress.

(b)

Courtesy of the Library of Congress.

Figure 1 **(a)** The underground workings of the Homestake Mine, Lead, South Dakota, 200-foot (61-meter) level. Two miners are standing atop of a wood beam structure, 1908. **(b)** Homestake Mine, 1900.

recovered 94% of the gold from the ore. The tailings were placed in a cyanide tank and a cyanide solution sprayed over the heaps of tailing, causing a chemical reaction in which the gold was liquefied. The gold was removed or purified from the solution with the addition of zinc and activated carbon. The zinc precipitated (separated) the gold from the ore. The gold was removed from the cyanide solution in a filter press (**Figure 2a**).

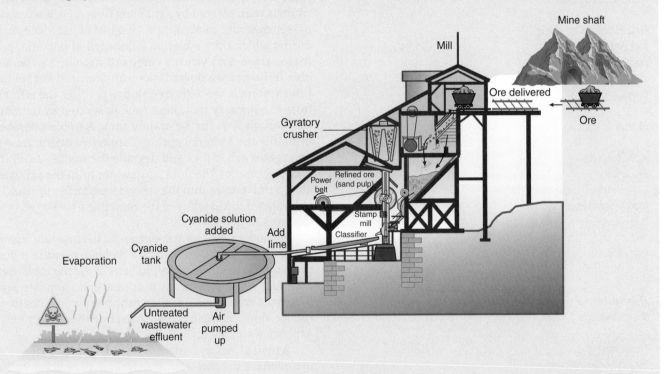

(a) Environment

Figure 2 **(a)** Simplified diagram of gold mining and milling operations before bioremediation efforts. The ore from the mine is delivered to a gyratory crusher. It is the first in a series of crushing machines used to process the ore. The solid rock ore chunks are crushed into 2-inch (5-centimeter) pebbles that are subsequently processed through the stamp mill. A total of 180 stamps each with a falling weight of 1,550 pounds (703 kilograms) and 300,000 tons (272,155 metric tons) of water are used in stamping the ore every 24 hours to pulverize the pebbles into sand pulp. The ore is passed through a screen into a rod mill followed by a ball mill that further crushes the ore into pulverized sand that is sifted through a classifier screen. The sand is raked out of the classifier into a cyanide tank. Lime is added to the sand pulp. The wet sand is allowed to settle and is raked flat for the next step in removing the gold flakes from the ore (the wastewater effluent is discharged into the environment). Air is pumped up through the raked sand and a cyanide solution is sprayed onto the sand that solubilizes or liquefies the gold. The liquid gold is separated from the sand and is refined by a series of purifications (which will create more effluent that is discharged into the environment). The sand is recycled back into the mine and is used to fill in/stabilize the tunnels where the ore was removed.

Whitewood Creek, killing everything. Whitewood Creek ran black for over 100 years.

In 1971, field surveys indicated contaminants from the Homestake gold mining and milling operations were discharged at a rate of 3,000 tons (2,721 metric tons) per day, impacting biota not only in Whitewood Creek but other receiving waters, such as Bull Run Creek, and the floodplains of the Belle Fourche and Cheyenne Rivers. The pollutants accumulated, tainting surface and ground waters, affecting the ecosystem's diversity of life in the soil, water, and plants. The hazardous substances included cyanide, arsenic, cadmium, chromium, copper, lead, manganese, mercury, selenium, nickel, silver, and zinc. The EPA gave Homestake Mine 1 year to remove the cyanide in 1983.

In 1984, "the microbes" were employed to clean up the contaminated environment. During the first stage of the bioremediation process, all of the mining effluent from the underground mine was filtered through tanks containing about 20,000 pounds (9 metric tons) of *Pseudomonas paucimobilis* (bacteria) that formed a slimy biofilm on rotating biological

contactors. The bacteria metabolically reduced the cyanide into carbonate, ammonia, and sulfate (**Figure 3**). Any freed contaminating metals in the effluent were absorbed by the bacterial biofilm. The filtrate from the first stage was pumped into a second set of biological contactors coated with a biofilm of nitrifying bacteria that converted ammonia to nitrate. Before the effluent from the second stage was released into the Whitewood Creek, trout maintained at the treatment plant were exposed to the effluent. If the trout were unaffected by the effluent, it was safe to dump the treated wastewater into Whitewood creek (**Figure 2b**). Each day, 4 million gallons (15 million liters) of clean effluent poured into the Whitewood Creek. Within 6 months of implementing the cleanup process, trout moved back into the Whitewood Creek and local streams.

Over a period of 126 years, more than 41 million ounces (1,162 metric tons) of gold and 9 million ounces (255 metric tons) of silver were pulled from the mine. Lead, South Dakota was once called the richest 100 square miles

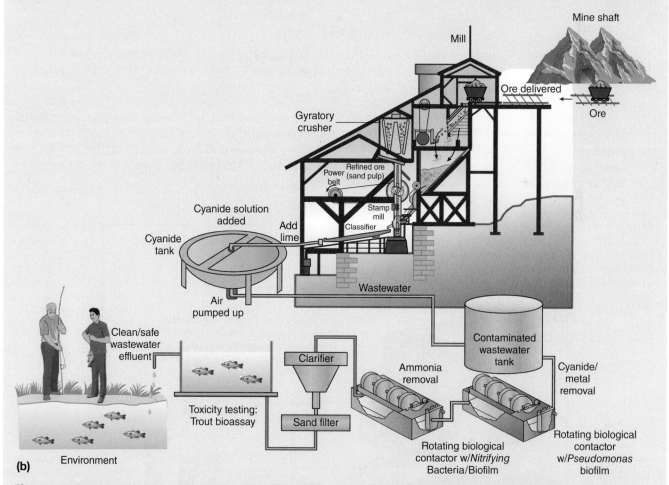

Figure 2 (b) Simplified diagram of gold mining and milling operations after bioremediation efforts. The gold mining process similar to that in (a) is used but the toxic effluent is treated through a series of bioreactors that removes the toxic chemicals used to extract the gold from the ore. It is a two-step process in which a series of 24 rotating biological contactors that are coated with a slimy biofilm of *Pseudomonas* bacteria break the cyanide down into ammonia, sulfate, and carbonate. The bacterial biofilm absorbs contaminating metals. The effluent from this step is transferred to a second bioreactor, which is a series of 24 rotating biological contactors coated with nitrifying bacteria that break down the ammonia. The effluent from this treatment process is passed through a clarifier and sand filter to remove any contaminating particles and dislodged bacteria from the bioremediation process. The filtered water to be discharged is released into a tank containing healthy trout. If the trout are not affected by the discharge, the water is free of hazardous chemicals, making it pure enough to be released into the environment.

(continues)

BOX 6.6 Microbes Clean Up Lead, South Dakota (continued)

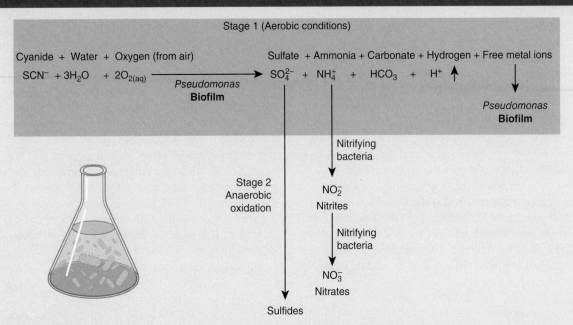

Figure 3 The chemistry behind Homestake's microbial feat. It required two stages to detoxify the waste generated by the gold mining and milling operations. Trillions and trillions of bacteria were employed every day to get the job done! See Figure 6.4 (the nitrogen cycle) for further details regarding stage 2 of this process.

(260 square kilometers) on Earth. The Homestake Mine reached a depth of 8,000 feet (1.5 miles, or 2.4 kilometers) and more than 370 miles (595 kilometers) of tunnels. The falling prices of gold, high production costs, and the lower than expected ore grades ultimately led to its closure in 2001.

Today, the caverns serve as laboratory space for world-leading research. Astrophysicists from all over the world are studying neutrinos (subatomic particles emitted from the sun) at a depth of 7,400 feet (2,255 meters) below the surface of the Earth.

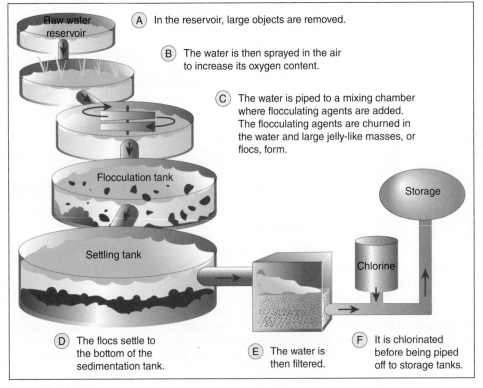

FIGURE 6.18 Steps in sewage treatment.

Summary

This chapter presents positive aspects of human association with the microbial world, a world whose presence is unseen but is manifest in many ways every day. Microorganisms constitute a major component of Earth's biomass, and their role in the cycles of nature underpins life itself.

Earlier civilizations learned by experience to harness the metabolic activities of microbes in the production of foods, including fermented milk products and alcoholic beverages. In more recent years, as the science of microbiology developed, the diversity of microbes has been further exploited. Microbes are used as research tools, and have spawned many disciplines in biology. They are factories for the production of many useful products, including products of genetic engineering, and as agents of cleaning by means of bioremediation. *The Microbes' Contribution to Biology* (1956) by A. J. Van Niel and C. B. Kluyver described the importance of microorganisms to the study of biology. If that text were updated today, many more pages would be necessary.

Because of the barrage of microbial threats, perhaps there is too much emphasis on strategies to wipe out, destroy, sanitize, or scrub away microbes. In so doing, useful, life-sustaining microbes are killed, thereby "throwing out the baby with the bath water." The following quotation serves as an appropriate conclusion to this chapter: "If you take care of your microbial friends, they will take care of your future."

KEY TERMS

abiotic
adenosine triphosphate (ATP)
anaerobic process
bacteriocins
bacteriophage
bioaugmentation
biofilm
biogeochemical cycles
bioinsecticides
bioremediation
biotic
carbon cycle
chemotherapy
Chlorella
climate change
colitis
consumers
cyanobacteria
decomposers
denitrifying
distilled spirits
ecosystem
enology
exotoxin
fecal microbiota transplantation
fermentation
gene therapy

genetic engineering
glioblastoma
Gram stain
herbivore
host
Human Genome Project
human growth hormone
human insulin
infectious diseases
leguminous plants
microbiome
microbiota
nitrification
nitrogen cycle
oncolytic viruses
pasteurization
pigmentation
prebiotics
predator
primary producers
probiotics
quorum sensing (QS)
recombinant DNA technology
starter culture
symbiosis
virotherapy
wort

SELF-EVALUATION

○ **PART I: Choose the single best answer.**

1. The greatest recyclers of all time are _____.

   a. plants
   b. microbes
   c. animals
   d. humans

2. Nitrification is characterized by the _____.
    a. conversion of nitrates into ammonia   c. conversion of ammonia into nitrates
    b. fixation of atmospheric oxygen   d. return of nitrogen to the atmosphere

3. The color of red wines is the result of _____.
    a. the addition of red dye to the wine   c. the type of cork used to bottle the wine
    b. the aging process   d. the fermentation of white moscato grapes

4. _____ is a bacterium used to clean up oil spills.
    a. *Streptococcus* b. *Serratia* c. *Pseudomonas* d. *Fusarium*

5. A sweet wine is characterized by _____.
    a. the color of grapes used   c. fermentation in a bottle
    b. little remaining sugar   d. a relatively high sugar concentration

6. Which series of chemical reactions is mediated by enzymes of a variety of strains of bacteria and yeasts that break down sugars to small molecules?
    a. Symbiosis   c. Photosynthesis
    b. Fermentation   d. Respiration

7. A gram of yogurt contains about 1 million _____.
    a. staphylococci   c. lactobacilli
    b. clostridia   d. mycobacteria

8. Which of the following plays a major role in sewage treatment?
    a. *Streptococcus*   c. *Zoogloea*
    b. *Clostridium*   d. *Bacillus*

9. Which of the following foods is *not* produced by using microbes?
    a. Kefir   c. Beer
    b. Soy sauce   d. Mayonnaise

10. Which of the following may be of value in preventing yeast infections during antibiotic therapy?
    a. Wine consumption   c. Gene therapy
    b. Probiotics   d. Starvation

○ PART II: Fill in the blank.

1. Plants with nodules along their root systems are called _____ plants.

2. Foods supplemented with _____ are called probiotics.

3. _____ is the genus and species of the yeast frequently used in making bread and alcoholic beverages.

4. _____ was genetically engineered to remove protein that initiates ice formation on plants and affords protection from frost.

5. Wines are produced from fruit juices, whereas beers are produced from fermentation of _____

6. _____ is the science of wine making.

7. _____ uses genetically engineered _____ to deliver genes into a patient's cells.

8. _____ are essential in the conversion of atmospheric _____ to organic compounds and back to the atmosphere.

9. _____ bacteria convert free nitrogen gas to ammonium.

10. _____ bacteria convert ammonium to nitrites.

○ PART III: Answer the following.

1. Distinguish between bioremediation and bioaugmentation.

2. Using specific examples, explain the expression "Microbes make the world go around."

3. Cite and briefly explain the ways microbes can be used in research and in industry.

4. Discuss the relationship between prebiotics and probiotics.

5. Cite examples of fermentation end products, and discuss whether any of them have commercial value.

6. Name the five basic components of wine used by wine tasters.

7. Explain the reasoning why fecal transplantation can cure *Clostridium difficile* infections in some patients.

8. What is quorum sensing, and what is its significance in the microbial world?

9. What causes bread to "rise"?

10. Discuss the importance of genetically engineering microbes.

BACTERIA

What marvels there are in so small a creature.

—Antonie
van Leeuwenhoek
(1862–1723)

Microscopy pioneer

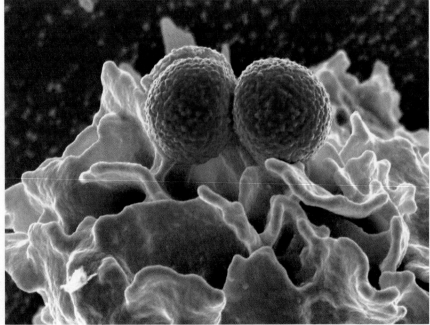

Courtesy of CDC/ the National Institute of Allergy and Infectious Diseases (NIAID).

Digitally colorized scanning electron micrograph of two blue-colored, coccus-shaped methicillin-resistant *Staphylococcus aureus* (MRSA) bacteria being phagocytized by an uncolored white blood cell. The strain depicted here is MRSA252, one of the leading causes of hospital-acquired infections in the United States and United Kingdom.

1. Detail the process of the Gram-staining procedure and note its importance in distinguishing between Gram (+) and Gram (–) bacteria.

2. Discuss how bacterial capsules and pili promote virulence.

3. Explain why endospores can remain viable even after hundreds of years have passed.

4. Draw and label the phases of an *E. coli* growth curve.

5. Discuss why it is not always possible to culture bacteria in artificial media.

6. Describe the advantages and disadvantages of rapid diagnostic tests.

7. List "atypical" bacteria and describe how they differ structurally from "typical" bacteria.

Case Study: Don't Touch That Armadillo!

Charles Brown was a healthy 55-year-old male who was born and raised in central Florida. One day while driving his car to the grocery store he accidentally hit an armadillo that was slowly crossing the road. As soon as he noticed what had happened, he got out and kicked the armadillo to the side of the road. When he kicked the armadillo, he got the remains of the armadillo on the soles of his shoes; the remains also splattered onto his left ankle and arm. He recalled the smell of it before wiping it off his arm with wipes he had in the car. Two years passed. Several large inflamed skin lesions appeared on his left ankle and thigh that would not go away. It prompted him to make an appointment with his primary care physician who referred him to a dermatologist.

Dermatologist Ann Weiss examined the lesions. She used a sterile safety pin to test skin sensitivity by pinpricking the center of each lesion. Charles did not have any feeling where the lesions were pricked, indicating sensory nerve loss (**Figure 1**). Skin punch biopsies of the lesions were sent to the medical technology laboratory for testing and analysis. Technicians stained preparations of skin smears and analyzed them microscopically. A specialized method, **acid-fast staining**, determined that there were red-stained, rod-shaped, bacteria indicative of the genus *Mycobacterium*.

The two medically important species of *Mycobacterium* are *M. tuberculosis* (causes tuberculosis) and *M. leprae* (causes **leprosy**). **Mycolic acids** are fatty acid components of the cell wall unique to mycobacteria that gives it a thick, waxy property, making it impermeable to most antibiotics. The mycolic acids of the cell wall resist staining by ordinary methods like

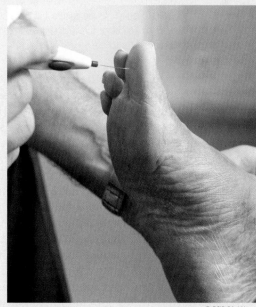
© BSIP SA / Alamy.

Figure 1 A doctor performing sensory testing.

Gram staining. An acid-fast staining procedure is used that involves steaming carbol fuchsin (a red dye) into the bacterial cell wall. The carbol fuchsin enters the mycobacteria but it resists decolorization with acid alcohol. Non-acid-fast bacteria are decolorized with acid alcohol and are counterstained with methylene blue (**Figure 2**). Mycobacteria grow very slowly, with a **generation time** of 12.5 days (compared to *E. coli*'s generation time of 20 minutes). This combination of traits is the reason why antibiotic treatment to cure patients takes 1 to 2 years to complete and patient compliance is challenging.

(continues)

Case Study: Don't Touch That Armadillo! *(continued)*

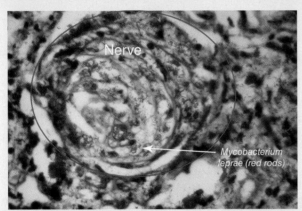

Courtesy of CDC/Arthur E. Kaye.

Figure 2 Microscopic view of an acid-fast stained skin smear from a patient with leprosy. *Mycobacterium leprae* are acid-fast (carbol fuchsin/red-stained rod-shaped bacteria) that resist decolorization with acid alcohol. The *M. leprae* in this preparation invaded a cutaneous nerve. *M. leprae* cannot be grown in bacteriological media or tissue culture, but it has been grown in the footpads of nude mice. Therefore, it is much harder to conduct research (e.g., growth curves) on this bacterium.

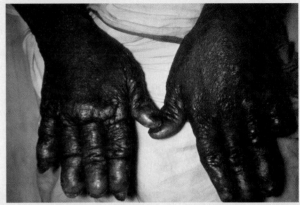

Courtesy of the CDC/Dr. Andre J. Lebraun.

Figure 3 The hands of a patient suffering from leprosy, also known as Hansen's disease. The digits of both hands eroded due to the course of the disease. Numerous skin nodules are present. The disease was named after the Norwegian physician Gerhard Armauer Hansen who in 1873 discovered rod-shaped bacteria, later determined to be *M. leprae*, present in the cells of leprous skin nodules but not in other skin cells. His reported observations contested the theory that leprosy was an inherited disease.

Dr. Weiss informed Charles that he was infected with *M. leprae*, causing leprosy. *M. leprae* is carried by the bloodstream into the peripheral nerves causing irreversible nerve damage. The upper respiratory tract, skin, and other tissues also are affected. Charles was quite defensive and questioned the diagnosis. He knew that leprosy existed as far back as biblical times. Lepers were stigmatized because of their appearance of deformities and the fear of **contagion**. He had read about lepers being thrown off ships by the coast of Molokai, Hawaii. If they survived the swim to the island, they lived their remaining years in the leper colony established there. Dr. Weiss reassured Charles that the disease is treatable and people are not segregated today. The disease is rare. About 95% of the world's population are naturally immune to leprosy and will never acquire the infection. Dr. Weiss explained that the nerve damage or lack of sensation makes leprosy patients very susceptible to being injured, which can result in serious, untreated infections, leading to gross deformities. If motor neurons are repeatedly damaged, it leads to weakness and paralysis of specific muscles, leading to deformities of the hands and feet such as wrist drop, claw hand, and foot drop (**Figure 3**). Loss of eye sensation leads to the loss of normal reflexes that prevent eye damage caused by the accidental entry of dust or other particles into the eye. The result is loss of vision/blindness. The bacterium grows best at 27°C to 30°C (80.6°F to 87°F); hence, it prefers cooler areas of the human body such as the extremities (skin, hands, and feet). This temperature, however, favors growth of *M. leprae* in armadillos, the only other known natural **reservoir** for *M. leprae* besides humans. Armadillos are mammals that sustain an average body temperature of 90°F.

Charles was further questioned whether he had any contact with armadillos. Researchers determined in the 1970s that armadillos could be naturally infected by *M. leprae*. With the advent of molecular biology techniques, subsequent DNA sequencing studies determined that the genome of the same strain of *M. leprae* from leprosy patients was essentially genetically identical to the strain present in infected wild armadillos. Charles said he had had several exposures to armadillos in the past 10 years in Florida but he was adamant that he had not been to Texas or Louisiana where wild armadillos infected with *M. leprae* were most prevalent. About 10% to 20% of armadillos in the Deep South carry *M. leprae*. Dr. Weiss explained that the incubation period for *M. leprae* is 2 to 20 years and that more recent evidence suggests that the geographic range of wild armadillos is expanding farther north (**Figure 4**), as well as the range of armadillos infected with *M. leprae*. In 2015, 178 new cases of leprosy were reported in the United States. Of these cases, 129 (72%) were reported in Arkansas, California, Florida, Hawaii, Louisiana, New York, and Texas. From 1985–2015, a total of 6,076 new cases of leprosy were reported in the United States.

Charles was not convinced that an armadillo was the cause because many people around the world have gotten leprosy in places where the nine-banded armadillo does not exist (**Figure 5**). Dr. Weiss cautioned Charles that it has not been proven but

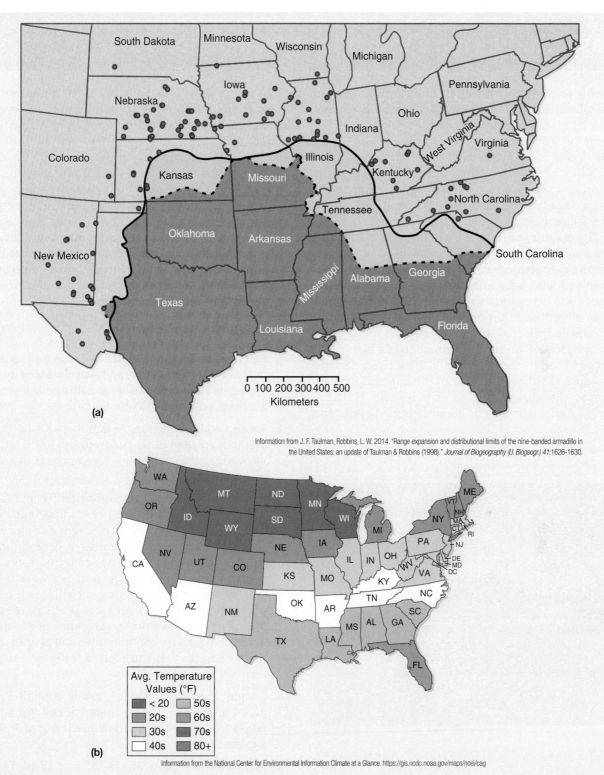

Information from J. F. Taulman, Robbins, L. W. 2014. "Range expansion and distributional limits of the nine-banded armadillo in the United States: an update of Taulman & Robbins (1998)." *Journal of Biogeography (U. Biogeogr.) 41*:1626-1630.

Information from the National Center for Environmental Information Climate at a Glance. https://gis.ncdc.noaa.gov/maps/ncei/cag

Figure 4 **(a)** Range and distribution of armadillos in the United States. Armadillo populations crossed the Rio Grande/Rio Bravo River (Mexican/Texas border) sometime during the 1820s. Populations of the wild armadillos spread north and eastward. Small groups of captive armadillos were released into south/central Florida in 1920s, and these populations spread north and westward, likely merging with the wild populations originating from Texas. Historically, the patterns of leprosy and distribution range of wild armadillos were linked. The region shaded in red indicates the distribution of armadillo populations documented in 1994. The yellow regions represent the current estimated range based on surveillance reports of 153 biologists in 2012. The aqua blue circles represent additional sightings of wild armadillos during 2003–2012. **(b)** Average daily temperature for the coldest month of the year, January 2017 (48 contiguous U.S. states). The survey results of the distribution of armadillos suggest that winter temperature extremes limit population expansion in the wild. Subfreezing temperatures do not limit the foraging of armadillos if forest leaf litter is available; persistent snow or ice cover limits armadillo foraging opportunities that would lead to eventual starvation. With the exception of Minnesota and Wisconsin, all states in the eastern half of the contiguous United States are potential regions for supporting established breeding armadillo populations in the wild.

(continues)

Case Study: Don't Touch That Armadillo! *(continued)*

that the route of transmission, if not through contact with wild armadillos infected with M. *leprae*, is thought to be by a respiratory route because the nasal discharge of leprosy patients contains high numbers of M. *leprae* cells, suggesting droplet transmission/inhalation of M. *leprae* aerosol. It is possible that the bacteria may enter through broken skin by direct, close person-to-person contact between an infected person and a susceptible person.

The majority of leprosy cases in the United States have been among individuals who lived or worked in leprosy-endemic locations outside of the United States and who were infected while abroad. Because Charles had no contact with anyone who had leprosy, it was more likely that he was infected after contact with a wild, infected armadillo. He also denied eating undercooked armadillo. Armadillo is a blue-collar delicacy, used in stews and chili. Most likely Charles was exposed to body fluids from the armadillo 2 years prior to the appearance of skin lesions. Unfortunately, Charles was genetically susceptible to infection by M. *leprae*.

© Roger Gasser.

Figure 5 Nine-banded armadillo taxidermy specimen. Armadillo in Spanish means "little armored one." Bony plates cover its legs, head, and body. It has nine flexible bands across its midsection, allowing it to be relatively nimble (like a short toy slinky). Armadillos are nocturnal, aloof, docile, and slow-moving mammals that have very poor eyesight, making them easy to trap in the wild. However, these same features mean that they are not very good at crossing the highway! Hence, armadillo are amongst the most common roadkill in Florida. No bait is needed to trap them in cages. They can dig deeply and quickly to evade predators. Armadillos mainly eat insects, grubs, and earthworms. Gardeners who live in areas with armadillos are not fond of the holes they dig during their nighttime shenanigans.

Charles was treated with a combination of antibiotics—dapsone (100 milligrams; is weakly **bactericidal** and has anti-inflammatory properties), minocycline (100 milligrams; is weakly bactericidal and has anti-inflammatory properties), and rifampin (600 milligrams; highly bactericidal, which should kill 99.999% of the bacteria within 3 months)—daily for 2 years. He responded well with no adverse effects to the chronic antibiotic therapy. The smaller lesions disappeared within a week of treatment. Larger lesions remained hyperpigmented and were insensitive to pain (**anesthetic**). A year after treatment was completed, Charles showed no signs of relapse. Charles avoided future contact with wild armadillos.

Questions and Activities

1. Read further into the chapter to the section "Culturing Bacteria: Diagnostics" and speculate why microbiologists have been unable to cultivate M. *leprae* in artificial bacteriological medium.

2. Besides acid-fast staining of skin smears, what other methods are used to diagnose leprosy? (Hint: https://www.cdc.gov/leprosy/health-care-workers/laboratory-diagnostics.html)

3. Because the **mode of transmission** for leprosy is not fully understood, researchers have not ruled out indirect modes of transmission such as contact with contaminated **fomites** or contaminated clothes and linens. Scientists collected soil samples in areas where leprosy patients were active (e.g., hospitals and resettlement villages for leprosy patients in India). An average of 35% of samples tested positive for the presence of M. *leprae*. Could contact with contaminated soil play an indirect role in the transmission of M. *leprae*, causing leprosy? Explain.

4. Google "armadillo recipes." How is armadillo prepared and eaten? Could undercooked armadillo play a role in foodborne transmission of M. *leprae*? Explain.

5. Create a world map. Shade/colorize all countries with endemic leprosy cases.

6. The development of antibiotic-resistant M. *leprae* is inevitable. Researchers are trying to develop antibiotics that would inhibit the biosynthesis of mycolic acids. Would this type of antibiotic be useful to treat M. *leprae* infections? Why or why not? Would this new antibiotic be bactericidal to E. coli or *Staphylococcus aureus* or to *Mycobacterium*

tuberculosis? Why or why not? Do you anticipate this type of antibiotic to be highly toxic to humans? Why or why not?

7. Leprosy can be caused through zoonotic transmission. What is a **zoonosis**? Besides leprosy, what are two other examples of zoonotic diseases? How are these diseases transmitted to humans?

8. The distribution of wild armadillos is expanding farther north. Could **climate change** play a role in this? Explain your answer. Provide an example of an **infectious disease** that has expanded into a region linked to climate change. What is the disease, and how is it transmitted?

Information from Domozych Renee, B. A. et al., 2016. "Increasing Incidence of Leprosy and Transmission from Armadillos in Central Florida: A Case Series." *JAAD Case Reports* 2:189–192.

Preview

The quotation on the first page of this chapter is the words of van Leeuwenhoek, a Dutch merchant by vocation and a lens grinder by avocation, in a report to the Royal Society of London in 1675. Van Leeuwenhoek's development of a simple microscope allowed him to see and describe "small **animalcules**." Microbes date back to the antiquity of life. A scientist at California Polytechnic Institute reported the isolation of ancient bacteria from a bee in a specimen of amber, a hardened resin from ancient pine trees, dating back 25 million to 40 million years. This account is reminiscent of the novel and movie *Jurassic Park*, in which a scientist extracted dinosaur DNA from entombed mosquitoes that had fed on dinosaurs.

The notion that bacteria are simple because they are unicellular is far from the truth; they may be small, but they are not simple. They are complex biological entities. The fact is that within a matter of only several hours after the invasion of a human body by some pathogenic bacteria severe illness and even death may result. This is equally true of infected elephants and whales, whose mass is measurable in tons, and yet they, too, fall victim to microscopic microbes. It would take a fantastic number of bacteria to equal 1 ton.

The anatomy of the bacterial cell as an organized and functional structure meeting the characteristics associated with life is described in this chapter. Under appropriate conditions some bacteria can undergo **binary fission** in as little as 20 minutes, resulting in huge populations in 24 hours. An appreciation of the bacterial growth curve is necessary to understand the dynamics of growth when only a few pathogens are introduced into the body.

Many known bacteria are easily grown in the laboratory, facilitating the diagnosis of bacterial infection. *Mycoplasma* spp., *Chlamydia* spp., and *Rickettsia* spp. represent bacterial genera that are atypical, "oddball" bacteria, and their unique characteristics are described later in the chapter. They cause a variety of diseases in humans.

Cell Shapes and Patterns

Examination of bacterial cells under a microscope reveals the presence of a variety of shapes (**morphology**) and patterns of arrangement (**FIGURE 7.1**). The majority are either rod-shaped, known as **bacilli** (singular, *bacillus* or *rod*); spherical, known as **cocci** (singular, *coccus*); or spiral shaped, known as **spirilla** (singular, *spirillum*). For the most part, rods tend to occur as single cells, but some species form chains. Characteristic groupings of cocci are more common and are useful in identification. **Streptococci** are chains of cocci resembling a string of pearls; **staphylococci** look like a bunch of grapes; **diplococci** occur in pairs; and **tetrads** are groupings of four. Undoubtedly, the terms *strep* and *staph* are familiar to you. Spiral or curved rods are

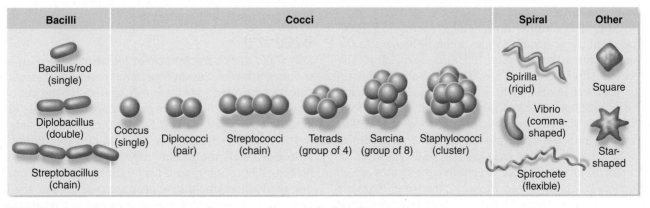

FIGURE 7.1 Bacterial cell shapes and patterns of arrangement observed using light microscopy.

categorized as **spirilla** (rigid spiral), **spirochetes** (flexible spiral), and **vibrios** (comma-shaped curved rods). Other bacteria have been described as star shaped, square shaped, or arranged in palisades. Microscopic determination of shape and pattern is often the first step in the identification of bacteria.

Naming Bacteria

Expectant couples spend hours searching through baby-naming books, conferring with family and friends, and otherwise agonizing over their choice of the perfect name for their expected child, particularly the firstborn. Biologists have established names for microbes and have also put considerable thought into doing so, because the names frequently identify important characteristics. You may feel overwhelmed by the long and difficult to pronounce names, but at least you can impress your family and friends.

Microbes are named in accordance with the **binomial system of nomenclature** established by Carolus Linnaeus in 1735. This system is not restricted to microbes but applies to all organisms. The names are Latin, and each organism carries two names (*binomial*), the first designating the **genus** and the second designating the **species**. For example, humans are *Homo* (genus) *sapiens* (species), and the fruit flies most commonly used in genetic studies are *Drosophila melanogaster*. Both the genus and species names are italicized. Note that the first letter of the genus name is always capitalized, but the species name is not.

Unfortunately for the student, microbiologists have not been consistent in assigning names. Some microbes are named in honor of the scientist responsible for first describing them, whereas others indicate the microbe's habitat, shape, associated disease, or combinations of these factors. A few examples make this clear. *Rickettsia prowazekii* (the cause of epidemic typhus) is named in honor of Howard T. Ricketts (1871–1910) and Stanislaus von Prowazek (1875–1915), both of whom died of the disease they were studying. *Legionella pneumophila* is named after an outbreak of pneumonia at an American Legion convention in 1976. *Escherichia coli* is named in commemoration of Theodor Escherich (1857–1911) and identifies the *coli* (large intestine) as the bacterium's habitat.

The use of descriptive morphological terms (e.g., *bacillus*) can lead to confusion. *Bacillus anthracis* belongs to the genus *Bacillus* (with a capital "B") and morphologically is a bacillus (with a lowercase "b") or rod-shaped bacterium. But *E. coli* is also bacillus or rod-shaped, belonging to the genus *Escherichia*. Many other bacilli do not belong to the genus *Bacillus*. So all *members of the genus Bacillus are bacillus (or rod-shaped), but not all bacilli are classified under Bacillus*. The same case can be made for streptococci and diplococci. *Diplococcus pneumoniae* strains are all diplococci (cocci arranged in pairs), but so is *Neisseria gonorrhoeae*. (The genus *Diplococcus* was later renamed *Streptococcus*.)

Anatomy of the Bacterial Cell

To appreciate the challenge posed by bacteria, it is necessary to be familiar with their structure and properties. The fact that they are procaryotic cells sets them apart from the other microbes. The structures that compose these cells are outlined in **TABLE 7.1** and illustrated in **FIGURE 7.2**; note that some anatomical features are not common to all bacteria.

Envelope

The bacterial **envelope** consists of the **capsule**, **cell wall**, and **cell (outer) membrane** [in Gram (–) bacteria]. A capsule is not present in all species, but the cell wall (with the exception of the mycoplasmas, to be described later in the chapter) and the cell (plasma) membrane are structures present in all bacteria.

○ Capsule

When present, the capsule is not integral to the life of the cell. In fact, the capsule can be easily removed by treating a culture with appropriate enzymes or by manipulating the presence of nutrients in the culture; in either case the cells grow just as well in the laboratory but lack capsules. The progeny of these noncapsulated cells will have capsules. In some species, the presence of a capsule promotes **virulence** (the capacity to produce disease), as, for example, in the bacillus that causes anthrax (*B. anthracis*). The streptococci responsible for strep throat and necrotizing fasciitis (flesh-eating streptococcal disease) are another example of encapsulated organisms. The presence of the capsule interferes with **phagocytosis**. Phagocytosis is an important body defense mechanism by which bacteria are ingested and killed by a variety of phagocytic cells. The capsular material may become so thick that it resembles a slime layer; it is like trying to grab onto a slippery fish. Occasionally, hunks of this mucuslike layer can be found in the broth used to grow the culture. A condition known as **ropy milk** is a nuisance to the milk industry because of the growth of the nonpathogen *Alcaligenes viscolactis* and the shedding of its slime layer into milk. The fact that this bacterium is not a disease producer is of little comfort as your tongue and palate unexpectedly encounter this ropy, mucuslike material.

○ Cell Wall

Cell walls are characteristic of all bacteria (again, with the exception of the mycoplasmas) and are a structure shared with plant, algae, and fungal cells. The cell wall is a rigid, corsetlike structure responsible for the characteristic shape of the cell. Further, it confers resistance to the cell from the inward diffusion of water. Without a strong cell wall, the plasma membrane would swell, and **lysis** (cell bursting) would occur. In the 1940s, the antibiotic **penicillin** came into widespread use despite the fact that its mechanism of action was not yet understood. Several years later scientists discovered that penicillin

TABLE 7.1 Anatomical Features of the Bacterial Cell

Structure	Function
Cell envelope	
Capsule[a] (glycocalyx, slime layer)	Promotes virulence
Cell wall (with the exception of mycoplasmas)	Confers structure and shape; provides tensile strength
Plasma membrane	Controls selectively permeable movement of molecules into and out of the bacterium; referred to as the "gatekeeper"
Cytoplasm[b]	
Nucleoid	DNA-rich area *not* enclosed by a membrane
Plasmids[a]	Circular, nonchromosomal DNA that confers properties such as antibiotic resistance
Endospores[a]	Confers extreme resistance to environmental factors
Ribosome	Protein synthesis
Chromosome	Determinant of genetic traits
Inclusion bodies	Storage and reserve supply of nutrient materials
Appendages	
Flagella[a]	Motility
Pili[a]	Adhesion to surfaces, bridge for transfer of DNA (conjugation)

[a] Structure not found in all bacterial cells.
[b] Area within plasma membrane; contains numerous structures.

interferes with the synthesis of **peptidoglycan** in bacterial cell walls, making bacteria subject to lysis. This is the mechanism by which the "wonder drug" penicillin and related antibiotics indirectly deliver a death blow to certain bacteria. A number of other antibiotics target bacterial cell walls, leading to the death of these cells. (Because human cells do not have cell walls, these antibiotics are, generally, of less toxicity.)

Bacteria are divided into two groups based on differences in the chemistry of their cell walls, namely, Gram positive (+) and Gram negative (–) as demonstrated by the Gram stain, a **differential staining procedure** introduced by Hans Christian Gram (1853–1938) in 1884. The Gram-staining procedure is an important first step in identifying the type of specific bacterium as the cause of infection, whether it is Gram (+) or Gram (–). A specimen from a culture source or from a swab is taken from an individual (e.g., skin, mouth) and spread onto the surface of a slide and then flooded with crystal violet, a purple dye. Both Gram (+) and Gram (–) cells stain a purple color after application (about 30 seconds) of the crystal violet. Iodine is the next reagent (about a minute), which acts as a **mordant**, followed by addition of alcohol (30 to 60 seconds)—the differential stage; the alcohol dehydrates a backbone layer of the Gram (+) cell wall, preventing the loss of the crystal violet iodine complex. In Gram (–) cells, the alcohol dissolves the relatively abundant outer membrane of the lipopolysaccharide bilayer and makes small pores in the inner layer component characteristic of Gram (–) cells through which the crystal violet-iodine complex diffuses outward [or *decolorizes* the Gram (–) bacteria]. At this stage these Gram (–) cells appear colorless. The addition of safranin, a pink-red counterstain in contrast to the initial purple crystal violet, results in these cells appearing as a pink-red color. In contrast, in Gram (+) bacteria the initial crystal violet-iodine complex is retained preventing the entrance of safranin, and hence these bacteria are a purplish-blue color (**FIGURE 7.3**).

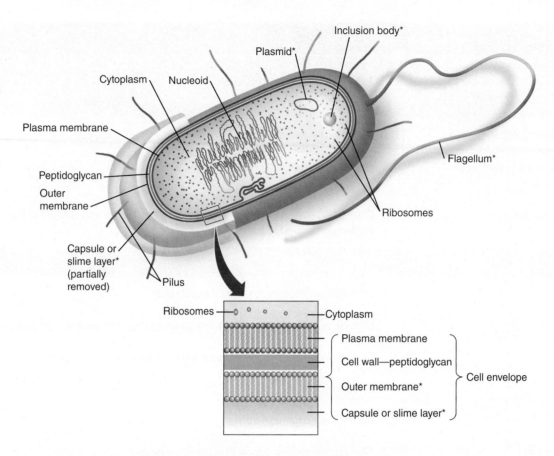

FIGURE 7.2 A "composite" Gram (−) bacterial cell. Note that if this were a Gram (+) bacterium, endospores would be produced from within the cytoplasm and the cell wall structure would not be the same. The asterisks indicate structures not present in all bacteria. *Structures not present in all bacteria.

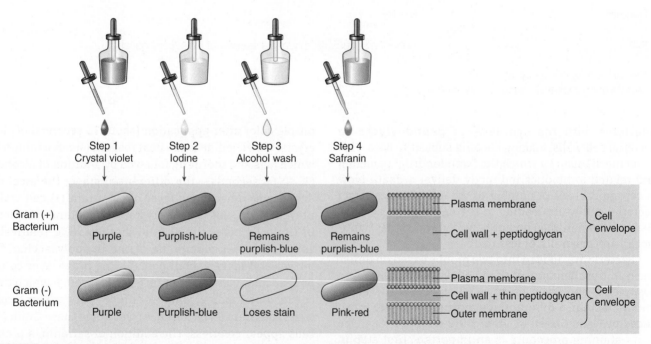

FIGURE 7.3 Gram-staining procedure. The color changes that occur during the staining of bacteria are shown in each step. When iodine is added, it acts as a mordant, creating a crystal violet-iodine complex that binds tightly to the peptidoglycan. Note this causes a slight color change from purple to purplish-blue or purplish-black. Due to the thin layer of peptidoglycan in the Gram (−) cell wall, the crystal violet stain can be washed out with alcohol (decolorizing the bacterium; it becomes colorless/clear). Therefore, a secondary safranin stain is used to counterstain the Gram (−) bacterium. The Gram (+) bacterium is not decolorized, retaining the crystal violet-iodine complex. The differences between a Gram (+) and Gram (−) bacterial cell wall are shown to the right of the Gram-staining procedure.

Perhaps you, or someone you know, have been diagnosed with a Gram (+) or a Gram (–) infection. *The distinction is important, because the choice of antibiotics is strongly influenced by the Gram-stain reaction of the causative bacteria* (**TABLE 7.2**). For example, penicillin and a variety of related antibiotics are effective against Gram (+) but not against Gram (–) bacteria; on the other hand, the antibiotic ciprofloxacin is effective against Gram (–) bacteria. Examples of **broad-spectrum antibiotics** that are effective against both Gram (+) and Gram (–) bacteria are the tetracyclines and the chloramphenicols. The differentiation between Gram (+) and Gram (–) cells is based on the chemistry of their cell walls.

Cell walls of Gram (+) and Gram (–) bacteria are composed of peptidoglycan, a backbone that gives the cell wall its characteristic rigidity. Some of the molecules associated with the bacterial cell wall are not found elsewhere in nature. Cell walls of Gram (+) bacteria have a thick peptidoglycan layer, whereas this layer is relatively thin in Gram (–) cells. Additionally, in Gram (–) cells there is an **outer membrane** external to the peptidoglycan layer that is associated with virulence. When Gram (–) bacteria undergo lysis, an **endotoxin** that causes fever and damage to the host is released from the outer membrane. Gram (+) bacteria, such as *Clostridium tetani*, which causes tetanus, produces an **exotoxin**. Endotoxins are incorporated into a Gram (–) cell wall in contrast to exotoxins, which are not. Exotoxins are formed within the cytoplasm of a Gram (+) bacterium and secreted to the outside of the cell.

Cell Membrane

The passage of molecules between the bacterial cell and its external environment is controlled by the bacterial plasma membrane. To understand this point, think of chains of streptococci residing on the surface of your pharynx (**FIGURE 7.4**). The surface is a warm, moist, and nutrient-rich environment that is optimal for bacterial replication. Oxygen and nutrients diffuse into the streptococci through their plasma membranes to allow for metabolic processes consistent with life and replication. During their growth and replication, as a natural part of their metabolism, certain bacteria secrete exotoxins. It is these **toxins** that produce the sore throat and fever associated with a "strep throat."

The bacterial plasma membrane is **selectively permeable**, meaning that not all molecules are able to pass freely into or out of the cell. The structure of the plasma membrane, inherent in its chemistry and physical arrangement, accounts for its "gatekeeper" function; it is double layered and has pores and transport molecules instrumental in the movement of materials between the bacteria and their environment. The physical processes of **diffusion** and **osmosis**, plus the size and the charge of the molecules, govern the selective permeability of the plasma membrane. The bacterial

TABLE 7.2 Infectious Diseases Produced by Gram (+) and Gram (–) Bacteria

Bacterium	Infectious Disease
Gram (+)	
Bacillus anthracis	Anthrax
Clostridium botulinum	Botulism
Clostridium tetani	Tetanus
Listeria monocytogenes	Listeriosis
Staphylococcus aureus	Skin infections, toxic shock, hospital-associated infections
Streptococcus mutans	Tooth decay
Streptococcus pyogenes	Streptococcal sore throat; scarlet fever, rheumatic fever, necrotizing fasciitis
Streptococcus pneumoniae	Pneumonia
Gram (–)	
Escherichia coli O157:H7	Bloody diarrhea
Elizabethkingia anophelis	Pneumonia, invasive bloodstream hospital-associated infections
Borrelia burgdorferi	Lyme disease
Bordetella pertussis	Whooping cough
Campylobacter jejuni	Gastroenteritis
Shigella spp.	Shigellosis (foodborne infections)
Salmonella spp.	Salmonellosis (foodborne infections)
Haemophilus influenzae	Pneumonia
Neisseria gonorrhoeae	Gonorrhea
Neisseria meningitidis	Meningococcal disease
Yersinia pestis	Bubonic, pneumonic, and septicemic plague (Black Death)
Vibrio vulnificus	Necrotizing fasciitis, bloodstream infections, severe gastroenteritis
Vibrio cholerae	Cholera

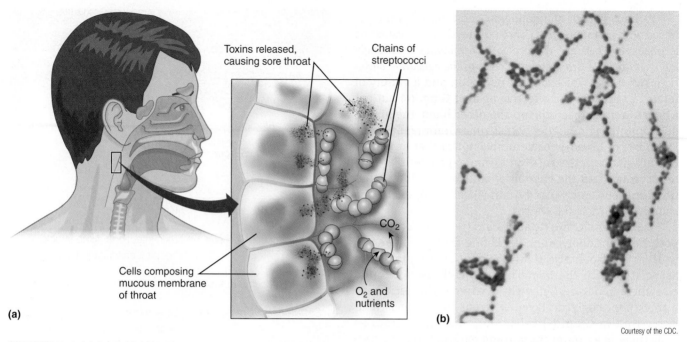

(a)

(b)

Courtesy of the CDC.

FIGURE 7.4 **(a)** Streptococci growing on the surface of the pharynx. **(b)** Gram stain of streptococci chains visualized by light microscopy at 1000X magnification.

plasma membrane is the site for energy-generating reactions of the cell and also contains enzymes that are instrumental in cell wall assembly and membrane synthesis. A plasma membrane is not unique to bacterial cells but is common to all cells, procaryotic and eucaryotic, and has a remarkably similar structure in all cells.

Cytoplasm

The cytoplasm is the part of the cell enclosed within the plasma membrane, and within it are numerous cellular constituents that function in cell growth and replication. Figure 7.2 illustrated the variety of structures found within this region, such as inclusion bodies, ribosomes, and plasmids.

◦ Nucleoid

The bacterial cell lacks a nuclear membrane; hence, it is a procaryotic cell. Within the cytoplasm is a DNA-rich area, demonstrated by staining procedures and transmission electron microscopy (TEM), which is referred to as a **nucleoid**; the area is not defined by a membrane. It is relatively easy to extract bacterial DNA from *E. coli* and some other bacteria. Most bacterial cells contain a single chromosome present as a circular, double-stranded stretch of DNA (dsDNA). (The bacterium *Vibrio cholerae* has two circular chromosomes. A few bacteria have been found to contain a linear chromosome, and some have both a linear and a circular chromosome.) This is in contrast to eucaryotic cells, in which the DNA is organized into "multiple" discrete linear structures, the chromosomes.

(Humans, for example, have 46 chromosomes organized into 23 pairs.) The mechanism of cell division in bacteria is binary fission, which, simply put, means "splitting in two." A **fission ring** directs the formation of a septum that divides the bacterium. After the **septum** is completed, the bacterium pinches in two, forming two bacteria (**FIGURE 7.5**).

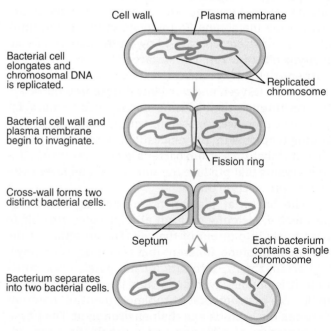

FIGURE 7.5 Binary fission of a Gram (+) bacterium. The process is the same in Gram (−) bacteria.

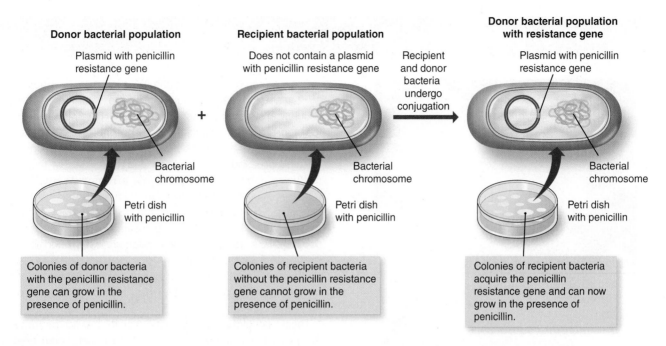

FIGURE 7.6 The infectious nature of plasmids. This diagram demonstrates how bacteria can develop resistance to the antibiotic penicillin. Bacteria containing plasmids with a penicillin-resistance gene can conjugate with bacteria that do not contain plasmids or that do not contain plasmids with a gene that codes for resistance to the antibiotic penicillin. During conjugation, the plasmid containing the penicillin-resistance gene is transferred to the recipient bacterium, allowing it to resist and survive the lethal effects of penicillin to bacteria. Note that bacteria contain more than one copy of a plasmid; therefore, the donor bacterium remains penicillin resistant to the effects of penicillin as well.

○ Plasmids

Plasmids are small circular molecules of nonchromosomal dsDNA that are found in the cytoplasm of some bacteria (Figure 7.2) and are independent of the chromosomal DNA. Under laboratory conditions plasmids are not essential to the cell; it is possible to "cure" (eliminate) them by laboratory manipulation without killing the bacterium. Plasmids may consist of only a few to a large number of genes, and they code for the production of proteins that have significant functions. Plasmids may carry genes that code for capsule or toxin synthesis conferring virulence to the bacterium. Plasmids may be transferred from one bacterial cell to another.

Other plasmids carry genes that confer **antibiotic resistance**; these plasmids are referred to as **R (resistance) factors**. Imagine, therefore, a population of bacteria in your intestinal tract in which, initially, only a few cells possess plasmids conferring resistance to three different antibiotics. It is possible that within only several hours plasmids may be transferred into other cells, resulting in most of the microbial population becoming resistant to multiple drugs (**FIGURE 7.6** and see **BOX 7.1**). Some plasmids allow for the exchange of DNA between bacterial cells. Because plasmids are self-replicating DNA molecules, the donor bacteria retain their plasmids while transferring copies to the recipient cell during conjugation.

BOX 7.1 Promiscuous Bubonic Plague Bacteria

In 1995, a 16-year-old male living in the Ambalavao District of Madagascar was suffering from chills, fever, and muscle pain (**myalgia**). The teenager sought medical attention. To the attending physician, malaria seemed to be the most logical diagnosis because it was common in this region. The teenager's symptoms were consistent with malaria. Many viral diseases also cause similar symptoms. Typically, the only treatment for a viral illness is supportive care. The boy was treated with quinine,

an effective inhibitor of *Plasmodium* spp., the protozoans that cause malaria.

Three days later, the boy's health deteriorated quickly. He had a high-grade fever that soared to 105.8°F (41°C). The teenager was delirious and extremely weak. A painful **bubo** (an inflamed lymph node) appeared in his right groin (**Figure 1**). As soon as the doctor noticed the bubo, he immediately suspected the teenager was suffering from **bubonic plague**. It is the most common form of the **plague**. The physician had experience treating patients

(continues)

BOX 7.1 Promiscuous Bubonic Plague Bacteria *(continued)*

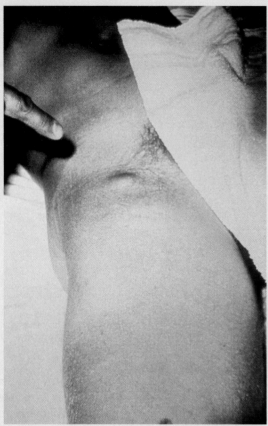

Courtesy of the CDC.

Figure 1 Bubonic plague patient with a swollen bubo in his groin.

who suffered from bubonic plague. In about 70% of cases, a bubo forms in the groin. Bubonic plague is caused by the bacterium *Yersinia pestis*. It is transmitted from rodent to rodent by the bites of fleas carrying *Y. pestis*. Bubonic plague occurs through rodent-to-human transmission by infected fleas biting/feeding on infected rodents that subsequently bite humans. Human-to-human transmission of **pneumonic plague**, the less common form of the disease, occurs by droplets containing the *Y. pestis* bacteria that are spread by a patient who is coughing and sneezing due to a lung infection. (See the CDC's Plague homepage for more details: https://www.cdc.gov/plague/index.html).

The bubo was punctured for culturing, and the diagnosis was confirmed as bubonic plague. *Y. pestis* is a Gram (–), rod-shaped bacterium that invades the bloodstream, liver, lungs, and other body sites. Hemorrhages occur under the skin, and the dried blood turns black, hence the term "**Black Death**" or "Black Plague." The mortality rate is 66% if untreated and 8–10% if treated with antibiotics. Plague can still be fatal, despite effective antibiotics. Human-to-human transmission is rare and usually requires direct contact with a patient suffering from pneumonic plague.

The recommendations of the World Health Organization (WHO) are to treat the bubonic plague with the following antibiotics: streptomycin, chloramphenicol, or tetracycline. The teenager's quinine treatment was discontinued and treatment with streptomycin was started. His body did not respond to the antibiotic, and his health continued to decline. The medical technology lab technicians performed **antibiotic sensitivity testing** on the bacteria isolated from the bubo. Doctors were shocked to discover that the bacterium was resistant to all of the WHO recommended antibiotics for treatment and **prophylaxis** (tetracycline and sulfonamides), in addition to the WHO's recommendation for alternative antibiotics to classic therapy such as ampicillin, kanamycin, spectinomycin, and minocylcine. The patient isolate was susceptible to cephalosporins and trimethoprim.

These antibiotic susceptibility testing results prompted the physician to immediately switch to trimethoprim treatment, which likely led to the teenager's recovery. The boy was extremely weak for more than a month after treatment. The doctor was alarmed by the multidrug-resistant traits of the newly isolated *Y. pestis* strain. The attending doctor contacted experts at the WHO, and the teenager's clinical isolate was sent to several national laboratory reference centers for further analysis.

Researchers were able to quickly ascertain that the teenager's *Y. pestis* strain, designated 17/95, contained four different plasmid DNAs. It had three plasmid DNAs commonly found in *Y. pestis* clinical isolates (named pFra, pPla, and pYV), but none of the common plasmids contained antibiotic-resistance genes. The 17/95 isolate harbored the common plasmids found in *Y. pestis* in addition to a large plasmid DNA, which was about 150,000 base pairs in length. The plasmid DNA encoded genes that conferred resistance to ampicillin, chloramphenicol, kanamycin, streptomycin, spectinomycin, sulfonamides, tetracycline, and minocycline. Typically, a bacterium harbors copies of only one type of plasmid DNA. The fact that the 17/95 isolate carried four different plasmids was also very unusual.

Researchers were concerned that the large plasmid DNA found in the 17/95 strain had the potential to be transferred to other types of bacteria in a natural setting (e.g., during the course of an infection in the human body). Typically, compatibility issues prevent the transfer of plasmids to bacteria of other genera. These same compatibility issues are associated with the number of different types of plasmids present in a bacterium. A special set of DNA sequences called an ORI (origin of replication) determines if a plasmid can replicate in a bacterium. The ORI is usually genus specific. For example, a plasmid DNA found in *E. coli* has an ORI that only allows the plasmid to be replicated in *Escherichia* but not in other bacteria, such as *Yersinia* or *Staphylococcus*. The researchers set up experiments to determine if the large plasmid DNA could be transferred to other gram-negative bacteria such as *E. coli* or to avirulent strains of *Y. pestis* through **conjugation** (bacterial sex). To their surprise, the large plasmid was easily transferable to more than one type of *E. coli* strain and to different avirulent *Y. pestis* strains.

The researchers were alarmed by the evolution of the multidrug-resistant *Y. pestis* 17/95 strain. Their findings were reported in the *New England Journal of Medicine*. They wondered how the large multidrug-resistant plasmid (named pIP1202) was originally transferred to *Y. pestis* 17/95. They analyzed the ORI of the plasmid and found that it was similar to ORIs of other plasmid DNAs transferrable to bacteria that are members of the *Enterobacteraceae* family. The bacterial members of this family are all Gram (–) rod-shaped bacteria that are typically found in the intestines of humans. *Y. pestis* circulates in the bloodstream, lymphatic vessels, spleen, liver, and sometimes lungs. How did *Y. pestis* Y95 manage to have sex (conjugate) with human intestinal bacteria? Could *Y. pestis* 17/95 conjugate with bacteria in the gut of a flea that ingested blood containing *Y. pestis* (recipient bacteria) and a donor gut bacterium that contained the large plasmid? Bacteria are promiscuous (**Figure 2**)!

The Black Death arrived on the island of Madagascar in 1898 by way of steamboats from India carrying rats with infected fleas. The plague has never disappeared. However, through improved housing and sanitation and the discovery of antibiotics and insecticides, plague was controlled by the 1950s. For the next 30 years, only 20 to 30 cases of plague were confirmed each year. However, since the 1990s, the number of confirmed plague cases has increased every year, from 800 to nearly 2,000 in 1996. In the past two decades, plague continues to be an endemic disease in Madagascar. An unprecedented outbreak of pneumonic plague (see **Figure 3**) started on August 1, 2017, which was contained by November 27, 2017. Patient zero was a 31-year-old man who was visiting Madagascar's central highlands when he developed malaria-like symptoms. Four days later, he died. Pneumonic plague is the most virulent form of the plague. It can kill patients within 12 to 24 hours. Immediate diagnosis and treatment are critical to survival. The 2017 outbreak saw approximately 2,417 cases and 209 deaths (9% case fatality rate). The majority of cases (1,854, or 77%) were the pneumonic form of the plague, 355 cases were bubonic plague, and 1 case was septicemic plague (infection in the bloodstream). Additionally, 207 cases were not classified clinically. None of the 81 healthcare workers who contracted the plague died. More than 4,770 individuals were involved in **contact tracing** supervised by medical doctors and students. A total of 7,318 contacts were identified during the outbreak who were required to take a course of prophylactic antibiotics to prevent person-to-person spread of the disease.

During the course of the epidemic, health officials were concerned about the spread of the plague through a centuries-old tradition called *famadihana*, also known as "the turning of bones," "body turning," or "dancing with the dead." The tradition involves exhuming the bones of deceased relatives from crypts, rewrapping them in fresh cloth, and dancing with the wrapped corpses before returning them to their graves (**Figure 4**). During *famadihana*, the bacteria can still be transmitted and contaminate whoever handles the deceased body. To stop the potential spread, dignified and safe burial protocols were finalized and submitted to national authorities. Approximately 90% of the population consulted was in favor of these measures. The plague victims were buried in anonymous mausoleums, not in tombs or crypts that could be reopened. Other key actions to prevent its spread outside of Madagascar involved increasing public awareness of plague and enhancing surveillance for the disease at airports and seaports.

Questions and Activities

1. Note that in Figure 2 (panel 1), the unknown gut bacterium (donor) that belongs to the *Enterobacteraceae* family has peritrichous flagella. What is the function of bacterial flagella?

2. Note that in Figure 2 (panel 2), a pilus is shown on *Y. pestis*. What is the function of a bacterial pilus?

3. Besides genes that code for antibiotic resistance, what are the functions of at least two other genes that can be carried by a plasmid?

4. Why was the isolation of the first multidrug-resistant strain of *Y. pestis* 17/95 in 1995 a cause for concern?

5. If the large plasmids (pIP1202) carried by *Y. pestis* 17/95 were discovered in a strain of *Salmonella* or *Klebsiella pneumoniae*, would you be concerned? Explain.

6. If the 2017 pneumonic plague outbreak had been caused by *Y. pestis* 17/95, do you anticipate there would be more or fewer confirmed cases? More or fewer plague deaths? Explain.

7. The **R-nought** at the height of the plague epidemic in October 2017 was estimated to be 1.73. Define R-nought. List the factors that affect R-nought.

8. Compare and contrast bubonic and pneumonic plague.

9. Create a world map and shade in locations of the world in which plague is endemic. (Hint: Download the WHO Fact Sheet on Plague, http://www.who.int /mediacentre/infographic/plague/en/)

10. Only one other clinical isolate to date has been shown to contain the large 150 base pair multidrug-resistant conjugative plasmid (pIP1202). It was isolated in 1997 from a plague patient in the Ampitana District, Madagascar, which is 120 kilometers (80 miles) from the 1995 patient in the Ambalavao District. Hypothesize why no other strains from clinical isolates have been found to carry pIP1202 in the past 20 years.

(continues)

BOX 7.1 Promiscuous Bubonic Plague Bacteria (*continued*)

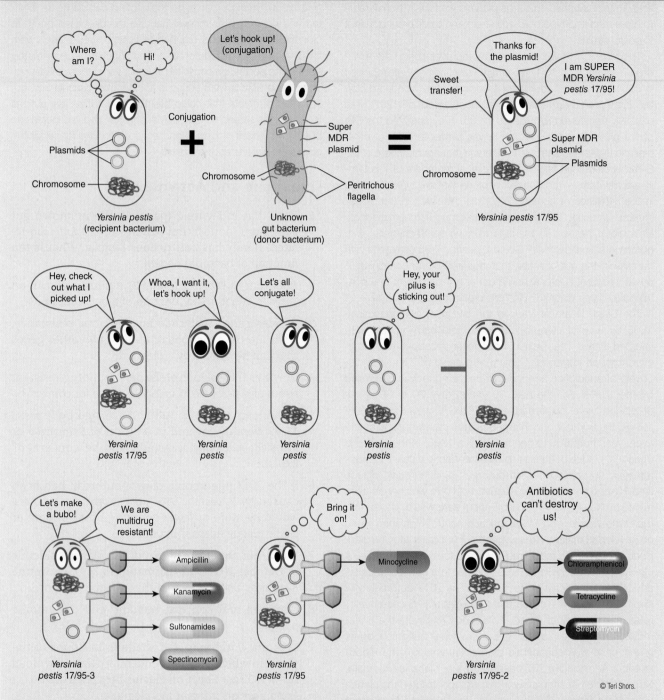

Figure 2 Cartoon depicting the development of the multidrug-resistant *Yersinia pestis* 17/95 strain. The smaller red, blue, and green plasmids (designated pFra, pPla, and pYV, respectively, by researchers) were known to replicate in *Y. pestis*. *Y. pestis* conjugated with an unknown bacterium that belongs to the *Enterobacteraceae* family. In doing so, the 150,000 base pair pIP1202 plasmid was transferred to *Y. pestis*. The bacterial clinical sample from the teenager suffering from plague that contained this plasmid was named *Y. pestis* 17/95. The multidrug-resistant plasmid pIP1202 contains eight different antibiotic-resistant genes. It encodes resistance to the effects of ampicillin, chloramphenicol, kanamycin, streptomycin, spectinomycin, tetracycline, monocycline, and sulfonamides. The *Y. pestis* 17/95 strain multiplies and can also transfer its conjugative plasmid to other strains of *Y. pestis*, resulting in populations of multidrug-resistant plague bacteria. Some of the structures of the cartoon bacteria labeled in panel 1 of the cartoon strip discussed in this chapter, such as flagella and pili, play a role in virulence.

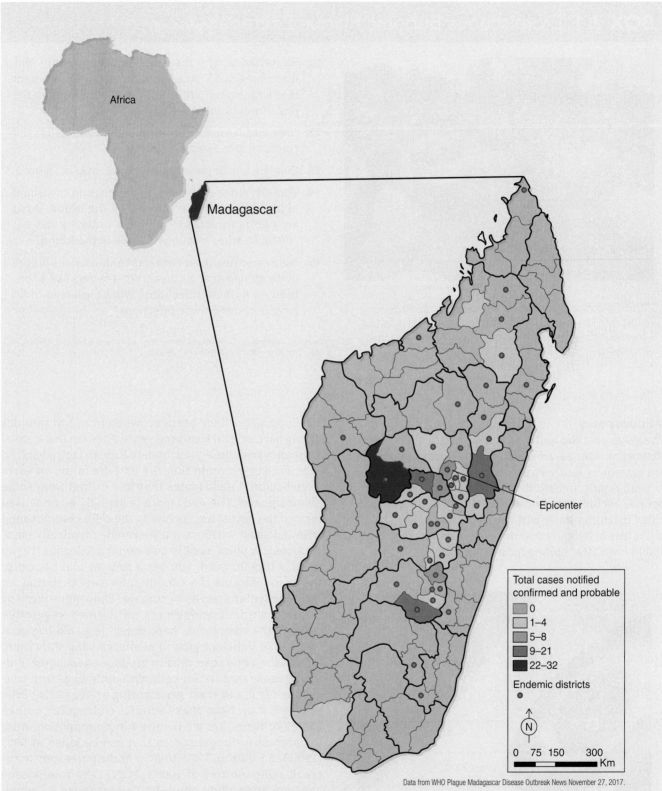

Figure 3 Geographic distribution of plague cases in Madagascar as of November 22, 2017. Madagascar is an island off the east coast of Africa. The **epicenter** of the outbreak was in the Ankazobe District. Madagascar is one of the poorest countries in the world, with more than 75% of the population living below the poverty line, surviving on less than $1.90 per day.

Data from WHO Plague Madagascar Disease Outbreak News November 27, 2017.

(continues)

BOX 7.1 Promiscuous Bubonic Plague Bacteria (*continued*)

© Travelib Madagascar / Alamy Stock Photo.

Figure 4 Antananarivo, Madagascar. People waiting to exhume corpses during the rite of *famadihana* in the highlands of Madagascar. If a pneumonic plague victim were exhumed, it could result in the spread of the disease.

11. Hypothesize why the same pIP1202 plasmid was discovered in a Y. *pestis* strain from a plague patient that was located 120 kilometers (80 miles) from the first strain isolated.

12. Create at least five more dialogue callouts for the comic strip in Figure 2.

13. Create a fourth panel for the comic strip in Figure 2.

14. The 2017 pneumonic plague outbreak on the island of Madagascar was contained on the island. What are two reasons as to why this outbreak did not spread to other locations outside of the island?

15. Eighty-one healthcare workers contracted the plague in the Madagascar outbreak. What factors likely contributed to these infections? What measures could have prevented these infections?

Information from Galimand M, et al. Multidrug resistance in *Yersinia pestis* mediated by a transferable plasmid. New England Journal of Medicine. 1997;337(10):677–680.

⊙ Endospores

The genera *Bacillus* and *Clostridium*, both of which contain pathogenic species, are endospore producers. **Endospores** are extremely hardy structures that are highly resistant to heat, drying, radiation, and a variety of chemical compounds, including alcohol (**FIGURE 7.7**). Most bacteria are killed in boiling water within only a few minutes, but this is not true of bacteria that make endospores; they remain viable even after boiling for a few hours.

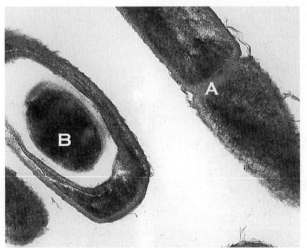

Courtesy of Dr. Sherif Zaki and Elizabeth White/CDC.

FIGURE 7.7 A transmission electron micrograph of *Bacillus anthracis* showing **(a)** vegetative cell division by binary fission and **(b)** an endospore forming within the middle of the vegetative cell (developed during "hard times"). Note the hard-shell endospore coat that provides protection for the endospore.

Endospores have been recovered from soil samples dating back several thousand years. They are in a state of dormancy with little or no metabolism. In fact, a handful of scientists maintain that the first life forms on Earth were bacterial endospores that had drifted from some distant planet. The endospore is actually a condensed form of the bacterium, including the cell's chromosomal DNA, encased within a multilayered, chemically complex coat. A plant seed is somewhat analogous. If you plant a tomato seed, you get a tomato plant because the seed contains the genetic information coding for that particular species of tomato. Endospore-forming bacteria without endospores are termed **vegetative cells**. In the laboratory, "hard times" (e.g., old bacterial cultures or deficient growth media, causing starvation) induce the vegetative cells to produce endospores. Providing these sporulating cells with optimal growth conditions results in their germinating as vegetative cells (**FIGURE 7.8**). Note that bacterial endospores, unlike spores of fungi, are *not* involved in reproduction. Also, in contrast to fungal spores that can be killed at 60°C (140°F), *Bacillus* and *Clostridium* endospores can withstand temperatures of 100°C (212°F). The rule is "one cell, one endospore, one cell." The organism *B. anthracis* is an endospore former. The endospore's hardiness and infectivity if the endospores are aerosolized and inhaled make it a top candidate for biological warfare and terrorism. You may have seen news accounts of threatened terrorism using the anthrax "spores"; some proved to be hoaxes, but others, unfortunately, have been the real thing.

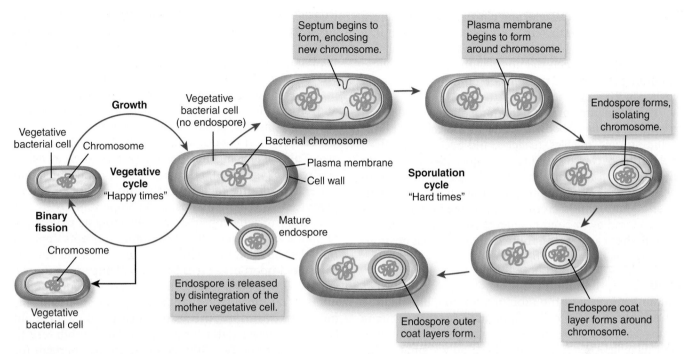

FIGURE 7.8 The vegetative *and* sporulation cycles of a Gram (+) bacterium that produces endospores. Remember that endospores are formed when "times are hard" (e.g., starvation), and vegetative cells multiply when all is copacetic.

The endospore-forming anaerobic genus *Clostridium* includes three species that cause serious and potentially life-threatening diseases of humans: tetanus (*C. tetani*; lockjaw), botulism (*C. botulinum*; a type of food poisoning), and gas gangrene (*C. perfringens*). The stage for tetanus is set when you step on a rusty nail carrying endospores and sustain a puncture wound, providing anaerobic conditions for germination of the endospores. (It does not have to be a nail, nor does the object need to be rusty. Have you had your tetanus booster shot?) Botulism is a serious and potentially lethal disease caused by *C. botulinum*. Those who are thinking of doing their own home canning with vegetables from their garden need to be aware that endospores are resistant to boiling. The food must be cooked in a pressure cooker after being packed into jars and sealed carefully. The point is that "home canners" need to be fully aware of what they are doing and cannot take shortcuts. Occasionally, commercially canned vegetables can be contaminated with growing *C. botulinum*; a "bulging" can is a telltale sign and should be immediately brought to the attention of the store manager.

Appendages

○ Flagella

Flagella confer **motility** (movement). They are composed of a protein called **flagellin** and are found in some species of bacteria. One or more flagella are arranged in clumps or are distributed all over the bacterium; the particular arrangement of the flagella is consistent for those species that have flagella (**FIGURE 7.9**).

Each flagellum is like a long tail measuring over 10 micrometers (μm), many times longer than the length of a single bacterium. They are fragile to stain and extremely thin, making it difficult to observe flagellar structures under a light microscope. The anatomy of flagella reveals that they are anchored by their

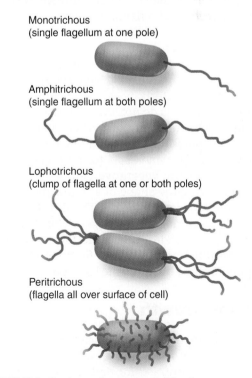

FIGURE 7.9 Structure and arrangement of flagella.

basal bodies to the cell wall and plasma membrane from which they extrude through the cell wall. The motility of live and unstained bacteria can be observed by suspending a drop of bacterial liquid culture on a slide and examining this special type of wet mount, called a *hanging drop*, with a light microscope.

The flagellum rotates like a propeller and is driven by rings in the basal body. Descriptive and colorful terms, including *runs*, *tumbles*, *twiddles*, and *tweedles*, are used to describe bacterial motility. Receptor sites are present within the cell resulting in directed movements—a process called **chemotaxis**—toward or away from a chemical stimulus. Some bacteria "swim" toward an attractant (glucose) and away from repellents (acids). It is amazing to think they can detect minute changes in their environment. This is remarkable behavior.

Pili

Pili are composed of the protein **pilin** and, like flagella, extrude through the cell wall and are present in many Gram (–) species. They are shorter, straighter, and thinner than flagella. A bacterium may have only a single pilus or up to several hundred; in some bacteria, including those that cause the sexually transmitted disease gonorrhea, pili serve as **adhesins** and anchor bacteria to the mucous membrane of the vagina or the penis. This is an early step in the **colonization** and subsequent establishment of disease.

Other pili, referred to as **sex pili**, function in bacterial conjugation, the transfer of DNA (i.e., plasmids) by forming a bridge between two bacterial cells through which plasmid DNA passes from donor to recipient. Conjugation plays a major role in the spread of antibiotic resistance genes among bacteria of different species.

Now that you have studied that anatomy of a typical bacterium, one can appreciate that there are some anatomical features that are different and some that are the same among pathogenic bacteria. For example, many pathogenic bacteria possess capsules, whereas nonpathogenic bacteria do not. Others possess lophotrichous or peritrichous flagella, and others do not. *Bacillus* spp. form endospores within the middle of a vegetative cell; in contrast, *Clostridium* spp. form an endospore at a terminal end of the vegetative cell, which is often described as having the shape of a tennis racket. Some pathogens produce large storage granules within or at the poles of a bacterial cell. Arrangements may be singular, chains, or palisades (Chinese letter arrangement). **FIGURE 7.10** illustrates different features of selected Gram (+) and Gram (–) pathogenic bacteria.

Bacterial Growth

For unicellular organisms, both procaryotic and eucaryotic, the terms *growth*, *multiplication*, and *replication* are used synonymously. In contrast, for multicellular organisms (consider the human), *growth* refers to an increase in the size of an individual as a unit and is the result of replication at the cellular level (mitosis), leading to an increase in the total number of cells in an individual, but not in the number of individuals. A population, microbial or otherwise, if allowed to grow unchecked, would take over the Earth and, ultimately, exterminate all other forms of life. Can you imagine a universe populated entirely by *E. coli*?

A system of checks and balances limits population size. This system includes abiotic factors, such as availability of oxygen (O_2), temperature, pH conditions, and water, and biotic factors, including suitable habitat, predator–prey relationships, and competition. A course in ecology usually considers these factors as they apply to plants and animals, but these same concepts apply to microbes as well. Interaction between bacterial species is just as important as it is between plants and animals. (Microbial ecology is a specialized course that might be of interest to a microbiology major.) For example, as Fleming observed in 1928, the chemical secreted by the mold *Penicillium rubens* (later termed *penicillin*) is inhibitory to various bacteria. Penicillin is a defense mechanism for the mold in the same sense that poisonous secretions of some jellyfish or the venom of certain snakes serve as defense mechanisms in their predator–prey relationships.

Bacterial growth provides insight into the dynamics of population growth as it occurs in nature, including in the body (**FIGURE 7.11**). A **growth curve** illustrates what occurs when *E. coli* incubating in the mayonnaise in the tuna salad at the beach on a warm day is ingested and the microbe multiplies in your intestinal tract. The worst is yet to come when 4 or 5 hours later on the way home, family members, usually children, might develop nausea, vomiting, and diarrhea; to add to this hypothetical scenario, consider that you might be in the middle lane of a congested highway with no opportunity to pull over.

To prepare a growth curve, samples of liquid cultures are collected at frequent intervals and then plated onto solid media in petri dishes. Population counts are performed by allowing the bacteria to develop into (**macroscopic**) **colonies** that can be counted (**FIGURES 7.12** and **7.13**). (One-hour intervals are convenient—except for the student who needs to culture-sit throughout the night and into the next day.) Based on these colony counts, a growth curve is constructed illustrating the four major phases.

Lag Phase

The **lag phase** is a "get ready for growth" stage and represents time of adaptation to the new environment; adaptation also occurs in actual infection. Under both circumstances there may be an initial drop in the count because of adjustment and survival of only a part of the microbial population. In infection, immune mechanisms of the body more readily kill those cells that are "less fit" for survival so only the fittest survive—a nice example of Darwinism at the microbial level.

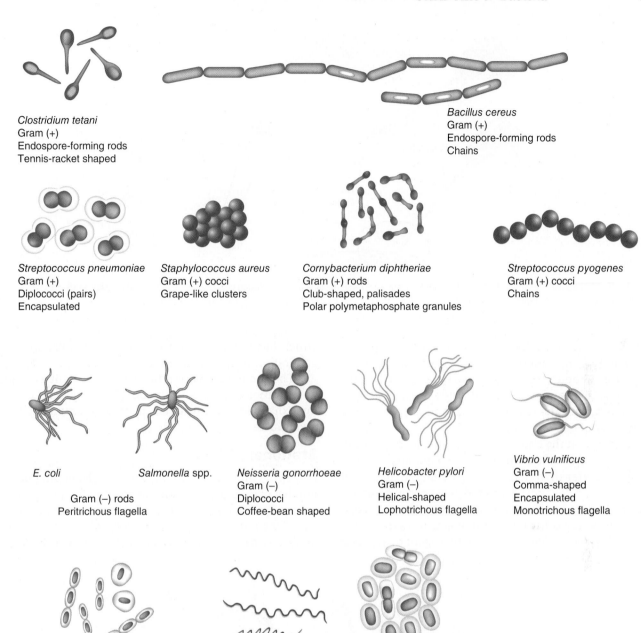

FIGURE 7.10 Illustration of different Gram (+) and Gram (–) pathogenic bacteria. Gram reactions, shape, and other anatomical characteristics are listed below each type of bacterium.

Log (Exponential) Phase

During the **log phase** the number of bacteria increases exponentially, governed by the generation time—the time it takes a bacterium to undergo **binary fission**, producing two "new cells." E. coli, for example, has a generation time of only 20 minutes, assuming optimal growth conditions. At the conclusion of each generation time, the population is doubled.

As an example, consider the dynamics of growth starting with a single E. coli rod. How many would you have in a 24-hour period during which 72 generations would have occurred? After the first 20 minutes there would be 2 bacteria, 20 minutes later (elapsed time of 40 minutes) there would be 4 bacteria, and after another 20 minutes (elapsed time of 60 minutes) there would be 8 bacteria. Although this may not be particularly impressive, consider

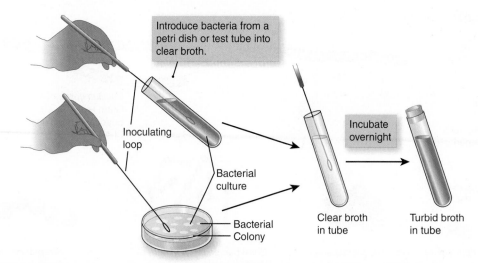

FIGURE 7.11 Overnight growth of *E. coli*. A sterile broth is inoculated with bacteria from an isolated colony on an agar plate. **Aseptic technique** is used to perform the procedure. The broth is incubated at 37°C (98.6°F). The bacteria multiply rapidly, causing the broth to look cloudy/turbid.

the solitary bacterium has now reached a population of 4,096 bacteria in 4 hours, or just 12 generations. What about after 10 hours, or after 24 hours, and so on? Mathematically, this can be expressed as 2^n, or 2^1, 2^2, and 2^3, respectively, with the exponents 1, 2, and 3 representing the number of generations.

Therefore, at the end of 24 hours the total population would be 2^{72}, and at 48 hours (2^{144}) the population's mass would be greater than that of Earth! Calculate the actual population and you will be amazed. Most of the common bacteria involved in human disease have short generation times. At the other end of the spectrum is the spirochete *Treponema pallidum*, the causative agent of syphilis,

which has a generation time of 33 hours. Examples of generation times are given in **TABLE 7.3**.

The conditions within the test tube present a number of limitations, as in any functioning ecosystem, including exhaustion of nutrients, space, and accumulation of toxic waste.

Stationary Phase

Sometime after about 10 or 12 hours, depending on the number of cells started with, adverse conditions set in and binary fission slows, leading to the **stationary phase** in which the number of new bacteria growing is equal to the number of bacteria dying. What are the adverse conditions

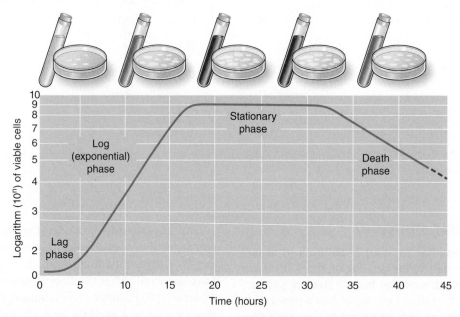

FIGURE 7.12 Bacterial growth curve. Growth of bacterial cultures in a tube containing broth medium were plated at various times during a 43-hour period after the broth was inoculated with *E. coli*. At the beginning of the experiment, visually, the broth will appear clear but then become turbid/cloudy over time due to the presence of high numbers of bacteria as a result of bacterial multiplication. A small volume of bacteria is also plated onto agar plates. Dilutions will be necessary in order to obtain isolated colonies on agar plates. One gets very good at doing dilutions in microbiology!

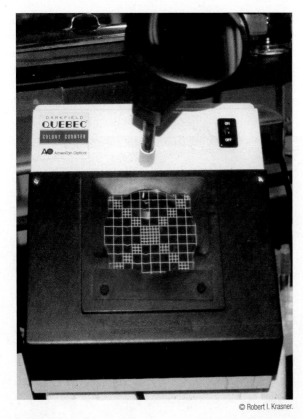

© Robert I. Krasner.

FIGURE 7.13 A colony counter. Agar plates with colonies are placed on the counting grid. A magnifying glass facilitates counting the colonies.

TABLE 7.3 Microbe Generation Times

Microbe	Generation Time[a]
Bacteria	
Escherichia coli	20 minutes
Staphylococcus aureus	30 minutes
Clostridium botulinum	35 minutes
Listeria monocytogenes	50 minutes
Mycobacterium leprae	12.5 hours
Mycobacterium tuberculosis	18 hours
Treponema pallidum	33 hours
Psychrobacter cryopegella[b]	39 days
Protozoans	
Leishmania donoviani	10 hours
Giardia lamblia	18 hours

[a] Times are approximate and growth is at 37°C (98.6°F body temperature).
[b] This is a Gram (−) rod-shaped nonpathogenic bacterium isolated from Siberian permafrost. The bacteria replicated at −10°C (14°F). The growth curve experiments to determine its generation time took months to do.

that arrest the growth of a culture? Having started with a single species, interspecies competition is not an issue, but there is competition among the bacteria for the depleting supply of nutrients and O_2. Toxic wastes, including carbon dioxide (CO_2), accumulate. Binary fission continues to occur but at a rate about equal to the death rate, and thus the culture has now reached a plateau, so there is no net increase in cell numbers. The culture has just about had it at this time, but as far as the individual bacterial cells are concerned, the worst is yet to come—the dreaded death phase!

Death Phase

The adverse conditions in the test tube become more and more pronounced. When the number of cells dying exceeds the rate of binary fission, the **death phase** has begun. It still might be possible to save the culture by simply transferring it into a larger tube with fresh medium or, alternatively, by taking a small sample of the culture that should contain some viable cells and inoculating a fresh tube of medium. If the bacterium is an endospore former (*Bacillus* or *Clostridium*), endospore formation may have resulted, and the cells may remain viable for years in the endospore stage.

Significance of Bacterial Growth

The dynamics of growth is of more than academic interest. What is the significance of all of this in terms of the natural state of infection? Think of the intestinal tract,

the ear, an area of skin, or the pharynx as a test tube representing a particular ecological niche, each with its own distinctive environment. Certainly, the "ecosystem" of the intestinal tract differs markedly from that of the middle ear. In most cases, symptoms of infection do not appear until there is a relatively large bacterial population. How do you apply this to the growth curve? Under most circumstances the initial introduction of bacteria into the body escapes notice, and this represents the lag phase. Within only several hours, depending on the particular organism and its generation time, a dramatic increase in the bacterial population takes place. It is during this exponential or log phase that you experience the symptoms associated with that infection, such as lethargy, abdominal pain, nausea, vomiting, diarrhea, headaches, and fever.

Intervention, by taking an antibiotic or other medicines as a supplement to the immune system, is an attempt to interrupt the log phase, enter the stationary phase, and, as quickly as possible, hasten the microbial death phase. In the absence of intervention, the body's immune system responds to the presence of the microbial pathogen by calling into play defense mechanisms, which attempt to destroy the invading microbes, thereby halting the continuation of the logarithmic phase. Should the body's defenses fail, the log phase continues, and the person will die of overwhelming infection.

Culturing Bacteria: Diagnostics

What happens to the throat swab that your physician or nurse practitioner uses to swab your pharynx, possibly when "strep throat" is suspected, or to the urine, fecal, or blood specimens that might be taken when bacterial infection is considered? (You cannot forget the throat swab! It is quite normal to gag, because, in fact, the swab does hit the spot that is responsible for the gag reflex. It is important to obtain a sample from your pharynx and not from the roof of your mouth.) The purpose of obtaining a specimen is to grow the suspect organism so that it can be identified. The swab or a sample of the specimen is placed into a tube of liquid nutrient broth and/or streaked across the surface of a Petri dish containing nutrients and agar (derived from seaweed). **Agar** produces a consistency slightly firmer than gelatin. Franny Hesse (1850-1934) is credited with the use of agar as a solidifying agent of bacteriological media in the late 1800s. The application of solid media remains an isolation technique used in microbiology research and diagnostic laboratories throughout the world.

Growth on agar allows the production of colonies with recognizable and identifiable properties (**FIGURE 7.14**).

Pure cultures are obtained by the streaking of a loopful of bacterial culture across the surface of an agar-containing Petri dish. Each isolated colony is a **"pure" culture** and is the result of the binary fission of a single bacterium (at least theoretically). Colony characteristics, including size and pigmentation (**FIGURE 7.15**), are valuable clues

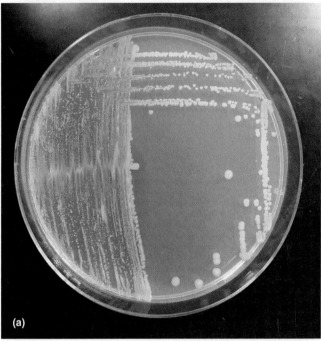

(a)

© Teri Shors and Amanda Prigan.

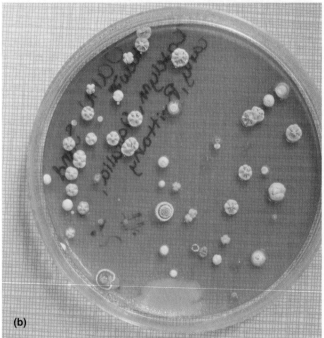

(b)

© Teri Shors and Amanda Prigan.

FIGURE 7.15 Colony characteristics. Species of bacteria display characteristic colony morphology, including pigmentation, texture, and size. **(a)** *Micrococcus luteus* colonies on nutrient agar. **(b)** Different soil samples were diluted with sterile water and plated onto R2A agar. Note the diverse types of colonies representing the diversity of soil bacteria present on the medium.

© Teri Shors and Amanda Prigan.

FIGURE 7.14 Isolation. The specimen is streaked onto the surface of an agar plate, which is then incubated to allow development of colonies. The same type of bacterial colonies may appear differently on different agar plates because the type of components in the medium will allow one to distinguish different information about the metabolism of the bacteria that may aid in diagnostics. **(a)** *E. coli* colonies present on MacConkey's agar appear red. **(b)** *E. coli* colonies present on nutrient agar appear white/cream. **(c)** *Staphylococcus aureus* colonies on mannitol salt agar cause the medium to turn yellow. **(d)** *S. aureus* colonies on nutrient agar.

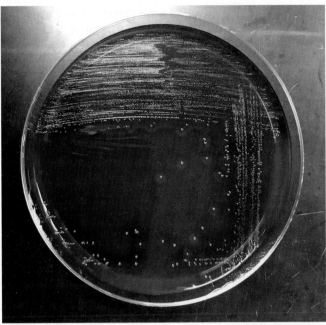

© Teri Shors and Amanda Prigan.

FIGURE 7.16 Diagnosis of strep throat. The streptococci responsible for strep throat are easily identified. A swab of the throat is streaked onto the surface of an agar plate enriched with sheep blood. The presence of clear zones around the areas of growth establishes the diagnosis.

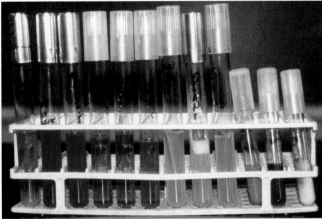

© Robert I. Krasner.

FIGURE 7.17 A variety of media in test tubes are used to determine metabolic properties of bacteria that are useful in identification.

in identification. For example, *Streptococcus pyogenes*, which causes strep throat, is easily identifiable on blood agar (agar to which sheep blood has been added). The streptococci secrete a protein (**hemolysin**) that destroys the sheep red blood cells surrounding the colonies. This results in clear zones because of the lysis of the red blood cells and confirms the diagnosis of a sore throat caused by streptococci (**FIGURE 7.16**).

In most cases, further studies are necessary to identify the organism. A variety of broth and agar media are available for growth and identification, and the particular medium selected depends on the source of the specimen. After all, the bacteria that are commonly associated with intestinal tract infections are different from those that cause skin, urinary tract, or respiratory tract infections. After incubation, the cultures are examined for characteristic properties (**FIGURE 7.17**). Gram stains are performed and are important tools to aid in identification. Further, the isolate is usually spread onto the surface of an agar plate, antibiotic disks are added (**FIGURE 7.18**), and the plate is incubated and examined after growth has occurred. **Zones of inhibition** (no growth) occur around those antibiotic disks to which the bacteria are sensitive, because the antibiotic leaches from the disk into the moist agar surface. This is an important tool for the physician in prescribing an appropriate antibiotic; other factors, including toxicity, cost, and duration of therapy, also are taken into account.

The process of identification may take only 24 hours, as in the case of streptococci; microbes with longer

generation times may require extended incubation to achieve sufficient growth before moving on to the next stage. In the case of tuberculosis, bacterial identification takes several weeks because of the long (18-hour) generation time characteristic of *M. tuberculosis* (Table 7.3).

Diagnostic microbiology has come a long way since the days of Frau Hesse. "Short-cut" techniques are available for diagnosing a variety of diseases caused by microbes or viruses, including urinary tract infections (UTIs), diarrhea (e.g., salmonellosis and shigellosis), strep throat, syphilis, malaria, schistosomiasis (worm), rabies, influenza, and hepatitis.

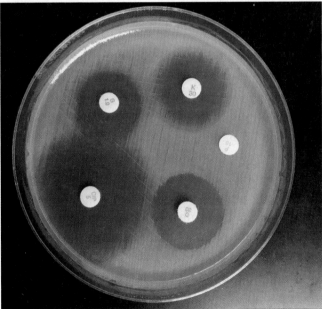

© Teri Shors and Amanda Prigan.

FIGURE 7.18 Determination of antibiotic sensitivity. The specimen is spread onto the surface of an agar plate, antibiotic disks are added, and the plate is incubated. Zones of inhibition (no growth) around a disk indicate the bacteria are sensitive. The disk is coded with the name and concentration of the antibiotic. Note that the spectrum of activity varies for each culture.

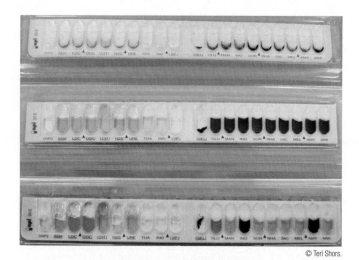

© Teri Shors.

FIGURE 7.19 API20E Gram (–) bacterial identification. Upper strip: uninoculated; middle strip: inoculated; lower strip: results after 24 hours of incubation at 37°C (98.6°F). Note the color changes in the cupules after incubation/growth of the microbe in the test strip. Similar testing is available to identify Gram (+) bacteria.

Certain bacterial infections caused by Gram (–) bacteria are identified using rapid and cost-saving multitest procedures. Instead of inoculating a number of separate test tubes to determine the biochemical test results of a pure culture (**FIGURE 7.19**), a plastic strip known as an **API-20E** test strip is inoculated with a bacterial suspension. It contains 20 mini test tubes (cupules) with dehydrated substrates (media). During inoculation, the media is reconstituted and the strips are allowed to incubate for 18 to 24 hours. The bacterial metabolism causes color reactions that are immediate or revealed with the addition of reagents. The reactions are read by sophisticated automated instrumentation that assigns a specific code, the Analytical Profile Index, from which the initials API are derived. The API identifies the microbe to its genus and species more quickly, leading to a rapid diagnosis and appropriate treatment (Figure 7.19). Persons suffering from Montezuma's revenge are grateful!

It is not always possible to culture the suspect organism, particularly if the infected individual is already receiving antibiotics. Further, as too frequently happens, the urine, throat swab, or blood specimen may have been left at room temperature too long, resulting in its drying out, or the time necessary for the organism to grow out may impose too long a delay. Fortunately, another avenue of diagnosis is available. Microbes leave telltale "footprints" of their presence by eliciting the production of specific antibodies by the host. **Antibodies** are produced in response to molecules called **antigens** that are part of the microbe or that are secreted (toxins). Because antibodies are specific and react only with the antigen that caused their production, the presence of antibodies helps to establish the identity of the bacteria involved.

The tools of molecular biology have opened research and diagnostics in microbiology and, particularly, in virology. The application of molecular biology has added precision and rapidity in a cost-effective manner to diagnostic procedures. Essentially, molecular biology as applied to diagnostics is focused on getting to the "heart" of the microbe (i.e., its genetic content). Diagnostic bacteriology has lagged somewhat in using this "new" approach because traditional culture methods and the more recent development of rapid tests have served well over the years.

Sore throats are common among kids. About 10% of all sore throats are caused by *Streptococcus pyogenes*. Strep throat is highly contagious and if untreated can increase the individual's risk for developing complications such as rheumatic fever, meningitis, and diseases affecting the heart and kidneys. Rapid strep antigen tests are used to quickly determine if *S. pyogenes* is the culprit, allowing for immediate antibiotic therapy if the test is positive. The throat and tonsils are swabbed to collect bacteria from the infected area for testing. The swab is placed in a plastic tube and drops of reagents are added to extract streptococcal antigens, followed by drops of antibodies against *S. pyogenes* that capture the antigens, resulting in a colored band or line in the reaction zone of a test strip (**FIGURE 7.20**). The test results are available within 10 to 15 minutes. A normal or negative test means that *S. pyogenes* may not be present. Following a negative result, an additional throat swab is taken and used to inoculate a blood agar plate (Figure 7.16). This is a safeguard against false-negative reactions (meaning someone actually has a strep throat infection even though the rapid strep antigen test result was negative). A positive result indicates an infection.

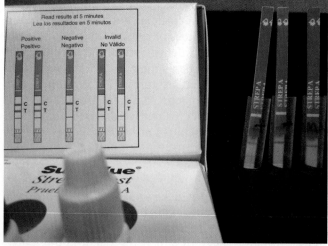

© Teri Shors.

FIGURE 7.20 Photograph of rapid strep antigen test results. The tube with the test strip on the left side is a positive control [positive result: *Streptococcus pyogenes* antigens are present in the sample. It contains a colored band in the upper test (or T) region and colored band in the lower control (or C) region]. A band in the test and control regions is a positive result. The tube in the middle represents a negative control; a colored band forms in the control region but not in the test region. The patient strip (M) on the right displays a band in the control but not the test region. It is a valid negative result.

Culture follow-up is not required, and the patient is treated with antibiotics. The next time a doctor or nurse swabs the back of your throat—an unpleasant experience—try hard to resist gagging as you anticipate your rapid strep test results!

Medical laboratory microbiologists need to be highly trained to identify suspected disease-producing bacteria properly. The techniques are simple, but interpretation of the laboratory findings can be complicated. The microbiologist must be thoroughly familiar with the normal bacterial inhabitants of the body to distinguish suspected disease producers. Many directors of today's clinical microbiology laboratories have a Ph.D. or M.D., or both.

Oddball (Atypical) Bacteria

The foregoing description applies to "typical" bacteria, but there are three important "oddball" groups of **atypical bacteria**, each of which possesses distinctive properties distinguishing it from the typical bacteria and from each other. These groups are significant atypical pathogens in humans. They are mycoplasmas, chlamydias, and rickettsias; their properties are summarized in **TABLE 7.4**.

Mycoplasmas

Mycoplasmas are bacteria lacking cell walls; they are the "smallest of the small." Perhaps you have heard of "walking pneumonia"; the causative agent is *Mycoplasma pneumoniae*. Because they lack cell walls, they are unaffected by common antibiotics that block cell wall synthesis such as penicillin, **cephalosporins**, and **bacitracin**.

Chlamydias

The **chlamydias** are slightly larger than the mycoplasmas and can be seen with a light microscope. These coccoid bacteria are **obligate intracellular parasites**, meaning that they can only grow within a cell, a property shared with rickettsias and viruses. Transmission is from person to person. *Chlamydia* spp. are responsible for a variety of diseases, including urethritis (inflammation of the urethra), trachoma (a common cause of blindness), and lymphogranuloma venereum (a sexually transmitted disease).

© Robert I. Krasner.

FIGURE 7.21 A memorial statue of Anne Frank in Amsterdam. Anne Frank died of typhus fever at the age of 15 in Bergen-Belsen, a German concentration camp. The Anne Frank Museum is in Amsterdam (http://www.annefrank.org/en/).

Rickettsias

Rickettsias are rod-shaped bacteria, the largest of the three groups of atypical bacteria, and can be seen with a light microscope. With the exception of Q fever, diseases caused by these bacteria are acquired through the bites of mosquitoes, ticks, fleas, lice, and other arthropods. Like the chlamydias they are obligate intracellular parasites. They are commonly grown in the yolk sac of (live) chicken embryos for laboratory study.

According to some historians, typhus fever (distinguished from typhoid) defeated Napoleon's army in the Russian invasion of 1812. Typhus fever is caused by bacteria (*Ricksettsia* spp.) that are transmitted to humans through biting fleas, lice, and chiggers. Perhaps you have read *The Diary of Anne Frank*, the true story of a teenage girl who hid from the Nazis with her family in a secret attic between 1942 and 1944. The family was betrayed and sent to Bergen-Belsen, a German concentration camp. Anne died there in 1945 of typhus fever at the age of 15, along with hundreds of others shortly before the camp was liberated by Allied forces (**FIGURE 7.21**).

TABLE 7.4 Comparison of Mycoplasmas, Chlamydias, and Rickettsias

Bacterial Group	Size (micrometers)	Cell Wall	Obligate Intracellular Parasite
Mycoplasmas	0.3–0.8	Absent	No
Chlamydias	0.2–1.5	Present	Yes
Rickettsias	0.8–2.0	Present	Yes

Summary

Bacteria are a distinct group of unicellular microbes that have existed, possibly, since the beginnings of life on Earth. They are small, but they are not simple. Bacteria display a diversity of shapes and patterns of arrangement. Most are rods and cocci. Cocci are subdivided into streptococci, staphylococci, diplococci, and tetrads.

Bacteria have a complex anatomy consisting of an envelope, cytoplasm, and appendages. The envelope includes the capsule, the cell wall, and the cell membrane. The cytoplasm is the region of the cell within the cell membrane and includes a nucleoid, plasmids, and endospores. Appendages are flagella and pili, both of which extrude through the cell wall. The capsule, plasmids, endospores, flagella, and pili are not present in all bacteria.

Bacteria exhibit considerable diversity in generation time—the time it takes a cell to undergo binary fission resulting in two new cells. Under optimal laboratory conditions, the generation time for E. coli is 20 minutes, whereas the generation time for *Treponema pallidum* is 33 hours. The four major phases of the curve are the lag phase, log (exponential) growth phase, stationary phase, and death phase. Bacterial growth curves provide insight into the dynamics of bacterial growth that occur in an infected individual. The fact that most bacteria can be readily cultured in the laboratory facilitates identification of the causative organism, allowing appropriate treatment. Mycoplasmas, rickettsias, and chlamydias are three groups of atypical bacteria, each of which includes pathogens (disease-producing microbes).

KEY TERMS

acid-fast staining
adhesins
agar
anesthetic
animalcules
antibiotic resistance
antibiotic sensitivity testing
antibodies
antigens
API-20E
aseptic technique
atypical bacteria
bacilli
bacitracin
bactericidal
basal bodies
binary fission
binomial system of nomenclature
Black Death
broad-spectrum antibiotics
bubo
bubonic plague
capsule
cell (outer) membrane
cell wall
cephalosporins
chemotaxis
chlamydias
climate change
cocci
colonies
colonization
conjugation

contact tracing
contagion
death phase
differential staining procedure
diffusion
diplococci
endospore
endotoxin
envelope
epicenter
exotoxin
fission ring
flagella
flagellin
fomite
generation time
genus
growth curve
hemolysin
infectious disease
lag phase
leprosy
log phase
lysis
macroscopic
mode of transmission
mordant
morphology
motility
myalgia
mycolic acids
mycoplasmas
nucleoid

obligate intracellular parasites

osmosis

outer membrane

penicillin

peptidoglycan

phagocytosis

pili

pilin

plague

plasmid

pneumonic plague

prophylaxis

pure culture

R (resistance) factors

rickettsias

reservoir

R-nought

ropy milk

selectively permeable

septum

sex pili

species

spirilla

spirochete

staphylococci

stationary phase

streptococci

tetrad

toxins

vegetative cells

vibrio

virulence

zones of inhibition

zoonosis

SELF-EVALUATION

○ **PART I: Choose the single best answer. Questions 1 to 4 are based on the following key:**

　　a. capsule

　　b. cell wall

c. plasma membrane

d. pili

1. Accounts for Gram-stain reaction

2. Responsible for gatekeeper function

3. A deterrent to phagocytosis

4. Important in adhesion to host cells

5. Assume that the organism *Robertus krasnerii* has a generation time of 20 minutes and that you are starting with one cell. How many generations will it take to achieve a population of 2,048 cells?

　　a. Cannot determine

　　b. 10

c. 11

d. 12

6. In examining a Petri dish culture with antibiotic sensitivity disks, the greatest sensitivity would be demonstrated by the disk with the _____.

　　a. most growth in the vicinity of the disk

　　b. smallest zone of no growth

c. largest zone of no growth

d. no zone of inhibition

7. The property of "hardiness" is associated with which of the following?

　　a. flagella

　　b. cell wall

c. capsule

d. endospore

8. Which group of bacteria is unique in lacking a cell wall?

    a. mycobacteria   c. rickettsias

    b. chlamydias   d. mycoplasmas

9. Gram (+) bacteria possess many layers of _____.

    a. starch   c. protein

    b. peptidoglycan   d. sialic acid

10. _____ are small, circular nonchromosomal DNA molecules that replicate independently inside of certain bacteria.

    a. Inclusion bodies   c. Plasmids

    b. Flagella   d. Pili

○ PART II: Fill in the blank.

1. Cocci that tend to form chains are called _____.

2. *Mycobacterium* spp. contain _____ in their cell walls that give them a property of being waxy and impermeable to Gram staining and antibiotic intake.

3. A defense mechanism of the body is the ability of certain immune cells to engulf and destroy bacteria. This process is known as _____.

4. _____ and _____ are examples of endospore-forming genera.

5. _____ is a rapid test used in the doctor's office to detect *Streptococcus pyogenes* antigens.

6. _____ is credited with the application of agar (replacing gelatin) in bacteriological media.

7. A bacterium that has flagella all over its surface is described as _____.

8. _____ is a form of cell division used by all bacteria.

9. It is best to treat a bacterial infection with antibiotics while it is in the _____ phase.

10. _____ are responsible for a variety of diseases, including urethritis, trachoma, and lymphogranuloma.

○ PART III: Answer the following.

1. Draw and label a "composite" Gram (+) bacterial cell. Indicate what structures are not found in all bacterial cells.

2. The term *bugs* is conventionally used to describe all microbes. If you are ill and diagnosed with a "bug," what are the limitations of this diagnosis?

3. Explain what causes deformities in individuals who have not been treated for leprosy.

4. Why is it necessary to use pure cultures of bacteria in the diagnostic laboratory?

5. Endospores are extremely hardy structures. Create a list of at least four environmental conditions or processes aimed to kill vegetative cells that will have no or little effect on bacterial endospores.

6. List the possible direct and indirect modes of transmission for leprosy. Speculate why there has been an increase in leprosy cases in the United States.

7. Draw and label six different bacterial shapes. Include the two shapes of the majority of bacteria and put an asterisk next to them.

8. Explain why a bacterial infection caused by *Mycobacterium leprae* or *M. tuberculosis* requires 1 to 2 years of antibiotic therapy, in contrast to a strep throat infection, which will require 3 to 10 days of antibiotic therapy.

9. Draw a typical bacterial growth curve and label the four major phases.

10. Explain why it is not always possible to culture a suspect microbe from an infected person.

VIRUSES AND PRIONS

We can't predict what a virus we've never seen will do.

—Marc Lipstitch

Professor of Epidemiology, Harvard School of Public Health

Ebola doesn't disappear. It just goes into hiding.

—David Quammen

Science, nature, and travel book author and journalist

Courtesy of CDC/James Gathany.

Centers for Disease Control and Prevention (CDC) intern Maureen Metcalfe is using a transmission electron microscope (TEM) to view a prepared sample of Variola virus (the virus that causes smallpox). An electron microscope uses a beam of electrons that pass through the thin-section of the specimen and onto an imaging surface. A digital camera is used to capture the image. The TEM is capable of revealing details that are tens of thousands of times smaller than can be seen with the highest-quality light microscope. Note that the thin-section preparations are dead/noninfectious; therefore, she does not need to wear gloves to use the TEM to observe the virus particles.

ShutterStock, Inc. / happykanppy.

LEARNING OBJECTIVES

1. Discuss why viruses are not considered to be "alive."

2. Explain how it was determined that viruses are smaller than bacteria.

3. Summarize the differences between viral genomes and the genomes of their cellular hosts.

4. Describe the basic structure of typical icosahedral and helical-shaped viruses and list examples of each.

5. Describe complex-shaped viruses and provide examples.

6. Compare and contrast giruses and virophages.

7. Summarize the steps in viral replication.

8. Explain why a virus requires a metabolically active host cell in order to multiply or replicate.

9. Detail why it is necessary to continue to develop more viral diagnostic tests even though there are very few antivirals available to treat infections caused by viruses.

10. Define *gene therapy*, *phage therapy*, and *virotherapy*, and explain how viruses are utilized in these therapeutic applications.

Case Study: Dr. Crozier's Puzzling Eye Color Change

Even when it's over, it's not over.

—Dr. Dan Bausch, Tulane School of Public Health and Tropical Medicine

The above quotation was spoken by Dan Bausch in reference to survivors and the often lingering complications of Ebola virus infections, or **"post-Ebola syndrome"** (**Figure 1**). An unprecedented Ebola epidemic erupted out of a small village in Guinea, West Africa, at the end of 2013. Five months into the outbreak, **Ebola virus disease (EVD)** spread to Conakry, a city with a population of 2 million people. It was the first time in human history that Ebola virus infections occurred in such a highly populated location. It was an infectious disease expert's nightmare. All prior epidemics occurred in or near remote small villages with no international airport. These EVD fires could be put out rather quickly using contact tracing. **Contact tracing** involves the identification and diagnosis of people who may have come into contact with an infected person. With EVD, contacts are closely monitored for 21 days since their last exposure with a sick individual. The incubation period for EVD is 2 to 21 days.

Healthcare workers from all over the world volunteered to care for EVD patients and to stop the spread of the epidemic in West Africa that raged from the end of December 2013 through the end of March 2016. The final

Courtesy of CDC Epidemiologist Rebecca Hall, M.P.H.

Figure 1 Survivor wall at the Bong Ebola Treatment Unit (ETU) located in Bong County, Liberia. At the time this photograph was taken, the Bong ETU had admitted more than 400 Ebola patients since it had opened in September 2014. Workers were overjoyed that the number of Ebola survivor handprints surpassed the initial space made available. Approximately 17,323 Ebola patients survived the largest Ebola epidemic in human history.

World Health Organization (WHO) situation report dated March 27, 2016, listed 28,646 cases and 11,323 deaths (**Figure 2**). Of the 881 healthcare workers who contracted EVD in the EVD-intensive countries of Guinea, Liberia, and Sierra Leone, 513 died. Healthcare workers were

(continues)

Case Study: Dr. Crozier's Puzzling Eye Color Change
(*continued*)

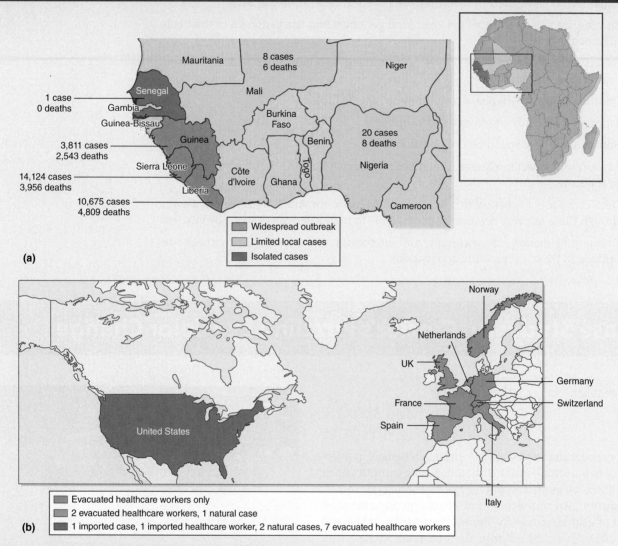

Figure 2 **(a)** Cumulative EVD cases and deaths in Africa during the 2014–2016 epidemic. Liberia, Sierra Leone, and Guinea were the EVD-intensive countries. The epidemic did cross borders but fortunately those cases were isolated and contained quickly. **(b)** The 2014–2016 Ebola epidemic spread outside of Africa. Most of the EVD cases were healthcare volunteers who contracted Ebola in West Africa and were then evacuated to other countries for treatment. Three natural cases occurred outside of Africa. All three cases were nurses tasked with the care of an EVD patient (two nurses in the United States and one nurse in Spain). EVD deaths also occurred outside of West Africa: two in the United States (one imported case and one healthcare evacuee who contracted EVD in West Africa) and two in Spain (two healthcare evacuees who contracted EVD in West Africa).

21 to 32 times more likely to be infected by Ebola virus than the general West African adult population.

Dr. Ian Crozier, an American infectious disease expert, was a volunteer for the WHO. He was infected while caring for Ebola patients in an Ebola Treatment Unit (ETU) in Kenema, Sierra Leone. The 43-year-old doctor was diagnosed on September 6, 2014, and was airlifted to **Emory University Hospital's Serious Communicable Disease Unit** located in Atlanta, Georgia, on September 9. Dr. Crozier underwent a brutal fight with Ebola virus. He was treated with an experimental drug, **TKM-Ebola**. The British nurse

and EVD survivor Will Pooley flew into Atlanta from England to donate blood. Dr. Crozier received a transfusion of several units of Pooley's convalescent plasma. Dr. Crozier's **viral load**, the number of viruses present in his bloodstream, was 100 times that of other EVD patients treated by the healthcare staff who had specialized training to care for patients in the Serious Communicable Disease Unit at Emory.

In the week following Dr. Crozier's evacuation to the United States, his condition deteriorated rapidly. He developed severe gastroenteritis, losing 8 to 10 liters (8.5 to 10.5 quarts) of diarrhea a day. Large amounts of intravenous fluid replacements were needed to keep his electrolytes and fluid volumes balanced. Symptoms of viral **encephalopathy** began. He was delirious and confused. Soon, there was an onset of acute severe respiratory failure. Dr. Crozier was intubated and placed on a mechanical ventilator for the next 12 days. His kidneys were failing, requiring the need for 24 days of dialysis. More symptoms emerged, such as highly elevated liver enzyme levels, muscle damage, and atrial fibrillation of the heart (which is an "old man's disease"). He was unconscious during this time period.

After 40 days of treatment at Emory, his ventilator was removed. His mental status was affected. He was unresponsive for a period of time and had prolonged **disconjugate gaze** (i.e., both eyes are not fixed on the same point, which causes double vision). Doctors feared that he would die or suffer from brain damage. After regaining consciousness, he was extremely weak and 14 kilograms (30 pounds) lighter. His hair had fallen out, and his muscles were atrophied, affecting his balance.

After 44 days, Dr. Crozier was a survivor. His blood tested negative for Ebola viruses; however, Ebola viruses were isolated from his semen. It was not uncommon for Ebola survivors to shed virus in their semen after infection. Advised to abstain from sex or to use condoms for at least 3 months, Dr. Crozier was discharged from the hospital and he returned home to be with his family in Phoenix.

As he began the recovery process, like so many other survivors, Dr. Crozier had symptoms referred to as post–Ebola syndrome. His short-term memory was affected, and he had difficulty finding the right words when he spoke. He struggled with exercise fatigue. Many of his symptoms were rheumatoid in nature, similar to an autoimmune disease that attacks the joints.

Less than 2 months after he was discharged, his eyes began to bother him. His eyes were red and inflamed,

sensitive to light with a burning sensation, and his near vision was fading in his left eye. After visiting ophthalmologist Steven Yeh back in Atlanta, it was discovered that his eye pressure was extremely high. Dr. Crozier was diagnosed with **uveitis** (inflammation that affects middle tissue in the eye wall, or uvea). His eyesight continued to decline. Dr. Yeh was concerned that because of Dr. Crozier's intense battle with EVD he was immunocompromised, which could make him vulnerable to eye infections caused by cytomegaloviruses or other herpesviruses. A sample of eye fluid was withdrawn from his left eye using a needle. The sample from his left eye contained twice as many Ebola viruses than he had had in his bloodstream when he was initially treated for EVD in Emory University's Serious Communicable Disease Unit. Because his blood and tears remained negative for Ebola virus, there was no risk to anyone who had casual contact with him.

Dr. Crozier was treated with high-dose steroids to reduce the inflammation, but the vision in his left eye continued to worsen from 20/40 to 20/100 to 20/2000 eyesight, and then to hand motions only (essentially he could see nothing). His eye pressure dropped to nothing. Oddly, the color of his left eye changed from blue to green (**Figure 3**). A patch was placed over his left eye and he was admitted to the hospital and placed on an antiviral and additional steroid treatment. Slowly, his eyesight came back. Eye complications represent an example of just one of the types of complications that some of the approximately 17,340 EVD survivors have experienced. An interview with Dr. Crozier in which he shares his poignant memories as a healthcare volunteer in Kenema and his personal battle with EVD is available at the following URL: https://www.youtube.com/watch?v=xi29GJZllts.

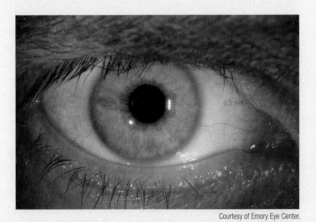 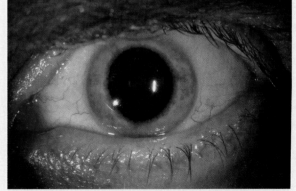

Courtesy of Emory Eye Center.

Figure 3 Photograph of Dr. Crozier's healthy eye (left photo) with uveitis (right photo). Uveitis was a complication caused by Ebola virus infection that he experienced after being discharged from Emory University's Serious Communicable Disease Unit. He was treated there for EVD after contracting the illness in Sierra Leone in 2014. Ebola virus was detected in Dr. Crozier's eye. The eye infection was so severe that it caused his eye iris color to change from blue to green (right photograph). Uveitis was reported as a complication experienced by survivors of this epidemic as well as previous Ebola outbreaks.

(continues)

Case Study: Dr. Crozier's Puzzling Eye Color Change
(*continued*)

Questions and Activities

1. All healthcare workers tasked with caring for Ebola patients must wear **personal protective equipment (PPE)**. Define PPE.

2. List at least five other situations in which a healthcare worker would be required to wear PPE.

3. What are the **modes of transmission** for Ebola virus?

4. Can an Ebola survivor be an infectious carrier of Ebola viruses? Explain.

5. Watch Dr. Crozier's interview. How was he initially infected by Ebola virus?

6. What two experimental treatments did Dr. Crozier receive?

7. Could a vaccine prevent Ebola virus infection? Why or why not?

8. List complications of EVD experienced by survivors. (*Hint*: Watch Dr. Crozier's interview and visit

https://www.cdc.gov/vhf/ebola/healthcare-us /evaluating-patients/messages-for-care-of-survivors -ebola-.html.)

9. Since this epidemic, at least one additional Ebola virus epidemic had occurred at the time of this writing. Where did this Ebola outbreak occur? How many cases were identified? How many deaths? What was the mortality rate?

10. The **R-nought** of the Guinea strain of Ebola virus in West Africa was 2.0. Calculate the R-nought for Ebola virus using the statistics for the U.S. cases (11 cases total, 2 deaths, 2 nurses infected after caring for imported case/Thomas Eric Duncan).

11. Dr. Crozier had more Ebola viruses in his body than any of the other patients treated at the Emory Ebola unit. He survived after a very challenging battle against infection. Why did he survive? (Review his interview).

Information from Varkey, J. B., et al., 2015. "Persistence of Ebola virus in Ocular Fluid During Convalescence." *New England Journal of Medicine 372*: 2423–2469.

Preview

Viruses are subcellular infectious agents responsible for a variety of **infectious diseases** in humans. Zika virus, Ebola virus, measles, and influenza are but a few examples of viruses that cause disease. It was not until the 1930s that viruses were described in terms of their biological and chemical properties. All viruses require a metabolically active **host cell** in order to replicate. Viruses are unique in that they have either DNA or RNA genomes, whereas cells contain only DNA genomes. The nucleic acid in viruses is wrapped in a protein coat, or **capsid**; some have an **envelope** around the coat; others have protruding spikelike structures. A generalized cycle of viral replication follows the sequence of adsorption, penetration, replication, assembly, and release, but the strategies differ depending on the specific virus. Because of their host cell requirement for replication, cell culture and fertile chicken eggs are used to cultivate viruses in the laboratory. The diagnosis of a specific viral disease is usually based on the patient's symptoms rather than on isolation of the virus.

Bacteriophages are viruses that infect bacterial cells; these viruses have served as models toward understanding the dynamics of virus–host cell relationships that occur in human viral diseases. The use of bacteriophages to treat bacterial infections, called **bacteriophage therapy**, has reemerged as a possible therapy of great importance given the increase in antibiotic-resistant bacteria. **Gene therapy** and **virotherapy** as cancer-fighting treatments are moving forward quickly into medical practice.

Prions are infectious proteins capable of causing fatal neurodegenerative diseases in animals and humans. They are subcellular agents unique in that they lack nucleic acids.

Viruses as Infectious Agents

What biological agents are smaller than a cell, not considered to be alive, able to be transmitted from person to person, potentially fatal, and have caused the largest recorded outbreak of disease on a worldwide scale (pandemic) in the history of humankind? A key word in the title of the chapter is a giveaway to the answer to this riddle: *viruses*. The pandemic referred to is the influenza (Spanish flu) **pandemic** of 1918, which killed 50 million people around the globe in only 1 year. HIV/AIDS, too, is a pandemic viral disease that began in the latter part of the 20th century and has had a devastating impact worldwide, but particularly in Africa. According to the director of the United Nations Children's Fund, "*by any measure the HIV-AIDS pandemic is the most terrible undeclared war in the world with the whole of sub-Saharan Africa a killing field.*"

Viruses are **subcellular** (simpler than a cell), and therefore can replicate (their sole claim to life) only when circumstances result in their gaining access to a metabolically active host cell. They have no life outside the cell and exist only as "freeloaders." The host cell is taken prisoner and turned into a virus-producing machine. They literally need to "get a life" from a host cell. On a positive note, the fact that viruses take over a cell is being exploited through **genetic engineering** and gene therapy by using "tamed" or **attenuated viruses** as vehicles for the delivery of "good" genes into cells to replace defective or nonexistent genes and as virotherapy to treat aggressive cancers.

Everyone has experienced numerous bouts of common viral infections, including colds, influenza, and gastroenteritis. Human immunodeficiency virus (HIV), Ebola virus, measles virus, and West Nile virus cause diseases that have been widely publicized. In 1997, a potential outbreak of avian influenza in Hong Kong, dubbed by the media as the "bird flu virus," caused a scare around the world. Subsequently, between 2003 and 2011 the WHO documented 566 cases with 332 deaths—a staggering mortality rate of 59%. Millions of chickens and other birds were slaughtered in an attempt to halt the spread.

Fortunately, that scare proved to be unwarranted, but other potential viral threats have reemerged. In 2015, an outbreak in Brazil caused by Zika virus unraveled a troubling link between Zika virus infection and birth defects. Infants were being born with a rare brain malformation known as **microcephaly**. Children with microcephaly have brains much smaller compared to the brain size of babies of the same sex and age. Zika viruses are transmitted by day-biting mosquitoes (*Aedes* spp.). Only one in five people infected with Zika virus exhibit symptoms. Expectant mothers became anxious as Brazilian authorities urged women not to get pregnant. Zika virus, and the resulting Zika virus disease (ZVD), has spread to Texas and Florida. The Centers for Disease Control and Prevention (CDC) is responsible for making the public aware of many diseases, including ZVD (**FIGURE 8.1**).

Ebola virus disease (EVD) garnered front-page headlines when it broke out in the Democratic Republic of the Congo in 1996, and reemerged in September 2007, causing 187 deaths out of 264 cases. It returned again in 2008 and 2011 in Uganda and the end of 2013 in Guinea, West Africa (see Case Study). In 2011, cases of mumps and measles reemerged in Canada, the United States, the United Kingdom, the European Union (i.e., Germany, Italy, Romania, and Greece), Australia, and New Zealand, spreading by air travel. In most of the aforementioned locations these were the first mumps or measles cases in 20 or more years. These examples attest to the occurrence of both **emerging** (new) and **reemerging infectious diseases**.

Smallpox is a terrible viral disease that once decimated large segments of the population. The disease was described as early as about 10,000 B.C. Examination of the body of Pharaoh Ramses V, who died in 1157 B.C., revealed pustule-like scars on the face, neck, and shoulders, which appeared to have been caused by Variola (the virus that causes smallpox). Shamefully, Variola virus was used in biological warfare long before its identity was known. Some historians claim that in the 1760s English troops, under the guise of appeasement, gave the Native Americans blankets from a smallpox hospital in hopes the Native Americans would contract the disease. The Native American population had not been previously exposed to Variola virus (the causative agent of smallpox) and therefore all Native Americans lacked immunity and were susceptible to contracting the disease. Smallpox devastated Indian Nations, wiping out 20-50% of the tribes.

Smallpox was the first disease for which **immunization** was available, thanks to Lady Mary Wortley Montagu's

Courtesy of the CDC.

FIGURE 8.1 CDC infographic that provides "need to know" information about Zika virus infections.

promotion of a procedure to prevent smallpox known as **variolation** during the 1720s and the pioneering work of vaccination by Edward Jenner in 1796. Smallpox is the first and, to date, the only infectious disease to be eradicated from the face of Earth. The word *eradicated* is to be emphasized as compared with the word *controlled*. The irony is that this triumph renders the world susceptible to the threat of smallpox because the virus still exists in a couple of government laboratories and could find its way into the hands of terrorists. **Vaccination** has ceased. Now, essentially almost all humans on Earth are susceptible to Variola virus infection if the virus were unleashed by a **bioterrorist** into the general population.

The term *virus* can be found in the mid-19th-century writings of Pasteur and others who preceded him long before the **Germ Theory of Disease** was established and the biology of viruses was known. *Virus* means "poison" or "slime" and has a negative connotation. By the 1940s viruses had been well described, and the term now has a biological basis. In more recent years, *virus* has become part of computer jargon and indicates malicious computer code that is easily spread (infectious) and destroys or steals data.

Based on the work of the Russian scientist Dmitri Ivanovsky in 1892, viruses were described as "filterable." Ivanovsky sifted extracts of tobacco leaves with symptoms of tobacco mosaic disease through filters that were known to retain bacteria. He then injected these extracts into healthy tobacco plants and observed symptoms of tobacco mosaic disease (**FIGURE 8.2**). It was clear that the infectious agent was, indeed, smaller than bacteria. Several years later, the Dutch microbiologist Martinus

Willem Beijerinck demonstrated that tobacco mosaic disease was caused by a filterable agent. The term *filterable* was eventually dropped. A breakthrough occurred in 1935 when Wendell M. Stanley, an American biochemist, purified and crystallized tobacco mosaic virus, opening the door for more sophisticated examination of a purified virus in an effort to further elucidate its structure. The invention of the **transmission electron microscope (TEM)** at about the same time allowed viruses to be observed.

All metabolically active cells are potential hosts for viruses. Even *Escherichia coli* and other bacteria serve as hosts. Bacterial viruses are known as *bacteriophages* (usually shortened to "phages"). Phages, because of the ease of growing the bacteria that serve as their hosts, have served as models for studying animal and plant viruses. Our current knowledge of viruses has been largely gained by studying bacterial viruses.

Virus Structure

Among infectious agents that infect humans and other mammals, viruses are the second smallest; prions are the smallest. Bacteria are described as **microscopic**, meaning they can be seen only with a **light microscope**. Viruses, however, are in a different realm, the realm of the TEM, and they typically range in size from about 20 to 350 nm (0.020 to 0.35 µm). Well over 1,000 phage particles can fit inside of an *E. coli* bacterium that is about 0.2 µm wide by 1.0 µm in length. It is important to compare the sizes of microbes and viruses to their host cells (**FIGURE 8.3**).

The 2003 discovery of **Mimiviruses** living inside of amoebas in a cooling tower changed the definition of virus size and size of a viral genome. Mimivirus was not only larger than previously seen viruses, but its genome was similar in size to atypical bacteria, such as mycoplasmas, chlamydias, and rickettsias. In 2008, a nearly identical but larger virus named Mamavirus was described, followed by the 2011 discovery of the largest viral relative named Megavirus (**BOX 8.1**). If there is a Mamavirus, will there be a Papavirus? What will the next enormous virus be called—Gigavirus or a Teravirus?

These larger viruses are now referred to as **giruses**. Giruses have been discovered in samples collected from cooling towers (ventilation and air conditioning systems), seawater, stagnant water, rivers, lakes, ponds, mountain and hypersaline soils, sewage treatment systems, hospital environments, stool samples from healthy and sick individuals, the internal and digestive organs of medicinal leeches, and as a contaminant of contact lens cleaning solutions. Giruses represent the exception to the traditional size criteria. This text will focus on the characteristics

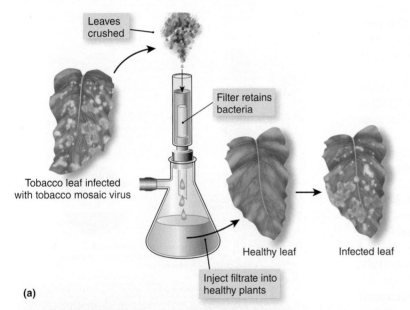

Leaves crushed

Filter retains bacteria

Tobacco leaf infected with tobacco mosaic virus

Healthy leaf   Infected leaf

Inject filtrate into healthy plants

(a)

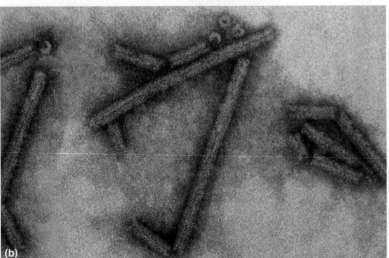

(b)

© Phototake/Alamy Images.

FIGURE 8.2 (a) Discovery of "filterable" viruses. Viruses pass through the pores of the filter with the liquid filtrate into the bottom of the flask. **(b)** A transmission electron micrograph of tobacco mosaic virus.

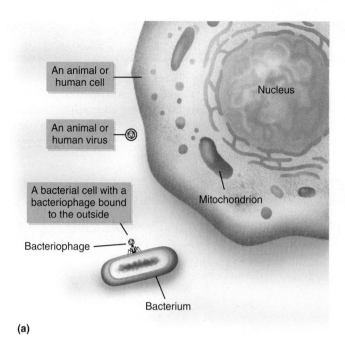

(a)

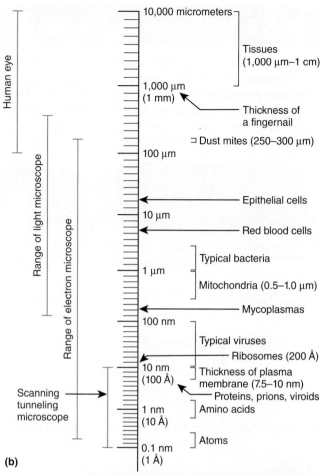

(b)

Information from Prescott, L. M., Harley, J. P., and Klein, D. A. Microbiology, Sixth Edition.
McGraw-Hill Higher Education, 2005.

FIGURE 8.3 **(a)** Viruses are smaller than their host cells. This illustration demonstrates size comparisons of an animal or human eucaryotic cell with its organelles compared to a virus and a bacterium compared to a bacteriophage. **(b)** The limits of microscopic and human eye resolution. Each major division represents a 10-fold change in scale. To the right of the scale are approximate sizes of molecules, cells, organelles, and tissues.

BOX 8.1 Big and Bizarre Viruses

A pneumonia outbreak occurred in Bradford, England, in 1992, caused by what was thought to be bacteria isolated from a cooling tower. Its identification as a bacterium was based on its large size and the fact that it stained Gram (+). The organism was named Bradford coccus (**Figure 1**). The researchers were all specialists in the study of oddball (atypical) bacteria, but they were not aware that they had stumbled upon a microbe that would change the classic definition of a virus.

As a part of their studies they were using water samples from a cooling tower in an effort to co-culture amoebas with Bradford cocci. They are symbiotic organisms that failed to culture separately. The investigators revealed that the Bradford cocci lived within the amoebas. Attempts to find and sequence a particular genetic marker, **16S rRNA gene**, used as a tool to help in distinguishing bacteria and viruses were fruitless. However, observing the coccus through a TEM, it

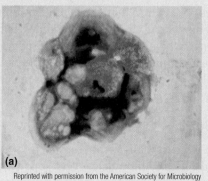

(a)

(b)

Figure 1 **(a)** Gram stain of Mimiviruses inside of an amoeba. **(b)** Transmission electron micrograph of Mimivirus. Note the dense hairlike projections (fibrils) covering the capsid.

(continues)

BOX 8.1 Big and Bizarre Viruses (continued)

looked like a hairy, large insect virus known as an *iridovirus*. The Bradford coccus today is known as a Mimivirus—short for "mimicking microbe." Similar hairy viruses, Mamavirus from a Paris water-cooling tower and Megavirus isolated from seawater at a marine station at Las Cruces, Chile, also have been identified.

Another giant virus was discovered in 2010, in the contact lens solution used by a 17-year-old French woman suffering from an aggressive eye infection known as **keratitis**. Doctors determined that the woman's eye was contaminated with two different types of bacteria, as well as an amoeba carrying a new giant virus named Lentille (Table 1).

TABLE 1 Size Ranges of Big and Bizarre Viruses

Name	Source	Isolated	Capsid Size ("hairs" or fibrils not included)
Shanvirus	Stool sample from a 17-year-old female in Tunisia with pneumonia who did not respond to antibiotics	Africa (Tunisia)	640 nm
Megavirus	Seawater	Chile	520 nm
Lentille	Contact lens solution of patient with keratitis	France	500 nm
Mamavirus	Cooling tower	France	450 nm
Longchamps	Decorative fountain	France	450 nm
Terra1	Soil	France	420 nm
Hirudovirus	Internal organs of *Hirudo medicinalis* leeches from a small stream	Africa (Tunisia)	410 nm
Mimivirus	Cooling tower	England	400 nm
Bus	Cooling tower	France	400 nm
Pointerouge1	Seawater	France	390 nm

and diseases caused by the vast majority of traditional viruses. Interestingly, a subviral agent was recently discovered *inside* of Mimivirus that was named a **virophage**, which means "virus eater" from the Latin *virus* and Greek *phagin* ("to eat"). It is essentially a virus that infects a girus!

As viewed through a TEM, viruses are simple in structure and may consist only of a nucleic acid genome that is wrapped in a protein coat, or capsid, whereas others may have an envelope around the capsid with additional protruding "spikes." Whereas many bacteria are typically rod or coccus shaped, the majority of viruses are icosahedral or helical (Figure 8.4a). Exceptions include the complex shape of larger viruses, such as Variola (smallpox) virus or Mimivirus, that overlap with the size of the smaller bacteria. A complete infectious viral particle is referred to as a **virion**.

Nucleic Acids

Viruses are unique in that they have either RNA or DNA **genomes**, but never both. Cells, whether procaryotic or eucaryotic, contain both DNA and RNA, but the cellular genome is only composed of DNA. In the RNA viruses, the RNA serves as the repository for genetic information. For example, the influenza A virus is an RNA virus that replicates

to produce more influenza A viruses. Hence, the genetic information must be encoded in the RNA. Further, some viruses are unique in that their genome is single-stranded DNA (ssDNA); others have double-stranded RNA (dsRNA). Other than in these viruses, DNA is double stranded and RNA is single stranded. Hence, in terms of nucleic acid content, four categories of viruses are possible: dsDNA, ssDNA, dsRNA, and ssRNA. This fact is important for the classification of viruses (**TABLE 8.1**), along with other morphological (shape) properties. Small viruses have genomes that contain only a few genes, whereas larger viruses have viral genomes that contain a few hundred genes. *E. coli*, by comparison, has approximately 4,000 genes, and human cells are estimated to have about 25,000 genes, possibly more.

Protein Coat

The protein coat is called the *capsid*, and the term **nucleocapsid** refers to the nucleic acid genome plus the protein coat. The capsid, in turn, consists of protein units called **capsomeres**. The three arrangements of these capsomeres result in the structural shape or morphology of a virus particle: **helical**, **icosahedral**, or **complex** (**FIGURE 8.4**).

TABLE 8.1 Type of Nucleic Acid Genome

dsDNA genome	ssDNA genome	dsRNA genome	ssRNA genome
Human papillomavirus (genital warts and cervical cancer) Epstein-Barr virus (mononucleosis) Adenoviruses (common cold) Herpes simplex virus (lip and genital herpes) Varicella-zoster virus (chickenpox and shingles) Variola virus (smallpox)	Parvovirus B19 (possibly slapped-cheek disease)	Rotavirus (gastroenteritis) Reovirus (may trigger celiac disease)	Hepatitis A virus (hepatitis) Poliovirus (poliomyelitis) Norovirus (gastroenteritis) Ebola virus (Ebola virus disease) Influenza A virus (influenza) Zika virus (Zika virus disease) Measles virus (measles) Heartland virus (tickborne illness) Powassan virus (encephalitis)

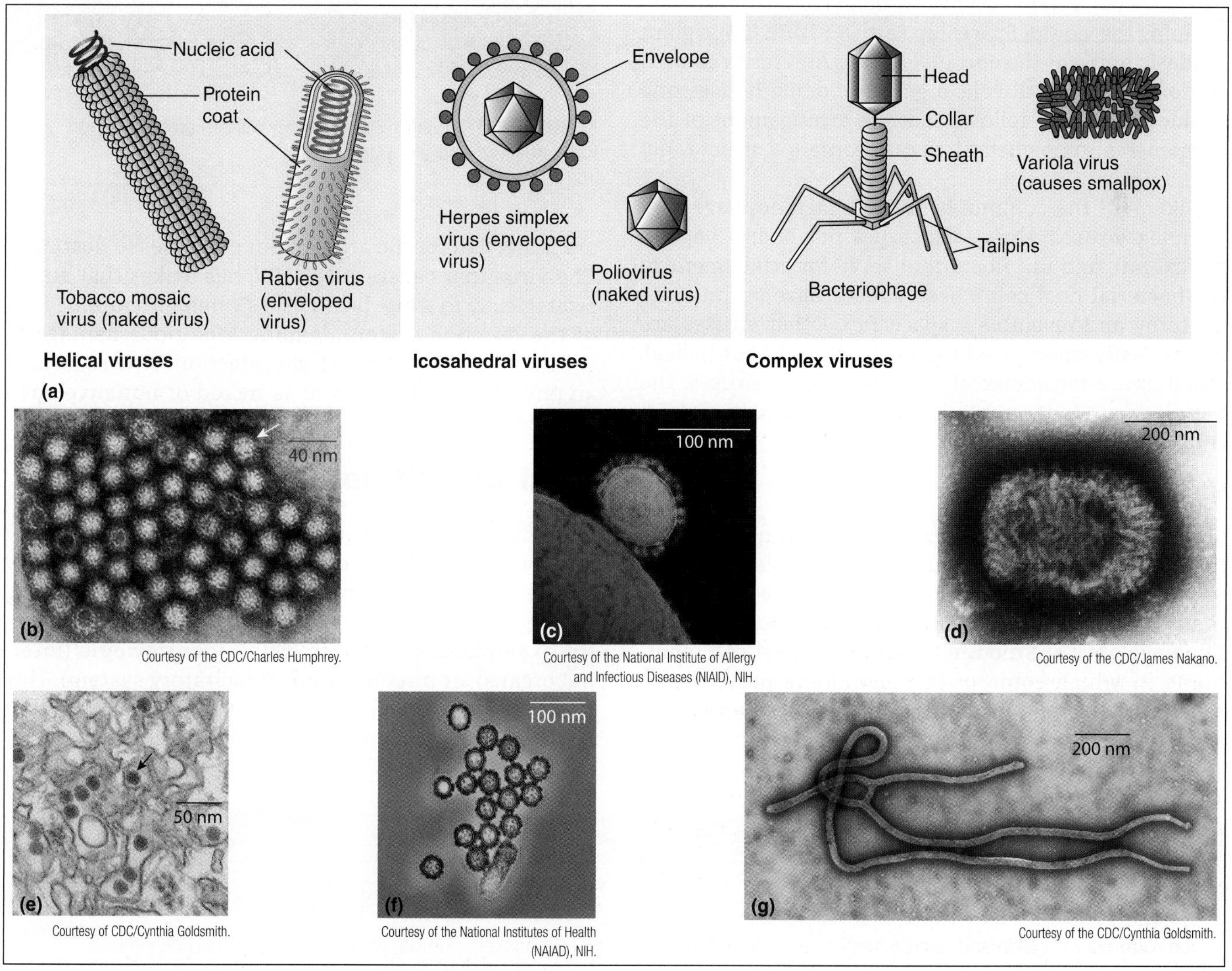

Courtesy of the CDC/Charles Humphrey.

Courtesy of the National Institute of Allergy and Infectious Diseases (NIAID), NIH.

Courtesy of the CDC/James Nakano.

Courtesy of CDC/Cynthia Goldsmith.

Courtesy of the National Institutes of Health (NAIAD), NIH.

Courtesy of the CDC/Cynthia Goldsmith.

FIGURE 8.4 **(a)** Basic virus structure. **(b)** Digitally colorized transmission electron micrograph of a cluster of noroviruses. The arrow points to a single virion. Noroviruses are icosahedral-shaped, naked viruses that are about 40 nm in diameter. Noroviruses are associated with gastrointestinal outbreaks on cruise ships, schools, and other crowded places. Outbreaks usually occur in winter. Hence, the term "winter vomiting disease" is often used in reference to norovirus outbreaks. **(c)** Highly magnified digitally colorized electron micrograph Middle East respiratory syndrome coronavirus (MERS-CoV). This image shows the ultrastructural details of a single virus. MERS-CoV is an enveloped icosahedral-shaped (spherical) virus with a diameter of approximately 100 nm. The virus was identified in 2012 in Saudi Arabia as the cause of severe respiratory illness in humans. Half of those infected died. **(d)** Highly magnified (310,000×) transmission electron micrograph of a negatively stained Variola virus particle. Variola virus is an enveloped, complex-shaped virus. It is sometimes referred to as being brick shaped. **(e)** Digitally colorized transmission electron micrograph of Zika virus (blue virions). Zika viruses are enveloped, icosahedral/spherical viruses that have a diameter of approximately 50 nm. Zika viruses are transmitted to people by mosquito bites. **(f)** Digitally colorized transmission electron micrograph of influenza A viruses (H1N1). Influenza A viruses are enveloped and icosahedral or **pleomorphic** in shape. The surface of the virions is depicted in black. **(g)** Digitally colorized transmission electron micrograph of Ebola virus. Ebola viruses are enveloped and helical, with a diameter of 80 nm and a length up to 1,200 nm.

Helical viruses consist of a series of rod-shaped capsomeres that during assembly form a continuous helical tube containing the nucleic acid characteristic of the particular virus. Tobacco mosaic virus is an example of a helical virus. Imagine eating a turkey wrap; the shell is like the protein coat, and the turkey is the nucleic acid.

In most polyhedral viruses, the capsomeres form icosahedrons—three-dimensional, 20-sided triangular structures that confer a geodesic-appearing shape to the virus. These icosahedral viruses, based on their appearance when visualized through a TEM, also are described as spherical viruses. Triangles, structurally stable geometric configurations in which the stress is evenly distributed, are used by engineers to create geodesic structures. On a trip to Epcot Center at Disney World in Florida, there is no escaping the view of Spaceship Earth (**FIGURE 8.5**), a giant geodesic dome at the entrance. At the Antarctic research station at the South Pole, a geodesic dome houses one of the facilities. It follows that the arrangement of the capsomeres in polyhedral viruses confers a structurally sound design.

Some of the commonly studied bacteriophages are complex viruses; they consist of a polyhedral head, a helical tail, and tail fibers that serve for attachment to the bacterial host cell. These viruses have an intricate anatomy and resemble a spacecraft. Other viruses are not as easily categorized as either polyhedral or helical, and thus are considered atypical or complex viruses. The Variola virus that causes smallpox is an example of a complex virus.

Viral Envelopes

Some viruses, during the process of exiting from the host cell, acquire a piece of the host cell's plasma membrane, which constitutes an additional layer covering the capsid. These viruses are called **enveloped viruses**. The envelopes are modifications of the host cell membrane in which some of the membrane proteins are replaced with viral proteins. Some of these appear as spikes because they protrude from the membrane and

© Songquan Deng/123RF.

FIGURE 8.5 Spaceship Earth at Disney World's Epcot Center is an example of an icosahedral geodesic dome.

are important for the attachment of viruses to host cells. The virus that causes AIDS, HIV, has spikes that attach strategically to white blood cells (**T helper lymphocytes**) of the immune system, leading to serious damage to the individual's ability to fight infection. Viruses lacking an envelope are referred to as **naked** or **nonenveloped viruses** (Figure 8.4b).

Viral Classification

Superficially, viruses are classified as plant, animal, or bacterial viruses in accordance with the host organism they infect. Clinically, they may also be classified according to the organ or organ system involved (**TABLE 8.2**), for example, **dermotropic** (skin), **viscerotropic** (internal organs), or **pneumotropic** (respiratory system). This latter classification, however, is limited by the fact that some viruses can infect more than a single organ or

TABLE 8.2 Clinical Classification of Viruses		
Category	Tropism (Tissue Affinity)	Infectious Diseases
Dermotropic	Skin and subcutaneous tissues	Chickenpox, shingles, measles, mumps, smallpox, rubella, herpes simplex, genital warts and cervical cancer (caused by papillomaviruses)
Neurotropic	Brain and central nervous system tissues	Rabies, Zika virus disease, La Crosse encephalitis, West Nile encephalitis, poliomyelitis, Powassan virus disease
Viscerotropic	Internal organs	Yellow fever, HIV/AIDS, hepatitis A and B, infectious mononucleosis, dengue fever, gastroenteritis (e.g., norovirus)
Pneumotropic	Lungs and other respiratory structures	Influenza, common cold, respiratory syncytial disease, severe acute respiratory syndrome (SARS), Middle East respiratory syndrome (MERS)

organ system. Current classification schemes are based on the biology of the virus, starting with its nucleic acid content. Despite the fact that viral **taxonomy** (classification) is complex, a workable scheme is based on the type of nucleic acid (ssDNA, dsDNA, ssRNA, or dsRNA), capsid structure (helical or icosahedral), number of capsomeres, and other chemical and physical properties. Over 4,400 viral species have been described by the **International Committee on Taxonomy of Viruses (ICTV)** and each has been assigned to 1 of 122 recognized families (https://talk.ictvonline.org/). Who knows how many viruses are out there?

Viral Replication

The fact that viruses require a host cell to replicate makes them difficult to cultivate and study in the laboratory. Although it is easy to grow bacteria in lifeless medium in test tubes or on Petri plates, the culture (growth and replication) of viruses requires the presence of living or metabolically active host cells; these techniques are described later in this chapter. This is particularly complicated in the case of animal and plant viruses but, nevertheless, is routinely done in laboratories specialized in these techniques. Bacteriophages, on the other hand, are much easier to study because their host cells are bacteria, which can be readily grown.

The system that has been most extensively investigated and that has served as a model for animal viruses is the T4 bacteriophage (a complex dsDNA phage) and *E. coli*. The T4 phages enter the *E. coli* cell and use the bacterium's metabolic machinery to produce a large number of new T4 phages, which are released by causing the **lysis** of their bacterial hosts. This is a **lytic cycle** of phage replication. The new virions are then capable of infecting other *E. coli* bacteria (**FIGURE 8.6**).

Bacteriophage T4 can also infect their *E. coli* host cells and the T4 phage genome is integrated into the genome of the bacterial cell. The integrated T4 phage genome is called a **prophage**. It replicates as a prophage with the cellular chromosomal DNA. No infectious virions are produced. This is called a **lysogenic cycle** (Figure 8.6).

The replication of T4 in *E. coli* is a model for the replication of viruses in general, although there is considerable variation in the strategies demonstrated by specific viruses. The replication cycle is divided into five stages (**TABLE 8.3**): **adsorption**, **penetration**, **replication**, **assembly**, and **release**. Depending on the particular virus, the time required to complete the replication cycle varies. For example, the length of the cycle for poliovirus is only 6 hours, whereas it is 36 hours for a herpesvirus. The replication cycle is of more than academic interest. This process occurs when you have a viral infection.

Adsorption

As an obvious prelude to replication, the virus must penetrate the host cell, a process that first requires adsorption, or attachment onto the host cell surface. Bacterial cells have **receptors** on their cell walls that serve as sites of attachment. Human or animal cells do not contain cell walls. Instead, the receptor molecules are embedded within their cellular plasma membranes. Receptors are involved in normal cellular activities. Viruses just happened to be able to attach to certain receptors. Viruses do not seek out a particular host cell; it is a matter of chance encounter. To view

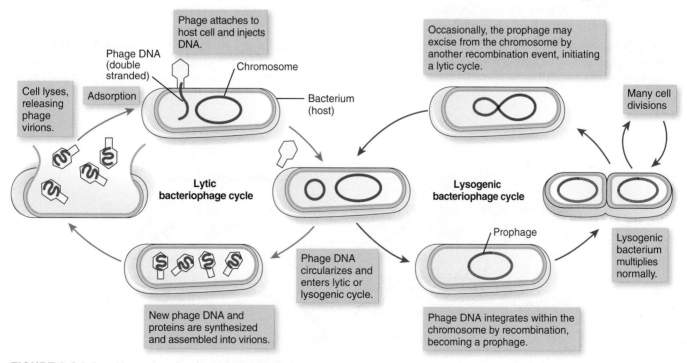

FIGURE 8.6 Lytic and lysogenic cycles of bacteriophage replication.

TABLE 8.3 Generalized Viral Replication Cycle*

Stage	Description
Adsorption (attachment)	Viruses attach to cell surface receptor molecules by spikes or fibrils, capsids, or envelope.
Penetration	Entire viral particle or only nucleic acid genome enters via endocytosis or by fusion with the cellular plasma membrane.
Replication	Process is complex, and details depend on the particular virus and its genome structure. Replication of the viral genome may occur in the nucleus or cytoplasm. Once replicated, genes are expressed, leading to production of viral components.
Assembly	Components are assembled into mature viruses.
Release (exit)	Viruses are extruded from host cell by budding (e.g., HIV) or lysis of host cell plasma membrane.

*Specific strategies vary with particular viruses.

it another way, bacteriophages must "dock" with a receptor molecule onto which it fits (**FIGURE 8.7**).

Consider a piece from a jigsaw puzzle; it can only fit into an appropriate complementary shape. Specificity is thereby required and establishes the **host range**. For example, E. coli strain B serves as the host for a particular bacteriophage, called T4, whereas E. coli strain C serves as the host for bacteriophage ΦX174. Phage T4 does not infect strain C, nor does bacteriophage ΦX174 infect strain B, despite the fact that these two E. coli strains are virtually identical. The strains differ in their receptor molecules, which are specific for the tail fibers of either phage T4 or phage ΦX174. Hence, the host for bacteriophage T4 is E. coli strain B, and the host for bacteriophage ΦX174 is E. coli strain C. Specificity is always the name of the game to some degree, but it may not always be as fussy as in the E. coli–bacteriophage system.

In animal and in human viruses, too, specificity establishes the host range. HIV has spikes on its surface that dock with the human CD4 receptor and a chemokine **coreceptor** present on the surface of a particular type of white blood cell (i.e., T helper lymphocyte), allowing, ultimately, the replication cycle of the virus to be completed. During this process, the T helper lymphocytes are destroyed, rendering the immune system severely impaired. The **hemagglutinin** spikes located on the surface of influenza A viruses bind to **sialic acid** receptor sites present on cells of the respiratory tract, including the lungs.

Penetration

At this point the virus is positioned on the surface of the host cell. This, in itself, is of little consequence because viruses must get inside of their hosts. Therefore, one way or another, penetration of the host cell is necessary for translation of viral proteins and replication of

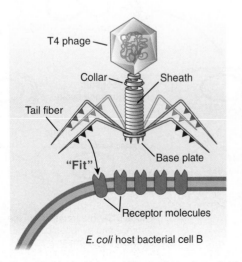
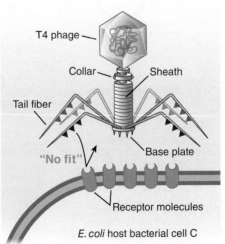

FIGURE 8.7 Bacteriophage "docking" with bacterial host cell receptor molecules.

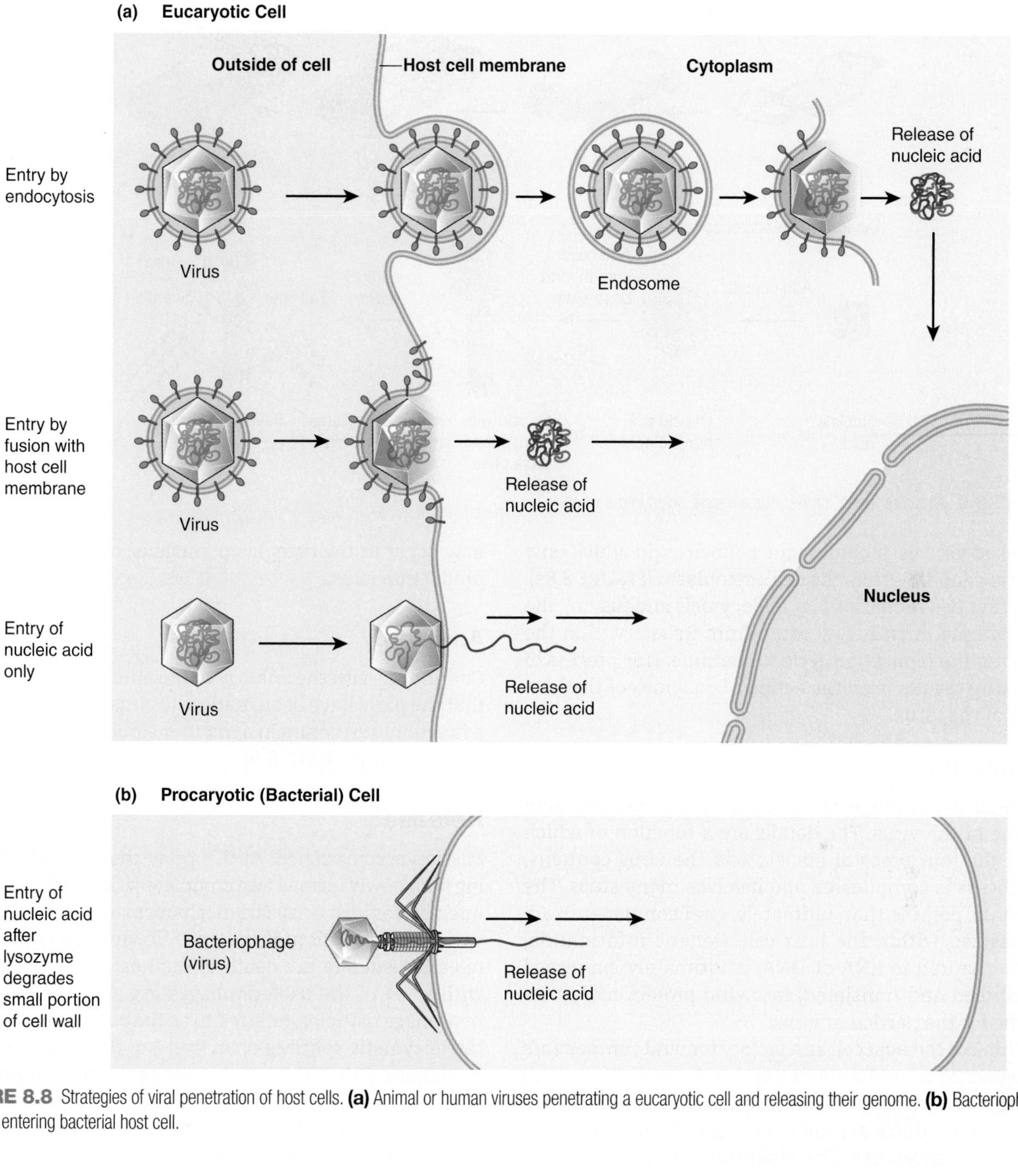

(a) Eucaryotic Cell

(b) Procaryotic (Bacterial) Cell

FIGURE 8.8 Strategies of viral penetration of host cells. **(a)** Animal or human viruses penetrating a eucaryotic cell and releasing their genome. **(b)** Bacteriophage genome entering bacterial host cell.

viral genetic material. Keeping in mind that it is the viral nucleic acid genome, whether ssDNA, dsDNA, ssRNA, or dsRNA, that carries the genetic message, the critical factor is that the viral nucleic acid genome must enter the host cell (**FIGURE 8.8**).

Perhaps the most intriguing mechanism of penetration is observed in some of the bacteriophages, including the T4 phage, of *E. coli* strain B. The T4 phage initially attaches to the bacterial cell wall by phage tail fibers and docks its base plate on the cell wall of its bacterial host. The base plate contains the enzyme lysozyme, which degrades a portion of the **peptidoglycan** of the cell wall. The phage DNA genome is transferred across the cell wall

into the bacterium by an unknown mechanism. The capsid remains on the outside of the bacterium and is of no further consequence.

In the case of animal or human viruses, the genome is not injected into the host cell. Rather, some enveloped viruses enter the cell by a process called **endocytosis** in which the complete virion is engulfed by the host cell and subsequently contained within a vesicle (Figure 8.8a). This is the case for Variola virus and other poxviruses. In other enveloped viruses, **fusion** of the envelope with the eucaryotic host cell plasma membrane occurs and the nucleocapsid enters, as, for example, with the mumps and measles viruses (Figure 8.8a). Still a third strategy exists for some of

Assembly of an automobile

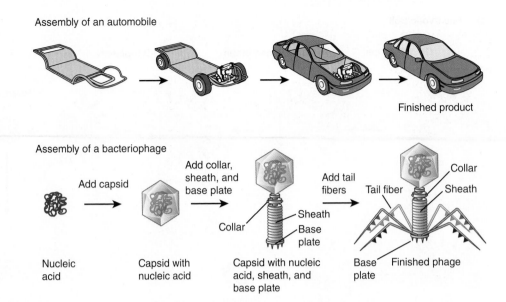

Finished product

Assembly of a bacteriophage

FIGURE 8.9 All of the "parts" of the virus are assembled into a virion.

the naked viruses, including the poliovirus, in which only the viral genome enters the cell's cytoplasm (FIGURE 8.8a). In the first two scenarios (i.e., endocytosis and fusion), the viral genome must be uncoated from its site within the capsid for the replication cycle to continue. This process of **uncoating** results from the action of enzymes of the host cell or of the virus.

Replication

The replication stage is directed by the nucleic acid genome of the virus. The details are a function of which one of the four types of nucleic acid the virus contains; the process is complicated and involves many steps. The important point is that, ultimately, viral components are synthesized within the host cell. Genetic information, whether stored in RNA or DNA, is ultimately expressed (transcribed and translated) into viral protein molecules specific for the particular virus.

Think of the host cell as a factory for viral components analogous to an automobile factory. Instead of making fuel injectors, dashboards, airbags, and steering mechanisms, the products are tail fibers, capsids, nucleic acids, spikes, and envelopes. The blueprint is the nucleic acid genome of the bacteriophage that has penetrated the bacterial host cell. Depending on the specific virus, production

may occur in the cytoplasm, nucleus, or possibly in both production sites.

Assembly

Continuing with the analogy of the automobile plant, now that the parts have been made, the virus is assembled into a functional structure in a manner similar to that of a production line (**FIGURE 8.9**).

Release

Mission accomplished! At this point the host cell is teeming with newly formed and complete virions, all potentially infective for host cells. The mechanism of release of virions varies as a function of the specific virus. In some cases release results in the death of the host cell. For example, with some of the bacteriophages (e.g., T4), the release of new phage particles, referred to as the **burst**, is the result of the enzymatic splitting open (lysis) of the cellular plasma membrane (**FIGURE 8.10**). As many as a few hundred new phage virions may be released. The entire process from adsorption to burst averages about 20 to 40 minutes in phages. In other cases death is not the outcome.

How is release accomplished with animal or in human viruses? Nonenveloped animal viruses are released from

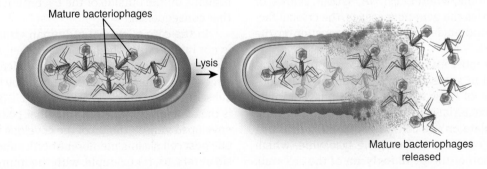

FIGURE 8.10 Lytic cycle of bacteriophage infection involves the lysis of the bacterial host cell by phages.

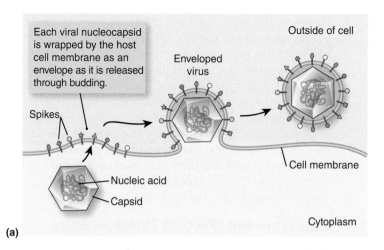

Each viral nucleocapsid is wrapped by the host cell membrane as an envelope as it is released through budding.

Outside of cell

Enveloped virus

Spikes

Cell membrane

Nucleic acid

Capsid

Cytoplasm

(a)

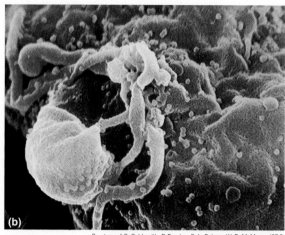

(b)

Courtesy of C. Goldsmith, P. Feorino, E. L. Palmer, W. R. McManus/CDC.

FIGURE 8.11 **(a)** Release by budding (of enveloped viruses). **(b)** Colorized scanning electron micrograph of HIV (spherical-shaped green viruses) budding from cultured lymphocytes (red). Multiple round, green bumps on cell surface represent sites of assembly and budding of virions.

the eucaryotic host cell by a process similar to the release of some bacteriophages involving lysis and death of the host cell. Release is somewhat more complicated in the case of enveloped viruses (**FIGURE 8.11**).

Generally, before release the expression of viral genes brings about the production of viral envelope components in some cases. **Budding**, or *extrusion*, is, essentially, the reverse of endocytosis. The virus pushes its way through the plasma membrane and pinches off a piece of the plasma membrane–spike complex, which now envelopes the nucleocapsid. Hence, the envelope consists of elements encoded by viral genes and by host cell membrane genes. Unlike lysis, budding is a gradual and continuous process and does not necessarily kill the cell. Ultimately, cell death is due to the takeover of the cell's machinery and accumulation of viral components. The number of mature virions released is enormous, ranging from a few thousand for poxviruses to more than a hundred thousand for poliovirus. Each of these virions is an infectious particle capable of invading a healthy host cell and causing damage symptomatic of the particular virus. All it takes is a few viral particles, which, by chance encounter with host cells bearing appropriate receptor molecules, initiate infection and within a short time cause symptoms, possibly death.

As stated earlier, all viruses, including bacteriophages, follow a similar sequence of events in their replication cycle (adsorption, penetration, replication, assembly, and release). The strategies differ; this is particularly true in the replication stage and depends on the nucleic acid content. It is remarkable that the minute amount of viral genome that gains access to the host cell directs the events leading to new virus particles. Whatever the uniqueness of the strategy, the significant point is that viral replication proceeds at a rapid rate, producing thousands of new virions. Considering that each virus is infectious, allowing for ongoing replication cycles involving more and more host cells, it is not difficult to understand the overwhelming effects that

occur when one acquires a viral infection. Antibacterial drugs (**antibiotics**) are ineffective against viruses because, unlike bacteria, viruses lack a cell wall and other structures unique to bacteria that are targeted by antibiotics. Some antiviral drugs have been developed, but they are not as effective against viruses as antibiotics are against bacteria.

Host Cell Damage

Imagine what it would be like inside a host cell that has been invaded by a virus. It would probably be an awesome experience to witness the devastation. In many cases the outcome is death or at least cell damage. The virus leaves its specific imprint, known as the **cytopathic effect (CPE)**, evidenced by cell deterioration. The particular CPE, viewed under an inverted light microscope, is helpful in the identification of many viral infections in cell cultures. **Cell culture** is a method of growing viruses that will be described in the next section of this chapter. The microscopic observation of virus-infected cells may be a characteristic CPE, such as the rounding of the cells, cell shrinkage, detachment, cell lysis and fusion, or syncytium formation. **Syncytia** are cells fused together into giant multinucleated cells that are believed to allow viruses to spread from cell to cell more readily. A number of different viruses cause syncytium formation during viral replication, including HIV and measles virus (**FIGURE 8.12**). In some cases, CPE is applicable to tissue specimens taken directly from a patient suffering from a viral infection or at autopsy.

Rabies is a viral disease that can be transmitted by a bite from certain wild rabid animals, including skunks, raccoons, bats, mongooses, and foxes, directly to humans or dogs or from dogs to humans. Consider the situation when a raccoon whose bizarre behavior is suggestive of rabies bites a dog or a human. After the raccoon is shot or captured and killed, brain tissue is examined for the presence of **Negri bodies**. Negri bodies are a CPE that is seen

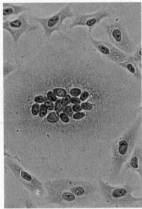

Courtesy of Shmuel Rozenblatt, Tel Aviv University, Israel.

FIGURE 8.12 Visualization of monkey kidney cell cultures infected with measles viruses using an inverted light microscope. A syncytium, cells fused together to form a giant multinucleated cell, is shown in this photograph. The plasma membranes of 17 cells infected with measles viruses fused together to form one syncytium.

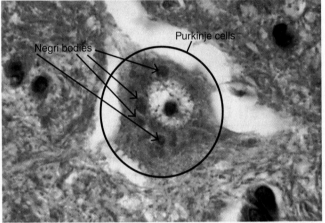

Courtesy of the CDC/Dr. Daniel P. Peri.

FIGURE 8.13 A photomicrograph of hematoxylin-eosin stained brain tissue taken during an autopsy after a patient died of rabies **encephalitis**. Negri bodies (depicted by the arrows) can be seen inside of the cytoplasm of a Purkinje cell (circled neuronal cell). Purkinje cells are some of the largest neurons in the cerebellum of a human brain. Hematoxylin stains nuclei deep purple, and eosin stains the Negri bodies dark pink.

in about 50% of brain tissues examined from rabid animals (**FIGURE 8.13**). The gold standard to diagnose rabies in animals is the **direct fluorescent antibody (dFA) test** used to detect rabies virus capsid proteins (**antigens**) in brain tissue. (This is why an animal suspected to be rabid should not be shot in the head!) Other CPEs are diagnostic for other particular viruses, including those that cause smallpox, herpes, and the common cold.

Cultivation of Viruses

Cultivation of bacteriophages is simple, based on the fact that easily grown bacteria serve as convenient host cells and meet the requirements for a metabolically active cell.

However, the replication of animal viruses is hindered by the necessity to use animal cells as host cells. In earlier years this presented a serious problem for virologists. Viruses were studied by injecting them into laboratory animals and observing their resulting symptoms; animals were killed so that the effects of infection could be observed at the organ and tissue levels, particularly the development of characteristic CPEs. The use of live animals was expensive and posed ethical issues, and these problems remain today. Animals are still used in certain circumstances, but **embryonated (fertile) chicken eggs** and cell cultures are the primary methods that have largely replaced animal use.

Embryonated (Fertile) Chicken Eggs

Some viruses can be grown in embryonated chicken (or other) eggs, a discovery made in the 1930s. The fertile chicken egg is a convenient and relatively inexpensive alternative to inoculation of live animals. The embryo is enclosed within the protective eggshell and, further, is covered by the shell membrane, providing a sterile and nutrient-rich environment for the developing embryo. Several different sites and membranes within the egg provide unique ecological niches that support the growth of particular viruses. With the use of sterile techniques, a hole is drilled through the shell, and the viral preparation is injected into the selected cavity, onto the appropriate membrane, or into the embryo itself, depending on the particular virus or suspected virus under study. Viral multiplication may be detected by the death of the embryo, pocks or lesions on membranes, or abnormal growth of the embryo. The viruses are recovered by harvesting fluids from the egg (**FIGURE 8.14**).

For the manufacture of influenza vaccines, influenza A and B viruses are grown in embryonated chicken eggs. As a result, the vaccine may contain egg proteins, prompting the question as to whether people who are allergic to eggs have the potential for an allergic reaction to the flu vaccine. Manufacturing an avian influenza A (H5N1 or H7N9) vaccine is more problematic because these influenza A viruses kill chicken embryos. For this reason, companies have developed cell culture lines that are permissive to infection by avian influenza A viruses.

Cell Culture

The application of cell culture techniques to the cultivation of viruses was widespread by the early 1950s. These techniques opened the door to the advancement of animal and human virus studies because the viruses could now be grown at lower cost, rapidly, and on a large scale. Cells are cultured in a variety of sterile laboratory containers, including flasks, bottles, and Petri dishes, containing a complex mixture of nutrients appropriate to the particular cells under culture. The viral suspension is then introduced into the culture flask; viral replication is detected by microscopic observation to detect CPEs and lysis of cells as evidence of a productive viral infection (**FIGURE 8.15**).

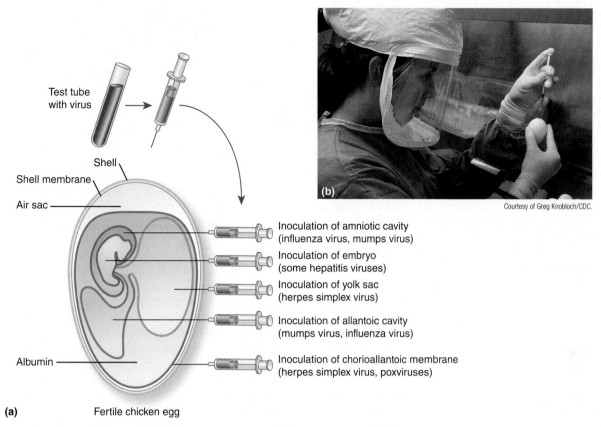

Test tube with virus

Shell

Shell membrane

Air sac

Inoculation of amniotic cavity
(influenza virus, mumps virus)

Inoculation of embryo
(some hepatitis viruses)

Inoculation of yolk sac
(herpes simplex virus)

Inoculation of allantoic cavity
(mumps virus, influenza virus)

Albumin

Inoculation of chorioallantoic membrane
(herpes simplex virus, poxviruses)

(a) Fertile chicken egg

Courtesy of Greg Knobloch/CDC.

FIGURE 8.14 (a) Cultivation of viruses in fertile chicken eggs. Several different sites and membranes within the egg can be inoculated with viruses. Different viruses will replicate in the site that supports their replication. For example, herpes simplex viruses are cultivated by inoculating the yolk sac. Herpes simplex viruses will not replicate in other locations, such as the amniotic cavity within the fertile egg. **(b)** Researcher inoculating fertilized eggs with influenza A viruses in order to prepare a vaccine to prevent influenza caused by the inoculated strain of influenza A virus.

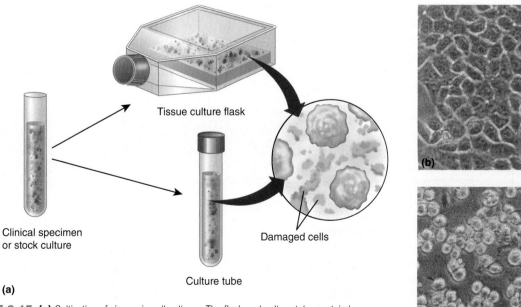

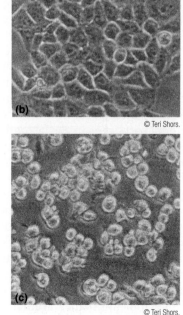

Tissue culture flask

Clinical specimen
or stock culture

Damaged cells

Culture tube

(a)

© Teri Shors.

© Teri Shors.

FIGURE 8.15 (a) Cultivation of viruses in cell cultures. The flask and culture tube contain layers of live host cells in nutrient broth to support viral growth. **(b)** Uninfected monkey kidney cells viewed by inverted light microscopy. The confluent monolayer of cells remains attached and intact to the bottom of a tissue culture dish. **(c)** Monkey kidney cells infected with vaccinia virus. CPEs include rounding and detachment of the cells from the tissue culture dish.

Diagnosis of Viral Infection

Although cell and embryonated egg techniques are available and routine in the hands of trained laboratory personnel, *there is usually no attempt to culture the viruses from those with suspected viral infections,* as is routinely done for bacterial infections. There are several reasons for this, including cost and limited viral laboratory facilities, but a prime reason is that results of cell culture take too long to be useful for clinical diagnosis. Nevertheless, definitive tests are available when indicated. The presence of **antibodies** in the patient's blood may be of diagnostic value; these molecules are produced in response to (foreign) antigens (e.g., viral capsid proteins), and their presence is an indication of current or past infection. Accurate identification of a virus may be essential for the protection of the public health, as in the 1999 encephalitis outbreak in New York, when tissue specimens taken at autopsy identified the virus not as the St. Louis encephalitis virus but rather as the West Nile encephalitis virus.

The majority of viral infections are based on the patient's symptoms, including fever, general aches and pains, weakness, nausea, and muscle fatigue. If there is a telltale rash, as in measles, or other characteristic signs, such as the skin lesions associated with chickenpox, the diagnosis is more apparent and more accurate. Usually, the patient has to settle for vague and meaningless terms and expressions like "it's a bug" or "24-hour grippe, flu, or intestinal virus." Recovery from the more common viral infections usually occurs without complications, thanks to the immune system.

In view of the fact that most acute viral infections are not identified using routine commercial tests, there remain reasons why definitive laboratory tests are needed and developed in terms of public health. Commercial test kits are used to screen blood donations for HIV, hepatitis B virus, hepatitis C virus, and West Nile viruses, reducing the spread of these viral diseases through blood transfusions. Viral testing is used to monitor the effectiveness of vaccination programs. **Rapid diagnostic tests** for viral diseases such as West Nile encephalitis enable authorities to initiate mosquito-control measures. However, most hospital clinical laboratories perform very few viral diagnostic tests.

Molecular technology is now routine in the form of commercialized kits in specialized virology diagnostics laboratories. Some approaches used to detect and identify bacterial pathogens in the laboratory are modified to diagnose infections caused by viruses. Techniques used to detect viruses from clinical specimens involve microscopy, detection of patient antibodies against a specific virus, viral antigen detection, viral genome detection and isolation, or detection of the virus suspect in cell cultures. Inverted light microscopy is used to visualize CPEs of virally infected cells, whereas TEM is used to observe individual virus particles. TEM is especially useful to detect viruses that cannot be cultured in the laboratory. **Enzyme-linked immunosorbent assays (ELISAs)** are designed to detect viral antigens or antibodies against viruses present in patient serum. ELISAs are based on antibodies binding to

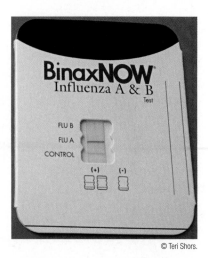

© Teri Shors.

FIGURE 8.16 Rapid influenza A and B test. This patient is suffering from an influenza A infection.

antigens and subsequently detected by a resulting color change. The enzyme reacts with its substrate to produce a color change. Like the rapid strep test, rapid influenza tests distinguish influenza A and B virus infections from a number of other viruses in about an hour (**FIGURE 8.16**).

Molecular diagnostics is the new gold standard for the diagnosis of viral encephalitis and viral **meningitis** and is performed on patient cerebrospinal fluid collected by a "spinal tap." **Polymerase chain reaction (PCR)** is used to amplify viral nucleic acid genomes and has replaced brain biopsies. Variations of PCR are also used to monitor viral loads of AIDS and hepatitis C patients in the management of their antiviral therapy. **TABLE 8.4** lists examples of viral pathogens and the most useful diagnostic approach.

As an anonymous alternative to testing in a clinic, home kits purchased online (via the Internet) are available to diagnose **sexually transmitted infections (STIs)**, such as the FDA-approved Home Access® Express HIV-1 Test System and OraQuick In-Home HIV test by OraSure Technologies (**FIGURE 8.17**). The FDA has also approved a number of other home kits for other viral diseases.

TABLE 8.4 Most Useful Diagnostic Procedures Used for Viral Infections

Diagnostic Approach	Virus
Microscopy	Epstein-Barr virus, poxviruses, norovirus, Varicella zoster virus
ELISA (to detect viral antigens or antibodies)	Influenza A and B viruses, hepatitis B and C viruses, papillomaviruses
PCR-based method	HIV, rabies virus, Ebola virus
Cell culture	Adenoviruses, influenza A and B viruses, mumps virus, rubella virus

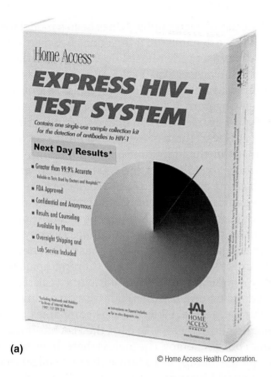

(a)

(b)

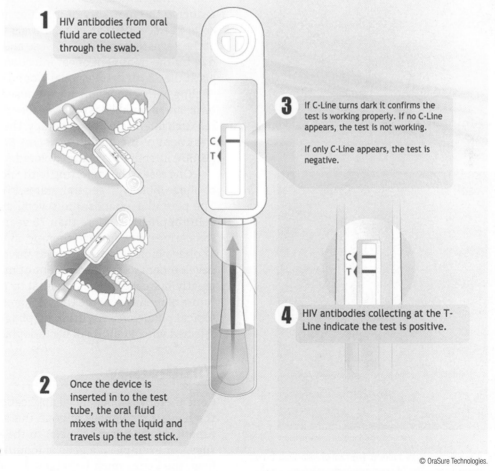

1 HIV antibodies from oral fluid are collected through the swab.

3 If C-Line turns dark it confirms the test is working properly. If no C-Line appears, the test is not working.

If only C-Line appears, the test is negative.

4 HIV antibodies collecting at the T-Line indicate the test is positive.

2 Once the device is inserted in to the test tube, the oral fluid mixes with the liquid and travels up the test stick.

(c)

FIGURE 8.17 Two types of in-home HIV-1 testing kits have been approved by the FDA for over-the-counter purchase by individuals who are 17 years or older. **(a)** The finger is pricked and a drop of blood is placed on a special filter paper in the Express HIV-1 Test System. The self-collected blood sample is express-mailed to the Home Access Health Corporation for laboratory testing. **(b)** OraQuick is a lateral flow test visually read by the user. No Laboratory needed. **(c)** The procedure for the OraQuick test kit involves self-collecting oral fluid by swabbing the gum line with a testing stick. The testing stick is inserted into a tube that contains testing reagents. Within 20 minutes, results are visually observed. Positive results must be confirmed in a medical setting.

Phage Therapy

Recall from earlier in the chapter the devastating effects of the T4 bacteriophage on *E. coli* (**FIGURE 8.18**). The phage can wipe out an entire *E. coli* culture. Hence, the possibility of using bacterial viruses to control bacterial infections may not seem so far-fetched and has been realized since their discovery. *Arrowsmith*, a 1926 Pulitzer Prize–winning novel by Sinclair Lewis, includes in the plot the discovery of a bacteria-destroying phage. Hundreds of bacterial species, including many pathogens, are subject to bacteriophage infections, frequently resulting in bacterial cell damage or death. Obviously, this interrupts the bacterial replication cycle.

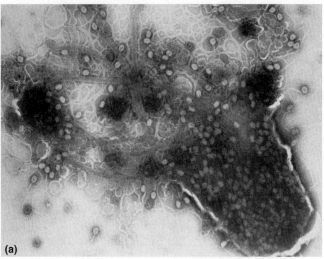

(a)

© Biozentrum, University of Basel/Science Photo Library/Photo Researchers, Inc.

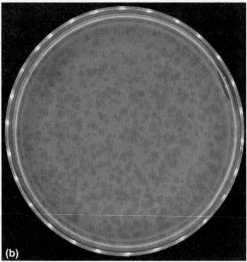

(b)

Courtesy of Giles Scientific Inc., www.biomic.com

FIGURE 8.18 The lytic activity of a phage. **(a)** TEM of lysis of an *E. coli* bacterium by T4 bacteriophages. The T4 (light ovals with stalks) infect the bacterium, using the cell's genetic machinery to code for their own replication. Crowded with T4 progeny, the cell's plasma membrane bursts, destroying the cell and releasing phages to infect other cells. 40,000×. **(b)** A lawn of *E. coli* with plaques (lighter-colored areas).

The use of bacteriophages has been a model for understanding the dynamics of interactions between animal and plant viruses and their hosts. However, the earlier interest in bacterial viruses was based on their use as therapeutic agents to fight bacterial infections. To put it another way, investigators realized the possible potential of phages to kill pathogenic bacteria. Phage therapy, although controversial, was increasing in popularity as a mode of treatment in the early part of the 20th century but was essentially abandoned in the Western world when antibiotics became available in the l940s. However, research and implementation of phage therapy continued, primarily in Eastern Europe. Georgia, formerly part of the Soviet Union, remains a stronghold of human phage therapy.

In the past several years there has been renewed worldwide interest in human phage therapy, sparked by the highly significant problem of antibiotic resistance. In the United States and in other countries, in addition to Eastern Europe, several biotechnology companies are actively pursuing phage therapy. To date, the FDA has not approved human phage therapy. The logic is there, however, and perhaps phage therapy has a future. Recent clinical trials indicate renewed interest in phage therapy, including a study that demonstrated the effectiveness of phage therapy in reducing the mortality of infection (*Pseudomonas aeruginosa*) in mice. Further, clinical trials using phage therapy at the Royal National Throat, Nose and Ear Hospital in London for the treatment of ear infections (otitis); clinical trials at the Southwest Regional Wound Care Center in Texas dealing with wound infections caused by a variety of bacteria; and a number of other studies serve as indicators of research interest in phage therapy. The use of phages in the treatment of **methicillin-resistant Staphylococcus aureus (MRSA)** infections is being explored.

One researcher, referring to the potential applications of phage biology in general, states, "*The best is yet to come.*" The potential extends into several areas, including agricultural practices. More than 70 years have passed since the early trials of phage therapy; since then, knowledge of phage biology garnered from the role of phages in the development and advancement of molecular biology has greatly increased their potential in human therapy. The future of medicine in the United States could include using topical phage creams, drinking phage cocktails, or being injected with a culture of bacteriophages!

Virotherapy

Gene therapy is a beneficial application of viruses. One of the biggest challenges of gene therapy is delivering the functional, or "good," gene(s) to the correct cells or tissues. For example, if a gene is required to function in the liver, the genes must be targeted to the liver cells without harming other types of cells. How can it be confirmed that a gene is targeted to the liver and not the big toe? The answer is that there are definitive, noninvasive **reporter genes** that track patient gene therapy.

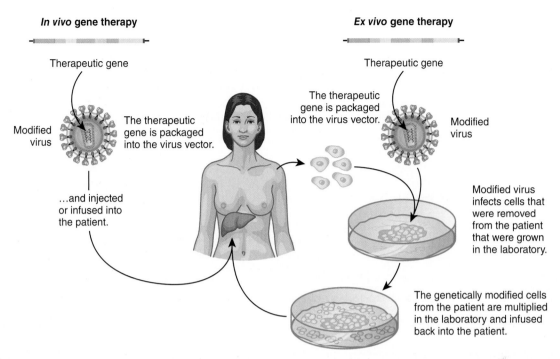

FIGURE 8.19 Comparison of *in vivo* and *ex vivo* gene therapy treatments using an engineered virus vector containing the "good gene" needed to repair or compensate for a defect within a cell.

Viruses are good gene delivery "vehicles," or vectors. They can be engineered to attach and enter specific types of cells and not cause harm to normal cells. With **in vivo gene therapy**, the engineered virus can be directly injected into the patient. With **ex vivo gene therapy**, the patient's cells are removed, cultured in the laboratory, incubated with the engineered virus, and then transfused back into the patient (**FIGURE 8.19**). If the virus can insert the correct gene into the nuclear DNA/genome of the cell, it functions indefinitely in the patient. If the virus does not insert the correct gene into a chromosome of the cellular genome, the effect of the good gene delivered is temporary. **TABLE 8.5** lists popular virus vectors used in gene therapy trials.

Cancer treatment is now at a crossroads. The traditional options to treat cancer are surgery, radiation, and chemotherapy. Even today, with most advanced cancers, radiation and chemotherapy are toxic and remission is brief because of the emergence of chemotherapy-resistant cells. Virotherapy is in development as a new treatment of incurable cancers. Virotherapy is the use of oncolytic (cancer-killing) viruses to destroy cancer cells. **Oncolytic viruses** replicate inside of cancer cells and kill them while sparing nearby healthy cells (**FIGURE 8.20**). Even before viruses were visualized with the TEM, physicians reported observations of malignant tumors regressing during a viral infection. One of the most-cited examples is a 1904 report by Dr. George Dock describing a 42-year-old woman with leukemia who went into remission following a "bout of influenza." In a 1953 report, a 4-year-old boy suffering from leukemia experienced a similar remission after the onset of chickenpox. Reports in the 1970s described

TABLE 8.5 Features of Popular Gene Therapy Viral Vectors

Virus Vector	Cell Target	Integration into Cell's Nuclear DNA?
Retrovirus	Infects dividing cells.	Yes, random integration
Adenovirus	Infects dividing and nondividing cells.	No
Adeno-associated virus	Infects dividing and nondividing cells. Requires a "helper" virus to replicate inside of cells.	Yes, 95% of the time integrates into chromosome 19
Herpes simplex virus	Can infect cells of the nervous system.	No, but stays in the nucleus for a long time
Vaccinia or related poxvirus	Can infect dividing cells.	No

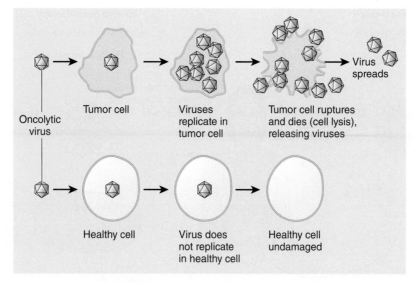

FIGURE 8.20 Oncolytic viruses replicate inside of cancer cells and kill them but do not harm the nearby healthy (noncancerous) cells.

remissions coinciding with measles infection. Over the next few decades the scientific community pursued the idea of using viruses to treat cancer. Most of the experiments involved curing mouse tumors, but, unfortunately lacked quality control.

Advances in genetic engineering of viruses, increased understanding of the immune system, new knowledge of virus–host cell interactions, and the isolation of many animal viruses that are harmless in humans but can infect and kill human cancer cells have pushed virotherapy forward.

The first oncolytic virus was approved in 2005 for the treatment of head and neck cancer. The number of recent research publications reflects the pace of virotherapy. During the 1990s, there were a handful of publications. In 2017, a PubMed search revealed an average of 19 scientific peer-reviewed publications per month. Without a doubt, virotherapy will be a part of future cancer therapies. A day may come when the viruses can be used to treat skin cancers, brain gliomas, and other cancers. Oncolytic viruses will be used to defeat one of modern man's most evasive and pernicious threats—cancer.

Biology of Prions

Name an infectious agent smaller than a virus that lacks both DNA and RNA and poses a potential threat to the world's beef supply. The answer is *prions*, a shorthand term for "proteinaceous infectious particles." They cause **bovine spongiform encephalopathy (BSE)**, or **mad cow disease** in cows, and the related human brain disease, **variant Creutzfeldt-Jakob disease (vCJD)**, which is an infectious form of the disease transmitted through the consumption of beef contaminated with prions that cause BSE in cows. A hereditary form of the human disease is known as **Creutzfeldt-Jakob disease (CJD)**. Related diseases appear in a variety of animals, such as

scrapie in sheep and **chronic wasting disease (CWD)** in deer.

Prions, in their *normal form*, are proteins found in the brains of mammals. Like all proteins, they are composed of specific sequences of amino acids, and their presence is the result of gene expression in the same way that eye color is a manifestation of protein pigments conferred by genes that control eye color. In their *misfolded* or *abnormal form*, prions are pathogenic, highly stable, and resistant to inactivation by freezing, drying, and heating at usual cooking temperatures; they are resistant to pasteurization and conventional sterilizing temperatures. **TABLE 8.6** presents the biological distinctions between bacteria, viruses, and prions. The *PrP* gene, located on human chromosome 20, encodes the normal cellular prion protein, PrP. PrP is also called PrP^c. PrP^{res} refers to the abnormal misfolded or *res*istant form responsible for neurodegenerative changes.

Stanley Prusiner received the 1997 Nobel Prize in Physiology or Medicine for his discovery of prions. According to the prion theory, BSE, vCJD, scrapie, and other transmissible spongiform encephalopathies (TSEs) are the result of abnormally folded prions that latch onto normal prions and convert them into altered and misfolded forms, a property that reveals their infectious nature (**FIGURE 8.21**). The expression "one bad apple spoils the barrel" is applicable. Robert Louis Stevenson's novel, *Strange Case of Dr. Jekyll and Mr. Hyde* (1886), can be used as another analogy. Dr. Jekyll is the "good guy," and Mr. Hyde is the "evil beast" living within Jekyll. The consequence of these abnormally folded prions in the brain is the loss of motor coordination, dementia, other neurological symptoms, and death. Autopsy reveals a Swiss cheese–like or spongelike brain full of holes. No treatment or cure is available.

The BSE–vCJD puzzle is not complete; not all the pieces of the puzzle are in place, nor have they even been

TABLE 8.6 Comparison of Bacteria, Viruses, and Prions

	Characteristics	Infectious?	Nucleic Acid	Immunogenic?	Resistance to Disinfection
Bacteria	Procaryotic cell	Yes	DNA and RNA	Yes	Moderate*
Viruses	Protein coat and nucleic acid, noncellular	Yes	DNA or RNA	Yes	Moderate*
Prions	Misfolded protein, noncellular	Yes	No DNA, No RNA	No	Resistant to all typical disinfection methods

*Depends on the particular test condition and the specific bacterial or viral species.

identified. The idea that abnormal prions are able to convert normal prions to an altered form is a revolutionary hypothesis. Nucleic acids are the building blocks of DNA, the hereditary material found in all microbes that is passed on from generation to generation. DNA makes

(a)

(b)

Neuron

PrP^c

PrP^{res}

A prion protein (PrP^{res}) interacts with a normal prion precursor (PrP^c).

The PrP^c is converted to PrP^{res}.

Additional PrP^{res} convert more PrP^c into PrP^{res}.

PrP^c throughout the neuron are converted into PrP^{res}. As the neurons die, vacuolar areas form in the grey matter.

(c)

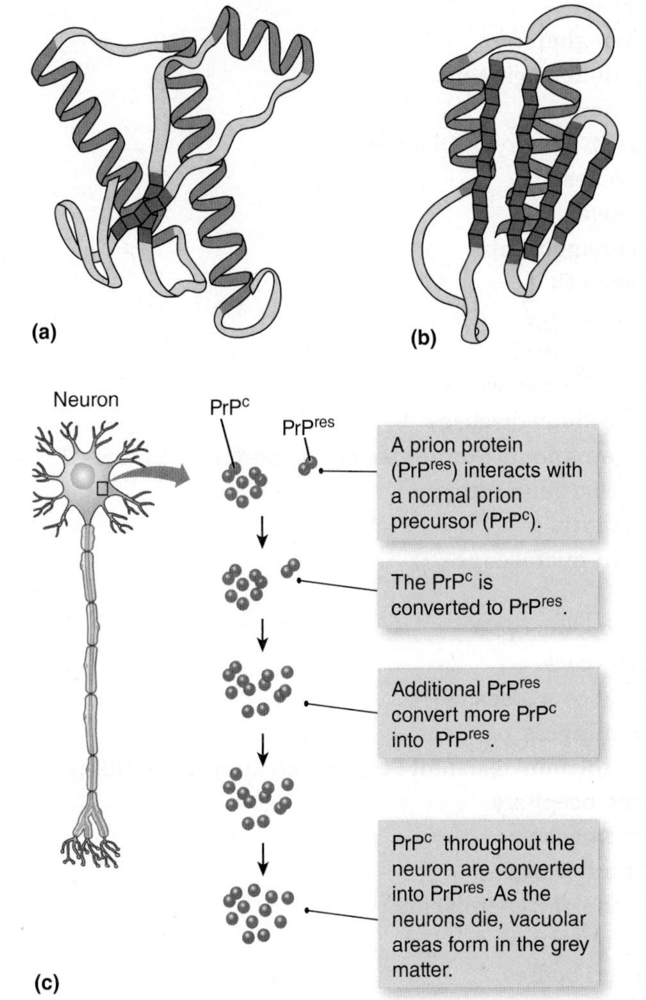

FIGURE 8.21 **(a)** Tertiary structure of a normal prion (PrP^c). **(b)** A misfolded prion (PrP^{res}). **(c)** Conversion of normal prions (PrP^c) to infectious prions (PrP^{res}). The misfolding allows the proteins to clump together and contribute to disease in ways that are not fully understood.

up the genes that encode proteins, which are essential for life's activities. Prions lack genes but are, nevertheless, infectious, representing a new agent of infection. Some skeptics of the prion hypothesis argue that there is no definitive proof that a slow virus is not the trigger necessary for initiation of conversion of proteins to a mutant infectious agent. Further, they assert that prions synthesized in a test tube, and thus free of viruses, have not been found to produce disease.

However, in the 1960s it was suggested that an infectious agent lacking genetic material might be responsible for disease. One landmark study indicated that brain tissue removed from sheep with scrapie, a neurological disease, remained infectious, even after radiation that would destroy DNA or RNA. The most significant missing piece of the prion puzzle has to do with the trigger that induces the normal protein to misfold. Many scientists believe that the prion protein alone is the infecting agent. Prusiner acknowledges the need for further experiments and the possibility of a "missing factor" that might chaperone PrP into an abnormal shape. Perhaps Prusiner's missing factor will turn out to be a virus yet to be discovered. The Nobel Committee was not bothered by the unanswered questions. The deputy chair stated, "*The details have to be solved in the future. But no one can object to the essential role of the prion protein in those brain diseases.*" Further, prions may be linked to Alzheimer's and Parkinson's diseases as well as to other neurological conditions.

Summary

Viruses are subcellular infectious agents. They subvert the host cell's metabolic machinery for their own replication. A unique aspect of viruses is that they contain either dsDNA, ssDNA, dsRNA, or ssRNA genomes, whereas cells contain DNA genomes.

Naked viruses consist of only a protein coat—the capsid—containing nucleic acid, whereas viral capsids are wrapped with a membrane to form an enveloped virus. Viruses are helical, icosahedral, or complex. They typically range in size from 20 nm to 350 nm, but there are exceptions (e.g., giruses such as Mimivirus). Bacterial viruses—bacteriophages—are models for understanding viral replication, a process that consists of five stages: adsorption, penetration, replication, assembly, and release. Viruses that cause human diseases exhibit these stages with variations in the strategies used, depending on the specific virus.

Viruses require a host for replication. Therefore, viruses cannot be grown on nonliving medium and so are cultured in embryonated chicken eggs or in cell cultures, both offering a supply of living cells allowing for their replication. Identification of the specific virus-causing infection is impractical and routinely not done.

Diagnosis, in most cases, is based on the symptoms of disease.

Treatment of bacterial diseases with bacteriophages sounds bizarre but is based on sound logic. The idea is attractive because of the growing problem of infections caused by antibiotic-resistant bacteria. Viruses are used as delivery vehicles for gene therapy. Virotherapy is a new field advancing toward the treatment of incurable cancers.

The misfolded form of prions are pathogenic proteins that are very stable in the environment. They cause neuro-degenerative diseases in animals such as BSE and scrapie and in humans, CJD, and variant CJD.

KEY TERMS

16S rRNA gene
adsorption
antibiotics
antibody
antigens
assembly
attenuated viruses
bacteriophage
bacteriophage therapy
bioterrorist
bovine spongiform encephalopathy (BSE)
budding
burst
capsid
capsomere
cell culture
chronic wasting disease (CWD)
complex
contact tracing
coreceptor
Creutzfeldt-Jakob disease (CJD)
cytopathic effect (CPE)
dermatropic
direct fluorescent antibody (dFA) test
disconjugate gaze
Ebola virus disease (EVD)
embryonated (fertile) chicken eggs
emerging infectious diseases
Emory University Hospital's Serious Communicable Disease Unit
encephalitis
encephalopathy
endocytosis
envelope
enveloped virus

enzyme-linked immunosorbent assays (ELISAs)
ex vivo gene therapy
fusion
gene therapy
genetic engineering
genome
Germ Theory of Disease
giruses
helical
hemagglutinin
host cell
host range
icosahedral
immunization
infectious diseases
International Committee on Taxonomy of Viruses (ICTV)
in vivo gene therapy
keratitis
light microscope
lysis
lysogenic cycle
lytic cycle
mad cow disease
meningitis
methicillin-resistant *Staphylococcus aureus* (MRSA)
microcephaly
microscopic
Mimiviruses
mode of transmission
naked virus
Negri bodies
nonenveloped virus
nucleocapsid
oncolytic viruses
pandemic

penetration
peptidoglycan
personal protective equipment (PPE)
pleomorphic
pneumotropic
polymerase chain reaction (PCR)
post–Ebola syndrome
prion
prophage
rapid diagnostic tests
receptors
reemerging infectious diseases
release
replication
reporter genes
R-nought
scrapie
sexually transmitted infections (STIs)

sialic acid
subcellular
syncytia
taxonomy
T helper lymphocytes
TKM-Ebola
transmission electron microscope (TEM)
uncoating
uveitis
vaccination
variant Creutzfeldt-Jakob disease (vCJD)
variolation
viral load
virion
virophage
virotheraphy
viruses
viscerotropic

SELF-EVALUATION

O PART I: Choose the single best answer.

1. Which of the following is a complication experienced by survivors of Ebola virus disease?
   - a. Uveitis
   - b. Chickenpox
   - c. Melanoma
   - d. Urinary tract infections

2. The "turkey wrap" analogy applies viruses with a _____ shape.
   - a. complex
   - b. rectangular
   - c. icosahedral
   - d. helical

3. The units of the viral capsid are called _____.
   - a. hats
   - b. tubules
   - c. capsomeres
   - d. envelopes

4. Viruses are known to infect _____.
   - a. plants
   - b. bacteria
   - c. humans
   - d. all organisms

5. The envelope of an animal virus is derived from the _____ of its host cell.
   - a. plasma membrane
   - b. cell wall
   - c. cytoplasm
   - d. nucleus

6. The general steps in a viral multiplication cycle include (1) release, (2) penetration, (3) adsorption, (4) replication, (5) assembly. Which of the following is the correct sequence of these steps?
   - a. 2, 3, 4, 5, 1
   - b. 2, 1, 3, 4, 5
   - c. 3, 2, 4, 5, 1
   - d. 1, 2, 3, 4, 5

7. Virotherapy converts viruses into _____-fighting agents.
   a. cancer
   c. cholesterol
   b. diarrhea
   d. fat

8. _____ are the infectious proteins that cause variant CJD.
   a. Viroids
   c. Prions
   b. Virophages
   d. Prunes

9. Cellular changes observed in cell cultures infected with viruses are called _____.
   a. clumping effects
   c. necrotic effects
   b. athogenic effects
   d. pathogenic effects

10. The new gold standard for diagnosis of viral encephalitis is _____ testing
    a. cell culture
    c. PCR
    b. ELISA
    d. animal

○ PART II: Fill in the blank.

1. Based on nucleic acid categories, there are _____ viral groups.

2. A complete viral particle is called a(n) _____.

3. Viruses can be grown in tissue culture or in _____.

4. The acronym PCR stands for _____.

5. _____ viruses replicate inside of cancer cells and kill them while sparing nearby healthy cells.

6. _____ are ineffective in treating viral infections.

7. The acronym ELISA stands for _____.

8. _____ are viruses that infect bacteria.

9. All viruses require a living _____.

10. A(n) _____ virus lacks an envelope and is relatively stable in the environment.

O PART III: Answer the following.

1. Describe the five stages of viral replication.

2. The genetic material of viruses is unique. Discuss this statement.

3. Why did the discovery and understanding of viruses lag about 40 years behind knowledge about bacteria?

4. How does the possibility of being treated for a bacterial infection with live bacteriophage strike you? Describe the logic of this strategy.

5. What do you think about the possibility of being treated for advanced cancer with an oncolytic virus? Describe the logic of this strategy.

6. What is the treatment for Ebola virus disease?

7. Explain how it was ruled out that Mimivirus was not an atypical bacterium.

8. Compare and contrast the following: bacteria, giruses, viruses, virophages, and prions.

9. Explain why the development of cell cultures to propagate viruses is such a boon to virology research.

10. Explain why viruses can be found everywhere on Earth.

BACTERIAL GENETICS

> "*Within one linear centimeter of your lower colon there lives and works more bacteria (about 100 billion) than all humans who have ever been born. Yet many people continue to assert that it is we who are in charge of the world.*"
>
> —Neil deGrasse Tyson
>
> *American astrophysicist, science communicator, and author*

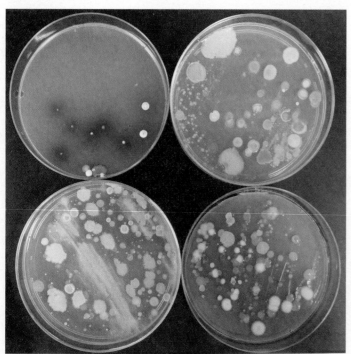

© Teri Shors.

Photograph displaying different types of colonies of soil bacteria growing on bacteriological media. Soil bacteria influence important processes in ecosystems and represent a large proportion of genetic diversity on Earth. The white powdery, elevated bacterial colonies on the plate in the upper left of this figure secrete a brown compound. These bacteria belong to the family *Actinomycetes*. Note that the brown color secreted in the medium surrounding several different colonies is likely an antimicrobial compound. Actinomycetes produce the majority of antibiotics that have been manufactured for clinical use.

LEARNING OBJECTIVES

1. Explain why microbial diversity is important to maintain healthy ecosystems.
2. Compare and contrast the chemical structure of DNA and RNA.
3. Summarize the molecular processes utilized by cells to interpret the language of DNA into a functional protein within a cell.
4. Define *transformation*, *conjugation*, and *transduction* and the association of these processes with bacterial diversity.
5. Describe an application of a microbiota-based therapy.
6. List four diseases associated with gut dysbiosis.
7. Summarize the potential of synthetic biology in maintaining human health in the future.

Case Study: Solving the Mystery of Why Vampire Bats Can Live on Blood

As pre-nursing major Diego Gonzales sat listening to a college instructor lecture about nutrition, his mind began to wander. The instructor was reviewing the types of foods consumed based on vegan, vegetarian, flexitarian, pescatarian, ketogenic, Mediterranean, and omnivorous diets. He was fascinated that there were so many popular diets being advertised in the mass media, such as Atkins, the Paleo diet, the Zone diet, and the HCG diet.

Until the age of 10, Diego lived in Brazil. The Amazon rainforest covers approximately 60% of Brazil. He remembered his parents telling him about a distant cousin who lived in a rural area near Portel who died of rabies after being bit by a hairy-legged vampire bat. The bats were losing their native habitat due to **deforestation** caused by gold mining. The hairy-legged vampire bats are **sanguivores**, living exclusively on blood, feeding mainly on the blood of wild birds, and occasionally on domesticated birds such as chickens.

Diego's cousin lived in a dwelling without windows. He was bit while sleeping at night. The bats use innate infrared-sensing capabilities to quickly identify accessible blood vessels, bite their prey with their razor-sharp incisors, and feed without any disturbance. Vampire bat saliva contains analgesic (pain-killing) substances and anticoagulant substances that thin the blood as it is drawn. The prey is unaware of its **blood meal**, a process that takes about 20 minutes.

Diego knew that hairy-legged vampire bats (**Figure 1**) could survive on about 2 tablespoons (2 milliliters) of blood per day and could not survive more than 2 or 3 days without a blood meal. As Diego daydreamed about the vampire bats, he raised his hand and asked the instructor how it was possible that vampire bats could survive on a blood diet.

The instructor took time to address his questions by discussing the nutritional value of blood and the physiological adaptations needed to digest and survive on a blood diet. The digestive tract, circulatory system, and kidneys of vampire bats are adapted to rapidly process

© Michael Lynch/ShutterStock, Inc.

Figure 1 Common vampire bat. Vampire bats are sanguivores, or "blood sippers."

(continues)

Case Study: Solving the Mystery of Why Vampire Bats Can Live on Blood (continued)

and digest blood. Blood consists of about 93% protein, 1% carbohydrates, low levels of vitamins and lipids, high salt, and even some **bloodborne pathogens**! Therefore, the diet is low in nutrients, lipids, essential amino acids, and vitamins and high in salt. The kidneys of the bats have special adaptations to handle the high-protein diet. Without these adaptations, the rapid digestion of the blood would cause an accumulation of urea, leading to kidney problems, high blood pressure, and fluid retention.

The instructor proceeded to describe a recent study published in *Nature Ecology & Evolution* in 2018. A large group of researchers collaborated to answer the question of how the vampire bat evolved from eating a diet of fruits (frugivorous bats) to a blood diet that is low in nutritional quality. To do this, they analyzed the hologenomes of the sanguivorous vampire bat and compared it to the hologenomes of carnivorous, insectivorous, and frugivorous bats. A **hologenome** is the entire set of genes of a host, in addition to the genes of its **microbiota**. The researchers sequenced the entire genomes of the different types of bats based on their diet and the genomes of the microbiota present in their bat droppings.

It was determined that the vampire bat genome contains a higher number of integrated **transposons** (also known as jumping genes) that can move around the genome. Many copies of a transposon named MULE-MuDR were present within the genes of the vampire bat genome that would normally function to challenge a blood diet. *The transposon insertions knocked out the functions of these genes, allowing the bats to survive on a blood-only diet without getting sick.* The knocked-out genes played a role in controlling the host's immune response toward viruses and antigen processing and lipid and vitamin metabolism.

The researchers also found that the microbiota of the vampire bat droppings (which represent its gut microbiota) contain unique microbial species not found in the gut microbiota of bats that survive on other diets. Surprisingly, the gut microbiota of vampire bats contained more than 280 bacterial species known to be pathogenic to mammals and an abundance of *Amycolatopsis mediterranei*. *A. mediterranei* bacteria have been shown to offer "protection" to their hosts because they

can produce antimicrobial and antiviral compounds that inhibit bacteriophages and poxviruses.

Diego was fascinated with the findings that the instructor shared with the class. The lecture hour passed by quickly, and it was time for him to get up and attend the next class on his schedule, Microbiology (which was also his favorite class).

Questions and Activities

1. Define *hologenome*, and explain how it was used to study the vampire bat's adaptation to a blood diet.

2. What are transposons, and how can they affect an organism's genome?

3. How does the gut microbiota of vampire bats differ from that of bats that feed on insects, meat, or fruit?

4. Speculate how it can be possible for the vampire bats to carry so many pathogens in their gut and not get sick.

5. When vampire bats take a blood meal from a human or a bird, it can ingest pathogens present in the blood. Why don't vampire bats get sick from feeding on blood?

6. Create a **hypothesis** to address why the hairy-legged bats prefer bird blood over blood from animals or humans. Explain how you would test your hypothesis.

7. Discuss how the **disease triangle model** relates to the rabies outbreak in Brazil.

8. What pathogen causes rabies?

9. Why didn't Diego's cousin wake up when a vampire bat took a blood meal from him?

10. What are the signs and symptoms of rabies? How is it treated?

11. Why did Diego's cousin die of rabies?

Information from M. Lisandra, Z. Xiong, M. Escalera-Zamudio, A. K. Runge, J. Theze, D. Streicker, . . . M. P. Thomas Gilbert. (2018). Hologenomic adaptations underlying the evolution of sanguivory and the common vampire bat. *Nature Ecology & Evolution, 2,* 659–668; M. C. Schneider, P. C. Romijn, W. Uida, H. Tamayo, A. Belotto, J. B. da Silva, & L. F. Leanes. (2009). Rabies transmitted by vampire bats to humans: An emerging zoonotic disease in Latin America? *Rev Panam Salud Publica/Pam American Journal of Public Health* 25:260–269.

Preview

It is estimated that more than 1 trillion (10^{12}) species of microbes exist. Only a small percentage of microbial species have been described because researchers have not been able to grow them as pure cultures in the laboratory. Both **mutation** and **recombination** by **transformation**, **transduction**, and **conjugation** are the mechanisms by which new genes and new recombinations of genes arise in different bacteria, accounting for their diversity. The **gene** is the basic unit of heredity and represents a DNA segment of a **chromosome** consisting of tightly coiled DNA; most bacteria have a single chromosome. The processes of **transcription** and **translation** of mRNA culminate in the synthesis of proteins. Sexual reproduction and mutation account for variation in eucaryotic microbes. Bacteria, however, are **procaryotes** and multiply through **binary fission**. During transformation, foreign ("naked") DNA, such as a **plasmid**, is taken up by bacteria; transduction is characterized by bacteriophage-mediated transfer of phage DNA, and in conjugation, plasmid or chromosomal DNA is transferred from donor to recipient bacterium during physical contact. These processes are some of the tools for **genetic engineering** and **biotechnology** applications. **Synthetic biology** is an advancement that allows for the manipulation of genes to create new and unnatural biological products or to manipulate existing biological products into systems that function in a natural way.

DNA Structure

This section and the sections leading up to Bacterial Genetics are intended as a review of the basic genetics typically covered in an earlier biology course. The general information is provided as a lead-in to the more specialized area of bacterial genetics.

Deoxyribonucleic acid (DNA) is, as the name implies, a nucleic acid (as is **ribonucleic acid**, or **RNA**) and consists of two chains of **nucleotides**. Each nucleotide is composed of three building blocks, namely, a phosphate, a sugar, and a nitrogen-containing base (**FIGURE 9.1**). The sugar is **deoxyribose**, a five-carbon sugar. The phosphate and the deoxyribose molecules are the same in all nucleotides; the variable part of each nucleotide is the nitrogenous base, of which there are four—adenine (A), guanine (G), cytosine (C), and thymine (T). Hence, nucleotides are referred to as A, G, C, and T. Adenine and guanine are similar in structure and are classified as **purines**; thymine and cytosine are closely related and are called **pyrimidines**. The sugar and phosphate molecules are the backbone of a sequence of chemically joined nucleotides; the nitrogenous bases poke inward.

The two strands of DNA twist to form a supercoiled double helix and are held together by hydrogen bonds between the nitrogenous bases of the nucleotides. Think

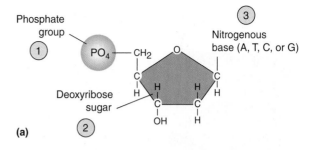

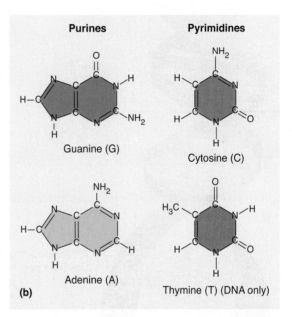

FIGURE 9.1 The structure of DNA. **(a)** The nucleotide. Each nucleotide is composed of (1) one to three phosphate groups, (2) a sugar, and (3) a nitrogenous base. **(b)** The four nitrogenous bases in DNA. Note that thymine is present in DNA; in RNA uracil is present in place of thymine.

of the DNA structure as a ladder in which the sides—the vertical aspect—are the sugar phosphate backbone and the rungs—the horizontal aspect—are the nitrogenous bases. Imagine now that the sides are twisted to establish a double helix. The two strands of DNA are **antiparallel**, meaning they "run" in opposite directions. As stated, the DNA strands are held together by the bases, but in a **complementary** way, meaning that A and T are a paired "fit," as are G and C, resulting in an equal number of A and T nucleotides and, similarly, an equal number of G and C nucleotides (**FIGURE 9.2**).

DNA Replication

Before bacterial cell division, **DNA replication** begins, resulting in the **progeny** receiving their "fair and equal share" of DNA. The starting point of replication is the enzymatic breakdown of the hydrogen bonds holding the two DNA strands together. The two DNA strands begin to unzip at specific sites, called the **origin of replication**,

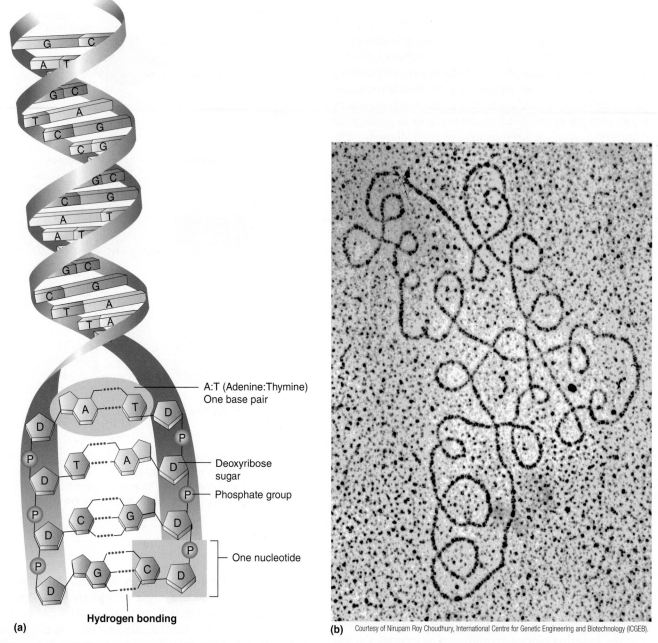

A:T (Adenine:Thymine)
One base pair

Deoxyribose sugar

Phosphate group

One nucleotide

Hydrogen bonding

(a)

FIGURE 9.2 (a) DNA is double stranded (dsDNA), helical, and antiparallel. (D = deoxyribose sugar; P = phosphate backbone; A, T, C, and G are the nucleotides.) The red dotted lines represent the hydrogen bonding between the nitrogenous bases of the nucleotides. Two hydrogen bonds occur between adenine and thymine, and there are three hydrogen bonds between cytosine and guanine. Note that in RNA uracil takes the place of thymine. **(b)** Electron micrograph of super-coiled chromosomal DNA from a lysed bacterial cell.

leaving exposed nitrogenous bases on each DNA strand to which free nucleotides in the "soup" of the bacterial **cytoplasm** attach, following complementary base pairing (A–T and G–C). As unzipping proceeds, each original DNA strand serves as a **template** for a new strand. The enzyme **DNA polymerase** synthesizes this new complementary DNA strand. The result is two double helixes, each consisting of an old and a new strand. The replication process is called **semiconservative** because each helix conserves

one old (original) DNA strand and one new DNA strand. In the event of an error, enzymatic-mediated correction, also known as **proofreading** usually occurs; if the error is not corrected, a mutation has arisen. The process is complex; the analogy of unzipping a jacket may help. As unzipping proceeds, the double-stranded DNA zipper separates into its two (single) components exposing the "teeth" upon which another single zipper (DNA strand) is built (see **FIGURES 9.3** and **9.4**).

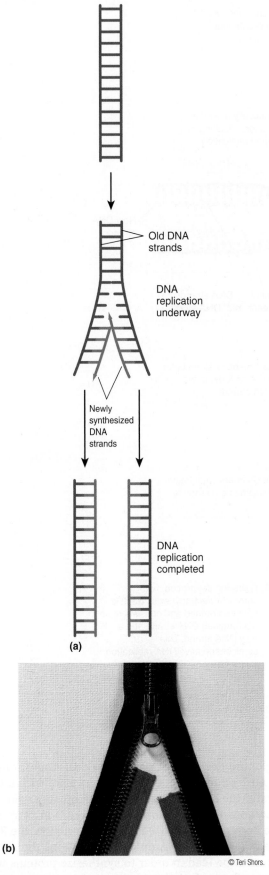

FIGURE 9.3 **(a)** Simplified illustration of DNA replication. **(b)** Zipper analogy. As unzipping of the two DNA strands occurs, a new separate zipper is built.

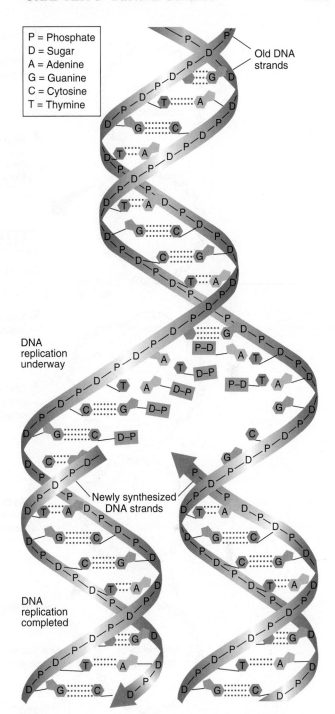

FIGURE 9.4 DNA replication illustrated at the molecular level.

The foregoing description of DNA replication is typical of eucaryotic cells in which there are multiple linear pairs of chromosomes. However, most bacteria have a single, looplike chromosome, resulting in some unique characteristics of DNA replication as compared with eucaryotic cells. The chromosome is enzymatically "nicked" and unzipped at the origin of replication into a V-shaped, two-pronged **DNA replication fork** along which new nucleotides are enzymatically assembled by DNA polymerase on each prong in a complementary fashion (**FIGURE 9.5**). The result is two double helixes, each of which ends up in a daughter cell during binary fission.

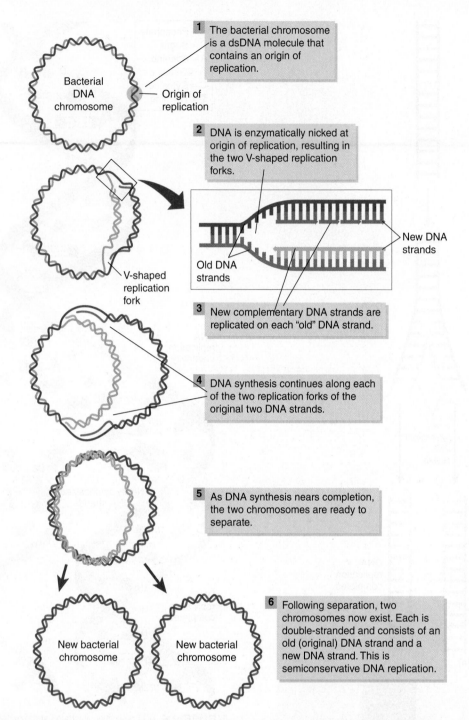

1 The bacterial chromosome is a dsDNA molecule that contains an origin of replication.

Bacterial DNA chromosome — Origin of replication

2 DNA is enzymatically nicked at origin of replication, resulting in the two V-shaped replication forks.

V-shaped replication fork

New DNA strands

Old DNA strands

3 New complementary DNA strands are replicated on each "old" DNA strand.

4 DNA synthesis continues along each of the two replication forks of the original two DNA strands.

5 As DNA synthesis nears completion, the two chromosomes are ready to separate.

New bacterial chromosome

New bacterial chromosome

6 Following separation, two chromosomes now exist. Each is double-stranded and consists of an old (original) DNA strand and a new DNA strand. This is semiconservative DNA replication.

FIGURE 9.5 Semiconservative replication. One bacterial chromosome is replicated into two double-stranded DNA chromosomes.

Transcription: DNA to mRNA

Having established the structure of DNA and its mechanism of replication, the process by which DNA, a series of nucleotides, is eventually converted into a series of amino acids, the building blocks of proteins, needs to be elucidated. The process of "sequence to sequence" is divided into the two steps of transcription and translation, both of which require ribonucleic acid (RNA). The structure of RNA is similar to that of DNA, with the exceptions that RNA is single stranded; the nucleotide uracil (U) is substituted for the nucleotide thymine (T) and base pairs with adenine (A); and the sugar, **ribose**, a five-carbon sugar similar to deoxyribose in DNA, has one more oxygen in its structure than does deoxyribose. Based on its function, the RNA is called **messenger RNA (mRNA)**. It contains the message that must be read in order to synthesize proteins within the cytoplasm of a bacterium.

Genetic information is transferred from DNA to mRNA during transcription (**FIGURE 9.6**). The process of

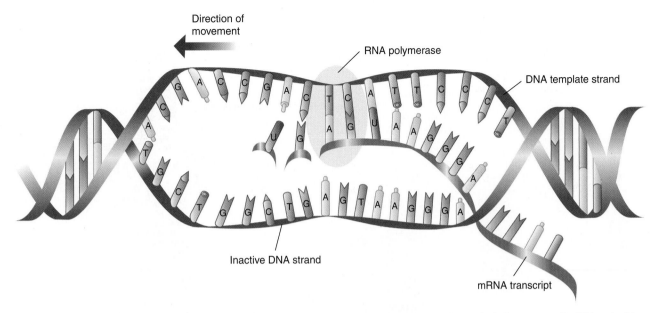

FIGURE 9.6 Transcription. The enzyme RNA polymerase moves along the DNA strand and unzips it into two strands. At the promoter site, RNA nucleotides complementary to the DNA template sequence are added to create the single-stranded mRNA transcript as unzipping proceeds.

transcription is similar to the mechanism by which DNA replicates in the sense that one strand of DNA serves as a template for the mRNA strand. A major distinction is that in DNA replication the whole molecule is copied, whereas in making an mRNA transcript only a short segment of single-stranded DNA is transcribed at a time. Transcription begins at a **promoter site** in a single strand of the DNA and ends at a **terminator sequence**. Within the segment between the promoter and the terminator the double-stranded DNA is unzipped (as in DNA replication), and one strand acts as a template for the enzyme **RNA polymerase** to make a complementary strand of mRNA. At the terminator sequence the enzyme and the new strand of mRNA detach from the DNA; the remaining single-stranded segment of DNA in the bacterium then converts back into a duplex.

The result is a strand of mRNA complementary to the template DNA strand (i.e., the blueprint). Transcription and translation occur simultaneously within the cytoplasm of a bacterium. The final stage of protein synthesis, translation, takes place on the ribosomes. Thus the mRNA carries the message of DNA and is appropriately called "messenger."

Translation: mRNA to Protein

Translation is the next and final stage in which the gene product, or **protein**, is synthesized. It is characterized by converting the genetic information encoded in mRNA (written in the four-letter alphabet of nucleic acids) into a protein (with its alphabet of 20 **amino acids**). Transcription, therefore, refers to accurately copying the genetic information from DNA to mRNA (both in the language of nucleic acids), whereas translation refers to converting

the genetic information encoded in mRNA into a "foreign" language (i.e., that of protein)—much as one would translate a menu from a French restaurant into English. The point of the central dogma is that the encoded information of the gene (DNA) flows into mRNA and then from the mRNA into the sequence of a protein.

Proteins are macromolecules and constitute about 50% of the dry weight of organisms. They are the most abundant compounds in the cytoplasm, function in a wide range of activities, and are extremely versatile. **Enzymes**, biological catalysts, and **antibodies** are all proteins, as are certain structural compounds of cell wall membranes and other cell structures. Proteins function in gene regulation and in numerous complex reactions in the bacterial cell. Amino acids are the building blocks of protein. Each amino acid consists of a central core (carbon) to which is attached an amino (NH_2) and a carboxy group (COOH), as well as a side chain, represented by an R group (**FIGURE 9.7**). The R group may be as simple as a single hydrogen atom, as in the amino acid glycine, or complex, as in phenylalanine. The distinctive part of each of the 20 naturally occurring amino acids is the chemistry of the R group. Amino acids are chemically joined together by **peptide bonds** to establish long chains of amino acids known as **polypeptides** (**FIGURE 9.8**).

Completion of the process of translation requires a second type of RNA similar to mRNA, called **transfer RNA (tRNA)** (**FIGURE 9.9**). This cloverleaf-shaped molecule transfers a specific amino acid molecule, one by one, onto the ribosome as a sequence of amino acids—a polypeptide—is chemically linked together through peptide bonds. The question arises as to how the tRNA molecules and their attached amino acids are directed in a particular sequence corresponding to the mRNA sequence.

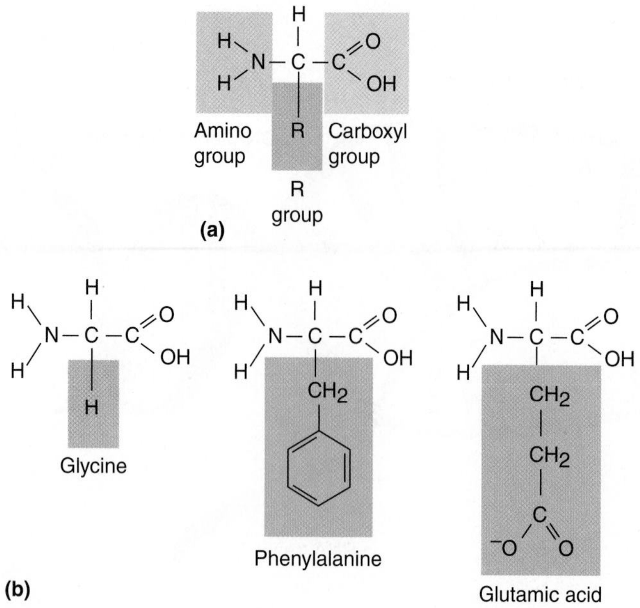

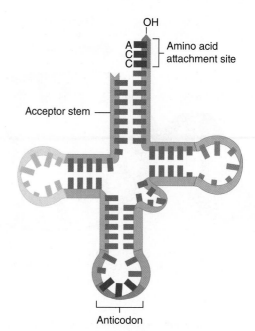

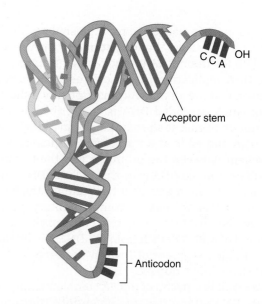

(a) Schematic configuration

(b) Natural configuration

FIGURE 9.9 (a) The "cloverleaf" structure of tRNAs. **(b)** A more correct or natural tRNA structure illustrating the folding of tRNA. In both (a) and (b) there are two binding sites—the anticodon and the amino acid attachment sites.

FIGURE 9.7 Amino acids. **(a)** A generalized structure for an amino acid, the building blocks of a protein. **(b)** Three examples of amino acid structures that contain different R groups.

The answer to this question goes back to mRNA. At a designated "start" point, the messenger RNA nucleotides are read on the ribosome in groups of three, referred to as **codons**. Each codon, in turn, relates to an **anticodon**, a series of three nucleotides on tRNA, in a complementary manner (A–U, G–C). Each tRNA molecule has a specific anticodon site to which is chemically bonded a specific amino acid. Hence, each tRNA molecule has two "binding sites." One is the anticodon region, consisting of three nucleotides that can attach to an mRNA codon in a base-complementary fashion (the language of nucleic acids). The second is the binding site for a particular amino acid (the language of proteins; Figure 9.9). Thus, the tRNAs carry the language and translate nucleic acid (mRNA) into protein, one amino acid at a time.

The codon dictionary is shown in **TABLE 9.1**. Note that the relationship is not always one codon–one amino acid. Consider the mathematics involved. If there are four nucleotide bases to choose from in building a codon of three bases, then 64 combinations ($4^3 = 64$) are possible. This would seem to present a problem because there are

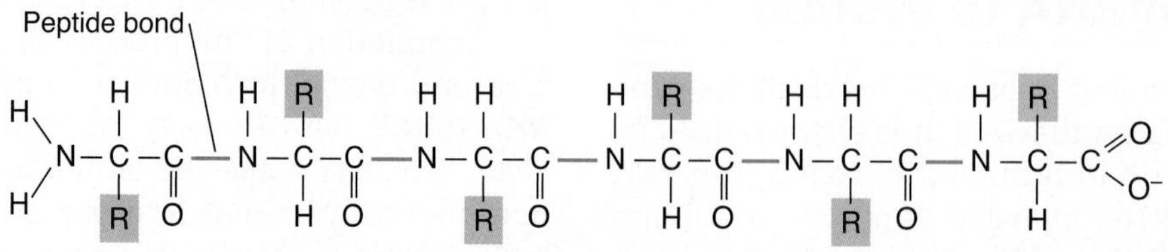

FIGURE 9.8 Amino acids are held together by the formation of peptide bonds (shown in red) during translation. The amino acids establish the primary structure of a polypeptide.

TABLE 9.1 The Genetic Code Decoder

The genetic code embedded in mRNA is decoded by knowing which codon specifies which amino acid. On the far left column, find the first letter of the codon; then find the second letter from the top row; finally, read up or down from the right-most column to find the third letter. In bacteria, AUG codes for N-formyl methionine, or ᶠmet.

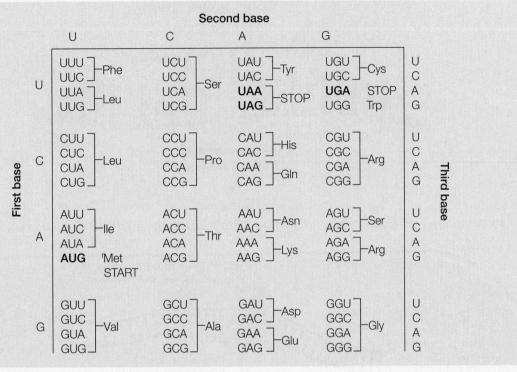

Ala = alanine; Arg = arginine; Asn = asparagine; Asp = aspartate; Cys = cysteine; Gln = glutamine; Glu = glutamic acid; Gly = glycine; His = histidine; Ile = isoleucine; Leu = leucine; Lys = lysine; ᶠMet = formylmethionine; Phe = phenylalanine; Pro = proline; Ser = serine; Thr = threonine; Trp = tryptophan; Tyr = tyrosine; Val = valine.

only 20 amino acids. However, examination of Table 9.1 shows that most amino acids are coded for by one or more codons, and hence the genetic code is referred to as being **redundant**. The amino acids serine and glycine have four codons and arginine—the record breaker—has six codons. Also, one codon (AUG, formylmethionine or ᶠmet) serves as a "start" (initiator) codon and three act as "stop" (terminator) codons (UAG, UAA, and UGA), as seen in the codon dictionary (TABLE 9.1), and do not code for any amino acids (nor is there a tRNA that binds to the sequences).

The sequence of amino acids in a polypeptide determines the particular protein that results out of the thousands of proteins that exist. Even very few mutations in the genetic material can have a profound impact on the final protein that forms. Consider sickle cell anemia, a life-threatening disease in humans, characterized by abnormal hemoglobin. The hemoglobin molecule consists of four polypeptide chains and a total of 584 amino acids. A mistake in amino acid number six is the source of this devastating condition. The number six amino acid is normally glutamic acid, but in sickle cell anemia valine is substituted because of a mutation in the gene that codes for hemoglobin. This is an example of a **point mutation**, in which one nucleotide is replaced with another nucleotide.

Because of the redundancy of the genetic code, there is some "wobble" room for amino acid flexibility, allowing for minor variation. Consider, as an example, the amino acid serine, and that if a mutation were to occur that changed the codon in the mRNA sequence from UCC to UCA, serine would still be the amino acid added to the polypeptide. However, if the mutation caused UCC to become UUC, then phenylalanine would replace serine in the protein (Table 9.1), leading to the production of an incorrect protein. Other types of mutations are addressed later in this chapter.

The basic process of translation is shown in **FIGURE 9.10**. *E. coli* has a **generation time** of 20 minutes. How is it that *E. coli* can divide so rapidly and contain all of its necessary living components? In procaryotes, the processes of transcription and translation are coupled together. In other words, transcription and translation occur simultaneously inside of the cytoplasm of a bacterium (**FIGURE 9.11**). In all eucaryotic organisms, transcription

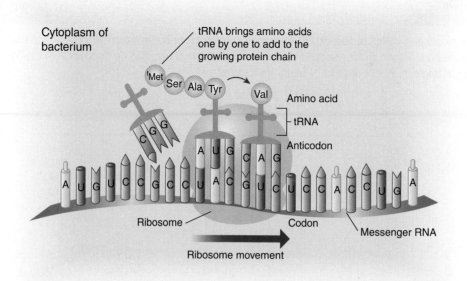

FIGURE 9.10 Diagram showing the process of translation in which mRNA is translated into protein in the cytoplasm of a bacterium.

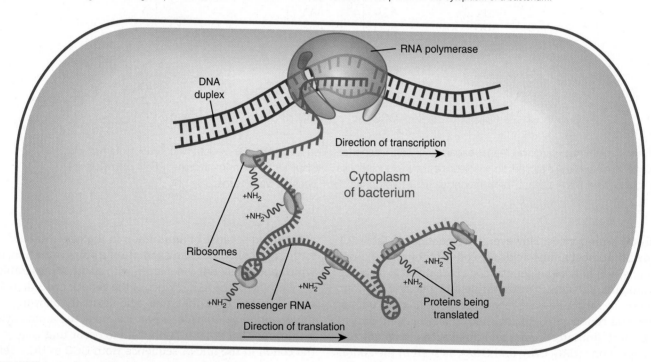

FIGURE 9.11 Transcription and translation are coupled processes that occur simultaneously within the cytoplasm of bacteria.

occurs in the nucleus, and the transcribed mRNAs are exported out of the nuclear pores and into the cytoplasm where they are translated by ribosomes into proteins.

Gene Expression

Both procaryotic and eucaryotic cells have a biochemical switchlike mechanism that allow genes to be turned on and off; it would be a waste of energy for genes to be constantly "turned on" when their protein products are not required. Major products of protein synthesis, as already

mentioned, are enzymes, which are biological catalysts. Some enzymes are constitutive enzymes, which are constantly produced because the gene switch is always in the "on" position. Other enzymes are called **inducible enzymes**, because their synthesis is the result of genes that can be turned on or off depending on the circumstances. The classical example of gene expression is in *E. coli* in which a particular enzyme (ß-galactosidase) is produced to break down the energy-rich lactose molecule into its two constituent sugars. As a matter of energy conservation, the gene that codes for the enzyme ß-galactosidase is

TABLE 9.2 Bacterial Pathogens and Their Genomes

Bacterial Pathogen	Infectious Disease	Genome* (Base Pairs)	Number of Genes*
Acinetobacter baumannii	Healthcare-associated infections	4,340,000	4,296
Bordetella pertussis	Whooping cough (pertussis)	4,086,186	3,816
Borrelia burgdoferi	Lyme disease	910,724	850
Chlamydia trachomatis	Trachoma (eye infection that can cause blindness)	1,044,459	911
Clostridium tetani	Tetanus	2,799,251	2,373
Escherichia coli	Enteritis	4,938,920	4,585
Helicobacter pylori	Ulcers	1,667,867	1,566
Listeria monocytogenes	Food poisoning	2,944,528	2,926
Mycobacterium tuberculosis	Tuberculosis	4,403,837	4,189
Staphylococcus aureus	Skin infections, food poisoning, healthcare-associated infections	2,870,000	3,017
Stenotrophomonas maltophilia	Serious respiratory infections in immune-compromised individuals	4,850,000	4,565
Vibrio vulnificus	Invasive bloodstream infections	4,950,000	4,790
Yersinia pestis	Plague	4,702,289	4,167

*Variation exists within species depending upon strain.

only turned on when lactose is present. Obviously, it would be a waste for the enzyme to be produced in the absence of lactose. In this case, the lactose acts as the **inducer** that turns the gene that codes for ß-galactosidase "on."

Another example of gene regulation is **repression**, or turning genes "off." The mechanism is similar to inducer mechanisms. Repression, however, is triggered not by the substrate of the enzyme, but rather by the end product of the reaction. A good example of this mechanism is the synthesis of the amino acid tryptophan; the presence of this amino acid switches off the genes for the further production of tryptophan. The on and off switches are under the control of a group of functionally related genes known as **operons**. *The major point of this brief description of gene expression is to emphasize that genes are not always in the "off" or "on" position, but rather are regulated.*

Chromosomes

Confusion exists regarding the terms *chromosome* and *gene*. A bacterium contains a circular **genome**, which usually consists of a single chromosome with from as few as 166,000 base

pairs to as many as 12,200,000 base pairs (**TABLE 9.2**). In most cases, only one copy of the chromosome is present, but there are exceptions; for example, *Epulopiscium fishelsoni* has as many as 100,000 copies of its chromosome. *E. fishelsoni*, a **gigantobacteria** that maintains a symbiotic relationship in the gut of sturgeonfish, has a length of 30 micrometers to more than 600 micrometers (in comparison to *E. coli*, which has a length of about 2 micrometers).

Mention the word *chromosome*, and Gregor Mendel, an Austrian friar noted for his pioneering work on genetics in the 1850s and 1860s, comes to mind. Mendel is best known for experiments on patterns of inheritance in pea plants earning him the title "father of modern genetics." His many and varied contributions were not fully recognized until the early years of the 20th century—some 60 years later.

A chromosome consists of long strands of DNA that are found in the nucleus of eucaryotic cells, whereas a *gene* is a piece or a segment of that DNA. The DNA, as has been noted, is, in turn, a sequence of nucleotide bases that, ultimately, through transcription and translation, prescribe a series of amino acids that are translated into proteins. The scenario is reminiscent of a portion of Jonathan Swift's "On Poetry: A Rhapsody" (1733):

> ... So, naturalists observe, a flea
> Has smaller fleas that on him prey;
> And these have smaller still to bite 'em;
> And so proceed ad infinitum. ...

The chromosome number is constant for each species, but does not differentiate each species; the number of chromosomes does not follow any pattern or particular logic. Consider a few examples; humans have 23 pairs of chromosomes, but so do guppies; mosquitos have 6, field horsetails have 216, and a particular type of fern is the record breaker, having 1,440 chromosomes (**TABLE 9.3**).

Bacterial Genetics

With this basic background, the remainder of the chapter focuses on bacterial (procaryotic) genetics. Both mutation and recombination by transformation, transduction, and conjugation are the mechanisms by which new genes and new recombinations of genes arise in procaryotic cells, accounting for their diversity. Mutation is random and haphazard and may produce genes that are adverse to survival; recombination is far more efficient because it deals with existing genes.

Mutations

As previously described, a mutation is a change in the nucleotide sequence. Some mutations are of little consequence to the microbe, whereas others have an adverse effect. Rarely, the mutation may have a beneficial effect. Considering the complex processes of transcription and translation, both of which involve a change from one language to another (DNA to mRNA and mRNA to protein), it is not surprising that errors—mutations—occur. Although cell mechanisms correct some of these mistakes, others slip through, and the error is passed on from generation to generation. Point mutations (discussed earlier) are the simplest mutations, whereas other mutations involve **insertions** or **deletions** of nucleotides. Mutations can occur spontaneously, but chemical and physical agents called **mutagens** can also induce mutations. Physical mutagens include ionizing radiation (X-rays and gamma rays) and nonionizing ultraviolet light. Chemical mutagens include nucleotide analogs (similar to nucleotides). It is possible for a mutation to revert back to its original state. Different types of mutations are summarized in **TABLE 9.4**.

Transposons, also called "jumping genes," in the genome are also responsible for mutations. Transposons are able to "jump" from one site on a chromosome to another or from a chromosome to a plasmid or from a plasmid to a chromosome. Plasmids are extrachromosomal, typically circular (some are linear) pieces of double-stranded DNA located in the cytoplasm of some bacteria that replicate independently of the chromosome. They carry a few to a large number of genes and are not

TABLE 9.3 Number of Chromosomes of Diverse Organisms*

Species	Number of Chromosomes
Typical bacterium	1
Mosquito	6
Rye	14
Pigeon	18
Tulip	24
Earthworm	36
Domestic cat	38
Bread wheat	42
Human	46
Guppy	46
Gorilla	48
Plum tree	48
Potato	48
Cow	60
Nine-banded armadillo	64
Chicken and domesticated dogs	78
Snowy owl	80
Goldfish	94
Field horsetail	216
Adder's tongue fern	1,440

*Variation exists within species.

essential for the survival of the cell. It is possible to "cure" (eliminate) them without interfering with the viability of the cell. Several kinds of plasmids have been identified, including those that code for antibiotic resistance and those that confer **virulence factors**. The fact that transposons are able to move into a plasmid opens up the possibility that they can jump from one bacterial cell to another and possibly even to a few eucaryotic cells. Transposons carry the genetic information, allowing them to be mobile.

TABLE 9.4 Types of Mutations in Bacteria

Type of Mutation	Description
Point	Substitution of one nucleotide for another.
Nonsense	One nucleotide change in an amino acid codon leading to a stop codon; e.g., UAU and UAC code for amino acid tyrosine, but UAA and UAG are stop codons.
Missense	One nucleotide changes in a codon, leading to a different amino acid; for example, GUU codes for amino acid valine and GCU codes for amino acid alanine.
Insertion	Addition of one nucleotide leads to a shift or change in the reading frame of codons; for example, the nucleotide sequence AGU CCA UUU ACC codes for the amino acid sequence of serine, proline, phenylalanine, and threonine. The addition of G in the second codon in front of A establishes the sequence AGU CCG AUU UAC, which codes for the amino acids serine, proline, isoleucine, and tyrosine.
Deletion	Deletion of one nucleotide leads to a shift or change in the reading frame (*frameshift mutation*); for example, the nucleotide sequence AGU CCA UUU ACG codes for the amino acids serine, proline, phenylalanine, and threonine. However, if the nucleotide A in the second codon is deleted, the sequence is now AGU CCU UUA CG, resulting in the amino acids serine, proline, and leucine.

Recombination

An important distinction needs to be made between the significance of the process of recombination in bacteria as procaryotic cells and in eucaryotes, including eucaryotic microbes. Recombination in sexually reproducing organisms (eucaryotes) is *vertical* and involves the fusion of male- and female-type gametes. Each gamete carries half the chromosome number (haploid) as the result of meiosis. Therefore, the offspring produced by gamete fusion have the full complement (diploid) number of chromosomes characteristic of that species. Human reproduction depends on the fusion of a sperm and an egg cell, each carrying 23 chromosomes, resulting in restoration of 46 chromosomes—the diploid number. Which one of the two identical chromosomes is selected from the 23 pairs of each parent is unpredictable but results in new **recombinants** based on the "genetic gamble." Obviously, recombination and reproduction are inextricably linked in eucaryotes.

But what about bacteria? *The strategy is very different: recombination and reproduction are separate and distinct events.* Bacteria reproduce by binary fission, an asexual process in which one bacterium splits into two, each of which splits into two, and so on, resulting in a population in which all of the bacteria are identical, or *clones* (barring mutation).

Recombination of bacterial genes is separate from reproduction. Recombination occurs through three distinct processes: transformation, transduction, and conjugation. Although these processes are distinct from each other, they have several key factors in common:

○ They are unidirectional.

○ Multiplication is not an outcome.

○ They are examples of *horizontal* gene transfer (the movement of DNA from one bacterium to another).

○ They are based on **homologous recombination** (integration of foreign donor DNA into host DNA).

○ They occur in nature, although described as laboratory phenomena.

○ They result in new genetic recombinations.

Transformation

Transformation is the uptake of "naked" fragments of DNA in the environment that were released by dead bacterial cells; the fragments are linear and in the order of only about 10 to 20 genes. The upshot is that the recipient (host) bacteria (those that have picked up and integrated the donor DNA fragments) are now recombinant cells. In a bacterial population, only about 10% to 20% of the bacteria are actually transformed; the major limitation is that the recipient bacteria must be able to take up DNA from the environment; that is, they must be **competent** for foreign DNA to enter through the cell wall and cell membrane.

Although some bacteria are naturally competent to take up DNA, most are not competent and can be induced to competency in the laboratory by techniques that create pores in their membrane. These include electroporation (electric shock), chemicals (such as calcium chloride), and hot and cold treatment. A large variety of Gram (–) and particularly Gram (+) cells are naturally competent, meaning they take up and integrate fragments of donor DNA into a complementary region of their chromosome (i.e., homologous recombination), occurring most frequently near the end of the logarithmic growth phase. Naturally competent cells take up DNA from a variety of sources, including bacteria belonging to other genera.

Entry of the foreign DNA into the recipient cell is not the end of the story. Although the donor DNA is now within the recipient bacterium, it must be integrated into the recipient DNA, displacing a piece of the bacterial chromosome. This is where homology comes into play; homologous sequences are the key to integration. Recall that A and T (or A and U) are complementary nucleotides, as are G and C. The donor DNA fragment is positioned next to a complementary DNA sequence of the recipient bacterium's chromosome. After enzymatic excision of the recipient host bacterial DNA fragment, the donor DNA replaces the excised fragment by "breakage and reunion." The process is complex and mediated by specific enzymes. Only a single strand of DNA enters into the recipient cell and is integrated into one strand of the host DNA, but in the next round of cell division one daughter cell retains the parent genotype and the other is the recombinant genotype. In general, the closer the relationship between the donor and the recipient bacterial strains, the more likely it is that there will be a greater number of homologous sequences, and hence a greater number of genetic recombinants. If there is no "fit" between the donor and recipient DNA, the donor DNA is disintegrated (**FIGURE 9.12A**).

Plasmid DNAs in the environment can also be picked up by a recipient bacterium through transformation (or by conjugation, which is discussed in a different section). Remember that plasmids are small, circular dsDNA molecules that contain an origin of replication, allowing them to replicate independently of the host chromosome (**FIGURE 9.12B**). Typically, plasmids only contain a few genes that are not essential to the daily survival of the bacterium. Instead, the genes can help the bacterium overcome challenging situations. Pathogenic bacteria often contain plasmids. Plasmids may contain antibiotic-resistance genes, genes that regulate capsule synthesis of the host, genes for fertility (involved in conjugation), or genes that code for toxins or antitoxins. Plasmids can be manipulated in the laboratory to contain *foreign genes* from other organisms for study or production in biotechnology applications (e.g., production of therapeutic proteins such as insulin that can be purified and used by diabetics).

Transduction

Transduction is another form of genetic exchange in which genes are transferred from donor host cells to recipient bacterial cells. The unique aspect of transduction is that the transfer agent, the **vector**, is a **bacteriophage**.

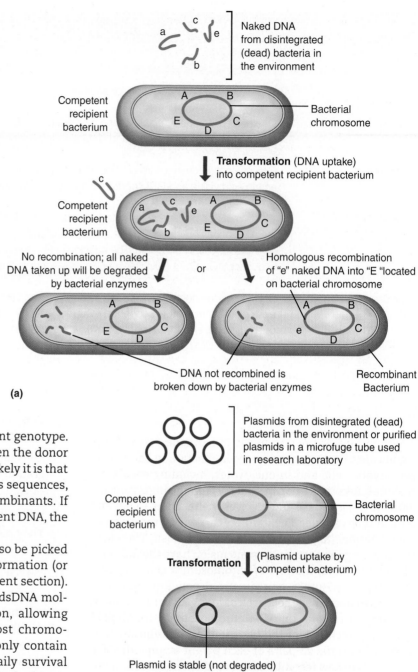

FIGURE 9.12 **(a)** Physiologically competent bacteria are able to take up naked or foreign DNA present in the environment from dead and degraded bacteria. A DNA fragment that shares the same sequences of DNA as the chromosome of the competent recipient bacterium may recombine and *replace* that section of DNA. This is called *homologous recombination*. The foreign DNA must be homologous. *Therefore, transformation only occurs between a few, close relatives.* The resultant recombinant bacterium will express the foreign genes it received and pass them along to the next generation of bacteria. Any DNA that is not incorporated into the chromosome of the bacterium will be degraded by bacterial enzymes within the cytoplasm of the bacterium. **(b)** Physiologically competent bacteria can also take up plasmid DNAs in the environment or plasmids used by scientists performing research. The plasmids are not integrated into the chromosome of the recipient bacterium and replicate independently. The plasmid will be passed on to the next generation of bacteria. Bacteria will typically only take up plasmids from a few close relatives.

Lytic bacteriophages are capable of infecting bacterial cells of appropriate specificity, in which case they follow the usual pattern of adsorption, penetration, synthesis, assembly, and release during lysis of the host. **Temperate bacteriophages** (as opposed to lytic bacteriophages) exhibit a **lysogenic cycle** that is the basis for two types of transduction: **generalized transduction** or **specialized transduction**. Both types of transduction are the result of the packaging of bacterial DNA into the bacteriophage protein coat, the **capsid** (to complete the head of the phage), resulting in "defective" phages. These phages are defective in that they contain only bacterial DNA or a combination of both bacterial and bacteriophage DNA. When the defective phage infects another recipient bacterial cell, the bacterial DNA from the first host bacterium becomes part of the genome of the second bacterial cell; homologous recombination by transduction has occurred.

In the case of generalized transducing lytic phages, enzymes break the bacterial chromosome into many fragments (**FIGURE 9.13**). During assembly, each DNA fragment has an equal chance of being packaged along with phage DNA into the capsid of the phage head. At the completion of the lytic replication cycle, most of the newly released bacteriophages are normal; that is, they contain only phage DNA, but a small percentage are defective and carry random bacterial genes in addition to their phage genes or the entire package of phage DNA is replaced by bacterial DNA from the host recipient bacterium. These bacteriophages with their accidentally packaged foreign bacterial DNA may infect other recipient bacteria and thereby transfer bacterial genes into a new host bacterial cell. *The hallmark of generalized transduction is that any bacterial chromosomal DNA fragment can be transferred.*

Specialized transduction, in contrast, is characteristic of temperate or **lysogenic bacteriophages** (**FIGURE 9.14**). In these phages the viral DNA (phage genome) is incorporated into the bacterial chromosome by homologous recombination. The incorporated phage DNA or a piece of it, called a **prophage**, represents a state of "peaceful coexistence" between the bacteriophage and the bacterial host, and the prophage is replicated with each round of bacterial cell division as though as it were an integral part of the host chromosome. At some point, either under natural circumstances or laboratory induction, the bacteriophage DNA is enzymatically excised from the bacterial chromosome of the host and enters into a lytic replication cycle, ultimately releasing large numbers of new bacteriophage particles. Most of the time excision is precise, and the resulting new phage particles that are released contain only their full complement of phage genes. Rarely, about one in a million times, the breakout is not "clean" and a few genes from either side of the attachment site adjacent to the prophage are attached to the phage DNA. When the defective phage infects a new bacterial recipient, these new bacterial genes (traits) are transferred. *Whether generalized or specialized transduction*

has taken place, the overarching significance is the creation of new combinations of genes contributing to bacterial diversity.

Conjugation

Bacterial conjugation is another mechanism of genetic recombination characterized by direct cell-to-cell contact between a mating pair of bacteria (**FIGURE 9.15**). Transmission is horizontal and in one direction. Conjugation requires that the donor bacterial cell harbors a plasmid in its cytoplasm, called an **F factor**, for **fertility factor**. Some F plasmids produce **F pili** (sex pili) that act as a bridge between donor and recipient bacteria. Those bacteria carrying the **F plasmid** are **F+ bacteria**, and those that lack the F factor are **F− bacteria**.

The conjugation process is initiated when contact is made between the sex pilus of the F+ donor bacterium and receptor sites on the F− recipient bacterium; the pilus shortens, drawing the two mating bacteria closer together. Within only a matter of minutes, one strand of the donor DNA plasmid is enzymatically nicked and a single strand begins to cross over into the F− recipient bacterium, a process that is completed in a few minutes (**FIGURE 9.16**). (Although it is convenient to think of the pilus as a channel for which the DNA passes from donor to recipient, there is doubt about this, and the actual mechanism of transfer is not known.) The plasmids in both the donor and recipient are now single stranded. Based on complementarity, a new strand is synthesized around the templates in both cells, resulting in double-stranded DNA in the donor and the recipient. The F− recipients have been converted to F+ status, and are now able to function as F plasmid donors.

The transfer of DNA is not limited to bacteria containing F plasmids; chromosomal DNA can also be transmitted from donor to recipient bacteria (**FIGURE 9.17**). If, as happens, the F plasmid in the donor bacterium is integrated into that same donor chromosome, a new cell type, called *Hfr*, results. These **Hfr bacteria** exhibit a much **higher frequency** of recombination with F− (hence the abbreviation Hfr). The integration of the F plasmid into the recipient bacterial chromosome resulting in Hfr bacteria is reversible, leading to a mixed population of Hfr and F+ bacteria. When Hfr and F− bacterial conjugation occurs, the first genes transferred are from the initiation site of the F plasmid, followed by chromosomal genes in a linear sequence along the single-stranded DNA.

The number of genes transferred is a function of the duration of conjugation and the fragility of the chromosome; only rarely is the entire chromosome transferred. The genes that dictate the fertility status of the donor bacterium are the last to be transferred; therefore, the F− cell will not usually become an F+ cell, but because it does carry some chromosomal genes it is now referred to as a *recombinant F− bacterium*. Although the details of conjugation may be complex, the significance is clear; conjugation

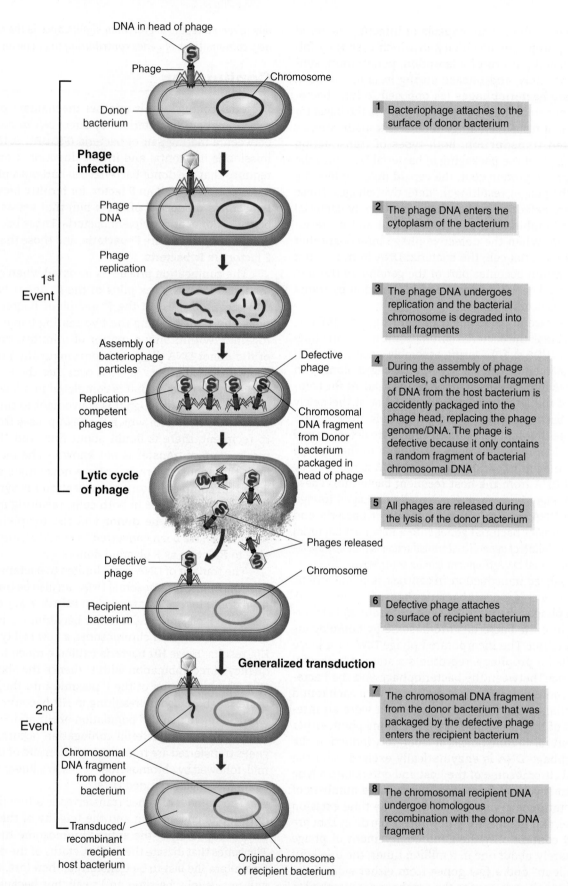

FIGURE 9.13 Generalized transduction is a genetic recombination process in which (1) a bacteriophage packages a fragment of chromosomal DNA during the course of a lytic infection of a host bacterium and (2) the resultant defective bacteriophage transfers the packaged host chromosomal DNA fragment into a recipient bacterium. The donor DNA fragment is integrated into the chromosome of the transduced/recombinant recipient host bacterium.

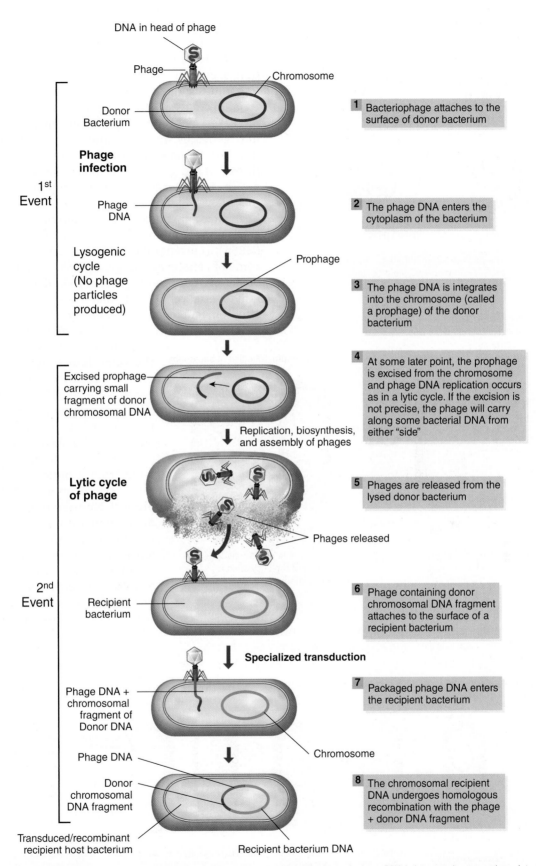

FIGURE 9.14 Specialized transduction is a genetic recombination process in which (1) a bacteriophage DNA is integrated as a prophage into the chromosome of a donor bacterium. (2) At some point in time, the prophage is activated into a productive/lytic replication cycle. New phages are released during lysis. On occasion, during break out or assembly and release of phage particles a phage will also package a small fragment of DNA from the chromosome that was flanking either side of the excised integrated prophage DNA. The bacteriophage then attaches to a new host recipient bacterium and transfers the packaged host chromosomal DNA fragment along with the phage DNA into the recipient host bacterium. The donor DNA fragment is integrated along with phage DNA into the chromosome of the transduced/recombinant recipient host bacterium.

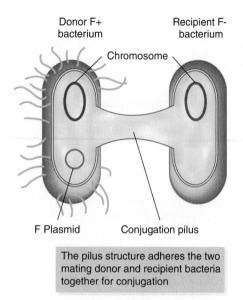

The pilus structure adheres the two mating donor and recipient bacteria together for conjugation

FIGURE 9.15 Bacterial conjugation is the direct transfer of DNA from an F⁺ donor bacterium to an F⁻ recipient bacterium.

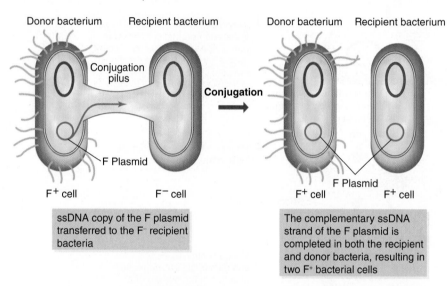

ssDNA copy of the F plasmid transferred to the F⁻ recipient bacteria

The complementary ssDNA strand of the F plasmid is completed in both the recipient and donor bacteria, resulting in two F⁺ bacterial cells

FIGURE 9.16 Bacterial conjugation between an F⁺ donor bacterium and an F⁻ recipient bacterium results in the transfer of the F⁺ plasmid to the F⁻ recipient bacterium and ultimately two F⁺ bacteria.

is a major genetic recombination event leading to bacterial diversity.

Whatever the mechanism of genetic recombination that occurs in bacteria, transformation, transduction, or conjugation, along with mutations, account for bacterial diversity (**TABLE 9.5**). Diversity, in turn, leads to increased ability to adapt to new environments. These new genes by mutation and new genetic recombinations are the raw material for Darwin's "survival of the fittest" at the genetic and molecular level. The "fittest" are those bacteria in a diverse population with genes that allow them to adapt, survive, and reproduce under hostile circumstances. As an example, consider that the emergence of methicillin-resistant *Staphylococcus aureus* (MRSA), a current major public health problem, is an outcome of genetic exchange. In 1955, a number of species of *Shigella*, a cause of dysentery, suddenly became resistant to tetracycline, sulfanilamide, chloramphenicol, and streptomycin in Japan, causing great concern. Researchers found that strains of *E. coli*, also isolated from the intestinal tract, were resistant to the same four drugs. Investigations proved that the antibiotic-resistant genes were transferred from *Shigella* spp. to *E. coli* via plasmids (conjugation). In 1995, a strain of *Yersinia pestis*, a cause of the **plague**, or **Black Death** (a term used in Medieval times), isolated from a patient in Madagascar was resistant to eight different antibiotics! Numerous other examples of the genetic transfer of antibiotic resistance as well as transfer of virulence factor genes by genetic recombination events have been identified. **FIGURE 9.18** provides an overview of how antibiotic resistance occurs as a result of genetic recombinations.

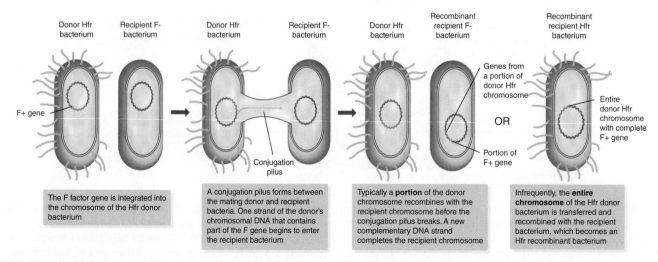

The F factor gene is integrated into the chromosome of the Hfr donor bacterium

A conjugation pilus forms between the mating donor and recipient bacteria. One strand of the donor's chromosomal DNA that contains part of the F gene begins to enter the recipient bacterium

Typically a **portion** of the donor chromosome recombines with the recipient chromosome before the conjugation pilus breaks. A new complementary DNA strand completes the recipient chromosome

Infrequently, the **entire chromosome** of the Hfr donor bacterium is transferred and recombined with the recipient bacterium, which becomes an Hfr recombinant bacterium

FIGURE 9.17 Mating of a donor Hfr bacterium with a recipient F⁻ bacterium usually results in a recombinant F⁻ recipient bacterium after conjugation.

TABLE 9.5 Comparison of Genetic Recombination Mechanisms of Bacteria

Genetic Recombination Process in Bacteria	Characteristics
Transformation	Uptake of naked DNA from the environment. Requires competent recipient bacteria.
Transduction	DNA carried by bacteriophage moves from donor to recipient bacteria. Generalized transduction is the result of bacterial donor chromosomal DNA fragments taken up by a bacteriophage following lysis of the bacterial host cell. Specialized transduction is the result of bacteriophage DNA carried along with excised bacterial donor DNA fragments following lysis of the bacterial host cell.
Conjugation	Mating contact between donor and recipient bacteria via F gene integrated into the donor chromosome and conjugation pilus. F⁺ to F⁻ Hfr to F⁻ (rarely Hfr to F⁺)

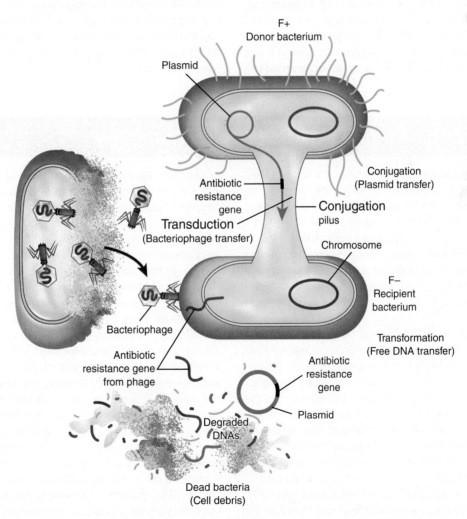

FIGURE 9.18 Overview of the different genetic recombination processes of bacteria. Antibiotic-resistance genes can be transferred to different bacteria through conjugation (plasmid transfer), transformation (free DNA in the environment is taken up by a bacterium), or transduction (bacteriophage transfer). Genes that code for the production of capsules or pili or the metabolic breakdown of different nutrients can also be transferred. The recombination process is not limited to the transfer of a specific gene or group of genes and ultimately is responsible for the diversity of bacteria.

Synthetic Biology

A knowledge of the genetics of microbes is of prime importance in exploiting and harnessing them by genetic manipulation for their beneficial use. A number of useful products made possible by genetic engineering in human medicine, animal husbandry, and agriculture are on the market. Understanding the genetics of bacteria also increases the opportunity to understand and to cope with microbes that are disease producers. In order to more effectively forestall epidemics and pandemics, DNA sequencing (i.e., determining the exact order of the nitrogenous bases—thymine, guanine, cytosine, and adenine) has contributed greatly to microbial genetics. To a large extent, biological warfare is based on the ability to engineer "designer bacteria" that produce large amounts of toxins that are better able to survive the hazards of the environment and that resist the effects of antibiotics.

Synthetic biology became the "new kid" on the block at the beginning of the 21st century, a discipline resulting from the rapid development of techniques to manipulate genetic material. **Next-generation DNA sequencing**, also called **high-throughput sequencing**, has enabled researchers to sequence the genetic code (DNA) of an entire organism or multiple genomes in a given sample at once, which is referred to as metagenomics. **Metagenomics** is the study of all genetic material present in environmental and biological samples such as feces, (refer back to the opening case study in which the hologenomic approach was used by researchers to study vampire bats and sanguivory), soil, plant roots, wastewater from a treatment facility, samples from extreme environments such as glaciers or deep mines, and even diseased tissues!

Metagenomics is used to identify the diverse microbes present in complex samples. Before these technological advancements, microbes had to be grown separately in the laboratory, the genetic material was extracted followed by sequencing. It has opened the door to many new and undiscovered diverse microbes that scientists were unable to grow in the laboratory. Metagenomics portrays a more realistic picture of the microbes or microbial communities present in the sample of interest.

The objective of synthetic biology is to produce unnatural biological molecules from "scratch," and possibly to create life or to assemble naturally occurring biological molecules into systems that function unnaturally as the result of modifying existing systems of genetic material. Synthetic biology (**BOX 9.1**) uses an engineering approach to biology and has numerous definitions. Some scientists view this discipline as a "new biological revolution." "Designer genes" may become as popular as "designer jeans."

BOX 9.1 Engineering Live Microbes as Therapies

Synthetic biology is exciting and has almost unlimited potential, but it presents both promises and perils that need to be considered. The public has the right to ask, "Do we really need it?" "Can it be contained?" or "What about the risk of creating a Frankenstein monster?" Bioengineers are already developing an inventory of *biobricks*; that is, biological parts containing genetic material. As an analogy, although not perfect, consider a group of children assembling a pile of Legos of assorted shapes and colors into a variety of forms from the same pile of pieces in diverse ways.

An article published in 2009 by Michael Rodemeyer of the Woodrow Wilson International Center for Scholars entitled "New Life, Old Bottles: Regulating First-Generation Products of Synthetic Biology" refers to the "golden dilemma" faced by regulators charged with developing constraints. ("Goldilocks and the Three Bears" is an old nursery rhyme. Goldilocks finds the porridge in the bears' house "too hot" or "too cold" until, finally, she pronounces one "just right.") If regulators are too hesitant and precautionary, they interfere with the benefits of synthetic biology, but if they are too lenient, their judgment could cause irrevocable damage. It is a juggling act and needs to be "just right."

In 2010, Craig Venter, a genomist highly recognized for his leading role on the **Human Genome Project**, again entered center stage with an announcement that his laboratory had created the first artificial life form, a bacterial cell that was able to produce proteins as an expression of its synthetic DNA and, thereby, opening a Pandora's box to the development of designer bacteria for the production of biofuels, pharmaceuticals, and chemical and agricultural products. Not all sequences can be synthesized in a "willy-nilly" sense; some take months to complete and other attempts have failed.

Through the use of metagenomics, we know even more about the complexities of the human **microbiome**. The most diverse and abundant microbes reside in the gut. The gut contains more than 1,000 different species of friendly bacteria. Gut microbiota play a strong role in stimulating the normal development and *education* of the immune system. Interest in the potential impact of gut microbiota on health and disease has surged. Scarcely any peer-reviewed scientific research reports can be found prior to 2000 using the search terms of "microbiota," "gut microbiota," or "oral microbiota." However, in the past 5 years the number of research papers on the topic has increased dramatically, with cumulative reports tallying 33,808 articles by the middle of 2017. Diseases such as obesity, inflammatory (irritable) bowel syndrome (IBS), diabetes, cancers, *Clostridium difficile* infections, Parkinson's disease, and numerous others have been reported to be associated with an imbalance of gut microbiota, referred to as gut **dysbiosis**. So far, observations by researchers and clinicians suggest that supplementing the body with "good bugs," or **probiotics**, can restore the gut to its natural healthy microbiota, alleviating symptoms of disease. To date, **fecal transplantation** is being used in the United States to successfully treat *C. difficile* infections. *C. difficile* infections are usually a complication of long-term antibiotic therapy. The antibiotics reduce the diversity of bacterial species of the gut microbiota, resulting in the overgrowth of *C. difficile* and debilitating chronic diarrhea.

In 2013, a microbiota-based therapy in combination with pre-biotics and antibiotics was used in a study conducted by medical researchers at Zhejiang University, China, to treat critically ill patients suffering from avian influenza A virus (H7N9) infections. Secondary bacterial infections are associated with infection by the H7N9 strain of influenza A virus, some of which are severe or even fatal. Infections by this strain of influenza A virus have about a 40% mortality rate. In their study, those patients with a mean age of 48 years who were treated with antibiotics in combination with capsules of probiotic bacteria containing *Clostridium butyricum* or *Bacillus subtilis*, and *Enterococcus faecium* and **prebiotics** stayed in the hospital an average of 13 days and did not suffer from secondary bacterial infections.

Today, synthetic biology has approached the frontier of therapeutics in which efforts are being made to engineer **GRAS (generally regarded as safe)** probiotic bacteria to restore the diversity of gut microbiota as a means to treat chronic diseases linked to gut dysbiosis and provide long-lasting benefits. GRAS probiotic bacteria are also being genetically modified to function as **live vaccines** to prevent infectious diseases.

Probiotic researchers have focused efforts on genetically engineering bacteria that have been used in foods and agriculture for hundreds of years (hence the term GRAS). They are genetically modifying different species of lactic acid–producing bacteria found in dairy products and other fermented foods such as *Lactobacillus lactis, L. gasseri, L. paracasei, L. plantarum*, and *Bifidobacterium longum*.

In addition to these food-producing probiotic bacteria, a nonpathogenic strain of *E. coli* that colonizes the human gastrointestinal tract but does not cause disease is being used in microbiota-based clinical applications. The name of the strain is *E. coli* Nissle 1917 (often referred to as *EcN*). It was isolated from a stool sample of a World War I soldier in 1917 by Dr. Alfred Nissle while working at a military hospital near Freiburg, Germany. He made the discovery that in comparison to other soldiers, the soldier did not suffer from diarrhea or any gastrointestinal diseases. At the time, soldiers were deployed in a location heavily contaminated with *Shigella*, causing soldiers to experience severe gastroenteritis.

In the laboratory, Nissle mixed pure cultures of *EcN* with pure cultures of pathogenic intestinal bacteria such as *Salmonella* and made the discovery that *EcN* was a strong antagonist of the pathogenic bacteria. Nissle applied for a patent, and a company began to manufacture and sell preparations of *EcN* as a medical probiotic marketed as Mutaflor®. Mutaflor is still sold around the world today as a medical probiotic to treat or prevent ulcerative colitis; chronic constipation; diarrhea in infants, toddlers, and children; IBS, Crohn's disease; and other inflammatory bowel diseases. *EcN* has been extensively studied for more than 100 years.

Attenuated strains of *Salmonella typhimurium* that cannot cause foodborne salmonellosis are also being developed for microbiota therapy. Attenuated or weakened strains of bacteria are no longer capable of causing disease due to mutations in their genetic makeup.

At the time of this writing, at least 170 GRAS probiotic bacteria have been genetically engineered as microbiota-based therapies (Table 1). The majority of them are in early phases of development. The main concern over the **genetically modified organisms (GMOs)** is their **biocontainment**, the

TABLE 1 Notable GRAS Probiotics Being Developed to Treat or Prevent Diseases

GRAS Genetically Engineered Probiotic Bacterium	Disease	Engineering/Mechanism
Escherichia coli Nissle 1917	Obesity and Inflammatory bowel disease (IBS)	Bacterium engineered to contain gene that codes for N-acetylphosphatidylethanolamines (NAPEs) to control food intake and weight gain.
Lactobacillus lactis	HIV vaccine	Bacterium engineered to contain the genes to produce *Streptococcus pyogenes* pili with the HIV p24 antigen fused to the ends of the pili, which will induce the body's immune response.
Lactobacillus lactis	Inflammatory bowel disease (IBS)	Bacterium engineered to contain anti-inflammatory cytokine genes (interleukin-10 or interleukin-27) to inhibit inflammation.

(continues)

BOX 9.1 Engineering Live Microbes as Therapies *(continued)*

GRAS Genetically Engineered Probiotic Bacterium	Disease	Engineering/Mechanism
Lactobacillus plantarum	High blood pressure (hypertension)	Bacterium engineered to contain gene that codes for angiotensin-converting enzyme inhibitory peptide (ACEIP), which helps blood vessels relax and decreases systolic blood pressure.
Salmonella typhimurium	Prevention of cholera	Bacterium engineered to contain the gene that codes for cholera toxin-B (CrxB) antigen, which induces immunity against *Vibrio cholerae*.
Lactobacillus paracasasei	Type 2 diabetes	Bacterium engineered to contain gene that codes for exendin-4 peptide, enhancing insulin production in β cells.
Salmonella enterica serovar Typhimurium	Cancer (melanoma)	Based on quorum sensing. When engineered bacteria reach high density (a quorum), the cells lyse and release antitumor proteins (pore-forming toxins called hemolysin E) to kill cancer cells.

interactions of the genetically engineered GRAS probiotic bacteria with commensal microbes in the body. **Figure 1** illustrates strategies used to treat infectious diseases or cancer using microbiota-based therapies.

It is not too far in the distant future when "intelligent" genetically engineered GRAS probiotic bacteria will be able to colonize the human gut, detect and eliminate pathogenic bacteria, and thereby prevent diseases at the earliest possible moment. The "intelligent" bacteria will be engineered to contain **synthetic gene circuits** that are involved in decision making (**Figure 2**). These "smart" engineered probiotic bacteria will sense changes within the community of the gut microbiota ecosystem and

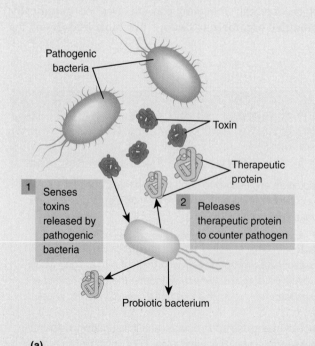

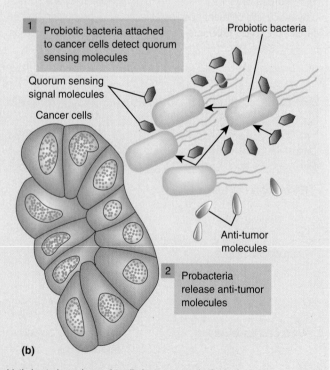

Figure 1 **(a)** Probiotic bacteria engineered to treat bacterial infections. **(b)** Probiotic bacteria engineered to eliminate cancers in the body.

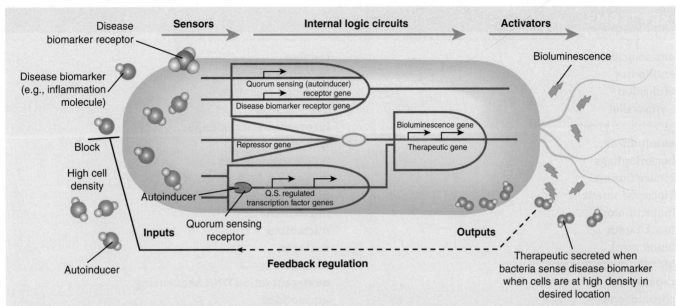

Figure 2 Simplified schematic illustrating the principles behind a genetically engineered bacterial therapy with synthetic biology for improving human health. These probiotic bacteria essentially contain synthetic genetic logic circuits that execute different tasks in a designed, logical manner. A probiotic bacterium is genetically modified to contain biosensor genes that encode a disease biomarker receptor. The disease biomarker proteins present in the body will bind to the receptor, allowing the therapeutic bacterium to sense or detect the presence of **disease biomarkers**, such as inflammation molecules. The programmed bacteria contain quorum-sensing (QS) genes. When the density of the engineered bacteria is high due to the accumulation of **autoinducer** molecules (encoded by QS genes) that are bound to QS receptors present on the engineered DNA that codes for specialized **transcription factors**, the genes that encode both a therapeutic protein and bioluminescence (light) will be turned on. The therapeutic proteins will be secreted and bind to the biodisease marker proteins, blocking their negative effects in the body (e.g., anti-inflammatory molecules that block inflammation or signal other proteins in the body to block inflammation). The production of the therapeutic protein and bioluminescence is coupled. Therefore, biomedical teams will be able to visualize the oscillating or blinking of bioluminescent light during the production of the drug in real time. A gene that codes for a **repressor** that blocks the production of transcription factors is also part of the synthetic gene circuit. The repressor is produced when the biodisease markers are not sensed or detected by the engineered bacteria. Ultimately, the bacterium is engineered to contain different genes that will function as biological sensors, enabling logical processing of its environment and activating the production of a therapeutic drug.

trigger the appropriate responses to correct dysbiosis before it can affect more distant sites in the body, such as the respiratory tract, heart, and brain. Imagine intelligent, genetically engineered probiotic bacteria that can produce and secrete an anti-inflammatory molecule that will stop inflammation in its tracks! Or imagine even further that smart probiotic bacteria could combat invading pathogens or detect and kill cancer cells or regulate mood and behavior. One day, microbiota-based therapies used in combination with personalized medicine will be the most effective approach to maintaining human health.

Summary

DNA is the repository of genetic information in procaryotic microbes. DNA is a double-stranded, supercoiled molecule. Genes are segments of DNA composed of the sugar deoxyribose; phosphate; and the nucleotides adenine (A), guanine (G), thymine (T), and cytosine (C). Mutations and genetic recombination events of transformation, transduction, and conjugation, all of which provide for new genes and new recombinations of genes, account for diversity in bacteria. Transformation is characterized by uptake of naked DNA by competent cells, phages are the vectors for the transfer of DNA from a recipient to a donor cell in transduction, and conjugation is characterized by physical contact between a bacterial mating pair. Bacterial genetics is the basis for genetic engineering, the cornerstone upon which biotechnology is built. Synthetic biology, a relatively new and developing area, allows for the bioengineering of unnatural life to create biological molecules and the assembly of natural biological molecules into systems that function unnaturally. Microbiota-based therapy is a recent development of synthetic biology that utilizes gene circuits involved in logical decision-making to treat chronic diseases in humans.

KEY TERMS

amino acids
antibodies
anticodon
antiparallel
attenuated strain
autoinducer
bacteriophage
binary fission
biocontainment
biotechnology
Black Death
blood meal
bloodborne pathogens
capsid
chromosome
codon
competent
complementary
conjugation
cytoplasm
deforestation
deletion mutation
deoxyribonucleic acid (DNA)
deoxyribose
disease biomarkers
disease triangle model
DNA polymerase
DNA replication
DNA replication fork
dysbiosis
enzymes
F (fertility) factor
F⁻ bacteria
F plasmid
F⁺ bacteria
fecal transplantation
gene
generalized transduction
generation time
genetic engineering
genetically modified organisms (GMOs)
genome
gigantobacteria
GRAS (generally regarded as safe)
Hfr bacteria
high-throughput sequencing
hologenome
homologous recombination
Human Genome Project
hypothesis

inducer
inducible enzymes
insertion mutation
live vaccines
lysogenic bacteriophage
lysogenic cycle
lytic bacteriophage
messenger RNA (mRNA)
metagenomics
microbiome
microbiota
mutagen
mutation
next-generation DNA sequencing
nucleotides
operon
origin of replication
peptide bonds
plague
plasmid
point mutation
polypeptides
prebiotics
probiotics
procaryotes
progeny
promoter site
proofreading
prophage
protein
purines
pyrimidines
recombinants
recombination
redundant
repression
repressor
ribonucleic acid (RNA)
ribose
RNA polymerase
sanguivores
semiconservative
specialized transduction
synthetic biology
synthetic gene circuits
temperate bacteriophages
template
terminator sequence
transcription
transcription factors

transduction
transfer RNA (tRNA)
transformation
translation

transposon
vector
virulence factors

SELF-EVALUATION

○ PART I: Choose the single best answer.

1. With reference to RNA, which is correct?
  a. A–T
  b. U–U
  c. U–T
  d. G–C

2. The uptake of naked DNA is called _____.
  a. transformation
  b. conjugation
  c. transduction
  d. DNA comingling

3. Which expression describes transcription?
  a. mRNA to protein
  b. DNA to mRNA
  c. DNA to protein
  d. DNA to mRNA to protein

4. Anticodons refer to:
  a. tRNA
  b. mDNA
  c. mRNA
  d. DNA

5. The integration of donor DNA fragments into a complementary region of a donor's chromosome is called
_____.
  a. operon switching
  b. homologous recombination
  c. transduction
  d. jumping genetics

6. The transfer structure of a bacterium used in conjugation is a(n) _____.
  a. pilus
  b. plasmid
  c. bacteriophage
  d. endospore

7. A(n) _____ is bacteriophage DNA incorporated into the chromosome of its host bacterium.
  a. prophage
  b. plasmid
  c. insertion mutation
  d. retrovirus

8. A _____ is the entire set of genes of a host, in addition to the genes of its microbiota.
  a. prophage
  b. chromosome
  c. dark DNA
  d. hologenome

9. Which of the following is not a characteristic of DNA?
  a. Double stranded
  b. Antiparallel
  c. Complementary
  d. Contains uracil

10. Which of the following is a macromolecule that constitutes about 50% of the dry weight of an organism?
  a. DNA
  b. RNA
  c. Proteins
  d. Lipids/membranes

O PART II: Fill in the blank.

1. _____ is a term used for protein synthesis.

2. A bacterium with the F plasmid integrated into its chromosome is called a(n) _____ cell.

3. _____ is the transfer of DNA into mRNA.

4. Bacterial cells must be _____ in order to take up foreign DNA through the cell wall via transformation.

5. Jumping genes are also called _____.

6. GRAS stands for _____ microbes for the development of microbiota-based therapies.

7. _____ is the hereditary material of a cell.

8. Bacterial _____ refers to a variety or different types of bacteria.

9. _____ allows for bioengineering of unnatural life through the assembly of natural biological molecules.

10. A(n) _____ microbe or virus is a weakened strain of a pathogen that can no longer cause disease but can be used in vaccine applications.

O PART III: Answer the following.

1. What are synthetic gene circuits, and how could they potentially be used to treat diabetes in the future?

2. What are the three major differences between RNA and DNA?

3. What is gut dysbiosis, and what diseases are associated with it?

4. Why are "defective" phages so named?

5. A stretch of mRNA is GCUUACCGAUAC.

   a.   What sequence of amino acids will be translated?

   b.   The above mRNA sequence is derived from a sequence of DNA. What is that DNA sequence?

   c.   Assume a deletion of the first U occurs. List the resulting amino acids sequence derived from the altered mRNA sequence.

6. Name and draw a tRNA molecule, clearly indicating the two "binding" ends.

7. List the different types of mutations. Which types are the most serious and why?

8. List the three main types of genetic recombinations or genetic material transfer processes. Discuss how a pathogenic bacterium could become antibiotic resistant through one of these processes (Hint: FIGURE 9.18.)

9. Compare and contrast generalized and specialized transduction.

10. Define F⁺, F⁻, and HFr cell types, and describe how they relate to bacterial mating.

PART 3

MICROBIAL DISEASE

CHAPTER 10

CONCEPTS OF MICROBIAL DISEASE

> "*Given enough time a state of peaceful coexistence eventually becomes established between any host and parasite.*"
>
> —René Dubos
> (1901–1982)
>
> *French-born microbiologist*

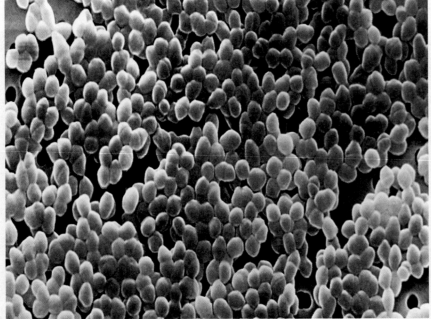

Courtesy of the CDC/Janice Haney Carr.

Digitally colorized scanning electron micrograph showing large numbers of *Enterococcus* spp. bacteria. *Enterococcus* spp. are a leading cause of hospital-acquired bloodstream infections, surgical wound infections, **urinary tract infections (UTIs)**, and catheter-related infections. *Enterococcus* spp. infections are becoming more challenging to treat due to the development of antibiotic resistance.

© Shutterstock / happykanppy.

LEARNING OBJECTIVES

1. Summarize the three types of microbial relationships associated with the symbiosis umbrella.

2. Explain how dysbiosis of gut microbiota may be involved in Parkinson's disease or obesity.

3. Define *quorum sensing* as it pertains to bacterial group behaviors.

4. Differentiate between and provide examples of offensive and defensive strategies used by pathogens to establish disease.

5. List at least five different bacterial species and two phyla that live on or inside of the human body, and discuss the importance of a healthy human microbiome in preventing disease.

6. Explain why antigenic variation is a significant microbial defense mechanism.

7. Define the three major factors that determine the severity of an infectious disease.

8. Summarize the defined periods that describe the course of microbial disease.

Case Study: The Smelly Chicken Factory Worker

Ian Langerhan was a 29-year-old male employed at a poultry-processing factory in Newport, Wales. One day while dressing chickens (**Figure 1**), he accidently pricked his finger with a chicken bone. He regretted that he was not wearing any gloves. Ian had a small puncture wound on his finger. His entire finger and part of his right hand became very red and inflamed over the course of a few days. He was concerned that he had **sepsis** (bacteria multiplying in his bloodstream causing infection and damage) or **bacteremia** (present but not multiplying in his bloodstream) and needed immediate medical attention. The other odd symptom was the fact that the injured finger smelled really bad.

Ian arrived at a hospital emergency room for treatment. A doctor carefully examined his hand and prescribed antibiotics. He assured Ian that the odor would disappear as soon as the infection was resolved by antibiotic treatment. The wound did heal and the inflammation and redness went away, but the odor continued to linger. It seemed to be emanating from his entire arm. Ian was so embarrassed. When Ian was confined to a small examination room at the clinic, medical staff found the odor to be intolerable. They had to use nose plugs while examining him.

Doctors assumed that Ian had contracted an unusual bacterium from the chicken bone that had punctured his skin that was ultimately the cause of Ian's horrific body odor. They needed to identify the bacteria that were causing the odor. Two months after antibiotic treatment, skin biopsies were sent to the clinical laboratory in order to isolate what bacteria were causing the odor.

(a)

© Andresr/ E+/Getty.

(b)

© Teri Shors.

Figure 1 **(a)** Workers at a factory dressing chickens for resale to grocery stores and restaurants. Dressing a chicken occurs after the chicken has been plucked and washed. It involves removing the feet and innards. **(b)** Typical dressed chicken sold in grocery stores.

(continues)

Case Study: The Smelly Chicken Factory Worker (continued)

Medical laboratory technicians were tasked with isolating the culprit. They isolated Gram (+) endospore-forming rods from skin biopsy samples taken from several sites along Ian's affected arm and other extremities. The bacteria were strict **anaerobes**. The technicians narrowed their identification down to one genus: *Clostridium*. Through additional metabolic biochemical testing, it was determined that there were three different species of *Clostridium* present on Ian's skin: *C. novyi*, *C. malenominatum*, and *C. cochlearium* (**Table 1**).

Immunofluorescent staining of skin biopsies determined that the sebaceous glands within Ian's skin harbored the clostridial bacteria (**Figure 2** and Table 1). All three species were known to be residents of normal **microbiota** in the oral cavity and gastrointestinal tract of humans. However, none of them were known to colonize *human skin*. Stool/fecal cultures from Ian repeatedly tested negative for these bacteria in the gastrointestinal tract. Saliva samples also tested negative. However, all three of the clostridial species were reported to exist in poultry. This evidence pointed to

TABLE 1 Patient Culture/Laboratory Test Results

Clinical Laboratory Testing	Rationale for Performing the Test	Results
Culture of skin biopsies	Isolation of bacteria producing the odor	Successfully isolated bacteria for further characterization and identification
Gram stain	Characterization of bacterial isolates from patient specimen	Gram-variable, endospore-forming rods
Additional characterization of bacterial isolates	Genus and species identification	Three strict anaerobes isolated: *C. novyi* type B variant, *C. malenominatum*, and *C. cochlearium*
Immunofluorescent staining of skin biopsies	Determination whether and where the smelly *Clostridium* bacteria were located on the skin	The sebaceous glands stained positive for *Clostridium* spp.
Stool cultures	To determine whether the *Clostridium* species were residents of Ian's gastrointestinal tract microbiota	*Clostridium* spp. not isolated from Ian's stool
Blood draw/serum analysis	Determination of whether Ian was immune suppressed, which would explain why his immune system was not clearing the *Clostridium* spp.	No immune deficiencies identified
Air extraction method and gas liquid chromatography on metabolites produced by the bacteria isolated from patient	Identification of the odiferous metabolites secreted by the bacteria isolated from Ian's skin	*Clostridium* spp. produced acetic, propionic, *n*-butyric, and 4-methylvaleric acids
Antibiotic susceptibility testing of *Clostridium* spp.	Determination of which antibiotics could be used to eliminate the odor-producing bacteria from Ian's skin	All *Clostridium* species identified were highly susceptible to antibiotics in a laboratory setting but failed to eradicate the clostridia from the patient even 5 years after the initial infection

Information from C. M. Mills, M. B. Llewelyn, D. R. Kelly, & P. Holt. (1996). A man who pricked his finger and smelled putrid for 5 years. *Lancet, 348* (9), 1282.

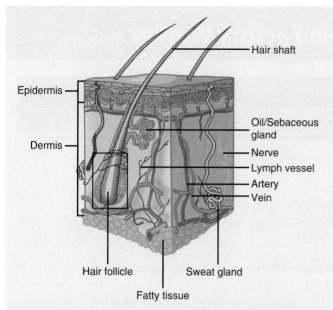

Figure 2 Diagram of human skin portraying the oil, or sebaceous, glands; the niche for the odor-producing clostridial bacteria isolated from the patient. Sebaceous glands are connected to hair follicles. They produce sebum, which lubricates the hair shaft and moisturizes the skin. Clostridial bacteria produce lipases, enzymes that can break down sebum as a nutrient source for the bacteria. The sebaceous glands may provide the anaerobic environment necessary to sustain multiplication of anaerobic bacteria. The skin has two types of sweat glands. Eccrine sweat glands are found all over the body and control temperature by excreting watery sweat. Apocrine sweat glands are concentrated in the armpits and groin. Both types of sweat glands excrete sweat that is odorless but produce odor when the sweat mixes with bacteria on the skin's surface. Sweat and sebum form an acidic layer (pH 4.5 to 5.6) over the skin that acts as a moisturizer and a barrier to bacterial infections.

Labels in figure: Hair shaft, Epidermis, Oil/Sebaceous gland, Dermis, Nerve, Lymph vessel, Artery, Vein, Hair follicle, Sweat gland, Fatty tissue

the chicken bone as the likely source of the bacteria that colonized Ian's skin.

Antibiotic sensitivity testing was used to determine which antibiotics would be effective in eradicating the bacteria from his skin. An air extraction technique was used to identify the volatile, odiferous clostridial **metabolites**. The bacterial isolates from Ian's skin produced the following putrid metabolites: acetic, propionic, *n*-butyric, and 4-methylvaleric acids. Acetic acid smells like vinegar; propionic acid smells like pungent body odor; *n*-butyric acid smells like rancid butter, Parmesan cheese, and vomit; and 4-methylvaleric acid has a cheesy odor.

This was a fascinating discovery because most human body odor is caused by different species of *Staphylococcus* and *Corynebacterium*. *Clostridium* infections in humans are usually severe. A team of medical researchers wondered why the anaerobic bacteria, in particular *C. novyi*, did not cause a severe infection such as **gas gangrene**. Three types of *C. novyi* have been identified: A, B, and C. All three types produce toxins

that can cause extreme edema, hemorrhaging, and fast-spreading necrosis (death) of tissues or gas gangrene. Further characterization determined that Ian's skin was colonized by *C. novyi* type B. Why wasn't Ian's infection more invasive and severe? *C. novyi* is difficult to cultivate in the lab due to its extreme anaerobic nature. It makes performing laboratory tests to answer questions about its physiology challenging.

Ian's quality of life was impacted by this strange complication associated with an infected puncture wound caused by a chicken bone. He became socially isolated. Besides antibiotics, many different approaches were used to eradicate the odor-causing bacteria (**Table 2**). Nothing worked. Ian became more and more depressed. He felt like a human guinea pig. Chronic antibiotic therapy wreaked havoc with his intestinal microbiota but didn't damper the body odor. Antibiotics, along with other drugs and procedures, caused some uncomfortable side effects.

Laboratory blood tests determined that Ian was not immune suppressed. Doctors wondered if he was a genetic anomaly in that his immune system had a "blind spot" that could not detect and clear the repulsive bacteria that had become *commensal* residents of his skin microbiota. After 5 years, when all attempts to eradicate the Ian's body odor were unsuccessful, a team of doctors working on his case at the Royal Gwent Hospital and the University Hospital of Wales threw up their hands. They published a case report in *The Lancet*, asking peers if any of them had successfully treated a similar case and for suggestions in either eradicating the skin bacteria or Ian Langerhan's body odor. This case puzzled the medical community. None of their interventions helped. After a few more years, Mr. Langerhan's mysterious odor spontaneously disappeared on its own!

Is it possible that the *Clostridium* species colonizing Ian's skin did not produce any toxins that attack human tissues/skin? Did the bacterial isolates only cause severe disease in poultry or farm animals? In other words, is tissue specificity essential for disease? Did clostridia waste products build up within the sebaceous glands, slowly killing off the bacteria, reducing odor as the number of bacteria declined? The dermis and epidermal layer of human skin consume oxygen. Oxygen is absorbed from the atmosphere into the layers of skin. Did Ian's good skin circulation over the course of time eventually kill the strictly anaerobic bacteria? Did the low pH of his skin inhibit clostridia lipases from breaking down the sebum in the sebaceous glands, harming its survival inside of the sebaceous glands? Enzymatic reactions are dependent upon certain pH conditions. Many unanswered questions about Ian's mysterious bout with body odor remain.

(continues)

Case Study: The Smelly Chicken Factory Worker (continued)

TABLE 2 Patient Treatment and Rationale

Treatment	Rationale for Therapy	Outcome
Flocloxacillin	Antibiotic (inhibits bacterial cell wall synthesis; attempt to inhibit odor-producing bacteria)	No impact on odor
Ciprofloxacin	Broad-spectrum antibiotic against Gram (+) and Gram (–) bacteria (inhibits bacterial DNA gyrase/replication; attempt to inhibit odor-producing bacteria)	No impact on odor
Erythromycin	Antibiotic (inhibits bacterial protein synthesis; attempt to inhibit odor-producing bacteria)	No impact on odor
Metronidazole	Antibiotic (inhibits nucleic acid synthesis; effective against anaerobic bacteria; attempt to inhibit odor-producing bacteria)	No impact on odor
Surgery	Explore tissue around puncture wound; removal of bone or other cause of odor	No chicken bone remnants, pus, or soft tissue damage
Hyperbaric oxygen treatment (2-month period)	Kill anaerobic bacteria such as *Clostridium* spp. that are causing the odor. Oxygen kills strict anaerobes	No impact on odor
Isotretinoin	Reduce inflammation, shrink sebaceous/oil glands (secrete less sebum/oily, waxy matter); reduce nutrient availability for lipase-producing clostridial bacteria, reducing odor	No impact on odor
Ultraviolet light treatment plus psoralen	Used to treat severe skin dermatitis; may alter skin microbiota, reducing odor	No impact on odor
Colpermin	Reduce sweating/odor	No impact on odor
Pro-banthene	Reduce sweating/odor	No impact on odor
Chlorophyll	Promote healing and reduce odors	No impact on odor
Antibiotic withdrawal	Allow restoration of skin normal microbiota to outcompete odor-causing bacteria, thereby reducing odor	No immediate impact on odor

Information from C. M. Mills, M. B. Llewelyn, D. R. Kelly, & P. Holt. (1996). A man who pricked his finger and smelled putrid for 5 years. *Lancet, 348*(9), 1282.

Questions and Activities

1. The bacteria that colonized Ian's skin were strict anaerobes. What is a strict anaerobe?

2. Explain how the skin could provide an anaerobic niche for *Clostridium* spp.

3. Define *commensalism*. How does it differ from mutualism and parasitism? (Hint: read the section on *Biological Associations*.)

4. Create a list of at least a handful of bacterial genera that are normal microbiota of skin.

5. What other bacterial genera that colonize the skin are known to produce odors? What are their Gram reactions and shapes? Do they produce endospores?

6. Create a list of at least a handful of bacterial genera that are normal microbiota of the gastrointestinal tract.

7. Create a list of at least a handful of bacterial genera that are normal microbiota of the oral cavity/mouth.

8. Explain how it is possible that the *Clostridium* bacteria were not killed or inhibited by antibiotic treatment but that in the laboratory the bacteria were sensitive to the effects of the antibiotics used in testing.

9. List the symptoms caused by *C. botulinum*, *C. perfringens*, and *C. tetani* infections in humans. What is a natural **reservoir** for these species of clostridia?

10. Compare and contrast the symptoms of infections caused by *C. botulinum*, *C. perfringens*, and *C. tetani* infections in humans with those experienced by Ian Langerhan.

11. Given that oxygen kills strict anaerobes, explain how it is possible that they can be cultured in the laboratory.

12. Could the bacteria isolated from Ian also be growing in the sweat glands of his skin? Why or why not?

13. Why was Ian's stool tested for the clostridial bacteria?

14. Besides antibiotic treatment, list three other treatments and their rationale to cure Ian's infection.

15. What are the symptoms of *C. difficile* infections in humans? Why or how do these infections occur in humans? Is it possible that the odor-causing bacteria flourished in Ian's skin in ways similar to *C. difficile* infections of the gastrointestinal tract? Explain.

16. Is it possible that endospore formation played a role or roles in the infection process? Explain.

17. Many questions were raised at the end of this case study. If it were possible to easily study the physiology of the clostridia isolates in the research laboratory, what experiments would you design to answer at least two of these questions?

18. Create a list of five questions you would ask Ian pertaining to this case if he were invited to class one day.

19. Create a list of five questions you would ask the team of medical doctors and researchers that worked tirelessly to treat Ian's body odor.

Information from C. M. Mills, M. B. Llewelyn, D. R. Kelly, & P. Holt. (1996). A man who pricked his finger and smelled putrid for 5 years. *Lancet, 348*(9), 1282.

Preview

"*No man is an island*" is a popular expression dealing with human interactions. Biologically, this expression can be extended to "no organism is an island." Wherever one looks in the biological world, associations between varieties of species are apparent. Even bacterial cells harbor viruses—**bacteriophages**. Humans are part of an invisible ecosystem of microbes and, like it or not, are a part of the food chain and serve as a microbial fast-food restaurant. A variety of microbes live in peace on or inside of their human **hosts**, as relatively permanent inhabitants that constitute the normal microbiota (also referred to as the human **microbiome**), or as transient microbial visitors just passing through (**BOX 10.1**). Bacteria communicate to each other through a chemical language known as **quorum sensing (QS)**. Many group behaviors achieved through quorum sensing are associated with bacterial survival.

BOX 10.1 You Need Guts to Survive: "Know Thyself"

"*Messieurs, c'est les microbes qui auront le dernier mot.*" (Gentlemen, it is the microbes who will have the last word.)

—Louis Pasteur (1822–1895), French microbiologist and chemist

In the 1970s, it was estimated and documented through numerous reports in the scientific community that 40 trillion bacteria, 10 times the number of human cells, colonized the human body. These numbers have recently been called into question. Scientists today believe that it is more likely a tie, with new estimates of about 3.7×10^{13} human cells and 4×10^{13} bacteria in the body. According to Norwegian social anthropologist Dr. Ole Bjorn Rekdal, "*The story of the ten-to-one ratio has all the*

characteristics of an academic urban legend." Regardless of the exact numbers, we live in a microbial world, and the bacteria are the rulers of our human "planet."

Many of the bacterial residents likely coevolved with their human hosts over millennia. At birth, humans are completely sterile, but then are immediately colonized by microbes from the local environment. As we age, the community of microbes that inhabit us, our microbiome, becomes more stable and complex in the absence of outside disturbances such as infection or antibiotics. The proportion of bacterial species present in the human microbiota vary from person to person; the microbiome of an individual is unique. (Refer to Figure 10.4 and Table 10.1 for examples of different microbiota that populate multiple locations in or on the human body; this feature focuses on gut microbiota.)

(continues)

BOX 10.1 You Need Guts to Survive: "Know Thyself" (continued)

TABLE 1 Diversity of Human Gut Microbiota Based on Dietary Practice

Dietary Practice	Type of Food Consumed or Restricted	Characteristics	Bacterial Gut Microbiota
Vegans[a]	High in vegetables and legumes, nuts No meat or meat products, such as gelatin; no fish or poultry; no dairy (milk) or eggs; no processed foods.	Low fat High fiber	Decreased *Bacterioides* spp., *Bifidobacterium* spp., *E. coli*, and bacteria belonging to *Enterobacteraceae*
Vegetarians[b]	High in fruits and vegetables No meat or meat products; may eat products that come from them such as milk and eggs.	Low fat High fiber	Increased proportions of *Bacteriodes/ Provotella* spp., *Bacteroides thetaiotaomicron*, *Clostridium clostridioforme*, and *Faecalibacterium prausnitzii*
Omnivores (Western diet)	Eat meat, including beef, pork, lamb, poultry, fish, and meat products such as eggs and milk; consume some fruits and vegetables, etc.	Low fiber	Increased *Alistipes* spp., *Bacteriodes* spp., *Bilophila* spp.; and butyrate-producing *Clostridium* spp. Decreased levels of fermicutes, such as *Roseburia* spp., *Eubacterium rectale*, and *Ruminococcus bromii*

[a]Many vegans refrain from eating sugar and consuming some wines. Some debate whether certain foods, such as honey, fit into a vegan diet. Some practice a raw vegan/ raw food diet that consists of unprocessed vegan foods that have not been heated above 46°C (115°F), based on the idea that foods cooked above these temperatures will lose a significant amount of their nutritional value and are harmful to the body.

[b]Besides not eating beef, pork, poultry, fish, shellfish or any animal flesh of any kind, vegetarians are further distinguished based on whether they eat dairy and/or eggs. Lacto-ovo vegetarians eat eggs and dairy products. Lacto-vegetarians do not eat eggs but eat dairy. Ovo-vegetarians eat eggs but do not eat dairy products.

Many factors influence the composition of gut microbiota. The anatomy of the human gut is defined by genetics. As mammals evolved, so, too, did their digestive tracts. **Carnivores** are meat eaters and depend on predation to acquire food. Their large intestines are not as complex as those of herbivores, because microbes are not required to digest or break down plant cellulose. In contrast, **herbivores** consume only vegetation. They depend on microbial fermentation in a more substantial large intestine to break down plant material. Humans are **omnivores**, consuming both animals and plants. Thus, the human gut requires some microbial fermentation in the large intestine. Researchers compared the anatomy of the human gut and its associated microbiota with that of 59 other mammals and determined that herbivores have more diverse gut microbiomes than carnivores. Overall, related mammalian species have similar gut microbiota. Table 1 lists some differences in the gut microbiota of humans who are vegan, vegetarian, or omnivorous. Even though there are differences in the bacterial species of gut microbiota, the number of total gut bacteria does not change.

Human genetics influences the composition of microbiota through immune system activity (which can determine predisposition to certain diseases), body mass index (BMI), diet, and lifestyle (**Figure 1**). Diet and lifestyle are responsible for gut microbiota variation between individuals and populations. Aging affects the richness and diversity of gut microbiomes. Before the age of 3 years, the human gut microbiome is "childlike" and less diverse. As we age, the diversity of the human gut microbiome increases.

Bacteria of the phyla Bacteroidetes and Firmicutes dominate *healthy* human gut microbiota. Bacteroidetes are Gram (−) anaerobic or aerobic rod-shaped bacteria. The Firmicutes, which comes from Latin, *firmus*, meaning "strong," and *cutis*, "skin," referring to the bacterial cell wall, consist mostly of Gram (+) bacteria. The gut microbiome plays important roles in host health and disease. When healthy microbiota are impaired due to disease or altered in composition through antibiotic treatment for a bacterial infection or other causes such as exposure to harmful chemicals in the environment, the term **dysbiosis** is used to describe the microbiota imbalance (**Figure 2**).

Healthy gut microbiota results in the establishment and maintenance of a robust community of microbes that is impervious to intestinal pathogens through various mechanisms such as the body's immune system responses. This property is called **colonization resistance**. The environment, host genetics, and the human microbiome interact to maintain balance or stability in the gut. Even though gut microbiota is highly variable from person to person, a growing number of studies are highlighting the fact that certain microbiota can be harmful to health. Dysbiosis is associated with an expanding list of

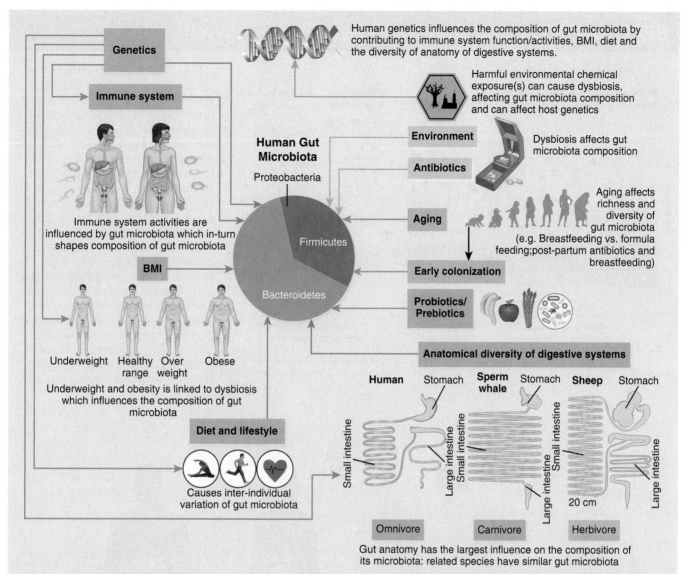

Figure 1 The composition of the human gut microbiome is influenced by many factors. The strongest influence on the makeup of gut microbiota is the evolution of the human gut's digestive anatomy and physiology. Host genetics will impact microbiota diversity through immune system activities, BMI, diet, and lifestyle. Other factors can independently impact diversity of the gut microbiota such as early colonization of the gut after birth (e.g., postpartum antibiotics or breastfeeding), aging, and the intentional consumption of probiotics/prebiotics. Dysbiosis, caused by antibiotic therapy or exposure to harmful chemicals in the environment, affects the composition of gut microbiota.

chronic diseases that includes obesity, inflammatory bowel disease (IBD), type 2 diabetes (adult onset), and **necrotizing enterocolitis** in newborns (**Figure 3**). Necrotizing enterocolitis is the most common and serious intestinal disease of premature babies, with a 25% mortality rate.

Researchers are using germ-free mice models to study associations between dysbiosis and obesity as well as neurodegenerative diseases such as Parkinson's disease (PD). For example, it is known that the gut microbiota of an obese individual differs from that of someone of a healthy weight (**Figure 4A**). When germ-free mice raised under sterile conditions received fecal microbiota transplants from obese donors, the germ-free mice became obese, suggesting that gut microbiota is involved in weight gain or weight loss. Furthermore, in clinical trials, when fecal microbiota from individuals of a healthy weight were

subsequently transplanted into an individual who was obese, the obese person lost weight.

Obesity also has a hereditary component. Germ-free mice received fecal microbiota from identical twins (**Figure 4B**). Gut microbiota of mice who received a fecal transplant from an obese twin donor gained weight. The obese mice had an abundance of *Blautia* spp. and low numbers of *Methanobrevibacter smithii*, *Christensenenella minuta*, and *Akkermansia mucinphila*. Mouse recipients of fecal microbiota from a lean identical twin maintained a lean weight, and their gut microbiota had an abundance of *M. smithii*, *C. minuta*, and *A. mucinphila* and low numbers of *Blautia* spp. The recipient mice were eating the same low-fat, high-fiber diet. Researchers mapped genes correlated with an increase in *M. smithii*, *C. minuta*, and *A. mucinphila* in their gut microbiota to genes involved with food intake and weight as well

(continues)

BOX 10.1 You Need Guts to Survive: "Know Thyself" (continued)

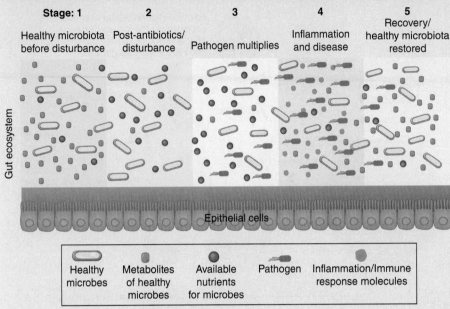

Stage: 1
Healthy microbiota before disturbance

2
Post-antibiotics/ disturbance

3
Pathogen multiplies

4
Inflammation and disease

5
Recovery/ healthy microbiota restored

Gut ecosystem

Epithelial cells

Healthy microbes | Metabolites of healthy microbes | Available nutrients for microbes | Pathogen | Inflammation/Immune response molecules

Information from J. A. Ferreyra, K. M. Ng, & J. L. Sonnenburg. (2014). The enteric two-step: Nutritional strategies of bacterial pathogens within the gut. *Cellular Microbiology*, *16*, 993–1003.

Figure 2 The stages of gut microbiota–pathogen interactions. In stage 1, a healthy gut ecosystem consisting of symbiotic microbes consuming nutrients, produces high levels of metabolites (metabolic waste), and fends off bacterial pathogens. In step 2, dysbiosis has occurred due to antibiotic treatment for a bacterial infection. The healthy gut microbiota has decreased in numbers. In stage 3, bacterial pathogens in the gut act as opportunists and cash in on the increase in available nutrients, which became available because fewer normal gut microbiota are available to consume them. The bacterial pathogen multiplies. In stage 4, the bacterial pathogens cause disease and changes in the gut (e.g., inflammation, immune response molecules), retarding the diversity and replication of the healthy gut microbiota. In stage 5, the healthy gut microbes recover, increasing metabolite output while outcompeting the pathogens. The diversity of healthy gut microbiota may or may not return to the initial healthy state it was in during stage 1.

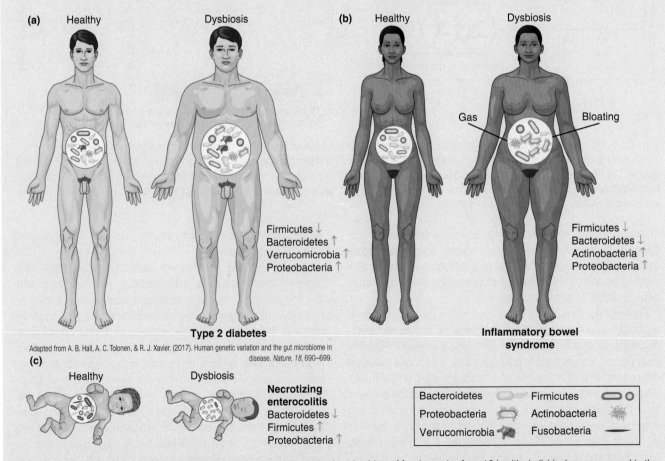

(a) Healthy   Dysbiosis

Firmicutes ↓
Bacteroidetes ↑
Verrucomicrobia ↑
Proteobacteria ↑

Type 2 diabetes

Adapted from A. B. Hall, A. C. Tolonen, & R. J. Xavier. (2017). Human genetic variation and the gut microbiome in disease. *Nature*, *18*, 690–699.

(b) Healthy   Dysbiosis

Gas   Bloating

Firmicutes ↓
Bacteroidetes ↓
Actinobacteria ↑
Proteobacteria ↑

Inflammatory bowel syndrome

(c) Healthy   Dysbiosis

Necrotizing enterocolitis
Bacteroidetes ↓
Firmicutes ↑
Proteobacteria ↑

Bacteroidetes | Firmicutes
Proteobacteria | Actinobacteria
Verrucomicrobia | Fusobacteria

Figure 3 Examples of dysbiosis associated with disease. In each example, the microbiota of fecal samples from 10 healthy individuals are compared to the microbiota of fecal samples from patients with **(a)** adult-onset type 2 diabetes, **(b)** inflammatory bowel disease (IBD), and **(c)** necrotizing enterocolitis. The gut microbiota of the patients is composed of different types and numbers of bacterial species than the gut microbiota from healthy individuals.

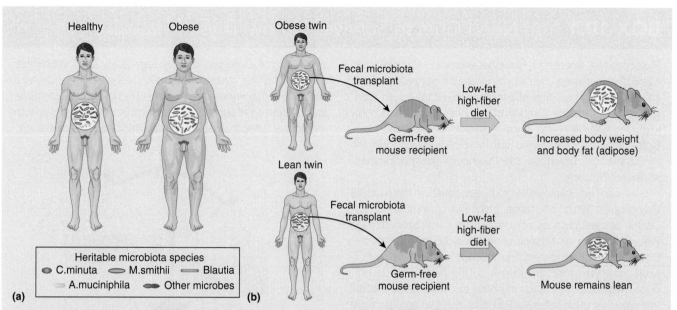

Figure 4 **(a)** Obesity is associated with an altered composition of gut microbiota such as an increase or abundance of *Blautia* spp. in an obese person compared to the gut microbiota of an individual of a healthy weight. An abundance of *Methanobrevibacter smithii*, *Christensenenella minuta*, and *Akkermansia mucinphila* are present in the gut microbiota of healthy weight individuals and not present in high numbers in the gut microbiota of obese individuals. **(b)** The bacterial species present in the gut microbiota of individuals of healthy weight or obese weight are inherited. This is shown through studies involving identical twins. When germ-free mice are transplanted with fecal gut microbiota from an obese identical twin, the mouse becomes obese. If the germ-free mouse receives a fecal microbiota from a lean identical twin, the mouse remains lean.

as those related to immunity. Will fecal microbiota transplants be the next big weight-loss fad? Will capsules of *M. smithii*, *C. minuta*, and *A. mucinphila* be prescribed for weight loss in the future?

Similarly, when germ-free mice raised in sterile conditions or mice depleted of microbiota by antibiotics were given a fecal microbiota transplant from a patient with Parkinson's disease, the mice had

symptoms of neurological deficits similar to that displayed with Parkinson's disease. Parkinson's disease is a neurodegenerative disease associated with gut dysbiosis. Recent studies have confirmed that those with Parkinson's disease suffer from gut microbiota dysbiosis. The gut microbiota of Parkinson's disease patients has abundant or increased *Blautia*, *Coprococcus*, *Roseburia*, and

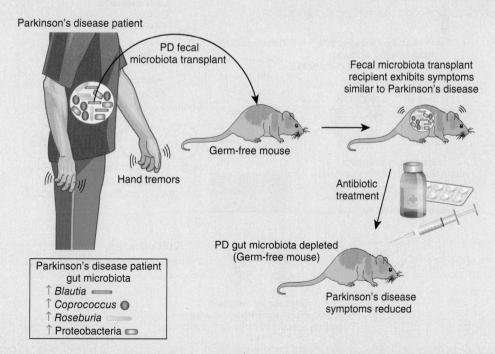

Figure 5 Germ-free mice receive a fecal microbiota transplant from a Parkinson's disease (PD) patient. Parkinson's is associated with gut dysbiosis. Mice recipients of the Parkinson's gut microbiota exhibited symptoms similar to Parkinson's disease. The mouse recipients were treated with antibiotics to kill the transplanted Parkinson's disease gut microbiota, and the Parkinson's symptoms in the mice were reduced.

(continues)

BOX 10.1 You Need Guts to Survive: "Know Thyself" (continued)

Proteobacteria. Species of *Faecalibacterium* and *Prevotellaceae* bacteria are decreased. Tremors or impaired motor movements in patients with Parkinson's have been correlated with pathogenic bacteria in the gut that are members of the *Enterobacteriaceae* family (e.g., *Proteus*). After the mice recipients of the fecal microbiota transplant were treated with antibiotics (which killed the transplanted gut microbiota), their Parkinson's-like symptoms disappeared (**Figure 5**).

How can the bacteria in your gut cause a disease like Parkinson's? What are these pathogenic microbes doing? Do the bacteria produce neurotransmitters that travel through the vagus nerve and cause central nervous system effects? Do the bacteria stimulate the production of inflammation molecules that then affect the brain? Do the gut microbiota produce metabolites that reach the brain to cause neural effects? Will treatment of diseases like Parkinson's use an approach that alters the composition of gut microbiota?

The idea that gut microbiota dysbiosis may be involved in causing disease is gaining the attention of medical researchers around the world (**Figure 6A**). *Because colonization resistance plays an important role during the development of gut microbiomes, therapies that restore and promote gut health will allow us to strengthen or boost colonization resistance.* Engineering gut microbiota for health benefits and therapies to prevent and treat diseases associated with gut microbiota dysbiosis could be observed across many disciplines in the near future. **Nanotechnology** is in development to monitor the human microbiome in real time. The current measures available to engineer gut microbiota are fecal microbiota transplantation, designing probiotic bacteria to inhibit quorum sensing (see further discussion of quorum sensing in chapter text),

engineering bacteriophages, pharmaceuticals (e.g., antibiotics), consumption of probiotics/prebiotics, and dietary changes (**Figure 6B**). These measures would be used to *intentionally change the composition of gut microbiota*, which, in turn, will interfere with microbial activities that are associated with specific disease effects.

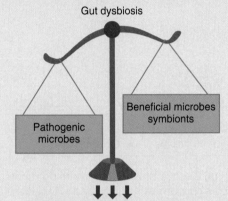

Inflammatory/Autoimmune diseases
 Irritable bowel syndrome, Multiple sclerosis, Psoriasis, Rheumatoid arthritis
Obesity
Type 2 diabetes
Cardiovascular diseases
Neurodegenerative diseases
 Parkinson's disease and Alzheimer's disease
Psychiatric diseases
 Anxiety and depression
Neurodevelopmental diseases
 Autism
Injury
 Spinal cord
Kwashiokor
Liver diseases
Cancer
 Colon

(a)

Fecal microbiota transplant (from healthy donors)
Restores composition of gut microbiota

Gut dysbiosis

Probiotics Prebiotics
Probiotics/Prebiotics
Changes composition of gut microbiota

Genetically engineer commensal probiotic bacteria
Inhibit quorum sensing behavior (produce bacteriocins and/or biofilm degrading molecules

Vegetarian Vegan Mediterranean Raw
Dietary Changes
Changes composition of gut microbiota

Is anyone out there?

Antibiotics
Changes composition of gut microbiota

Genetically engineer bacteriophages/Phage therapy
Target(kill) specific pathogenic gut microbes

(b)

Figure 6 (a) Gut dysbiosis occurs when pathogenic bacteria in the gut have multiplied, causing an imbalance in gut microbiota and damage to the gut or other parts of the body. Dysbiosis is associated with the chronic diseases listed in this diagram. **(b)** Current types of technologies or strategies available to restore or engineer gut microbiota after dysbiosis occurs. Methods include use of biological interventions such as fecal transplants or engineered bacteria or bacteriophages to reduce the pathogenic microbes in the gut or nonbiological tools such as antibiotic therapy, probiotics/prebiotics, or dietary changes in order to shift the microbiota composition toward a healthy balance.

Some microbes almost always cause damage and disease to their host on contact, whereas others, including the normal microbiota, may cause disease only when the host immune defense mechanisms are compromised, as is the case in individuals with HIV/AIDS. Whatever the circumstance, the underlying factor is that there is a dynamic interplay between the number of microbes, quorum sensing, and microbial virulence mechanisms, as described in this chapter, and the immune mechanisms of the host. The **course of a disease** can be characterized by a sequential series of five stages.

Biological Associations

The term **symbiosis** means "*living together*" and describes an association between two or more species. According to parasitiologist Clark Phares Read (1921–1973):

> *Symbiosis, particularly parasitism, is frequently regarded in distasteful terms by the hygiene-conscious citizen. We have a tendency to think of it as a peculiar and abnormal association of some lower organism with a higher one. There is an element of snobbishness in such a view, which must be quickly abandoned when a discerning look is taken of the living world. In nature there is probably no such thing as a symbiote-free organism. The phenomenon of symbiosis is quite as common as life itself.*

Symbiosis is like an umbrella with three possibilities: mutualism, commensalism, and parasitism (**FIGURE 10.1**). **Mutualism** may be the "best of all worlds" and is like a blissful marriage in which both members of the association enjoy benefits. A classic example of mutualism is the association between termites and the protozoans that constitute a part of the normal microbiota of the termite's intestine. Termites can cause severe damage to a house, a fact exploited by pest control companies with creative slogans such as "Bug off!" and other catchy phrases (**FIGURE 10.2**). The termites' ability to digest wood (cellulose, a carbohydrate component of plant cell walls) is based on enzymes secreted by its protozoan residents. In turn, the protozoans have it made; they live in a stable environment with plenty of food passing through.

© Robert I. Krasner.

FIGURE 10.2 An attention-getting landmark atop a building on Route 95 in Providence, Rhode Island.

Consider *E. coli*, a part of the human microbiome and an inhabitant of the human large intestine. These bacteria enjoy the comforts of home while producing vitamin K, a factor required for blood clotting. Lichens are commonly seen and represent a mutualistic relationship between two microbial forms, a fungus and an algae or cyanobacterium (**FIGURE 10.3**).

Commensalism is a relationship between two or more species in which one benefits and the other is indifferent, that is, neither benefited nor harmed. Consider, for example, most of the normal microbiota of the body, organisms that reside in or on the body and take their energy from the metabolism of host secretions and waste products but make no obvious contribution to the partnership with their host, nor do they damage their host. A humorous poem by W. H. Auden summarizes the commensal relationship enjoyed by the normal microbiota and the host (**BOX 10.2**).

© Teri Shors.

FIGURE 10.3 Lichens are a symbiotic association of a fungus and an algae or cyanobacterium. Each organism provides benefits to the other.

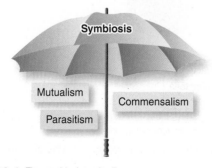

FIGURE 10.1 The symbiosis umbrella.

BOX 10.2 A Poem About the Human Microbiome

For creatures your size I offer
   a free choice of habitat,
so settle yourselves in the zone
   that suits you best, in the pools
of my pores or the tropical
   forests of armpit and crotch,
in the deserts of my forearms,
   or the cool woods of my scalp

Build colonies: I will supply
   adequate warmth and moisture,
the sebum and lipids you need,
   on condition you never
do me annoy with your presence,
   but behave as good guests should
not rioting into acne
   or athlete's-foot or a boil.

Reproduced From W. H. Auden, Excerpted from "A New Year Greeting," *Epistle to a Godson and Other Poems.* New York: Random House, 1972.

As pointed out in the poem, the human body offers a wide variety of ecological niches, each of which has a population of particular bacterial species (**FIGURE 10.4**; **TABLE 10.1**). Microbes are not the only commensals to enjoy our bodies. Scrapings taken from the landscaped furrows of your forehead just below the hairline and examined microscopically might reveal mites crawling about. Mites are distantly related to spiders and are small enough to take up residence in sebaceous (sweat) glands and hair follicles. These examples illustrate that humans are part of the microbial food chain.

Whereas mutualism is at one end of the spectrum, **parasitism**—a relationship in which the parasite lives at the expense of its host—is at the other end. Parasitism is

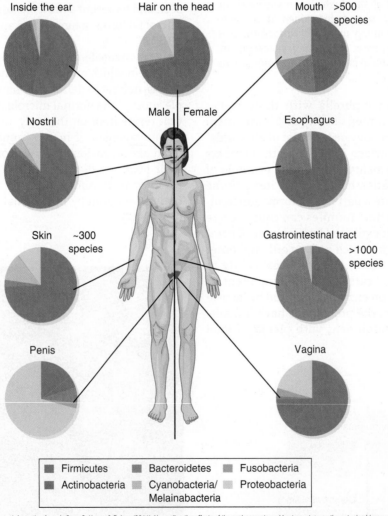

Information from A. Spor, O. Koren, & R. Ley. (2011). Unraveling the effects of the environments and host genotype on the gut microbiome. *Nature Reviews Microbiology, 9,* 279–290.

FIGURE 10.4 We are not alone! The dominant diverse bacterial phyla that represent normal microbiota in various body locations of a human: mostly commensal microbes. Examples of bacterial genera within the different phyla are listed in Table 10.1.

TABLE 10.1 Human Microbiome Bacteria

Phylum	Gram Reaction	Examples of Genera
Firmicutes	Gram (+)	*Clostridium, Streptococcus, Staphylococcus, Bacillus, Listeria, Eubacterium, Lactobacillus, Ruminococcus, Veillonella**
Bacteroidetes	Gram (–)	*Bacteroides, Prevotella*
Fusobacteria	Gram (–)	*Fusobacterium, Streptobacillus*
Acintobacteria	Gram (+)	*Corynebacterium, Micrococcus, Rothia, Proprionibacterium, Bifidobacterium, Gardnerella, Actinomyces*
Cyanobacteria/Melainabacteria	Gram (–)	Unclassified genera
Proteobacteria	Gram (–)	*Escherichia, Enterobacter, Proteus, Serratia, Salmonella, Hemophilus, Moraxella, Klebsiella, Pseudomonas, Campylobacter, Helicobacter, Acinetobacter, Neisseria*

*Veillonella is Gram (–) but still belongs in the Firmicute phylum, which are mostly Gram (+) bacteria.

an association in which one species, the parasite, lives at the expense of the other, the host. There are at least as many definitions of the word *parasitism* as there are letters in the word itself, but they all point to harm or death of the host. The host (in this text) is the human, and microbes are the parasites. The term *parasite* is used to encompass all microbes, viruses, and worms that produce disease. On the other hand, some biologists restrict the term to protozoans (animal-like) and worms, and other biologists restrict it to worms.

The distinctions between mutualism, commensalism, and parasitism are not airtight. The borders between them are hazy and are frequently crossed. Think of these relationships as being like a slippery seesaw (**FIGURE 10.5**).

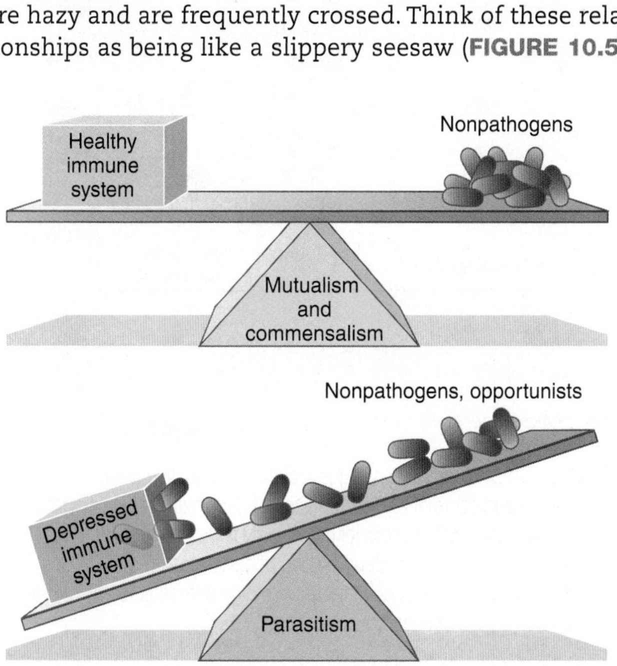

For example, in HIV/AIDS and other conditions resulting in a compromised immune system, nonpathogenic microbes, including the normal microbiota, are responsible for potentially fatal infections; these microbes are appropriately referred to as **opportunistic pathogens**. A major cause of death in AIDS is a fungal infection caused by the opportunist *Pneumocystis jirovecii*. Burn victims are at major risk of serious bacterial infection resulting from nonpathogenic microbes, including their own normal microbiota. Further, organisms that are part of the normal microbiota can take the slippery slide from commensalism to parasitism. For example, when *E. coli*, a typical resident of the intestine, gains access to another area of the body—as might happen if the intestinal wall is pierced during surgery—a potentially fatal condition called **peritonitis** may result (**FIGURE 10.6**).

Here's a bizarre relationship that truly escapes categorization as has been described; choose the one you think best! *Toxoplasma gondii* is a protozoan parasite in rats that blocks the neuronal circuitry that makes rats terrified of

FIGURE 10.5 The seesaw of symbiosis—from mutualism to parasitism.

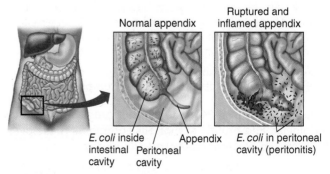

FIGURE 10.6 Peritonitis: Microbes in the "wrong" place.

cats, but, even more strange, these rats become sexually attracted to the scent of cat urine. So the cats eat the rats and, in so doing, acquire the parasite.

Parasitism: A Way of Life

The term *parasite* has a negative connotation and is sometimes used to describe human interactions. To be called a parasite is hardly a compliment. (You may have had the unpleasant experience of being part of a group project, perhaps a student presentation, in which an individual in the group "parasitizes" the others, meaning that he or she makes little contribution and takes advantage of everybody else's effort.)

Parasitism is the basis for microbial disease and for the challenge between microbes and humans. The parasite lives at the expense of the host and, in so doing, causes damage to the host. According to Lewis Thomas (1913–1993), an American physician, author, educator, and administrator known for his popular essays in biology and medicine, *"Disease usually results from inconclusive negotiations for symbiosis, an overstepping of the line by one side or the other, a biological misinterpretation of borders."* Thomas refers to the "slippery slide" previously described. Parasitism is a constant and delicate dance of biological adaptation in the Darwinian sense of survival of the fittest as both host and parasite coevolve.

Rats, Lice, and History, Hans Zinsser's intriguingly titled 1935 classic text, speaks of parasitism in a somewhat philosophical sense:

> *Nature seems to have intended that her creatures feed upon one another. At any rate, she has so designed her cycles that the only forms of life that are parasitic directly upon Mother Earth herself are a proportion of the vegetable kingdom that dig their roots into the sod for its nitrogenous juices and spread their broad chlorophyllic leaves to the sun and air. But these—unless too unpalatable or poisonous—are devoured by the beasts and by man; and the latter, in their turn, by other beasts and by bacteria. . . . The important point is that infectious disease is merely a disagreeable instance of a widely prevalent tendency of all living creatures to save themselves the bother of building, by their own efforts, the things they require. Whenever they find it possible to take advantage of the constructive labors of others, this is the direction of the least resistance, the plant does the work with its roots and green leaves. The cow eats the plant. Man eats both of them; and bacteria (or investment bankers) eat the man. . . . That form of parasitism, which we call infection, is as old as animal and vegetable life."*

(Reprinted from Hans Zinsser, *Rats, Lice, and History,* 1935, courtesy of Little, Brown and Company.)

Hans Zinsser (1878–1940) was an American physician, bacteriologist, and prolific author. His comments made over 80 years ago are remarkably true to this day.

Bacteria *Talk* to Each Other: Quorum Sensing

Up until the late 1970s, microbiologists assumed that pathogenic bacteria lived a life of solitude. If a bacterium encountered a suitable host, it could colonize it and establish residency without causing harm, or acquire nutrients, multiply, and move on to a new host, or it could invade a host, causing damage directly to it during the course of an infection.

Through the course of bacteriology research, we now know that bacteria are social creatures that communicate with each other by producing intracellular signal molecules called **autoinducers**. The autoinducers are secreted into the surrounding environment and accumulate. When the autoinducers reach a certain concentration, other bacteria in the vicinity *sense*, or recognize, that they are not alone. The bacterial population size has reached a *quorum*, signaling the bacteria to perform a socially synchronized, genetically preprogrammed response. The response is the expression of certain genes that are involved in group behaviors (**FIGURE 10.7**). This chemical signaling communication among bacteria through the exchange of autoinducers is called *quorum sensing*.

You may have heard of term *quorum* if you have ever served on committees that are charged with making decisions. No decisions are valid unless a quorum present. A quorum is the minimum number of members required to conduct committee decision-making. Many group behaviors tasked through quorum sensing are associated

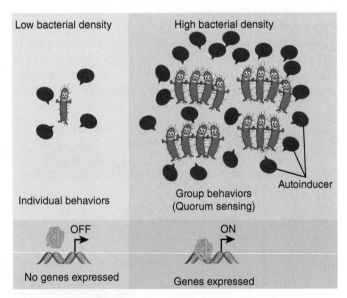

FIGURE 10.7 Quorum sensing is a chemical signaling system used among bacteria to coordinate the regulation of gene expression and group behavior. Left panel: When bacterial cell density is low, the autoinducers diffuse away and are not detectable; group behavior does not take place. Right panel: The autoinducer concentration is high and detected by quorum-sensing bacteria, enabling group behavior.

TABLE 10.2 Group Behaviors of Quorum-Sensing (QS) Bacteria

Group Behavior	Function/Task
Bioluminescence (light production)	Survival within host (e.g., bioluminescent *Vibrio fischeri* are symbionts living in the light organs of Hawaiian bobtail squids. The bacteria produce light at night, which camouflages the squids that are feeding nocturnally, making them undetectable by predators, allowing both the bacterial symbionts and their hosts to survive.)
Extracellular polymeric substances (EPS) production	Group production of EPS by QS bacteria allows them to stick together or to surfaces, forming a biofilm (e.g., on a catheter, river rock, etc.)
Fermentation	Metabolism; growth or multiplication of QS bacteria (e.g., fermentation of foods)
Antibiotic synthesis	Survival/defensive action of QS bacteria; competitive advantage in acquiring nutrients
Bacteriocin production	Survival/defensive action of QS bacteria; competitive advantage in acquiring nutrients
Pigmentation	QS bacteria defensive mechanism to avoid predation by higher organisms
Sporulation	Survival of QS bacteria when nutrients are in short supply
Bacterial–plant interactions	Symbiotic or disease-causing behavior of QS bacteria with plant hosts
Production of toxins or enzymes	Virulence factors produced by QS bacteria are used to invade a host, causing infection and damage to the host
Swarming/motility	Survival; allows QS bacteria to respond rapidly to environmental changes within the host and/or its environment
Conjugation	Survival; increases genetic diversity of a QS bacterial population

with bacterial survival. Examples of group behaviors signaled by quorum sensing are listed in **TABLE 10.2**.

Each bacterial species produces an autoinducer that has a unique chemical structure, or *chemical language.*

Communication occurring among the same species of bacteria is called **intraspecies communication** and that between different species is **interspecies communication** (**FIGURE 10.8**). *Quorum sensing allows bacteria to be*

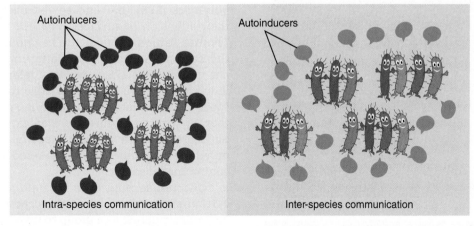

FIGURE 10.8 Left panel: Intraspecies quorum sensing is a process by which bacteria of the same species secrete autoinducers and only those bacteria of the same species can detect and respond to it. Right panel: Interspecies quorum sensing occurs when a bacterial species sends and receives signals to other species in a mixed population of bacteria that will detect and respond to it.

TABLE 10.3 Host Environments of Quorum-Sensing Bacteria

Host Environment	Specific Location of Quorum-Sensing (QS) Bacteria
Marine animals	Gastrointestinal tracts and light organs (e.g., QS bacteria in Hawaiian bobtail squid)
Humans	Biofilms in lungs of patients with cystic fibrosis or tuberculosis, on urinary catheters and prosthetic devices, or oral biofilms associated with periodontitis
Agriculture/plants	Symbiosis; root nodules of legumes by nitrogen-fixing symbiont *Bradyrhizobium japonicum* Plant diseases (crown gall)
Aquaculture/aquatic organisms	Disease/pathogens of farmed fish and other aquatic organisms
Soils	Biofilms within the rhizosphere of forest soils
Rivers	Biofilms on river rocks
Food fermentation	Sourdough bread, kefir, yogurt, wines

multicellular, similar to higher organisms. Quorum-sensing bacteria have been isolated from a variety of environments (**TABLE 10.3**). Have you ever observed a bag of expired greens in the refrigerator? The greens turn brown and slimy. This is probably due to plant pathogens such as *Pseudomonas* spp. or *Erwinia* spp. responding to quorum-sensing autoinducers (**FIGURE 10.9**).

Quorum Sensing in Bacterial Infections

"It is not in numbers, but in unity, that our great strength lies; yet our present numbers are sufficient to repel the force of all the world."

—Thomas Paine, Common Sense (1776)

© Teri Shors.

FIGURE 10.9 Bagged greens have an expiration date for a reason. In this photograph, expired romaine lettuce has wilted, become slimy, and turned brown. Quorum-sensing bacteria present on the romaine lettuce are the likely causes for its physical changes resulting in food spoilage.

Bacterial pathogens use quorum sensing to coordinate their assault and invasion into a host. Many clinically relevant Gram (+) and Gram (−) bacteria coordinate gene expression to produce multiple **virulence factors** and associated group behavior such as **biofilm** formation once a cell density threshold is reached (**TABLE 10.4**). Virulence factors, such as enzymes and toxins, allow pathogenic bacteria to invade, multiply, and cause damage to host tissues. (See section "Microbial Mechanisms of Disease" for more details about virulence and bacterial pathogenicity.)

Biofilms are communities of bacteria permanently attached to a wet surface such as the lining of the colon, the lungs of cystic fibrosis patients, and medical devices such as catheters, stents, and heart valves. Quorum-sensing bacteria produce **extracellular polysaccharide polymers (EPS)** that encase the bacteria, facilitating their adhesion to surfaces (**FIGURE 10.10**). Similar to Thomas Paine's quote at the beginning of this section, the strength of pathogenic quorum-sensing bacteria lies in unity. The EPS provides a defense or barrier that individual bacteria find impossible to achieve. Biofilms cause chronic infections in humans. Prolonged antibiotic therapy is often required to eradicate these infections but regrettably even this strategy fails. Sometimes the only long-term cure is surgical removal of the biofilm-contaminated source.

Quorum sensing does create a social dilemma in certain situations. Such is the case when pathogenic bacteria cooperate as a group to invade a host. Quorum-sensing bacteria detect autoinducers, and they cooperate by producing their own chemical signals that regulate the expression of certain genes that code for virulence

TABLE 10.4 Quorum-Sensing Pathogenic Bacteria		
Pathogenic Bacterium	**Gram Reaction**	**Type of Infection/Disease**
Pseudomonas aeruginosa	Gram (–)	Biofilms in lungs/complications of cystic fibrosis patients
Staphylococcus aureus	Gram (+)	Abscesses and endocarditis (inflammation of the heart)
Escherichia coli	Gram (–)	Gastroenteritis
Vibrio cholerae	Gram (–)	Cholera
Stenotrophomonas maltophilia (emerging pathogen)	Gram (–)	Meningitis, soft tissue infection, endocarditis, invasive infections caused by the bacteria colonizing a patient's intravenous and irrigation fluids
Aeromonas spp. (emerging healthcare-associated pathogen)	Gram (–)	Gastroenteritis, urinary tract infections, sepsis, peritonitis

factors. If a group of quorum-sensing bacteria turn on the expression of a gene that codes for an enzyme that is a **protease**, the proteases damage the host by degrading proteins within the host tissues, allowing them to invade the host. In doing so, the degraded host tissues are freely available as nutrients (food) for the quorum-sensing bacteria and **cheater bacteria**. Cheaters respond to quorum-sensing signals but do not help the group (they do not turn on genes that code for proteases). The cheaters show up and use the nutrients made available through the group activity (turning on protease gene expression, which results in the degradation of host tissues, a delightful food source for cheaters) even if they did not participate by producing proteases to help the group (**FIGURE 10.11**). In doing so, cheaters can quench or inhibit quorum-sensing group activities. The formation of biofilms by quorum-sensing bacterial pathogens restricts cheaters. Quorum sensing regulates the expression of genes that are involved in the production of a thick, protective layer of EPS that secures the nutrient supply and outcompetes individual freeloader, cheater bacteria.

Quorum Quenching to Treat Superbug Infections

Antibiotics have lost their effectiveness in treating infections caused by superbugs or multidrug-resistant bacteria. A recent report by the UK government estimates that by 2050 multidrug-resistant bacteria could cause an additional 10 million human deaths annually and cumulative world economic losses of $100 trillion. The development of new and innovative approaches for better and more effective treatments of multidrug-resistant infections is urgently needed to address this worldwide public health threat.

A strategy that quenches quorum sensing will block the expression of genes that code for virulence factors and group activities of pathogenic quorum-sensing bacteria but will not kill the pathogen. It is theorized that the host will not be invaded or damaged by infection. Blocking communication would stop group behavior such as the formation of biofilms, forcing quorum-sensing bacteria to live as individuals fending for themselves. Chronic infections caused by *Pseudomonas aeruginosa* in cystic

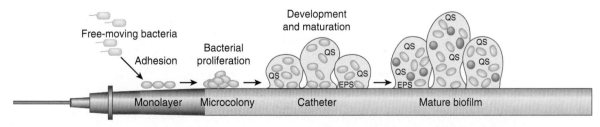

FIGURE 10.10 Illustration of the formation of a biofilm on a catheter. Quorum-sensing bacteria produce EPS during development and maturation of the biofilm. Biofilms may contain communities of a single or multiple species of bacteria. It protects the bacteria encased by the biofilm from attack by the body's immune system and effective antibiotic treatment.

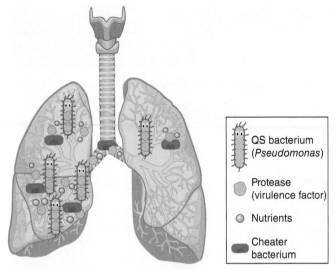

FIGURE 10.11 Quorum-sensing bacteria cooperate when responding to autoinducers. As a group, they turn on the expression of a gene that codes for a protease. Proteases degrade (damage) lung tissues, allowing the bacteria to invade the lining and tissues of the lungs. Protease activity results in tissue breakdown into nutrients (food) for the quorum-sensing bacteria and nearby cheater bacteria. Cheater bacteria respond to the autoinducing signals but do not cooperate by producing proteases. Instead, they take advantage of the nutrients made available through quorum-sensing bacteria group behavior.

fibrosis patients are an attractive target because of the lack of effective treatment options available. Will therapies that block quorum sensing be effective against all bacterial pathogens? Could the combination of antibiotics and anti-quorum–sensing therapies result in more effective treatments and less development of microbial antibiotic resistance?

Three different research strategies are being employed to study and develop therapies to quench quorum sensing:

1. Inhibiting the synthesis or production of autoinducers so that no autoinducers accumulate or can be detected

2. Directly degrading or destroying the autoinducers after they are synthesized (making them undetectable by quorum-sensing bacteria)

3. Inhibiting the ability of quorum-sensing bacteria to detect autoinducers (**FIGURE 10.12**)

Even though the field of quorum research continues to acquire new knowledge, to date, no **quorum-sensing inhibitors (QSIs)** are available to treat superbug infections. It will be many years before QSIs are approved by the U.S. Food and Drug Administration (FDA) as clinically safe for real-life applications. More research is needed to further understand the complexities of quorum sensing in diverse natural environments such as within a healthy human gut versus dysbiosis of the human gut in order to drive innovations toward mainstream therapies.

Microbial Mechanisms of Disease

Here we note the distinction between the terms *infection* and *disease*. Some microbiologists use these two terms interchangeably, and that is the case in this text. Others use the term **infection** to describe the multiplication of microbes on or in the body without producing definitive symptoms; in this sense, the normal microbiota cause infection. **Disease** is a possible outcome of microbial invasion, whether by normal microbiota or exogenous (from

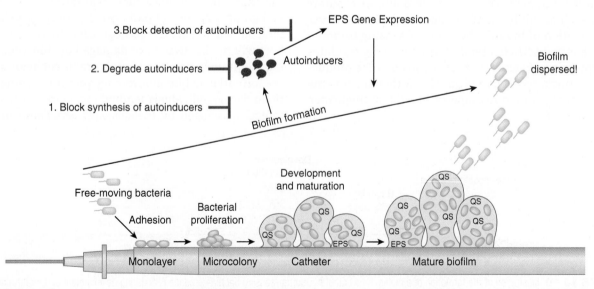

FIGURE 10.12 Quenching quorum sensing by blocking synthesis of autoinducers, degrading autoinducers, or blocking the detection of autoinducers by quorum-sensing bacteria theoretically should disrupt biofilm formation.

the outside) microbes, resulting in impairment of health to some degree. Again, the "slippery slide" principle exists, as it frequently does in biology, between the terms *infection* and *disease*. Early symptoms of many infectious diseases include lethargy and a sense of ill health, or **malaise**. As experience indicates, **infectious diseases** run the gamut of severity from mild to moderate to severe to lethal. Some diseases are **subclinical** or **asymptomatic**, as evidenced by the presence of telltale antibodies against diseases that were never diagnosed because the individual had no signs or symptoms of disease. With a number of infections only a certain percentage of infected individuals experience signs and symptoms of an infection. For example, only about 20% of those infected with Zika virus experience symptoms.

A growing body of evidence indicates that some diseases traditionally thought to be noninfectious, non-microbial diseases, such as some cardiac diseases, most cancers, metabolic diseases, mental disorders, and nutritional deficiency disorders, may well be caused by microbes. But how exactly do microbes cause infections? The chance of acquiring a particular infection and the severity of the accompanying symptoms depend on three major factors:

1. **Dose (*n*)**: The number of microorganisms to which the potential host has been exposed.

2. **Virulence (*V*)**: Microbial mechanisms or weapons.

3. **Resistance (*R*)**: The host immune system.

A dynamic battle rages between disease-producing microbes and the resistance of the host that can be viewed as a tug-of-war or a seesaw. This struggle can be summarized by the equation $D = nV/R$. D represents the severity of infectious disease, the numerator refers to the microbes (*n* = number of organisms, *V* = virulence factors), and the *R* in the denominator refers to resistance (immunity). Collectively, these factors determine whether disease occurs and, if so, its severity. Particularly significant, in most cases, are the early hours of challenge; if the microbes gain the upper hand over resistance, then they continue to multiply and damage the host, establishing disease.

Pathogenicity and Virulence

The number of infectious agents (*n*) gaining access to (in or on) the host plays a crucial role in determining whether disease will take place. For most bacteria a minimal dose, the **infective dose (ID)**, is necessary to establish infection. Highly virulent microbes have smaller IDs, and in some cases fewer than a hundred bacteria are needed. Those of low virulence can establish infection only when the number of infectious agents is large. It only takes about 10 bacteria to cause tuberculosis, and as few as 18 noroviruses are necessary for infection to

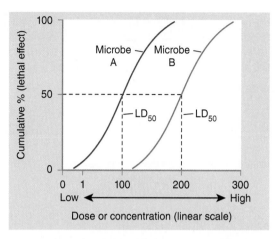

FIGURE 10.13 An LD_{50} dose–response curve. The higher the LD_{50} value, the less virulent the microbe is because it requires a higher number of microbes, or infective dose, to kill its host.

occur. In contrast, as many as 1 million cholera bacteria may be needed to establish infection. The **50% lethal dose (LD_{50})** is a laboratory measurement of virulence determined by injecting animals with graded doses (varying concentrations) of microbes; the dose that kills 50% of the injected animals in an established time is the LD_{50} (**FIGURE 10.13**). See **TABLE 10.5** for examples of infective doses for a variety of infectious diseases.

The term *pathogen*, previously introduced, refers to the ability of microbes to produce disease. Some groups of microbes are considered to be highly pathogenic, including those that cause tetanus, botulism, plague, AIDS, and Ebola virus disease. They almost always cause disease. (It should be kept in mind that the vast majority of bacteria are not pathogens.) Virulence is a measure of pathogenicity and encompasses those specific factors (e.g., toxins) that enable pathogens to overcome host defense mechanisms and to multiply and cause damage. Even within the same species of bacteria variation in virulence exists as demonstrated by LD_{50} determinations.

Microbes gain entry into the body through portals of entry that include the skin and mucous membranes of the mouth, nose, genital tract, and blood. Microbes that are able to establish disease have adaptive **defensive strategies** that allow them to escape destruction by the host immune system and **offensive strategies** (**TABLE 10.6**)

Krasner's Notebook

Frequently, to make a point, the authors have asked in their respective classes for a show of hands as to "Who feels tired and not so great?" Just about every hand goes up (including the author's). Is this the beginning of an infectious disease or conditions related to academic life, including preparing papers, studying (or lack of it), boredom, lack of sleep, and . . . (you can fill in the rest)?

TABLE 10.5 Infective Doses of Selected Human Pathogens

Pathogen	Infectious Disease	Type of Pathogen	Mode of Transmission*	Infective Dose
Influenza A virus	Influenza	Virus	Airborne or droplet transmission	1,000 to 2,000 viruses
Vibrio cholerae	Cholera	Bacterium	Oral–fecal	10,000 to 1 million
Hepatitis B virus	Hepatitis B or serum hepatitis	Virus	Bloodborne	0.00001 to 0.00000001 milliliter of blood containing hepatitis B virus
Norovirus	Winter vomiting disease	Virus	Oral–fecal	≤ 18 viruses
Measles virus	Measles	Virus	Airborne or droplet transmission	1 virus
Mycobacterium tuberculosis	Tuberculosis	Bacterium	Airborne or droplet transmission	10 bacteria
Shigella spp.	Shigellosis	Bacterium	Oral–fecal	< 10 bacteria
Salmonella spp.	Salmonellosis	Bacterium	Oral–fecal	> 100,000 bacteria

*Main mode of transmission. Some diseases may have other possible modes that occur at lower rates.

TABLE 10.6 Bacterial Mechanisms of Virulence

Virulence Factor	Function	Bacterial Pathogen
Defensive strategies		
Adhesins	Fixation to host cell surfaces and linings	*Neisseria gonorrhoeae* (attachment to urethral lining); *Streptococcus mutans* (causes tooth decay and sticks to surface of teeth)
Capsules	Interfere with uptake of bacteria by host immune defenses (phagocytosis)	Bacteria that cause anthrax, plague, or streptococcal diseases (e.g., scarlet fever, "strep" throat, pneumococcal infection)
Miscellaneous	Interfere with uptake of bacteria by host immune defenses (phagocytosis)	"Waxy coat" of *Mycobacterium tuberculosis* (causes tuberculosis) and *Mycobacterium leprae* (causes leprosy); M protein in streptococcal cell walls
Offensive strategies		
Enzymes	Destroy integrity of tissue structure	Hyaluronidase (breaks down hyaluronic acid of connective tissue), hemolysins (bring about lysis of red blood cells), collagenases (break down collagen in connective tissue), leukocidins (destroy white blood cells)
Exotoxins	Specific activities that interfere with vital host functions	Botulinum toxin (one of the most potent neurotoxins known; interferes with transmission of nerve impulse, resulting in flaccid paralysis); tetanus neurotoxin (interferes with transmission of nerve impulses, resulting in irreversible muscle contraction)
Endotoxins	Produces shocklike symptoms, chills, fever, weakness	Structural components of all Gram (–) cell walls

that result in damage to the host. The distinction between these two strategies is not always clear, because some factors act both defensively and offensively. Think of it as being like a basketball team—to win the game (analogous to establishing infection), the team (pathogens) must have effective defensive strategies to counter its opponent (the host immune system), but that is not enough. The team must also have effective offensive strategies to shoot baskets to score points. Each group of pathogens has evolved unique defensive and offensive virulence factors, many of which are present in more than one group. New molecular biology techniques have increased an understanding of the mechanisms of bacterial virulence.

Defensive Strategies

O Bacterial Adhesins

Many pathogens possess cell surface molecules called **adhesins** (**FIGURE 10.14**) by which they adhere to **receptor** molecules at a **portal of entry** in a Velcro-like manner. A clear example of this is represented by the causative agent of gonorrhea, *Neisseria gonorrhoeae*; it produces adhesins by which bacteria hold fast and colonize on the warm, moist membranes of the urogenital tract despite the periodic and forceful flow of urine. The bacteria that are primarily responsible for dental decay, *Streptococcus mutans*, remain attached to the surface of teeth by adhesins despite the constant irrigation of the teeth by saliva. Bacteria lacking adhesins are subject to being washed or swept away and are not able to colonize the surface of the host cells lining the portal of entry.

O Capsules and Other Structures

Some bacteria contain **capsules** surrounding their cell walls. In many cases, the presence of the capsule material interferes with the process of **phagocytosis**, an important resistance mechanism of the host defenses by which bacteria are engulfed by scavenger immune cells of the body. It can be experimentally demonstrated that removal of the capsule from encapsulated bacteria renders them more susceptible to phagocytosis (**FIGURE 10.15**). A number of bacteria, including those that cause plague, anthrax, pneumococcal pneumonia, and strep throat, contain capsules. The presence of a capsule does not necessarily confer virulence, nor does the lack of a capsule exclude virulence. It is to be emphasized that *bacterial virulence is due to a combination of virulence factors.*

The streptococci responsible for streptococcal sore throat and a variety of other diseases have a protein in their cell walls termed the **M protein**. Those streptococci containing the M protein display resistance to phagocytosis. Mutant strains not producing this factor and strains in which the M protein has been experimentally removed are readily engulfed by phagocytic cells. Note that the streptococci possess both capsules and M protein—a double whammy against phagocytosis. *Mycobacterium tuberculosis*, the bacterium that causes tuberculosis, and a related bacterium,

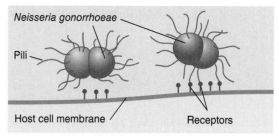

(a) Pili

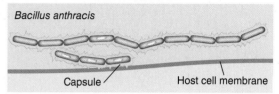

B Capsules

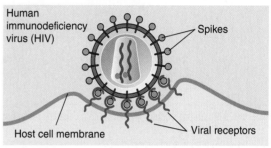

(c) Spikes

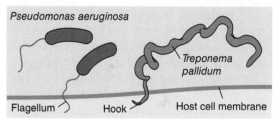

(d) Hooks or flagella

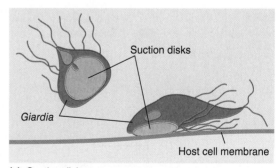

(e) Suction disks

FIGURE 10.14 Examples of adhesins produced by bacteria: **(a)** *Neisseria* produce pili, **(b)** *Bacillus anthracis* produces capsules **(c)** HIV contains glycoprotein spikes **(d)** *Pseudomonas aeruginosa* produces a flagellum and **(e)** *Giardia* produces suction disks that act as adhesions.

Mycobacterium leprae, that causes leprosy, are characterized by a distinctive "waxy" cell wall that allows the bacteria not only to resist digestion within phagocytic cells but, further, to multiply inside them. These are notable examples of structural components that act as virulence factors.

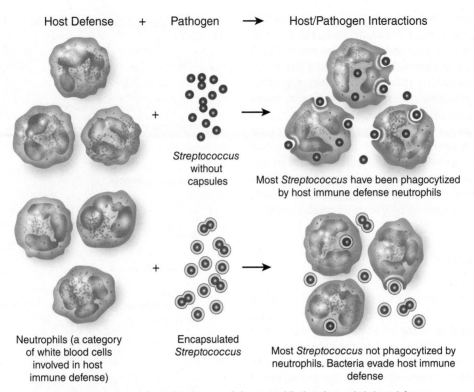

Host Defense   +   Pathogen   →   Host/Pathogen Interactions

Streptococcus without capsules

Most *Streptococcus* have been phagocytized by host immune defense neutrophils

Neutrophils (a category of white blood cells involved in host immune defense)

Encapsulated *Streptococcus*

Most *Streptococcus* not phagocytized by neutrophils. Bacteria evade host immune defense

FIGURE 10.15 *Streptococcus* spp. that contain capsules resist phagocytosis by neutrophils that play a role in host defenses.

Antigenic Variation

Antigenic variation is a strategy exhibited by some microbes that allows them to evade the immune system of their host, another remarkable example of microbial evolution and adaptation. They are "masters of disguise." Influenza A viruses are well-known pathogens that "change their coat" by changing their surface antigens (**FIGURE 10.16**). A major function of the human immune system is to defend the body against microbial infection; certain cells of the immune system recognize components of microbes, called **antigens**, as foreign and produce **antibodies** specifically targeted against these antigens, marking the microbes for destruction. The upshot is that antibodies fail to recognize and to target these new coats. The antigenic changes of influenza A viruses can occur within the course of a single

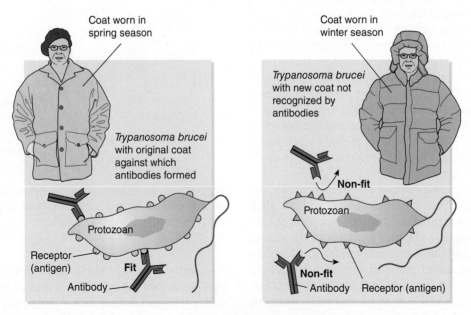

Coat worn in spring season

Trypanosoma brucei with original coat against which antibodies formed

Protozoan

Receptor (antigen)

Antibody

Fit

Coat worn in winter season

Trypanosoma brucei with new coat not recognized by antibodies

Non-fit

Protozoan

Non-fit   Antibody   Receptor (antigen)

FIGURE 10.16 Changing one's coat. *Trypanosoma brucei* can form new surface antigens not recognized by host defense antibodies. This is an important immune defense evasion strategy. This is analogous to people wearing different coats to withstand weather conditions.

year, explaining why **vaccination** (immunization) to prevent viral infection needs to be repeated each year. *Trypanosoma brucei* are a species of protozoans responsible for African sleeping sickness, have genes for as many as a thousand distinct surface antigens, and switch from one gene to another, resulting in disguise—quite a trick. The causative agents of relapsing fever and gonorrhea, both bacterial diseases, also are masters of antigenic disguise. Antigenic variation is an effective strategy that allows microbes to evade the host's antibody defense system, an important component of the immune system.

○ Enzyme Secretion

One of the most remarkable examples of defensive strategies is exhibited by *Helicobacter pylori*, the causative agent of certain peptic ulcers. An unusual adaptation allows the bacteria to grow in the extremely acid environment of the stomach. The hydrochloric acid in the stomach has a pH of 1 to 4; neutral pH is 7. This acid environment is intolerable for most microbes. How does *H. pylori* counteract the acidity? *H. pylori* manufactures the enzyme **urease**, which splits the normal component urea into ammonia and carbon dioxide. The ammonia helps to reduce the acidity, protecting the microbe in its own limited environment. *H. pylori* has been confirmed as the cause of most cases of peptic (gastric and duodenal) ulcers. *H. pylori* also secretes toxins (offensive strategy) that contribute to the formation of ulcers (**FIGURE 10.17**).

Offensive Strategies: Extracellular Products

A variety of products are produced and released by bacterial cells as they grow; these factors are referred to as *extracellular substances*. Some are classified as **toxins**, others are classified as **enzymes**. Some exhibit both enzymatic

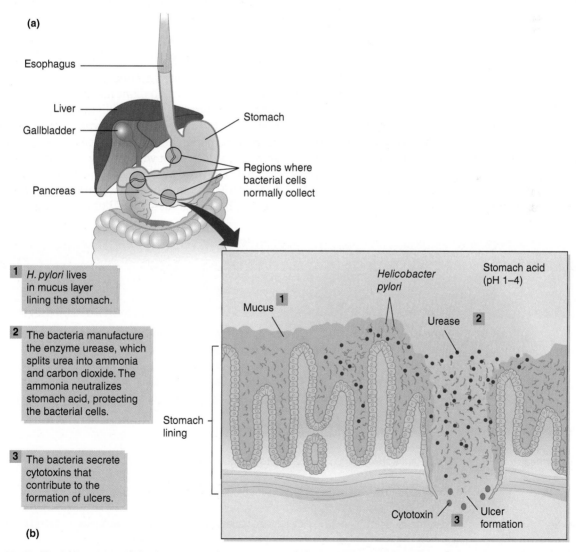

FIGURE 10.17 The **(a)** location and **(b)** progression of peptic ulcers associated with *Helicobacter pylori* infection. *H. pylori* produce toxic virulence factors such as urease and cytotoxins that act at different levels of infection. Urease neutralizes stomach acidity by degrading urea present on epithelial cells, whereas cytotoxins are involved in the development of ulcers.

and toxic activities. The distinction between toxins and enzymes is not always clear, so it is best to think of them collectively as extracellular products.

Exoenzymes

Some pathogens secrete **exoenzymes**, or *spreading factors*, which foster the spread of invading bacteria throughout the tissues by causing damage to host cells in their immediate vicinity and, in so doing, break down tissue barriers. Think of this action as being like that of a plow clearing a field so that seeds can be planted in the exposed soil and crops can grow. **Hyaluronidase** is an exoenzyme equivalent of a plow; it breaks down the "ground" substance, the hyaluronic acid content of connective tissue, reducing its viscosity, fostering the spread and penetration of microbes deeper into the tissues.

Collagenase is another example of an exoenzyme spreading factor; it breaks down the structural framework of collagen, a vital part of connective tissue, resulting in gas gangrene and massive areas of necrotic (dead) tissue. Extensive tissue **debridement** (cutting away of necrotic tissue) or amputation may be necessary to halt infection and avoid death.

Hemolysins are exoenzymes that destroy red blood cells through destruction of cell membranes. **Kinases** break down clots, allowing entrapped bacteria to spread. (One kinase—**streptokinase**, a product of streptococci—is now an important therapeutic tool for dissolving blood clots in patients suffering from heart attacks and has proved to be quite effective in preventing strokes and deaths.) **Coagulase** forms a network of threads around bacteria, affording protection against phagocytosis, and **leukocidins** destroy white blood cells.

Exotoxins

The word *toxin* conjures up thoughts of substances that are poisonous to the body. Toxins are major virulence factors for many pathogenic microbes; they are classified as either exotoxins or endotoxins. These products differ from each other in their chemical composition, modes of action, and nature of their release.

Exotoxins are protein molecules that are synthesized within the microorganism and released ("exo") into the host tissues during the growth and metabolism of the microbes. Exotoxin production is primarily associated with Gram (+) bacteria; the ability to produce toxins is called **toxigenicity**. Exotoxins are readily soluble in body fluids, are rapidly transported throughout the body, and may act at sites quite distant from the site of colonization of the organism. They are highly toxic and specific in their activity. Exotoxins are grouped into three principal types:

1. **Cytotoxins**, which kill or damage host cells
2. **Neurotoxins**, which interfere with transmission of nerve impulses
3. **Enterotoxins**, which affect the cells lining the gastrointestinal tract, leading to diarrhea

TABLE 10.7 lists a number of bacterial toxins and their effects on the host. Exotoxins are among the deadliest poisons. It is estimated that 1 milligram of botulinum neurotoxin (which causes botulism, a type of food poisoning) can kill half the population of a major city. This may be an exaggeration, so let's say it will kill only one-fourth of the population. What's the difference? Whatever the numbers, the potency of this neurotoxin is remarkable! **Toxoids** are toxins that have been detoxified but retain their antigenicity.

TABLE 10.7 Disease Manifestations by Bacterial Exotoxins		
Disease	Bacterial Pathogen (Exotoxin-Producer)	Exotoxin Activity and Result
Botulism	*Clostridium botulinum*	Neurotoxin: prevents transmission of nerve impulse, resulting in (flaccid) limp paralysis
Tetanus	*Clostridium tetani*	Neurotoxin: prevents transmission of inhibitory nerve impulse, resulting in rigid contractions of skeletal muscles (lockjaw)
Clostridial food poisoning	*Clostridium perfringens*	Enterotoxin: diarrhea
Traveler's diarrhea	*E. coli* (enterotoxigenic)	Enterotoxin: diarrhea
Gastroenteritis	*E. coli* O157:H7	Cytotoxin: bloody diarrhea; kidney damage
Scarlet fever	*Streptococcus pyogenes*	Cytotoxin: causes damage to blood capillaries, resulting in a red rash
Whooping cough (pertussis)	*Bordetella pertussis*	Cytotoxin: kills ciliated epithelial cells in the respiratory tract

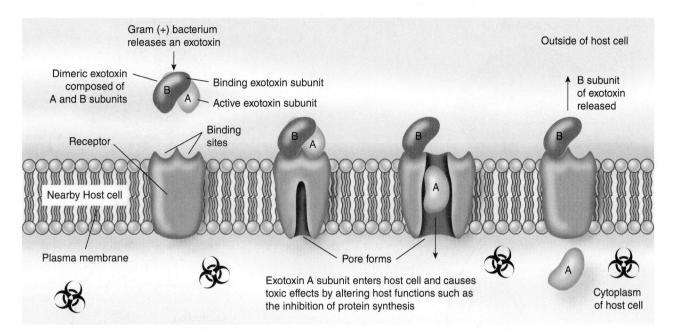

FIGURE 10.18 The AB model of exotoxin activity.

They are used in immunization against those bacterial infections that are primarily a result of exotoxin activity. In addition to botulism, a number of potentially fatal bacterial diseases are characterized by the production of specific exotoxins, including diphtheria, tetanus, cholera, and pathogenic *E. coli* infections. These toxemic diseases are largely a manifestation of the toxin and not the presence of the bacteria; *injection of toxin by itself into laboratory animals mimics the symptoms of the disease.*

Although each toxin has a specific mechanism of activity (Table 10.7), the AB model (**FIGURE 10.18**) has been proposed to explain the general mode of activity of some toxins. According to this model, exotoxins are composed of two subunits referred to as the A (active) fragment and the B (binding) fragment. Isolated A fragments have been shown to be enzymatically active but lack the ability to bind and to enter cells; isolated B fragments are able to bind to target cells but lack toxicity. Hence, the AB complex is necessary for exotoxin activity; the activity of the A subunit follows the activity of the B subunit. The specific nature of the A fragment activity is a function of the particular exotoxin.

In the early years of microbiology, the effect of toxins on human patients and on laboratory animals was observed, but little was known about how they worked. With the advent of molecular biology, toxins have been studied in great detail to determine their mechanism of activity. **FIGURE 10.19** illustrates the activity of the botulinum and tetanus neurotoxins.

Bacteria may be infected by bacteriophages. In some cases the phage DNA becomes incorporated into the bacterial chromosome in a latent or dormant state and does not bring about lysis of the bacterium.

The integrated bacteriophage genome into the host bacterium chromosome is termed a **prophage**, and the bacterium is said to be **lysogenized**. During bacterial cell division the prophage is replicated as a part of the chromosome, and the resulting daughter cells continue to harbor the phage DNA, conferring new properties, possibly including toxin production, in the lysogenized bacteria. This phenomenon is referred to as **lysogenic phage conversion** (**FIGURE 10.20**). For example, presence of the *tox* gene in lysogenized *Corynebacterium diphtheriae* results in production of an exotoxin that is responsible for the symptoms of diphtheria. When the bacteria are "cured" (the prophage is removed), diphtheria exotoxin production no longer takes place.

Endotoxins

Endotoxins are quite different from exotoxins, as summarized in **TABLE 10.8**. Unlike exotoxins, endotoxins are not usually released during bacterial growth and metabolism; they are structural components of the outer membrane of Gram (−) cells. They are released as these bacteria undergo disintegration, although some endotoxin material may be released during bacterial multiplication. **Endotoxins** are not proteins but are molecules known as lipopolysaccharides (LPS) that are a component of all Gram (−) cell walls. As indicated in Table 10.7, each exotoxin has a specific action; in contrast, all endotoxins, regardless of their source, produce the same host response, characterized by shocklike symptoms, chills, fever, weakness, formation of small blood clots, and possibly death. They are considerably less toxic than exotoxins but can cause severe symptoms during infections by Gram (−) bacteria. Ironically, individuals suffering from

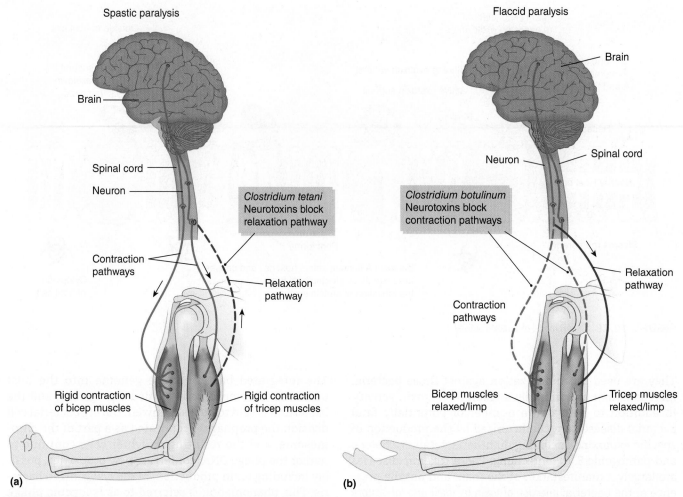

FIGURE 10.19 Involvement of botulinum and tetanus bacterial neurotoxins in spastic (contracted) or flaccid (limp) paralysis. *Clostridium tetani* causes tetanus and *C. botulinum* causes botulism. **(a)** Contracted/spastic paralysis. **(b)** Flaccid paralysis.

Gram (–) infections may experience exaggerated symptoms during antibiotic treatment because of the release of endotoxins from dead bacteria.

Virulence Mechanisms of Nonbacterial Pathogens

O Viruses

Recall that viruses are completely dependent on their host cells for survival and replication. On the defensive side, the fact that viruses replicate inside of cells of their host allows them to hide from components of the host immune system. The ability of some viruses, for example, the influenza A virus, to change their surface antigens affords a measure of defense against host defense antibodies.

On the offensive side, death of the host cell may result from the lysis of the cell membrane caused by:

- Production of large numbers of replicating viruses
- Disruption of the host cell's ability to synthesize proteins
- Damage to the host's cell membrane
- Inhibition of the host cell's metabolism

Some viruses have attachment molecules enabling them to dock with specific target host cells, allowing for adsorption and penetration. For example, consider that HIV docks with strategic cells (T lymphocytes) of the immune system, causing the infected individual to become severely immunocompromised. Other viruses produce **cytopathic effects** that kill the host cell. Some viruses cause adjacent cells to fuse and form a **syncytium** (a network of cells fused together).

Some viruses leave telltale fingerprints called **inclusion bodies**. The bodies are "viral debris" consisting of viral parts (genomes, protein capsids, envelopes) discarded in the process of viral assembly. These inclusion bodies are important in diagnosis, because they can be identified microscopically and associated with a particular virus.

O Eucaryotic Microbes (Including Helminths)

Many protozoans, fungi, and helminths are pathogens, but their mechanisms of virulence are not well defined. Many fungi secrete toxins and enzymes that cause damage to host cell tissues and aid in their invasion. *Giardia,*

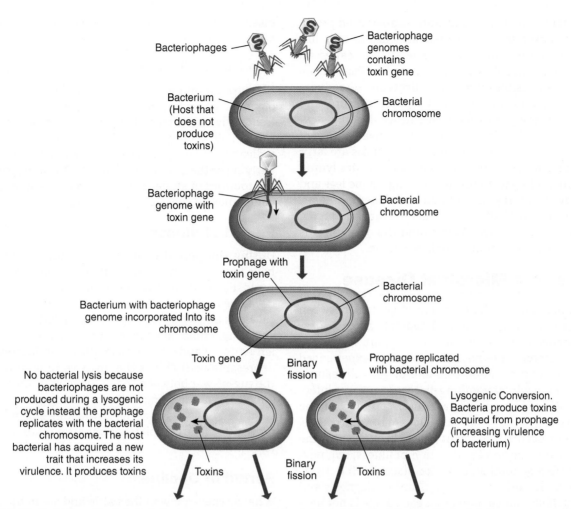

FIGURE 10.20 Mechanism of bacteriophage lysogenic conversion. A bacterium is infected with a bacteriophage. The bacteriophage genome contains a toxin gene. During lysogenic phage conversion, the prophage is incorporated into the bacterial host chromosome. The bacteria have acquired a new trait; becoming a toxin-producing bacteria. The production of toxins will increase bacterial virulence. Not all bacteriophage genomes carry toxin genes. Bacteriophage genomes can also contain genes that code for other virulence factors. When the bacteriophages exit a lysogenic cycle, its genome may pick up additional genes from the host, including genes that code for toxins or other virulence factors.

TABLE 10.8 Characteristics of Bacterial Exotoxins and Endotoxins

Property	Exotoxins	Endotoxins
Site	Released from bacteria during growth and metabolism	Retained (for the most part) within outer membrane and released when bacterium undergoes death by lysis
Cell source	Primarily Gram (+) bacteria	Gram (−) bacteria
Activity	Specific for each toxin	Essentially similar for all endotoxins
Chemical nature	Protein	Lipopolysaccharide (LPS)
Toxicity	High toxicity	Minimal toxicity
Heat stability	Unstable; usually destroyed at about 60°C (140°F)	Stable; can withstand temperatures of 100°C (212°F)
Examples of bacterial diseases	Tetanus, scarlet fever, diphtheria, gas gangrene, botulism	Meningococcal meningitis, typhoid fever, salmonellosis

an important and widespread water-transmitted protozoan that causes severe diarrhea, attaches to cells lining the small intestine by a virulence factor called an adhesive or sucking disk (Figure 10.14). The *Plasmodium* protozoan responsible for **malaria** infects and reproduces within host red blood cells, causing their rupture, and the trypanosome protozoan that causes sleeping sickness exhibits antigenic variation, a defensive strategy of virulence, as do influenza A viruses. Helminths are large extracellular parasitic worms that can obstruct lymph circulation, leading to grotesque swellings in the legs and other body structures, a condition called **elephantiasis**. *Ascaris* worms can "ball up," obstruct the intestinal tract, and migrate into the liver. Waste products of helminths can cause toxic and immunologic reactions.

Course of a Microbial Disease

As explained earlier, the outcome of contact between a pathogen and a host (human) depends on the number of pathogens and their virulence factors versus the immune system of the host. Everyone has experienced the miseries of a common cold and of gastrointestinal disturbances of winter vomiting disease (norovirus infections) manifested by vomiting and diarrhea or (perish the thought!) both. In most cases the disease runs its course, and recovery occurs. Treatment may help to alleviate the symptoms and shorten the duration of illness, but, nevertheless, five general stages take place (**TABLE 10.9**), and their characteristics are significant in diagnosis and treatment. It should be noted that each stage is not necessarily distinct.

Incubation Period

The **incubation period** is the time between the pathogen's access to the body through a portal of entry and the display of signs and symptoms. During this time, although the infected individual is not necessarily aware of the presence of a pathogen, the microbes may be spread to others. The incubation time is quite consistent in some diseases but variable in others. The incubation times for several diseases are listed in **TABLE 10.10**.

Prodromal Period

The **prodromal period** is relatively short and not always obvious. The symptoms are vague and mild and are frequently characterized by tiredness, headache, muscle aches, and "feeling lousy." (These symptoms may not be indicative of a disease at all, but the result of too much partying or the stress associated with exams, papers, and the routines of student life.) In some infections, the individual may be contagious during this stage.

Period of Illness

During the period of illness you might feel like you have been hit in the head with a hammer and wish you could die but aren't that lucky. In this period, the disease develops to the most severe stage, accompanied by typical signs and symptoms that may include fever, nausea, vomiting, chills, headache, muscle pain, fatigue, swollen lymph glands, and a rash. (What an impressive and depressing list!) This is the invasive time during which the tug-of-war between the pathogen's virulence factors and the host's immune system is taking place. It is a critical time in that, as in a tug-of-war, one side wins. Either recovery will be complete, or impairment or death of the host will result.

Period of Decline

What a relief! You won the battle and are in the **period of decline**. The signs and symptoms begin to disappear, the body returns to normal, and life is worth living again.

Convalescence Period

During the **convalescence period**, recovery takes place, strength is regained, damaged tissues are repaired, and rashes disappear. In some cases healthy and chronic carrier states develop, as might happen with cholera and typhoid fever; the carrier state can exist for years.

TABLE 10.9 Course of Microbial Disease	
Period	Description
Incubation	Period between initial infection and appearance of symptoms; considerable variation among diseases
Prodromal	Period in which early symptoms appear; usually short and not always well characterized
Illness	Period during which the disease is most acute and is accompanied by characteristic symptoms
Decline	Period during which the symptoms gradually subside
Convalescence	Period during which symptoms disappear and recovery ensues

TABLE 10.10 Incubation Period of Infectious Diseases

Pathogen	Type of Pathogen	Infectious Disease	Incubation Period
Rhinovirus or *adenovirus*	Virus	Common cold	12 to 72 hours (usually 24 hours)
Clostridium botulinum	Bacterium	Botulism	12 to 36 hours
Varicella zoster virus	Virus	Chickenpox	2 to 3 weeks (average is 14 to 16 days)
Mumps virus	Virus	Mumps	12 to 26 days (average is 18 days)
Staphylococcus aureus	Bacterium	Food poisoning	2 to 4 hours
Hepatitis B virus	Virus	Serum hepatitis (hepatitis C)	45 to 180 days (average is 60 to 90 days)
Hepatitis A virus	Virus	Hepatitis A	14 to 28 days
Hepatitis C virus	Virus	Hepatitis C	2 weeks to 6 months
Mycobacterium leprae	Bacterium	Leprosy	1 to 20 years (average is 5 years)
Ebola virus (Guinea strain)	Virus	Ebola virus disease	2 to 21 days (average is 8 to 10 days)
Norovirus	Virus	Winter vomiting disease	12 to 48 hours
Zika virus	Virus	Zika virus disease	3 to 12 days
Measles virus	Virus	Measles	10 to 14 days until rash appears
Influenza A virus	Virus	Influenza	1 to 4 days (average is 2 days)
Rabies virus	Virus	Rabies	2 to 8 weeks or longer (average is 18 to 21 days)
Middle East respiratory syndrome coronavirus	Virus	Middle East respiratory syndrome (MERS)	2 to 4 days

Summary

This chapter considers microbial pathogens and their disease-causing mechanisms. "No organism is an island" is an expression implying that all organisms live in symbiotic association with other organisms. These associations can take the form of mutualism, commensalism, or parasitism, and can slide from one to the other. Parasitism is the basis for microbial disease; the parasite lives at the expense of the host and causes damage to the host. Bacteria communicate to each other through a chemical language known as quorum sensing. Many group behaviors tasked through quorum sensing are associated with bacterial survival. A healthy microbiome plays a role in preventing disease. Research has shown that dysbiosis of gut microbiota is associated with obesity, Parkinson's disease, diabetes, and autoimmune diseases. Future treatment of dysbiosis may involve the reengineering of gut microbiota.

The mechanisms of microbial disease have been described as a tug-of-war between (1) the number, or dose, of microorganisms to which a potential host has been exposed and their virulence factors, and (2) the immune mechanisms of the host. The outcome of the host–parasite dynamic determines the severity of infection, varying from subclinical to death. Microbes vary considerably, even within a species, in their pathogenicity, as a function of their virulence factors. Pathogenic microbes have evolved defensive strategies by which they are able to withstand the immune system of the host and offensive strategies that cause damage to the host. Although there is considerable variation in the manifestation of illness, depending on the specific microbe and the immune system of the host, the general pattern of infectious diseases can be described as the incubation, prodromal, illness, decline, and convalescence periods.

KEY TERMS

adhesins
anaerobes
antibodies
antigenic variation
antigens
asymptomatic
autoinducers
bacteremia
bacteriophages
biofilm
capsules
carnivores
cheater bacteria
coagulase
collagenase
colonization resistance
commensalism
convalescence period
course of a disease
cytopathic effects
cytotoxins
$D = nV/R$
debridement
defensive strategies
disease
dose (n)
dysbiosis
elephantiasis
endotoxins
enterotoxins
enzymes
exoenzymes
exotoxin
extracellular polysaccharide polymers (EPS)
50% lethal dose (LD$_{50}$)
gas gangrene
hemolysins
herbivores
hosts
hyaluronidase
immunofluorescent staining
inclusion bodies
incubation period
infection
infectious disease
infective dose (ID)

interspecies communication
intraspecies communication
kinases
leukocidin
lysogenic phage conversion
lysogenized
M protein
malaise
malaria
metabolites
microbiome
microbiota
mutualism
nanotechnology
necrotizing enterocolitis
neurotoxins
offensive strategies
omnivores
opportunistic pathogens
parasitism
period of decline
peritonitis
phagocytosis
portal of entry
prodromal period
prophage
protease
quorum sensing (QS)
quorum-sensing inhibitors (QSIs)
receptor
reservoir
resistance (R)
sepsis
streptokinase
subclinical
symbiosis
syncytium
toxigenicity
toxins
toxoids
urease
urinary tract infection (UTI)
vaccination
virulence (V)
virulence factors

SELF-EVALUATION

○ **PART I: Choose the single best answer.**

1. Endotoxins are associated primarily with _____.
   - a. viruses
   - b. Gram (+) bacteria
   - c. Gram (–) bacteria
   - d. fungi

2. Pili are _____.
   - a. adhesions
   - b. exotoxins
   - c. proteases
   - d. phagocytic

3. Which of the following is *not* an example of a virulence factor?
   - a. Hyaluronidase
   - b. Collagenase
   - c. Hemolysins
   - d. Ribosomal RNA

4. Microbial diseases are the result of a biological association, which can best be described as _____.
   - a. commensalism
   - b. symbiosis
   - c. mutualism
   - d. parasitism

5. Bacterial _____ interfere with phagocytosis.
   - a. M proteins
   - b. adhesions
   - c. capsules
   - d. flagella

6. _____ is a fatal condition that can happen if the intestinal wall is pierced during surgery, allowing bacteria of the intestines to gain access to another part of the body.
   - a. Sinusitis
   - b. Peritonitis
   - c. Tinnitus
   - d. Senioritis

7. Which strategies involve the production of extracellular products that foster the spread of bacteria throughout the tissues by causing damage to host cells?
   - a. Offensive
   - b. Defensive
   - c. Feeding
   - d. Evasion

8. Which strategies allow microbes to escape destruction by the host immune system?
   - a. Offensive
   - b. Defensive
   - c. Feeding
   - d. Evasion

9. *Corynebacterium diphtheriae* harbors a(n) _____ that contains a toxin gene responsible for the symptoms of diphtheria.
   - a. prophage
   - b. pigment
   - c. enzyme
   - d. plasmid

10. The _____ is the time between the pathogen's access to the body through a portal of entry and the display of signs and symptoms.
   a. period of illness
   b. prodromal period
   c. incubation period
   d. convalescence period

PART II: Fill in the blank.

1. E. coli, a resident of the intestinal tract that produces vitamin K, demonstrates a type of symbiotic association called _____.

2. _____ interfere with transmission of nerve pulses.

3. _____ affect the cells lining the gastrointestinal tract, leading to diarrhea.

4. _____ kill or damage host cells.

5. A relationship between two or more species in which one benefits and the other is indifferent is called _____.

6. _____ is a process by which bacteriophage DNA is incorporated into the bacterial chromosome and confers new properties on the bacterial host.

7. Bacterial communication through autoinducers is called _____.

8. The disruption of gut microbiota is also called _____.

9. _____ is a relationship between two microbes that benefit from each others' presence.

10. Enzymes called _____ are produced by bacteria and break down clots, allowing entrapped bacteria to spread.

PART III: Answer the following.

1. Explain how quorum sensing allows bacteria to be multicellular, similar to higher organisms.

2. The equation $D = nV/R$ expresses the severity of an infectious disease. Identify each term (D, n, V, and R) in the equation.

3. Regarding symbiotic associations, a "slippery seesaw" exists. What does this mean? Give examples.

4. Explain the AB model that has been proposed to explain the general mode of activity of some toxins.

5. Define *biofilm*, and explain why pathogenic bacteria that form biofilms are difficult to eradicate from the body.

6. Antigenic variation is a significant microbial defense mechanism. Provide an example.

7. What does the term LD_{50} mean?

8. Name a "spreading factor" and describe its action.

9. List the five periods assigned to the course of a microbial disease and describe what happens at each period.

10. Define *infectious dose* and *incubation period*. What does each measure?

EPIDEMIOLOGY AND CYCLE OF MICROBIAL DISEASE

> "*If you think you are too small to make a difference, try sleeping with a mosquito.*"
>
> —Dalai Lama

© Teri Shors.

At the beginning of the spring 2018 semester, a norovirus, or winter vomiting disease, epidemic hit the University of Wisconsin–Oshkosh. Approximately 600 college students became ill. The majority of students affected lived in the dormitories on campus. The students hit hardest lived in a dormitory located near the campus daycare center, where many of the students in this dormitory also worked. An online self-reporting tool was used to track the number of norovirus cases on campus. Signs were posted outside and inside of every campus building reminding students to wash their hands and to take other precautions to prevent the spread of norovirus infections. Self-serve food services located in different classroom buildings were closed for a week during the height of the epidemic. Quick foods to go such as donuts and bagels were individually wrapped for purchase. Self-serve hot dogs and salad bars were closed for 1 to 2 weeks. Sick students were asked to remain in their dorm rooms until they were symptom-free for 48 hours. Other norovirus epidemics were happening simultaneously in the United States and abroad, including an epidemic at the Winter Olympics, held in South Korea.

© ShutterStock, Inc. / happykanppy

1. State the relationship between herd immunity, R-nought, and epidemics.

2. Review the cycle of microbial infection.

3. Provide examples of infectious diseases transmitted through indirect and direct transmission.

4. Describe control measures that can be used to prevent diseases transmitted by arthropods.

5. Identify the top five healthcare-associated infections in the United States.

6. Summarize the One Health concept.

7. Explain why epidemics can cause fear.

8. Define *zoonoses*, and explain their association with human and environmental health.

Case Study: You Can Get Fatal Herpes from a Monkey!

The clear waters created by some of the world's largest artesian springs located in Silver Springs, Florida, led to the beginning of the Glass-Bottom Boat Jungle Cruise. In the 1920s, a nature theme park with animal exhibits, rides, and a water park was created to accompany the boat cruises. Visitors could observe fish, turtles, and alligators beneath the glass-bottom boats and a myriad of birds above it. Tourists saw wild boars, foxes, bobcats, black bears, white-tailed deer, fox squirrels, wild turkeys, coyotes, alligators, and gopher tortoises along the gin-clear river. Many consider the operation to be Florida's first tourist attraction. Silver Springs merged into Silver River State Park, creating Silver Springs State Park in 2013.

Local folklore suggested that rhesus macaques escaped in Silver Springs during the filming of the 1939 movie *Tarzan Finds a Son*, starring Johnny Weissmuller. According to a report by the Department of Wildlife Ecology and Conservation, UF/IFAS Extension, this was not the case. No macaques appeared in the film. In an effort to increase tourism and revenue, and not realizing that macaques were good swimmers, Colonel Tooey, manager of the Glass-Bottom Boat Jungle Cruise operation, intentionally released six rhesus macaques on an island located in the Silver River during the 1930s. The macaques swam off the island to surrounding forests. Boat operators lured the macaques with food to the shore to entertain tourists. Again, in 1948, Colonel Tooey released six more macaques along the north shore of the Silver River.

From the 1930s through 1978, three different species of monkeys were intentionally released throughout Florida; rhesus macaques, squirrel monkeys, and

vervet monkeys (**Figure 1**). Rhesus macaques live in groups and thrive in human-dominated areas. The free-ranging macaques have gained a reputation as an **invasive species**. They cause environmental damage, such as the consumption of over 50 plant species in Silver Springs, resulting in shoreline erosion along the river. The macaques affect other native species, such as by eating shorebird eggs, causing a decline in the shorebird population. By the 1980s, Silver Springs had about 400 macaques. From 1998 to 2012 a private trapper was permitted by the state of Florida to capture rhesus macaques between Silver Springs and the Ocklawaha River. About 830 macaques were captured, and 700 of them were sold to biomedical research facilities. This generated public controversy, halting the practice.

The nonnative rhesus macaques were introduced into Florida during a time when little was known about viruses. The transmission electron microscope (TEM) was invented in 1931. The first report in the scientific literature to include clear electron micrographs of virus particles, ectromelia virus, or mousepox, and vaccinia virus was published in 1938. After the macaques were released, Captain Tooey didn't anticipate they would swim off the island. This unexpected turn of events created situations in which humans had closer contact with the macaques. Tourists began feeding the macaques, heightening the opportunity for zoonoses to occur.

Infectious diseases that spread between animals and humans are known as **zoonotic diseases**, or **zoonoses**. The intentional domestication of animals for farming or agriculture approximately 10,000 years ago was likely

(continues)

Case Study: You Can Get Fatal Herpes from a Monkey!
(continued)

Silver Springs State Park
1930s, 1948 Rhesus Macaques
1970s Squirrel monkeys

Masterpiece Gardens
Lake Wales
1960s
Squirrel monkeys

Titusville 1976
Rhesus Macaques

Naples
1960s
Squirrel monkeys

Florida Atlantic University
Boca Raton
1970 Squirrel monkeys

Key Lois and Raccoon Key
1973 and 1978
Rhesus Macaques

Bannet House Museum
and Gardens, Fort Lauderdale
1940s or 1970s
Squirrel monkeys

Dania Beach
1950s
Vervet monkeys

(a)

Rhesus monkeys: © Yoyochow23/ShutterStock, Inc.
Squirrel monkeys: © Meunierd/ShutterStock, Inc.
Vervet monkeys: © Alta Oosthuizen/ShutterStock, Inc.

(b)

© Brian Parker/Alamy.

Figure 1 (a) Since the 1930s, three different species of monkeys were intentionally introduced into Florida for tourism or conservation purposes. Rhesus macaques may pose risks to human health. Recent studies have found that the rhesus macaques in Silver Springs State Park test positive for herpes B virus. Herpes B virus can infect humans. Without treatment, infections results in fatal encephalitis with a 50% mortality rate. **(b)** The glass-bottom boat cruises are a tourist attraction even today in Tijuana, Mexico.

the beginning of zoonotic disease transmission through the close and frequent contact of humans with new zoonotic pathogens. Human measles likely arose from a rinderpest-like virus of sheep and goats. *Mycobacterium bovis* of cattle, which is indistinguishable from *M. tuberculosis* in humans, causes at least 3% of all human tuberculosis cases today through the consumption of unpasteurized dairy products. Scientists estimate that six out of every known infectious disease in people are spread from animals. Every year, tens of thousands of Americans will get sick from harmful infectious agents spread between animals and people.

Wildlife, or free-roaming animals, has also been an important source, or **reservoir**, of infectious disease transmission to humans. Well-known zoonoses in human history prior to 1900 were rabies and the bubonic plague, also known as the Black Death. Hiking, hunting, camping, logging, and close contact with animals living in zoos or safari parks are activities that may represent risk factors for acquiring certain zoonoses with a wildlife reservoir.

The first human case of herpes B virus infection occurred in 1932. Dr. William Bartlett Brebner was using macaques for poliovirus vaccine research. He was accidently bit on his fingers by a healthy rhesus macaque. Three days later, he noticed pain, swelling, and redness at the bite sites. Six days after the bite, he was admitted to the hospital. He had a temperature of 101.4°F (38.5°C). The bites were red and associated with **lymphadentitis** (inflammation of the lymph nodes). He was given a tetanus antitoxin injection. He appeared to be improving for the next few days during which time herpeslike blisters formed at the bite sites. Seven days after his admission to the hospital, he suffered from abdominal cramps that lasted for 2 days. A couple of days after the abdominal cramps disappeared, his health declined quickly. Brebner's kidneys began to fail, and he suffered from rapidly ascending **myelitis** (inflammation of the spinal cord) that resulted in **flaccid paralysis** of his lower extremities. Eighteen days after the monkey bite, he died of respiratory paralysis at the young age of 29. For a detailed description of his final days and a very interesting read, see lead author and famous poliovirus researcher Dr. Albert Sabin's 1933 report titled "Acute Ascending Myelitis Following a Monkey Bite, with Isolation of a Virus Capable of Reproducing the Disease" in the *Journal of Experimental Medicine* (volume 59, pages 115–136), published 1 year after Dr. Brebner's death.

After Dr. Brebner passed away, Dr. Frederick P. Gay and Margaret Holden obtained brain tissue from Dr. Albert Sabin. They isolated an ultrafilterable agent that was similar to human herpes simplex virus-1 (oral herpes) that they named "W" virus. Drs. Albert Sabin and Arthur Wright independently isolated the same virus from Brebner's tissues. They named it "B" virus, after Dr. Brebner. Later, names such as herpes B virus were used.

By 1959, 17 more herpes B cases had been identified. Only five survived. The cases described were all biomedical employee–associated exposures to monkeys during the production of the poliomyelitis vaccine, either preparation of monkey kidney cell cultures for the cultivation of polioviruses or for safety testing of the finished vaccine. To date, worldwide, a total of 50 cases (21 fatal) of herpes B have been confirmed. All of the documented cases were associated with occupational exposures through monkey bites, scratches, or exposures to monkey tissues. Herpes B viruses cause a very mild or asymptomatic infection in rhesus macaques. The monkeys are a natural reservoir of the virus.

From 1990 to 1992, there were 28 medical reports of *nonoccupational* macaque bites in the United States. In 2017, news reports of unwanted monkey interactions with Silver Springs State Park guests made newspaper headlines. A family recorded a movie that portrayed aggressive monkeys and shared it on social media. One homeowner photographed 50 monkeys crouched around a backyard birdfeeder. At least 18 confirmed reports had been made of park-goers who had received monkey bites and scratches since the macaques had been introduced into the park.

In February 2018, a report was published in the Centers for Disease Control (CDC) publication, *Emerging Infectious Diseases* in which researchers presented results of two studies to determine if the rhesus macaques in Silvers Springs State Park carry herpes B virus and if they shed herpes B viruses in their saliva or feces. The report described a study conducted during 2000–2012 in which blood samples were collected from 317 macaques captured along the Silver and Ocklawaha Rivers. About 25% of the blood samples contained antibodies to herpes B virus (which means these macaques are infected, healthy carriers of herpes B virus). The second study was conducted during 2015–2016. The researchers collected saliva and fecal samples from the macaques in the park to determine if they shed herpes B viruses. About 2.5% of the saliva samples and none of the fecal samples contained herpes B viruses. Therefore, they determined that the monkeys in the park shed herpes B viruses, putting humans at risk for contracting a highly fatal pathogen.

Coinciding with the 2018 report, the Florida Fish and Wildlife Conservation Commission ruled that feeding wild monkeys in the park was prohibited. The rule went into effect on February 11, 2018. The commission also supported the removal of the rhesus macaques from the environment to reduce the public health threat they pose. It's safe to say that there will be no more monkeying around in Silver Springs State Park!

(continues)

Case Study: You Can Get Fatal Herpes from a Monkey!
(continued)

Questions and Activities

1. How is herpes B virus transmitted from a rhesus macaque to a human?

2. List at least three other types of herpesviruses that infect humans, and identify what diseases they cause in humans.

3. Define *invasive species*, and list three examples of invasive species that have been introduced into Florida.

4. What are natural predators of monkeys in Silver Springs State Park? Provide reasons why the monkeys were able to breed and flourish in the park.

5. What is the treatment for herpes B virus infections in humans? (Hint: http://www.cdc.gov/herpesbvirus/index.html.)

Information from S. M. Wisely, K. A. Sayler, C. J. Anderson, C. L. Boyce, A. R. Klegarth, and S. A. Johnson. (2018). Macacine herpesvirus 1 antibody prevalence and dDNA shedding among invasive rhesus macaques, Silver Springs State Park, Florida, USA. *Emerging Infectious Diseases, 24*, 345–350.

Preview

Microbes are a potential threat to the world population. This chapter describes the biology of microbes and the mechanisms by which they cause disease from a public health aspect and focuses on the factors responsible for infectious diseases in populations. Basic concepts of epidemiology are presented so that the occurrence and prevalence of disease in a particular environment at a specific time can be better understood. The existence of microbial disease requires a chain of linked factors that constitute the cycle of disease. These factors are reservoir, transmission, portal of entry, and portal of exit.

In hospitals and in long-term healthcare facilities, all factors involved in the cycle of infectious diseases are present in a concentrated way. These facilities are hotbeds for the transmission of microbes among hospital personnel, visitors, and patients.

Concepts of Epidemiology

Epidemiology is an investigative methodology designed to determine the source and the cause of diseases and disorders that produce illness, disability, and death in human populations. Epidemiologists have been dubbed "disease detectives"; they are among the first group to be dispatched by the CDC when the threat of an **outbreak** occurs anywhere in the world. Their sleuthing is directed at understanding why an outbreak of a disease is triggered at a particular time and place. Epidemiologists consider the age distribution of the population, sex, race, personal habits, geographical location, seasonal changes, modes of transmission, and other factors. These parameters are used to design public health strategies for control and prevention of future outbreaks. Historically, epidemiology is based on an understanding of the causes and distribution of infectious diseases, but modern epidemiology has branched out to other public health problems, including heart disease, obesity, smoking, diabetes, Alzheimer's

disease, concussions, alcohol and drug abuse, alcohol and pregnancy, cancer, pesticide-related illness and injury, chronic obstructive pulmonary disease (COPD), mental health, "road rage" and other acts of violence, carbon monoxide poisoning (CO), and exposure to lead paint.

Epidemiologists focus on the frequency and distribution of diseases in populations and classify diseases as sporadic, endemic, epidemic, and pandemic. **Sporadic infectious diseases** are those that occur only occasionally and at irregular intervals in a random and unpredictable fashion. Typhoid fever, eastern equine encephalitis, and tetanus are examples. Diseases that are continually present at a steady level in a population and pose little threat to public health are **endemic infectious diseases**. The common cold, herpes simplex virus-1 infections, HIV/AIDS, chronic hepatitis, and foodborne illnesses are endemic across the United States and can occur on a year-round basis; Zika virus disease is endemic in Texas, Florida, and a number of U.S. territories (Puerto Rico, American Samoa, and the U.S. Virgin Islands). A disease is said to be **epidemic** when there is a sudden increase in the **morbidity** (illness rate) and the **mortality** (death rate) above that normally seen, causing a potential public health problem. Throughout history, epidemics have resulted in more deaths than those caused by wars, and they have influenced the course of history. Plague, or the "Black Death," has bedeviled humankind at least since the reign of Emperor Justinian in the 6th century; the 14th-century epidemic was particularly devastating. Smallpox, cholera, and typhus fever are other examples of past epidemics. Epidemics may arise from an explosion of sporadic or endemic diseases or, it would seem, from out of nowhere. **Pandemic infectious diseases** are those that spread across continents, and possibly worldwide; HIV/AIDS is a pandemic. Cholera has been responsible for pandemics on several occasions. In 1918, what was then referred to as "Spanish influenza" was perhaps the greatest pandemic of all time. The term *epidemic* has been borrowed to indicate a variety of conditions unrelated to infectious diseases that are present beyond the norm. For example, college officials talk of smartphone usage as a problem of "epidemic" proportions, and the term is often attributed to school violence, obesity, smoking, and cancer.

Epidemiologists use the term **epicenter** to define the point of origin or location of an infectious disease outbreak. Guinea, West Africa, was the epicenter of the 2014–2016 Ebola virus disease (EVD) epidemic. **Patient zero**, or the index case, represents the first infected individual in a new disease outbreak of an epidemiological investigation. Investigators trace the start of the epidemic retrospectively to the first person who became ill in the disease outbreak.

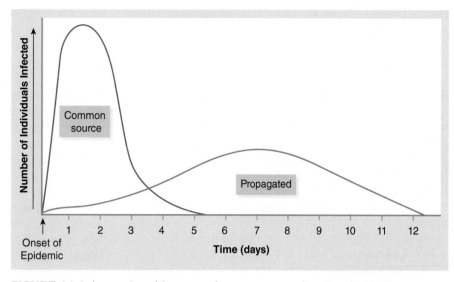

FIGURE 11.1 A comparison of the courses of common-source and propagated epidemics.

Epidemiologists describe the source and spread of epidemics as common-source epidemics or propagated epidemics. **Common-source epidemics** arise from contact with a single contaminated source and are usually associated with feces-contaminated food and water. Typically, a large number of people become ill quite suddenly, and the disease peaks rapidly in the population. **Winter vomiting disease**, or norovirus infection, is an example of a common-source epidemic (refer to the chapter opener figure about the University of Wisconsin–Oshkosh norovirus outbreak in the 2018 spring semester). A **propagated epidemic** is the result of direct person-to-person contact; the microbe is spread from infected individuals to susceptible noninfected individuals. As compared with common-source epidemics, the number of infected individuals rises more slowly and decreases gradually. Syphilis (sexual transmission) and HIV infections (transmitted through the sharing of

needles for drug injection use) are examples of propagated epidemics. **FIGURE 11.1** illustrates the courses of common-source and propagated epidemics.

The number of individuals in a population who are immune (nonsusceptible or resistant) to a particular disease as compared with those who are nonimmune (susceptible) is an important factor in the occurrence of epidemics. Immunity can be the result of having had a particular infection (natural immunity) or of having been vaccinated. The term **herd immunity** (group immunity) refers to the proportion of immune individuals in a population. Infectious disease can only be spread to susceptible individuals; therefore, the smaller the number of susceptible individuals, the less opportunity for contact between them and infected individuals. Public health officials strive to maintain high levels of herd immunity against communicable diseases to minimize their progressing to epidemic status (**FIGURE 11.2**).

FIGURE 11.2 Depiction of herd immunity. **(a)** Cows vaccinated against parainfluenza-virus 3 are protected from infection. **(b)** Cows not vaccinated to prevent infection by parainfluenza virus are infected and become sick.

Hence, vaccination is required in the elementary grades against a variety of infectious diseases; proof of an up-to-date immunization history is required for college admission.

The basic reproduction rate, denoted as **R-nought**, or R_0, is a measure of the potential for transmission, that is, the mean number of secondary cases occurring in a susceptible population in the wake of a particular infection. The population density and the duration of contagiousness and other factors need to be considered. For an infection to spread, the R_0 value must be greater than 1; if it is less than 1, then the infectious disease will die out. The greater the R_0 value, the greater the chance of spread, making it more difficult to establish measures of control. Pertussis (whooping cough), for example, has a high R_0 value of between 12 and 17. **TABLE 11.1** lists R_0 values for several infectious diseases. If you watched the movie *Contagion* (2011), recall that the infectious agent MEV-1 had an R_0 of 4 to 6 on the sixth day of the pandemic, and cases increased exponentially over time.

As described, the epidemiologist's toolbox contains a number of strategies to deal with endemic diseases, epidemics, to pandemics, and to predict their course as components of epidemic theory. A decrease in herd immunity can lead to **reemerging infectious diseases**. A case in point is the epidemic of diphtheria that occurred in the newly independent states of the former Soviet Union in the early 1990s. A decline in the public health infrastructure resulted in fewer children receiving diphtheria vaccination and a decline in herd immunity. When the disease was introduced into the susceptible (unvaccinated) population, possibly by returning military personnel, diphtheria reached epidemic proportions.

Frequency and distribution with attention to age, gender, diet, lifestyle, and other factors describing a disease-afflicted population is a starting point in investigative epidemiology in order to establish the strategy necessary to halt an outbreak. Epidemiologists use a variety of graphs, charts, tables, and maps to establish the parameters of an outbreak (**FIGURE 11.3**). These figures include real-life examples of infectious disease statistics published by the CDC.

Surveillance of disease outbreaks and of factors that could trigger outbreaks is an important mission of public health organizations throughout the world, including the **World Health Organization (WHO)**, the CDC, the **Center for Disease Research and Policy (CIDRAP)**, and agencies at the state and local levels. To keep track of infectious diseases and conditions in the United States, physicians are required to report cases of certain diseases, referred to as *notifiable diseases*, to their local health departments; these are then reported to the CDC using the National Electronic Disease Surveillance System (NEDSS) Base System (NBS). In 1994, 49 diseases were listed as nationally notifiable; 100 diseases were reportable in 2018 (**TABLE 11.2**).

The specific diseases are decided on at an annual meeting involving state departments of health and the CDC. An increase or a decrease in the number of notifiable diseases does not necessarily reflect a change in the health status but may be the result of reorganization each year. To further assist public health and medical personnel, the CDC publishes the journal *Emerging Infectious Diseases*, as well as the *Morbidity and Mortality Weekly Report*, which contain data organized by states on morbidity and mortality of particular infectious diseases in the United States and throughout the world.

Surveillance information is also gathered through online reporting tools such as **ProMED-mail** (http://www.promedmail.org). It assists in disseminating outbreak information (such as location or confirmed disease) as rapidly as possible. It reaches more than 60,000 subscribers in at least 185 countries, and thousands more refer to its website. It has been successful in alerting local, national, and international organizations about new outbreaks. **HealthMap** (www.healthmap.org) is another example of a free online informal global resource created in 2006 by a team of researchers at Boston Children's Hospital to monitor disease outbreaks and provide **real-time surveillance** of emerging public health threats. The HealthMap app for iPhone and Android smartphones provides the latest real-time disease outbreak information. HealthMap aggregates content from ProMED-mail, the WHO, GeoSentinel, the World Organization for Animal Health (OIE), the Food and Agriculture Organization of the United Nations (FAO), Eurosurveillance, Google News provided by Google, Moreover provided by Verisign, Wildlife Disease Information Node, Baldu News, and SOSO Info (**FIGURE 11.4**).

TABLE 11.1 Basic Reproduction Rate (R_0) Within Human Populations

Disease	Type of Causative Agent	R_0 Range
Measles	Virus	12–18
Pertussis	Bacterium	12–17
Diphtheria	Bacterium	6–7
Poliomyelitis	Virus	5–7
Rubella	Virus	5–7
Mumps	Virus	4–7
HIV/AIDS	Virus	2–5
Influenza A	Virus	2–3
Ebola virus disease	Virus, Guinea strain	2
Rabies	Virus	<1

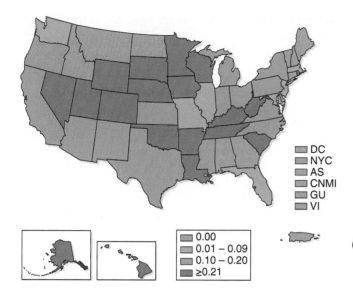

DC
NYC
AS
CNMI
GU
VI

0.00
0.01 – 0.09
0.10 – 0.20
≥0.21

Abbreviations:
DC: District of Columbia
NYC: New York City
AS: American Samoa
CNMI:Commonwealth of Northern Mariana Islands
GU: Guam
VI: U.S. Virgin Islands

(a)

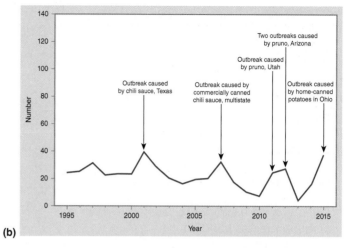

(b)

FIGURE 11.3 Graphs, charts, and maps used by epidemiologists to illustrate infectious disease frequency and distribution. **(a)** United States and U.S. territories: incidence of pediatric influenza-associated deaths expressed as rates per 100,000 population, 2015. **(b)** Number foodborne botulism cases in the United States, 1995–2015. **(c)** Reported cases of *Chlamydia* spp. infections by age group and gender in the United States during 2015 expressed as rates per 100,000 population. Information from the August 11, 2017, edition of the CDC's *Morbidity and Mortality Weekly Report, 64*(53), 1–143.

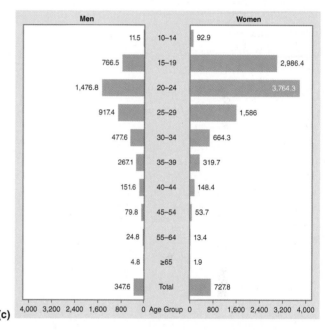

(c)

TABLE 11.2 Infectious Diseases Designated as National Notifiable, United States, 2018	
Disease	
Anthrax	
Arboviral neuroinvasive and nonneuroinvasive diseases	California serotype virus diseases
	Chikungunya virus disease
	Eastern equine encephalitis virus disease
	Powassan virus disease
	St. Louis encephalitis virus disease
	West Nile virus disease
	Western equine encephalitis virus disease

(continues)

TABLE 11.2 Infectious Diseases Designated as National Notifiable, United States, 2018 (*continued*)

Disease	
Babesiosis	
Botulism	
Brucellosis	
Campylobacteriosis	
Carbapenemase-producing carbapenem-resistant Enterobacteradeae (CP-CRE)	CP-CRE, *Enterobacter* spp.
	CP-CRE, *Escherichia coli*
	CP-CRE, *Klebsiella* spp.
Chancroid	
Chlamydia trachomatis infection	
Cholera	
Coccidioidomycosis	
Cryptosporidiosis	
Cyclosporiasis	
Dengue virus infections	Dengue
	Dengue-like illness
	Severe dengue
Diphtheria	
Ehrlichiosis and anaplasmosis	*Anaplasma phagocytophilum* infection
	Ehrlichia chaffeensis infection
	Ehrlichia ewingii infection
	Undetermined human ehrlichiosis/anaplasmosis
Giardiasis	
Gonorrhea	
Haemophilus influenzae, invasive disease	
Hansen's disease (leprosy)	
Hantavirus infection, non-Hantavirus pulmonary syndrome	

Disease	
Hantavirus pulmonary syndrome	
Hemolytic uremic syndrome, postdiarrheal	
Hepatitis A, acute	
Hepatitis B, acute	
Hepatitis B, chronic	
Hepatitis B, perinatal virus infection	
Hepatitis C, acute	
Hepatitis C, chronic	
Hepatitis C, perinatal infection	
HIV infection (AIDS has been reclassified as HIV Stage III)	
Influenza-associated pediatric mortality	
Invasive pneumococcal disease	
Legionellosis	
Listeriosis	
Lyme disease	
Malaria	
Measles	
Meningococcal disease	
Mumps	
Novel influenza A virus infections	
Pertussis	
Plague	
Poliomyelitis, paralytic	
Poliomyelitis, nonparalytic	
Psittacosis	
Q fever	Q fever, acute
	Q fever, chronic

(continues)

TABLE 11.2 Infectious Diseases Designated as National Notifiable, United States, 2018 (*continued*)

Disease	
Rabies, animal	
Rabies, human	
Rubella	
Rubella, congenital syndrome	
Salmonellosis	
Severe acute respiratory syndrome-associated coronavirus (SARS-CoV disease)	
Shiga toxin-producing *Escherichia coli* (STEC)	
Shigellosis	
Smallpox	
Spotted fever (rickettsiosis)	
Streptococcal toxic shock syndrome	
Syphilis	Syphilis, primary
	Syphilis, secondary
	Syphilis, early nonprimary, nonsecondary
	Syphilis, unknown duration or late
Tetanus	
Toxic shock syndrome (staphylococcal)	
Trichinellosis	
Tuberculosis	
Tularemia	
Typhoid fever	
Vancomycin-intermediate *Staphylococcus aureus* (VISA)	
Vancomycin-resistant *Staphylococcus aureus* (VRSA)	
Varicella (deaths only)	
Varicella (morbidity)	
Vibriosis	

Disease	
Viral hemorrhagic fever	Crimean-Congo hemorrhagic fever virus
	Ebola virus
	Lassa virus
	Lujo virus
	Marburg virus
	New World arenavirus, Guanarito virus
	New World arenavirus, Junin virus
	New World arenavirus, Machupo virus
	New World arenavirus, Sabia virus
Yellow fever	
Zika virus disease and Zika virus infection	Zika virus disease, congenital
	Zika virus disease, noncongenital
	Zika virus infection, congenital
	Zika virus infection, noncongenital

Information from the CDC. (2018). 2018 National Notifiable Conditions. Retrieved from https://wwwn.cdc.gov/nndss/conditions/notifiable/2018/

©Teri Shors.

FIGURE 11.4 The smartphone HealthMap app uses informal online disease outbreak sources to monitor and carry out real-time surveillance of emerging public health threats.

Cycle of Microbial Disease

For infectious diseases to exist at the community level, a chain of linked factors needs to be present, somewhat reminiscent of a parade of circus elephants linked trunk to tail. These factors are reservoirs, modes of transmission, portals of entry, portals of exit, susceptible host, and infectious agent (**FIGURE 11.5**). An understanding of these factors is imperative to attempt to break the cycle somewhere along the path. For example, if insects are involved in transmission, then controlling their population is a target; for those microbes transmitted by drinking water, providing safe drinking water is a goal. Shrinking the reservoir (where the microbes exist in nature) is a potential target for other diseases. In some instances, a combination of targets is preferable.

For a particular microbial disease to exist there has to be a pathogen as the causative agent and a host in which the pathogen takes up residence. The potential for disease to occur and its outcome are a result of the complex interaction between the number of invading microbes and their virulence and the host immune system. **Communicable diseases** are **infectious diseases** in which the pathogen can be transmitted directly or indirectly from its reservoir to the host portal of entry.

Reservoirs of Infection

A **reservoir** is a site in nature in which microbes survive (and possibly multiply) and from which they may be transmitted. All pathogens have one or more reservoirs, without which they could not exist. Knowledge and identification of these reservoirs are important, because the reservoirs are prime targets for preventing, minimizing, and eliminating existing and potential epidemics. The fact that humans are the only reservoir of Variola virus, which causes smallpox, and that person-to-person transmission of smallpox takes place were key factors in the eradication of this disease.

Additionally, humans are the only known reservoir for the infectious agents that cause gonorrhea, measles, and poliomyelitis. Animals, as well as plants

and nonliving environments, also serve as reservoirs. In some cases the source of the pathogen is distinct from the reservoir and is the immediate location from which the pathogen is transmitted. For example, in typhoid fever the reservoir may be an individual with an active case of the disease who sheds *Salmonella typhi* bacteria in feces; the immediate source would be water or food contaminated with fecal material. In contrast, in most sexually transmitted diseases the human body serves as both reservoir and source.

Active carriers are those individuals who suffer from an infectious disease, whereas **healthy carriers** have no symptoms and unwittingly pass the disease on to others. Typhoid Mary (Mary Mallon), a cook and healthy lifetime carrier of the bacteria that cause typhoid fever, was responsible for about 10 outbreaks, 53 cases, and 3 deaths due to typhoid fever during her lifetime. **Chronic carriers** are those who harbor a pathogen for long periods after recovery, possibly throughout their lives, without ever again becoming ill with the disease. In the case of chronic (and healthy) carriers of typhoid fever, removal of the gallbladder may be effective in eliminating the carrier state; intensive therapy with antibiotics works in other cases. Tuberculosis is another disease in which carriers play a significant role. Depending on the particular infection, carriers discharge microbes through portals of exit, including respiratory secretions, feces, urine, and vaginal and penile discharges.

Domestic and wild animals serve as reservoirs for pathogens that infect humans. About 75% of recently emerging infectious diseases affecting humans are of animal origin. An infectious disease in humans caused by an animal pathogen is referred to as a *zoonosis* or *zoonotic disease* (**TABLE 11.3**).

Microbes of animals that are most closely related to humans have the greatest chance of making the "species leap" to humans, or erasing the **species barrier**. Consider, for example, HIV (the virus that causes AIDS), which is thought to have a reservoir in chimpanzees and is now a human pathogen. **Prions** cause both mad cow disease and its human counterpart, variant Creutzfeldt-Jakob disease (vCJD); prions jumped from cattle to humans. Perhaps all infectious diseases in humans originated in other species and jumped the species barrier. Monkeys are reservoirs for the microbes that cause malaria, yellow fever, herpes B (see opening case study), and numerous other diseases in humans. The reservoirs for the spirochete bacteria that cause Lyme disease, a major problem in the Midwest and Northeast United States, are deer and small rodents such as voles and mice. Hantavirus pulmonary syndrome, a relatively new disease in the United States, uses a variety of rodent species, particularly the deer mouse, as reservoirs. Many mammals, including dogs, raccoons, skunks, foxes, mongooses, and bats, serve as reservoirs for rabies.

Eradication of zoonotic diseases is particularly challenging because it is, ultimately, dependent on eradicating the reservoirs. Malaria is a protozoan disease

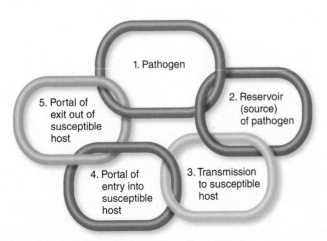

FIGURE 11.5 The cycle, or chain, of microbial infection.

1. Pathogen

2. Reservoir (source) of pathogen

3. Transmission to susceptible host

4. Portal of entry into susceptible host

5. Portal of exit out of susceptible host

TABLE 11.3 Zoonotic Diseases

Infectious Disease	Pathogen (type)	Usual Mode of Transmission	Main Reservoirs
Anthrax	*Bacillus anthracis* (bacterium)	Direct contact (e.g., ingesting contaminated meat)	Livestock, wild animals, environment (e.g., contaminated soil)
Salmonellosis*	*Salmonella* spp. (bacterium)	Ingestion/foodborne (oral–fecal)	Poultry, cattle, sheep, pigs
Tuberculosis (bovine)	*Mycobacterium bovis* (bacterium)	Ingestion (e.g., unpasteurized milk or dairy products)	Cattle
Cat scratch fever	*Bartonella henselae* (bacterium)	Direct contact (scratch or bite of cat)	Cats
Listeriosis*	*Listeria monocytogenes* (bacterium)	Ingestion of contaminated food, direct contact (mother to child in womb or during birth)	Cattle, sheep, soil
Lyme disease	*Borrelia burgdorferi, Borrelia mayonii* (bacterium)	Deer tick bite	Small mammals, deer in the wild
Ehrlichiosis	*Ehrlichia* spp., *Neorickettsia* spp., *Anaplasma* spp. (bacterium)	Deer tick bite	Small mammals, deer in the wild
Leprosy (Hansen's disease)	*Mycobacterium leprae* (bacterium)	Direct contact with bodily fluids of an infected animal	Armadillos, soil contaminated by infected armadillos in the wild
Plague*	*Yersinia pestis* (bacterium)	Flea bite	Rats and their fleas
Rabies*	Rabies virus or Lyssavirus	Direct contact with saliva from infected animal into break in skin or on mucous membranes/bite	Dogs, bats, skunks, foxes, raccoons, mongooses (Caribbean)
Avian influenza*	Avian influenza A virus	Inhalation, direct contact of nasal secretions	Poultry, wild water fowl
Swine influenza*	Swine influenza A virus	Inhalation, direct contact of nasal secretions	Pigs
West Nile encephalitis	West Nile virus	Mosquito bite	Wild birds and some reptiles, mosquitoes
Monkeypox	Monkeypox virus	Bites, close contact with pets	Exotic pets such as prairie dogs, Gambian giant pouched rats, dormice
Hantavirus syndromes	Hantavirus	Inhalation of aerosols of rodent saliva, urine or feces	Rodents in the wild (e.g., deer mice)
Ebola virus disease*	Ebola virus	Contact with bodily fluids of bats or preparation of bushmeat	Bats, **bushmeat**

(continues)

TABLE 11.3 Zoonotic Diseases (continued)

Infectious Disease	Pathogen (type)	Usual Mode of Transmission	Main Reservoirs
Middle East respiratory syndrome (MERS)*	MERS-coronavirus (MERS-CoV)	Contact with camels or drinking raw camel milk	Dromedary camels
Toxoplasmosis	*Toxoplasma gondii* (parasite)	Ingestion of oocysts, or "eggs" (fecal–oral, undercooked meat)	Cats
Giardiasis (beaver fever)*	*Giardia intestinalis*, *Giardia lamblia*, or *Giardia duodenalis* (parasite)	Ingestion of contaminated water (oral–fecal)	Wildlife, humans
Cryptosporidiosis*	*Cryptosporidium parvum* (parasite)	Ingestion of oocysts, or "eggs," shed in water (fecal–oral)	Cattle, sheep, pets
Taeniasis	*Taenia* spp. (parasitic tapeworms)	Ingestion of undercooked beef, pork, or raw fish (sushi)	Cattle, pigs, wild-caught fish (e.g., Alaskan salmon)
Ringworm of the skin or dermatophytosis (Tinea corporis)	*Trichophyton rubrum* and *Microsporum canis* (fungi)	Direct contact	Cats, dogs, cattle, many animal species

*Can be transmitted person to person.

transmitted by mosquitoes. Intensive spraying with the pesticide DDT (dichlorodiphenyltrichloroethane) in the 1940s markedly reduced the mosquito population and the number of malaria cases. The mosquitoes developed resistance to DDT eventually, leading to a reemergence of malaria.

O Nonliving Reservoirs

Some organisms are able to survive and multiply in nonliving environments. Soil and water are the major nonliving reservoirs of infectious diseases. The rod-shaped bacteria that cause tetanus and botulinum, both members of the same bacterial genus, *Clostridium*, are **endospore** formers and thus can survive for many years in soil. These bacteria are part of the normal **microbiota** of horses and cattle and are deposited in their feces onto the soil. The use of animal fertilizers contributes to their distribution. Certain helminth (worm) parasites (e.g., hookworms) deposit their eggs onto the soil, establishing a reservoir for human infection (**FIGURE 11.6**).

Contaminated drinking water and foods are major reservoirs for many microbes that cause **gastroenteritis**, ranging from mild to severe to fatal. CDC experts estimate that 1 in 6 Americans get sick from eating contaminated foods or beverages and about 3,000 die. The list of culprits includes bacteria, viruses, and protozoa (**TABLE 11.4**). Because of the potential for an outbreak of waterborne and foodborne illnesses, local departments of public health devote considerable attention to sanitary

measures designed to minimize risks; their activities include monitoring food service establishments, beaches, and swimming pools and certifying food handlers. The CDC Food Safety Home Page (www.cdc.gov/foodsafety) is an excellent resource for information about different foodborne illnesses, foodborne outbreaks in real-time, surveillance, and educational content about food safety (**FIGURE 11.7**).

Courtesy of Alaine Kathryn Knipes; Parasitic Disease Branch (DPDx); Division of Parasitic Diseases and Malaria.

FIGURE 11.6 Soil can be a reservoir for microbes and helminth eggs. Inongu Mtumba is in Tanzania shown here preparing fecal specimens and diagnosing helminth infections in the field.

TABLE 11.4 Water and Food as Reservoirs of Infection

Type of Microbe	Examples of Waterborne and Foodborne Infections
Bacteria	Salmonellosis; shigellosis; campylobacteriosis; cholera; anthrax; food poisoning (*Clostridium perfringens* infections); *E. coli* O157:H7; *E. coli* 0145, 0121, and 026 infections
Viruses	Hepatitis A, poliomyelitis, winter vomiting disease
Protozoa	Giardiasis, amebiasis, cryptosporidiosis
Helminths (worms)	Taeniasis, ascariasis, trichinellosis

TABLE 11.5 New Food Vehicles Associated with Multistate Outbreaks in the United States

Food Category	
Meats, poultry, fish	Kosher broiled chicken livers, scraped tuna product
Fruits, vegetables, and greens	Bagged spinach, broccoli powder (on snack food), canned chili sauce, carrot juice, cucumbers, hot peppers, whole fresh papayas
Products with nuts	Hazelnuts, peanut butter, raw cashew cheese
Spices	Pepper
Other	Dog food, frozen pot pies and meals, raw, prepackaged cookie dough

CDC Food Safety: Foods Linked to Foodborne Illness https://www.cdc.gov/foodsafety/foods-linked-illness.html

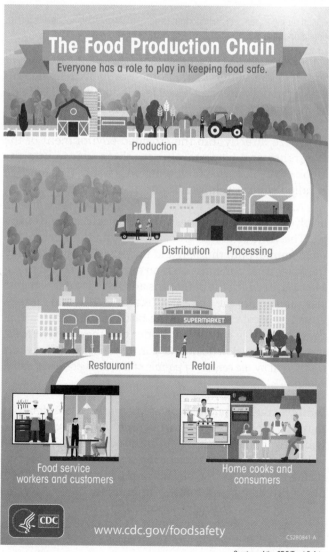

FIGURE 11.7 CDC food safety infographic.

Courtesy of the CDC/Food Safety.

PulseNet is a national network of public health and food regulatory agency laboratories that identify and stop foodborne diseases. It prevents an estimated 270,000 infections every year in the United States from the most common causes of foodborne illness: *Salmonella* spp., *E. coli* 0157, and *Listeria monocytogenes*. **Food Safety News** (www.foodsafetynews.com) was created in 2009 to provide news coverage for anyone who cares about the safety of the food supply. **TABLE 11.5** contains a list of food vehicles that were reported as new or not previously associated sources of multistate foodborne outbreaks in the United States since 2006.

Modes of Transmission

The next link in the cycle of disease is transmission, the bridge between reservoir and the portal of entry (**FIGURE 11.8**). The **portal of entry** is the site at which the microbe enters the host. The **mode of transmission** is the mechanism by which an infectious agent is spread through the environment to another person. More simply put, transmission answers the question, "How do you get the disease?" Several modes of transmission are possible, and they can be grouped into two major pathways: direct and indirect. Each of these, in turn, can be subgrouped into three categories (**TABLE 11.6**).

◗ Direct Transmission

The most common type of **direct transmission** is person-to-person contact, in which the infectious agent

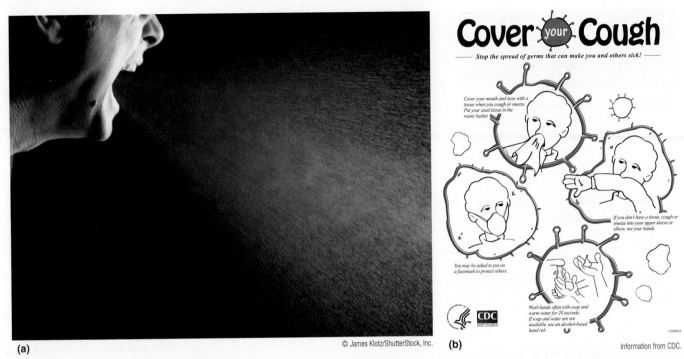

(a)
© James Klotz/ShutterStock, Inc.
(b)
Information from CDC.

FIGURE 11.8 (a) Droplet transmission. As many as 20,000 droplets may be produced during a sneeze. It is important to carry a handkerchief or tissue and to cover your nose and mouth when sneezing. **(b)** Cough etiquette and influenza prevention infographic.

is directly and immediately transferred from a portal of exit to a portal of entry. The **portal of exit** is the site from which microbes leave a host and may infect another host. Sexual contact, kissing, and touching are the most common examples. Transmission is facilitated by contact of the warm, moist mucous membranes of one individual with the warm, moist mucous membranes of another, as occurs in sexually transmitted diseases. Sexual contact is an example of **horizontal transmission** (i.e., transmission from one person to another). **Droplet transmission** is also direct and horizontal and involves the projection of infected aerosol from coughing, sneezing, talking, and laughing onto the conjunctiva of the eyes or onto the mucous membranes of the nose or mouth. Influenza, whooping cough, and measles are spread by droplets.

Droplets are about 10 micrometers (0.0003 inches) or greater in diameter and travel less than 1 meter (3.2 feet); as many as 20,000 droplets may be produced in a sneeze (Figure 11.8A). Think about that the next time you cough or sneeze directly into the crowded environment of the classroom and be certain to cough or sneeze into the crook of your elbow and not into your hand. This practice is called **cough etiquette** (Figure 11.8B).

A second type of direct and horizontal transmission involves animal bites, rabies being the most common example. The virus is directly transmitted from the saliva of the rabid animal onto the skin and underlying tissues. Finally, transplacental transmission is an example of **vertical transmission**, in which the pathogens are passed from mother to offspring across the placenta (Zika virus, HIV, and measles virus), in breast milk, or in the birth canal (bacteria that cause syphilis or gonorrhea). Notice that these categories of direct transmission do not have intermediates. *Microbes are transferred by the contact of portals of exit with portals of entry.*

O Indirect Transmission

Indirect transmission involves the passage of infectious material from a reservoir or source to an **intermediate vehicle** and then to a host. The intermediate vehicle can be living or nonliving. **Vehicleborne transmission** is accomplished by food, water, biological products (organs, blood, blood products), and **fomites** (inanimate objects) as, for example, desk surfaces, doorknobs, or escalator rails.

Waterborne transmission is a serious problem throughout the world and is a major cause of death in many developing countries as a result of fecal–oral transmission, in which pathogens are transmitted from the

TABLE 11.6 Modes of Transmission	
Direct	Indirect
Contact (e.g., kissing, sneezing, coughing, singing, sexual contact)	Vehicles (fomites, e.g., doorknobs, eating utensils, toys, facial tissue)
Animal bites	Airborne (e.g., through aerosols created by shaking bedsheets, sweeping, mopping)
Transplacental	Vectors (e.g., mosquitoes, ticks, fleas)

©Teri Shors. ©Teri Shors. ©Teri Shors. ©Teri Shors. ©Teri Shors. ©Teri Shors.

FIGURE 11.9 We encounter many different kinds of fomites in our daily lives. A few examples are shown here: **(a)** water faucet; **(b)** doorknob in a residence; **(c)** bathroom fomites, such as toilet paper dispensers, toilet seats, feminine product waste containers, and flush handles; **(d)** items at a self-service food counter; **(e)** cell phone and coffee cup; and **(f)** computer keyboard and mouse.

feces of one individual to another by hand-to-mouth transfer. Public, semiprivate, and private water supplies must all be carefully monitored for the presence of fecal pathogens.

Fomites play a significant role in the transmission of pathogens. The list of fomites is seemingly endless and includes objects in common use, such as doorknobs, telephones, faucets, computer keyboards, and exercise equipment. Toys are fomites and contribute to illness in children wherever the toys are shared. Surgical instruments, medical equipment (e.g., catheters, intravenous equipment, and syringes), bedding, and soiled clothing also are fomites. An interesting study involving soiled saris (garments worn by Indian women) as fomites, conducted in 51 slum areas in Dhaka, Bangladesh, revealed a positive correlation between the number of misuses of dirty saris and episodes of childhood diarrhea (**FIGURE 11.9**).

The list of fomites and their role may cause you some concern, but there is at least a partial solution to the problem. The simple act of frequent handwashing has been shown to markedly reduce hand-to-mouth (and nose and eye) infection. Frequent wiping of tabletops and counters with disinfectants is effective and a sign of good hygiene in a sanitation-conscious restaurant. In health and fitness centers it has become a widespread practice to wipe down exercise equipment with disinfectant after use. Athletes are at higher risk of contracting **community-acquired methicillin-resistant *Staphylococcus aureus* (CA-MRSA)** infections. The CDC has created several posters to remind high school and college athletes not to share personal items that may have had contact with infected skin (**FIGURE 11.10**).

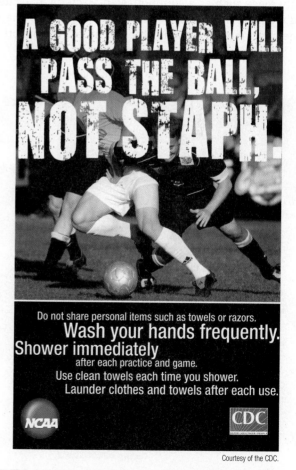

Courtesy of the CDC.

FIGURE 11.10 Poster created as a collaboration between the CDC and the National Collegiate Athletic Association (NCAA) to educate college athletes about the prevention of CA-MRSA infections.

Airborne transmission by **aerosols** is the second type of indirect transmission. Aerosols are suspensions of tiny water particles and fine dust in the air; they are distinct from droplet nuclei, as they are smaller than 4 micrometers (0.0002 inches), travel more than 1 meter (3.2 feet), and are small enough to remain airborne and "hang" for extended periods. Aerosols cause outbreaks of Legionnaires' disease, Valley fever or coccidioidomycosis, and psittacosis, or parrot fever. Psittacosis is a respiratory infection caused by *Chlamydia psittaci* from infected pet birds such as parrots and cockatiels and poultry. The bacteria can infect people who breathe in the dust of bird droppings while caring for and cleaning up after infected birds (**FIGURE 11.11A**). The birds do not have symptoms or seem sick. Aviary and pet shop employees, veterinarians, poultry workers, and bird owners are at increased risk of infection. Valley fever occurs in parts of the southwestern U.S., including Arizona, and parts of Texas, New Mexico, and California. Coccidioidomycosis is acquired by inhaling the fungal spores of *Coccidioides immitis* during a dust storm.

The microbes in aerosols may not come directly from humans, birds, or animals, but they may be present in dust particles where they can survive for months. Most hantavirus pulmonary syndrome infections can be traced back to when the victim cleaned out mouse droppings from a dusty place such as a summer cabin. Bacteria, parasites, and viruses can be disseminated by changing bed linens, sweeping, mopping, scraping the kitty litter box, and other activities (**FIGURE 11.11B**). Hospital personnel are keenly aware of this, as reflected in the practice of using wet mops and damp cloths to wipe surfaces.

The third type of indirect transmission is by **vectors**, living organisms that transmit microbes from one host to another. The term *vector* is sometimes more broadly used to cover any object that transfers microbes, but this is, strictly speaking, incorrect usage. Ticks, flies, mosquitoes, lice, midges, and fleas are the most common vectors, and they belong to the same biological phylum, the Arthropoda, along with lobsters and crabs. (It may be difficult to understand what flies, fleas, ticks, and lobsters have in common—but the edibility of lobsters certainly sets them apart!) **FIGURE 11.12** is a diagram illustrating examples of direct versus indirect transmission.

Arthropods are invertebrate animals with jointed appendages; *arthro* means "joint," as in *arthritis* (inflammation of joints), and *pod* means "foot," as in *podiatrist* ("foot doctor"). Further, they all have segmented bodies and a hardened exoskeleton. The arthropods are members of the largest phylum and consist of many diverse species that are divided into four subphyla (**TABLE 11.7**). They are considered to be the most successful of all living animals in terms of the huge number of species and their distribution. **FIGURE 11.13** provides photographs of the different types of arthropods that transmit pathogens to humans.

Spiders, ticks, and mites hatch from eggs as six-legged larvae and undergo metamorphosis to eight-legged adults

(a) Courtesy of the CDC/Eric Grafman.   (b) ©Teri Shors.

FIGURE 11.11 Pets can be a reservoir for infectious agents. **(a)** Care should be taken to avoid inhaling aerosols of dried secretions from infected pet birds, such as this beautifully colored Jenday conure, also known as a Jandaya parakeet. Thoroughly wash your hands with soap and water after contact with birds or their droppings. **(b)** Pregnant women should not change the kitty litter in order to avoid contact with *Toxoplasma gondii*, the parasite that causes toxoplasmosis.

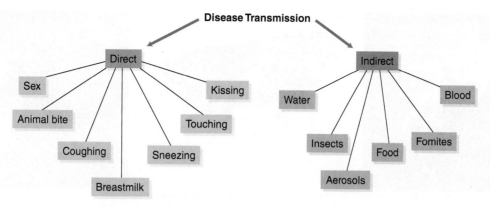

FIGURE 11.12 Direct versus indirect transmission of infectious agents that cause disease in humans.

TABLE 11.7 The Phylum Arthropoda Contains Disease Vectors

Subphylum Chelicerata	Subphylum Hexapoda
Scorpions	Insects (many subgroups)
Chiggers	Midges
Daddy longlegs spiders	Fleas*
Mites*	Mosquitoes*
Horseshoe crabs	Lice*
Deer ticks* (also called blacklegged ticks)	Sand flies*
Lone star ticks*	Black flies*
Wood tick*	Tsetse flies*
Dog ticks*	House flies*
	Blow flies*
	Kissing bugs*
	Cockroaches*
Subphylum Crustacea	**Subphylum Myriapoda**
Water fleas	Millipedes
Isopods	Centipedes
Fairy shrimp	
Crabs	
Copepods*	
Lobsters	
Barnacles	
Shrimp	

*Arthropod vectors that transmit pathogens that cause disease in humans.

with two body segments and mouth parts adapted for the sucking of blood. Ticks transmit a variety of infectious diseases, including Lyme disease, Rocky Mountain spotted fever, Colorado tick fever, Powassan encephalitis, Heartland virus disease, Bourbon virus disease, babesiosis, anaplasmosis, and ehrlichiosis. In addition to their role as vectors, some ticks are important reservoirs because they exhibit **transovarial transmission** (the passage of microbes into their eggs).

Insects are an extremely large and diverse group of arthropods with well over 1 million species. You may think of them as pests because (depending on the species) they bite, eat our crops, are bizarre-looking, some carry obnoxious odors, and are associated with uncleanliness. Insects have three body segments (the head, the thorax, and the abdomen) and six legs; some have one or two pairs of wings while others have two. Some have mouth parts adapted for puncturing the skin and sucking blood. "Kissing bugs" suck blood from their hosts and are vectors of Chagas disease, which is endemic in Central and South America.

Because many arthropods play a significant role in the cycle of infectious diseases, officials at public health departments know that arthropod control can lead to disease control. Mosquito abatement programs have been carried out on numerous occasions to control malaria. All the lower 48 states are threatened by West Nile virus, a mosquito-borne virus, resulting in **insecticide** spraying to control the mosquito population. Several other viruses that cause encephalitis (brain swelling and other neurological damage) belong to a group called **arboviruses**, and are so named because they are arthropodborne (shortened to "arbo").

Arthropods can be either mechanical vectors or biological vectors. **Mechanical vectors** transmit microbes passively on their feet and other body parts; the microbes do not invade, multiply, or develop in the vector. Houseflies, for example, feed on exposed human and animal fecal material and then transfer microbes on their feet to food and eating utensils. Typhoid fever, shigellosis, and other gastrointestinal diseases characterized by diarrhea or dysentery may be spread in this way. Covering of

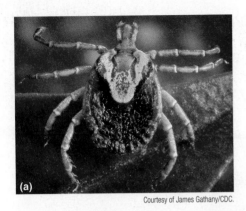

Courtesy of James Gathany/CDC.

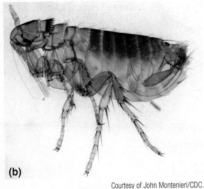

Courtesy of John Montenieri/CDC.

Courtesy of James Gathany/CDC.

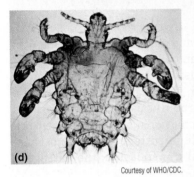

Courtesy of WHO/CDC.

Courtesy of WHO/CDC.

Courtesy of Peggy Greb/USDA ARS.

FIGURE 11.13 A "bug" parade. **(a)** Tick. **(b)** Flea. **(c)** Mosquito. **(d)** Louse. **(e)** Kissing bug. **(f)** Tsetse fly.

human and animal waste to avoid exposure to flies is an obvious answer, but this is not always possible in poverty-stricken areas, under wartime conditions, in refugee camps, and in other circumstances involving large groups of people when it is difficult to maintain good sanitation.

Even in the best of circumstances, flies have access to dog feces and can mechanically transmit microbes to kitchen areas. It is disturbing to think that a fly that has just lunched on dog feces in your backyard or in a neighboring park may walk across the chicken salad that you prepare for a picnic. Cockroaches also serve as mechanical vectors; remember this when you see them marching across a kitchen counter. In December 2000, Chinese newspapers reported that Beijing was in the grip of a roach menace. Roaches were invading restaurants, hotels and motels, and even hospitals. Roaches carry more than 40 kinds of bacteria, some of which are pathogens. Cheap rundown motels are sometimes referred to as "roach motels."

Biological vectors, unlike mechanical vectors, are necessary components in the life cycles of many infectious disease agents and are required for the multiplication and development of the pathogen; transmission by biological vectors is an active process. As an example, when a mosquito picks up a *Plasmodium* spp. (the protozoan that causes malaria) while taking a blood meal from an infected person, the protozoa are not at an infective stage. Further development in the mosquito's body results in protozoa that are now infective for hosts. Depending on the particular vector, protozoa may be carried in the saliva and injected into the tissue while biting; other vectors have the nasty habit of regurgitating infectious secretions into and around the bite; and others defecate infectious material onto the bite area. Itching usually results, and scratching facilitates entry of the protozoan. Mosquitoes, fleas, lice, and ticks are common biological vectors.

Mosquitoes can rightly be considered as "public health enemy number one" based on their transmission of bacterial, viral, protozoan, and helminthic (worm) diseases (**TABLE 11.8**). In a textbook about mosquitoes, the author Dr. Andrew Spielman (1930–2006) wrote:

> She doesn't aerate the soil, like ants and worms. She is not an important pollinator of plants, like the bee. She does not even serve as an important food item for some other animal. She has no "purpose" other than to perpetuate her species. That the mosquito plagues humans is really, to her, incidental. She is simply surviving and reproducing.

Ticks are not only significant vectors of disease but also direct sources of disease. Tick paralysis is an example and is characterized by ascending flaccid paralysis resulting from a toxin in tick saliva; the paralysis usually disappears within several days. A 2006 CDC report cited and described a cluster of four cases in Colorado from May 26 to May 31, 2006. Tick populations depend largely on the number of deer in the area; more deer per square kilometer means more ticks.

Fleas are the biological vectors of the *Yersinia pestis* bacterium, and rats are the reservoirs. Rats are important reservoirs for other microbial diseases, including Lassa fever and, in the southeastern United States, hantavirus pulmonary syndrome. Rats are a serious public health concern around the world (**BOX 11.1**).

TABLE 11.8 Diseases Transmitted by Arthropod Bites

Disease	Type of Pathogen	Genus of Pathogen	Arthropod Vector	Distribution
Plague	Bacterium	*Yersinia pestis*	Fleas	Southeast Asia, Central Asia, South America, western North America
Lyme disease	Bacterium	*Borrelia* spp.	Deer tick	Europe, North America, Australia, Japan
Rocky Mountain spotted fever	Bacterium	*Rickettsia* spp.	Wood and dog ticks	United States
Murine typhus or endemic typhus	Bacterium	*Rickettsia typhi*	Fleas	Worldwide
Louse-borne typhus or epidemic typhus	Bacterium	*Rickettsia prowazekii*	Lice	Eastern Europe, Asia, Africa, South America
Zika virus disease	Virus	Flavivirus	Mosquito (*Aedes* spp.)	Africa, Asia, the Caribbean, North America, Central America, South America, Pacific Islands
Powassan encephalitis	Virus	Flavivirus	Deer tick	Great Lakes region of the United States, Canada
Colorado tick fever	Virus	Coltivirus	Dog tick	Rocky Mountain states, United States
Heartland virus disease	Virus	Phlebovirus	Lone star tick	United States
La Crosse encephalitis	Virus	Bunyavirus	Mosquito (*Ochlerotatus* spp.)	United States
West Nile encephalitis	Virus	Flavivirus	Mosquito (*Culex* spp.)	Africa, Caribbean, North America, South America, Central America, India, Australia, Middle East, Russia, Europe, Southeast Asia
Dengue fever	Virus	Flavivirus (Dengue 1, 2, 3, and 4)	Mosquito (*Aedes* spp.)	India, Southeast Asia, Pacific Islands, Mexico, South America, Caribbean, United States (Texas/ Mexico border)
Yellow fever	Virus	Flavivirus	Mosquito (*Aedes* spp.)	Tropical South America, Africa
Chikungunya virus disease	Virus	Alphavirus	Mosquito (*Aedes* spp.)	Africa, Southeast Asia, Philippines
Malaria	Protozoan	*Plasmodium* spp.	Mosquito (*Anopheles* spp.)	Africa, Southwestern Pacific, South America, Southeastern Asia, India
Babesiosis	Protozoan	*Babesia* spp.	Ticks	United States, Europe

(continues)

TABLE 11.8 Diseases Transmitted by Arthropod Bites (*continued*)

Disease	Type of Pathogen	Genus of Pathogen	Arthropod Vector	Distribution
American trypanosomiasis (Chagas disease)	Protozoan	*Trypanosoma* spp.	Kissing bug	South and Central America
African trypanosomiasis (sleeping sickness)	Protozoan	*Trypanosoma* spp.	Tsetse flies	West, Central, and East Africa
Leishmaniasis	Protozoan	*Leishmania* spp.	Sand flies	Central America, South America, Africa, India, and other parts of Asia, Europe
Lymphatic filariasis or elephantiasis	Helminth	*Wuchereria* spp., *Brugia* spp.	Mosquito (*Culex* spp., *Anopheles* spp., *Aedes* spp., *Mansonia* spp.)	Central America, South America, Africa, India, and other parts of Asia
Onchocerciasis	Helminth	*Onchocerca* spp.	Black flies	Central America, tropical South America, Africa

BOX 11.1 Oh, Rats!

The title of this box is a familiar expression that you have heard or used. Rats transmit infectious diseases by direct contact (biting), by serving as reservoirs for diseases transmitted by arthropods (most notably plague-carrying fleas), and by contamination of food or water with urine and feces. Many species of rats can be found throughout the world. The Norway rat (*Rattus norvegicus*) is the most common rat in the United States; it is also known as the brown rat, sewer rat, street rat, as well as other nonaffectionate names. As Shakespeare wrote in *Romeo and Juliet* (1660), "*A rose by any other name would smell as sweet.*" In her 1913 poem titled "Sacred Emily," poet Gertrude Stein wrote, "*A rose is a rose is a rose*"; let it now be said that "*A rat is a rat is a rat.*"

The Norway adult rat measures about 23 centimeters (9 inches) long and has a blunt nose, small ears, and heavy-looking body. These rats have an active sex life, producing four or five litters per year, totaling about 20 young. The young become sexually mature at about 3 to 4 months. Norway rats are socially active and live in colonies. They are more than just pests, eating just about anything, ruining a large part of the world's food supply. They destroy buildings and household furnishings, attack domestic animals, and gnaw on electrical wires, causing fires. Furthermore, they produce about 50 droppings a day in which *Salmonella* spp. bacteria, a pathogen for humans, thrive.

Control is based on minimizing their food supply and invoking good sanitation measures. When a labor strike involving garbage workers hits a large population, and mountains of foul-smelling trash bags pile up along the sidewalks and streets, the rat population increases, as does the potential for ratborne infectious diseases. Controlling the rodent population and disposal of trash is a major and ongoing problem in developing countries. The following is a partial list of major infectious diseases transmitted to humans by rats and/or rat fleas:

- Salmonellosis
- Leptospirosis
- Hantavirus pulmonary syndrome
- Rat bite fever
- Murine typhus
- Tularemia
- The plague (Black Death, see **Figure 1**)

Courtesy of the CDC.

Figure 1 A public health expert is setting traps for rats in order to study the role of this disease vector in plague. During the Middle Ages, homes and workplaces in Europe were inhabited by flea-infested rats carrying the bacterium *Yersinia pestis* (causing the plague).

In the Middle Ages, the rat population was so severe around the world that people lived in fear of the plague and other diseases. Numerous books, poems, and essays have been written about the plague and about rats. The town of Hamelin was particularly besieged and brought in the services of the Pied Piper to rid them of their rats. A delightful poem in Robert

Browning's (1812–1889) *The Pied Piper of Hamelin, Germany: A Children's Story* follows:

Verse 2 (partial)
Rats!
They fought the dogs and killed the cats,
And bit the babies in the cradles,
And ate the cheeses out of the vats,
And licked the soup from the cook's own ladles,
Split open the kegs of salted sprats,
Made nests inside men's Sunday hats,
And even spoiled the women's chats. . . .

Verse 7 (partial)
And out of the houses the rats came tumbling.
Great rats, small rats, lean rats, brawny rats,
Brown rats, black rats, grey rats, tawny rats,
Grave old plodders, gay young friskers,
Fathers, mothers, uncles, cousins,
Cocking tails and prickling whiskers,
Families by tens and dozens,
Brothers, sisters, husbands, wives—
Followed the Piper for their lives.
From street to street he piped advancing,
And step for step they followed dancing,
Until they came they followed dancing,
Until they came to the river Weser
Wherein all plunged and perished. . . .

Vectorborne diseases are as numerous and varied as are their vectors, and they are emerging and reemerging throughout the world. Factors responsible for their emergence or reemergence include genetic changes in both vectors and pathogens, resulting in resistance to insecticides and drugs. In the first half of the 20th century, considerable progress was made in the fight against vectorborne diseases. Most of these diseases were brought under control, and by the 1960s their threat, except in Africa, was greatly diminished (**TABLE 11.9**). In fact, malaria was eliminated from many countries of the world. However, no country is immune to the potential threat and spread of vectorborne diseases. A case in point is West Nile virus (named after the Nile River in Africa), which emerged in 1999 in the state of New York. This was the first appearance of the virus in the United States; it had been previously reported only in Africa and Asia.

In 1989, in response to the growing problem of vectorborne diseases, the CDC established what is now known as the Division of Vector-Borne Diseases, presently located in Fort Collins, Colorado. The division is responsible for surveillance, prevention, and control of vectorborne diseases. It is charged with the investigation of national and international epidemics of bacterial and viral diseases transmitted to humans by arthropods, primarily mosquitoes, ticks, and fleas. To prevent and control these diseases, biologists in the division work with the three populations involved: the pathogen, the host, and the vector.

Portals of Entry

The next step in the cycle of disease involves access into (or onto) the body through portals of entry. Some microbes have a single portal, but others have more than one. **Body orifices** (openings to the outside), including the mouth, nose, ears, eyes, anus, urethra, and vagina, and penetration of the skin make it possible for microbes to gain access. To some extent, human behavior influences the portal of entry; the transmission of the virus that causes AIDS is a case in point. The most common portal of entry for sexually transmitted diseases is the urethra in males and the vagina in females, but the throat and the rectum may also serve for entry.

The portal of entry is an important consideration in the outcome of host–parasite interactions. Bubonic plague results from the bite of a plague-infected flea, but if the *Yersinia pestis* bacteria gains entry into the lungs through the respiratory tract, the result is the more lethal pneumonic plague. Anthrax, also a bacterial disease, is another example. The three varieties of anthrax are (1) cutaneous anthrax, which results when the skin is the portal of entry; (2) gastrointestinal anthrax, which occurs as the result of oral ingestion of the bacteria; and (3) inhalation anthrax, which results from the organisms entering through the respiratory tract. **TABLE 11.10** and **FIGURE 11.14** summarize the portals of entry by anatomical site.

TABLE 11.9 Successful Vectorborne Disease Control

Infectious Disease	Location	Year(s)
Yellow fever	Cuba	1900–1901
Yellow fever	Panama	1904
Yellow fever	Brazil	1932
Anopheles gambiae infestation	Brazil	1938
Anopheles gambiae infestation	Egypt	1942
Louseborne typhus	Italy	1942
Malaria	Sardinia	1946
Yellow fever	Americas	1947–1970
Yellow fever	West Africa	1950–1970
Malaria	Americas	1954–1975
Malaria	Global	1955–1975
Onchocerciasis	West Africa	1974–present
Bancroftian filariasis	South Pacific	1970s
Chagas disease	South America	1991–present
Dengue*	Torres Straight between mainland Australia and Papua New Guinea	2005–2016

Information from D. J. Gubler. (1998). Resurgent vectorborne diseases as a global health problem. *Emerging Infectious Diseases*, *4*(3), 442–450.

*Dengue information from M. O. Muzari, G. Devine, J. Davis, B. Crunkhorn, A. van den Hurk, . . . S. Ritchie. (2017). Holding back the tiger: Successful control program protects Australia from *Aedes albopictus* expansion. *PLOS Neglected Tropical Diseases*. doi: 10.1371.journal.pntd.0005286

TABLE 11.10 Infectious Disease Cycle: Portals of Entry and Exit

Portal Type	
Portals of entry	**Pathogens**
Respiratory tract	*Streptococcus pneumoniae, Mycobacterium tuberculosis*, influenza A virus, rhinovirus, *Bordetella pertussis*, measles virus, Middle East respiratory syndrome-coronavirus (MERS-CoV)
Gastrointestinal tract	*Vibrio cholerae, Salmonella* spp., *E. coli*, norovirus, hepatitis A virus, *Shigella* spp., *Dracunculus medinensis, Giardia lamblia, Listeria monocytogenes*
Urogenital tract	*Neisseria gonorrhoeae, Chlamydia* spp., HIV, papillomaviruses, Herpes simplex virus-2
Skin (hair follicles, sebaceous glands, wounds, arthropod bites)	*Staphylococcus aureus, Proprionibacterium acnes, Bacillus anthracis*, rabies virus, papillomaviruses, *Plasmodium* spp.

Portals of entry	Pathogens
Blood (transfusion, blood products, arthropod bites, placental transfer)	*Treponema pallidum*, HIV, rubella virus, Zika virus, *Toxoplasma gondii*, cytomegalovirus, *Trypanosoma cruzi*, hepatitis C virus, *Elizabethikingia* spp., *Listeria monocytogenes*

Portals of exit	Pathogens
Respiratory tract	*Streptococcus pneumoniae*, *Mycobacterium tuberculosis*, influenza A virus, rhinovirus, *Bordetella pertussis*, measles virus, Middle East respiratory syndrome-coronavirus (MERS-CoV)
Gastrointestinal tract	*Vibrio cholerae*, *Salmonella* spp., *E. coli*, norovirus, hepatitis A virus, *Shigella* spp., *Dracunculus medinensis*, *Giardia lamblia*, *Listeria monocytogenes*
Urogenital tract	*Neisseria gonorrhoeae*, *Chlamydia* spp., HIV, papillomaviruses, herpes simplex virus-2
Skin	*Staphylococcus aureus*, *Proprionibacterium acnes*, papillomaviruses, herpes simplex virus-1, *Candida albicans*
Blood (transfusion, blood products, arthropod bites, placental transfer)	*Toxoplasma gondii*, cytomegalovirus, hepatitis B and C viruses, *Treponema pallidum*, *Toxoplasma gondii*, *Listeria monocytogenes*

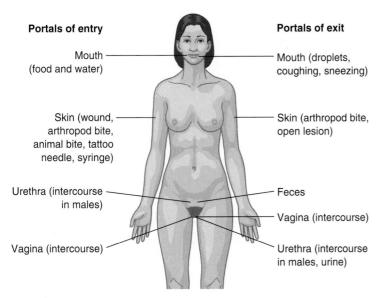

FIGURE 11.14 Pathogens enter the human body through preferred portals of entry and are shed through portals of exit.

Portals of Exit

Once microbes have gained access into the body, whether or not disease results is determined by the interaction between the number of pathogens and their virulence and the immune system of the host. To complete the cycle of infectious disease and to allow the spread of disease into the community, pathogens require a portal of exit (Table 11.10 and Figure 11.14). In some cases the portal of exit relates to the area of the body that is infected. This is particularly true for organisms that cause diseases of the respiratory tract (such as colds and influenza). However, this is not always the case. For example, the spirochete-shaped bacterium that causes syphilis uses the urogenital tract as the portal of exit but can invade the skin and nervous system. The eggs of some disease-producing worms exit the body in fecal material, survive in soil, and remain infectious for long periods of time. HIV and Ebola virus exits the body through semen and vaginal discharges as well as through the blood. Diseases transmitted by arthropods enter the body through the bites of insects, and these insects also serve as avenues of exit. A mosquito biting an individual with malaria will become infected by the *Plasmodium* spp. parasite present in the individual's blood.

Healthcare-Associated Infections

Healthcare-associated infections (HAIs), formerly referred to as *nosocomial (hospital-acquired) infections*, are infections acquired by patients during their hospital stay or during their confinement in other long-term healthcare facilities; infections acquired by hospital personnel are also considered HAIs. People are hospitalized because they are ill and require treatment beyond what home care can provide. HAIs include **central line–associated bloodstream infections (CLABIs), catheter-associated urinary tract infections (CAUTIs), surgical site infections (SSIs), ventilator-associated pneumonia (VAP)**, laboratory-identified hospital-onset methicillin-resistant *Staphylococcus aureus* (MRSA) bacteremia (bloodstream) events, and laboratory-identified hospital-onset ***Clostridium difficile* (CDI)** events. Ironically, on any given day, about 1 in 25 hospital patients in the United States has at least one HAI. The types of infections that occur in healthcare settings are listed in **TABLE 11.11**.

Hospital Environment as a Source of Healthcare-Associated Infections

What are the factors unique to the hospital environment that place patients and hospital staff at an increased risk for acquiring infections? To begin with, the patient population consists of ill individuals who may have

TABLE 11.11 Pathogens in Healthcare Settings

Pathogen	Pathogen Type	Infectious Disease
Acinetobacter baumannii	Gram (–) bacterium	Bloodstream or open wound infections; especially tracheostomy sites
Burkholderia cepacia	Gram (–) bacterium	Serious respiratory infections in patients with weakened immune systems or chronic lung diseases
Clostridium difficile	Gram (+) endospore-forming bacterium	Colitis (overuse of antibiotics is the most important risk for infection by this bacterium)
Clostridium sordellii	Gram (+) endospore-forming bacterium	Pneumonia, endocarditis, arthritis, peritonitis, myonecrosis
Klebsiella spp.	Gram (–) bacterium	Bloodstream infections, wound or surgical site infections, and meningitis
Legionella pneumophila	Gram (-) bacterium	Pneumonia
Methicillin-resistant *Staphylococcus aureus* (MRSA)	Gram (+) bacterium	Skin and bloodstream infections
Mycobacterium abscessus	Gram (+) bacterium	Infection of skin, soft tissues and lung infections in patients with various lung diseases
Pseudomonas aeruginosa	Gram (–) bacterium	**Sepsis**, pneumonia
Mycobacterium tuberculosis (TB)	Gram (+) bacterium	Tuberculosis
Vancomycin-intermediate and vancomycin-resistant *Staphylococcus aureus* (VISA or VRSA, respectively)	Gram (+) bacterium	Can cause serious infections such as sepsis, **pneumonia**, **endocarditis**, and **osteomyelitis**
Vancomycin-resistant *Enterococci* (VRE)	Gram (+) bacterium	Urinary tract infections, bloodstream infections, wound infections associated with catheters or surgical procedures
Hepatitis A, B, and C	Viruses	Hepatitis A, B, or C
Human immunodeficiency virus (HIV)	Virus	HIV/AIDS
Influenza A and B viruses	Virus	Influenza
Norovirus	Virus	Severe or prolonged gastroenteritis

Information from CDC (2018). Healthcare-associated infections. Retrieved from https://www.cdc.gov/hai/index.html

a compromised (weakened) immune system. A weak immune system increases a patient's susceptibility to **opportunistic pathogens**, including the patient's own microbiota. Antibiotics are heavily used, but also mis-used, in hospitals to treat or to prevent infections, fos-tering the development of antibiotic-resistant strains of bacteria. Drugs to purposely suppress the immune sys-tem (as in organ transplantation or treatment of autoim-mune disorders), prolonged bedrest, and restrictive diets are necessary components of treatment but are traumatic to the body and counterproductive to the maintenance of a healthy immune system.

Diagnostic and treatment protocols frequently involve extensive surgery and the use of invasive procedures, including the insertion of catheters into the urethra, swallowing of tubes, puncturing of veins with needles for intravenous therapy, and placement of nasal tubes. Ther-mometers, bedpans, urinals, eating utensils, and night table surfaces are only a few of the many fomites that pose potential risk. Hence, the equipment and devices involved in patient care contribute to transmission. The hospital staff, including physicians, nurses, laboratory techni-cians, and maintenance workers, may become complacent and unwittingly (and carelessly) transmit microbes from patient to patient; some may be healthy carriers.

All the factors involved in the cycle of infectious diseases are pres-ent in a concentrated way in hospitals and in long-term healthcare facilities, establishing these environments as reservoirs of pathogens. A relatively small number of bacterial species are responsible for most healthcare-associated infections but these are species common to the environment. Some sites in the body are more prone to healthcare-associated infections than others; the urinary tract, surgi-cal sites, and the respiratory tract are most susceptible.

The CDC has prioritized work-ing toward the elimination of five healthcare-associated infections that patients can get while receiving medi-cal treatment:

o CLABSIs

o CAUTIs

o MRSA bacteremia

o SSIs

o CDIs

CLABSIs occur when a tube is not put in correctly into a large vein or kept clean. It becomes a way for microbes to enter the body and cause deadly bloodstream infections. CAUTIs occur when catheters are not put into the patient correctly, are not kept clean, or are left in the patient for so long that microbes travel through the cath-eter and cause urinary tract infections (UTIs) in the blad-der and kidneys.

Methicillin-resistant *Staphylococcus Aureus* (MRSA) infections are spread by contaminated hands in the hos-pital setting and can cause serious bloodstream infections. SSIs occur when microbes enter a surgical incision. Some-times infections involve only the skin and sometimes they can involve tissues under the skin, organs, or implants. The most common SSIs occur after abdominal or colon sur-gery. CDIs occur in patients who are taking antibiotics. The antibiotics compromise the patient's microbiota, allowing *Clostridium difficile* to flourish in the small intestine, causing potentially deadly diarrhea, which can spread in health-care settings. **FIGURE 11.15** contains a summary of CDC surveillance of healthcare-associated infections based on data gathered in 2014. Despite the risk of healthcare-asso-ciated infections, be assured that the advances in medicine far outweigh the risks of hospitalization.

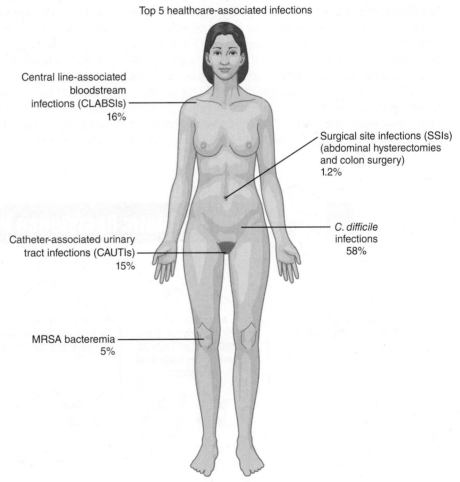

Top 5 healthcare-associated infections

Central line-associated bloodstream infections (CLABSIs) 16%

Surgical site infections (SSIs) (abdominal hysterectomies and colon surgery) 1.2%

C. difficile infections 58%

Catheter-associated urinary tract infections (CAUTIs) 15%

MRSA bacteremia 5%

FIGURE 11.15 Body site distribution and percentage of the top five healthcare-associated infections in U.S. hospitals based on 2014 data.

Control Measures

HAIs are a serious problem in hospitals and other medical facilities in terms of mortality, morbidity, and financial burden, and every hospital has strategies of prevention and control. The fact that the frequency and spectrum of antibiotic-resistant organisms are on the rise contributes to the problem (**BOX 11.2**). All hospitals are required to have an **infection control officer** and an infection control committee to maintain accreditation by the American Hospital Association.

BOX 11.2 *Iraqibacter*

David (last name intentionally omitted) was a young soldier whose military vehicle was hit by an Iraqi pipe bomb, throwing him into the air. He suffered numerous injuries, including shrapnel piercing his skin and deep cuts, and his right leg was barely attached. Thanks to the quick attention of a military rescue team, he was airlifted to an army field hospital in Germany. His leg was dangling, his blood pressure was low, and he was suffering from internal bleeding. Despite the poor odds of survival, his life was saved by the experienced medical personnel at the hospital, at the expense of losing his leg at the knee.

It is hard to believe that the worst was yet to come, a scenario that brought him even closer to death and could only be treated by amputation of his left leg, also just below the knee. The second amputation happened, not on the battlefield, not in a war zone, nor in foreign territory, but in a major military hospital in the United States resulting from massive infection by a bacterium—*Acinetobacter baumanii.* It is ironic that people can survive long periods of time without food, massive wounds, loss of limbs, and unimaginable hardships to be felled the unseen—microorganisms. David's ordeal is not uncommon, although the details vary. What happened to David and to the hundreds of other survivors that suffered a similar fate? What's the link to healthcare-associated infections in hospitals and other long-term medical care facilities?

The answers to these questions are not totally clear. Initially, physicians in the military thought the *Acinetobacter* spp. were soil organisms carried into the battlefield wounds. It was logical, but disproven; bacteria isolated from the wounds and from the soil were not the same. Actually, wound contamination in the military with this bacterium was noted back in 2003, and its frequency in returning service members led to the name *Iraqibacter*. Continued sleuthing revealed widespread contamination of field hospitals by service members and by equipment along the long route home and, ultimately, to the general population.

A. baumanii is not a "new" bacterium; it is an opportunistic Gram (–) rod-shaped bacterium that is multidrug resistant and listed as a serious threat in a report by the CDC outlining the top 18 drug-resistant threats to the United States. It is resistant to at least three different classes of antibiotics. Each year, approximately 12,000 *Acinetobacter* infections are reported in the United States. Of these, about 7,300 are multidrug-resistant infections that result in about 500 deaths. Its drug resistance is the result of a large section of drug-resistance genes acquired by conjugation with *Legionella pneumophila* (**Figure 1**).

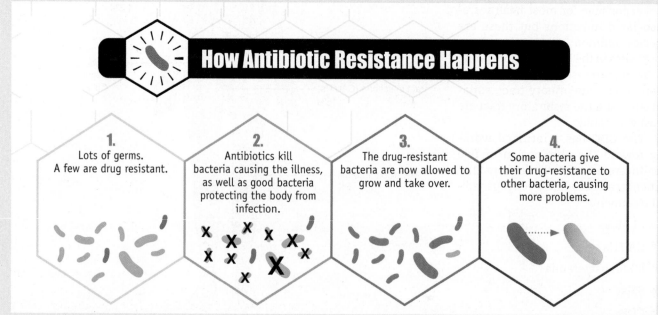

How Antibiotic Resistance Happens

1. Lots of germs. A few are drug resistant.

2. Antibiotics kill bacteria causing the illness, as well as good bacteria protecting the body from infection.

3. The drug-resistant bacteria are now allowed to grow and take over.

4. Some bacteria give their drug-resistance to other bacteria, causing more problems.

Courtesy of the CDC/Melissa Brower.

Figure 1 The four steps in the development of antibiotic resistance. In step 4, *Acinetobacter* acquired drug-resistance genes from the bacterium *Legionella pneumophila.*

Table 1 indicates that *A. baumanii* is a cause of significant healthcare-associated infections that some microbiologists fear will continue to increase. It is not a threat to healthy people, but it is to older adults; organ transplant patients taking immune-suppressive medications and certain other drugs; those suffering from autoimmune disorders on immune-suppressive drugs; children; and surgical patients. The bacterium has been recovered from a variety of surfaces, including respirators, catheters, and other devices used in patient care. It can cause a host of problems, such as pneumonia, urinary tract infections, bloodstream infections, and infections in other parts of the body.

TABLE 1 The CDC's Top 18 Drug-Resistant Threats in the United States

Category of Threat	Pathogens
Urgent	*Clostridium difficile*, carbapenem-resistant *Enterobacteriaceae* (CRE), *Neisseria gonorrhoeae* (cephalosporin resistance)
Serious	Multidrug-resistant *Acinetobacter*, drug-resistant *Campylobacter*, fluconazole-resistant *Candida*, extended spectrum *Enterobacteriaceae* (ESBL), vancomycin-resistant *Enterococcus* (VRE), multidrug-resistant *Pseudomonas aeruginosa*, drug-resistant nontyphoidal *Salmonella*, drug-resistant *Salmonella* serotype *typhi*, drug-resistant *Shigella*, methicillin-resistant *Staphylococcus aureus* (MRSA), drug-resistant *Streptococcus pneumoniae*, drug-resistant *Mycobacterium tuberculosis* (XDR-TB)
Concerning threats	Vancomycin-resistant *Staphylococcus aureus* (VRSA), erythromycin-resistant Group A *Streptococcus*, clindamycin-resistant Group B *Streptococcus*

Information from CDC (2018). Antibiotic/antimicrobial resistance. Retrieved from http://www.cdc.gov/drugresistance/

Hospitals spend considerable time and money to minimize the possibility of microbial contamination in all aspects of the hospital environment. The infection control officer is responsible for training of hospital personnel in basic infection control procedures, including isolation procedures, proper techniques of disinfection and sterilization, and the surveillance and reporting of cases of infectious diseases in both patients and staff. The infection control officer and the infection control committee are also responsible for insect control; good housekeeping; and safe practices for the disposal of feces, urine, bandages, dressings, and other potentially contaminated materials.

Education emphasizing the importance of the simple act of handwashing is vital. (Semmelweis talked about handwashing 160 years ago!) Numerous studies have demonstrated that this single simple procedure is the most important practice in minimizing HAIs. In some studies, shockingly low rates (well under 50%) of handwashing by healthcare workers, including physicians and nurses, have been revealed.

One Health

CDC experts recognize that the health of humans is connected to the health of animals and the environment they share together. Six out of every 10 infections in humans are spread by animals causing zoonoses. Each year, around the world, it is estimated that zoonoses cause 2.5 billion human cases of infectious diseases and 2.7 million deaths. *The health of wildlife in our environment can serve as a warning sign for potential illness in people.* For example, birds infected with West Nile virus often die. Bird die-offs are a warning sign that West Nile virus is circulating, and humans are hosts located within the same environment for the West Nile virus, warning us that potential infections and disease can occur (**FIGURE 11.16A**).

As the human population increases, human activities such as industrialization (the manufacturing of goods) and political struggles related to geographical factors cause global changes that affect trade and result in the migration of humans and other species in the same environment. Global changes are associated with decreased biodiversity and migration resulting in habitat deterioration. Ultimately, global changes favor the emergence or reemergence of infectious and noninfectious diseases (**FIGURE 11.16B**).

An example that best illustrates the impact of global changes resulting in disease is the **deforestation** practices of humans in forested areas of the Amazon region in Brazil. Trees are removed for purposes of mining, dam construction, logging, and petroleum exploration. Deforestation results in loss of habitat for thousands of common vampire bats. Vampire bats are common carriers of rabies virus. They feed on blood. They rarely feed on humans, but increased deforestation meant that the vampire bats,

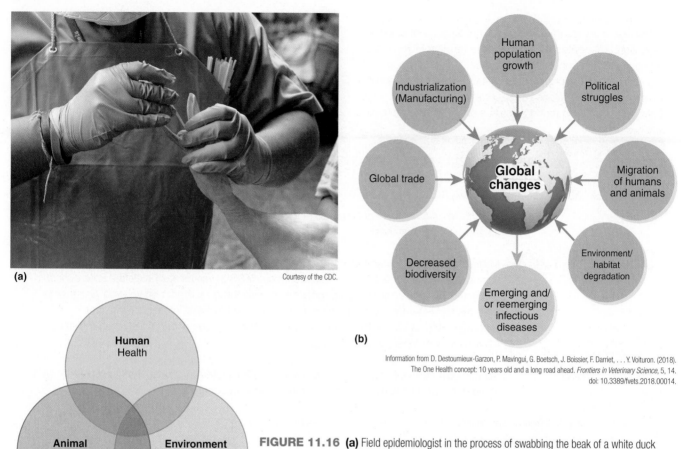

(a)

(b)

Information from D. Destoumieux-Garzon, P. Mavingui, G. Boetsch, J. Boissier, F. Darriet, . . . Y. Voituron. (2018). The One Health concept: 10 years old and a long road ahead. *Frontiers in Veterinary Science*, 5, 14. doi: 10.3389/fvets.2018.00014.

(c)

FIGURE 11.16 **(a)** Field epidemiologist in the process of swabbing the beak of a white duck during an infectious disease investigation. Approximately 60% of infectious diseases in humans are zoonoses. **(b)** Global change favors infectious disease outbreaks. **(c)** One Health in action. This diagram depicts the overlapping activities from concept to action. It is a holistic, transdisciplinary approach to health. Collaborative relationships between human, animal (domestic and wildlife), and environmental health partners are fostered. Disease surveillance activities are coordinated and communication between the different health sectors are developed as a uniform message to the public.

which normally live almost exclusively in dark places such as hollow trees and caves, ventured into the homes of people who lived near an abandoned gold mine in Maranhao in 2005. At the time, a generator failed, leaving people in villages without electricity for weeks in the dark at night. Their houses were quite simple dwellings that had many openings that were vulnerable to the entrance of bats. The vampire bats bit several hundred people at night, causing at least 23 human rabies deaths in a rabies epidemic.

The CDC has adopted a **One Health** approach that involves designing and implementing programs, legislation, and research to respond to and prevent zoonoses, food safety problems, and other public health emergencies (**FIGURE 11.16C**). The CDC's One Health Office works in the United States and around the world to protect the health of humans, animals, and the environment. It is a collaborative, transdisciplinary approach working at the local, regional, national, and global levels. The WHO works closely with the Food and Agriculture Organization of the United Nations and the World Organization for Animal Health to reduce the risks of food safety hazards, zoonoses, and other public health threats at the humans–animals–environment interface.

Epidemiology of Fear

The fear of epidemics can reach epidemic proportions, but the threat of infectious disease, according to some experts, is out of proportion. Zika virus disease, Ebola virus disease, norovirus, measles outbreaks, Powassan encephalitis, avian influenza (H5N1, H7N9, and H7N4 subtypes), Middle East respiratory syndrome (MERS), *Vibrio vulnificus* flesh-eating bacteria, Alaskan salmon tapeworms in raw sushi, and antibiotic-resistant bacterial infections have captured the public's attention and led to an explosion of television programs, popular books, and movies.

At the end of 2013, the worst Ebola virus disease epidemic in human history began to kindle in West Africa. The Ebola fire burned for over 2 years. It took leadership, adaptation, and innovation to extinguish the large-scale Ebola epidemic. Tireless healthcare workers finally saw the epidemic come to a close in 2015. CDC Director Tom Frieden said, *"We've made great progress, but we can't let our guard down. There will continue to be cases and clusters of Ebola, but an epidemic of this kind we've had in the past year never has to happen again."*

In 2015, Zika virus emerged to infect humans in the Western hemisphere for the first time. Zika virus is transmitted through *Aedes* spp. day-biting mosquitoes. Brazil was the epicenter for a large epidemic, with Zika virus infecting thousands of people in South America. Zika virus was considered a harmless virus until this time. The Brazilian Ministry of Health declared a public health emergency because of an unusual increase in microcephaly in newborns associated with mother's infected with Zika virus. Only one in five people infected with Zika virus experience symptoms. Expectant mothers were understandably anxious. Brazilian doctors were urging women not to become pregnant. Evidence determined that Zika virus could be transmitted through sexual intercourse.

The WHO declared the Zika virus disease outbreak a Public Health Emergency of International Concern in February 2016. The announcement occurred just before the 2016 Summer Olympic Games, which were to be held in Rio de Janeiro. Some athletes chose to withdraw over safety concerns, and tourism suffered as a result of the Zika alarm. Zika virus disease traveled north to U.S. territories and parts of Texas and Florida. In May 2016, WHO Director-General Dr. Margaret Chan said, *"The rapidly evolving outbreak of Zika warns us that an old disease that slumbered for 6 decades in Africa and Asia can suddenly wake up on a new continent to cause a global health emergency."*

During August 2017, experts in nine countries issued health warnings about the potential spread of bubonic and pneumonic plague emanating from an epidemic in Madagascar. Plague is endemic in Madagascar. The same plague that had wiped out nearly two-thirds of Europe's population in the 13th century caused an outbreak in Madagascar that was much larger in scale and spread faster than past outbreaks. It was finally contained in December 2017, with cumulative tallies of 2,417 cases and 209 deaths (9% mortality). A total of 7,318 contacts were identified and provided with prophylactic antibiotics.

In 2017, the United States was hit with three consecutive hurricanes. The Strategic National Stockpile responded and military experts were deployed in order to deliver supplies to the affected areas. Six federal medical stations were shipped to Puerto Rico. Vaccines for hepatitis A and B, tetanus, pneumococcal pneumonia, rabies, and influenza, along with other medicines and supplies, were shipped to Puerto Rico (**FIGURE 11.17**). **Climate change** is not only increasing hurricane activity, but also the occurrence of infectious disease outbreaks. For example, flooding due to hurricane activity results in more breeding areas for mosquito vectors that carry pathogens such as Zika virus, West Nile virus, Chikungunya virus, and Dengue viruses. Contaminated floodwaters have been associated with skin infections, including infections caused by *Vibrio vulnificus* and other flesh-eating bacteria.

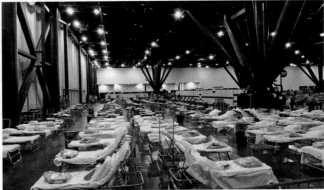

Courtesy of CDC/Kamelya Hinson.

FIGURE 11.17 This photograph shows the interior of the George R. Brown (GRB) Convention Center in Houston, Texas, after it was converted into an emergency shelter in response to Hurricane Harvey. This site was made ready within 12 hours and accepted 160 people displaced by the hurricane and in need of nonemergency medical care.

In March 2013, for the first time, a second highly lethal new **avian influenza A virus (H7N9)** was infecting people in China. The first deadly avian influenza A virus (H5N1) made its presence in Hong Kong in 1997. It startled experts because the virus jumped directly into humans from birds and it had a 60% mortality rate. Previous influenza A virus subtypes infected pigs and subsequently were transmitted to humans. Fortunately, the Hong Kong influenza A virus (H5N1) did not spread easily from person to person and it fizzled out. But this second avian influenza A virus (H7N9) has experts even more alarmed. CIDRAP Director Dr. Michael Osterholm said *"when it comes to influenza, expect the unexpected."* Concerns were raised that the virus had the ability to mutate and adapt into a strain that could readily infect people and spread more easily from person to person.

Sporadic influenza A (H7N9) cases continue to occur through contact with infected poultry or contaminated environments such as **live bird markets**. As of October 2017, 1,565 laboratory confirmed cases of human infections caused by the Asian H7N9 avian influenza A virus had been documented. The virus causes severe pneumonia and has about a 39% mortality rate. Most cases occur as a result of occupational exposure to live poultry on a farm. In February 2018, the WHO announced the first human case of influenza caused by a third emerging Asian **avian influenza A virus (H7N4)**. The patient was a 68-year-old woman from Jiangsu Province in China who had preexisting heart disease and hypertension. With the recent marking of the 100th anniversary of the 1918 Spanish flu, public health experts continue to raise concerns over the need for more pandemic preparedness.

In his first inaugural address on March 4, 1933, President Franklin Delano Roosevelt spoke eloquently of the danger of fear. His often-quoted words were, *"So first of all let me assert my firm belief that the only thing we have to fear is fear itself—nameless, unreasoning, unjustified terror which paralyzes needed efforts to convert retreat into advance."* His words, focused on World War I, are applicable to the spread of infectious diseases. The take-home message is that awareness; surveillance; development of new diagnostic tools, antimicrobials, and vaccines; commonsense precautions; and calmness are paramount. Fear can be paralyzing.

Summary

Epidemiologists classify disease as sporadic, endemic, epidemic, or pandemic, depending on its frequency and distribution. These categories are not absolute; a particular infectious disease can slide from one classification to another. Common-source epidemics arise from contact with a single contaminant, resulting in a large number of people becoming ill suddenly; the disease peaks rapidly. Propagated epidemics are characterized by direct person-to-person (horizontal) transmission, a gradual rise in the number of infected individuals, and a slow decline.

A chain of linked factors is required for infectious diseases to exist and to spread through a population. These factors are reservoirs of disease, transmission, portals of entry, and portals of exit. Understanding the characteristics of microbes and the diseases they cause is necessary to break the cycle somewhere along its path. The reservoir is the site where pathogenic microbes exist in nature and from which they can be spread. Active carriers, healthy carriers, and chronic carriers are reservoirs, as are wild and domestic animals. Nonliving reservoirs include contaminated water, food, soil, and surfaces.

Transmission is the bridge between reservoir and portal of entry. Person-to-person contact is the most common type of horizontal direct transmission and allows for the immediate transfer of microbes. Vertical transmission is another type of direct transmission and is categorized by the passage of pathogens from mother to offspring across the placenta, in the birth canal during delivery, or in breast milk. In direct transmission there are no intermediaries. Indirect transmission involves the passage of materials from a reservoir or source to an intermediate vehicle and then to a host. The intermediate vehicle can be nonliving or living. Water, food, fomites, and aerosols are significant nonliving vehicles of indirect transmission. Vectors are living organisms that transmit microbes from one host to another. Mechanical vectors passively transfer microbes on their feet or other body parts, whereas biological vectors are required for the multiplication and development of the pathogen within the vector.

Portals of entry are the next consideration in the cycle of infectious diseases. Some pathogenic microbes have a single preferred portal of entry into the body, whereas others have more than one. Body orifices, including the mouth, nose, ears, eyes, anus, urethra, and vagina, are portals of entry; the skin can be penetrated and is another portal of entry. For the cycle of infectious disease to continue in a population, pathogens must exit from the body. In many cases, the portals of entry and the portals of exit are the same.

Healthcare-associated infections are a serious problem in hospitals and other medical facilities in terms of mortality, morbidity, and financial burden, and every hospital has strategies of prevention and control. The fact that the frequency and spectrum of antibiotic-resistant organisms are on the rise contributes to the problem. The public should not be paralyzed by the fear of infection. Awareness, surveillance, commonsense precautions, and calmness are the best preventive measures. The CDC's One Health concept is aimed to integrate multiple, transdisciplinary approaches in order to ultimately reach a healthy equilibrium between humans, animals, and the environment.

KEY TERMS

active carrier
aerosols
arboviruses
arthropods
avian influenza A virus (H7N4)
avian influenza A virus (H7N9)
biological vectors
body orifices
bushmeat
catheter-associated urinary tract infection (CAUTI)
Center for Disease Research and Policy (CIDRAP)
central line–associated bloodstream infection (CLABSI)
chronic carrier
climate change
Clostridium difficile (CDI) event
common-source epidemic

communicable diseases
community-acquired methicillin-resistant *Staphylococcus aureus* (CA-MRSA)
cough etiquette
deforestation
direct transmission
droplet transmission
endemic infectious disease
endocarditis
endospore
epicenter
epidemic
epidemiology
flaccid paralysis
fomites
Food Safety News

gastroenteritis
healthcare-associated infections (HAIs)
HealthMap
healthy carrier
herd immunity
horizontal transmission
indirect transmission
infection control officer
infectious diseases
insecticide
insects
intermediate vehicle
invasive species
live bird markets
lymphadentitis
mechanical vectors
methicillin-resistant *Staphylococcus aureus* (MRSA)
microbiota
mode of transmission
morbidity
mortality
myelitis
opportunistic pathogens
One Health
osteomyelitis
outbreak
pandemic infectious disease
patient zero

pneumonia
portal of entry
portal of exit
prions
ProMED-mail
propagated epidemic
PulseNet
real-time surveillance
reemerging infectious disease
reservoir
R-nought (R_0)
sepsis
species barrier
sporadic infectious disease
surgical site infection (SSI)
surveillance
transovarial transmission
vectors
vectorborne diseases
vehicleborne transmission
ventilator-associated pneumonia (VAP)
vertical transmission
wildlife
winter vomiting disease
World Health Organization (WHO)
zoonoses
zoonotic disease

SELF-EVALUATION

○ PART I: Choose the single best answer.

1. Which of the following are involved in the worldwide surveillance of infectious diseases?
   a. CIDRAP
   b. ProMED-mail
   c. HealthMap
   d. A–C are involved in surveillance.

2. Morbidity refers to _____.
   a. the death rate
   b. the illness rate
   c. person-to-person contact
   d. a zombie state

3. Ebola virus disease outbreaks in Africa are best described as _____.
   a. sporadic
   b. endemic
   c. vehicle transmission
   d. pandemic

4. Which of the following arthropods is *not* a disease vector?
   a. Fly
   b. Tick
   c. Centipede
   d. Mosquito

5. Which of the following is *not* spread by direct transmission?
   a. Gonorrhea
   b. HIV/AIDS
   c. Mononucleosis (the "kissing disease")
   d. Chikungunya virus disease

6. _____ refers to a proportion of immune individuals to a particular infectious disease within a population.
   a. Herd immunity
   b. Passive immunity
   c. Horizontal transfer
   d. Vertical transmission

7. Six out of every 10 infections in humans are spread by _____.
   a. fomites
   b. animals
   c. arthropods
   d. humans

8. _____ are an example of an intermediate vehicle that may be contaminated with microbes and serve in their transmission.
   a. Reservoirs
   b. Fomites
   c. Vectors
   d. Portals

9. Which of the following infectious diseases is *not* transmitted by a mosquito?
   a. West Nile encephalitis
   b. Dengue fever
   c. Zika virus disease
   d. Measles

10. The biological vectors of the bacterium *Yersinia pestis* are _____.
    a. fleas
    b. mosquitoes
    c. deer ticks
    d. house flies

○ **PART II: Fill in the blank.**

1. A worldwide outbreak of a disease is called a(n) _____.

2. Arthropods can either be _____ vectors or _____ vectors.

3. _____ is the only disease that has been eradicated.

4. Pathogens access the body through _____.

5. All hospitals are required to have a(n) _____ who is responsible for disease prevention in the hospital.

6. Typhoid Mary was a healthy _____ of the bacterium that causes typhoid fever.

7. The most common healthcare-associated infection is caused by _____.

8. Malaria is caused by _____.

9. R_0 is a measure of _____.

10. The higher the R_0 value, the _____.

○ **PART III: Answer the following.**

1. Distinguish between vertical and horizontal transmission. Give examples of each.

2. What is meant by zoonoses? Give three examples of zoonoses.

3. List the top five types of healthcare-associated infections in the United States.

4. How can vectorborne diseases be prevented? Why are vectorborne diseases a growing problem?

5. Would you consider avian flu or malaria a greater threat to human survival? Explain.

6. List the human body's portals of entry and portals of exit for pathogens.

7. Discuss how rats (rodents) can cause disease. Name two diseases associated with rat populations.

8. Why is it difficult to control rabies in the wild?

9. Explain why hospitals and long-term healthcare facilities are hotbeds of infections for patients.

10. List microbes that are often the contaminants of water and food, resulting in water or foodborne infections.

BACTERIAL DISEASES

> "*It is much more important to know what sort of a patient has a disease than what sort of disease a patient has.*"
>
> —Sir William Osler
> (1849–1919)
>
> *Canadian physician and founding professor of Johns Hopkins Hospital*

Courtesy of the CDC/Amanda Mills.

A sign in the interior of a bathroom reminding employees to wash their hands. Handwashing saves lives.

LEARNING OBJECTIVES

1. Explain why the spread of emerging antibiotic-resistant strains of pathogenic bacteria in the hospital environment is difficult to control from a public health perspective.

2. List bacterial infections that can occur after natural disasters and catastrophic events.

3. Discuss the impact of multistate foodborne outbreaks on the food industry and the public health responses in the United States and Europe.

4. List at least three types of bacterial toxins and discuss the symptoms they induce and the damage they cause inside of the human body during the course of an infection.

5. Summarize why certain bacterial infections may increase in incidence at a world-wide level due to climate change.

6. Explain bacterial generation time, and how it affects the length of antibiotic therapy when treating patients.

7. Identify those groups of people who are most at risk for developing serious complications of bacterial infections, such as sepsis, pneumonia, necrotizing fasciitis, or meningitis, and explain why they are more susceptible to such complications.

Case Study: From Sea to Sepsis

Two individuals sought emergency medical care after exposure to seawater in the Gulf of Mexico. The first patient was a 77-year-old man, "Jim," who had been crabbing in the Gulf of Mexico the day before he sought medical care. The second individual, 31-year-old "Diego," had gotten a new tattoo on his right calf 8 days before he went swimming in the Gulf of Mexico. He arrived for urgent medical care 3 days after he went swimming.

Jim was pricked multiple times by crab claws while crabbing. He developed painful **bullae** (watery blisters) on his right arm. He also suffered from malaise, bouts of fever and chills, dizziness, and vomiting. Jim had no pre-existing conditions (e.g., diabetes, liver disease, immuno-deficiency, or skin conditions). Two days before he sought medical attention, Diego began suffering from fever and chills. His tattoo and other areas on both legs were red and inflamed. Diego had a history of **liver cirrhosis**, and he said he averaged a six-pack of beer every day.

Within 2 to 3 hours of Diego's admission to the hospital, bullae rapidly erupted on the affected areas of skin that progressed to **necrotizing fasciitis**, also referred to as the "flesh-eating" disease (**Figure 1**). His blood pressure was dangerously low. Doctors suspected he was suffering from a bacterial infection and immediately treated him with antibiotics. Deep tissue biopsies

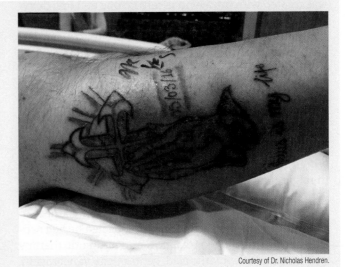
Courtesy of Dr. Nicholas Hendren.

Figure 1 Patient's tattoo site with progressing bullae.

taken from his legs and blood cultures confirmed their suspicions. The cultures that came back were Gram (–), comma-shaped (curved rods), oxidase-positive bacteria and identified as *Vibrio vulnificus*. Within 24 hours of admission, Diego went into **septic shock** and acute

(continues)

Case Study: From Sea to Sepsis (continued)

kidney failure. His bloodstream and soft skin tissues were teeming with bacteria. Over the following 2 weeks doctors aggressively **debrided** the dead soft tissues, cared for his wounds, and continued antibiotic treatment. Unfortunately, Diego's preexisting conditions (e.g., liver disease from alcohol consumption) complicated his treatment. Two months later he died.

Jim also went into septic shock caused by severe **cellulitis** and soft tissue infection. Cellulitis is a potentially serious bacterial skin infection in which the affected skin is red and swollen, warm, and painful to the touch. The doctors suspected that Jim had a bacterial infection and immediately began treating him with three antibiotics: vancomycin, doxycycline, and ceftriaxone. Blood cultures and clinical specimens from ruptured bullae were ordered. *V. vulnificus* was cultured from his blood. After 3 days, he was moved out of an **intensive care unit (ICU)**. His antibiotic regimen was changed to doxycycline and ceftriaxone. He underwent aggressive wound care in which Silvadene cream was applied twice daily to affected areas. **Silvadene** contains 1% silver sulfadiazine, which is an antimicrobial agent. It is used only as a topical medication on skin.

Jim was discharged after an 11-day hospital stay. For the next 3 weeks, antibiotic therapy was continued. He was treated with intravenous ceftriaxone and oral doxycycline. For the next 3 months, he had right arm pain. Intravenous antibiotic therapy was prolonged. He was admitted twice into the hospital for recurrent cellulitis. He suffered from **sepsis** caused by an *Enterococcus* spp. His antibiotic therapy was changed to cephalexin. Three months later, his arm finally healed. Doctors instructed Jim to wear gloves at all times if he handled seafood.

Just like the expression "everything comes in threes," a third patient, 38-year-old "Larry," arrived in the ER with symptoms of fever 102°F (39°C), chills, nausea, and **myalgia** (pain). He had two necrotic lesions on his left leg. Upon taking his medical history, Larry reported he had insulin-dependent diabetes and consumed three to six cans of beer every day. Doctors suspected he had sepsis. He was transferred to an ICU and antibiotic therapy was initiated: clavulanic acid, gentamicin, vancomycin, and ceftazidime. A blood sample was drawn and sent to the medical technology lab for culture and analysis. Larry died 3 days later.

V. vulnificus was cultured from his blood. A follow-up investigation by environmental health inspectors revealed that Larry had eaten raw oysters harvested from Galveston Bay, Texas, (Gulf of Mexico), 4 days before being admitted to the hospital. No oysters were available for analysis. However, it is known that *V. vulnificus* is present in up to 50% of oyster beds with the water conditions that occur in the Gulf of Mexico during warm months [i.e., temperatures greater than 68°F (20°C) during May through October and salinity is less than 16 parts per thousand]. Larry had no other possible exposure to *V. vulnificus*, such as ingesting other raw shellfish or skin exposure to seawater or shellfish.

V. vulnificus causes sepsis, wound infections (including necrotizing fasciitis requiring amputation), and gastroenteritis. Transmission occurs through the consumption of raw or undercooked seafood, especially oysters, or through contamination of an open wound, sores, cuts, or by seawater or seafood drippings. *About one in seven people who experience a* V. vulnificus *wound infection dies.* Individuals who are at high risk for developing fatal sepsis with a 50% mortality rate suffer from one or more of the following conditions:

- Liver disease as a result of **hepatitis**, **cirrhosis**, alcoholism, or cancer
- **Hemochromatosis** (iron overload)
- Diabetes
- Kidney disease
- Cancer (including **lymphomas**, **leukemia**, **Hodgkin's disease**)
- Stomach disorders
- Any illness or treatment that weakens the body's immune system, including HIV/AIDS

Transmission also occurs when open wounds, cuts, burns, or sores (including recent tattoos) are exposed to brackish water where *V. vulnificus* naturally occurs throughout the world. *V. vulnificus* is considered one of the most dangerous waterborne pathogens. It is highly sensitive to water temperature and salinity. It resides in high numbers in filter-feeding shellfish (oysters, clams, and mussels). It was first recognized as a human pathogen in the 1970s.

In 1996, an unexplained outbreak of *V. vulnificus* infections causing severe soft tissue infections and sepsis occurred in Israel among fish market workers and fish consumers. The fish were cultivated on inland brackish fish farms. Prior to this, no *V. vulnificus* infections were reported in Israel. The origins of this **emerging bacterial disease** were not fully understood at the time. The strain was sent to the **Centers for Disease Control and Prevention (CDC)**.

The number of *V. vulnificus* infections reported is steadily increasing. Between 1996 and 2005, the number of cases increased by 41% in the United States. Outbreaks were reported in locations with no previous incidence in diverse climate zones (**Table 1**). *Reports of*

TABLE 1 Worldwide *Vibrio vulnificus* Outbreaks*

Location	Year(s)	Number of Cases	Number of Deaths	Percentage Resulting in Sepsis	Percentage with Wound Infections	Percent Mortality
Florida	1981–1993	141	50	53%	33%	56%
23 U.S. states	1988–1996	422	143	43%	45%	38.4%
United States	1997–2006	428	62	N/A	66%	17%
California	1991–2010	88	39	N/A	N/A	44.3%
Denmark (hospital)	1994	11	1	9.1%	36.4%	9.1%
France, New Caledonia	2008	3	3	N/A	N/A	100%
Germany, Mecklenburg-Vorpommem	2006	3	0	N/A	100%	0%
Greece	1997–2003	9	2	N/A	100%	22%
Israel	1996–1997	62	0	N/A	100%	0%
Japan, Ariake Sea Coast	1984–2008	37	24	N/A	0%	64.9%
Japan	1999–2003	94	58	72.3%	22.3%	61.7%
South Korea	2001–2010	588	285	N/A	N/A	48.5%
Taiwan, Chi Mel Hospital	1998–2011	121	35	N/A	N/A	29%
Taiwan, National Medical Center	1996–2011	140	18	58%	78%	18%

*Based on reports in primary literature.
N/A: data not available.
Information from S-P. Heng. V. Letchumanan, C. Y. Deng, N. S. Ab Mutalib, T. M. Khan, L. H. Chuah, . . . L. H. Lee. (2017). *Vibrio vulnificus*: An environmental and clinical burden. *Frontiers in Microbiology, 8*, 997.

epidemics have been linked to extreme weather events such as heatwaves and heavy precipitation in temperate regions within the past century. Eighteen wound infections were reported in the affected region following hurricane Katrina in 2005. Five of the cases (28%) died; three deaths were associated with V. *vulnificus* infections and two were associated with V. *parahaemolyticus* infections. During the extremely warm summers of 1994, 2003, and 2006, a plethora of case reports appeared in the medical literature describing *Vibrio*-associated wound infections correlated with recreational swimming and bathing in the Baltic Sea. The most intense heatwave ever experienced in Scandinavia, in 2014, was associated with the highest yearly total of *Vibrio* infections in Finland and Sweden.

Global temperatures have increased nearly 0.8°C (1.4°F) since the late 19th century and by approximately 0.2°C (0.35°F) per decade over the past 25 years. Temperatures are predicted to increase by 1.8°C to 5.8°C (3.1°F to 10.1°F) by the end of the 21st century. *The anticipated global warming and reduced salinity of coastal regions will provide new areas for pathogenic strains of* Vibrio *species to naturally occur.* Despite the increase in the reporting of V. *vulnificus* cases, conclusive evidence linking the emergence of V. *vulnificus* with **climate change** remains contentious. Perhaps, *Vibrio* species represent an important and verifiable weather gauge of climate change in marine environments. To date, the CDC estimates that about 205 cases of V. *vulnificus* infection occur in the United States every year.

(continues)

Case Study: From Sea to Sepsis (continued)

© Teri Shors.

Figure 2 Refractometer for measuring salinity.

After a hurricane or storm surge causes flooding, individuals can be exposed to brackish coastal waters. Recreational fisherman (e.g., along the U.S. Gulf Coast) or those swimming or diving in potentially brackish waters should monitor the salinity of waters using a **refractometer** (**Figure 2**). Knowing the salinity of the water can save lives. Refractometers measure how much light bends or refracts when it enters a liquid. The more salts dissolved in water, the more resistance the light will meet, and the more the light will bend or refract. The salinity of freshwater is nearly zero parts per thousand (ppt). The salinity of the ocean is typically 35 ppt. Brackish water is a mixture of seawater and freshwater in an estuary (i.e., locations in which a freshwater source, such as a river, meets and mixes with the ocean) or seawater and freshwater can mix after a hurricane or storm surge causes flooding. *Brackish water usually has a salinity of 0.5 to 34 ppt.* V. vulnificus *prefers a habitat with a water temperature range of 9°C to 31°C (48.2°F to 87.8°F) and salinities between 15 and 25 ppt.* Individuals should cover open wounds, cuts, sores, recent tattoos, and burns and wear protective gear (e.g., boots, gloves, eyewear) to prevent contact with seawater and seafood containing V. vulnificus.

Questions and Activities

1. List the three **modes of transmission** by which an individual can be infected by V. vulnificus.

2. Define sepsis, and explain why it is often associated with high mortalities.

3. Salinity of water can be measured using a hydrometer or refractometer. Which method is most accurate? Why is it important to get an accurate reading if you are planning on spending time in freshwater near coastal areas of Florida or Texas that may contain seawater after recent floods?

4. Besides potential infection by V. vulnificus, what other infections could you get if you ate raw oysters or uncooked seafood?

5. Explain why individuals who have liver disease or are immune compromised are more at risk to infection and complications caused by V. vulnificus infection.

6. Most pathogens that infect humans flourish in media that contains very little salt. In contrast, V. vulnificus is a pathogenic **halophile**. Explain how a pathogenic halophile differs from a **halotolerant** pathogen.

7. V. vulnificus is an **opportunistic pathogen**. Define opportunistic pathogen. List five human bacterial pathogens, the diseases they cause, and whether the pathogens on your list are halophiles, halotolerant, or **halointolerant**.

8. Create a hypothesis about what type of **virulence factors** V. vulnificus would need in order to cause an infection in a person who ingested raw, contaminated oysters. What about infection through an open wound? Design an experiment to test your hypothesis in a research laboratory.

9. Sea salts are used to make foods such as artisan cheeses and dry-cured meats. Sea salt has been widely perceived to be a chemically pure and sterile food ingredient, but recent investigations have determined that many different brands of seas salts contain microbial contaminants. Perform an Internet search to identify the microbes that have been found in sea salts. Can they cause food spoilage? Disease in humans?

10. List five countries that have had outbreaks of V. vulnificus in recent decades.

11. V. vulnificus is not the only bacterial species that can cause necrotizing fasciitis. List other bacteria that can also cause necrotizing fasciitis.

12. A 2015 study determined that swimmers in the Chesapeake Bay, Maryland, were exposed to V. vulnificus. Cases were reported in Alabama in 2018. Follow up using HealthMap (http://www.healthmap.org/en/) to research how many new cases of V. vulnificus have occurred in the past 6 months in the United States. List all cases, locations, and mode of transmission.

13. *V. parahaemolyticus* and *V. cholerae* also cause disease. Compare and contrast these two bacterial species with *V. vulnificus* (e.g., disease symptoms, treatments, modes of transmission).

14. What type of environmental conditions does *V. vulnificus* thrive in?

15. What types of catastrophic disasters have *V. vulnificus* infections been associated with? Why did these disasters increase the incidence of *Vibrio* infection?

16. List any preexisting conditions an individual might have that would increase his or her likelihood of developing a severe *V. vulnificus* infection that could potentially result in death.

17. Some experts suggest that *Vibrio* species represent an important and verifiable weather gauge of climate change in marine environments. Explain their rationale.

18. Which antibiotics are most commonly used to treat *V. vulnificus* infections?

19. In 2018, a heatwave occurred in parts of North America and Europe. Research CDC information on the number of reported infections, the mode of transmission, and the number of deaths caused by *V. vulnificus* and *V. parahaemolyticus* in the United States and Eurosurveillance reports (https://www.eurosurveillance.org) for cases in Europe.

Information from C. Baker-Austin, J. Trinananes, N. Gonzalez-Escalona, & J. Martinez-Urtaza. (2017). Non-cholera *Vibrios*: The microbial barometer of climate change. *Trends in Microbiology, 25,* 76–82; Centers for Disease Control and Prevention (CDC). (1996). *Vibrio vulnificus* infections associated with eating raw oysters—Los Angeles, 1996. *MMWR, 45,* 621–624; Centers for Disease Control and Prevention (CDC). (2017). *Vibrio vulnificus* infections and disasters. Fact Sheet. https://www.cdc.gov/disasters/vibriovulnificus.html; N. Hendren, S. Sukumar, & C. S. Glazer. (2017). *Vibrio vulnificus* septic shock due to a contaminated tattoo. *BMJ Case Report.* doi:10.1136/bcr-2017-220199; A. J. Sheer, K. P. Kline, & M. C. Lo. (2017). From sea to bloodstream: *Vibrio vulnificus* sepsis. *American Journal of Medicine, 130,* 1167–1169.

Preview

This chapter presents a long list of bacterial diseases, but the diseases discussed represent only a small fraction of those that are known. Only the major bacterial diseases of personal and public health significance are presented. The diseases are organized according to their route of transmission: healthcare-associated infections (HAIs), foodborne and waterborne, airborne, sexually transmitted, contact, soilborne, and arthropodborne. Note that in some cases transmission may be accomplished by more than one route. For example, listeriosis may be foodborne or airborne, and anthrax may be airborne or acquired by direct contact. To the extent possible, statistics indicating the current incidence of these diseases, primarily from the U.S. Centers for Disease Control and Prevention (CDC), is included as is available, but the actual numbers are secondary to the status of the disease. In most cases, the numbers vary depending on the source of information and are underreported as the result of inadequate surveillance, failure to report, misdiagnosis, and underuse of laboratory testing. These factors are of greater significance in developing countries. Attention is paid in varying degrees to the practical considerations of transmission, pathogenesis, prevention, and, to some lesser extent, treatment. Most of these diseases are responsive to antibiotic therapy, although antibiotic resistance is an increasingly greater problem. Preventive vaccines are available in many cases, and others are the subject of active research.

Some of the diseases presented are ones not usually expected to occur in the United States or in other developed countries; however, transmission can happen because of an increase in immigration and in international travel, including eco-travel. Humanitarian purposes, too, require an understanding of public health.

Healthcare-Associated Infections (HAIs)

On any given day in the United States, about 1 in 25 patients is suffering from at least one **healthcare-associated infection (HAI)**. HAIs are infections that patients acquire while they are receiving health care for another illness. HAIs can occur in any healthcare setting, including hospitals, long-term care facilities or nursing homes, physician offices, **urgent care clinics**, and **ambulatory surgical centers**, also referred to as outpatient or same-day surgery centers. It seems ironic that while receiving health care one can get more than they bargained for by acquiring another infection—but it does happen. Healthcare facilities are akin to a storage facility for human pathogens. The worst pathogens will be there. HAIs result in tens of thousands of deaths and cost the U.S. healthcare system billions of dollars every year. **TABLE 12.1** lists the estimated number of HAIs in U.S. hospitals in 2011.

Reducing HAIs is a priority. Significant progress has been made in preventing some types of infections, but more needs to be done. The following factors raise the patient's risk of acquiring an HAI:

- Catheters (bloodstream, endotracheal, and urinary)
- Surgeries
- Injections of medication
- Overuse or improper use of antibiotics
- Infectious diseases spreading between patients and healthcare workers
- Healthcare facilities that are not properly cleaned and disinfected

TABLE 12.1 Healthcare-Associated Infections (HAI) in U.S. Hospitals, 2011	
Type or Body Site of HAI	Estimated Number of HAIs
Pneumonia	157,500
Surgical site during inpatient surgery	157,500
Gastrointestinal illness	123,100
Other types of infections	118,500
Urinary tract infections	93,300
Primary bloodstream infections	71,900
Estimated Total Number of Infections in Hospitals	**721,800**

Information from CDC (2018). Healthcare-associated infections. Retrieved from https://www.cdc.gov/hai/index.html

The following are common HAIs that patients develop in hospitals:

- Bloodstream infections
- *Clostridium difficile* infections
- Methicillin-resistant *Staphylococcus aureus* (MRSA) infections
- Surgical site infections
- Urinary tract infections

FIGURE 12.1 shows the common HAIs by body site and the bacterial pathogens involved. Besides the bacterial infections covered in this section, less common HAIs are caused by other bacteria (e.g., *Legionella* spp.), viruses, and fungal pathogens. Steps can be taken to prevent HAIs by making a conscious effort to work toward improving care and protecting patients (e.g., effectively managing water systems, because many of these pathogens can be found in tap water).

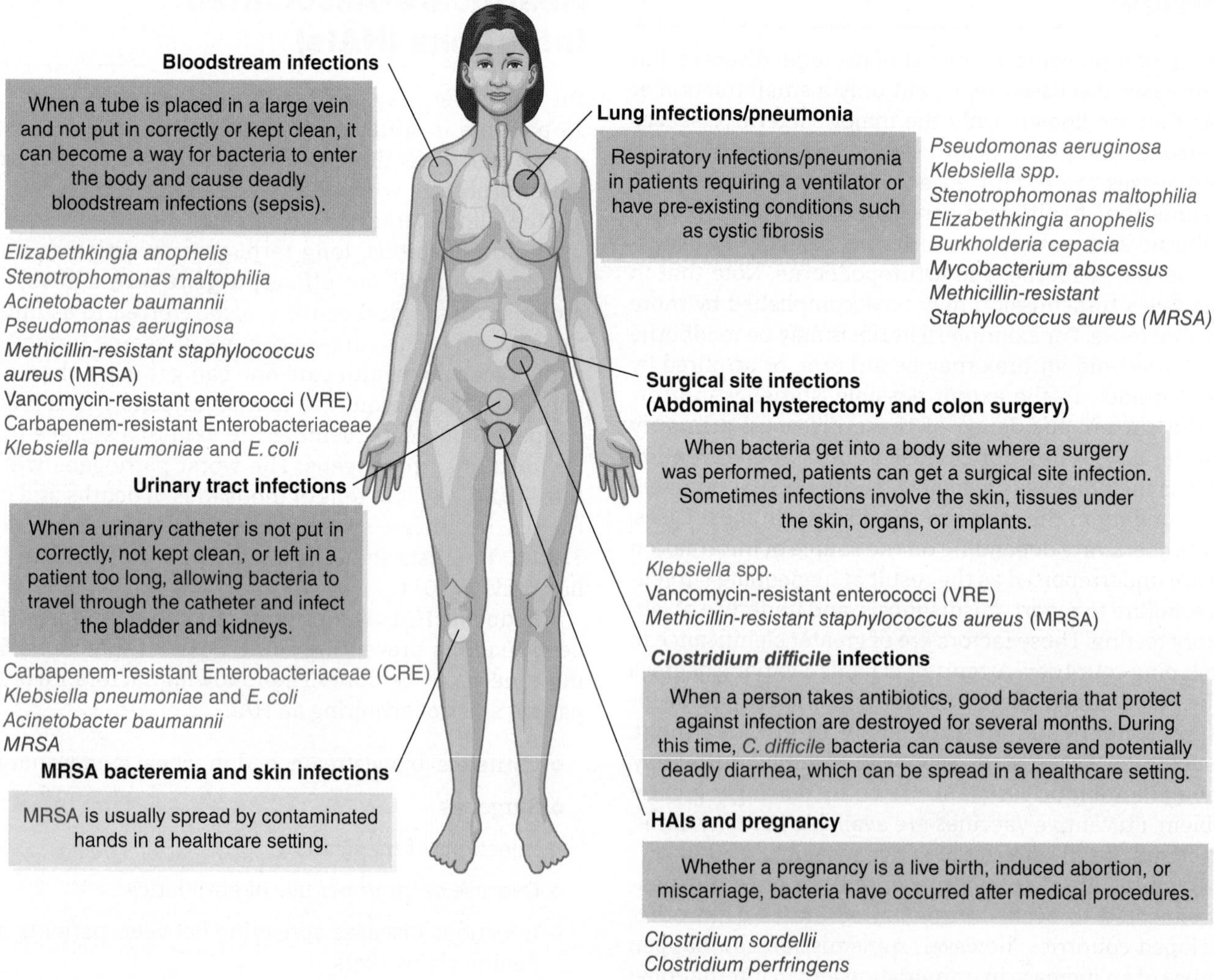

Bloodstream infections

When a tube is placed in a large vein and not put in correctly or kept clean, it can become a way for bacteria to enter the body and cause deadly bloodstream infections (sepsis).

Elizabethkingia anophelis
Stenotrophomonas maltophilia
Acinetobacter baumannii
Pseudomonas aeruginosa
Methicillin-resistant *staphylococcus aureus* (MRSA)
Vancomycin-resistant enterococci (VRE)
Carbapenem-resistant Enterobacteriaceae
Klebsiella pneumoniae and *E. coli*

Urinary tract infections

When a urinary catheter is not put in correctly, not kept clean, or left in a patient too long, allowing bacteria to travel through the catheter and infect the bladder and kidneys.

Carbapenem-resistant Enterobacteriaceae (CRE)
Klebsiella pneumoniae and *E. coli*
Acinetobacter baumannii
MRSA

MRSA bacteremia and skin infections

MRSA is usually spread by contaminated hands in a healthcare setting.

Lung infections/pneumonia

Respiratory infections/pneumonia in patients requiring a ventilator or have pre-existing conditions such as cystic fibrosis

Pseudomonas aeruginosa
Klebsiella spp.
Stenotrophomonas maltophilia
Elizabethkingia anophelis
Burkholderia cepacia
Mycobacterium abscessus
Methicillin-resistant
Staphylococcus aureus (MRSA)

Surgical site infections
(Abdominal hysterectomy and colon surgery)

When bacteria get into a body site where a surgery was performed, patients can get a surgical site infection. Sometimes infections involve the skin, tissues under the skin, organs, or implants.

Klebsiella spp.
Vancomycin-resistant enterococci (VRE)
Methicillin-resistant *staphylococcus aureus* (MRSA)

***Clostridium difficile* infections**

When a person takes antibiotics, good bacteria that protect against infection are destroyed for several months. During this time, *C. difficile* bacteria can cause severe and potentially deadly diarrhea, which can be spread in a healthcare setting.

HAIs and pregnancy

Whether a pregnancy is a live birth, induced abortion, or miscarriage, bacteria have occurred after medical procedures.

Clostridium sordellii
Clostridium perfringens

FIGURE 12.1 HAIs by body site and type of infection. The names of the bacterial pathogens printed in purple are Gram (+) bacteria and the names of the bacterial pathogens printed in red are Gram (−) bacteria.

HAIs Caused by Gram-Negative Bacteria

Almost one-third of all HAIs and 60% of HAIs in intensive care units (ICUs) are caused by Gram-negative (–) bacteria. The Gram (–) bacteria causing the most HAIs are:

- *Stenotrophomonas maltophilia*
- *Pseudomonas aeruginosa*
- *Acinetobacter baumannii*
- *Burkholderia cepacia*
- Carbapenem-resistant *Enterobacteriaceae* (CRE)

Gram (–) bacteria are becoming increasingly resistant to antibiotics. When patients are exposed to the hospital environment multiple times, they may be at greater risk of acquiring or developing antibiotic-resistant Gram (–) bacterial infections.

O Stenotrophomonas maltophilia

Stenotrophomonas maltophilia (formerly known as *Pseudomonas maltophilia*) is an emerging opportunistic pathogen. An opportunistic pathogen is a microbe that normally does not cause an infection unless a host has a weakened immune system. It can cause a variety of infections but most commonly causes pneumonia in patients with chronic lung diseases such as **cystic fibrosis (CF)** (**TABLE 12.2**) and causes death in 14% to 69% of patients. Cystic fibrosis (CF) is a genetic disease caused by mutations in the cystic fibrosis conductance regulator (CFTR) gene. Patients with CF cannot effectively clear mucus in their airways. Mucus clogs the airways and traps bacteria, leading to lung damage and, ultimately, respiratory failure.

S. maltophilia is an environmental microbe commonly found in water, including tap water and bottled water. Odds are that it could be isolated from your kitchen sink or showerhead! It is often isolated from soda machine fountains, washed salads, ice machines, nebulizers, endoscopes, and a myriad of other sources (Table 12.2; **FIGURE 12.2**). *Stenotrophomonas* has the ability to adhere

© Teri Shors.

FIGURE 12.2 Waterborne bacteria such as *S. maltophilia, Pseudomonas aeruginosa*, and other Gram (–) bacteria have been isolated from sink drains, tap water, reverse osmosis systems, showerheads, and bottled water. They have been identified as sources of emerging opportunistic bacterial pathogens that cause HAIs. This kitchen sink contains both tap water (left faucet) and a reverse osmosis system (right faucet) to further purify drinking water. The reverse osmosis system forces contaminants in water through a fine membrane, leaving the contaminants behind, which are then flushed down the drain. The system should be maintained regularly (filter changes). *Always remember that tap water is not sterile!* The microbes found in tap water do not cause disease in healthy people but can be deadly for people with weakened immune systems.

well to plastics, forming **biofilms**. Biofilms have been identified on the surfaces of intravenous catheters, prosthetics, dental waterlines, and nebulizers. Infections with bacterial biofilms have emerged as a major public health concern because the biofilm-growing bacteria are highly resistant to antibiotics and the body's immune defenses. Experts are concerned that climate change could potentially impact the spread of waterborne pathogens such as *S. maltophilia* that can flourish at warmer temperatures, resulting in more infections and challenges to the safety of the drinking water supply.

TABLE 12.2 Sources of *Stenotrophomonas maltophilia* and Its Associated Infections

Types of *S. maltophilia* infection	Pneumonia,* sepsis, soft tissue and skin infections, cellulitis, osteomyelitis, meningitis, endocarditis, keratitis of the eye, urinary tract infection
Clinical Sources of *S. maltophilia*	Suction tubing, endoscopes, electronic ventilator sensors and circuits, nebulizers, dental suction hoses, contaminated disinfectants and handwashing soaps, tap water, water fountain drains, sink drains, cystic fibrosis patient coughs (aerosolized), ice machines, water faucets, showerheads
Environmental Sources of *S. maltophilia*	Showerheads, tap water and bottled water, washed salads, soda fountain machines, water treatment plant process, sinkholes, river water, water fountains and sink drains, home-use nebulizers of cystic fibrosis patients, contact lens solution, plant rhizosphere, snakes, goats, buffalo, deep sea vertebrates

*Most common infection.
Information from J. S. Brooke. (2012). *Stenotrophomonas maltophilia*: An emerging global opportunistic pathogen. *Clinical Microbiology Reviews, 25*, 2–41.

Pseudomonas aeruginosa

Pseudomonas aeruginosa infections usually occur in patients in the hospital who are on breathing machines or have devices such as catheters and/or with weakened immune systems. Respiratory infections caused by *P. aeruginosa* are the leading cause of infection and death in CF patients. Patients with wounds from surgery or from burns also are highly susceptible to serious, life-threatening *P. aeruginosa* infections. *Pseudomonas* can be spread on the hands of healthcare workers or by equipment that becomes contaminated and is not properly cleaned. CDC experts estimate that approximately 51,000 *P. aeruginosa* HAIs occur every year in the United States. Of these, more than 6,000 (13%) are multidrug resistant. *P. aeruginosa* infections are responsible for roughly 400 deaths each year.

P. aeruginosa HAIs have been linked to hospital water sources (it forms biofilms in water systems). *P. aeruginosa* can be isolated from tap water, sinks, drains, toilets, and showers (**FIGURE 12.3A**). Healthy individuals can develop mild

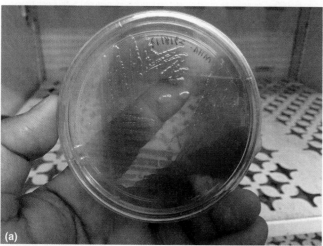

(a)

© Mohd Firdaus Othman/Shutterstock.

(b)

© Artazum/Shutterstock.

FIGURE 12.3 **(a)** Some strains of *Pseudomonas aeruginosa*, such as the colonies shown here, produce a green fluorescent pigment on Mueller-Hinton agar, a solid medium used to screen bacteria for antibiotic sensitivity in the medical technology laboratory. **(b)** Proper maintenance of hot tubs prevents the growth of contaminating bacteria such as *P. aeruginosa*, which can cause skin rashes, and Legionnaires' disease, caused by *Legionella*, a Gram (–) bacterium. Very serious *Pseudomonas* infections occur in people with cystic fibrosis or other chronic illnesses and weakened immune systems.

diseases caused by *P. aeruginosa* infections, especially after exposure to water. For example, *P. aeruginosa* has been found to cause ear infections in children, generalized rashes in people exposed to hot tubs or swimming pools that are not adequately chlorinated (**FIGURE 12.3B**), and, on occasion, eye infections in persons using extended-wear contact lenses.

Acinetobacter baumannii

Acinetobacter baumannii accounts for about 80% of the *Acinetobacter* infections that occur in ICUs and healthcare facilities with severely ill patients. *Acinetobacter* infections rarely occur outside of healthcare settings. It causes an opportunistic infection of the skin, soft tissues, lungs, urinary tract, and bloodstream. It has been dubbed as "Iraqibacter" because *A. baumannii* infections were suddenly reported among veterans and soldiers who served in Iraq and Afghanistan. Multidrug-resistant strains have spread to civilian hospitals.

Worldwide, *A. baumannii* accounts for up to 20% of all infections in ICUs. The bacterium can be found in water and isolated from patient's sputum, respiratory secretions, wounds, and urine (**FIGURE 12.4**). All strains are inherently resistant to multiple antibiotics.

Burkholderia cepacia

Burkholderia cepacia can be found in soil and water. These Gram (–) bacteria rarely cause infections in healthy people but can cause serious respiratory infections, including pneumonia, in CF patients and patients with weakened immune systems. Transmission of *B. cepacia* has also occurred through the use of contaminated mouthwash. In 2004, a *B. cepacia* outbreak occurred at a hospital in Texas. None of the patients had CF but all of them had been exposed to contaminated sublingual probes used to monitor carbon dioxide levels in tissues. Voluntary recalls of over-the-counter nasal sprays have occurred due to contamination with *B. cepacia*. *B. cepacia* can be spread through person-to-person contact, contact with contaminated surfaces, and environmental exposure. It is also associated with multidrug-antibiotic resistance.

Courtesy of the CDC/James Gathany; Jana Swenson.

FIGURE 12.4 CDC microbiologist and electron microscopist Janice Carr takes a scanning electron micrograph of *Acinetobacter baumannii*.

Courtesy of the National Institutes of Health.

FIGURE 12.5 Aerial view of the Clinical Center, NIH campus, Bethesda, Maryland. It was ground zero for a KPC outbreak that began in 2011. Medical researchers sequenced the DNA genomes of the KPC strains collected from patient specimens to trace the spread of the outbreak.

❍ Carbapenem-resistant *Enterobacteriaceae* (CRE)

Carbapenem-resistant *Enterobacteriaceae* **(CRE)** are a family of Gram (–) rod-shaped bacteria that cause difficult-to-treat infections because they are resistant to many antibiotics. Healthy people do not get CRE infections. CRE infections occur in immunocompromised patients on ventilators or who have intravenous catheters or are on a long course of certain antibiotics. Examples of bacteria in this family that cause HAIs are *Klebsiella* spp. and *E. coli*. Members of this family of bacteria are typical residents of the human gut that can become resistant to antibiotics. Roughly 15% of infections in hospital ICUs are caused by *K. pneumoniae*.

K. pneumoniae is a CRE known as **Klebsiella pneumoniae carbapenemase (KPC)**. **Carbapenemase** is an enzyme produced by certain bacterial pathogens that inactivate antibiotics such as penicillins, cephalosporins, monobactams, and **carbapenems**, making the bacteria resistant to the effects of the antibiotics. Infections caused by carbapenem-resistant bacteria have few treatment options and are associated with mortalities upwards of 50%. KPC is insidious; it silently colonizes the gastrointestinal tract without causing any signs or symptoms. Patients can be asymptomatic for long periods of time. They are silent **carriers** and **reservoirs** of KPC, spreading the bacteria to other patients and healthcare staff, making outbreaks difficult to control and stop. KPC can survive on the hands of hospital workers for several hours, facilitating HAIs.

Probably the most famous outbreak of KPC infections began in 2011, at the U.S. National Institutes of Health (NIH) Clinical Center located in Bethesda, Maryland (**FIGURE 12.5**). One might not expect an outbreak at such a prestigious place. The NIH is the largest medical research center in the United States, complete with a clinical hospital that opened its doors in 1953. To date, more than 500,000 patients have volunteered in clinical research studies. The first gene therapy treatments occurred at the NIH in 1990.

An outbreak began at the NIH Clinical Center after patient zero arrived on June 13, 2011, from an ICU in a New York City hospital. **Patient zero**, or the first patient of the outbreak, was discharged from the hospital 3 weeks before hospital personnel became aware that KPC was independently transmitted from patient zero to three additional patients. A total of 18 patients were infected by KPC, and 11 died. Of the 11 patients that died, 6 died as a result of KPC infection. The outbreak investigation was featured in the *Frontline* program "Hunting the Nightmare Bacteria" that aired on PBS television on October 22, 2013 (https://www.pbs.org/wgbh/frontline/film/hunting-the-nightmare-bacteria). Medical researchers sequenced the entire genomes of the KPC strains isolated from patients to piece together the origin and spread of the epidemic within the ICU wards at the NIH Clinical Center.

Foodborne and Waterborne Bacterial Diseases

After studying this topic you may become overly anxious about the safety of the foods you eat and the sources of the water you drink. Pathogens may be lurking in the kitchen, and it is certainly wise to take precautions and to use common sense; it is not necessary, however, to become paranoid! There are two levels of protection: (1) the immune system and (2) a system of surveillance practiced by national, state, and city health departments throughout the country. The U.S. Department of Agriculture (USDA), through its Food Safety Inspection Service (FSIS), is charged with ensuring the safety of the nation's commercial supply of meat, poultry, and egg products. The U.S. Food and Drug Administration (FDA) Food Safety Modernization Act (FSMA) was signed into law by President Obama on January 4, 2011. **FIGURE 12.6** illustrates the anatomy of the human digestive system, the usual target for foodborne and waterborne bacteria, resulting in **gastroenteritis** (vomiting and diarrhea).

Foodborne and waterborne infections have a significant impact on health in the United States and around the world; increasingly, more of the pathogens responsible for these infections are becoming resistant to antibiotics. According to CDC estimates, 1 in 6 Americans becomes sick from contaminated foods or beverages each year, and 3,000 die. The USDA estimates

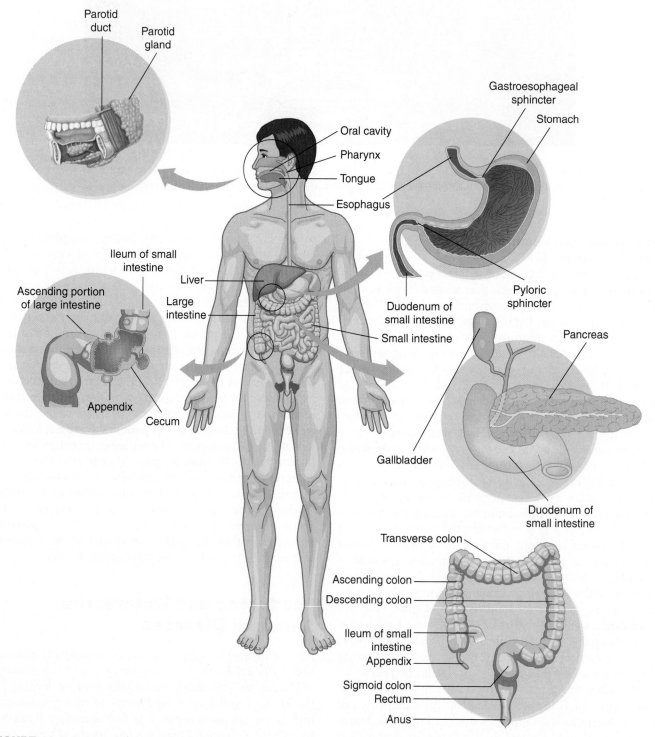

FIGURE 12.6 Anatomy of the human digestive system.

that foodborne illnesses cost the U.S. economy more than $15.6 billion each year. Many cases are undiagnosed and are sometimes passed off as a "stomach virus" (a nonexistent disease), signifying that these numbers are only the tip of the iceberg.

Product recalls in the United States and United Kingdom are increasing. Companies spend as much as $10 million in direct costs alone. The most common cause of food recall products is biological/microbial contamination. It is not uncommon for a major food preservation (canning or freezing) company or meat, chicken, or fish processing plant to issue an urgent warning regarding consumption of a particular food item.

Perhaps the most widely known outbreak in the last 25 years was the 1993 Jack-in-the-Box episode in the state of Washington. Patrons of the fast-food restaurant consumed undercooked hamburgers contaminated with *Escherichia coli* O157:H7. Three young children died, almost 70 children were hospitalized, and a total of 500 people became ill. More recently, in February 2007, ConAgra Foods recalled certain brands of peanut butter found to be contaminated with *Salmonella* bacteria; according to a CDC report over 628 people were sickened from consumption of the product in 47 states. More than 1 million pounds of ground beef were recalled in October 2007, by Cargill because of possible contamination with *E. coli*. In 2018, a multistate outbreak of *E. coli* O157:H7 infections occurred, with the source of the bacteria being romaine lettuce harvested in Yuma, Arizona. Contaminated canal water containing *E. coli* O157:H7 (present in human feces) was used to water the romaine lettuce growing in the field. It is unclear how the canal became contaminated. The 36-state outbreak resulted in 210 people confirmed infected. Of those affected, 96 were hospitalized and 27 developed **hemolytic uremic syndrome (HUS)**, a severe complication that causes kidney failure (**uremia**). Five deaths were reported from Arkansas, California, Minnesota, and New York.

Also in 2018, a multistate outbreak of *V. parahaemolyticus* infections occurred that was linked to fresh crab meat imported from Venezuela. The CDC recommended that consumers not eat, restaurants not serve, and retailers not sell fresh crab meat imported from Venezuela at that time. At least 12 people who ate fresh crab meat were infected; 4 were hospitalized. Illnesses were reported from Maryland, Louisiana, Pennsylvania, and Washington, D.C.

The riskiest foods and most common bacterial pathogens causing multistate outbreaks from 2010 to 2014 are provided in **TABLE 12.3**. Over the 5-year period, 120 outbreaks were recorded, with an average of 24 multistate outbreaks each year. A median of six states were involved in each outbreak. Salmonellosis was the largest cause of foodborne illness in the United States (63 cases, or 53%, of the total outbreaks in the 5-year period). The number of cases, hospitalizations, and deaths associated with the foodborne outbreaks are listed in **TABLE 12.4**.

Travelers to foreign countries need to be particularly careful about ingestion of food and water because the many problems of infrastructure, particularly in developing countries, do not ensure an optimal level of safety. It is easy to forget that ice cubes come from tap water or that thoroughly washing fruits and vegetables is to no avail if the water is contaminated. People who live in an area develop a level of immunity that travelers do not have and hence are less prone to get these diseases. Tourists' plans are frequently curtailed, and sightseeing becomes limited to the hotel room with an adjoining bathroom. A summary list of major foodborne and waterborne infections is presented in **TABLES 12.5** and **12.6**. Note that the types of foods and the symptoms are common denominators. "Boil it, cook it, or forget it" is an excellent rule.

The following are the "10 commandments" for reducing the risk of foodborne and waterborne infections:

1. Wash hands frequently, particularly after using the bathroom and after changing a baby's diaper.

2. Cook raw beef and poultry products thoroughly.

3. Avoid unpasteurized milk and juices and any food made from unpasteurized milk.

4. Properly wash food utensils, cutting boards, and raw fruits and vegetables, and frequently wash kitchen sponges in the dishwasher.

5. Beware of "double-dippers" (those who dip their cracker, celery, carrot stick, etc., into a spread, take a bite, and then dip the same item into the spread again).

6. Avoid eating at disreputable joints.

7. Do not eat raw shellfish.

8. Avoid food and drinks sold by street vendors, particularly when traveling abroad.

9. Do not share eating utensils, water bottles, or other such items.

10. Do not swim in contaminated waters.

Food Intoxication (Food Poisoning)

A distinction needs to be made between **food intoxication** (also called **food poisoning**) and **foodborne infection**. In the popular jargon, all food-related illnesses are falsely lumped together as food poisoning. Food intoxication refers to the ingestion of already-produced bacterial **toxins**; the bacteria may, in fact, no longer be present. Foodborne infection is the result of the ingestion of bacterial-contaminated foods and subsequent bacterial growth in the intestinal tract, secretion of bacterial toxins, and, possibly, invasion of the intestinal tract. The toxin is referred to as an **enterotoxin** (a toxin that affects the intestinal tract) and is responsible for the common, and very unpleasant, symptoms of nausea, vomiting, diarrhea, and, possibly, bloody stools and fever.

TABLE 12.3 Multistate Foodborne Outbreaks in the United States, 2010–2014

Food Category	Bacterial Pathogen				
	Salmonella	*E. coli* (producing Shiga toxin)	*Listeria*	*Vibrio*	Other Bacterial Pathogens
Fruits	13		3		1
Vegetable row crops	1	14			
Beef	5	8			
Sprouts	7	2	1		
Seeded vegetables	9				
Dairy		2	6		
Nuts/seeds	8	1			
Shellfish				6	1
Chicken	4				1
Fish	3	1			
Turkey	3				
Eggs	1				
Wild game		1			
Oils/sugars	1				
Pork	1				
Other	7	5	2		2
Total outbreaks	**63 (53%)**	**34 (28%)**	**12 (10%)**	**6 (5%)**	**5 (4%)**

Reproduced from S. J. Crowe, B. E. Mahon, A. R. Vieira, & L. H. Gould. (2015). Vital signs: Multistate foodborne outbreaks—United States, 2010–2014. *MMWR, 64,* 1221–1225.

TABLE 12.4 Number of Cases, Hospitalizations, and Deaths Associated with Multistate Foodborne Outbreaks in the United States, 2010–2014

Bacterial Pathogen	Cases		Hospitalizations		Deaths	
	Number	Percent	Number	Percent	Number	Percent
Salmonella	6,530	82%	952	65%	8	12%
E. coli (producing Shiga toxin)	636	8%	178	12%	1	2%
Listeria	271	3%	244	17%	57	86%
Vibrio	89	1%	6	<1%	0	0
Other	403	5%	80	5%	5	0
Total	**7,929**	**100%**	**1,460**	**100%**	**66**	**100%**

Reproduced from J. Crowe, B. E. Mahon, A. R. Vieira, & L. H. Gould. (2015). Vital signs: Multistate foodborne outbreaks—United States, 2010–2014. *MMWR, 64,* 1221–1225.

TABLE 12.5 Foodborne and Waterborne Bacterial Diseases Causing Food Intoxication (Food Poisoning)

Disease	Incubation Period (Time Before Onset of Symptoms)	Bacterial Pathogen	Source of Contaminated Food	Symptoms	Duration of Illness
Botulism	8–24 hours	*Clostridium botulinum*	Store-purchased contaminated foods, organic honey, home-canned products, illegal prison brew (pruno)	Double vision, blurred vision, drooping eyelids, slurred speech, difficulty swallowing, and muscle weakness that moves down the body	Variable
Clostridial food poisoning	8–16 hours	*Clostridium perfringens*	Meat, poultry, gravy, dried or precooked foods such as beans	Severe cramping, abdominal pain, watery diarrhea	24 hours
Staphylococcal food poisoning	1–6 hours	*Staphylococcus aureus*	Unrefrigerated or improperly refrigerated meats, potato and egg salads, cream pastries	Sudden onset of nausea and vomiting and abdominal cramps; diarrhea and fever may be present	24–48 hours
Fried rice (emetic syndrome)	1–5 hours	*Bacillus cereus*	Raw rice, fried rice at restaurants (storing boiled rice at room temperature or inadequate refrigeration above 4.4°C (40°F) or fried rice in buffet not maintained at 60°C (140°F)	Nausea and vomiting	12–24 hours
Bacillus cereus (diarrheal syndrome)	10–16 hours	*Bacillus cereus*	Meats, stews, gravies, vanilla custard sauce	Abdominal cramps, watery diarrhea, nausea	24–48 hours

Information from U.S. Food and Drug Administration. (n.d.). Foodborne illness-causing organisms in the U.S.: What you need to know. Fact Sheet. Retrieved from https://www.fda.gov/downloads/food/foodborneillnesscontaminants/ucm187482.pdf

TABLE 12.6 Foodborne and Waterborne Bacterial Diseases Causing Foodborne Infection

Disease	Incubation Period (Time Before Onset of Symptoms)	Bacterial Pathogen	Source of Contaminated Food	Symptoms	Duration of Illness
Salmonellosis*	6–48 hours	*Salmonella enterica* serotype *Enteritidis* and other species	Undercooked or raw eggs, poultry, meat, unpasteurized juice or milk, cheese, contaminated fruits and vegetables	Vomiting, abdominal cramps, diarrhea, fever	4–7 days

(continues)

TABLE 12.6 Foodborne and Waterborne Bacterial Diseases Causing Foodborne Infection (continued)

Disease	Incubation Period (Time Before Onset of Symptoms)	Bacterial Pathogen	Source of Contaminated Food	Symptoms	Duration of Illness
Shigellosis	4–7 days	*Shigella* spp.	Raw produce, contaminated drinking water, uncooked foods and cooked foods not reheated after contact with an infected food handler	Projectile diarrhea, fever, abdominal cramps, blood and mucus in stool	24–48 hours
Cholera	24–72 hours	*Vibrio cholerae*	Raw shellfish, water contaminated with fecal material of infected person	Severe and large volumes of diarrhea causing dehydration in a 24-hour period, muscular cramps, wrinkling of the skin	3–4 days if treated, but if untreated mortality rates as high as 50–60%
Traveler's diarrhea	1–3 days	*Escherichia coli* producing toxin	Water or food contaminated with human feces	Watery diarrhea, abdominal cramps, some vomiting	3–7 or more days
E. coli O157:H7 infection	1–8 days	*Escherichia coli*	Undercooked beef hamburgers, unpasteurized milk and juice, raw fruits and vegetables (e.g., sprouts, romaine lettuce, spinach), contaminated water	Severe, often bloody diarrhea, abdominal pain and vomiting, can lead to HUS/kidney failure	5–10 days
Campylobacterosis	2–5 days	*Campylobacter jejuni*	Raw and undercooked poultry, contaminated water, unpasteurized milk	Diarrhea, cramps, high fever, bloody stools	2–10 days
Listeriosis	9–48 hours for gastrointestinal symptoms, 2–6 weeks for invasive disease	*Listeria monocytogenes*	Vegetables, meats, cheese, ice cream, unpasteurized milk	Sore throat, fever, muscle aches, diarrhea; pregnant women may have flulike symptoms that lead to miscarriage or premature delivery; elderly may develop **bacteremia** or **meningitis**	Variable

Disease	Incubation Period (Time Before Onset of Symptoms)	Bacterial Pathogen	Source of Contaminated Food	Symptoms	Duration of Illness
Vibriosis	2–48 hours	*Vibrio parahaemolyticus*	Undercooked or raw seafood such as shellfish	Watery or bloody diarrhea, nausea, headaches, vomiting, fever, abdominal cramps	2–8 days
Vibriosis	1–7 days	*Vibrio vulnificus*	Undercooked or raw seafood such as shellfish, especially oysters	Healthy persons: vomiting, diarrhea, abdominal pain; High-risk individuals: sudden chills, fever, shock, bullae/skin lesions; can be fatal in persons with liver disease and weakened immune systems.	Variable

*Also associated with handling reptiles (turtles and iguanas). A multistate outbreak was associated with pet guinea pigs, 2015–2017.
Information from U.S. Food and Drug Administration. (n.d.). Foodborne illness-causing organisms in the U.S.: What you need to know. Fact Sheet. Retrieved from https://www.fda.gov/downloads/food/foodborneillnesscontaminants/ucm187482.pdf

O Botulism

Botulism—the most dangerous food intoxication—is a rare but serious disease caused by *Clostridium botulinum*, a Gram (+), anaerobic, **endospore**-forming, rod-shaped bacteria commonly found in soil. The **neurotoxin** produced by this organism is extremely deadly; it interferes with the passage of nerve transmitters (acetylcholine), resulting in muscle **paralysis**, including the diaphragm (a muscle) and the muscles of the ribs, leading to respiratory paralysis, that is, the inability to breathe. Death occurs within a day or two. Antibiotics are of no value in treatment because the problem is not one of infection but of intoxication. Treatment consists of **antitoxin therapy**, and in some cases mechanical ventilators are used. The number of cases in the United States is low, and the associated death rate in untreated cases is 70%. The telltale signs of *C. botulinum*–contaminated canned foods are bulging of the can and offensive odor and taste. The best defense is to heat foods for 10 minutes at 121°C (250°F), a process that destroys the bacterial neurotoxin.

An illegally produced alcoholic brew caused botulism outbreaks between 2004 and 2012 in prison inmates who consumed the brew while serving time in maximum security facilities located in Arizona, Utah, and California (**TABLE 12.7**). The prison moonshine is also called **pruno**,

TABLE 12.7 Botulism Outbreaks in Prisons Linked to Drinking Pruno, 2004–2012*

Year	State	Number of Cases	Age Range (Years)	Number Hospitalized	Number Intubated
2004	California	4	19–35	4	2
2005	California	1	30	1	1
2011	Utah	8	24–35	8	3
2012	Arizona	4	27–33	4	1
2012	Arizona	8	20–35	8	7

*No deaths were reported.
G. Briggs, K. K. Komatsu, S. Anderson, & S. Yasmin. (2012). Botulism from drinking prison-made illicit alcohol—Arizona, 2012. *MMWR, 62*(5), 88.

hooch, juice, jump, raison jack, chalk, and buck. The prisoners smuggled unpeeled potatoes (which contained soil with *C. botulinum* endospores that germinated in the brew) from the kitchen, along with apples, old peaches, jelly, and ketchup as a source of sugar. Sources of natural yeast include bread, corn, rice, and, in this case, potatoes. The mashed ingredients were added to hot water and placed in a plastic bag. Natural yeasts fermented the mixture of foods into alcohol and other bacterial end products such as carbon dioxide. The pruno was described as magenta in color, "*smelled like baby poop and tasted like vomit.*"

A quick Internet search will turn up many recipes for pruno. This was the first time that unpeeled potatoes were used to make pruno. It could be that crackdowns to prevent stealing of food from the prison kitchen drove inmates to search for alternative ingredients. Their desperation could have killed them! The potatoes were identified as the source of *C. botulinum* or the source of the botulinum neurotoxin. Prison kitchens no longer serve baked potatoes. All potatoes must be mashed. All inmates who drank the brew were hospitalized and treated with botulinum antitoxin. Wound botulism caused by injecting drugs using dirty and shared needles is also on the increase.

Although home-canned string beans, peppers, asparagus, sausage, cured pork and ham, smoked fish, and canned salmon are the most common sources of intoxication, commercially available products also can be a source. Four cases of botulism occurred in Florida and Georgia in September 2006; the cause was identified as Bolthouse Farms carrot juice. In July 2007, four cases of botulism—two in Texas and two in India—were reported after ingestion of chili sauce from Castlebury's Food Company. *Organic honey can contain spores of* C. botulinum *and should not be fed to children younger than 1 year of age.*

Ironically, minute doses of botulinum toxin, called **Botox**, have been used to treat a variety of common disorders associated with muscle overactivity, including strabismus (crossed eyes), stuttering, and uncontrolled blinking. Botox is advertised as "*from toxin to therapeutic agent.*" In May 2001, the journal *Neurology* reported on a study of a small number of patients in which it was claimed that treatment with Botox helped sufferers of chronic low back pain. Botox is also used cosmetically by injection into the skin to relieve wrinkling. Unfortunately, unscrupulous practitioners can cause serious problems, as happened to four persons who were injected with an unlicensed botulinum toxin; their symptoms were consistent with those of naturally occurring botulism. The unlicensed **botulinum toxin** prescription was almost 3,000 times the estimated human lethal dose.

Clostridium perfringens is associated with gas gangrene, a condition that devastates soldiers wounded in battle because of wound contamination with particles of soil containing spores. This bacterium is also an important cause of food poisoning (but some consider it a food infection). It is commonly associated with endospore contamination of meat, poultry, and beans. Recovery usually occurs within 24 hours, and there is no specific therapy.

In 1993, "wearing of the green" celebrations ended with outbreaks of gastroenteritis associated with contaminated corned beef served at St. Patrick's Day meals in Virginia and Ohio. You may recall that the genus *Clostridium* is an endospore former, and that endospores survive the high temperatures used in cooking and germinate as multiplying bacteria during cooling.

Staphylococcal Food Poisoning

Staphylococcal food poisoning is the most common type of food poisoning, and many cases remain undiagnosed. You probably have heard of this condition as "staph food poisoning" or by an older and incorrect term, **ptomaine poisoning**. *Staphylococcus aureus,* a Gram (+), coccus-shaped bacteria arranged in grapelike clusters, secretes an enterotoxin that produces intestinal tract symptoms, including abdominal cramps, nausea, vomiting, and diarrhea (**FIGURE 12.7**). The symptoms may be so severe that you wish you were dead, but don't have the luxury. There is hardly anyone who has not suffered from this; all you can do is wait it out and you will feel better within several hours. The list of potential food sources is large and includes coleslaw, potato salad, fish, dairy products, cream-filled pastries, and spoiled meats. How does contamination occur? The main **reservoir** for *S. aureus* is the nose, but the bacterium also causes boils, abscesses, or pimples on the skin, and these lesions are the usual source of seeding the staphylococci into food. A few hours under the sun in an improperly iced cooler is all it takes to allow for multiplication of the bacteria and enterotoxin production.

Bacillus cereus Fried Rice (Emetic) Syndrome and Diarrheal Syndrome

Bacillus cereus is a Gram (+) rod that produces endospores that can cause foodborne intoxication associated with more than one type of enterotoxin. Illness associated with *B. cereus* is likely underestimated because the duration of

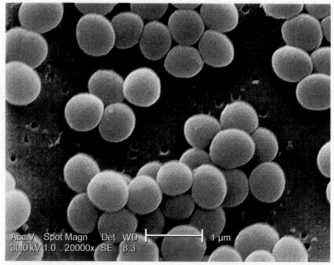

FIGURE 12.7 Colorized scanning electron micrograph of *Staphylococcus aureus.*

© Teri Shors.

FIGURE 12.8 *Bacillus cereus* can be isolated from raw rice and is the cause of fried rice emetic syndrome and diarrheal syndrome. The photograph shows one of the most commonly contaminated foods associated with emetic syndrome, chicken fried rice. Outbreaks have occurred when boiled rice was stored at room temperature for subsequent preparations of fried rice or fried rice was not appropriately maintained at the correct temperature on buffets.

symptoms on average is only 24 hours. *B. cereus* causes two types of foodborne illness: emetic and diarrheal syndromes. The emetic syndrome is caused by a heat- and acid-stable enterotoxin. An individual who consumes food, typically contaminated chicken fried rice at a Chinese restaurant buffet or by takeout or delivery of food ordered from restaurants, will experience nausea and vomiting 30 minutes to 5 hours after eating (**FIGURE 12.8**). The emetic syndrome resembles symptoms of those that occur with *S. aureus* intoxication or food poisoning.

Fried rice diarrheal syndrome is caused by a different enterotoxin that are produced during the multiplication of *B. cereus* in the small intestine. Symptoms include abdominal pain, watery diarrhea, seldom vomiting, and no fever. The diarrheal syndrome is more common in the United States and in Europe, in contrast to **fried rice emetic syndrome**, which is more common in Japan (where rice is a food staple in the diet of most Japanese families). The infective dose associated with *B. cereus* food poisoning ranges from 10,000 to 10^{11} bacteria.

Nonfoodborne illnesses have also been associated with *B. cereus*. Cases of intravenous drug users who have injected heroin contaminated with *B. cereus* endospores have been documented. It has caused bacteremia and **endocarditis** (inflammation of the heart) in injection drug users.

Foodborne and Waterborne Infection

O Salmonellosis

How do you like your eggs? If you answered over easy or sunny side up or if you like your Caesar salad dressing made the old-fashioned way with raw eggs, you are at risk for **salmonellosis** caused by *Salmonella enteritidis*–infected eggs.

Salmonella species are Gram (–) rods that are members of the *Enterobacteriaceae* family. Salmonellosis is a significant public health problem, in that it is a serious and potentially fatal infection that afflicts about 40,000 people; however, the actual numbers may be 30 times greater (1.2 million people) because milder cases are not diagnosed. Six hundred deaths from salmonellosis are reported each year in the United States. Undoubtedly, this figure can be increased by a factor of three because many cases are not reported.

The popularity of raising backyard chickens has been linked to salmonellosis outbreaks. In 2018, the CDC investigated multistate outbreaks linked to contact with chicks and ducklings purchased from specific hatcheries. In July 2018, 138 cases were reported from 44 states that were related to raising backyard chickens and ducks. Chickens and other poultry may carry *Salmonella* spp. that get inside of eggs before the shells are formed. Eggs can also become contaminated from poultry droppings. It is not recommended to wash eggs that have been collected because colder water can pull the *Salmonella* bacteria into the egg.

Salmonella infection is also associated with contaminated meats (especially chicken), eggs, seafood, and unpasteurized fruit juice. Lately, fresh produce has become an increasingly significant source of salmonellosis (and other foodborne bacteria), indicating a need for better growing and harvesting practices on the farm. Lettuce, raspberries, tomatoes, cantaloupes, precut watermelon, and alfalfa sprouts contaminated with *Salmonella*, *E. coli*, or other microbes have made it onto store shelves.

If you purchase a reptile or rodent as a pet you may be buying trouble as well. Iguanas and a variety of lizards, snakes, turtles, and guinea pigs are popular pets, but they can be asymptomatic or healthy carriers of a variety of *Salmonella* species (**FIGURES 12.9** and **12.10**). In 2008, an outbreak involving 103 cases distributed across 33 states was traced back to exposure to turtles.

One recommendation of the CDC is that pet store owners advise potential reptile owners of the risk of salmonellosis, particularly in children. Human salmonellosis has been associated with dog and cat pet treats, such as beefy munchies, bully sticks, ground turkey pet food, kitten grind, and beef toppers in the United States and Canada for several years. The common denominator is that in all the cases the persons infected had handled salmon or beef dog treats. In 2017, the CDC investigated a multistate outbreak of *Salmonella enteritidis* linked to

Krasner's Notebook

At about the age of 2 my daughter became ill and required hospitalization. At the time I was on a sabbatical leave at Georgetown University School of Medicine in Washington, D.C. As crazy as it may sound, my wife and I were relieved when the diagnosis was salmonellosis infection and, therefore, that the prognosis was excellent. The likely culprit was a small pet turtle that we had purchased a few weeks earlier. My wife, in a moment of intense relief, flushed the turtle down the toilet before I had the opportunity to culture from it—a foiled epidemiological study!

Courtesy of the CDC/Janice Haney Carr.

Courtesy of the CDC/Eric Grafman.

© Teri Shors.

© Teri Shors.

FIGURE 12.9 **(a)** Colorized scanning electron micrograph of *Salmonella* species. 10,400× magnification. Many species of birds, animals, and reptiles are carriers of *Salmonella* spp. **(b)** Chickens harbor *Salmonella* spp. in their intestines; the bacteria are then shed in droppings or feces. **(c)** Crows are carriers of *Salmonella* spp. **(d)** Bird feeders can become contaminated by bird droppings of birds that carry *Salmonella*, causing salmonellosis among some of the feeding birds. Wash your hands after handling bird feeders and bird baths!

pet guinea pigs. A total of 9 cases were reported from eight states dating back to 2015, with one hospitalization and no deaths. *Children younger than 5 years, the ones most desirous of having a pet, are the group most likely to acquire infection.*

Crows, those big, black birds, have increased in population in recent years, and it is now known that crow

feces carry *Salmonella*. (Think about that when a flock flies overhead, and wear a cap!) The most common infectious disease of birds that eat at bird feeders is salmonellosis. Healthy birds get infected when they eat at crowded feeders where bird carriers contaminate the feeder with their droppings. *Wash your hands thoroughly after touching a bird feeder or birdbath (Figure 12.9)! Remember that salmonellosis*

Courtesy of the CDC/Eric Grafman.

Courtesy of the CDC/Christine Prue.

© Teri Shors.

FIGURE 12.10 Pets have been associated with *Salmonella* outbreaks among children. **(a)** Reptiles such as turtles carry *Salmonella*. This photograph shows turtles on display at a pet store. **(b)** Amphibians such as the African dwarf frogs depicted in this photograph caused a multistate outbreak of salmonellosis in 2009. The frogs were carriers of *Salmonella*, serotype *typhimurium*. **(c)** Guinea pigs were linked to a multistate outbreak of salmonellosis caused by *S. enteritidis* infections. The outbreak began in 2015. A guinea pig in a pet store is pictured here. Children younger than 5 years, older adults, or people with weakened immune systems should use caution when handling or touching pets. These groups are at higher risk for infection and serious illness. Remember that pets can be healthy carriers, showing no symptoms of illness.

is a foodborne infection and requires the presence of bacteria that multiply in the intestinal tract, whereas staphylococcal and clostridial food intoxication, although incorrectly described as food poisoning, results from ingestion of already-formed toxins.

Salmonella typhi causes typhoid fever. The organism is able to survive in sewage, water, and certain foods, and gains access into the body through the ingestion of fecally contaminated food products; hands contaminated with fecal material, including some imported foods; and, rarely, sexual activities involving the anal area. Salmonella is transmitted by flies and fomites. After ingestion the organism invades the lining of the small intestines, causing ulcers and the passage of blood in the feces, accompanied by fever, possibly delirium, and the presence of rose-colored spots on the abdomen as a result of hemorrhaging in the skin.

Once an important disease worldwide, its prevalence has markedly decreased in developed countries as the result of sanitary engineering and sewage treatment facilities, luxuries not available in developing countries. In the early 1900s, **typhoid fever** was a major killer, with over 20,000 cases occurring annually in the United States. Approximately 300 cases are now reported annually to the CDC, about 75% of which are acquired during overseas travel. Some 22 million people and 200,000 deaths in developing countries become infected with S. typhi each year.

About 5% of those infected do not become ill. These healthy carriers of S. typhi present a public health problem, although antibiotic therapy usually renders the carrier free of the bacteria. In some cases surgery to remove the gallbladder, a site where the bacilli sometimes take up residence, may be performed to eliminate the carrier state. "Typhoid Mary" (Mary Mallon) was a healthy carrier of S. typhi.

Shigellosis

Shigellosis, an acute infection of the lining of the intestinal tract caused by various species of Shigella, a Gram (−), rod-shaped bacteria that is a member of the Enterobacteriaceae family. The infection is manifest by the usual gastroenteritis symptoms with the possible addition of dysentery, which may be bloody. Sources include eggs, shellfish, dairy products, vegetables, and water. Transmission is via the fecal–oral route. The **infective dose** of just 10 to 200 bacteria is enough to cause disease. As in all diseases involving loss of large volumes of water, dehydration is the major cause for concern and can lead to death via circulatory collapse. Treatment involves rehydration, either oral or intravenous, and possibly antibiotics. Estimates are that approximately 500,000 reported cases appear each year in the United States, but the actual figure may be as much as 20 times more because of unreported cases. **World Health Organization (WHO)** statistics indicate that in the developing world approximately 165 million cases and 1 million deaths occur annually.

Day care centers, because of their very young population, are potentially prone to outbreaks, necessitating that personnel be particularly diligent in handwashing and diapering procedures as measures of both prevention and control. Three outbreaks occurred in 2006 associated with day care centers in Kansas, Kentucky, and Missouri. In most cases the Shigella isolates were antibiotic resistant. In the three reported outbreaks the median ages were 7, 6, and 4 years, respectively; the case total was 994.

Cholera

Vibrio cholerae, a Gram (−) curved or comma-shaped rod, is the cause of **cholera** (FIGURE 12.11). This bacterium, like so many other intestinal tract pathogens, secretes an

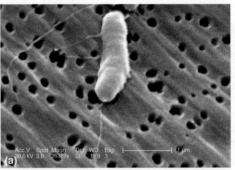

Courtesy of the CDC/Janice Haney Carr.

Courtesy of the CDC.

FIGURE 12.11 (a) Colorized scanning electron micrograph of *Vibrio vulnificus* (*Vibrio cholerae* looks similar). It is a curve-shaped rod that has a single polar flagellum. Magnification 26,367×. **(b)** Poverty, overcrowding, and primitive sanitation are a dangerous trio favoring the emergence and dissemination of this waterborne disease. This photograph shows a Trianon, Haiti, community member detaching buckets from the rear of his motorcycle to collect and transport water from a spring in the village. The water will be tested by National Directorate for Drinking Water and Sanitation (DINEPA) technicians to determine if it is safe to drink. DINEPA was created to address the challenges to lack of water regulations for rural areas in Haiti.

exotoxin that causes diarrhea. As much as 1 liter of fluid can be lost per hour for several hours. A liter is approximately equal to 1 quart, and a human has a total of 5 to 6 liters of blood, which is about 80% water. Unless rehydration is instituted quickly, death results in only a few hours; the mortality rate reaches 70% in untreated cases. The massive dehydration causes the eyes to take on a sunken appearance, the skin becomes dry and wrinkled, and muscular cramps in the legs and arms are a common complaint. Infection is acquired by the ingestion of

water or food contaminated by fecal material from *Vibrio*-infected persons.

Cholera reaches epidemic proportions in developing countries (BOX 12.1). Fortunately, the terrible loss of life from cholera throughout the world can be prevented through oral rehydration therapy. Prepackaged mixtures of sugar and salts are available at a very minimal cost, are readily transported to remote areas, and negate the use of intravenous injection and the associated costs that make such therapy prohibitive in countries where it is most needed.

BOX 12.1 Double Disaster Strikes Haiti

It was all over in less than a minute on January 20, 2010. The Earth trembled, trees swayed and snapped, buildings shook, houses were devastated, and people scrambled seeking refuge. An estimated 1.5 million Haitians were suddenly left homeless and 300,000 were killed by the earthquake that struck Haiti, an underdeveloped and impoverished nation.

The only good that arose from the rubble of the earthquake disaster was the outpouring of generosity displayed by private citizens and international agencies from around the world. Nevertheless, large numbers of Haitians fled to camps for **internally displaced persons**, or **IDPs**, and as so often happens, relatively few dollars trickled to those who needed it the most, namely, the inhabitants of the camps. Hygienic conditions were almost nonexistent and were an embarrassment to the world as well as a violation of human rights. Several months after the earthquake about half of the camps were still without basic services such as water and toilets. People slept under tarpaulins, sheets, or cardboard or out in the open; tents were a limited luxury that few possessed.

The sanitary conditions in the country set the stage for an outbreak of infectious disease that was bound to happen, and it did. In late October 2010, 9 months after the earthquake, cholera was confirmed in the country; within only several months thousands had to be hospitalized and 5,000 people died. The Haitian government feared that there would be more than 400,000 cases within a year's time.

This was the first cholera outbreak in Haiti in decades, leading to speculation as to the source of cholera bacteria. On this point, the scenario becomes contentious and sticky; no country or persons wants to be identified as the source of this fecally transmitted pathogen. One explanation is that United Nations peacekeeping troops from Nepal were quartered in an area of Haiti that had a leaky sewage system, and that the Nepalese troops were carriers of *Vibrio cholerae* bacteria present in their feces. The human feces was discharged into the drinking water through the leaky sewage systems. Nepal had recently suffered three waves of cholera. In this circumstance it would be like throwing kindling into a box of red hot coals. The Nepalese government was outraged and denied the allegation despite the fact that the cholera strain isolated from Haitians was much like the one isolated in Nepal. Other epidemiologists posed that the cholera bacteria had been dormant in waters for many years,

and that perhaps the breakdown in public health infrastructure resulting from the earthquake was responsible.

The crisis regarding the source of the cholera was defused with the attitude "It is what it is," and attention focused on recovery and prevention. The actual source of the *V. cholerae* bacteria was never fully determined. Three years after Haiti's devastating earthquake, technicians from the National Directorate for Drinking Water and Sanitation (DINEPA) were testing rural drinking water supplies and mapping water resources for future use (Figure 1).

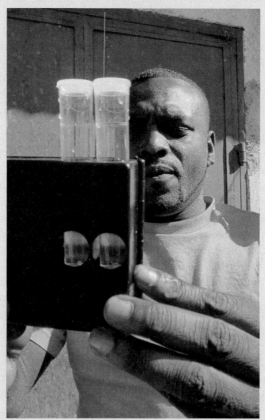

Courtesy of the CDC.

Figure 1 DINEPA technician testing water samples. The technician is looking for a color change in the water that would indicate the level of chlorine in it. Chlorine was added to rural water supplies to disinfect it for use as potable water.

Pandemics of cholera have been documented for hundreds of years, and it seems that no continent has escaped. In 2004, worldwide, more than 100,000 cases were reported and approximately 2,300 deaths. In the United States and in other industrialized countries, because of advances in water and sanitary engineering, cholera is rare. Outbreaks occurring in the United States are usually the result of tourists returning from Africa, Southeast Asia, or South America; in some cases the source has been contaminated seafood brought back to the United States. Occasionally, notices are posted in newspapers citing the closing of shellfish beds and beaches because of fecal pollution and the danger of *V. cholerae* and other fecal pathogens.

The genes for the enterotoxin produced by *V. cholerae* were identified in 1992 by scientists at the University of Maryland. Removal of these genes renders the bacteria nonpathogenic and candidates for a vaccine. The FDA approved a single-dose live, oral cholera vaccine called Vaxchora in the United States. The vaccine is intended for 18- to 64-year-olds who are traveling to an area of active cholera transmission. No country or territory currently requires vaccination to prevent cholera as a condition for entry. Vaxchora reduces the risk of severe diarrhea by 90% at 10 days after vaccination and by 80% at 3 months after vaccination. Three other vaccines—Dukoral (manufactured in Sweden), ShanChol (manufactured in India), and Euvichol-Plus/EuvicholR (manufactured in South Korea)—are available and prequalified by the WHO, but they are not yet available in the United States. The vaccines are not 100% effective in preventing cholera.

Noncholera Vibriosis

Besides being Gram (–) curved rods with a single polar flagellum, *Vibrio* species contain two circular chromosomes. Recall that most bacteria possess a single circular chromosome composed of double-stranded DNA. Over 100 species of *Vibrio* have been described. About a dozen of these cause disease in humans. Besides *V. cholerae*, three important species complete the "big four" human *Vibrio* pathogens: *V. parahaemolyticus* and *V. vulnificus* (see the opening case study and Figure 12.11A) and *V. alginolyticus*. **Vibriosis** causes an estimated 80,000 illnesses and 100 deaths in the United States every year. Most people are infected by consuming raw or undercooked seafood or through exposure to seawater during the months of May through October when water temperatures are warmer.

V. vulnificus is a common inhabitant of estuaries and coastal waters. It can cause foodborne illness, necrotizing wound infections, and sepsis. It is an important foodborne pathogen, and it is the leading cause of seafood-related mortality. Most mortalities occur in people with liver disease and immune suppression.

V. parahaemolyticus is the most prevalent foodborne pathogen associated with seafood consumption (especially shellfish) that causes acute gastroenteritis. Common symptoms are abdominal cramps, nausea, headaches, diarrhea, fever, and chills (Table 12.6). In recent decades *V. parahaemolyticus* has caused numerous large-scale outbreaks worldwide. In 2018, a multistate outbreak was linked to fresh crab meat contaminated with *V. parahaemolyticus* imported from Venezuela.

V. alginolyticus is an emerging pathogen. It is ubiquitous in seawater. It is not a foodborne pathogen. It tends to cause wound and ear infections that are resolved through antibiotic therapy. Epidemiological data suggest that the rapid increase of these infections in the United States and Europe may be related to climate change.

Escherichia coli

Escherichia coli, a Gram (–) rod, is a member of the *Enterobacteriaceae* family. It is frequently a news item because of an enterotoxin-producing strain called *E. coli* O157:H7, first recognized in the United States in 1982 (**FIGURE 12.12**). It is a highly significant cause of foodborne illnesses, and a number of known outbreaks have occurred. What is *E. coli*? The genus is named after German-Austrian pediatrician Theodor Escherich (1857–1911), who described the bacterium in the 1880s. The species (*coli*) indicates the location. The bacteria are residents of the normal **microbiota** of the large intestines (colons) of humans and other animals. Bacteria, particularly *E. coli*, account for much of the weight of fecal material; these normal inhabitants of the intestinal tract are beneficial because they produce vitamins that are absorbed into the body.

So why the bad rap for *E. coli* O157:H7, just one of the more than 100 strains of *E. coli* in the intestines of cattle and other animals? Somewhere and somehow, *E. coli* picked up the genes responsible for the production of an enterotoxin known as **Shiga toxin**, which produces severe damage, possibly to the point of death. Shiga toxin is encoded by a gene in *Shigella dysenteriae* that managed to find its way into the chromosome of certain strains of *E. coli*. More than half the states in the United States require that isolation of *E. coli* O157:H7 be reported to the

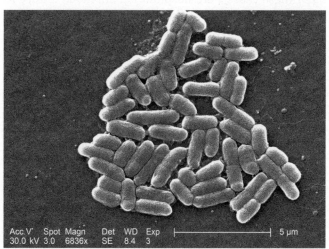

Courtesy of Janice Haney Carr/CDC.

FIGURE 12.12 Colorized scanning electron micrograph of *E. coli* O157:H7.

state health department for monitoring purposes in an effort to prevent outbreaks.

The consumption of ground beef is particularly dangerous, because contamination of the meat with fecal material may occur during the slaughtering process. The bacteria are mixed throughout large batches of meat obtained from many cattle when it is ground, in contrast to steaks, each of which is from a single source. With a little imagination the expression "One bad apple spoils the barrel" can be applied. The best advice is to avoid rare meat, particularly hamburgers. Many fast-food chain restaurants now specify the cooking time and temperature as a part of their policy and will not serve rare hamburgers. In most cases recovery from E. coli infection occurs within 5 to 10 days without specific treatment. Children 5 years and younger are at a greater risk of death due to E. coli O157:H7 due to enterotoxin destruction of red blood cells, low platelet counts, and kidney failure, a condition known as hemolytic uremic syndrome (HUS).

Note that ground beef and other meats are not the only carriers. Germany was under the gun in the spring and summer of 2011, when an outbreak of a new strain of E. coli, designated O104:H4, caused an outbreak of gastroenteritis. The strain is similar to O157:H7; it produces the Shiga toxin that can cause kidney, circulatory, and neurological symptoms. Initially, the source of the outbreak was an epidemiologist's nightmare. Cucumbers from Spain were first thought to be the source, much to the chagrin of that country, but, ultimately, the cause was pinpointed to contaminated sprouts. How the sprouts became contaminated is not clear. The end result was that in Germany 3,785 became sickened, causing hospitals to be overwhelmed; 45 people died. Spread of the bacteria to other European countries was limited to 91 confirmed cases and 1 death.

The consumption of contaminated fresh spinach caused a particularly serious outbreak in September 2006, resulting in illness in 183 persons in 26 states and 1 death; 95 of these cases were hospitalized. Shredded lettuce in Taco Bell restaurants in the Northeastern United States was responsible for 71 cases of E. coli O157:H7 distributed over five states; 53 persons were hospitalized, and 8 cases of HUS resulted.

In addition to O157:H7, characterized as enterohemorrhagic E. coli, several other strains of E. coli exhibit variation in the mechanism by which they cause diarrhea. The enterotoxigenic E. coli strains are one example. This group is the most common cause of **traveler's diarrhea**.

When you hear that a particular beach is closed because of fecal pollution, you now know that it is probably because of the presence of large numbers of E. coli. This bacterium is easy to detect, making it useful as an indicator organism in water quality, and its presence signifies that intestinal pathogenic bacteria may be present. Too bad about O157:H7 damaging the reputation of the usually nonpathogenic workhorse E. coli, probably the most widely investigated species of life. An E. coli culture (nonpathogenic) can be purchased for only several dollars from a variety of biological supply houses and is widely used in laboratories in high schools and colleges.

Campylobacteriosis

Campylobacter is a relatively new cause of intestinal disease in humans, although it has a long history of association with disease in animals, causing abortion and enteritis in sheep and cattle. It is the major cause of bacterial diarrhea in the United States and causes more cases of diarrhea than *Salmonella* and *Shigella* combined; the *Campylobacter* species responsible for this disease is *Campylobacter jejuni*, a Gram (–) spiral-shaped rod (**FIGURE 12.13**). Poultry, cattle, drinking water, and unpasteurized milk have been identified as potential sources; even one drop of "juice" from raw chicken meat is sufficient to establish infection. (Remember that when you are preparing a barbecue.)

The disease is usually self-limiting and lasts about 1 week. The infective dose is low, with fewer than 500 bacteria being enough to cause illness. The October 20, 1998, issue of *The New York Times* shocked the public with a report that 70% of chicken meat sampled from supermarkets was

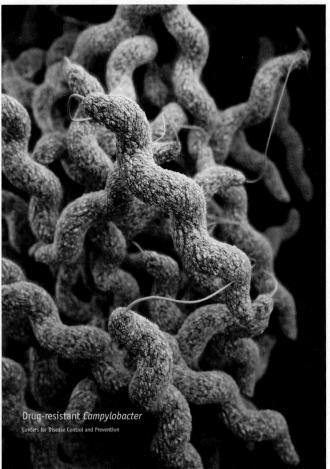

Drug-resistant *Campylobacter*
Centers for Disease Control and Prevention

Courtesy of the CDC/James Archer.

FIGURE 12.13 Three-dimensional computer-generated image of a cluster of spiral-shaped *Campylobacter* bacteria. The artist recreation was based upon a scanning electron micrograph image of the bacterium.

contaminated with *Campylobacter*. To make matters worse, many isolates were antibiotic resistant because of the widespread use of antibiotics in feed. The CDC estimates that 1.3 million cases of **campylobacteriosis** occur each year in the United States. Most are associated with eating raw or undercooked poultry, or eating something that touched it. Some cases are due to unpasteurized (raw) milk.

O Listeriosis

Listeriosis is caused by the Gram (+) rod *Listeria monocytogenes* (**FIGURE 12.14**) and is an increasingly significant cause of foodborne infection. The CDC estimates that about 1,600 people get listeriosis each year and about 260 die in the United States. The infection is most likely to sicken pregnant women and their newborns, older adults, and persons with compromised immune systems. The bacterium is ubiquitous in the environment. It has been isolated from soil, water, manure, plant materials, and the intestines of healthy animals (including humans, birds, and fish). The disease is associated with the ingestion of *Listeria*-contaminated foods, including luncheon or deli meats; hot dogs; soft cheeses, such as brie, and Hispanic-style cheeses made from pasteurized milk, such as queso fresco; ice cream; cantaloupe; raw (unpasteurized) milk; and, in 2007, in pasteurized milk. Processed foods, such as cold cuts, hot dogs, and soft cheeses may become contaminated during or after processing and serve as the source of epidemics or threatened epidemics. *Listeria* can multiply in foods inside of refrigerators, contributing to the problem.

Listeriosis is usually mild or subclinical in healthy adults and children and is manifested by nonspecific symptoms, including sore throat, fever, and diarrhea. *Pregnant women are approximately 20 times more susceptible than nonpregnant women.* Newborns can be infected as a result of their mothers' eating contaminated foods containing only a few *Listeria* bacteria and may become acutely ill

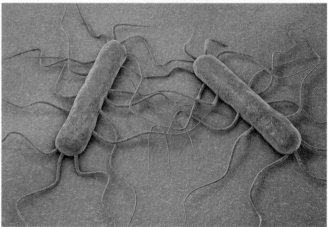

© Kateryna Kon/Science Photo Library/Getty.

FIGURE 12.14 Artist's rendition of *Listeria monocytogenes*. Each rod-shaped bacterium contains peritrichous flagella anchored into its cell wall, which are used to adhere to surfaces during biofilm formation. Listeriosis is rare but can cause serious illness and even death in newborns.

with meningitis (nervous system involvement); the mortality rate is about 60%, and antibiotics given promptly to pregnant women can often prevent infection of the fetus or newborn. *Pregnant women should avoid the following foods because of potential exposure to* Listeria monocytogenes:

- Deli meats
- Hot dogs
- Smoked fish
- Jerky
- Soft cheeses (brie, Camembert, Roquefort, feta, Gorgonzola, and Mexican-style cheeses such as queso blanco and queso blanco)
- Unpasteurized milk

A wrongful death suit against the Sara Lee Corporation was filed on February 2, 1999, after the death of a woman who consumed *Listeria*-contaminated hot dogs. The USDA reported that maintenance of the company's air-conditioning system and the resulting dust "*could have caused widespread contamination of meat products through the plant.*" Fifteen adults died and three miscarriages occurred in 16 states. In the summer and fall of 2002, in the northeastern United States, an outbreak of listeriosis affecting more than 53 people, most of whom were hospitalized, resulted in 8 deaths and 3 miscarriages or stillbirths. Precooked, sliceable deli meat was the culprit.

A dairy farm described as a "mom and pop" operation had to close down in February 2008, when contaminated milk caused listeriosis in four people. The cases included three men over the age of 70, all of whom died after June 2007, and a miscarriage in a 30-year-old woman. According to the owners, of the 330 cows on the farm about 130 were milked daily and produced 1,000 gallons of milk each day. How the *Listeria* contaminated the milk remains a mystery, because all the farm's milk was pasteurized.

Most outbreaks of listeriosis are associated with meat products and, to some lesser extent, with milk and milk products. However, starting in August 2011, a multistate outbreak of listeriosis was traced to cantaloupes grown at Jensen Farms in Colorado. The cases spread to 28 states, resulting in 30 deaths and 146 confirmed cases; it is considered to be the worst foodborne outbreak to have occurred in the United States in 100 years. Those who died ranged from 48 to 96 years.

In 2016, frozen vegetables were recalled due to *Listeria* contamination in the United States. By 2018, 107 countries, including the United States and Canada, recalled frozen vegetables contaminated with *L. monocytogenes*. At least 47 people became sick in five European countries and 9 people died. The estimated cost of the recall was $35 million. The problem is complex, because some countries reprocess and re-export some of their products, such as frozen vegetables. The frozen vegetables recalled were produced in Greenyard's Hungarian facility in Baja between August 13, 2016, and June 20, 2018.

○ *Clostridium difficile*

Clostridium difficile (**FIGURE 12.15**) is a Gram (+) endospore-forming rod closely related to *C. botulinum, C. perfringens,* and *C. tetani.* It is a cause of HAIs that result in severe **colitis** (inflammation of the colon). For some individuals, *C. difficile* colitis symptoms become severe and diarrhea becomes debilitating, resulting in extreme weight loss. The condition has generally been associated with hospital patients on chronic or prolonged antibiotic therapy. It can be spread in healthcare environments through contact with contaminated surfaces. Healthcare workers are required to wear gloves and gowns when caring for *C. difficile* patients. Remember that hand sanitizer will not kill *C. difficile!* The CDC estimated that nearly half a million *C. difficile* infections occurred in the United States in 2011, and that 29,000 died within 30 days of diagnosis. Deaths occurred in people 65 and older.

This infection has jumped from hospital and hospital-like environments to the community; the emergence of methicillin-resistant *Staphylococcus aureus* is a similar phenomenon. Generally, *C. difficile* is inhibited by the normal microbiota, but in some individuals on antibiotics the composition of the normal microbiota is altered to the degree that *C. difficile* may "grow out," causing mild to severe diarrhea and possibly death from inflammation and spread into the bloodstream. Alternative therapies, including probiotics and yogurt, are sometimes successful.

Fecal transplants are practiced by gastroenterologists in the United States (who have gotten over the "yuck" factor) as a safe, inexpensive treatment option for patients with reoccurring severe *C. difficile* colitis. About 80% of patients are cured. The remaining 20% usually require a second fecal transplant. The treatment involves introducing saline-diluted fecal matter from a donor (from a "poop" bank) into a patient via an enema or through a special catheter that is inserted down the throat, through the small intestines, and into the colon. Would you get a fecal transplant? The procedure is more common in Canada and Europe but is becoming a more common practice in the United States because specialized fecal banks have been established that collect and screen human fecal samples for pathogens to be used specifically for fecal transplant.

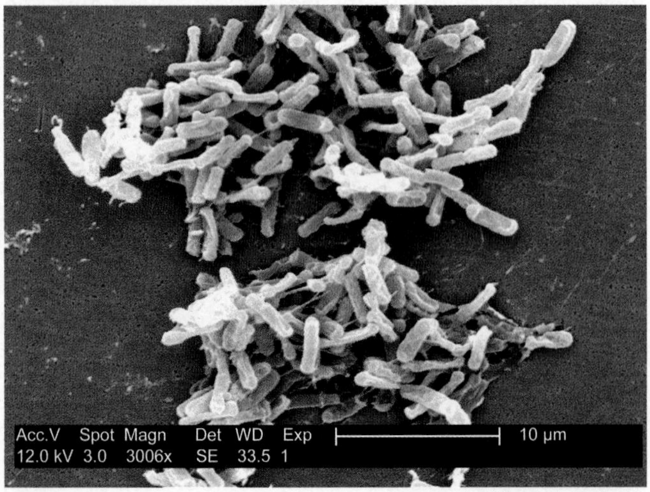

Airborne Bacterial Diseases

Airborne bacterial diseases are respiratory tract infections transmitted by air droplets or by contact with contaminated **fomites** (inanimate objects such as facial tissues, countertops, toys at a day care center, and cell phones); in some cases, the disease spreads beyond the respiratory tract to other areas of the body (e.g., *Mycobacterium tuberculosis,* the cause of tuberculosis, can spread to bones and organs). The humidity and temperature in air allow bacteria to remain viable for extended periods of time; hence, air serves as an excellent vehicle for transmission.

Air is a passive transmitter because the bacteria do not replicate in the air, as they do with foodborne intoxication. Respiratory infections generally respond well to antibiotic therapy, although antibiotic resistance is an increasing problem. Vaccines are available to prevent some respiratory infections but may not be practiced in many countries because of problems of availability, cost, and public health infrastructure. The upper respiratory tract contains the tonsils and pharynx, and the lower respiratory tract includes the trachea, larynx (voice box), bronchi, bronchioles, and lungs (**FIGURE 12.16**). **TABLE 12.8** provides a list of major airborne bacterial diseases categorized into upper and lower respiratory tract infections.

Courtesy of Lois S. Wiggs and Janice Carr/CDC.

FIGURE 12.15 Colorized scanning electron micrograph of *Clostridium difficile.*

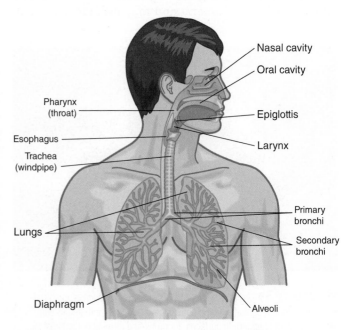

FIGURE 12.16 Anatomy of the human respiratory system.

TABLE 12.8 Airborne Bacterial Diseases*

Upper Respiratory Tract			
Infectious Disease	Incubation Period	Bacterial Pathogen	Symptoms
Pertussis (whooping cough)	7–10 days	*Bordetella pertussis*	Respiratory blockage causing violent and persistent coughing
Streptococcal infections (group A) and sore throat, fever (pharyngitis, tonsillitis strep throat, scarlet fever, flesh-eating strep, toxic shock syndrome, childhood fever)	2–5 days	Streptococci (group A)	Coughing, severe headache, red rash, strawberry tongue, multiorgan failure, destruction or "eating" of the flesh
Meningitis	1–3 days	*Neisseria meningitidis*	Coldlike symptoms, fever, delirium, stiffness of the back and neck
Meningitis	5–6 days	*Haemophilus influenzae* type B	Fever, stiff neck, altered mental status
Lower Respiratory Tract			
Disease	Incubation Period	Bacterial Pathogen	Symptoms
Legionellosis	2–10 days	*Legionella pneumophila*	Fever, muscle pain, cough, pneumonia
Tuberculosis	Several weeks	*Mycobacterium tuberculosis*	Fever, weight loss, cough, fatigue

*These diseases are treatable with antibiotic therapy; antibiotic-resistant strains pose a problem in some cases.

Before the advent of vaccination and antibiotic therapy, throughout the United States and other developed countries of the world so-called communicable disease hospitals were established to care for those with contagious, communicable, or **infectious diseases**. For example, **TB sanatoriums** were established during the late 1800s to treat tuberculosis patients. After antibiotics were discovered and could be used to cure patients, sanatoriums were no longer needed, and most closed during the 1950s.

Although sanatoriums were equipped to manage the constant threat of bacteria being introduced into the air by their victims through coughing and nasal secretions, death rates were very high. Young children were especially susceptible to these diseases. High fever, delirium, racking coughs, labored breathing, and horrible rashes were associated symptoms. Little could be done other than supportive therapy. It was a "wait, see, and pray" attitude. Frequently, the young and hapless victims were not hospitalized but were confined to their homes; **quarantine signs** or placards were placed on the doors by local public health officials to discourage visitors. These were the days of the horse-and-buggy doctors, and their selflessness was more than admirable as they sat by the bedsides and could do little more than apply cold towels to the patients burning with raging fevers. Because of advancements in sanitation, hygiene, vaccination, and antibiotics, specialized hospitals to care for patients with communicable diseases are no longer needed. Virtually all hospitals are able to invoke "isolation" strategies.

Upper Respiratory Tract Infections

○ Pertussis (Whooping Cough)

Whooping cough, also called **pertussis**, is a highly infectious and potentially lethal infectious disease caused by the Gram (–) coccobacillus-shaped bacterium known as *Bordetella pertussis* (**FIGURE 12.17**). Pertussis is a particular threat to infants and children younger than 4 years of age. Children who have whooping cough are the primary reservoir of the disease; there is no nonhuman reservoir. The bacteria are spread by talking, coughing, sneezing, and laughing.

The coccobacilli bind to ciliated epithelial cells that line the upper respiratory tract and secrete **exotoxins** that

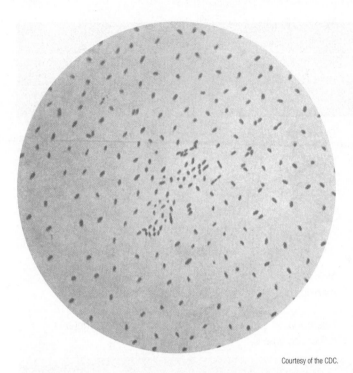

Courtesy of the CDC.

FIGURE 12.17 Gram stain of a cultured specimen containing Gram (–) coccobacilli-shaped *Bordetella pertussis* bacteria.

cause damage to the cilia lining the **upper respiratory tract**, causing airways to swell so that mucus cannot be cleared from the air passages. The net effect is a buildup of a thick gluelike mucus, promoting symptoms of a common cold followed by paroxysms (spasms) of violent, hacking (that sounds like "whooping"), persistent, and recurrent coughing, usually about 15 to 20 coughs, in an attempt to cough up the mucus. Coughing fits can last up to 10 weeks or more; pertussis sometimes is known as the "100-day cough." These episodes result in a need for oxygen, triggering deep and rapid inspirations throughout the partially obstructed passages that are responsible for the characteristic "whoop." The coughing may be so violent as to cause vomiting, hemorrhage, and even brain damage.

Between 1940 and 1945, before the advent of a pertussis vaccine, whooping cough was rampant. In the United States there were as many as 147,000 cases per year and more than 8,000 deaths. In 1991, the pertussis component, consisting of live B. *pertussis* bacteria, was modified to the use of killed bacterial cells, referred to as **acellular (aP)**, to minimize side effects. It is usually used in conjunction with the diphtheria and tetanus vaccines; this multiple vaccine is known as **DTaP (diphtheria, tetanus, pertussis) vaccine**.

Pertussis is considered a **reemerging infectious disease**; 17,972 cases occurred in 2016, a 55% decrease from 2014, in which there were 32,971 cases. In 2016, those most affected were children younger than 11 months of age and 11- to 19-year-olds; the seven deaths that occurred were all in children younger than 1 year of age.

An outbreak of pertussis occurred in an Amish community in Kent County, Delaware, between September 2004 and February 2005, in which 345 cases occurred, primarily in preschool-aged children. Although the Amish religious doctrine allows vaccination, childhood vaccination is low in many Amish communities. These statistics underscore the need to administer vaccination in all populations and to reevaluate the immunization strategy for pertussis. When the incidence of whooping cough increased in 2012 among 13- and 14-year-olds, it suggested to experts that immunity from vaccination was waning. The CDC has recommended increased vaccination coverage efforts, especially for pregnant women, adolescents, and adults.

One possible factor in the resurgence of whooping cough in the United States is that the original pertussis vaccine was controversial and accused of causing brain damage, although this has never been conclusively proved. Nevertheless, the bad press accounted for fewer children being vaccinated and a resultant increase in the disease. Two new and improved vaccines were licensed in 2005. Pertussis vaccine does not produce lifelong immunization, nor does recovery from whooping cough, contrary to popular belief. Persistent coughs in older children and adults may well be whooping cough, usually present in a milder form.

Whooping cough remains a devastating disease in unvaccinated populations. Experts estimate that each year worldwide there are an estimated 24.1 million pertussis cases and about 160,700 deaths of children younger than age 5. The largest proportion of cases (33% of cases and 58% of deaths) occur in African nations.

● Streptococcal Infections

Streptococci are a large group of Gram (+) cocci arranged in long chains of bacteria, the most significant one of which is *Streptococcus pyogenes* (**FIGURE 12.18**). The particular manifestation of disease is related largely to the specific *Streptococcus* and the toxin it produces. Infections range from generally mild, the most common of which is "strep throat," to a variety of life-threatening conditions. The bacteria reside in the nose and throat and are transmitted by respiratory droplets from infected persons or by contact with wounds and sores on the skin.

Shors' Notebook

My mother used to tell me about one of my siblings who got very sick as a little girl (this occurred before I was born). In 1960, my 2-year-old sister Mary contracted whooping cough. A doctor advised my mother that she would die. My mother continued to nurse and care for her like Florence Nightingale. She survived, beating the odds. I was born a year later. Even though my mother was not well educated, this experience imprinted the idea in her mind that vaccinations were necessary, and as childhood vaccines became available she made sure I was vaccinated.

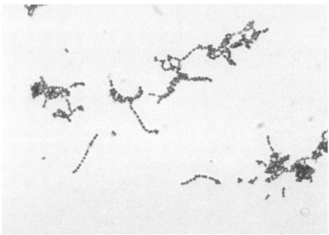

FIGURE 12.18 Gram stain showing Gram (+) Chains of *Streptococcus pyogenes*, the bacterium that cause scarlet fever. Magnification 1,000×.

Strep throat, more correctly termed streptococcal pharyngitis or tonsillitis, is a common and usually mild infection predominantly found in children 5 to 15 years of age. It is characterized by a red and sore throat, fever, and headache. Definitive diagnosis is by culturing a specimen from a throat swab or by a rapid test procedure. In some cases, scarlet fever results and is caused by a streptococcal strain producing **erythrogenic toxin**. The disease is characterized by a red rash and **strawberry tongue** (**FIGURE 12.19**).

FIGURE 12.19 Characteristic "strawberry tongue" of a patient with scarlet fever. Children with this disease frequently have an inflamed tongue along with a characteristic red rash.

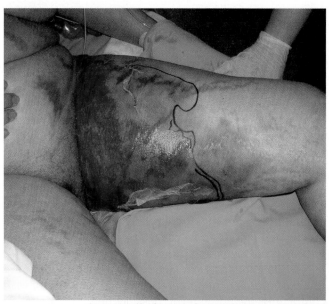

FIGURE 12.20 Necrotizing fasciitis.

Some streptococci are invasive and cause serious and life-threatening infections, including streptococcal **toxic shock syndrome**, characterized by multiorgan failure, and necrotizing fasciitis, more dramatically called flesh-eating disease. *The disease is a rarity*; it is horrendous and is the result of the bacterial infection of a minor wound with an invasive strain of streptococci that produces flesh-destroying enzymes and toxins and can spread through human tissue at a rate of about 3 centimeters (0.4 inches) per hour (**FIGURE 12.20**).

In 1994, several outbreaks of this rapidly progressing disease, dubbed by the media as the "*flesh-eating bacteria*," occurred in England and in the United States; one story was headlined "*the bacteria that ate my face*." It is unclear *exactly* how many necrotizing fasciitis cases occur each year in the United States. The CDC estimates about 650 to 850 cases of necrotizing fasciitis caused by *S. pyogenes*. Another 700 cases each year are caused by other bacteria, such as *Vibrio vulnificus* (see opening case study), *Staphylococcus aureus*, *Klebsiella* spp., *Clostridium* spp., *Bacteroides* spp., *E. coli*, and *Aeromonas hydrophilia*. The number of annual infections does not appear to be rising. Of those affected by necrotizing fasciitis, about one in five dies.

Treatments include intravenous (IV) administration of antibiotics and may include skin graft surgeries and amputations. **Hyperbaric oxygen therapy (HBOT)** is an emerging therapy in combination with antibiotic treatment. HBOT involves breathing pure oxygen in a pressurized room or tube. The air pressure is three times higher than normal air pressure. Under these conditions the lungs can gather more oxygen than would be possible breathing pure oxygen at normal air pressure. The blood carries oxygen throughout the body that helps to fight bacteria. It also stimulates the release of growth factors and stem cells to promote healing.

Glomerulonephritis, a kidney disease, and **rheumatic fever**, a condition involving the heart and joints, are possible long-term complications of repeated early childhood streptococcal infections. The incidence of these diseases has decreased drastically as a result of the advent of antibiotics, but unfortunately antibiotic-resistant strains are of increasing concern.

Bacterial Meningitis

Meningitis is an inflammation of the meninges, the three membranes covering the spinal cord and the brain. This condition is potentially serious because pathogens invade the nervous system. The key to survival is early diagnosis; several hours can make the difference between life and death. Meningitis can be caused by a variety of bacteria, viruses, protozoans, and fungi; this section deals only with bacterial meningitis. Early signs of bacterial meningitis are flulike symptoms that progress quickly to more definitive complaints, including fever, possibly delirium, and stiffness in the neck and back.

A number of bacteria can infect the meninges, but the two most common species are *Neisseria meningitidis*, a Gram (–) diplococcus, and *Haemophilus influenzae* type b, referred to as Hib, a Gram (–) coccobacillus-shaped bacterium. About 10% of the population are healthy carriers and harbor meningococci in the back of the nose and throat. These healthy carriers, along with infected individuals, spread the bacteria between people by respiratory droplets through coughing, sneezing, kissing, and sharing eating utensils. There is no animal reservoir. The meningococci are very fragile outside the body, which limits the incidence of the disease. The bacteria in droplets enter the nasal pharynx and may cross the epithelial cell barriers lining the pharynx and invade the bloodstream, from which they enter the meninges. Between 2003 and 2007, an estimated 4,100 cases and 500 deaths from bacterial meningitis occurred in the United States.

Hib meningitis occurs primarily in children younger than 5 years of age. Once the leading cause of bacterial meningitis in the United States, its incidence has dramatically decreased since the introduction of the Hib vaccine in the early 1990s. Although it is a less serious form of meningitis, it should not be regarded lightly. Three Hib vaccines are available in the United States for infants as young as 6 weeks of age: ActHIB, Hiberix, and PedvaxHIB.

Meningococcal disease and meningococcal septicemia is caused by the Gram (–) diplococcus-shaped bacterium *Neisseria meningitidis* (**FIGURE 12.21**). This bacterium lives in the throats of 5% to 10% of healthy people. It rarely causes meningococcal disease or meningococcal septicemia. However, when it does produce symptoms it affects all age groups but has a higher incidence in adolescents and young adults. Five serotypes of *N. meningitis* cause meningococcal disease: A, B, Y, C, and W. About 40% of meningococcal disease cases in the United States are caused by serotype B. Symptoms of **meningococcal disease** first appear as flulike illness (fever, headache, stiff neck) but rapidly worsen to nausea, vomiting,

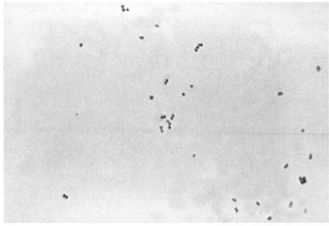

Courtesy of the CDC/Dr. Brodsky.

FIGURE 12.21 Gram stain of *Neisseria meningitidis*. It is a Gram (–) diplococcus (cocci arranged in pairs). Magnification 1,000×.

photophobia (eyes become sensitive to light), and confusion. Fortunately, the infection responds to antibiotic therapy, although some individuals are left with severe, irreversible damage. However, as public health officials point out, the risk of bacterial meningitis causing serious illness in those unvaccinated is about 10 times greater than those who are vaccinated.

Sometimes *N. meningiditis* bacteria cause sepsis or septicemia, a bloodstream infection that is referred to as **meningococcal septicemia** or **meningococcemia**. Bacteria enter the bloodstream, multiply, and damage the walls of blood vessels, causing bleeding to skin and organs. Symptoms include fever, fatigue, vomiting, cold hands and feet, myalgia, rapid breathing, diarrhea, and, in late stages, a purple rash (**FIGURE 12.22**). Despite antibiotics, some patients develop serious sequelae, necessitating amputation of limbs, fingers, and toes in an attempt to halt the spread of infection, neurological damage, and deafness.

College students living in dormitories appear to be particularly susceptible to meningococcal disease.

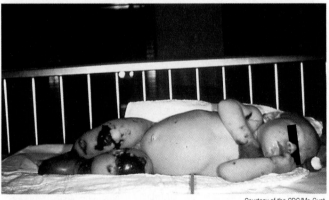

Courtesy of the CDC/Mr. Gust.

FIGURE 12.22 This 4-month old child is suffering from meningococcal septicemia caused by *Neisseria meningiditis*. She developed the purple rash and tissue necrosis.

A NOVA episode, "Killer Disease on Campus," aired on PBS television in 2002. The program included interviews of survivors and parents of children who had suffered from meningococcal disease or meningococcal septicemia. Quadrivalent vaccines have been approved to prevent infection by serotypes A, C, W, and Y: Menactra in 2005 and Menveo in 2010. The CDC recommends that all preteens and teens be vaccinated with a quadrivalent vaccine. Two vaccines that protect individuals against serotype group B are Trumenba, which was approved by the FDA in 2014, and Bexsero, which was approved in 2015. The CDC recommends that all 16- to 18-year-olds be vaccinated with the group B vaccine.

College-bound students should be vaccinated against this potentially deadly disease in view of their future close-quarter living conditions. Many states have mandated that students attending colleges or universities be vaccinated with the quadrivalent vaccine. The **Immunization Action Coalition** provides information on which states mandate vaccination for college or university students at http://www.immunize.org/laws/menin.asp.

Lower Respiratory Tract Infections

The **lower respiratory tract** consists of the larynx (voice box), trachea (windpipe), bronchial tubes, and lungs (Figure 12.16). The tissue of the lungs terminates in millions of air sacs named **alveoli**. The lungs are covered by a membrane called the pleura. The terms laryngitis, tracheitis, and bronchitis refer to their anatomical location; -*itis* means "inflammation." **Pleurisy** is an inflammation of the pleura, and **pneumonia** is an inflammation of the lungs. Lower respiratory tract infections are caused by a variety of bacteria, viruses, fungi, and protozoans.

○ Legionellosis

In 1976, the American Legion, a U.S. military veterans' association, held its 58th national convention at the historic Bellevue-Stratford hotel in Philadelphia; this convention was a particularly significant one, because it marked the 200th anniversary celebration of the founding of the United States of America (**FIGURE 12.23A**). However, the occasion was marred by a mysterious pneumonia that afflicted 182 Legionnaires, of whom 147 were hospitalized and 29 died. After an exhaustive investigation lasting over a year, it was discovered that the disease was caused by bacteria and was disseminated through the air-conditioning ducts of the hotel. The bacterium was later identified and named *Legionella pneumophila*, a Gram (–) rod (**FIGURE 12.23B**). The publicity that ensued over the next few years was dramatic and inspired musician and composer Bob Dylan in 1981 to write a song titled "Legionnaires' Disease."

Legionella spp. can cause Legionnaires' disease or Pontiac fever, collectively known as **legionellosis**. **Pontiac fever** is a milder upper respiratory infection than Legionnaires' disease. Symptoms resemble influenza and are primarily fever and muscle aches that begin a few hours to a few days after exposure to a number of different species

Courtesy of the CDC/Dr. Gilda Jones.

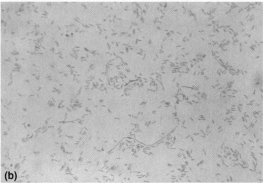

Courtesy of the CDC/Dr. Gilda Jones.

FIGURE 12.23 (a) The façade of the Bellevue-Stratford Hotel in Philadelphia, Pennsylvania, where the 1976 disease outbreak occurred and was given the name Legionnaires' disease. **(b)** Gram stain of *Legionella* showing the presence of Gram (–) rods. Magnification 1,000×.

of *Legionella*, such as *L. pneumophila*, *L. micdadei*, *L. anisa*, *L. longbeachae*, and *L. feeleii*. It resolves on its own within 2 days, often going undiagnosed. The mortality rate is zero. The disease is named after Pontiac, Michigan, where the first case was recognized in 1968.

Legionnaires' disease can be caused by a few different species of *Legionella* but 91% of cases are caused by *L. pneumophila*. Patients with Legionnaires' disease develop a fever, cough, chills, headache, and severe pneumonia. The infections can be acquired by persons of any age, but middle-aged and older persons are the most susceptible, as are cigarette smokers, those with chronic lung diseases, or people on steroids (which suppresses the immune system). When death occurs, it is usually attributable to shock and kidney failure.

Transmission rarely involves person-to-person contact (there has been one report of human-to-human transmission to date). Inhalation of aerosols that come from a water source contaminated with *L. pneumophila* is the **mode of transmission**. *Legionella* spp. are ubiquitous in lakes, streams, air-conditioning cooling towers, hot water heaters, hot tubs (whirlpools), spas, showerheads, indoor decorative fountains (e.g., in malls and hotel lobbies), drinking fountains or "bubblers," vegetable misters in grocery stores, and showers. Community-acquired outbreaks continue to occur throughout the world (**FIGURE 12.24**). The bacteria grow well in the warm and stagnant waters afforded by these environments and are then aerosolized.

Ironically, hospital outbreaks have occurred on frequent occasions because the temperature in the hot water lines is kept relatively low for safety reasons, fostering bacterial multiplication. Furthermore, *Legionella* spp. *are more resistant than most bacteria to disinfection by chlorine and hot water temperatures*. From 1985 to 1988, 26 cases of Legionnaires' disease were linked to shower use in a hospital. Ten of the patients died, most of whom had health risks such as smoking, steroid use, or chronic lung disease. More recently, during a CDC investigation for the surveillance of waterborne disease outbreaks associated with drinking water in the United States from 2013 to 2014 found that the most common cause of disease was *Legionella* spp. (57% of outbreaks). A total of 35 outbreaks resulting in 130 cases, 109 hospitalizations (88%), and 13 deaths were caused by *Legionella* spp. in 10 different states. The majority of outbreaks were community acquired; the drinking water sources were surface water impoundments (river/streams or water reservoirs) or wells pumping groundwater that was deficient in management of water treatment. Drinking water is defined as potable water for human consumption. Within this definition, drinking water is used for drinking, bathing, showering, handwashing, teeth brushing and oral hygiene, food preparation, and dishwashing. Fortunately, home and automobile air-conditioning units have never been implicated in causing legionellosis outbreaks.

Legionnaires' disease is a cause for concern because it causes HAIs. Approximately 76% of people have acquired Legionnaire's disease from a healthcare facility, and 25% of infected patients die. In 2016, 6,100 cases of Legionnaires' disease were reported in the United States, but this number is probably much lower than the actual number of cases because many cases are not diagnosed or are misdiagnosed. Infection can be acquired any time of the year, but the summer and early months of the fall are the seasons when most cases occur. Prevention of Legionnaires' disease can be accomplished by improvements in the design and maintenance of public drinking water, cooling towers, whirlpools, spas, hot tubs, and

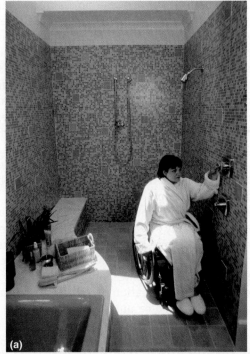

(a)

Courtesy of the CDC/Richard Duncan.

(b)

Courtesy of the CDC/Amanda Mills.

(c)

© Teri Shors.

FIGURE 12.24 *Legionella* spp. live in water environments. More recently, *Legionella* spp. have been implicated in outbreaks associated with drinking water that is aerosolized through **(a)** showering and **(b)** drinking water fountains. **(c)** Vegetable misters in grocery stores are also a potential source of *Legionella* spp.

other sources of warm, stagnant waters in which there is the potential for aerosolization.

Tuberculosis

Consumption, **white plague**, and **white death** are 19th-century terms for **tuberculosis (TB)**. These terms reflect the fact that the victims were excessively thin, pale, and weak and appeared to be consumed by their illness. TB is not, by any means, a disease of the past. In the United States and other industrialized countries, it is a reemerging infectious disease, yet only 40 years ago it was on the brink of extinction. March 24th is **World TB Day** (http://www.who.int/campaigns/tb-day/2018/event/en/), which commemorates the date in 1882, when Dr. Robert Koch announced the discovery that *Mycobacterium tuberculosis* causes tuberculosis. World TB Day provides awareness about TB-related problems. The U.S. theme for the 2018 observance was "*Wanted: Leaders for a TB-Free United States. We can make history. End TB.*"

Paintings from tombs in Egypt and examination of mummies dating back to 4000 B.C. indicate the antiquity of TB. The disease emerged in Neolithic times as human populations increased in size, settled down, and domesticated cattle. Presumably, human TB may have arisen from bovine TB. TB is as much a social disease as it is a microbial disease; in the 18th and 19th centuries its development as an urban plague was associated with poverty, poor housing, crowding, inadequate nutrition, and unemployment, all spawned in the wake of the Industrial Revolution. It is ironic that progress fueled the social conditions that allowed TB and other diseases to flourish. Before the discovery of the tubercle bacillus by Robert Koch in 1882, the environment of crowded tenements was associated with the cause of the disease.

The impact of TB was so profound in the 19th century that some feared it would bring about the end of European civilization. In many cities of America and Europe, TB was the leading cause of death, accounting for as many as 15% to 20% of fatalities. The treatment of TB centered on fresh air, bed rest, and good nutrition with plenty of fresh eggs, milk, and cream, all of which were provided in the TB sanatoriums (as hospitals for the care of people with TB were called) around the country. Porches and decks were prominent in these facilities to allow the patients to be in fresh-air environments, even during the winter. In some cases, surgical intervention to collapse a lung was practiced. The rationale was that in a "rested" state an infected lung would heal more quickly. A tube was inserted into the chest cavity to allow air to enter and collapse the lung.

TB declined as social conditions improved during the first few decades of the 20th century, accompanied by the development of immunization in the 1920s and antibiotics in the 1940s. Streptomycin, an antibiotic effective against the tubercle bacillus, was discovered in the 1940s, followed by the introduction in 1952 of **isoniazid**, a drug still widely used to treat TB.

TB: The Disease

The causative agent of tuberculosis (TB) is *Mycobacterium tuberculosis*. TB is an infectious disease of the lower respiratory tract. The bacilli or rod-shaped bacteria are transmitted by inhalation of infected droplets aerosolized by *M. tuberculosis*–infected individuals during coughing, sneezing, singing, talking, or laughing. Large numbers of bacteria are coughed up by individuals suffering from TB, and transmission is efficient because the waxy bacterial cell wall protects the microbe from drying.

The people at the greatest risk are those who spend relatively long periods with an infected person, including family members, friends, and coworkers. *Anyone can be infected by* M. tuberculosis. It is not an infectious disease that only affects people who have chronic preexisting conditions.

Before **pasteurization**, milk was significant in the transmission of TB. Less commonly, the bacteria can be acquired through the skin, and cases have been reported in laboratory personnel who have handled specimens containing *M. tuberculosis* and in people who have received tattoos. An embalmer contracted TB from an infected corpse while preparing the body for burial.

The first exposure to *M. tuberculosis* results in primary infection; in most cases there are no symptoms, and the individual is not even aware that infection has occurred. Cell-mediated immunity walls off the bacteria in lesions known as **granulomas**. This is the usual response in about 90% of individuals with primary infection. In the other 10% the immune response is not adequate, resulting in the escape of bacteria from the granulomas. The course of TB is illustrated in **FIGURE 12.25**. The bacteria cause symptomatic primary TB manifested by fevers, night sweats, weight loss, fatigue, and the coughing up of blood-tinged sputum. A classic sign of active TB is a blood-stained handkerchief. Symptomatic primary infection is more likely to occur in children, in the elderly, and in the immunocompromised, especially those with HIV/AIDS.

In most cases the primary infection is in the lungs, but infection can occur in the brain, spinal cord, kidney, bone, or cutaneous (skin) tissue (**FIGURE 12.26**). Although most cases of TB infection do not progress to acute disease, **reactivation TB**, or secondary TB, can result because the bacteria can remain in a **dormant state** for years and be reactivated decades later. If this happens any of the anatomical sites seeded during the primary infection may be affected, including primarily the lungs but also the bones and joints. It is likely that the fictional character Quasimodo, the hunchback in Victor Hugo's classic tale *The Hunchback of Notre Dame*, was a victim of spinal TB.

Diagnosis and Screening

All *Mycobacterium* species contain **mycolic acids** in their cell wall, which gives these bacteria a waxy physical characteristic making them impossible to **Gram stain**. *Mycobacterium* spp. cell walls resist or repel **crystal violet** or **safranin dyes** used in Gram staining. A method was developed in 1883, called the **Ziehl-Neelson staining procedure**

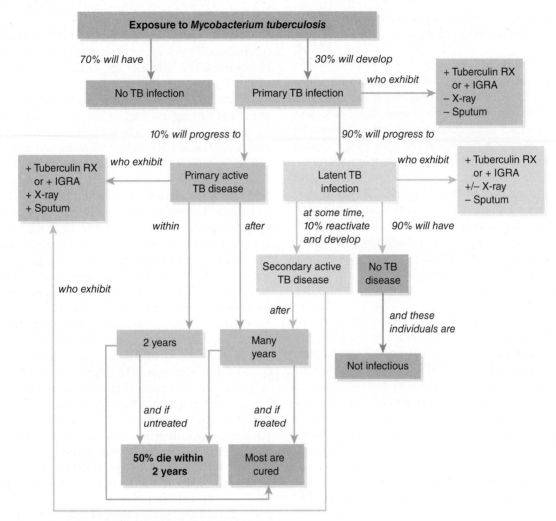

FIGURE 12.25 Algorithm for the diagnosis and course of TB.

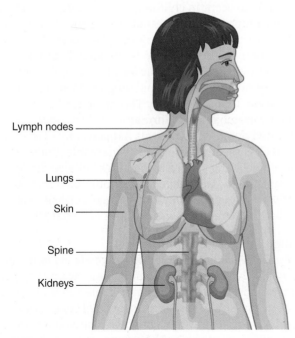

FIGURE 12.26 Common sites of TB infection.

or **acid-fast staining procedure** to identify **acid-fast bacteria**. The method was described by German bacteriologist Franz Ziehl (1859–1926) and German pathologist Friedrich Neelsen (1854–1898). Carbol fuchsin dye is used to stain heat-fixed, smear preps of **sputum** samples on glass slides. The slide of the fixed sputum sample is flooded with carbol fuchsin and steamed. The steaming facilitates the penetration of the lipoidal, waxy cell wall, allowing carbol fuchsin to enter and stain the cytoplasm of *Mycobacterium* (all other bacteria present in the fixed sputum sample will also stain with carbol fuchsin).

Subsequently, acid alcohol is added to the fixed sputum sample containing the carbol fuchsin–stained *Mycobacterium* and other bacteria (from human microbiota). *Mycobacterium* spp. remain stained by the carbol fuschin dye but other bacteria are decolorized. *Mycobacterium* is described as being acid-fast bacteria because the microbes resists decolorization with acid alcohol, whereas those cells that lose the dye upon addition of acid alcohol are **non-acid-fast bacteria**. The decolorized bacteria can be counterstained with methylene blue (**FIGURE 12-27**). This

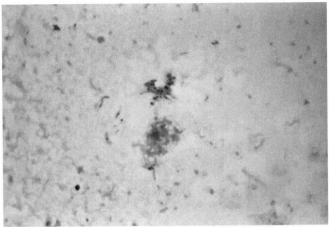

Courtesy of the CDC.

FIGURE 12.27 Photograph showing the presence *Mycobacterium tuberculosis* in a sputum smear. The Ziehl-Neelsen staining procedure was used to stain the acid-fast bacteria (fuchsia) with carbol fuchsin and the non-acid-fast bacteria (blue) are counterstained with methylene blue. Magnification 1,000×.

staining procedure allows one to *differentiate* between bacteria that are acid fast (carbol fuchsin stained), such as *Mycobacterium*, and those that are non-acid fast (most normal microbiota that stain with methylene blue).

The early symptoms of pulmonary TB resemble those found in a plethora of infectious diseases. Two tests are used to determine if a patient is infected with TB bacteria: the **Mantoux tuberculin skin test** (**FIGURE 12.28**) or a TB blood test called the **interferon gamma release assay (IGRA)**, which measures how strongly a person's immune system reacts to TB bacteria. In addition, **nucleic acid amplification tests (NAATs)** are available in certain laboratories to aid diagnosis and to detect antibiotic resistance. The NAAT is performed on a sputum sample from a patient.

The Mantoux tuberculin skin test is performed by the intradermal injection of a minute amount of purified protein derivative (PPD) from *M. tuberculosis*; the test is harmless because it does not use whole bacteria. After 48 to 72 hours the injection site is visually examined for the presence of an induration (a red, raised lesion); a lesion

measuring 5 millimeters (0.2 inches) or more in diameter constitutes a positive skin test.

Those testing positive may falsely jump to the conclusion they have TB, *but a positive skin test only indicates previous exposure to the organism—a condition referred to as **TB infection**, not **TB disease***. In TB infection, the TB bacteria are walled off in the granulomas; the person has no symptoms. Active symptomatic disease is the result of bacilli escaping from granulomas and multiplying inside of the body; the immune system cannot stop the TB bacteria from replicating. Many people in the United States have come from countries with high rates of TB and have contracted TB. Although they may have completely recovered, they may continue to manifest a positive skin test. Further, they may have received a TB vaccine (described later) as a preventive measure and will test positive, probably for their entire lifetime.

If a person has been determined to be infected with TB bacteria, additional tests are needed to determine if the patient has TB disease, such as a chest x-ray (**FIGURE 12.29**); microscopic examination of sputum for the presence of tubercle bacilli or acid-fast rod-shaped

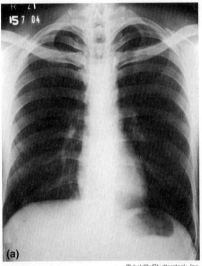

(a)

© hald3r/Shutterstock, Inc.

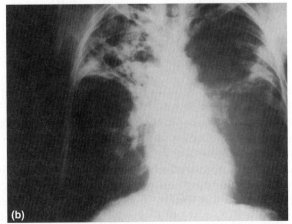

(b)

Courtesy of the CDC.

FIGURE 12.29 **(a)** A normal chest x-ray. **(b)** Chest x-ray of patient diagnosed with TB disease. The upper left and lower right lobes show infiltrated bacteria. This diagnosis was "far-advanced" tuberculosis.

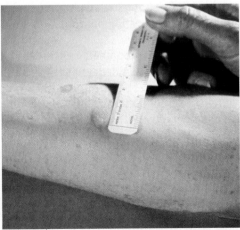

Courtesy of CDC.

FIGURE 12.28 Positive tuberculin skin test.

bacteria; and culture of sputum specimens for the demonstration of *M. tuberculosis*. It takes 6 to 8 weeks for colonies of *M. tuberculosis* to appear on solid agar plates. Culturing the bacteria from the patient remains the **gold standard** (definitive test) for laboratory confirmation of TB disease. If a patient does not have TB disease, but is infected with TB bacteria, the decision for treatment will be based on the patient's chances of developing TB disease.

○ Antibiotic Therapy and DOTS

The FDA has approved 10 drugs for the treatment of TB disease. TB disease is treated by a 6- to 9-month course of antibiotics along with supportive measures of adequate rest, a good diet, and oxygen therapy. *M. tuberculosis* has a slow **generation time** of between 15 and 20 hours. By way of comparison, *E. coli* has a generation time of 20 minutes. Hence, lengthy antibiotic treatments must be prescribed due to the slow growth and death rate of the TB bacteria. Of the FDA-approved drugs, the first line of antibiotics to treat TB disease are a cocktail regime in which the patient is prescribed all four of the following antibiotics listed or two of those listed for an even longer time period: isoniazid, rifampin, ethambutol, and pyrazinamide.

Toxicity and antibiotic resistance need to be considered in choosing the appropriate combination of antibiotics. The length of time needed for treatment can create a serious problem of noncompliance; further, the antibiotics are expensive. These factors severely hamper the treatment of TB in both developing and developed countries. This is particularly true in the homeless population, many of whom have TB and/or HIV/AIDS as well as drug or alcohol addiction. Even if the antibiotics are provided free of charge and are readily available, noncompliance remains a major issue. Attempts to control TB by supplying infected individuals with take-home medicines have been unsuccessful because the medications often are taken in a haphazard fashion, fostering the development of multiple antibiotic-resistant strains.

DOTS, or **direct observational therapy short course**, is a simple and effective method adopted by the WHO in 1992 to combat noncompliance and complacency. It has saved thousands of lives and minimized the emergence of drug-resistant strains. The DOTS strategy combines the five elements of (1) political commitment, (2) microscopy services, (3) drug supplies, (4) surveillance and monitoring, and (5) direct observation. Once an individual's sputum shows tubercle bacilli, healthcare workers must watch the patient swallow the full course of the prescribed anti-TB drugs on a daily basis (**FIGURE 12.30**). The sputum-smear test is repeated after 2 months and at the end of the 6- to 8-month treatment schedule.

The drugs used in the DOTS system are not new and when used correctly have a cure rate approaching 100%. Within the first 2 to 4 weeks of treatment, patients become noninfectious. The DOTS program is designed for large, poor populations. DOTS does not require hospitalization, a luxury not available for the countries hardest hit by TB. Further, the program is cost-effective. The World Bank rates DOTS "*one of the most cost-effective of all health interventions.*"

Courtesy of WHO/TDR/Andy Crump. Used with permission.

FIGURE 12.30 Patient in a DOTS program taking her TB medicine.

The tuberculosis antibiotic success story has been tarnished by the emergence of **multiple drug-resistant tuberculosis (MDR-TB)** and **extremely drug-resistant tuberculosis (XDR-TB)**. MDR-TB bacteria are resistant to at least two of the best antibiotics to treat TB, isoniazid and rifampin, both first-line drugs. XDR is a relatively rare type of MDR-TB and is resistant to isoniazid and rifampin and at least three of the second-line drugs. Misuse of antibiotics is the major cause. The regimen of TB therapy is long and expensive, frequently resulting in noncompliance, which, as noted earlier, fosters the development of antibiotic resistance.

○ Prevention: Vaccine Development

Public health strategies focus on prevention of disease. The high worldwide incidence of TB affords top priority to the development of new and effective vaccines against the disease. The **bacillus Calmette-Guérin (BCG) vaccine** for tuberculosis is available, but its use is controversial. The vaccine consists of attenuated (weakened) live TB bacteria. The protection rate is about 80% in children and less than 50% in adults, with the duration of protection ranging from 5 to 15 years.

The BCG vaccine is not used in the United States because the incidence of TB does not justify its use, except among high-risk groups, including health professionals charged with the care of patients with TB and military personnel serving in areas with a high rate of TB. Further, those receiving the BCG vaccine will have a positive tuberculin reaction for life, necessitating repeated chest x-rays for monitoring purposes.

○ TB: Current World Status

Tuberculosis is one of the world's deadliest infectious diseases. In 2016, 10.4 million people around the world had TB disease and 1.7 million died of TB. Over 95% of TB deaths occur in low- and middle-income countries. Seven countries, listed in order from highest to lowest, account for 64% of the total cases:

1. India
2. Indonesia
3. China
4. Philippines
5. Pakistan
6. Nigeria
7. South Africa

In 2016, 1 million children were diagnosed with TB disease and 250,000 children died of TB (including children with HIV-associated TB). TB is the leading killer of HIV-positive people; in 2016, 40% of HIV deaths were due to TB.

In 2016, 9,272 cases of TB were reported in the United States, the lowest case count on record. The CDC estimates that about 14% of cases are recent transmissions. Tuberculosis remains a current and dangerous public health crisis and health security threat made worse by the mushrooming occurrence of MDR-TB and XDR-TB strains of *M. tuberculosis*.

Mycobacterium Avium Complex (MAC)

Mycobacterium avium complex (MAC) refers to infections caused by two species of **nontuberculous mycobacteria (NTM)**, that is, mycobacteria that do not cause TB: *M. avium* and *M. intracellulare*. These bacteria are ubiquitous in the environment at a worldwide level. They can be found in soil (including potting soil), household dust, and water, especially drinking water. These bacteria are opportunistic pathogens that can cause life-threatening pulmonary infections in humans, other mammals, and birds. Infections occur in immunodeficient (e.g., HIV/AIDS or lymphoma patients), immunosuppressed (e.g., cancer patients on chemotherapy, transplant recipients on immunosuppressive therapy), and nonimmunosuppressed patients with risk factors for infection [e.g., smokers or people with lung conditions such as chronic obstructive pulmonary disease (COPD) or cystic fibrosis (CF) patients], and older adults whose immune systems are weakening due to the aging process.

NTM, especially *M. avium* and *M. intracellulare*, are contaminants in drinking water because they can grow as biofilms in household plumbing, including hot water systems. *M. avium* was detected in water aboard the Russian space station Mir. If disinfectants and soaps are flushed down water pipes, the disinfectants kill or inhibit the fast-growing competing bacteria, leaving the NTMs to grow and persist at their slow rate using minimal nutrients in the absence of the competing fast-growing normal microbiota in the plumbing system.

Because the NTMs form biofilms attached to the plumbing surfaces, the bacteria are not detected in water samples; however, they can be isolated from water faucet taps and showerheads (**FIGURE 12.31A**). The water

(a)

© Teri Shors.

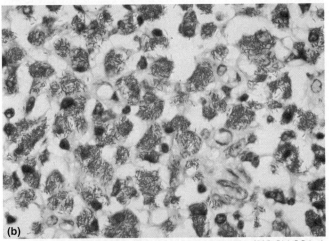

(b)

Courtesy of the CDC/Dr. Edwin E. Ewing Jr.

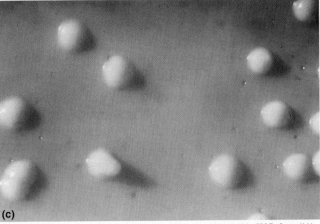

(c)

Courtesy of the CDC/Dr. George Kubica.

FIGURE 12.31 (a) Showerheads have been associated with *Mycobacterium avium* and *Mycobacterium intracellulare* (MAC) infections. The mycobacteria can form biofilms in household plumbing that are aerosolized deep into the lungs, causing opportunistic infections in people with compromised immune systems or underlying lung disease. **(b)** Photomicrograph of an acid-fast-stained lymph node taken from an HIV/AIDS patient suffering from an *M. intracellulare* infection. Note the many fuchsia-stained mycobacteria among the plump, blue histiocytes. **(c)** Colonies of *M. avium* on solid media shown here are smooth/waxlike, raised and opaque.

heater temperature and water source does affect the presence of NTMs in household plumbing. NTMs were less frequently isolated by researchers from water heater temperatures that were set at 55°C (131°F) or higher. Note that the manufacturer default setting for hot water heaters is usually 60°C (140°F) but to save energy costs recommendations are to reduce the temperature to 49°C (120°F). This is considered safe for the majority of the population but may not be such a good idea for those at risk for pulmonary infections.

Pulmonary MAC is the most common form. The incidence of MAC infections is increasing. The most common cause is thought to be aerosol inhalation of the bacteria. Cases have been associated with showering. Showers provide the ideal niche for bacterial biofilm formation, enabling MAC to accumulate in large numbers and then be dispersed into the passing water and aerosolized by the showerhead. The MAC are inhaled deep into the lower respiratory tract (*Legionella* bacteria can cause infections this way as well). The onset of symptoms is gradual, leading to coughing, weight loss, fatigue, and night sweats. It can sometimes progress to **lymphadenitis** (swollen lymph nodes primarily on one side of the neck) in children ages 1 to 4 years or HIV/AIDS patients. It is likely not contagious from person to person. Diagnosis is based on symptoms, culturing, and acid-fast staining (**FIGURE 12.31B**) and culturing mucus samples (**FIGURE 12.31C**), chest, or lung x-ray. Treatment includes antibiotic therapy but sometimes surgery is required to remove the affected lymph nodes.

Sexually Transmitted Infections (STIs)

Unlike most other infectious diseases, the **sexually transmitted infections (STIs)**, formerly called **venereal diseases (VDs)**, carry with them the stigma of "human wickedness" because they are associated with personal sexual behavior. **Syphilis** and **gonorrhea** may be passed from mother to child in utero. Ironically, these miserable diseases, recognized in the 1600s, were named after Venus, the Roman goddess of love. Before the **germ theory of disease** it was believed that some females passed onto their male partners a "noxious substance" that resulted in an STI (a flagrant example of sexism).

Bacteria that cause STIs are directly transmitted from the warm, moist mucous membranes of one individual to the warm, moist membranes of another individual during sexual practices, including vaginal and anal intercourse and oral sex. Semen and vaginal discharges are the major sources. The male and female genitourinary tracts are diagrammed in **FIGURE 12.32**. Promiscuous individuals are particularly prone to STIs because of their sexual behavior. Unfortunately, one STI is not exclusive of the others, so individuals may have more than one STI simultaneously. **TABLE 12.9** summarizes STIs.

The advent of public education programs, particularly during World War II, and the appearance of penicillin in 1943, led to a dramatic decrease in the incidence of syphilis and gonorrhea; predictions were that they would largely become diseases of the past. Unfortunately, these predictions did not come true. In recent times STIs have become a serious public health problem, particularly because of the emergence of antibiotic-resistant strains. The current concern with HIV/AIDS overshadows the continuing problem of other STIs. Vaccines are not available; prevention focuses on abstinence, monogamy, and safe sex.

Syphilis

Girolamo Fracastoro (1478–1553), an Italian poet and physician, wrote a poem in 1530, *Syphilis sive morbus gallicus*, about a shepherd boy named Syphilus who unknowingly offended Apollo, the sun god. Apollo vented his wrath on the boy by inflicting him with a "loathsome" disease. Ultimately, this new disease became known as *syphilis sive morbus gallicus*. The causative agent of syphilis is the Gram (−) spirochete-shaped bacterium *Treponema pallidum* (**FIGURE 12.33**).

T. pallidum frequently coinfects with pathogens that cause other STIs, including HIV/AIDS. Humans are the only known reservoir. In 2016, more than 27,814 cases of primary and secondary syphilis were reported in the United States, representing a 17.6% increase since 2015. A total of

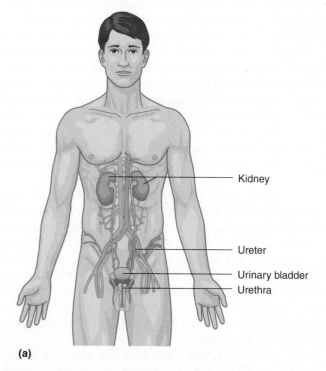

(a)

FIGURE 12.32 **(a)** Anatomy of the human genitourinary system. (continues)

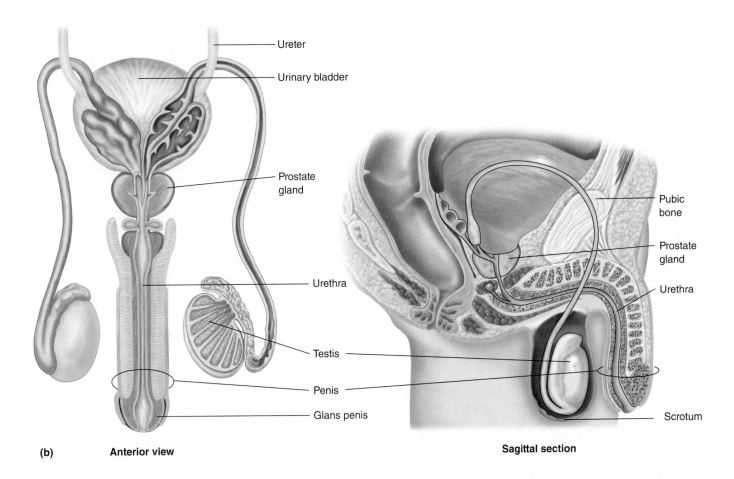

(b)   **Anterior view**   **Sagittal section**

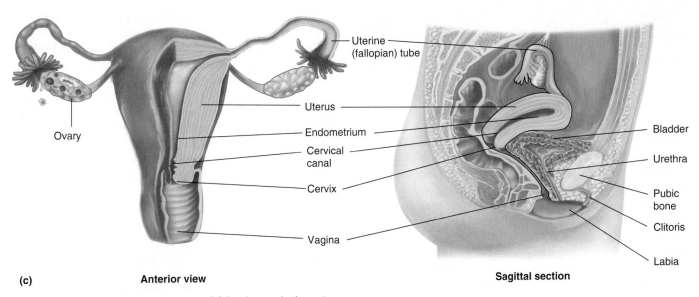

(c)   **Anterior view**   **Sagittal section**

FIGURE 12.32 (continued) **(b)** Male and **(c)** female reproductive systems.

90% of primary and secondary syphilis cases were males. Men who have sex with men account for 89.6% of cases in the United States. Higher numbers of cases were recorded in males in the 20- to 29-year-old age group. However, syphilis among men having sex with women continues to be a problem. The disease is sometimes referred to as "the great imitator" because, in its later stages, it mimics other diseases.

Syphilis progresses through a series of three stages: primary, secondary, and tertiary (**FIGURE 12.34**). **Primary**

TABLE 12.9 Sexually Transmitted Bacterial Diseases*

Sexually Transmitted Infection	Incubation Period	Bacterial Pathogen	Symptoms
Syphilis	16–28 days (possibly as long as 10 weeks)	*Treponema pallidum*	Painless sores on the penis or cervix, rashes on the palms and soles, and eventual paralysis and insanity
Gonorrhea	2–7 days	*Neisseria gonorrhoeae*	Burning urination, cervical and urethral infection, abdominal pain, sterility
Chlamydia	2–6 weeks	*Chlamydia trachomatis*	Ectopic pregnancy, infertility, inflammation of the testicles
Lymphogranuloma venereum	1–4 weeks	*Chlamydia trachomatis*	Sores on the penis and vagina that develop into painful buboes
Chancroid	3–5 days	*Haemophilus ducreyi*	Painful ulcers on the penis, labia, or clitoris

*These diseases are treatable with antibiotic therapy; antibiotic-resistant strains pose a problem in some cases.

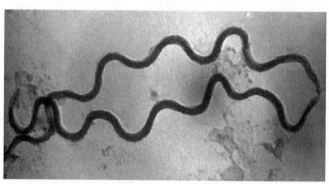

Courtesy of Joyce Ayers/CDC.

FIGURE 12.33 A photomicrograph of two spirochete-shaped *Treponema pallidum* bacteria.

syphilis is manifested by the appearance of painless **chancres** (sores) on the penis or on the cervix, which may be undetected. These chancres shed *T. pallidum* continuously (**FIGURE 12.35A**). The chancres disappear in about 4 to 6 weeks and give false hope of recovery to those infected. In untreated individuals, over approximately 5 years, symptoms of a rash, particularly on the palms and soles, appear and disappear. **Secondary syphilis** is systemic, meaning that the bacteria multiply and spread throughout the body (**FIGURE 12.35B**).

A single intramuscular injection containing 2.4 million units of long-acting benzathine penicillin G will cure a person who has primary, secondary, or early late syphilis. Three doses of intramuscular injections containing 2.4 million units of long-acting benzathine penicillin G per dose is enough to cure late latent syphilis or syphilis of unknown duration.

About one-third of those untreated progress to **tertiary syphilis**, an advanced stage that develops over 40 years, during which numerous organs and tissues, particularly those of the cardiovascular and nervous systems, show degenerative changes. Before the antibiotic era individuals with tertiary syphilis constituted a significant portion of the population of mental institutions. A poem by an anonymous author described the progression of syphilis. In the poem the term **gummas** refers to tumorlike lesions that develop during the tertiary stage; these gummas destroy nerve or skin tissue. Neurosyphilis is particularly disabling and can result in a shuffling walk (**tabes**), paralysis (**paresis**), and insanity.

Although sexual contact is the usual mode of transmission of syphilis, the bacteria can be passed in saliva and pose a threat to dentists, dental hygienists, and "deep kissers." Of particular concern is that infants can acquire **congenital syphilis** as the result of spirochetes that pass across the placenta from mother to baby. In 2016, 628 cases of congenital syphilis (a 27.6% increase since 2015) were reported in the United States. These infants develop serious problems, including **saber shins**, a condition in which the shinbone develops abnormally.

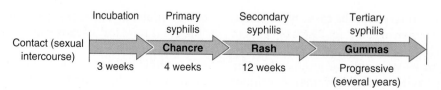

FIGURE 12.34 Stages and time line of syphilis. Times are approximate and subject to considerable individual variation.

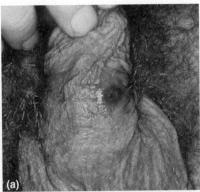

Courtesy of Dr. Gavin Hart and Dr. N. J. Fiumara/CDC.

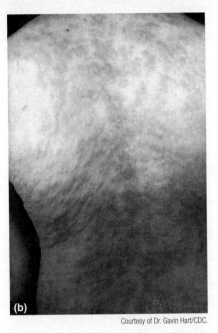

Courtesy of Dr. Gavin Hart/CDC.

FIGURE 12.35 Characteristic lesions of syphilis. **(a)** Chancre of primary syphilis on penis. **(b)** Lesions of secondary syphilis on the back; these lesions may occur on other body surfaces.

In March 2018, public health officials in Milwaukee, Wisconsin, were dealing with at least 127 individuals, including young teenagers, who had contracted HIV, syphilis, or both. This was a "cluster" event, meaning that those diagnosed could be connected to each other and had had recent contact. Most of the affected individuals were men, and 45% were HIV positive. The age range of those diagnosed was between 14 and 24 years of age. In response to this concerning outbreak, a 414ALL campaign to lower STIs and teen pregnancy rates in Milwaukee was formed. The campaign makes condoms more available, accessible, and acceptable for teens (https://www.414allmke.org/).

Considering the high incidence of syphilis on a worldwide basis, it is not surprising that there have been intensive efforts to develop a vaccine; thus far, efforts have been futile. This has been particularly disappointing because a genetic map of *T. pallidum* was completed in 1998, opening up new avenues of vaccine development.

Campaigns aimed at educating the public to practice safe sex remain vital; the military still plays an important role in this effort.

Gonorrhea

Gonorrhea, sometimes referred to as "clap" or "drip," is caused by *Neisseria gonorrhoeae*, a Gram (−) diplococcus bacterium. Like syphilis, humans are the only reservoir. During 2015–2016, the rate of reported gonorrhea cases increased by 22.2% among men and 13.8% among women, but this is an underestimate given that only about half of all cases are reported to the CDC. In 2016, the southern states had the highest number of gonorrhea cases, with an infection rate of 166.8 per 100,000 people. *N. gonorrhoeae* may be transmitted by vaginal, oral, or anal sex. **Pili** enable the bacteria to attach to cells lining the urethra, allowing them to hang on during the passage of urine. Gonorrhea is the second most prevalent STI in the United States (after chlamydia).

A major problem of controlling gonorrhea is that healthy carriers often remain without symptoms after infection but continue to transmit the bacteria for more than 10 years. As many as 10% of males and 30% of females may not show symptoms. In males the disease is characterized by a pus-containing drip from the penis and burning during urination (**FIGURE 12.36**). In females the disease may remain hidden, and the cervix and urethra are the most common sites of infection. Furthermore, *N. gonorrhoeae* is now on the list of **superbugs** (antibiotic-resistant bacteria that cause infections that are very hard to treat). In 2016, 44.1% of *N. gonorrhoeae* isolates were resistant to penicillin, tetracycline, ciprofloxacin, or some combination of those antibiotics. The proportion of patients treated with an alternative antibiotic, ceftriaxone, increased from 84% in 2011 to 96.9% in 2016.

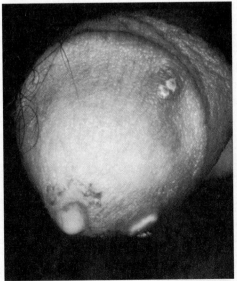

Courtesy of the CDC.

FIGURE 12.36 Male penis with discharge due to *Neisseria gonorrhoeae* infection.

Pelvic inflammatory disease (PID) occurs in about 50% of untreated females and is characterized by abdominal pain and, possibly, sterility. *N. gonorrhoeae* can be transmitted into the eyes of newborns during delivery, causing corneal damage and possibly blindness, a condition referred to as **ophthalmia neonatorum**. Erythromycin or another antibiotic ointment is placed into the eyes of newborns as required by law in most states.

Chlamydia

Chlamydia, caused by *Chlamydia trachomatis*, a Gram (–) coccobacillus, is the most common STI, as well as the most commonly reported nationally notifiable disease. Most people are not even aware of its existence; it is sometimes referred to as "the silent epidemic." In 2016, 1,598,354 cases were reported in the United States.

This disease is particularly prevalent in young adults and teenagers. Even those with the disease may be unaware that they are infected because approximately 70% of infected females and 30% of infected males lack symptoms. Females are particularly vulnerable to complications, including PID (as in gonorrhea), abnormal pregnancy, and infertility. Males are subject to inflammation of the testicles and infertility. The silent nature of the infection is a major obstacle in prevention; education and alertness to this disease is a major preventive measure. The CDC recommends sexually active females 25 years old and younger and women 25 years and older at risk for infection (e.g., women with new or multiple sex partners) be tested each year.

As in gonorrhea, newborns whose mothers are infected have a 50-50 chance of developing an eye-threatening condition; therefore, antibiotics are administered at birth. Chlamydia is the easiest STI to treat, and a single dose of antibiotic usually cures the disease in 1 week.

A silent epidemic of chlamydia presented a serious problem in the United Kingdom in 2005. It was the most common STI, affecting more than 10% of sexually active young men and women. In reality, the figures were probably very much higher, because many cases are not reported. The pharmaceutical company, Boots, with locations spread across the country, inaugurated a program in 2005 to encourage young people to be screened and

© Robert I. Krasner.

FIGURE 12.37 Chlamydia is rampant in the United Kingdom, prompting Boots, a pharmaceutical company with retail stores throughout the United Kingdom, to sponsor an awareness program. Banners like this are prominently displayed in drug stores.

to seek treatment if necessary (**FIGURE 12.37**). More than 908,488 chlamydia tests were done in England from January 1 to September 20, 2011, of which 7.4% tested positive.

Contact Diseases (Other Than STIs)

A number of other contact diseases are common in human populations, but only three are discussed here (**TABLE 12.10**): peptic ulcer, leprosy, and *Staphylococcus aureus* infections. Two uncommon HAIs are listed in Table 12.10 but are not discussed.

Peptic Ulcers

Students sometimes complain that the stress of exams, papers, (boring) lectures, and all other facets of academic life are "giving them ulcers." It is true that student life is stressful (so is a professor's!), but it won't give you ulcers. The popular myth that ulcers are directly associated with stress is no longer relevant. Take a look at **TABLE 12.11**, and you may be surprised to learn that those engaged in stressful occupations are no more at risk of getting ulcers than those in occupations considered nonstressful. Ulcers were also once thought to be the result of eating spicy foods. Treatment focused on hospitalization, bed rest, bland diet, and consuming large quantities of milk, cream, and antacids to coat the lining of the stomach and the duodenum (the first part of the small intestine). In some cases, surgery was performed to remove the diseased area. But relief was only temporary.

In 1982, two Australian physicians, J. Robin Warren and Barry Marshall, claimed that *Helicobacter pylori*, a comma-shaped Gram (–) bacterium, not stress or diet, was the cause of **peptic ulcers**. This idea knocked the medical

TABLE 12.10 Contact Diseases (Other Than STIs)*

Infectious Disease	Incubation Period	Bacterial Pathogen	Mode of Transmission	Symptoms
Leprosy/Hansen's disease	1–20 years (average of 5 years)	*Mycobacterium leprae*	Skin contact; air droplets	Development of lepromas; neurological damage
Staphylococcal skin infections	3–4 days	*Staphylococcus aureus*	Skin contact	Pimples, abscesses, and peeling on skin; systemic symptoms in toxic shock syndrome
Healthcare-associated infection causing severe toxic shock syndrome and infection of umbilical stump in newborns	Unknown	*Clostridium sordellii*	Healthcare-associated infection/transmission between persons or from the environment unknown	Nausea, vomiting, diarrhea, abdominal pain without fever
Healthcare-associated infection affecting skin and soft tissues	Unknown	*Mycobaccterium abscessus*	Healthcare-associated infection/transmission between persons or from the environment unknown	Skin is red, tender, warm to the touch, boils and pus-filled vesicles form. Fever, chills, and muscle aches may occur.
Peptic ulcers	Unknown	*Helicobacter pylori*	Unknown, but probably person to person, possibly food and contaminated water	Burning or gnawing in epigastrium, particularly when stomach is empty; possibly bleeding; nausea, vomiting

*These diseases are treatable with antibiotic therapy; antibiotic-resistant strains pose a problem in some cases.

TABLE 12.11 Stress and Ulcers: A Myth

Most Stressful Jobs in the United States in 2018	Least Stressful Jobs in the United States in 2018
1. Enlisted military personnel at E3 level or those with at least 6 years of experience	1. Diagnostic medical sonographer
2. Firefighter	2. Hair stylist
3. Commercial airline pilot	3. Audiologist
4. Police officer	4. University professor
5. Event coordinator	5. Medical records technician
6. Newspaper reporter	6. Compliance officer
7. Broadcaster	7. Jeweler
8. Public relations executive	8. Pharmacy technician
9. Senior corporate executive	9. Operations research analyst
	10. Medical laboratory technician

Do you believe that stressful jobs lead to ulcers? Which of these people listed above are most likely to have an ulcer? The answer may surprise you. All the workers on this list are just as likely to get an ulcer as any others you can imagine. A firefighter's stomach is no more likely to be riddled with ulcers from the stress of dealing with fires than a jeweler's stomach. Although stress and diet can irritate an ulcer, they do not cause it. Ulcers are caused by the bacterium *H. pylori* and can be cured with a 1- or 2-week course of antibiotics, even in people who have had ulcers for years.
Information from CareerCast.com.

establishment off its feet. Marshall went to the extreme of drinking cultures of *H. pylori* to prove his claim to a skeptical medical community, and, sure enough, he developed ulcers. Finally, in 1994, a National Institutes of Health Consensus Development Conference affirmed a strong association between *H. pylori* and peptic ulcers and recommended antibiotics for treatment.

How does *H. pylori* survive in the harsh acidic environment of the stomach and in the duodenum— environments hostile to most microbes? The answer is a remarkable example of biological adaptation. The bacterium produces the enzyme **urease**, which breaks down the compound urea, a metabolic product of the body, into carbon dioxide and ammonia, an alkaline compound that neutralizes the acidity. This clever survival strategy protects *H. pylori*.

The mechanism of transmission of *H. pylori* is not clear, nor is there an explanation why some persons infected with the bacterium develop ulcers but others do not. Direct person-to-person contact appears to be the most plausible route; humans are the primary reservoir. The possibility also exists that food and water are involved in the mode of transmission.

The diagnosis of peptic ulcers is based on several methods, including the presence of antibodies specific for *H. pylori* and the culturing of tissue, obtained by **biopsy**, to demonstrate the presence of the bacterium. A test called the **breath test** is about 94% to 98% accurate and is based on bacterial production of urease. For this test the patient drinks a preparation containing isotopically labeled urea. If present, *H. pylori* breaks down the urea, resulting in the formation of carbon dioxide, which is exhaled. Measurement of the labeled carbon dioxide in the breath determines the presence or absence of *H. pylori*.

H. pylori infection can be cured in 2 to 3 weeks with appropriate antibiotics and other medications. The recurrence rate after antibiotic treatment is about 6%, compared with 80% when only antacids are used. The bottom line is that ulcers are a bacterial infection and not caused by stress.

H. pylori infection is common throughout the world. It is estimated that 70% of the population in developing countries and 30% to 40% of the population of the United States and other developed nations are infected with *H. pylori* but only about 10% develop ulcers during their lifetime. Infection typically occurs during childhood and may persist lifelong unless treated. Studies have indicated that long-term *H. pylori* infection is associated with the development of gastric cancer, the second most common cancer worldwide.

Leprosy

Among all those diseases carrying a social stigma, **leprosy** is at the top of the list (**BOX 12.2**). Historical accounts of "affliction" point to leprosy; through the ages the disease conjured up dreadful images of rejection and exclusion from society, disfiguring skin lesions, and missing fingers and toes (**FIGURE 12.38A**). People suffering from the disease were required to carry and ring bells and to shout "unclean" as they approached others (**FIGURE 12.38B**). Fear in the minds of healthy individuals led to terrible tales of cruelty. The story of a Belgian Roman Catholic priest Father Damien de Veuster (born Jozef De Veuster) is one of great dedication and compassion. In 1870, Father Damien established a **leper colony** on the island of Molokai in Hawaii as a refuge for those with the disease. He spent his

BOX 12.2 Politics of the Leprosy Bacillus

In August 2001, I [Robert I. Krasner] visited the Leprosy Museum at St. Jørgens Hospital in Bergen, Norway (http://www .bymuseet.no/vaare-museer/lepramuseet/). This might appear to be an unusual activity; most tourists are attracted to the other museums and cultural sites that Norway has to offer. (In fact, not to my surprise, I was the only tourist there!) The hospital or **leprosarium** was founded around 1411 and stayed in operation continuously until 1946, providing a home for thousands of Norwegian lepers (**Figure 1**). Leprosy was common in the western parts of Norway into the late 19th century.

While visiting the museum I met Sigurd Sandmo, its curator, and spent several hours with him over a 2-day period. Sigurd is a medical historian with a vast amount of knowledge. I was so impressed with this young man that I invited him to write the following piece about leprosy:

Today Norway is a fully developed country and among the wealthiest nations in the world. But 150 years ago, one

Figure 1 Leprosy Museum in Bergen, Norway.

would consider Norway to be a poor outpost in Europe, with many of the problems to be found only in developing countries today. One of the most conspicuous problems was leprosy, a disease that was looked upon with great interest and concern by Norway's physicians and authorities. In the second half of the 19th century, the city of Bergen had the highest concentration of leprosy patients in Europe, and the city was an international center for leprosy research.

For most other European countries, where the disease had ceased to be a problem around 1500, leprosy had during the 19th century more or less become a subject for missionaries. In the reports from work among African and Asian lepers, the Levitical meaning of the disease was now reproduced. An almost forgotten medieval disease had renewed its religious and racial actuality. In Europe only Norway and Iceland experienced a second bloom of leprosy in the 19th century. The Norwegian distribution of leprosy was rather local, and the disease was mainly found among poor fishermen and peasants in the western parts of the country. With an incidence of more than 3% of the population in some districts, this was a shame for the young Norwegian state. In the 1870s, a Swedish physician referred to the problem as "the waves of shame on the shores of Norway."

The national authorities offered the medical circles in Bergen both economic freedom and political influence to have the problem solved. In return they expected visible results. From our modern point of view, we may see several scientific breakthroughs. In 1847, Daniel C. Danielssen published the first symptomatology of leprosy, which represents the birth of modern leprosy research. In 1856, Ove G. Hoegh founded the Norwegian Leprosy registry, probably the first national disease registry in the world. But the value of these events was hard to recognize at the time. Much more spectacular was Armauer Hansen's discovery of the leprosy bacillus in 1873. Hansen's publication from 1874, Preliminary Contributions to the Characteristics of Leprosy, is the earliest description of a microorganism as the cause of a chronic disease.

Hansen's discovery represented the kind of breakthrough the public health authorities were waiting for. It was the microscope against the Old Testament and the scientist's rationalism against the Levitical myth. And ever since, health workers have used the story of the discovery to fight the myth where it has survived. Hansen has become a symbol of dignity, humanism, and a scientific approach toward patients who are still suffering from Hansen's disease. When we come across Hansen stamps from Thailand or a Hansen monument in the Vietnamese jungle, we easily recognize the impact Hansen's discovery once had. The authorities, the mob, the church, the peasants, and the patients could all see how Hansen's scholarly approach challenged the 2,000-year-old stigma.

However, Hansen's discovery had immediate political implications too. At the time, Norway did not have any leprosy legislation. Admission to all hospitals was voluntary, as the common opinion was that the disease was either inheritable or a nonspecific condition caused by bad living conditions. During the first years after the discovery, Hansen's theory of a contagium vivum was met with a certain amount of skepticism from his colleagues, both in Norway and abroad. When Hansen presented his proposal for a new act requiring isolation of leprosy patients in Norway, many of his colleagues felt that this was an inhumane law, branding the patients as criminals. Hansen defended himself as a pragmatic scientist: How can we act humanely toward persons who have a contagious disease? His opponents read Hansen's contributions to the discussion with regret. However, the law was passed in 1885, and served as a model and inspiration for leprosy legislation and public action in disease colonies in several other countries, such as Greece, Hawaii, and Japan. Today the remains of colonies in Spinalonga, Greece, and in Molokai are sad sights for visitors. In Japan surviving patients now receive compensation for their sufferings caused by harsh legislation.

In the discussions concerning the public health work and legislation in Norway in the 1870s and 1880s, we can see how Hansen's work gradually became politicized. The discovery itself was internationally accepted during the 1880s, but many of Hansen's opponents and colleagues saw that the political answer to the leper question was complicated and that the new legislation in a way revitalized the old stigma. In one of Norway's medical journals, the discussion went on for several years. After years of heated debate with irritated colleagues, Hansen might have considered whether his fundamentalistic positivism and rationalism were, after all, suitable for discussing public health work.

Several elements from the discussion on leprosy legislation in the 1880s were repeated with regard to another disease 100 years later in Norway, as in most other countries; the AIDS discussions of the 1980s were surprisingly similar. Again, the possibility of establishing disease colonies was discussed and society's rights and responsibilities with respect to infected individuals were presented as a question of yes or no. And, again, the evidence of contagion was used by different groups with different interests to justify their political and religious opinions. Once again, scientific discoveries of microorganisms turned out to be powerful but unpredictable political weapons.

We often think of biological and medical discoveries as nonpolitical statements. And when we look at the single participant, we are usually right. But like the discovery of the leprosy bacillus, scientific breakthroughs are usually caused by priority given by political actors, and scientific results will often take on a political dimension once they leave the laboratory. Both the discovery of the leprosy bacillus and the discovery of human immunodeficiency virus remind us how vulnerable and risky this dimension can be.

Sigurd Sandmo, curator, Leprosy Museum, St. Jørgens Hospital, Bergen, Norway.

Courtesy of National Library of Medicine.

© The British Library/age fotostock.

FIGURE 12.38 (a) A patient with leprosy, in this medieval drawing, is examined by a physician, who is consulting other physicians. **(b)** A historic painting from the 14th century showing a female leper with missing limbs ringing a bell to announce her presence.

life there and ultimately died of leprosy (**FIGURE 12.39**). In 2009, he was the 10th American to be canonized as a saint by the Roman Catholic Church. The *Catholic Encyclopedia* refers to Father Damien as the *"Apostle of the Lepers."*

Today, the term "leper," because of its stigma, is not socially acceptable, and the disease is now known as **Hansen's disease**, after Gerhard Henrik Armauer Hansen (**FIGURE 12.40A**), the physician who first observed *Mycobacterium leprae*. *M. leprae* is the causative agent of Hansen's disease; note that it belongs to the same genus, *Mycobacterium*, as the tubercle bacillus (*Mycobacterium tuberculosis*). The disease is transmitted by skin contact and respiratory droplets over a long period of time; it is not considered to be highly infectious. The **incubation period** is long (1 to

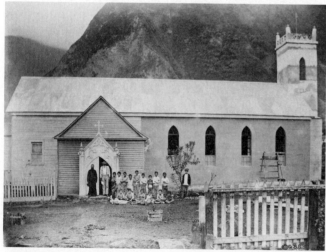

Courtesy of the Library of Congress.

FIGURE 12.39 Father Damien (1840–1889) standing with an adult and 16 boys outside the church at the leper colony at Kalaupapa on Molokai Island, Hawaii. Today, Kalaupapa is a National Historical Park of Hawaii (https://www.nps.gov/kala/index.htm).

© Robert I. Krasner.

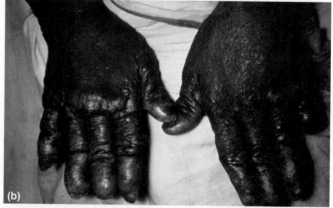

Courtesy of CDC/Dr. Andre J. Lebraun.

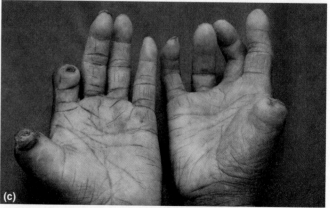

© Medicshots/Alamy Stock Photo.

FIGURE 12.40 (a) A statue of Gerhard Henrik Armauer Hansen. **(b)** With leprosy, tumorlike lesions appear on the skin. **(c)** A patient with leprosy is likely to have had the disease for 30 years or more. This patient has lost fingers because of the disease.

20 years; average is about 5 years), and it takes about 3 to 6 years of close contact with an individual with the disease for it to be acquired. Children appear to be more susceptible.

Hansen's disease is characterized by a variety of physical manifestations (**FIGURES 12.40B** and **C**), including disfiguring tumorlike skin lesions (**lepromas**) and neurological damage to the cooler peripheral areas of the body, such as the hands, feet, face, and earlobes. This may lead to curling of the fingers (**claw hand**), thickening of the earlobes, collapse of the nose, and possibly blindness. The individual may lose the ability to perceive pain in the fingers and toes, and accidental burns may occur, resulting in serious deformities. Fortunately, antibiotics are now available to arrest the disease, and patients with Hansen's disease can look forward to a normal life.

It is estimated that each year about 150 to 250 people are diagnosed with leprosy, and a total of about 3,000 people need care for leprosy each year in the United States. The cases are typically new immigrants from Central and South America or Asia or people in contact with infected armadillos in southern states (e.g., Florida and Texas). The national Hansen's Disease Program in Baton Rouge, Louisiana, is the United States' main care and research facility and operates under the auspices of the U.S. Department of Human Health and Services.

At one time more than 400 residents lived at the original Louisiana Leprosarium in Carville, which dated back to the 1890s (**FIGURE 12.41**). Closing the hospital caused

Krasner's Notebook

I visited the hospital for Hansen's disease in Carville, Louisiana, in 1962. The atmosphere in this modern hospital, located in a parklike setting, was remarkably upbeat, as was the attitude of the people with the disease.

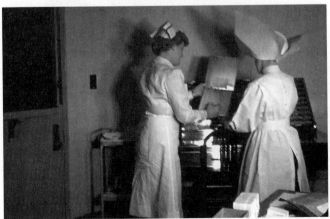

Courtesy of CDC/Elizabeth Schexnyder, ©National Hansen's Disease Museum, Curator.

FIGURE 12.41 Public health nurse (left) and a Sister of Charity nurse (right) are in the process of exchanging patient information in the "chart room" at the Louisiana Leprosarium during its operation. At the time 21 Sisters of Charity were on staff. The act of quarantining patients for leprosy remained law in the state of Louisiana until 1957. Today the Carville Louisiana Leprosarium is a national museum honoring leprosy patients (https://www.hrsa.gov/hansens -disease/museum/index.html).

great anxiety among patients who had spent their lives there. These residents were referred to specialized outpatient facilities throughout the country and were offered financial subsidies by the government. The leprosy center in Molokai has also been closed. Persons having completed or under appropriate antibiotic therapy are considered free of active infection. The WHO reports that the worldwide cumulative number of Hansen's disease new cases registered early in 2016 was close to 216,108. The number of new cases reported each year continues to decline.

Staphylococcus aureus Infections

Bacteria of the genus *Staphylococcus*, frequently called "staph," a Gram (+) group of cocci (arranged in grapelike clusters), of which there are over 30 types that cause a multitude of infections, are normal microbiota of the skin, mouth, nose, and throat; the skin is the largest organ in the human body (**FIGURE 12.42**). Usually, these bacteria

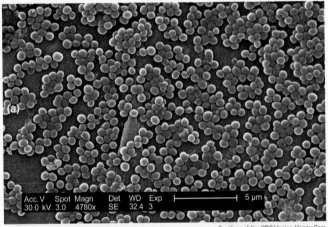

| Acc.V | Spot | Magn | Det | WD | Exp | | 5 µm |
| 30.0 kV | 3.0 | 4780x | SE | 32.4 | 3 | | |

Courtesy of the CDC/Janice Haney Carr.

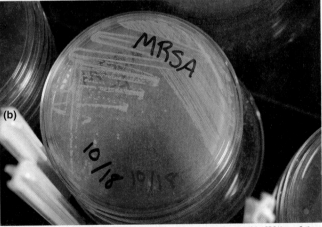

Courtesy of the CDC/James Gathany.

FIGURE 12.42 (a) Colorized scanning electron micrograph of *Staphylococcus aureus*. Note that the coccus-shaped bacteria clump together like clusters of grapes. Magnification 4,780×. **(b)** Methicillin-resistant *S. aureus* growing on mannitol salt agar plates. This medium is used to distinguish *S. aureus* from nonpathogenic species such as *S. epidermidis*. *S. aureus* ferments mannitol present in the medium into acidic metabolic end products, resulting in a pH change that causes the medium to turn yellow. *S. epidermidis* does not ferment mannitol into acidic metabolic products and the medium will remain pink.

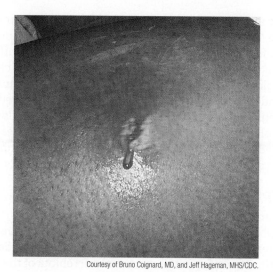

Courtesy of Bruno Coignard, MD, and Jeff Hageman, MHS/CDC.

FIGURE 12.43 Staphylococcal infection. Note discharge of pus.

present no problem. However, when the skin is broken, as can result from a wound, burn, or other circumstance, its normal integrity as a barrier is lowered, making possible the invasion of opportunistic strains of staphylococci. Certain strains of *S. aureus* are the most frequent staphylococci involved in infections. Healthcare-associated infections are a major problem.

Staphylococci are frequent causes of localized skin infections, occurring as pimples, abscesses, and inflamed lesions filled with a core of pus (**FIGURE 12.43**). **Abscesses** can progress to produce **boils**, which can develop into carbuncles. **Carbuncles** are larger and deeper lesions, and can reach baseball size. These lesions are extremely painful and are dangerous because they can progress into systemic (bloodborne and widespread) infections throughout the body.

A relatively common manifestation of staphylococcal (and streptococcal) infection is **impetigo**. This is a superficial infection of the skin that usually occurs around the mouth in the form of blisters that ooze a yellowish liquid. More frequently, impetigo occurs in very young children, particularly after a runny nose, which sets up irritation in the surrounding tissue. Impetigo is annoying but not particularly serious, other than the fact that it is spread easily from child to child. Impetigo can also be caused by streptococci. **Scalded skin syndrome** is also caused by staphylococci and tends to be more prevalent in children with infection of the stem of the umbilical cord. In these cases the skin can become blistery as a result of the production of an exfoliative toxin that peels away the skin to expose a red layer.

HAIs are a major problem, and **methicillin-resistant Staphylococcus aureus (MRSA)** is one of the leading causes of HAIs in the United States. Everyone is at risk. MRSA is widespread and is no longer only hospital acquired. MRSA infections are defined as **community acquired MRSA (CA-MRSA)** if the person infected with MRSA was not hospitalized within 2 years before the date of the MRSA

infection. These infections are usually skin infections that appear in the form of boils or sores. About 30% of people carry MRSA in their noses but show no signs and symptoms.

CA-MRSA infections are associated with athletic facilities, dormitories, military barracks, correctional facilities, and day care centers. The news media has focused attention on CA-MRSA strains that cause outbreaks among professional, college, and high school athletes participating in highly physical sports (person-to-person contact) such as football, wrestling, rugby, and soccer. MRSA infections have developed in athletes with preexisting cuts and abrasions who then shared bars of soap or towels with a MRSA-infected person or carrier. This is another good reason not to share hygiene or personal products! Today athletic locker rooms and health clubs are filled with posters educating the public about MRSA and personal hygiene, including the sharing of personal items and equipment (**FIGURE 12.44**).

Vancomycin-intermediate *Staphylococcus aureus* **(VISA)** and **vancomycin-resistant** *Staphylococcus aureus* **(VRSA)** are antibiotic-resistant strains causing HAIs in patients who have conditions such as diabetes and kidney disease, use catheters, have had recent MRSA infections, and have had recent exposure to vancomycin and other antibiotics. These staphylococcal bacteria can cause serious infections in patients such as sepsis, pneumonia, endocarditis, and **osteomyletis** (bone infection).

Practice good hygiene: Do not share personal items, such as towels or razors. Wash your hands frequently. Shower immediately after every practice and game. Use clean towels each time you shower. Launder clothes and towels after each use. Your health matters.

Courtesy of the CDC.

FIGURE 12.44 NCAA educational poster on community-acquired MRSA infections.

Toxic shock syndrome came to light in the late 1970s. A major outbreak occurred in 1980 and caused considerable fear, particularly among menstruating females who used a certain brand of tampon. The outbreak led to a recall of the tampons in question. Some strains of *S. aureus* secrete an exotoxin that causes a high fever, nausea, vomiting, peeling of the skin (particularly on the palms and the soles), and a dangerous drop in blood pressure that leads to life-threatening shock. The infection carries about a 3% fatality rate. Toxic shock syndrome can also occur in men and in nonmenstruating females and may be associated with surgical wound infections. It can also be caused by streptococci.

Soilborne Diseases

Anthrax, tetanus, and leptospirosis are a few of the most common soilborne bacterial diseases (**TABLE 12.12**). Anthrax and tetanus are caused by endospore-forming rod-shaped bacteria and survive in soil for many years. Bacterial endospores are extremely resistant to many environmental stresses.

Anthrax: Inhalation, Cutaneous, Gastrointestinal

Anthrax, a potentially deadly disease caused by *Bacillus anthracis*, a Gram (+) rod, is considered by security experts as one of the most probable weapons for biological warfare because it produces endospores that can be readily disseminated by missiles and bombs (**FIGURE 12.45**).

Anthrax is primarily found in large, warm-blooded animals, mostly sheep, cattle, and goats, and is acquired by the ingestion of spores during grazing. Infection can also occur in humans and is most prevalent in agricultural regions of the world. The three forms of anthrax disease vary by the mode of transmission and subsequent symptoms.

Inhalation anthrax is an occupational hazard for humans exposed to contaminated dead animals and animal parts (e.g., tanners and sheep shearers), who, in

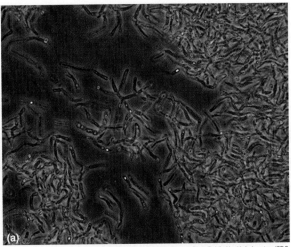

Courtesy of Larry Stauffer, Oregon State Public Health Laboratory/CDC.

Courtesy of Larry Stauffer, Oregon State Public Health Laboratory/CDC.

FIGURE 12.45 (a) Phase contrast photomicrograph of *Bacillus anthracis*. Arrows indicate chains of vegetative rod-shaped bacteria and refractile "points of light" endospores. **(b)** Mucoid *B. anthracis* colonies growing on heart infusion agar. The mucoid characteristic of growth is due to the production of a capsule that surrounds the bacterium.

TABLE 12.12 Soilborne Bacterial Diseases*				
Infectious Disease	Incubation Period	Bacterial Pathogen	Mode of Transmission	Symptoms
Anthrax	1–15 days	*Bacillus anthracis*	Contact with endospores in soil and in occupations involving handling of wool, hides (e.g., drum-making with hides), and meats	Cutaneous form produces skin lesion, headache, nausea, fever
Tetanus	4 days to several weeks	*Clostridium tetani*	Contact with spores from soil, animal bites, gunshot wounds	Lockjaw, muscle stiffness, and spasms due to production of tetanospasmin
Leptospirosis	7–13 days	*Leptospira interrogans*	Urine, soil, or water contaminated with urine from dogs, cats, sheep, rats, mice	Flulike with possible complications in liver and kidney

*These diseases are treatable with antibiotics; antibiotic-resistant strains pose a problem in some cases.

the course of their work, are most likely to inhale endospores; this condition is also called **woolsorter's disease**. Inhalation anthrax is the most severe form of the disease and is the greatest threat with the deadliest consequences in terms of biological warfare. The incubation period is 1 to 6 days. All it takes is a millionth of a gram (8,000 to 10,000 endospores) of B. anthracis to constitute a lethal dose; 1 kilogram (2.2 pounds) of B. anthracis endospores could kill billions of people. Early symptoms of disease are coldlike and progress to severe breathing problems within several days, followed by death 1 to 2 days later (**BOX 12.3**).

A second form of anthrax is **cutaneous anthrax**, also acquired by contact with B. anthracis endospores present in wool, hides, leather, or hair products (**FIGURE 12.46**). About

---

BOX 12.3 1979, The Year of the "Biological Chernobyl"

In April 1979, people, sheep, and cows began to die in Sverdlovsk (now called Yekaterinburg), a city with over 1 million people located in the former Soviet Union. Reports appeared in the Western media that the mysterious illness killing people in Sverdlovsk was anthrax. Russian officials said those affected had eaten black-market meat that was contaminated with *Bacillus anthracis*, the cause of anthrax. More deaths continued into May 1979.

City authorities wearing protective clothing and masks washed buildings with disinfectant and hosed down the streets without providing any explanation to city residents. Mud roads were paved. Residents were given mass vaccinations and antibiotics to ward off anthrax (which Russians referred to as **Siberian ulcer**). Anthrax was **endemic** in the Soviet Union, however, what they were experiencing was of a much larger scale and severity. Bodies of the deceased victims were placed in coffins containing chlorinated lime and buried in the same section of a city cemetery. At least 68 people died in a 2-month period.

Soviet doctors questioned the authorities and U.S. officials voiced their suspicions to the Soviet Union government. The outbreak caused intense international debate and speculation as to whether it was natural or accidental. It took over 8 years, but finally the answers surfaced. All of those who became ill had been outdoors in the evening or early morning of the day a maintenance person at Military Compound 19 forgot to replace a critical filter in a vent of the laboratory where B. anthracis was being secretly produced to weaponize missiles for a biological attack on the United States. It resulted in the release of a plume of B. anthracis endospores.

People in the path of the lethal plume inhaled the endospores, some falling ill even 6 weeks after the endospores were released. As investigations continued years after the outbreak, details surfaced about the wind direction on the day the endospores were released outside of Compound 19. It matched the trail of human deaths in Sverdlovsk and sheep deaths in small towns downwind

from the military facility (**Figure 1**). In 1992, post–Soviet Russia President Boris Yeltsin confirmed the accidental release of the deadly endospores being prepared to strike American cities. Their intentions backfired, resulting in the release of B. anthracis endospores into Russia's backyard. It was the largest outbreak of human inhalational anthrax in recorded history.

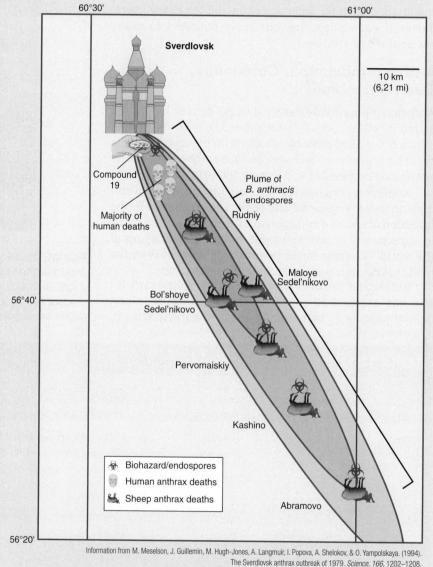

Information from M. Meselson, J. Guillemin, M. Hugh-Jones, A. Langmuir, I. Popova, A. Shelokov, & O. Yampolskaya. (1994). The Sverdlovsk anthrax outbreak of 1979. *Science, 166*, 1202–1208.

Figure 1 The majority of human inhalational anthrax deaths occurred within 5 kilometers (3.1 miles) downwind of Compound 19. Six villages where sheep died of anthrax were along the plume of aerosolized endospores. The villages ranged from 10 to 60 kilometers (6.2 to 37.2 miles) downwind from Compound 19 in Sverdlovsk.

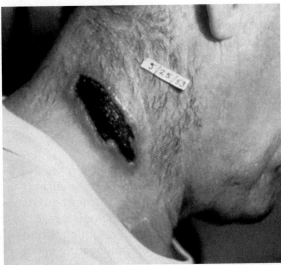

FIGURE 12.46 A cutaneous anthrax lesion.

20% of untreated cases result in death; death is rare with appropriate antibiotic therapy. A case of cutaneous anthrax was identified in a musician and a family member in Connecticut, presumably from animal skins used to make drums.

A third form, **gastrointestinal anthrax**, results from the ingestion of inadequately cooked meat contaminated with B. anthracis, leading to acute inflammation of the gastrointestinal tract characterized by abdominal pain, vomiting of blood, and severe diarrhea. Approximately 25% to 60% of untreated cases result in death. Person-to-person transmission does not occur.

Antibiotics are effective against all three forms of anthrax, but early intervention is necessary. One vaccine licensed by the FDA, Biothrax, is manufactured by the Emergent BioDefense Operations Lansing, LLC. The vaccine is available for people aged 18 to 65 years who are at increased risk for exposure, such as those who work with animals or animal products, laboratory professionals, or those who may be involved in responding to a potential outbreak.

Tetanus

Tetanus, also called **lockjaw**, conjures up terrible images and is usually associated with stepping on a rusty nail. Tetanus is a noncommunicable disease because it does not require human-to-human or animal-to-human contact. It is acquired by exposure to endospores of *Clostridium tetani*, a Gram (+) rod. The tetanus bacterium produces **tetanospasmin**, a neurotoxin considered to be the second most deadly bacterial toxin; estimates are that as little as 100 micrograms can kill an adult. (The C. botulinum neurotoxin is considered the most potent bacterial toxin.)

C. tetani is present worldwide and abundant in soil, manure, and dust. These bacteria are normal inhabitants of the intestinal tracts in horses and cattle and, more rarely, in humans. The fact that these bacteria produce endospores and the hardiness of the endospores explain their longevity and abundance in the soil. Tetanus develops when endospores gain access into the body through wounds. These endospores then germinate into vegetative cells and multiply in deep wounds that do not tend to bleed a lot (puncture wounds). Hence, in the "stepping on a rusty nail" scenario, the nail produces a relatively deep puncture wound, which in itself is of little concern; "rusty" implies that the nail has been in the soil for a long time. The crevices in the nail provide for the adherence of minute amounts of soil that potentially contain large numbers of endospores. Gunshot wounds, animal bites, knife wounds, and wounds caused by the prongs of a pitchfork all serve as grisly mechanisms of puncture wounds, possibly allowing the entrance of tetanus spores. The incubation period ranges from 3 days to 3 weeks but is usually about 8 days.

C. tetani, unlike most bacteria dealt with in this chapter, has no invasive ability. During growth, tetanus neurotoxin is produced that interferes with the relaxation phase of muscle contraction, resulting in uncontrollable muscular contraction. Stiffness in the jaw (lockjaw) is an early symptom resulting from the contraction of facial muscles; as more neurotoxin is produced the neurotoxins continue to spread, leading to contraction in other muscles, particularly in the limbs, stomach, and neck. Tetanus survivors report excruciating pain resulting from the spasms of muscle contractions, which can be strong enough to break bones. The extreme contractions in the back and rib muscles cause the body to arch severely to the extent that only the victim's head and heel are in contact with the surface, a position referred to as **opisthotonos** (**FIGURE 12.47**).

Death occurs by suffocation as a result of contraction of the muscles involved in breathing. Can you imagine taking a very deep breath and not being able to blow out to take another breath? In the United States, schoolchildren are required to be immunized against tetanus, as are children in day care facilities. The number of tetanus cases in the United States is now about 30 cases per year, a sharp decline from the 500 cases per year in the 1940s, and 5 deaths. Nearly all cases of tetanus are among people who have never received a tetanus vaccine or adults who have not stayed up-to-date on their 10-year booster shot. The occurrence of tetanus in developed countries is primarily due to absent or inadequate immunization. The tetanus vaccine does not provide lifelong protection and needs to be repeated at about 10-year intervals. Several vaccines are available to prevent tetanus in children, adolescents, and adults, including **DT (diphtheria, tetanus) vaccine**, **DTaP (diphtheria, tetanus, pertussis) vaccine**, **Td vaccine** (against tetanus and diphtheria), and **Tdap vaccines** (against tetanus, diptheria, and pertussis).

Krasner's Notebook

The photograph in Figure 12.47A was taken in Central America. The boy was approximately 10 years old and was brought into the hospital near death; there was no possible intervention. He died several hours later, a horrible outcome from, presumably, a minor wound. Immunization would have prevented this.

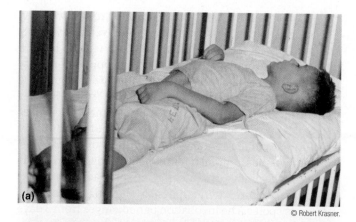

(a)

© Robert Krasner.

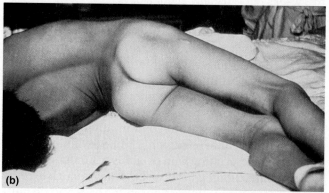

(b)

Courtesy of CDC.

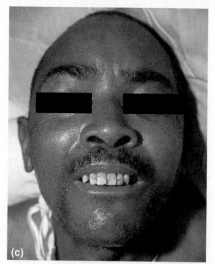

(c)

Courtesy of Dr. Thomas F. Sellers, Emory University/CDC.

FIGURE 12.47 Tetanus. **(a)** The *Clostridium tetani* neurotoxin causes muscle contraction and rigidity. **(b)** The back and rib muscles cause the body to arch severely to the extent to which the victim's head and heel are in contact with the surface, a position referred to as opisthotonos. **(c)** Risus sardonicus (sardonic grin) or facial tetany.

Tragically, **neonatal (newborn) tetanus (NT)** is a common manifestation of tetanus in the first month of life in developing countries, and the course of this disease follows the same pattern as tetanus. Symptoms occur within 2 weeks after birth; the infant fails to suck properly on the mother's breast, becomes irritable, and has convulsions. The disease is prevalent in the poorest nations of the world and occurs when an infant is delivered under unsanitary conditions to a nonimmunized mother.

Imagine a scenario in which a nonimmunized woman delivers a baby on the dirt floor of a small hut in a remote village in an impoverished area. An untrained birth attendant, unaware of the need for sanitary conditions, cuts the umbilical cord with an unclean razor blade, thereby introducing tetanus. In some traditions the baby's umbilical stump is treated with cow dung, ash, mustard oil, or other unsterilized "folk" products.

Neonatal tetanus remains a killer in the developing world. In 2015, about 34,000 died of the disease, a 96% reduction from the estimated 787,000 newborns who died of the disease in 1988. Neonatal tetanus is rare in countries with tetanus vaccine programs. Prevention is centered on vaccination of all women of childbearing age (because their protection is passed to their newborns), improved delivery and postdelivery practices, and education.

Leptospirosis

Leptospirosis, also known as **swamp fever**, is caused by the Gram (−) corkscrew-shaped bacterium *Leptospira interrogans* (**FIGURE 12.48A**). The reservoir of these

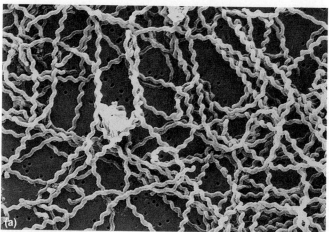

(a)

Courtesy of the CDC/Janice Haney Carr.

(b)

Courtesy of Evi Susanti Sinaga, Indonesia.

FIGURE 12.48 (a) Colorized scanning electron micrograph of corkscrew-shaped *Leptospira* spp. bacteria. **(b)** Photograph that depicts an investigation of a leptospirosis outbreak in Nogosari, Boyolali, Indonesia in June 2014. Farmers who work in the rice paddies from dawn until dusk without wearing any footwear are at high risk for contracting leptospirosis. The farmers photographed were interviewed by interns of the Indonesian Field Training Program (FETP).

bacteria are nonhuman hosts, including dogs, rodents, and a variety of wild animals, and is spread through their urine. Rats appear to be the most significant source of disease, probably because there is greater opportunity for contact with rat urine than with that of other animals. In fact, there is concern that inner-city residents may be at particular risk for leptospirosis, along with farmers, sewer workers, and workers in other occupations that may be exposed to infected rat urine. Additionally, bodies of water, riverbanks, and vegetation may become contaminated as a result of urine runoff from surrounding soil (**FIGURE 12.48B**).

The bacteria penetrate the human skin, enter the bloodstream, and rapidly invade virtually all organs. The infection can also be acquired by swallowing water from sources where rats may have urinated. It is not transmitted from human to human. Outbreaks can occur after hurricanes, floods, or heavy rains; anyone who has been in contact with floodwater or contaminated freshwater (rivers and streams) or soil can be at risk for infection. Epidemics among triathlon participants have been reported. A triathlon is a multisport involving the completion of three continuous endurance events, typically swimming, biking, and running components. Swimming in natural waters is the first leg of the race. (Neither author has ever considered participating in a triathlon but is accustomed to marathon writing as in preparation of this text.)

Leptospirosis can be a serious and potentially fatal disease, with symptoms including **jaundice**, fever, headache, nausea, and chills; those affected may require hospitalization. Diagnosis is based on blood tests. It is treatable with antibiotics such as penicillin and doxycycline, which should be given early in the course of disease.

Leptospirosis is most common in tropical countries. It is the most widespread zoonotic illness and has been classified as an emerging and infectious disease by the WHO and CDC. In 2015, a total of 10 states reported 96 leptospirosis cases. Puerto Rico reported the highest number of leptospirosis cases (45), followed by Hawaii (22) and Guam (11). The risk of acquiring leptospirosis can be greatly reduced by not swimming or wading in water that might be contaminated with animal urine and staying away from infected animals.

Arthropodborne Diseases

Arthropods play a major role in the transmission of infectious diseases. Arthropods are invertebrate animals such as insects, spiders, or crustaceans. Vector control is vitally important in breaking the complicated transmission cycle of these diseases. Numerous species of fleas, mosquitoes, flies, ticks, and lice are responsible for a variety of devastating diseases that have plagued humankind for thousands of years and have influenced the course of civilization. **TABLE 12.13** summarizes the arthropodborne bacterial diseases presented in this section.

Plague: Bubonic, Pneumonic, Septicemic

Plague, also known as the **Black Death**, conjures up terrible images of disease throughout the course of history. The term "plague" is frequently used in a general sense to describe an explosive outbreak with a high death rate. More strictly, however, plague refers to "the plague," caused by the Gram (–) rod-shaped bacterium *Yersinia pestis*, one of the most virulent bacteria known (**FIGURE 12.49**). All it takes is a single bacillus to establish infection.

TABLE 12.13 Arthropodborne Bacterial Diseases*

Infectious Disease	Incubation Period	Bacterial Pathogen	Mode of Transmission	Symptoms
Plague		*Yersinia pestis*	Fleas	
Bubonic	2–6 days			Hemorrhages under skin (Black Death), fever, buboes
Pneumonic	2–3 days			Pneumonia
Septicemic	1–6 days			Pneumonia
Lyme disease	3 days to 1 month	*Borrelia burgdorferi*	Ticks	Possibly expanding bull's-eye rash, flulike symptoms (headache, fatigue, fever, muscle pain); in later stages, arthritis-like neurological impairment, heart inflammation
Ehrlichiosis	7–10 days	*Ehrlichia chaffeensis* and *E. canis*	Ticks	Like a "bad flu" with fever, chills, headache, muscle pain, nausea; rash only rarely present

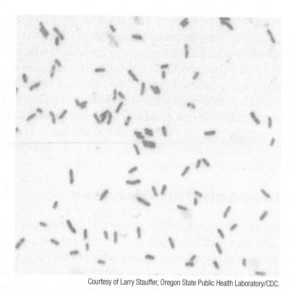

Courtesy of Larry Stauffer, Oregon State Public Health Laboratory/CDC.

FIGURE 12.49 Photomicrograph of Gram-stained *Yersinia pestis* as single Gram (–) rods. Magnification 1,000×.

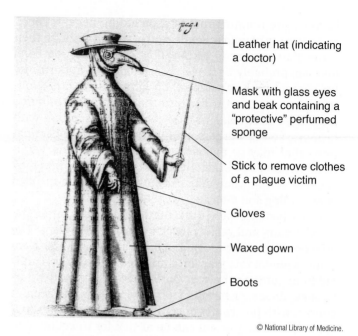

Leather hat (indicating a doctor)

Mask with glass eyes and beak containing a "protective" perfumed sponge

Stick to remove clothes of a plague victim

Gloves

Waxed gown

Boots

© National Library of Medicine.

FIGURE 12.50 Costume that was once thought to be protective against plague.

Medieval Europe was beleaguered with Y. *pestis* from 1347 to 1351, when homes were inhabited by fleas carrying the plague bacteria. In the short span of only 4 years one-third of the population, estimated at 25 million people, were ravaged and died of the Black Death. The dead were piled up and fed upon by rats, resulting in even more infected rats whose fleas spread the disease as the plague swept through populations. At the peak of the epidemic in the 1300s, as many as 800 people died each day in Paris, cutting the population in half. A major plague epidemic that killed over 12 million people began in Asia in 1890, and was carried to San Francisco by rat-infested ships at the turn of the 19th century. It was once believed that carrying sweet-smelling herbs and flowers and holding them to the nose protected against the poisonous vapor of plague. **FIGURE 12.50** depicts a person in protective clothing; the birdlike mask contains the plant material.

Ironically, plague, a devastating disease with a long and terrible history, gave rise to the children's rhyme and playground game, "Ring Around the Rosie," that has been around since the 1800s, and maybe even before. The actual wording varies by country and region, but it is characterized by children standing in a circle chanting the words. The American version is as follows:

> *Ring around the rosy,*
> *A pocketful of posies.*
> *Ashes, ashes,*
> *We all fall down!*

The significance of these words is open to interpretation, but many historians link the rhyme to the Great Plague. The first line refers to the rosy rash, an alleged symptom of the plague, posies refers to herbs carried as protection, ashes (or, in other versions, hush-a, a-shoo, or a-tishoo) refers to the victims' sneezing fits or to the burning of the

bodies, and, finally, the last line is the outcome. A modern version might read as follows:

> *Symptoms of serious illness,*
> *Flowers to ward off the stench.*
> *We're burning the corpses,*
> *We all drop dead!*

Plague is considered a reemerging infectious disease because of its increase throughout the world. Plague resurfaced in 1994 in Surat, India, about 643 kilometers (400 miles) from Mumbai, and resulted in the infection of several hundred people and about 50 deaths. In the 14th century, the Black Death required 17 years to cross the trade routes from China to Iceland. Today, it takes only about 10 hours to fly from Mumbai to London's Heathrow Airport. This is less than the incubation time for an infected person to make the trip without any symptoms of illness. If treatment is started early enough, the patient should be isolated and contacts should be notified. Antibiotic therapy, typically streptomycin or gentamycin, should begin as soon as possible.

Plague was first introduced into the United States in 1900 by rat-infested steamships that sailed from endemic areas of plague, mainly Asia. Today, **sporadic cases** occur in two regions of the United States: (1) northern New Mexico, northern Arizona, and southern Colorado, and (2) California, southern Oregon, and far western Nevada.

According to the CDC an average of seven cases of plague in humans are reported annually (**FIGURE 12.51**). The WHO reports that from 2010 to 2015, 3,248 plague cases were reported worldwide, including 584 deaths. The most endemic countries with human cases of plague are the Democratic Republic of Congo, Madagascar, and Peru.

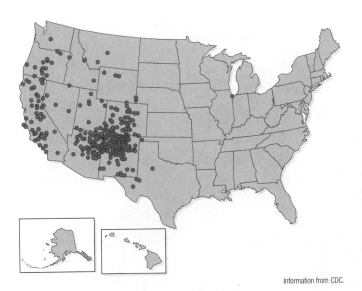

Information from CDC.

FIGURE 12.51 Since the mid-20th century, plague cases have typically occurred in the rural western United States. The circles on the map represent cases of human plague in the United States from 1970 to 2016.

Plague is a **zoonosis**; more than 200 species of mammals, primarily rodents (including gophers, ground squirrels, mice, and wild rats), serve as reservoirs. Rat fleas are the usual vectors, and the mode of transmission occurs primarily from infected rats to other animals, including humans, as a result of contact with a dead plague-infected animal or, more usually, being bitten by an infected flea. In **bubonic plague**, the best-known form of the disease, the bacteria localize in the lymph nodes, particularly in the nodes of the groin, armpits, and neck, causing the nodes to swell to the size of eggs. These enlarged nodes, known as **buboes**, are hard, red, and painful (**FIGURE 12.52**).

The bacteria invade the bloodstream, liver, lungs, and other sites. Hemorrhages occur under the skin, and the dried blood turns black, hence the term "Black Death."

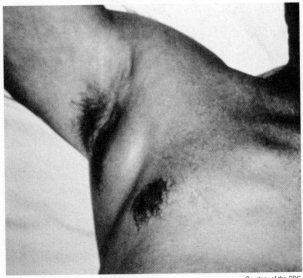

Courtesy of the CDC.

FIGURE 12.52 Bubo in the armpit of a person with the bubonic plague.

The mortality rate in untreated cases is over 50%. Disease control, as in all zoonotic diseases, centers on interrupting the transmission between reservoirs and humans. Bubonic plague, although it may be severe enough to kill its victims, is not normally infectious from person to person.

A second form of the plague, known as **pneumonic plague**, occurs when individuals with the bubonic form develop pneumonia and transmit the bacteria to others by coughing and through saliva. The bacteria invade the victims' lungs, which become filled with a frothy, bloody fluid. Pneumonic plague approaches 100% fatality without early detection and treatment. Currently, no plague vaccines are available in the United States. New vaccines are in development and are expected to be commercially available in the near future. **Septicemic plague** is a third form of the disease; it results from the spread of infection from the lungs to other parts of the body, but it can also be acquired by direct contact of contaminated hands, food, or objects with the mucous membranes of the nose or throat. This form of plague is considered to be 100% fatal.

Lyme Disease, or Borreliosis

The residents of the small New England town of Old Lyme, Connecticut, are probably not thrilled that a zoonotic disease bears its name because it was first described in their community in 1975. This tickborne disease is now present throughout the United States, particularly in the Northeast, the upper Midwest, and the Pacific Coast, and its incidence has dramatically increased for at least the last 15 years in the United States. In 2015, **Lyme disease**, or **borreliosis**, was the sixth most common nationally notifiable disease, and 95% of Lyme disease cases were reported in 14 states: Connecticut, Delaware, Maine, Maryland, Massachusetts, Minnesota, New Hampshire, New Jersey, New York, Pennsylvania, Rhode Island, Vermont, Virginia, and Wisconsin. Lyme disease is also found in Europe, Australia, the former Soviet states, China, and Japan.

The biology of Lyme disease is particularly complex, because five closely interrelated organisms must be present at appropriate times: (1) the Gram (–) spirochete *Borrelia burgdorferi* or *B. mayonii* in the United States (in Europe and Asia *B. afzelii* and *B. garinii* cause Lyme disease); (2) the **deer tick** (also known as the **blacklegged tick**, or *Ixodes scapularis*) that serves as the **vector**; (3) deer that serve as hosts for the maturation of tick eggs to the adult stage; (4) small rodents such as mice or voles that allow the further development of the tick; and (5) humans or other final hosts. The cycle requires 2 years to complete. It is important to understand the life cycle so as to target mechanisms that break the transmission cycle (**FIGURE 12.53**).

Adult ticks feed (take a **blood meal**) and mate on large mammals, particularly deer, in the fall and early spring. Female ticks become engorged with blood, fall off the deer, and lay their eggs on the ground; by summer the eggs hatch into **larvae**. The larvae feed (take a blood meal) on small mammals and birds in the summer and early fall and are inactive until the next spring, when

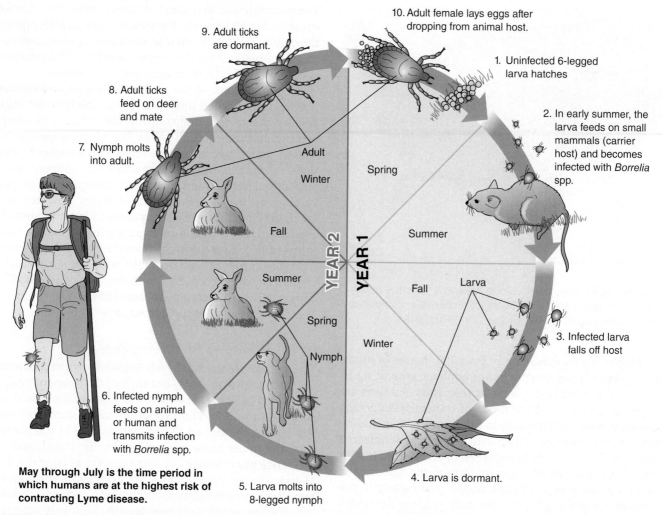

9. Adult ticks are dormant.

10. Adult female lays eggs after dropping from animal host.

8. Adult ticks feed on deer and mate

1. Uninfected 6-legged larva hatches

7. Nymph molts into adult.

2. In early summer, the larva feeds on small mammals (carrier host) and becomes infected with *Borrelia* spp.

Adult

Winter

Spring

Fall

Summer

YEAR 2

YEAR 1

Summer

Fall

Larva

Spring

Winter

Nymph

3. Infected larva falls off host

6. Infected nymph feeds on animal or human and transmits infection with *Borrelia* spp.

4. Larva is dormant.

May through July is the time period in which humans are at the highest risk of contracting Lyme disease.

5. Larva molts into 8-legged nymph

FIGURE 12.53 The cycle of Lyme disease is complicated and involves several stages over the course of 2 years. The larval stage takes place in steps 1 through 4, the nymph stage occurs during steps 5 and 6, and the development stage into adults take place in steps 7 and 8. Human cases of Lyme disease are most prevalent from May through July each year.

they molt into **nymphs**. The nymphs feed on small rodents and other small mammals and birds in the late spring and summer and mature into adults in the fall, thereby completing the 2-year cycle. The larvae and nymphs become infected with the spirochetes as they take a blood meal from their infected **intermediate hosts**, particularly white-footed mice; the spirochetes remain in the deer tick during the tick's developmental stages. The nymph is the major source of transmission to humans and other animals. Two or more days of feeding are required by the nymphs, during which the ticks transmit the spirochetes. The virtually undetectable poppy seed size of the nymphs allows them to escape being removed. Adults (**FIGURE 12.54**) are about the size of sesame seeds.

In its early stages the disease causes flulike symptoms, including fever, headache, and muscle and joint pain. Some individuals have a spreading **bull's-eye skin rash** (**FIGURE 12.55**). These are the "lucky" people because the vague flulike symptoms apply to many diseases, making the early diagnosis of Lyme disease difficult. Unfortunately, the rash may not occur until a month or more

© Teri Shors and Amanda Prigan.

FIGURE 12.54 The different life cycle stages of a deer tick, the vector of Lyme disease: larva, nymph, female and male adult ticks. Deer ticks are about the size of a poppy seed.

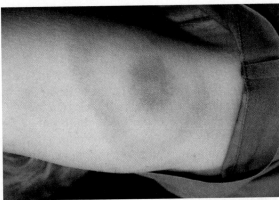

FIGURE 12.55 Bull's-eye rash of Lyme disease. This rash may appear about 1 month or more after the tick bite, but some patients never develop a rash.

Courtesy of James Gathany/CDC.

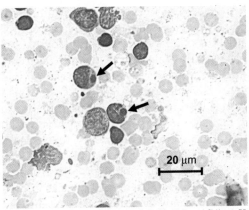

Reproduced from Williams CV, Van Steenhouse JL, Bradley JM, Hancock SI, Hegarty BC, Breitschwerdt EB. Naturally occurring *Ehrlichia chaffeensis* infection in two prosimian primate species: ring-tailed lemurs (*Lemur catta*) and ruffed lemurs (*Varecia variegata*). *Emerg Infect Dis.* 2002 December; 8(12): 1497–1500.

FIGURE 12.56 *Ehrlichia*, the rickettsia that cause ehrlichiosis, hide in white blood cells. The *Ehrlichia*-infected cells in the figure are stained purple.

after the tick bite, or it may never occur; other symptoms, including arthritis and numbness, may occur years later. Diagnosis includes history of tick exposure, symptoms, and blood tests to look for the presence of the antibodies to the spirochetes. A not unusual story illustrating the problems of Lyme disease diagnosis is that of a man who lived with a faulty diagnosis of multiple sclerosis for 9 years before being correctly diagnosed with Lyme disease. Infected individuals treated early with appropriate antibiotics usually recover completely and quickly. Those diagnosed with late and chronic Lyme disease may experience arthritis-like symptoms and neurological difficulties.

Intervention is based on avoiding tick-infested areas, particularly in May through July, when ticks are active and looking for a blood meal. People should take note of this and exercise the proper preventive measures. Measures of protection include use of an insect repellent containing DEET or other insecticidal products, appropriate clothing (pull socks up over pant legs), personal inspection for and removal of ticks after outdoor activity, and removal of leaves (dry clothes at the highest heat in a dryer for 10 to 15 minutes to kill ticks on clothing—washing clothes is not enough to kill ticks present on clothing). Ticks live on the tips of grasses and shrubs, so it is important to clear brush and tall grasses near houses, particularly if there are young children in the area. No vaccine is currently available against Lyme disease. Lymerix, introduced in 1998, was the first Lyme disease vaccine to be approved by the U.S. FDA. Disappointingly, the product was withdrawn by the manufacturer in 2002, presumably due to poor sales. Updates on Lyme disease testing and diagnosis and other information are available at https://www.cdc.gov/lyme/index.html.

Ehrlichiosis

Ehrlichiosis (previously known as human monocytic ehrlichiosis, or HME) is an emerging tickborne infection, similar to Lyme disease, caused by at least three different species of *Ehrlichia*, which are Gram (−) intracellular rod-shaped bacteria (**FIGURE 12.56**). *Ehrlichia* spp. infect white blood cells and circulate in the bloodstream; therefore, these pathogens may pose a risk of being transmitted through blood transfusions. The three species that cause ehrlichiosis in the United States are *E. chaffeensis*, *E. ewingii*, and *E. muris eauclairensis*; they are transmitted to humans through the bite of an infected **lone star tick** (**FIGURE 12.57**). The majority of cases are due to infection with *E. chaffeensis*.

Courtesy of the CDC/James Gathany.

FIGURE 12.57 Adult female lone star tick. Note the "lone star" mark on its surface. If infected, it can transmit *Ehrlichia* spp. that cause ehrlichiosis.

The disease was first reported in humans in 1986, although it has been known for years to exist in horses, dogs, cattle, and other mammals. Some biologists have suggested that the ticks may be moving from dogs to humans. Approximately 200 cases were reported in 2000, but in 2016 there were 1,377 cases, mostly in the eastern half of the United States. The majority of cases occur during June and July; the highest incidence of cases occurs in Oklahoma, Kansas, Missouri, Arkansas, Kentucky, Tennessee, Virginia, New Jersey, New York, Maryland, and Rhode Island.

People who spend time outdoors where deer are common should be diligent to check for ticks and be aware of and respond to any symptoms. The disease is characterized by fever, headache, chills, general malaise, nausea, vomiting, diarrhea, confusion, **conjunctivitis**, and, in some cases, a rash (in up to 60% of children, less than 30% adults). Diagnosis is based on clinical symptoms and can be later confirmed using specialized confirmatory laboratory tests. Doxycycline is the first line of treatment for adults and children of all ages.

Summary

This chapter presents a variety of diseases caused by bacteria organized on the basis of their mode of transmission; namely, healthcare-associated, foodborne and waterborne, airborne, sexually transmitted, contact, soil, and arthropodborne. Some bacterial diseases are diseases of antiquity, but others are new or emerging or reemerging. Some bacteria have caused major epidemics and pandemics throughout the centuries, and their death tolls have influenced the course of history.

The information presented makes it clear that despite advances in hygiene, immunization, and drug therapy in the 20th century, humans all over the world remain threatened by pathogenic bacteria. Foodborne and waterborne intoxications and infections are characterized by gastroenteritis and are a huge burden manifested by morbidity and mortality, particularly in less developed countries with an inadequate public health infrastructure. Despite the watchdog efforts in the production process and by different government entities, they continue to have a significant effect on the health of Americans.

Airborne bacteria cause diseases that are categorized into upper and lower respiratory tract infections. Immunization and antibiotic therapy have been instrumental in reducing the morbidity and mortality in those countries that are able to implement these practices.

STIs were on the decline after the appearance of penicillin in the 1940s, but the sexual revolution in the 1960s resulted in an increase in their prevalence. Contact diseases (other than those transmitted sexually) remain a problem; the necessity for direct contact limits their spread. The relatively recent discovery that most gastrointestinal ulcers are caused by bacteria, and not by stress as previously believed, is a major breakthrough in the treatment of ulcers. Direct person-to-person contact is thought to be the most plausible route of transmission.

Pathogenic bacteria are present in the soil; those that are endospore formers survive for very long periods of time. Circumstances leading to their transmission into the body result in infection. Arthropods are vectors in the transmission of several bacterial diseases.

KEY TERMS

abscesses
acellular (aP)
acid-fast bacteria
acid-fast staining procedure
alveoli
ambulatory surgical centers
anthrax
antitoxin therapy
arthropods
bacillus Calmett-Guérin (BCG) vaccine
bacteremia
biofilms
biopsy
Black Death
blacklegged tick
blood meal
boils
borreliosis
Botox

botulinum toxin
botulism
breath test
buboes
bubonic plague
bullae
bull's-eye skin rash
campylobacteriosis
carbapenemase
carbapenem-resistant *Enterobacteriaceae* (CRE)
carbapenems
carbuncles
carriers
cellulitis
Centers for Disease Control and Prevention (CDC)
chancres
chlamydia
cholera
cirrhosis

claw hand
climate change
colitis
community-acquired MRSA (CA-MRSA)
congenital syphilis
conjunctivitis
consumption
crystal violet
cutaneous anthrax
cystic fibrosis (CF)
debrided
deer tick
dormant state
direct observational therapy short course (DOTS)
DT (diphtheria, tetanus) vaccine
DTaP (diphtheria, tetanus, pertussis) vaccine
ehrlichiosis
emerging bacterial disease
endemic
endocarditis
endospore
enterotoxin
erythrogenic toxin
exotoxin
extremely drug-resistant tuberculosis (XDR-TB)
fecal transplants
fomites
food intoxication
food poisoning
foodborne infection
fried rice diarrheal syndrome
fried rice emetic syndrome
gastroenteritis
gastrointestinal anthrax
generation time
germ theory of disease
glomerulonephritis
gold standard
gonorrhea
Gram stain
granulomas
gummas
halointolerant
halophile
halotolerant
Hansen's disease
healthcare-associated infection (HAI)
hemochromatosis
hemolytic uremic syndrome (HUS)
hepatitis
Hodgkin's disease
hyperbaric oxygen therapy (HBOT)
interferon gamma release assay (IGRA)

Immunization Action Coalition
impetigo
incubation period
infectious diseases
infective dose
inhalation anthrax
intensive care unit (ICU)
intermediate hosts
internally displaced persons (IDPs)
isoniazid
jaundice
Klebsiella pneumoniae carbapenemase (KPC)
larva
Legionnaires' disease
legionellosis
leper colony
lepromas
leprosarium
leprosy
leptospirosis
leukemia
listeriosis
liver cirrhosis
lockjaw
lone star tick
lower respiratory tract
Lyme disease
lymphadenitis
lymphomas
Mantoux tuberculin skin test
meningitis
meningococcal disease
meningococcal septicemia
meningococcemia
methicillin-resistant *Staphylococcus aureus* (MRSA)
microbiota
mode of transmission
multiple drug-resistant tuberculosis (MDR-TB)
myalgia
Mycobacterium avium complex (MAC)
mycolic acids
necrotizing fasciitis
neonatal (newborn) tetanus (NT)
neurotoxin
non-acid-fast bacteria
nontuberculous mycobacteria (NTM)
nucleic acid amplification test (NAAT)
nymph
ophthalmia neonatorum
opisthotonos
opportunistic pathogen
osteomyletis

pandemics
paralysis
paresis
pasteurization
patient zero
pelvic inflammatory disease (PID)
peptic ulcers
pertussis
photophobia
pili
plague
pleurisy
pneumonia
pneumonic plague
Pontiac fever
primary syphilis
pruno
ptomaine poisoning
quarantine signs
reactivation TB
reemerging infectious disease
refractometer
reservoir
rheumatic fever
saber shins
safranin dyes
salmonellosis
scalded skin syndrome
secondary syphilis
sepsis
septic shock
septicemic plague
sexually transmitted infections (STIs)
Shiga toxin
shigellosis
Siberian ulcer
Silvadene
sporadic cases

sputum
strawberry tongue
strep throat
superbugs
swamp fever
syphilis
tabes
TB disease
TB infection
TB sanatoriums
Td vaccine
Tdap vaccines
tertiary syphilis
tetanospasmin
tetanus
toxic shock syndrome
toxins
traveler's diarrhea
tuberculosis (TB)
typhoid fever
upper respiratory tract
urease
uremia
urgent care clinic
vancomycin-intermediate *Staphylococcus aureus* (VISA)
vancomycin-resistant *Staphylococcus aureus* (VRSA)
vector
venereal diseases (VDs)
vibriosis
virulence factors
white death
white plague
whooping cough
woolsorter's disease
World Health Organization (WHO)
World TB Day
Ziehl-Neelson staining procedure
zoonosis

SELF-EVALUATION

○ PART I: Choose the single best answer.

1. Families with pet iguanas or turtles are particularly at risk for _____.
   a. *Campylobacter* infection
   b. listeriosis
   c. salmonellosis
   d. shigellosis

2. One of the most potent known neurotoxins is associated with _____.
   a. *Salmonella*
   b. botulism
   c. staphylococci
   d. cholera

3. _____ caused a "100-day cough."

    a. *Salmonella*   c. *Mycobacterium*

    b. *Bordetella*   d. *Legionella*

4. The beautiful fountain in a shopping mall may be a source of _____.

    a. listeriosis   c. Legionnaires' disease

    b. leptospirosis   d. leprosy

5. Which of the following bacteria is *not* a source of HAIs?

    a. *Legionella*   c. *Listeria monocytogenes*

    b. *Bacillus cereus*   d. *Staphylococcus epidermidis*

6. Which of the following is an upper respiratory tract infection?

    a. Diphtheria   c. Legionellosis

    b. Tetanus   d. Tuberculosis

7. Which of the following is *not* a location associated with community-acquired MRSA?

    a. Football fields   c. Dormitories

    b. Hospitals   d. Weight rooms

8. Leprosy is caused by _____.

    a. *Staphylococcus*   c. *Bacillus*

    b. *Mycobacterium*   d. *Yersinia*

9. Which of the following is *not* associated with arthropod transmission?

    a. Lyme disease   c. Black Death

    b. Ehrlichiosis   d. Anthrax

10. Which of the following bacteria is not a Gram (–) pathogen?

    a. *Clostridium botulinum*   c. *Pseudomonas aeruginosa*

    b. *Bordetella pertussis*   d. *Haemophilus influenzae*

❍ PART II: Fill in the blank.

1. _____ are infections that patients acquire while they are receiving health care for another illness.

2. _____ is the arthropod vector for Lyme disease.

3. *Mycobacterium* spp. contain _____ in their waxy, lipoidal, cell walls.

4. Patients with _____ (hint: a type of lung condition) are especially vulnerable to *Pseudomonas aeruginosa* infections.

5. Pregnant women should avoid foods like hotdogs, deli meats, and soft cheeses that may be contaminated with _____, a foodborne pathogen that would have serious negative effects on a developing fetus.

6. The generation time of *Mycobacterium tuberculosis* is _____.

7. The largest outbreak of human _____ occurred in Sverdlovsk in 1979.

8. _____ is the most common STI.

9. _____ is found in brackish water and can cause necrotizing fasciitis if it infects a sore or open wound.

10. *Legionella* spp. and *Mycobacterium avium* can be isolated from _____.

○ **PART III: Answer the following.**

1. Distinguish between food poisoning (intoxication) and food infection. Name two diseases in each category.

2. Explain the following terms:
   - a. Oral rehydration therapy
   - b. DOTS
   - c. Opportunistic infection
   - d. Debriding tissues

3. You are about to travel to a foreign country where sanitation is at a lower level than you are accustomed to. What precautions will you take during your stay?

4. List measures you can use to prevent tickborne diseases.

5. What is meningitis? List the bacteria that can cause meningitis.

6. List three bacterial pathogens found in the soil and the diseases they cause. How can these diseases be prevented?

7. Describe the three stages of syphilis. How is syphilis treated?

8. Explain why *Helicobacter pylori* can withstand the harsh pH of stomach acid.

9. Explain what it means if you call someone a "Typhoid Mary."

10. List at least four bacterial pathogens that are multidrug resistant. Explain why there is cause for concern if a pathogen is resistant to antibiotic therapy.

CHAPTER 13

VIRAL AND PRION DISEASES

"An efficient virus kills its host. A clever virus stays with it."

—James Lovelock

Independent scientist, environmentalist, and futurist

"Prions are 'almost immortal.'"

—Paul Brown

Prion expert, NIH research scientist

Samaritan's Purse medical missionary Dr. Kent Brantly survived Ebola virus disease (EVD). He was infected by Ebola virus while treating and caring for patients in Liberia during the 2014 Ebola epidemic in West Africa. Dr. Brantly is shown testifying at the U.S. House Foreign Affairs subcommittee hearing on "Global Efforts to Fight Ebola," held in Washington, D.C., on September 17, 2014. Dr. Brantly was one of the "Ebola fighters" honored as *Time* magazine's Person of the Year in 2014. The title is given to an individual or a group who has had the biggest impact on the world and news over the course of the year.

1. List three viral pathogens that cause gastroenteritis and their modes of transmission, and describe what public health measures are needed to control and prevent epidemics caused by these viruses.

2. Explain the relationship between chickenpox and shingles.

3. Compare and contrast the common cold and influenza, addressing the viral pathogens involved, mode(s) of transmission, signs and symptoms, treatment, and prevention.

4. Define *antigenic drift* and *antigenic shift* as it applies to influenza viruses.

5. List the natural reservoirs for MERS-CoV, influenza A viruses, rabies virus, Ebola virus, variola virus, and measles virus.

6. Evaluate which cancers associated with viral infections can be prevented through vaccination.

7. Explain why the number of new HIV infections continues to increase given that we know how HIV is transmitted and how HIV infections can be prevented.

8. List three viruses that cannot be inactivated by hand sanitizer and three viruses that can be inactivated by hand sanitizer and note the structural differences between the viruses that can be inactivated and the ones that cannot be inactivated.

9. List at least five reasons why the 2014–2016 Ebola epidemic in West Africa was unprecedented, and identify the measures that were needed to control and eventually halt the epidemic.

10. Compare and contrast Epstein-Barr virus and cytomegalovirus with regards to mode of transmission, signs and symptoms, treatment, and prevention.

11. Explain why it can be difficult to accurately diagnose Zika virus diseases, Chikungunya virus disease, and dengue fever.

12. Summarize the severe complications of infections caused by measles virus, cytomegalovirus, and Zika virus in newborns.

13. Evaluate the disease process of rabies versus transmissible spongiform encephalopathies (TSEs).

14. Review the differences between hepatitis B and C infections.

15. Generalize why coronaviruses are a source of emerging infections in humans.

Case Study: Surviving Ebola

Thirty-three-year old Dr. Kent Brantly (pictured in the chapter opener figure) arrived with his wife and children to begin medical missionary work in Monrovia, Liberia, in October 2013. One month after he arrived in Liberia, the worst Ebola epidemic in human history took hold in West Africa. In April 2014, Dr. Brantly and his team converted their hospital into an **Ebola virus disease (EVD)** care unit. Ebola patients began arriving in the unit for treatment on June 22, 2014.

On July 23, Dr. Brantly awoke with a fever and felt fatigued. On day 4 of his illness, he tested positive for Ebola virus. He was experiencing the **dry phase** of EVD characterized by sudden fever, intense weakness, pain, and headache. On day 6 of his illness, a **petechial rash** (clusters of pinpoint round spots on the skin due to capillaries bleeding, leaking blood under the skin) developed on his arms and chest, and his fever spiked to 104.5°F (40.3°C). The **wet phase** symptoms then began. He suffered from malaise, abdominal pain, and

(continues)

Case Study: Surviving Ebola (continued)

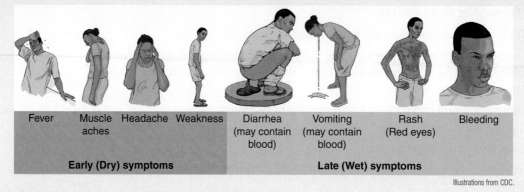

Fever	Muscle aches	Headache	Weakness	Diarrhea (may contain blood)	Vomiting (may contain blood)	Rash (Red eyes)	Bleeding
Early (Dry) symptoms				**Late (Wet) symptoms**			

Illustrations from CDC.

Figure 1 The two phases of infection caused by Ebola virus. Transmission of the disease occurs through exposure to the blood or secretions of an infected individual. The risk of infection is high during the burial of a diseased EBV patient.

profuse diarrhea. The wet phase of EVD is characterized by the development of vomiting and diarrhea as EVD progresses (**Figure 1**). The rash developed into a **maculopapular rash** (larger, flat red areas on the skin) that covered his body from his legs to his face. His stool was bloody/black due to bouts of **melena**, the production of dark, sticky feces following internal bleeding or swallowing of blood.

Later that same day, Dr. Brantly received a unit of **convalescent blood** from an Ebola survivor. Hospital staff continued supportive treatment, including oral hydration. On day 9, he was given an intravenous dose of **ZMapp**, an experimental **monoclonal antibody** cocktail. After a long plane flight across the globe, on day 10 he was transferred to the **Serious Communicable Disease Unit** at Emory University Hospital in Atlanta, Georgia.

Dr. Brantly was discharged from the hospital on day 30. He was a survivor of one of the most horrible and deadly infectious diseases known to humans. An NBC News reporter interviewed him about his experience, which was aired on September 5, 2014, in a segment titled: "*Saving Dr. Brantly: The Inside Story of a Medical Miracle.*"

Questions and Activities

1. Watch Dr. Brantly's interview. How does he believe he was infected with Ebola virus?

2. What is the mode of transmission of Ebola virus?

3. During what phase of EVD is a patient most contagious, and why?

4. At the time of this epidemic, a vaccine and antivirals were not available for EVD. Summarize reasons why this was likely the case.

5. EVD survivors suffer from complications of EVD. Perform research and list the types of complications Ebola survivors experience.

6. What precautions would you use in caring for a patient suffering from EVD?

7. In May and August 2018, two different outbreaks of Ebola were reported in the Democratic Republic of Congo (DRC), which is in the middle of a long civil war. How do such situations pose challenges for medical aid teams?

8. Follow up on the 2018 Ebola outbreaks in the DRC. How many EVD cases and deaths have occurred? Are the outbreaks over?

9. What additional resources, if any, are available to fight EVD in Africa that were not available in 2014 (e.g., vaccine, antivirals, medical partnerships, aid, etc.)?

Information from: G. M. Lyon, A. K. Mehta, J. B. Varkey, K. Brantly, L. Plyler, A. K. McElroy, . . . Emory Serious Communicable Diseases Unit. (2014). Clinical care of two patients with Ebola virus disease in the United States. *New England Journal of Medicine, 371*, 2402–2409; Saving Dr. Brantly [interview]. (2014, September 6). NBC News. Retrieved from https://www.nbcnews.com/feature/saving-dr-brantly

Preview

This chapter presents the major viral and prion diseases of humans. The field of virology changes quickly. Emerging viruses, including Ebola virus, Zika virus, Chikungunya virus, Powassan virus, avian influenza A viruses, new norovirus strains in Australia, human metapneumovirus, and Middle East respiratory syndrome coronavirus, to mention only a few, seem to have appeared suddenly from nowhere and threaten our existence.

For each viral disease presented, particular attention is paid to its **mode of transmission** and other factors involved in the disease: such as pathogenicity, incubation time, treatment, and (in some cases) immunization. With many bacterial diseases the affected area of the body frequently correlates with the mode of transmission and the **route of entry** of the bacteria into the body. For example, foodborne bacteria primarily infect the digestive system, and airborne bacteria primarily infect the respiratory tract. In the case of viruses there may be little connection between the route of entry and the particular organs and tissues of the body involved. The measles virus and varicella zoster virus (which causes chickenpox), for example, are airborne and enter the body through the respiratory tract, but the skin is their major target. Rabies is acquired through a break in the skin from the bite of a rabid animal, but rabies virus attacks the nervous system. In many cases the virus may have more than one route of entry and/or mode of transmission, making many viral diseases difficult to pigeonhole.

Immunization is available to prevent some viral diseases but not others. Keep in mind that antibiotics are not effective against viruses, but there are a few antiviral drugs (not antibiotics) that are used in specific cases.

As you study each of these viral diseases, recall the five stages of viral replication: adsorption, penetration, replication, assembly, and release. Remember that these stages occur in infected cells, causing damage to the host, which may be anywhere from subclinical to mild to severe to fatal. Frequently, the diagnosis of a particular viral disease is based on the clinical symptoms, which may be readily apparent, such as a swollen jaw in mumps and the characteristic rash of measles or chickenpox.

Laboratory diagnosis, however, based on demonstration of viral antigens or antibodies or culture of the specific virus, may be the only way to definitively diagnose many viral diseases. Because of the reporting time necessary and the expense involved, these procedures are not always done routinely. Frequently, it is assumed that the condition will simply run its course.

Prions are infectious proteins that cause a group of diseases of the brain and nervous system in humans and other animals called transmissible spongiform encephalopathies (TSEs). TSEs are rare, incurable, neurodegenerative fatal diseases. Creutzfeldt-Jakob disease (CJD) and variant CJD (caused by eating prion-contaminated beef products) are the most common TSEs in humans.

Foodborne and Waterborne Viral Diseases

The major foodborne and waterborne viral diseases contribute to the overall public health problem of diseases and cause significant morbidity and mortality. The Centers for Disease Control and Prevention (CDC) estimates that in the United States foodborne diseases (bacterial, parasitic, and viral) average 48 million cases annually, resulting in 128,000 hospitalizations and 3,000 deaths each year.

The United States has one of the safest public drinking water supplies in the world. Over 286 million Americans get their tap water from a community system in which the water is treated to be free from microbial or viral contamination. The drinking water quality must meet standards set forth by the U.S. Environmental Protection Agency (EPA) to ensure that if there are any chemicals and pollutants present that they are at safe and acceptable levels of concentration.

All of the viruses discussed in this section are **naked viruses** (**TABLE 13.1**). These viruses do not contain a lipid bilayer envelope that surrounds the capsid of the virus particle. The viral surface is composed of a protein shell, increasing its stability in the environment. *Hand sanitizer will do nothing to prevent infections caused by noroviruses, hepatitis A and E viruses, and polioviruses.* Washing your hands is the best way to prevent and control the spread of viruses that cause the foodborne and waterborne illnesses associated with gastroenteritis!

When decontaminating surfaces (e.g., countertops, sink faucets, door knobs, toilets), it is necessary to use a disinfectant with a short **dwell time** and one that is not caustic or damaging to surfaces. Dwell time is the length of time recommended for a disinfectant to sit on a surface in order to decontaminate it before being removed and the

TABLE 13.1 Foodborne and Waterborne Viral Diseases

Viral Illness	Incubation Period	Virus(es)	Symptoms	Immunization and Comments
Gastroenteritis	2–10 days	Noroviruses, rotaviruses, enteroviruses	Stomachache, diarrhea, dehydration, vomiting	RotaTeq and Rotarix vaccines available against rotavirus infections since 2010
Hepatitis	3–8 weeks	Hepatitis A and E viruses	Jaundice, abnormal liver function in tests	Vaccine available against hepatitis A virus for those at risk
Poliomyelitis	7–14 days	Poliovirus	Usually asymptomatic but can cause lifelong paralysis	Vaccine available since 1950s; disease close to eradication

surface dried. Typically, disinfectants contain the active ingredients of hydrogen peroxide and benzyl alcohol (e.g., **Oxivir TB**). While 5% to 10% bleach can also be used, it is caustic and will damage furniture and other surfaces.

Gastroenteritis and Noroviruses

Gastroenteritis is characterized by stomach and abdominal pain, diarrhea, vomiting, and abdominal cramps as a result of inflammation of the stomach and intestinal tract. The term "stomach flu" is popular but meaningless. An exact diagnosis of gastroenteritis is difficult because these symptoms are caused by a variety of viruses, bacteria, and protozoans. The symptoms usually last a few days, and recovery is uncomplicated. Dehydration is a potential problem, particularly in infants and in older adults because of fluid loss resulting from excessive diarrhea and vomiting.

The mode of transmission of the viruses that cause gastroenteritis is the **fecal–oral route**. Raw or undercooked shellfish from contaminated waters are also a potential threat. Outbreaks, following a 1- to 2-day incubation period, are most likely to occur in schools, day care centers, cruise ships, dormitories, nursing homes, military barracks, refugee camps, and disaster relief camps with their close quarters and shared toilet facilities.

Noroviruses (also known as Norwalk and Norwalk–like viruses, named after Norwalk, Ohio) are frequent causes of gastroenteritis, or "winter vomiting disease." Noroviruses are the leading cause of vomiting and diarrhea due to acute gastroenteritis among people of all ages in the United States. People living or gathering together can be easily infected for several reasons. First, fewer than 10 viruses are required for infection. Second, noroviruses persist in the environment, and continue to be shed after recovery. Third, these viruses are highly stable and resistant to inactivation by hand sanitizer. Good old-fashioned handwashing with soap and water is the best way to prevent the spread of noroviruses. Contaminated surfaces must be disinfected with a dilute concentration of bleach or a hydrogen peroxide–based disinfectant.

Shors' Notebook

At the beginning of the spring 2018 semester, a norovirus outbreak struck the University of Wisconsin–Oshkosh campus. Students living in the dormitories were hit particularly hard. Of course, these were students in my class! Custodial services staff clocked extra hours to disinfect classrooms. Certain food services were shut down. As a virologist, knowing the stability of noroviruses and the fact that the **infective dose** (or **infectious dose**) is 10 or fewer noroviruses to cause violent vomiting and diarrhea, I feared getting sick. I knew from conversations that it was not a fun illness. The eyes of a few students who contracted norovirus in my class were bloodshot because they experienced such violent bouts of vomiting.

One day, I walked into the restroom and noticed it looked like someone had been sick. I was so afraid to use the bathroom that I left, got in my vehicle, and drove home (a 7- to 10-minute drive from campus) to use the bathroom at my house! I never did get sick during this outbreak. It pays to be paranoid and to wash your hands religiously. The outbreak helped students gain firsthand knowledge about a viral disease. Many interesting discussions about noroviruses took place in my courses that semester.

Cruise ships, some of which carry over 3,000 passengers and 1,000 crew, are natural hotspots for noroviruses. The short incubation period was more than enough time for Australian passengers and crew on the cruise ship *Sun Princess,* which affected eight consecutive voyages from December 5, 2016, to February 6, 2017, to be clinging to the rails or "heading for the head" as they suffered from the nausea, vomiting, and diarrhea associated with norovirus. Hundreds of passengers got sick. More than 16,000 passengers were potentially exposed to noroviruses on journeys to New Zealand, Indonesia, Fremantle, Papua New Guinea, Margaret River, and the South Pacific.

Cruise ships sailing from a foreign port that participate in the CDC's **Vessel Sanitation Program** report gastroenteritis cases confirmed by the ship's medical staff to the CDC before entering a U.S. port. If you plan on taking a cruise, check out cruise ship inspection score at https://wwwn.cdc.gov/InspectionQueryTool/InspectionSearch.aspx

You don't need to ride the waves to become debilitated by this highly contagious virus. Over the past few years, norovirus outbreaks have been associated with restaurant buffets, food service workers, hospitals, even celebrants at family reunions. Unfortunately, because there are many different strains of norovirus, it is difficult to develop immunity.

Children younger than 5 years and older adults are particularly vulnerable to infection by noroviruses. The CDC estimates that in the United States norovirus infections account for 19 to 21 million norovirus illnesses, 400,000 emergency room visits, about 56,000 to 71,000 hospitalizations, and as many as 1.7 to 1.9 million outpatient visits annually. Noroviruses are the leading cause of acute gastroenteritis among U.S. children 5 years of age and younger that seek medical attention. By the age of 5, 1 in 278 children will be hospitalized, 1 in 14 will visit an emergency room, and 1 in 6 will receive outpatient care for norovirus illness. Your estimated lifetime risk of contracting a norovirus infection based on CDC estimates is shown in **FIGURE 13.1**.

Hepatitis A and E

Hepatitis is an inflammation of the liver and is *commonly* caused by one of four viruses designated hepatitis A, B, C, and E. **Jaundice**, a yellowish discoloration of the skin and eyes caused by increased levels of bilirubin in the blood, frequently occurs in hepatitis virus infections, along with abnormal liver function test results (**FIGURE 13.2**). Hepatitis B and C are presented later based on their mode of transmission.

Hepatitis A, formerly called **infectious hepatitis**, is usually a mild and self-limiting disease with an abrupt onset. Its average incubation time is approximately 30 days, but the incubation period may be as long as 2 months. Unlike infections caused by other hepatitis viruses, recovery is usually complete without chronic infection. Hepatitis A viruses are found in feces, and transmission is by the

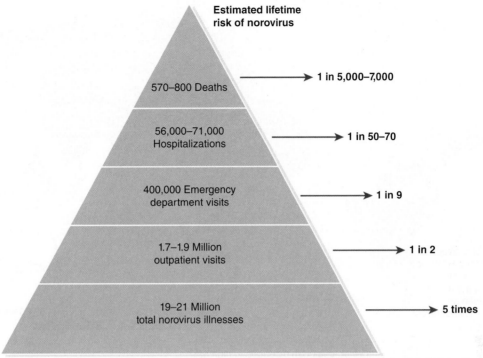

Estimated lifetime risk of norovirus

570–800 Deaths → 1 in 5,000–7,000

56,000–71,000 Hospitalizations → 1 in 50–70

400,000 Emergency department visits → 1 in 9

1.7–1.9 Million outpatient visits → 1 in 2

19–21 Million total norovirus illnesses → 5 times

Reproduced from CDC (2018). Burden of norovirus illness in the U.S. Retrieved from https://www.cdc.gov/norovirus/trends-outbreaks/burden-US.html.

FIGURE 13.1 This pyramid represents the estimated number of norovirus illnesses annually in the United States across all age groups.

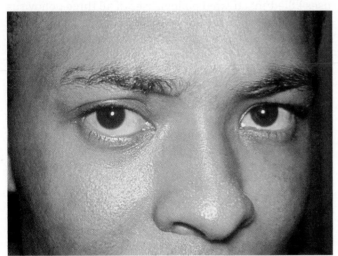

Courtesy of Dr. Thomas F. E. Sellers/Emory University/CDC.

FIGURE 13.2 Jaundice.

fecal–oral route, most frequently resulting from contamination of food and drinking water. Close personal contact, including oral sex, also contributes to transmission. Approximately 50% of cases are subclinical.

The CDC estimates that approximately 2,007 new hepatitis A infections occurred in the United States in 2016. Rates of hepatitis A infections were on the decline in the United States until 2012. Since 2012, the number of cases has begun to increase. Increases have been associated with multistate outbreaks linked to raw scallops in Hawaii and imported contaminated foods (e.g., pomegranate arils

from Turkey, frozen strawberries from Egypt). Good personal hygiene, emphasizing handwashing and good sanitation, is the most effective means of control.

Diagnosis is based on laboratory detection of antibodies in the blood serum; no specific treatment is available. The CDC recommends that all children in the United States be vaccinated against hepatitis A virus between their first and second birthdays. Additionally, it is recommended for use in high-risk individuals, including those traveling to foreign countries where the disease is endemic (**FIGURE 13.3**). Injections of blood products rich in hepatitis A virus antibodies (**immune globulin**; **GamaSTAN S/D**) are also used before and after exposure under defined conditions. In countries with poor sanitary conditions and hygienic practices, 90% of children have been infected with hepatitis A virus before the age of 10 years.

Hepatitis E virus, like hepatitis A virus, is transmitted by the fecal–oral route, most commonly through contaminated drinking water in developing countries. In developed countries, sporadic cases have occurred following the consumption of uncooked or undercooked pork or dry deer meat and raw shellfish. It is uncommon in the United States but is endemic in Asia, the Middle East, Africa, and Central America. No vaccine is available to prevent infection caused by hepatitis E virus.

Poliomyelitis

Two generations back the word **poliomyelitis** (usually shortened to polio), also known as **infantile paralysis**,

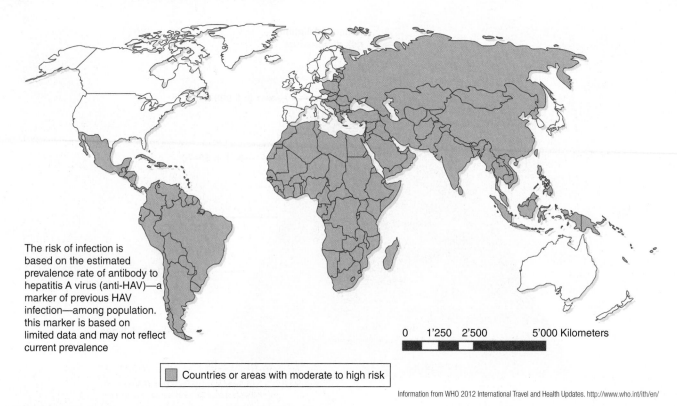

The risk of infection is based on the estimated prevalence rate of antibody to hepatitis A virus (anti-HAV)—a marker of previous HAV infection—among population. this marker is based on limited data and may not reflect current prevalence

☐ Countries or areas with moderate to high risk

Information from WHO 2012 International Travel and Health Updates. http://www.who.int/ith/en/

FIGURE 13.3 Countries or areas with moderate to high hepatitis A risk.

struck fear into the hearts of parents the world over. Now polio is expected soon to join smallpox, the first eradicated disease, in extinction. Immunization in most parts of the world has greatly reduced the incidence of the disease. The last case of naturally occurring polio in the United States was in 1979. Since that time, there have been a few cases of unvaccinated immunocompromised children who contracted vaccine-derived polio in the United States. From January 1 to August 8, 2018, 12 cases of wildtype polio were reported in Pakistan and Afghanistan and 21 cases of circulating vaccine-derived cases of polio in several African countries. To follow the progress of the **Polio Global Eradication Initiative**, go to their interactive surveillance website, which is updated in real time: http://polioeradication.org/polio-today/polio-now/. All this has taken place since the introduction of the first poliovirus vaccine, the Salk vaccine, in the 1950s, followed a few years later by the Sabin vaccine. The **Salk vaccine**

Krasner's Notebook

I remember the panic that ensued during my childhood when polio, or even suspected polio, threatened a neighborhood. Theaters, swimming pools, playgrounds, and all other places where young children might congregate were closed; children were kept indoors. Throughout much of the year the March of Dimes campaign to support the National Foundation for Infantile Paralysis was relentless in fundraising to support research toward combating the crippling disease of polio. In movie theaters containers bearing the label "March of Dimes" were passed around, and theater patrons were encouraged to drop in coins to support research aimed at developing a vaccine against polio.

uses killed/inactivated polioviruses of all three poliovirus strains. The **Sabin vaccine** is an oral vaccine composed of live but attenuated (weakened) polioviruses. In the United States and in other countries, the Salk vaccine has been replaced by the Sabin vaccine; in some countries the Sabin vaccine continues to be used.

Poliomyelitis is a highly transmissible infectious disease. Most cases of polio are asymptomatic, and only a small number result in paralysis, as happened with President Franklin Roosevelt (**FIGURE 13.4**). Transmission is from person to person by direct fecal–oral contact, by indirect contact with infectious saliva or fecal material, or by contaminated sewage or water. Poliovirus replicates in the tonsils and **Peyer's patches** (lymph nodes in the walls of the intestine near the colon), and it may invade the blood and be disseminated to the nervous system. Replication in spinal cord nerve cells causes **paralytic poliomyelitis**, resulting in severely deformed limbs (**FIGURE 13.5**). Paralysis occurs in about 7 to 21 days from the time of initial infection; in about 2% to 10% of cases death may result.

Bulbar poliomyelitis is an extremely dangerous form of polio that can paralyze the respiratory muscles, causing difficulty in swallowing and breathing. In the past, patients spent long periods, perhaps their entire lives, in **iron lungs** (**FIGURE 13.6**), which are airtight metal tanks that enclose the entire body, except for the head, and force respiration. These tanks have a diameter slightly larger than that of an oil drum and a length approximately equal to that of two drums.

In the 1950s, iron lungs were replaced by positive pressure ventilators, and they are used by less than a handful of people in the United States today. Electrical failures

Courtesy of Franklin D. Roosevelt Presidential Library and Museum.

Courtesy of Franklin D. Roosevelt Presidential Library and Museum.

FIGURE 13.4 (a) Franklin Delano Roosevelt (FDR) at Warm Springs. FDR fell victim to polio in 1921 at age 39, leaving this future president (1933–1945) a paraplegic. Throughout his 4 years as governor of New York and his 12 years as president of the United States, his legs were in heavy metal braces. This courageous and highly visible individual lent his name and influence to the fundraising campaigns of the National Foundation for Infantile Paralysis and to the March of Dimes to combat polio. **(b)** The press portrayed Roosevelt as a robust, physically strong leader.

Courtesy of NIP/Barbara Rice/CDC.

FIGURE 13.5 The crippling effects of polio. A man with a weak, withered leg.

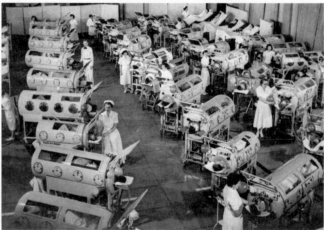

Courtesy Rancho Los Amigos National Rehabilitation Center, L.A. County Department Health Services. Used with permission.

FIGURE 13.6 Life in an iron lung in 1953. Many victims of polio had to spend their lives in an iron lung because they could not breathe on their own.

immunity, but only to the particular viral type (of which there are three) that caused the infection. Humans are the only reservoir for poliovirus, as is the case with smallpox and measles, making possible the eradication of these diseases. Higher nonhuman primates can also be infected by polioviruses and a potential reservoir of poliovirus but it is unlikely they are reservoirs in nature.

were a nightmare; attendants had to mechanically pump the bellows of an iron lung to keep a patient breathing. Dianne Odell died in 2008, at the age of 61 as the result of a power failure in her home after having lived in an iron lung for 51 years. She was one of the nation's oldest survivors of polio to be living in an iron lung.

After exposure to poliovirus, more than 90% of susceptible contacts become infected and acquire lifetime

Krasner's Notebook

In the summer of 2006, I visited the Jonas Salk Institute in La Jolla, California, to talk with Peter Salk, MD, about his father, Jonas. The two main cement laboratory buildings facing the Pacific enclose a spacious courtyard. The inscription leading up to the courtyard presents a quote from Jonas Salk: "*Hope lies in dreams, in imagination, and in the courage of those who dare to make dreams into reality.*"

Airborne Viral Diseases

The primary source of viral airborne diseases is respiratory droplets from an infected person to a susceptible person. Hence, having an "infectious laugh" may not be a desirable trait. A summary of the major airborne viral diseases is presented in **TABLE 13.2**. Almost all of the viruses that cause respiratory diseases listed in Table 13.2 are enveloped (surrounded by a lipid bilayer membrane). This means that alcohol-based hand sanitizers sprayed on hands or surfaces will dissolve or damage the external lipid membrane of the enveloped virus, reducing its infectivity and spread of disease. The exception to the rule are the adenoviruses, rhinoviruses, and enteroviruses that cause mild respiratory diseases such as the common cold. These are naked viruses, which are much more stable, like the viruses that cause gastroenteritis.

TABLE 13.2 Airborne Viral Diseases

Viral Disease	Incubation Period	Symptoms	Immunization and Comments
Common cold (caused by rhinoviruses, adenoviruses, enteroviruses, and coronaviruses)	1–3 days	Runny nose, sore throat, coughing, sneezing	No vaccine
Influenza	1–2 days	Chills, fever, muscle aches, sneezing, sore throat, fatigue	Annual flu vaccine for older adults and others at high risk; Relenza, Tamiflu, and Rapivab antivirals available
Respiratory syncytial virus infection	4–5 days	Runny nose, decrease in appetite, coughing, sneezing, fever, wheezing, irritability; can cause severe complications such as pneumonia in children younger than 1 year	No vaccine; life threatening; most prevalent cause of respiratory infection for infants younger than 6 months
Human metapneumovirus infection	3–6 days	Cough, nasal congestion, shortness of breath; can progress to bronchitis or pneumonia	No vaccine available, no specific antiviral treatment; new virus discovered in 2001, genetically related to RSV
Middle East respiratory syndrome (MERS)	2–14 days	Diarrhea, nausea, and vomiting followed by pneumonia, kidney failure (3 out of 10 people die)	No vaccine
Severe acute respiratory syndrome (SARS)	2–10 days	Fever, headache, cough, body aches, acute respiratory failure	No vaccine
Measles (rubeola)	10–21 days	Coldlike, Koplik's spots, rash	MMR vaccine
Mumps	10–20 days	Swollen jaw	MMR vaccine
German measles (rubella)	12–32 days	Fever, rash	MMR vaccine, prenatal transmission
Chickenpox and shingles	10–23 days (chickenpox); few to many years (shingles)	Chickenpox: Rash Shingles: Rash and intense pain along nerves, postherpetic neuralgia (PHN)	Vaccine now available*; prenatal transmission Shingles vaccines (Zostavax approved in 2006 and new two-dose Shingrix vaccine approved in 2017 has greater protection and lasts longer)

Viral Disease	Incubation Period	Symptoms	Immunization and Comments
Hantavirus pulmonary syndrome	1–3 days	Early symptoms: flulike Late symptoms: shortness of breath, coughing, lungs fill with fluid	No vaccine; transmitted by aerosolized fecal material, saliva, or urine from infected rodents; high fatality rate
Smallpox	7–19 days (average is 10–14 days)	Fever, headache, body aches centrifugal pustular rash and scabs	Effective vaccine; disease eradicated in 1977
Monkeypox	5–21 days (average is 6–16 days)	Fever, respiratory symptoms, pustular rash and scabs, lymphadenopathy	Smallpox vaccine prevents infection by monkeypox

*A "combination" vaccine called MMRV, which contains both chickenpox and MMR vaccine, is available to children younger than 12 years.

Common Cold

The common cold is the most frequent infection, with the highest loss of workdays and schooldays, presenting an economic burden. Each year in the United States there are millions of cases of the common cold. Adults have an average of two to three colds every year, and children have even more. Hence, although not considered life threatening, a cold is "nothing to sneeze at" (pun intended!). Next time you are in a drugstore or a supermarket take a look at the dozens of remedies marketed to relieve the symptoms of the common cold. Manufacturers of these remedies are ever wary to avoid use of the term "cure" for fear of legal liability, but they come very close to the edge in their claims.

The symptoms of sneezing, coughing, sore throat, stiffness, and that general "blah" feeling are all too familiar. Colds are popularly described as "summer colds," "head colds," "chest colds," and "winter colds." These expressions have no medical basis, but they do indicate a seasonality and hint at the variety of symptoms (**FIGURE 13.7**). Colds are caused by a large variety of distinctive groups of viruses. Within each group there may be over 100 different strains; this makes it difficult to establish immunization despite the fact that vaccine technology exists. Which strain, or strains, should be chosen for development of a vaccine? Although a common denominator might exist within all strains, one has not been identified.

Cold viruses are transmitted not only by respiratory droplets but also, to a large extent, by hands or **fomites**, including doorknobs, faucets, furniture, and toys. The incubation period is about 1 to 3 days. Some individuals with colds carry a handkerchief or tissues into which nasal and cough secretions and saliva are discharged. Others sneeze or cough directly into their hands and

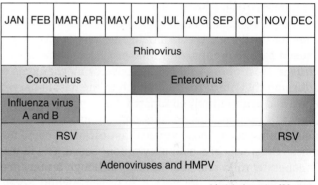

Information from various CDC sources.

FIGURE 13.7 A variety of viruses that cause respiratory illnesses circulate during different seasons of the year. Enteroviruses typically cause gastrointestinal and respiratory illness in the summer and autumn. Rhinoviruses cause respiratory illness from March through October. Adenoviruses and human metapneumonia viruses (HMPV) circulate and cause disease throughout the year, but more so in winter through late spring. Influenza A and B viruses cause illness from November through April. Respiratory syncytial virus (RSV) infection patterns are somewhat similar to influenza viruses. Typical coronavirus infections (not including SARS-CoV and MERS-CoV) usually peak from December through April.

have no reluctance in offering their hand for a friendly and infectious handshake. Three-ply tissues were introduced in the 1990s; the moisture-activated middle ply was impregnated with citric acid and sodium lauryl sulfate designed to kill the cold virus. Citric acid is a flavoring agent in soft drinks and sodium lauryl sulfate is a surfactant commonly found in many shampoos, shaving foams, toothpastes, and detergents. However, for a variety of reasons, including cost, efficiency, and the abrasiveness of the tissue, these tissues were not popular with the public but did not totally disappear from store

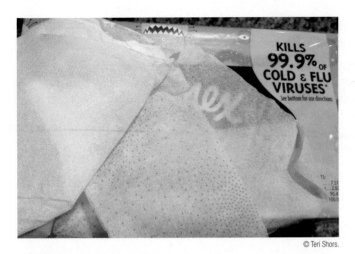

© Teri Shors.

FIGURE 13.8 Three-ply antiviral tissues with a special middle layer containing blue dots of chemicals that kill cold and flu viruses upon contact with mucus.

shelves. The "antiviral" Kleenex is still sold today, advertised to kill 99% of cold and flu viruses, including H1N1 influenza virus (**FIGURE 13.8**).

As mentioned, a large variety of viruses are responsible for the common cold, but about half of all colds are caused by two groups, the rhinoviruses and the adenoviruses. Most of the cold viruses possess mechanisms of adhesion that allow them to attach to the nasal pharynx and avoid being entirely eliminated during vigorous coughing and sneezing. This permits replication and spread to neighboring cells. Those viruses that escape become airborne. Diagnosis of a cold is symptom based. The illness is generally mild and self-limiting, although a nuisance. Laboratory tests are not necessary or worthwhile, except for epidemiological purposes.

Respiratory Syncytial Virus (RSV) and Metapneumovirus Infections

The respiratory syncytial virus (RSV) is highly contagious and endemic worldwide. RSV is the common cause of two serious respiratory diseases—**bronchiolitis** (an inflammation of the small airways of the lung) and **pneumonia** (infection of the lungs)—and is particularly life threatening to infants younger than 6 months of age. In fact, it is the most widespread cause of respiratory infections in this age group, and by the age of 2 years most children have had RSV infections. RSV has an incubation period of 4 to 5 days. The symptoms of the infection are nonspecific, which makes diagnosis difficult; they include fever, runny nose, ear infection, and pharyngitis. The virus is so named because in the respiratory tissues a **syncytium**— a network of large, fused, abnormal cells with multiple nuclei—may be present. The infection can progress to a serious lower respiratory tract infection, including obstructed airways.

For children younger than 5 years, RSV infections result in 2.1 million outpatient visits and 57,527 hospitalizations

in the United States each year. Virtually all children have had one RSV infection by age 3. In older children and in adults, the infection is manifested as a common cold. Outbreaks are a threat in pediatric wards and in nurseries, and the results can be devastating. Older adults are also adversely affected. Each year RSV infections cause 177,000 hospitalizations and 14,000 deaths among adults older than 65 years. Frequent and careful handwashing decreases the transmission of RSV. Treatment is largely supportive; additional oxygen with **intubation** and mechanical ventilation may be needed to help the patient breathe in some cases. No antivirals or vaccines are available to combat RSV.

Through molecular diagnostic testing, human metapneumovirus (HMPV), a relative of RSV, was discovered in 2001. It causes lower and upper respiratory infections in people of all ages but especially children, older adults, and people with weakened immune systems. Because it is a recently recognized respiratory virus, healthcare professionals may not routinely consider or test for HMPV. It commonly circulates in winter and lasts until or through spring. RSV and HMPV can circulate simultaneously during the respiratory virus season.

Influenza

The term "flu" is certainly familiar to you and is used somewhat loosely to describe coldlike respiratory symptoms accompanied by muscle pain (**myalgia**), but in most cases this is probably not the "true" flu. "Stomach flu" is another vague term used to describe symptoms associated with gastroenteritis; influenza viruses do not cause gastroenteritis. **Influenza** is caused by a virus that is as distinct from viruses that cause the common cold and those that cause gastroenteritis as dogs are from cats. So the term "flu" is frequently used inappropriately (particularly by students on exam days!), as are the "grippe" and the "24-hour" viruses.

The real flu or influenza is characterized by exaggerated coldlike symptoms, including headache, high fever or feverish/chills, myalgia (particularly in the back and legs), sore throat, cough, and fatigue. Influenza really wipes you out, whereas you are typically back to normal after a few days with a common cold. Influenza usually disappears in a week or two, and treatment, including bed rest and fluids, is aimed at relieving the symptoms. Some influenza viruses infect not only humans but other species, including dogs, pigs, seals, and birds (particularly ducks and other wild waterfowl, which are healthy **carriers** of influenza A viruses that are shed in their feces).

The three types of influenza viruses are designated A, B, and C. Influenza A virus causes **epidemics**, and occasionally **pandemics**, and is associated with animal **reservoirs**, particularly wild waterfowl and pigs. Influenza B virus is less severe, causing only epidemics, and there is no animal reservoir, whereas influenza C virus does not cause epidemics and produces only mild respiratory illness in humans. Influenza control is directed against types A and B.

Influenza is acquired from droplets and aerosols; fomites play a secondary role. The incubation time is 24 to 48 hours. Young children are particularly significant reservoirs of influenza viruses because of their relatively unhygienic practices. Conditions of colder temperatures (5°C or 41°F), low humidity (20%), crowding, and close intermingling favor transmission, as in theaters, classrooms, nursing homes, college dormitories, and military barracks. Influenza cases usually peak from about the middle of December through early March, the months in temperate climates when "fresh air" indoors is not possible in **temperate zones** or regions. Tropical areas do not exhibit seasonal variation.

Influenza Strains and Vaccines

Each year, in early October, preparation for the flu season begins; those 65 years and older and other high-risk groups, including people with immunodeficiencies and, in some cases, pregnant women, are particularly urged to get a flu shot (**FIGURE 13.9**). Why is this necessary each year? Lifetime, or at least long-term, prevention is usually the case in immunization against diseases such as polio, measles, and mumps—but not so with influenza.

The explanation lies in the biology of the influenza A virus. Protruding through the **viral envelope** are **hemagglutinin (H) spikes**, of which there can be as many as 500 on the surface of each virus, and **neuraminidase (N) spikes**, of which there are about 100 on the surface of each virus (**FIGURE 13.10**). Both of these serve as **virulence factors**.

The H spikes of influenza virus attaches to a host **receptor**, in this case, **sialic acid**, present on the epithelial cells that line the respiratory mucosa and aid viral penetration into these cells. The N spikes play a role in the last stage of viral replication, namely, release of new **virions**, allowing for spread to other cells. Both of these spikes are **antigens**, which means they provoke the

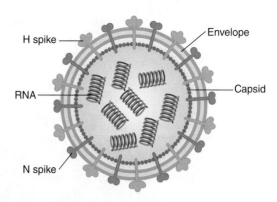

FIGURE 13.10 Influenza A virus H and N spikes located on the outside surface of the virus.

production of specific **antibodies** as part of the body's defense mechanism.

Now, here is the reason why influenza vaccination must be given on a yearly basis. These spikes undergo variation in their antigenic structure. The different forms are assigned numbers, for example, H1, H2, H3, N1, N2, and N3. There are 17 types of H and 9 types of N, allowing for 153 possible combinations. Hence, antibodies produced against H1 react against H1 spikes but not against H2 spikes. Think of it as a lock-and-key arrangement—a key made against lock H1 does not fit into the modified H2 lock but only into the original H1 lock (**FIGURE 13.11**).

The full nomenclature of a virus reflects the virus type (A, B, or C), when and where the virus was isolated, and the antigenic structure of the H and N spikes. For example, reports in the news media may state that the A/Philippines/82/H3N2 influenza virus is circulating during flu season. In this case, A is the type of influenza virus. It was first discovered in 1982 in the Philippines, and it contains H3N2 spikes.

Influenza virus H and N antigens continually undergo two types of changes, known as antigenic shift and antigenic drift. **Antigenic shift** occurs only with influenza type A virus. The shift is a major, abrupt antigenic change in the H or N spikes that results from the recombination of genetic material from different viral strains, creating a new influenza A virus strain (**FIGURE 13.12**). Consider that if a pig were to be coincidentally infected by both a human influenza A virus strain and a bird influenza A virus strain, a new hybrid or recombinant influenza A virus strain might develop that could then infect a human; the pig served as a **blender** (or recombination vessel), somewhat like a mixer creating a new hybrid influenza A viruses.

Flu pandemics commonly originate in China, where millions of pigs, birds, and people live in close quarters, allowing for new recombinations of influenza A virus strains and enormous opportunity for crossing the **species barrier**. Possibly, the deadly 1918 influenza arose as a result of antigenic shift (**BOX 13.1**). **Antigenic drift**, on the other

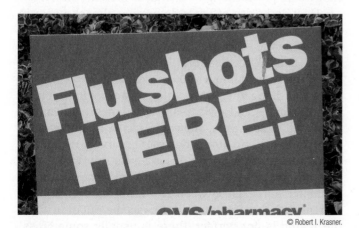

© Robert I. Krasner.

FIGURE 13.9 Individuals are urged to get their flu shots at local clinics; this sign is posted at a pharmacy that provides flu shots.

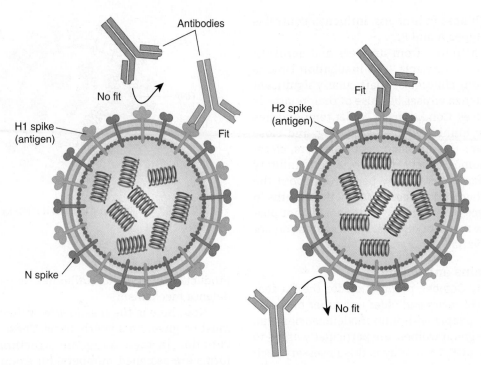

FIGURE 13.11 Lock-and-key arrangement of antigen and specific antibodies.

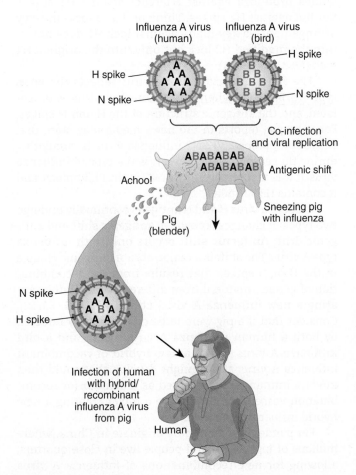

FIGURE 13.12 Antigenic shift occurs when influenza A viruses coinfect the same cell. During replication, an influenza A virus acquires a new H or N gene, resulting in a new or recombinant hybrid influenza A virus strain that could cause a pandemic.

hand, is a minor change in the H and N spikes occurring over a period of years (**FIGURE 13.13**). Both influenza A and B viruses can undergo antigenic drift. Because influenza B viruses only infect humans, antigenic shift of influenza B viruses does not occur.

The upshot of both antigenic shift and antigenic drift is that they enable influenza viruses to evade the antibody defense mechanisms of the host by "changing their coats" (surface H or N spikes). Because these changes continually occur, the influenza A and B virus strains that make up the vaccine year to year need to be adjusted. Antibodies against last year's viral strains are effective only to the extent that some of these strains may, coincidentally, be the same. If the changes in influenza A and B viruses did not occur, a flu shot would provide long-term, maybe lifetime, immunity.

How do health officials determine almost a year in advance of the flu season which viral strains are to be included in the vaccine cocktail for the coming year? The World Health Organization (WHO), in cooperation with other agencies, has over 100 surveillance sites around the world that gather information regarding the circulating influenza A and B viruses. These data are assembled in the fall of each year, allowing vaccine-manufacturing companies to begin production in February for the following flu season. Considering this is a "year in advance" process, it is not surprising that, depending on the year, some mismatch may exist between the viral strains anticipated and those that are included in the vaccine, but even when the match is not perfect, there is usually some cross-protection. Equally significant is the serious problem that some of the viral strains can mutate and become resistant to the antiviral drugs used to treat influenza patients.

BOX 13.1 The Coming Flu Pandemic?

Historically, influenza A virus was responsible for the occurrence of pandemics. Antigenic shifts in influenza A viruses occur periodically, and people all over the world do not have antibody protection against novel influenza A virus strains, leaving them vulnerable when one emerges. A study of influenza epidemics and pandemics that have occurred over the centuries indicates a cyclical pattern of influenza A outbreaks.

In 1918, the world was engaged in the horrors of World War I. As if that were not tragic enough, along came the influenza A virus, a biological agent knowing no political or national loyalties. Influenza took a toll on human life greater than the ravages of the war itself; the pandemic that it created is regarded as one of the most devastating of all time (**Figure 1**).

Little was known about viruses; it was not until about 20 years later that viruses were studied in earnest. Several million people died in the short span of only 120 days, including an estimated 500,000 to 675,000 in the United States. Estimates are that influenza A virus killed 50 to 100 million people worldwide in a single year, including 17 million in India. The virus killed rapidly; a person could be fine in the morning, feel sick in the afternoon, and be dead by nightfall! Bodies piled up, and burying of the dead became a major public health problem. The overworked, underpaid, and at-risk gravediggers, perhaps in an effort to keep their sanity during those terrible times, developed their own jargon, stories, and poems. One poem that surfaced was the following:

I lost a little bird
And its name was Enza
I opened the window
And in-flu-enza!

The 1918 influenza pandemic remains a mystery in terms of understanding why this virus was so virulent. Many influenza victims died because their lungs filled with fluid, causing them to drown. Generally, the senior members of the population are usually more prone to die from influenza, but an unusual aspect of the 1918 pandemic is that the disease struck 20- to 40-year-old adults the hardest. Typical influenza season deaths

are represented graphically by **U-shaped mortality curves** in which peak mortalities caused by influenza occur among the very young and older adults (**Figure 2A**). The 1918 influenza pandemic resulted in a **W-shaped mortality curve**. Death rates peaked during October 1918 (**Figure 2B**). So many deaths occurred among the childbearing population that there was a drop in U.S. **life expectancy** in 1918 (**Figure 2C**). Because influenza pandemics appear to be cyclical, for the sake of future readiness it is important to understand why the 1918 influenza pandemic was so devastating.

To this end, Johan Hultin, a 73-year-old pathologist from San Francisco, traveled to the Alaskan tundra to exhume the bodies of Inuits who died during the 1918 pandemic. Hultin brought back lung tissue specimens for study to determine the genes responsible for the unique virulence of the influenza A virus to provide valuable information about thwarting another potential pandemic (**Figure 3**). Hultin's samples were analyzed at the Armed Forces Institute of Pathology, and the evidence indicates it to be an H1N1 influenza A virus, but questions remain as to why this virus was so deadly.

During the 20th century major influenza pandemics occurred in 1918, 1957, and 1968. In fact, public health experts feared that we were overdue for an epidemic, possibly a pandemic. Contrary to what many virologists expected, the first pandemic of the 21st century started in North America, not in Southeast Asia or China. Mexico became the **epicenter** of the first pandemic of the 21st century (**Figure 4**). From the beginning of the outbreak in April 2009 through December 5, 2009, 208 countries reported cases. The influenza A virus was identified as a new H1N1 swine influenza A virus strain.

A striking observation of the 2009 pandemic was that more than 75% of infected people were younger than 30 years of age. The most affected group was 10- to 19-year-olds. Less than 3% of cases were persons over the age of 65. A plausible explanation was that older people born before 1957 still had some cross-reacting neutralizing antibodies that protected them from infection by the 2009 **H1N1 swine influenza A virus** strain; it was the same strain that dominated the other circulating strains during the pandemic in Mexico.

Response to a pandemic in the 21st century has technological advantages over responses to previous pandemics. The media responded with public service announcements on cough etiquette, handwashing reminders, school closings, and reminding sick individuals to stay home to minimize spreading the flu.

Flyers and posters were hung on walls in schools and other public places (**Figure 5**). Website healthmaps and smartphone applications were created to track influenza cases. Users could track outbreaks in their region and alert others by email when new cases were reported in their area. The WHO and the CDC's Emergency Response Team played a vital role in assessing, guiding, and optimizing resources to reduce the impact of the pandemic on businesses, hospitals, schools, and the community. The "wildfire" was manageable this time around.

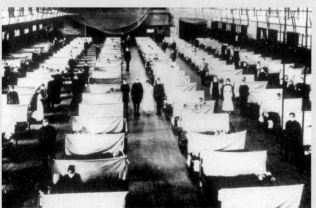

Courtesy of the National Library of Medicine.

Figure 1 A ward of patients with influenza. The 1918 influenza pandemic was catastrophic and killed millions of people all over the world.

(continues)

BOX 13.1 The Coming Flu Pandemic? (*continued*)

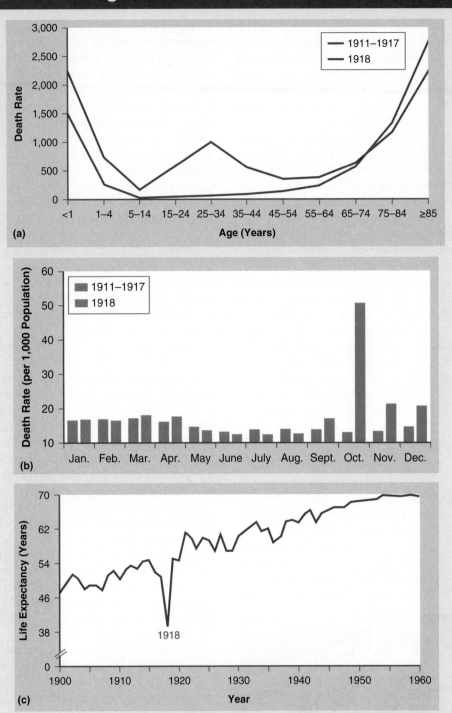

(a and c) Information from: J. K. Taubenberger. (1999). Seeking the 1918 Spanish influenza virus. *ASM News, 65,* 473–478.

Figure 2 **(a)** The graph compares the age distribution of influenza deaths in the United States during the 1918 pandemic compared to seasonal influenza deaths that occurred from 1911 through 1917. The 1918 influenza pandemic resulted in a U-shaped curve, whereas influenza deaths from 1911 to 1917 show a W-shaped curve, with mortality peaks among the very young and the elderly. **(b)** During the 1918 influenza pandemic, the month of October (which is typically the start of flu season) had the highest number of influenza deaths in the United States. **(c)** U.S. life expectancy dropped in 1918.

An **H3N2 swine influenza A virus** strain was circulating in the United States during the summer of 2012. More than 200 human cases in over 10 states were reported. Most cases were children showing hogs for 4H or FFA projects at county fairs. Fairgoers were advised to wash their hands and to avoid touching the pigs. The variant strain spread directly from pigs to humans but was not spreading from human to human. Dr. Michael Osterholm, director of the University of Minnesota's Center for Infectious Diseases Research and Policy commented, "*It's time to take what likely would be a very unpopular*

Courtesy of Dr. Johan Hultin.

Figure 3 Johan Hultin collecting tissue samples in 1997 from 1918 flu victims buried in an Alaskan mass grave.

© Dreamshot/Dreamstime.com.

Figure 4 Newspaper headlines in April 2009, announcing that influenza deaths had risen to 236 in Mexico.

© Teri Shors.

Figure 5 Poster on handwashing etiquette on a wall in the Halsey Science Center at the University of Wisconsin–Oshkosh.

step—tell organizations this year, pigs should stay home from the fair. These pigs shouldn't be at fairs." Per his warning followed by human flu cases at other county fairs in Ohio, the Cuyahoga County Fair Board in Ohio banned the pig exhibits. Several other county fairs closed their pig barns following human influenza cases. For the next few years, the swine flu viruses were quiet in the human population. A small blip appeared on the radar screen in August 2018 when fairgoers who went to the San Luis County Fair in California who had extended contact with pigs tested positive for a variant swine influenza A virus.

The world currently faces the threat of a potential pandemic of avian (bird) flu caused by H5N1 and H7N9 influenza A virus strains. Most epidemiologists and public health personnel would agree that it is not "if" but "when" an outbreak of avian flu will happen.

What is bird flu, and what are H5N1 and H7N9? All species of animals (and plants) are hosts for a variety of microbes and viruses. Wild waterfowl are the natural reservoir for influenza A

viruses, along with many other microbes, in their intestinal tract as normal **microbiota**, usually without signs of illness. The influenza A viruses are shed into their feces, nasal secretions, and saliva and can be transmitted to other birds that might have contact with these materials or contact with contaminated surfaces. Similarly, domestic birds (poultry) can also pick up microbes or viruses. The avian influenza A virus is highly pathogenic for poultry and has a mortality rate approaching 100%; it can destroy a flock of chickens in only a few days.

There are many strains of influenza A viruses, dictated by the combination of H and N spikes present on the surface of the virus. Rarely do these influenza A viruses infect humans, but human infection did happen in Hong Kong in 1997, when H5N1-infected poultry caused influenza in humans, making the spread of **H5N1 avian influenza A virus** into the human population. The species leap—chickens to people— had occurred. Once again, this time in China in 2013, an **H7N9 avian influenza A virus** crossed the species barrier into humans.

While the H5N1 and H7N9 avian influenza A viruses do not spread easily from person to person, they do cause severe infections in humans. From 2003 through 2018, the H5N1 virus was reported in 16 countries, with 860 cases and 454 deaths—a 53% mortality rate. From 2013 to 2018, a total of

(continues)

BOX 13.1 The Coming Flu Pandemic? *(continued)*

1,564 laboratory-confirmed cases had a 39% case-fatality rate. At the time of this writing, all cases were reported in China. The World Organization for Animal Health (formerly the Office International des Epizooties, OIE) maintains up-to-date statistics on animal and avian influenza.

Many species of waterfowl that carry influenza A viruses migrate long distances between their breeding grounds and nonbreeding locations. As natural reservoirs of influenza A viruses, waterfowl and shorebirds could spread highly pathogenic avian influenza A viruses, such as the H5N1 and H7N9 strains mentioned earlier, to human populations during migration. Migration patterns are well documented.

Migratory routes of birds are grouped together as flyways. A **flyway** is the entire range of a migratory bird species through which it travels on a yearly basis from breeding to nonbreeding grounds, including intermediate resting and feeding places. The **Pacific flyway** is a major north–south flyway for migratory birds in North America, extending from Alaska to Patagonia (**Figure 6A**). North America has four flyways: Atlantic, Mississippi, Central, and Pacific. The Pacific flyway is thought to be the most likely location for the introduction for highly pathogenic avian influenza A viruses into humans in the United States and Canada. Will this be the route for the coming flu pandemic in North America (**Figure 6B**)? Or will a new coronavirus emerge?

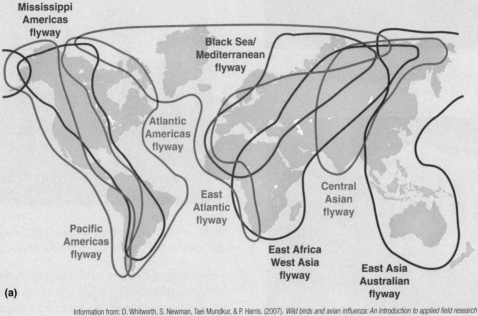

(a)

Information from: D. Whitworth, S. Newman, Taei Mundkur, & P. Harris. (2007). *Wild birds and avian influenza: An introduction to applied field research and disease sampling techniques.* FAO Animal Production and Health Manual. Rome: Food and Agriculture Organization of the United Nations.

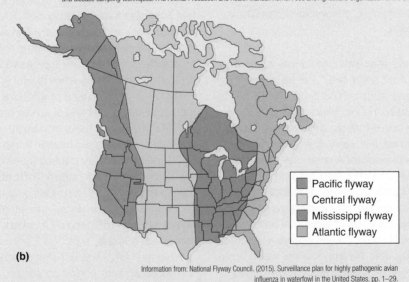

■ Pacific flyway
□ Central flyway
■ Mississippi flyway
□ Atlantic flyway

(b)

Information from: National Flyway Council. (2015). Surveillance plan for highly pathogenic avian influenza in waterfowl in the United States. pp. 1–29.

Figure 6 **(a)** Flyways of migratory birds. Some species of birds have very long annual migrations, breeding in the high latitudes of the Arctic during the northern summer and then traveling to more hospitable middle or southern latitudes as far south as South America. Some migratory species spread low pathogenic strains of avian influenza A viruses but also have the potential to spread highly pathogenic avian influenza A viruses—possibly a coming flu pandemic! **(b)** North America has four flyways. The flyway boundaries are defined based on known bird migration routes but are not geographically fixed or sharply defined.

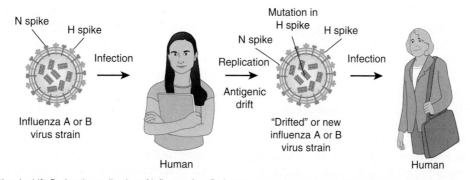

FIGURE 13.13 Antigenic drift. During the replication of influenza A or B viruses, a mutation or change occurs in the H or N antigens, resulting in a new strain. Antigenic drift is occurring all of the time among influenza viruses that infect humans. Antigenic drift results in influenza epidemics. Influenza A and B viruses are always changing. No universal vaccines are available to prevent all strains of influenza A or B. This is why a different seasonal flu vaccine is needed every year.

Influenza outbreaks are costly. The CDC estimates that a staggering number of lost work days occur each year due to influenza. The cost of vaccine preparation, vaccine administration, and loss of productivity runs into the billions of dollars in the United States alone. Additional statistical and epidemiological facts put together by the CDC are shown in **FIGURE 13.14**.

Many individuals, including those aged 65 and older, who are the most susceptible, are reluctant to be vaccinated against influenza. Their refusal of the vaccine is risky, because **secondary bacterial infections**, possibly leading to a fatal pneumonia, are more likely to occur in the older population. This is why many physicians prescribe antibiotics for their older patients who have been diagnosed with influenza.

The U.S. Food and Drug Administration (FDA) has approved three antivirals that specifically act against influenza A and B viruses: zanamivir (**Relenza**), oseltamivir (**Tamiflu**), and peramivir (**Rapivab**). The antiviral drugs do not prevent infection, nor are they cure-alls. If taken within about 48 hours after the appearance of symptoms, however, they can shave off a few days of illness and reduce the severity of a bout with influenza. They are not intended as substitutes for vaccines, and, as with all medicines, they have side effects. Furthermore, individuals need to recognize very early flu signs, which mimic those of many other microbial diseases, and visit their physician almost immediately to get a prescription for these drugs.

SARS and MERS

Coronaviruses are a diverse group of viruses that infect humans and a wide range of animals. A high frequency of recombination and mutations occur between different coronaviruses, which may be why they seem to be able to adapt to new hosts and changing environments.

As an animal pathogen, coronaviruses cause a variety of disease symptoms such as respiratory, gastrointestinal, liver, and neurological disorders resulting in mild disease, severe sickness, and even death in some cases. In humans, coronaviruses cause mild upper respiratory infections. However, when SARS-CoV emerged in 2002 and MERS-CoV emerged a decade later, that paradigm was challenged.

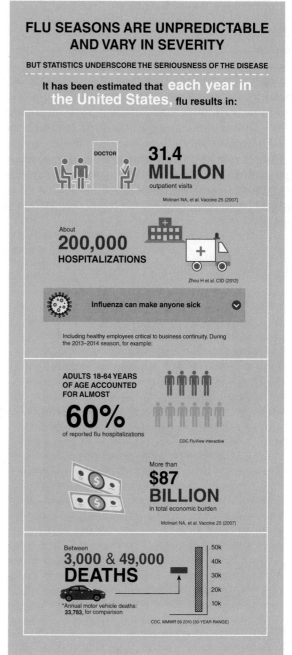

Courtesy of CDC.

FIGURE 13.14 CDC influenza prevention and epidemiology statistics/facts.

O SARS

Between November 2, 2002, and July 3, 2003, it appeared that the world was on the brink of a pandemic of severe acute respiratory syndrome (SARS) caused by **SARS-CoV**, a coronavirus. Approximately 8,096 cases and 774 deaths occurred over that relatively short period of time. Overall, about 10% of SARS-CoV patients developed severe complications and died. The outbreak first occurred in a rural area in Guangdong Province in China in 2002, but the People's Republic of China covered up the extent of the problem, thus delaying public health intervention. Not until February 2003 when an American businessman flying from China to Singapore became ill with respiratory distress, causing the plane to land in Hanoi, Vietnam, where the person died, was a public health response put into place.

This event was widely publicized, and the world was now aware of the cover up in China and a threatening pandemic. In short order SARS spread rapidly across continents into some 37 countries. Images of deserted streets and subways and people wearing face masks were common; thermal checkpoints for the detection of passengers with elevated temperatures were recommended by WHO at selected airports with international flights—all in an effort to halt the further march of the disease (**FIGURE 13.15**).

SARS is spread by close personal contact. Respiratory droplets propelled within about a meter (or yard) from an infected person onto the mucous membranes of the eyes, mouth, and nose of another person can establish infection. Transferring the coronavirus by touching fomites contaminated with SARS-CoV and transference onto the mucous membranes can also establish infection. Diagnosis can be difficult because of the appearance of the usual flulike symptoms—cough, fever, myalgia, sore throat, and shortness of breath—within usually a 2- to 10-day window. As is true of viral diseases in general, no specific treatment is available, limiting the care of those with SARS to supportive treatment. In the United States only eight people, all of whom had traveled to countries with SARS, had laboratory-confirmed evidence of the disease.

Epidemiologists have determined that bats are the natural reservoir of SARS-CoV. Humans were infected through an **intermediate host**, palm civet cats (**FIGURE 13.16A**), which likely had contact with infected bats or bat saliva or guano (**FIGURE 13.17**). By 2004, the CDC issued a "Notice of Embargo of Civets" because SARS-CoV was isolated from civets captured in areas of China where the SARS outbreak originated. In southern China, civet cats are a delicacy that were sold at **live animal markets** for consumption. The CDC also banned the importation of civets. The civet is a mammal with a catlike body, long legs, long tail, and a masked face resembling a raccoon or weasel. Antibodies against SARS-CoV were detected in blood samples drawn from handlers of civets.

Continued surveillance by the WHO, CDC, and other public health interests is imperative due to the rapid dissemination of the virus and its high morbidity and mortality rates. Further, the socioeconomic spinoffs were

© Akkharat Jarusilawong/ShutterStock, Inc.
© Rweisswald/ShutterStock, Inc.

FIGURE 13.16 Natural reservoirs for highly pathogenic human coronaviruses. **(a)** Palm civet cat is an intermediate host and reservoir for SARS-CoV. **(b)** Dromedary camels, including their milk, meat, feces, or urine, are a reservoir for MERS-CoV.

JOEL NITO/AFP/Getty Images.

FIGURE 13.15 Airport thermal scanners screening passengers arriving in Manila were re-introduced as a measure to prevent the spread of SARS from China into the Philippines.

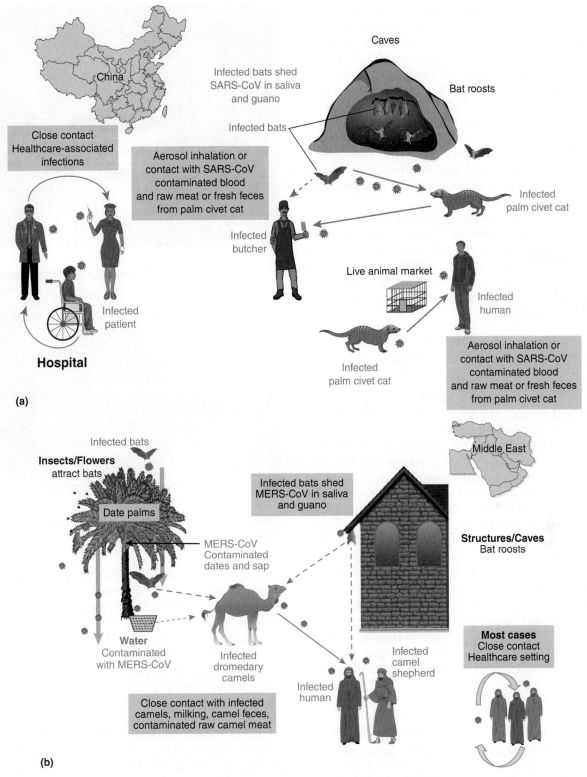

FIGURE 13.17 **(a)** Proposed model for SARS-CoV and **(b)** MERS-CoV zoonotic infections. Solid arrows indicate known mode of transmission. Dashed arrows are speculated modes of transmission. Many questions remain regarding the epidemiology of these highly pathogenic coronaviruses.

tremendous in this near pandemic, including ethnic and racial stereotyping and declines in tourism. The SARS epidemic and its sociological and ethnic impact serves as an excellent example of the broad parameters of public health.

○ MERS

Ten years after the SARS pandemic, another new human coronavirus emerged. In June 2012, a novel coronavirus was isolated from a patient who died from severe pneumonia and multiorgan failure in Saudi Arabia. The isolated coronavirus was eventually named Middle East respiratory syndrome coronavirus, or **MERS-CoV**. Unlike SARS-CoV, it has not caused thousands of cases over a short period of time. From September 2012 through August 10, 2018, 2,229 laboratory-confirmed cases had been noted, associated with 791 deaths (35% mortality rate). Cases have been reported in 27 countries. MERS-CoV is not spread easily from person to person unless there is close contact, such as in a healthcare setting in which the caregiver is not correctly using **PPE** (**personal protective equipment**; i.e., protective clothing, face masks, gloves etc.). **Healthcare-associated infections (HAIs)** have occurred in Saudi Arabia, the United Arab Emirates, and South Korea.

The mode of transmission is not fully understood, but dromedary camels (**FIGURE 13.16B**) are a major reservoir host for MERS-CoV and an animal source of infection in humans. MERS-CoV strains isolated from dromedary camels are identical to strains isolated from MERS patients. A model for MERS-CoV mode of transmission to humans and animals is shown in Figure 13.17.

Measles (Rubeola), Mumps, and German Measles (Rubella)

Measles, mumps, and German measles are considered childhood diseases. In 1968, a **measles**, **mumps**, **rubella vaccine (MMR)** was introduced. MMR consists of a mixture of live attenuated (weakened) measles, mumps, and German measles viruses. The vaccine has drastically reduced the incidence of these diseases. The low incidence does not mean people are unlikely to contract the diseases and therefore do not need to be vaccinated. The public needs to be aware of this, because a false sense of security can lead to failure to immunize, resulting in a nonimmunized and vulnerable population. If vaccination was halted and the population's **herd immunity** waned, many new cases would result. Although the triple vaccine is composed of live attenuated viruses, complications are rare, and the risks of acquiring one or more of these diseases is much greater than the risks of complications. The vaccine is given at 12 to 15 months of age and again at 4 to 6 years of age. The three diseases have no specific treatment, and prevention by immunization is the key.

○ Measles

Measles, also known as rubeola, is distinct from what is commonly known as German measles, discussed below. The measles virus is the most contagious human pathogen; estimates are that there is more than a 98% chance of becoming infected if exposed directly to someone with measles. *It has an infective dose of one virus.* The mechanism of transmission is by respiratory droplets, and the disease is fostered by overcrowding, low levels of immunity in the population, malnutrition, and inadequate medical care. Humans are the only reservoir for this disease (as is true of smallpox, chickenpox, and mumps), and therefore it could be eradicated. The symptoms of measles are

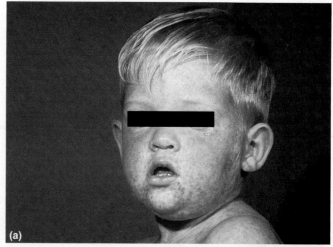
Courtesy of CDC.

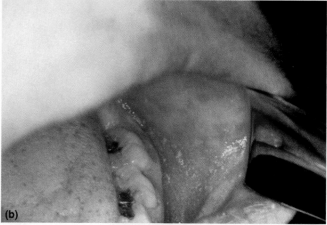
(b)   Courtesy of CDC.

FIGURE 13.18 Measles (rubeola). **(a)** Characteristic diffuse red rash covering the face and shoulders. **(b)** Koplik's spots inside the mouth.

similar to a cold, with the development of characteristic **Koplik's spots** in the mouth early in the disease, followed by a red rash (**FIGURE 13.18**) that starts on the face and spreads to the extremities and over most of the body. The disease is usually mild and self-limiting, but 1 in 500 children with measles develops potentially serious and even fatal complications, including pneumonia, ear infections, brain damage, and seizures.

Worldwide, measles is a significant disease and is a frequent cause of death, particularly in developing countries, despite the availability of an effective vaccine. In developing countries and in populations lacking immunization, more than 15% of children who contract measles die of the disease or its complications.

Malnourished children are particularly vulnerable. The Measles Initiative, launched in 2001, is led by the American Red Cross, the United Nations Foundations, the CDC, UNICEF, and the WHO. Since 2001, the Initiative has been active in 80 countries, delivering more than 1 billion doses of measles vaccine, raising measles vaccination coverage to 85% and reducing measles deaths by 74%.

In 1998, Dr. Andrew Wakefield, a British gastroenterologist, was the lead researcher and main author of a study published in *The Lancet* that described a new autism syndrome speculated to be triggered by MMR vaccination. The study lowered parental confidence in vaccination programs and created an MMR vaccination crisis in the United Kingdom and sparked questions about vaccine safety in North America. Evidence mounted that there was no increased autism risk associated with the MMR vaccine.

The paper was retracted 12 years later (**FIGURE 13.19**). Britain's General Medical Council ruled that the data presented by Dr. Wakefield were fraudulent, and he was found guilty of unethical medical practice and scientific misconduct. Despite the retraction, parents and autism advocacy groups continue to support Dr. Wakefield. Dr. Paul Offit, Chief of Infectious Diseases at Children's Hospital of Philadelphia, said, "*This retraction by The Lancet came far too late. It's very easy to scare people; it's very hard to unscare them.*"

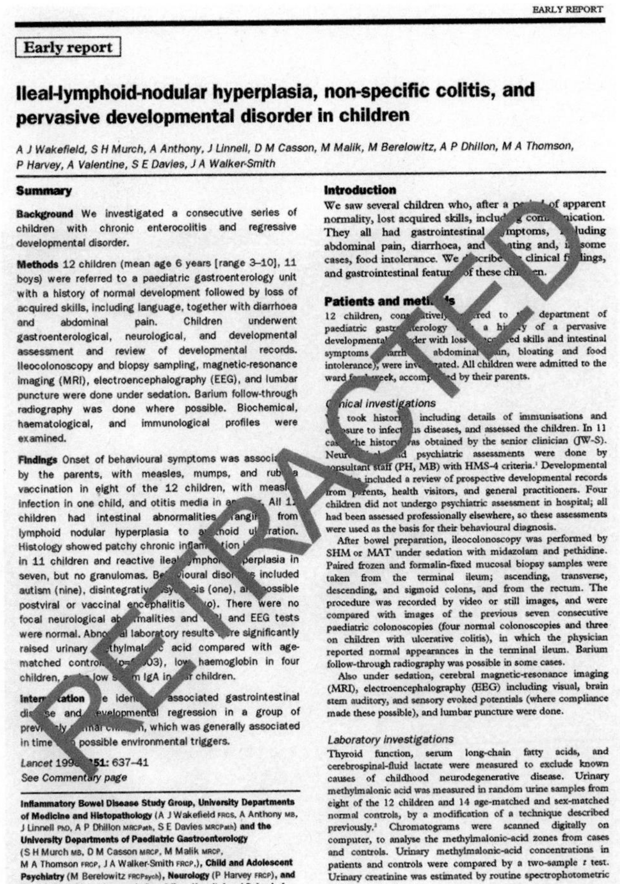

Hundreds of independent studies conducted over the past 15+ years have come to the same conclusion: vaccines do not cause autism. The scientific community agrees that neither vaccines nor their ingredients are linked to the development of autism.

The acquired immunity resulting from having had the disease is considered to be lifelong for survivors, whereas the MMR vaccination affords approximately a 20-year protection. "Susceptibles" slip through the regimen of immunization because of a false sense of security that these diseases are gone, a proposed link to autism, or because of the burden of healthcare costs, particularly in less developed countries. Therefore, outbreaks of measles will continue to occur, as happened in Fiji during February to May 2006, resulting in 132 cases.

Europe and the United States are no exception. From January 2016 to March 2018, 57 measles deaths were reported in the European Union. Between February 2017 and January 2018, 14,732 cases of measles were reported to the European Surveillance System by 30 countries. Most cases were reported in Romania (35%), Italy (34%), Greece (9%), and Germany (6%). This continued spread of measles across Europe is due to suboptimal vaccination coverage (87% of the cases were unvaccinated individuals). In 2000, measles was declared eradicated in the United States, but there continues to be a resurgence of outbreaks. Anti-vaxxers, those who choose to forgo vaccinations, have targeted immigrant parents with antivaccine propaganda. In 2008, antivaccine advocates Andrew Wakefield and the Organic Consumers Association began targeting Somali Americans in Minnesota with misinformation about autism and vaccines. The anti-vaxxer movement likely contributed to the largest measles outbreak in Minnesota in nearly 30 years with 79 cases; most cases were unvaccinated Somali American children.

A recent study analyzing the antivaccine movement in the United States reported that there are several metropolitan measles hotspots with high percentages of **nonmedical exemptions**. In other words, parents are not vaccinating their children based on reasons that are not medical (e.g., antivaccine beliefs). The metropolitan hotspots include: Spokane, Washington; Portland, Oregon; Salt Lake City and Provo, Utah; Houston, Fort Worth, Plano, and Austin, Texas; Troy, Warren, and Detroit, Michigan; Kansas City, Missouri; and Pittsburgh, Pennsylvania. Several U.S. cities located in less populated areas are also measles hotspots based on their higher nonmedical exemption rates (**TABLE 13.3**).

In 2015, the United States experienced a large multistate outbreak linked to Disneyland in Anaheim, California. It likely started from a traveler who became infected overseas with measles and subsequently visited the amusement park. In 2014, 23 different measles outbreaks occurred in the United States, including a large outbreak of 383 cases among unvaccinated Amish children in Ohio. Many of the cases were associated with cases brought in from the Philippines, which had

EARLY REPORT

Early report

Ileal-lymphoid-nodular hyperplasia, non-specific colitis, and pervasive developmental disorder in children

A J Wakefield, S H Murch, A Anthony, J Linnell, D M Casson, M Malik, M Berelowitz, A P Dhillon, M A Thomson, P Harvey, A Valentine, S E Davies, J A Walker-Smith

Summary

Background We investigated a consecutive series of children with chronic enterocolitis and regressive developmental disorder.

Methods 12 children (mean age 6 years [range 3–10], 11 boys) were referred to a paediatric gastroenterology unit with a history of normal development followed by loss of acquired skills, including language, together with diarrhoea and abdominal pain. Children underwent gastroenterological, neurological, and developmental assessment and review of developmental records. Ileocolonoscopy and biopsy sampling, magnetic-resonance imaging (MRI), electroencephalography (EEG), and lumbar puncture were done under sedation. Barium follow-through radiography was done where possible. Biochemical, haematological, and immunological profiles were examined.

Findings Onset of behavioural symptoms was associated, by the parents, with measles, mumps, and rubella vaccination in eight of the 12 children, with measles infection in one child, and otitis media in another. All 12 children had intestinal abnormalities, ranging from lymphoid nodular hyperplasia to aphthoid ulceration. Histology showed patchy chronic inflammation in the colon in 11 children and reactive ileal lymphoid hyperplasia in seven, but no granulomas. Behavioural disorders included autism (nine), disintegrative psychosis (one), and possible postviral or vaccinal encephalitis (two). There were no focal neurological abnormalities and MRI and EEG tests were normal. Abnormal laboratory results were significantly raised urinary methylmalonic acid compared with age-matched controls (p=0.003), low haemoglobin in four children, and a low serum IgA in four children.

Interpretation We identified associated gastrointestinal disease and developmental regression in a group of previously normal children, which was generally associated in time with possible environmental triggers.

Lancet 1998; **351:** 637–41

See Commentary page

Introduction

We saw several children who, after a period of apparent normality, lost acquired skills, including communication. They all had gastrointestinal symptoms, including abdominal pain, diarrhoea, and bloating and, in some cases, food intolerance. We describe the clinical findings, and gastrointestinal features of these children.

Patients and methods

12 children, consecutively referred to the department of paediatric gastroenterology with a history of a pervasive developmental disorder with loss of acquired skills and intestinal symptoms (diarrhoea, abdominal pain, bloating and food intolerance), were investigated. All children were admitted to the ward for 1 week, accompanied by their parents.

Clinical investigations

We took histories, including details of immunisations and exposure to infectious diseases, and assessed the children. In 11 cases the history was obtained by the senior clinician (JW-S). Neurological and psychiatric assessments were done by consultant staff (PH, MB) with HMS-4 criteria.[1] Developmental histories included a review of prospective developmental records from parents, health visitors, and general practitioners. Four children did not undergo psychiatric assessment in hospital; all had been assessed professionally elsewhere, so these assessments were used as the basis for their behavioural diagnosis.

After bowel preparation, ileocolonoscopy was performed by SHM or MAT under sedation with midazolam and pethidine. Paired frozen and formalin-fixed mucosal biopsy samples were taken from the terminal ileum; ascending, transverse, descending, and sigmoid colons, and from the rectum. The procedure was recorded by video or still images, and compared with images of the previous seven consecutive paediatric colonoscopies (four normal colonoscopies and three on children with ulcerative colitis), in which the physician reported normal appearances in the terminal ileum. Barium follow-through radiography was possible in some cases.

Also under sedation, cerebral magnetic-resonance imaging (MRI), electroencephalography (EEG) including visual, brain stem auditory, and sensory evoked potentials (where compliance made these possible), and lumbar puncture were done.

Laboratory investigations

Thyroid function, serum long-chain fatty acids, and cerebrospinal-fluid lactate were measured to exclude known causes of childhood neurodegenerative disease. Urinary methylmalonic acid was measured in random urine samples from eight of the 12 children and 14 age-matched and sex-matched normal controls, by a modification of a technique described previously.[2] Chromatograms were scanned digitally on computer, to analyse the methylmalonic-acid zones from cases and controls. Urinary methylmalonic-acid concentrations in patients and controls were compared by a two-sample *t* test. Urinary creatinine was estimated by routine spectrophotometric assay.

Children were screened for antiendomyseal antibodies and boys were screened for fragile-X if this had not been done

Inflammatory Bowel Disease Study Group, University Departments of Medicine and Histopathology (A J Wakefield FRCS, A Anthony MB, J Linnell PhD, A P Dhillon MRCPath, S E Davies MRCPath) **and the University Departments of Paediatric Gastroenterology** (S H Murch MB, D M Casson MRCP, M Malik MRCP, M A Thomson FRCP, J A Walker-Smith FRCP,), **Child and Adolescent Psychiatry** (M Berelowitz FRCPsych), **Neurology** (P Harvey FRCP), **and Radiology** (A Valentine FRCR), **Royal Free Hospital and School of Medicine, London NW3 2QG, UK**

Correspondence to: Dr A J Wakefield

© Teri Shors.

FIGURE 13.19 *The Lancet* retracted this controversial paper 12 years after it was published. The author suggested a link between the MMR vaccine and autism. Many parents in the United Kingdom lost confidence in vaccination because of this research study.

TABLE 13.3 Top 10 Measles Hotspots in the United States

Rank	State	Percentage of Nonmedical Exemptions: 2016–2017	County	Largest City
1	Idaho	26.7%	Camas	Fairfield
2	Idaho	19.7%	Bonner	Sandpoint
3	Idaho	18.2%	Valley	McCall
4	Idaho	17.1%	Custer	Challis
5	Idaho	16.1%	Idaho	Grangeville
6	Wisconsin	15.7%	Bayfield	Washburn
7	Idaho	15.6%	Boise	Horseshoe Bend
8	Idaho	14.9%	Kootenai	Coeur d'Alene
9	Idaho	14.6%	Boundary	Bonners Ferry
10	Utah	14.6%	Morgan	Morgan

Information from J. K. Olive, P. J. Hotez, A. Damania, M. S. Nolan. (2018). The state of the antivaccine movement in the United States: A focused examination of nonmedical exemptions in the states and counties. *PLoS Medicine.* https://doi.org/10.1371/journal.pmed.1002578

experienced a large measles outbreak. This demonstrates that travelers need to make sure their vaccinations are up-to-date! These outbreaks underscore the necessity to maintain a high level of immunity in the population as a preventative measure.

○ Mumps

The hallmark of **mumps** is a large swelling on one or both sides of the face, resulting from infection of the parotid gland, one of three pairs of salivary glands, which is located at the junction of the upper and lower jaw (**FIGURE 13.20**). Humans are the only natural hosts for this disease, which has an incubation period of 10 to 20 days. In temperate climates the disease usually occurs in late winter and early spring, most commonly in children younger than 15 years. Many children are asymptomatic; that is, they show no symptoms but nevertheless are rendered immune.

Symptoms other than swelling of the parotid glands (which may not occur) include fever, nasal discharge, and muscle pain. The virus can spread to other structures, including the testes, ovaries, meninges, heart, and kidneys. Complications are unlikely, but there has been some concern that in young males inflammation of the testes (orchitis) could lead to sterility. This has not proved to be the case, although this temporary complication is painful. Rarely, permanent deafness occurs and is usually confined to one ear.

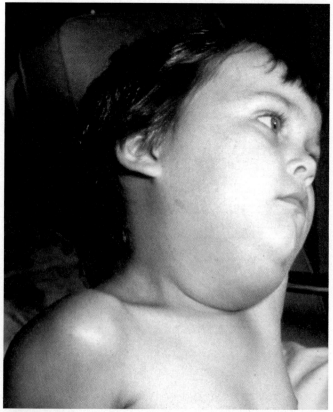

Courtesy of NIP/Barbara Rice/CDC.

FIGURE 13.20 Mumps. This child has swollen parotid glands.

Before the MMR vaccine, estimates are that 50% of the children in the United States became infected with mumps. The decline, due to the vaccine, is dramatic; 2,000 cases occurred in 1964, whereas 291 cases occurred in 2005. To the surprise of public health officials, outbreaks of mumps occurred in college students in 1990, despite them having had a second dose of the vaccine as recommended. In 2006, the United States experienced multistate outbreaks involving 6,584 mumps cases, affecting mainly students at colleges in the Midwest. These outbreaks were dubbed the "two-dose vaccine failure." The largest U.S. mumps outbreak since 2006 occurred beginning in July 2009. The outbreak was traced to an 11-year-old boy who returned from the United Kingdom, where there was an ongoing mumps outbreak.

Rubella (German Measles)

Of the several viral diseases that cause a rash, **rubella** is the mildest and has an incubation period of 12 to 32 days. In fact, for generations, rubella, also called "three-day measles," was thought to be a mild form of measles (rubeola). In 1829, a German physician recognized that these were two different diseases; hence, the term German measles came into use. The disease is endemic worldwide and is highly infectious; it is spread largely through respiratory secretions, but urine from infected individuals can also transmit rubella. Some individuals are asymptomatic and transmit the virus without knowing it. The characteristic rash starts on the face and progresses down the trunk and to the extremities; it resolves in about 3 days (**FIGURE 13.21**).

In itself, rubella is not regarded as a serious disease. *However, prenatal transmission can occur, particularly if the mother is infected during the first trimester of pregnancy, even if she is asymptomatic.* Major consequences include cardiac lesions, deafness, ocular lesions resulting in blindness, mental and physical challenges, and glaucoma. Approximately 15% of those exposed may escape infection during their childhood years, posing a risk that females may enter their childbearing years without having had measles. Therefore, it is particularly important that females be immunized against rubella. Immunization has been very successful in decreasing the incidence, leading to the CDC declaring in 2004 that rubella had been eliminated from the United States. In 2012, The Measles Initiative was combined with rubella control and elimination efforts, and thus now called the Measles and Rubella Initiative.

Chickenpox and Shingles

The expression "brothers under the skin" aptly describes chickenpox and shingles, two seemingly different diseases that cause skin eruptions. Historically, they were thought to be caused by different viruses, but it has long been established that they are manifestations of the same herpesvirus. The term "pox" is in the disease name, which can be confusing because chickenpox is not caused by a poxvirus, such as the virus that causes smallpox (variola virus).

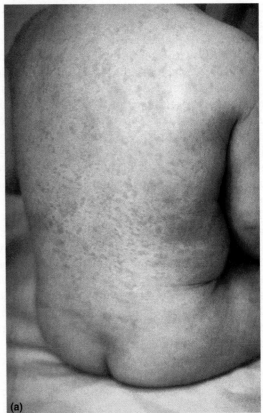

(a)

Courtesy of CDC.

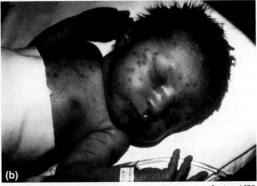

(b)

Courtesy of CDC.

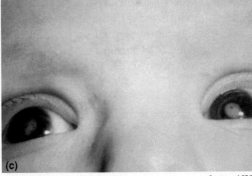

(c)

Courtesy of CDC.

FIGURE 13.21 Rubella (German measles) strikes both adults and children and is very dangerous to infants. **(a)** Child with rash on his torso. **(b)** Infant born with rubella. **(c)** Newborn with thickening of the lens of the eye that causes blindness.

Chickenpox is caused by a **herpesvirus**. The word *herpes* means "to creep or crawl" in reference to the herpes lesions or blisterlike pustules filled with fluid that form a rash. When you are infected with a herpesvirus, it remains with you for life. Your immune system does not clear it from your body. After your first exposure/infection, the wily herpesvirus goes into a dormant state during which it is inactive or "sleeping" (no infectious viruses are produced at this time) until it is triggered to wake up or be reactivated, causing disease symptoms and an infection in which the varicella zoster viruses begin to replicate and produce more viruses. Shingles is the reactivated form of chickenpox.

The herpesvirus that causes chickenpox is called **varicella zoster virus. Chickenpox** is a disease usually associated with childhood; **shingles** is the reactivated form of a varicella zoster infection, which is typically triggered after about the age of 60, and the risk of reactivation increases with age in some individuals who had chickenpox in their earlier years, but it has been reported in children as young as 8 years. Shingles is also more common in people with compromised immune systems.

Humans are the only hosts for chickenpox and shingles. Chickenpox has an incubation period of 10 to 20 days and is transmitted by airborne droplets or by contact with the fluid in the blisterlike skin lesions (vesicles) that develop. The disease is highly contagious, especially before the emergence of the rash. Chickenpox is caused by the only herpesvirus that can be spread person to person by airborne transmission. Like other herpesviruses, it is also spread by close contact.

Early signs include fever, headache, and generalized aches and pains, followed in a few days by an itchy rash with blisterlike, fluid-filled vesicles that are concentrated on the scalp, face, and trunk (**FIGURE 13.22A**). More than 250 to 500 vesicles can appear on the body at any one time; usually, adults with chickenpox develop more vesicles than do children. The vesicles tend to occur in a succession of crops over a 2- to 4-day period, and an individual may have a combination of newly emerging and old, "crusty" vesicles.

Chickenpox exists worldwide, and nearly all unvaccinated children are infected during their early years. Chickenpox, with rare exceptions, confers lifelong immunity, eliminating the need for vaccination. Like measles virus, varicella zoster virus can cross the placenta and cause

Shors' Notebook

I grew up at a time before there was a chickenpox vaccine. One day when I was 15 years old, I specifically remember my mother telling me play at my neighbor's house. They had 10 boys in their family—hence, we always had enough to play all kinds of games. Turns out that their 2-year-old nephew Jamie Fisher had the chickenpox. I had never had chickenpox. We played flag football all afternoon and Jamie was all over me. About 2 weeks later, I noticed an itchy bump on my forehead. Being 15, I thought it was a zit (pimple), so I squeezed it. Well—to make a long story short, I had chickenpox.

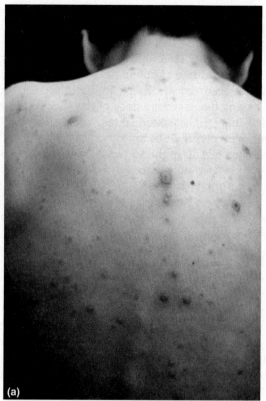

(a)
Courtesy of CDC.

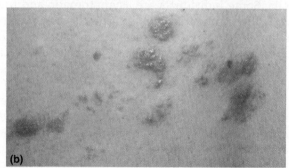

(b)
© Stephen VanHorn/ShutterStock, Inc.

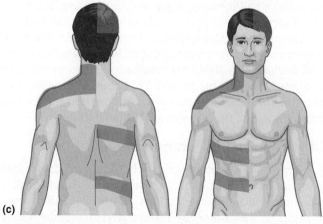

(c)

FIGURE 13.22 Chickenpox and shingles. These two diseases are caused by varicella zoster virus and are manifested by characteristic rashes. **(a)** Chickenpox. **(b)** Shingles. This painful condition is caused by the presence of the virus along sensory nerves, where they replicate and produce painful skin eruptions along the path of the nerve. **(c)** Body locations of shingles rashes. The spinal nerve, which girdles the trunk at about the belt line, is frequently involved.

serious fetal damage. Each year there are about 400 million cases of chickenpox, about 10,600 hospitalizations, and 100 to 150 deaths. In 1995, the FDA approved a live attenuated chickenpox vaccine. In 2010, an **MMRV (measles, mumps, rubella, varicella) vaccine** became available as a childhood vaccine. Two doses of the vaccine are about 90% effective in preventing chickenpox. Some people who are vaccinated against chickenpox may still get the disease, but it is usually milder with fewer blisters and little or no fever. The CDC recommends that children receive two doses of the vaccine, the first dose at 12 to 15 months of age and a second dose at 4 to 6 years of age (before beginning school).

Some parents purposely promote "chickenpox parties" to get their children naturally infected with varicella zoster virus. Their belief is that something "natural" (the disease) is better than something artificial (the vaccine) or that immunity from the disease will be more permanent than that from the vaccine. In 2011, state officials cracked down on a Tennessee woman whose Facebook page "Find a Pox Party Near You" was advertising the sale of lollipops licked by children with chickenpox (**FIGURE 13.23**). It is a federal crime to send "infectious diseases" or human pathogens across state lines. Chickenpox is not spread through oral secretions. A child licking one of the tainted pops will not get chickenpox. Consequently, the child could pick up bacteria or other viruses, spreading other infectious diseases.

Shingles is not life threatening, but it is miserable with painful complications occurring in some individuals (**FIGURE 13.22B**). About one in five adults with a history of chickenpox will develop shingles. The condition is described as causing one of the most intense pains of any

© Teri Shors.

FIGURE 13.23 In 2011, some parents who feared the chickenpox vaccine purchased lollipops prelicked by children suffering from chickenpox to try to infect their children the "natural" way during "pox parties." Health and legal authorities warned parents against this practice.

disease, which is not surprising because varicella zoster virus infects nerve fibers. During the initial viral infection, varicella zoster viruses take up residence, in a latent state, in collections of nerve cells (ganglia) located along the spinal column or in nerves supplying the face. The spinal nerve, which girdles the trunk at about the belt line, is frequently involved, resulting in this area being a common site for shingles (**FIGURE 13.22C**). These viruses remain in a **latent state** for years but may be triggered into an active replicating state by a variety of agents, including x-rays, certain drugs, and immunodeficiency, causing a painful eruption on the skin along the path of the nerve.

An individual with shingles is producing infectious varicella zoster viruses. They are contagious to individuals who have not had chickenpox; thus, people who have not had chickenpox can be infected with varicella zoster virus and contract chickenpox if they have had close contact with the person who has shingles. Shingles is caused only by varicella zoster virus that has been dormant since an individual acquired chickenpox. Shingles can cause a long-lasting intense burning sensation or pain called **postherpetic neuralgia (PHN)**.

The good news is that **Zostovax**, a shingles vaccine consisting of live varicella zoster viruses, was licensed by the FDA in 2006, and in 2017, the FDA approved **Shingrix**, a recombinant shingles vaccine. The CDC recommends healthy adults older than 50 years receive two doses of Shingrix 2 to 6 months apart. It provides strong protection against shingles and PHN.

Hantavirus Pulmonary Syndrome

The first known U.S. outbreak of **hantavirus pulmonary syndrome (HPS)** occurred in 1993, in the Four Corners area (where New Mexico, Colorado, Utah, and Arizona meet), with 24 cases of a severe flulike respiratory illness complicated by respiratory failure and, in some cases, death. Hantaviruses have been present in Africa since the 1930s but were associated with hemorrhagic (bleeding) symptoms; these were referred to as Old World hantaviruses. Many of the hantaviruses attack the kidneys, but the hantavirus that struck the Four Corners area was named the **Sin Nombre strain**, a New World hantavirus that attacks the lungs.

The fatality rate for HPS is at least 60%. The disease primarily strikes young, healthy adults, and death occurs in only several days after infection because of accumulation of fluid in the lungs, which interferes with oxygen diffusion. Oxygen therapy is frequently required in an attempt to avert death.

The Sin Nombre virus is carried by rodents, especially the deer mouse (**FIGURE 13.24**) and the cotton rat, both inhabitants of the Southwest. Because of the explosive nature of HPS, the mouse carrier has been referred to as "the mouse that roared." Transmission of the hantavirus to humans is the result of exposure to dried and aerosolized fecal material, saliva, or urine from infected rodents. It is also possible to contract the disease when fresh or contaminated rodent droppings get into the skin, eyes, food,

Courtesy of James Gathany/CDC.

FIGURE 13.24 A deer mouse. This mouse species is a vector of the hantavirus that causes HPS; it sheds the virus in urine, feces, and saliva.

or water or as the result of a rodent bite. The Sin Nombre hantavirus cannot be transmitted from human to human.

Prevention is based on avoiding contact with rodents and their excreta. Because the hantavirus is primarily spread by airborne contaminated droplets, entering barns or other buildings that are frequently closed for extended periods may be hazardous to one's health, as is disturbing rodent-infested structures. Diagnosis of HPS is difficult, because the early signs and symptoms of infection are characteristic of a variety of viral and bacterial diseases. Definitive diagnosis requires laboratory procedures.

Since the first case in 1993, a total of 728 cases of HPS have been reported in 36 states as of January 2017. More than 96% of cases have occurred in states west of the Mississippi River (**FIGURE 13.25**). The very first case occurred on May 14, 1993; a Navajo marathon runner in rural New Mexico known to be in excellent health collapsed and died of respiratory failure at an Indian Health Service hospital emergency room. Days before his collapse he visited a doctor twice complaining of flulike symptoms and shortness of breath, but his chest x-ray was normal. He was treated with antibiotics and acetaminophen. Two days before he fell ill, his fiancée died of the same mysterious respiratory illness. Both of them died from fluid buildup in their lungs. More cases with similar symptoms were reported, and by June 11, 1993, a total of 24 cases had been confirmed.

The most recent hantavirus outbreak in the United States occurred in 2017. It involved the investigation of 17 infected people with a hantavirus named the Seoul virus. All the individuals were linked to ratteries (rat-breeding facilities) with confirmed hantavirus infections. A total of 31 rattery facilities in 11 states had hantavirus infections.

The other notable hantavirus outbreak in the United States was a cluster of cases during the summer of 2012. Campers in Yosemite National Park, California, were infected while staying in tent cabins. Infected deer mice were living in the walls of the tent cabins (**FIGURE 13.26**). Three of 10 people infected died. The California Department of Public Health worked with Yosemite Park workers to increase routine measures to reduce the risk of hantavirus exposure to park visitors. Efforts included regular inspection and cleaning of cabins, rodent-proofing buildings, maintaining sanitation levels to discourage rodent infestations, and public education.

Smallpox and Monkeypox

Smallpox is caused by variola virus, and it was a worldwide scourge for thousands of years. Dr. Michael T. Osterholm, epidemiologist and director of the Center for Infectious Disease Research and Policy (CIDRAP) at the University of Minnesota School of Public Health, has called smallpox *"the lion king of infectious diseases."* Smallpox killed about 500 million people in the 20th century—more than then 320 million deaths caused by wars, the 1918 influenza pandemic, and HIV/AIDS combined.

Survivors of smallpox were often left blind from corneal ulcerations and badly scarred by pockmarks. Smallpox was so pervasive that it was considered unusual if someone did not have pockmarks on their face. President George Washington in 1751, at the age of 19, suffered from a severe case of smallpox. He had pockmarks on his face, yet portrait artist Edward Savage portrays Washington siting in uniform, his facial complexion smooth, free of smallpox scars (**FIGURE 13.27A**).

While reciting one of the most famous speeches in 19th-century U.S. history, the Gettysburg Address,

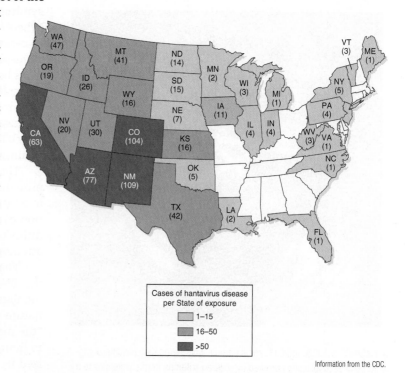

Cases of hantavirus disease
per State of exposure

☐ 1–15
▨ 16–50
■ >50

Information from the CDC.

FIGURE 13.25 Cumulative cases of hantavirus disease in the United States, May 1993 through January 2017.

President Abraham Lincoln was feeling the early symptoms of smallpox. Upon his return to Washington, D.C., the doctor told him he had a touch of varioloid, the old-fashioned name for smallpox.

The early signs of smallpox include high fever, fatigue, headache, and backache, followed in a few days by a rash concentrated on the extremities called a **centrifugal rash** (**FIGURE 13.28**). The rash is characterized by flat red lesions that become filled with pus and crust over in the second week; the lesions break out at the same time. (The rash of

© Eunika Sopotnicka/iStockPhoto.

FIGURE 13.26 Tent cabins with bear lockers in Yosemite National Park at Camp Curry were infested with deer mice carrying the Sin Nombre hantavirus in 2012.

© Everett Historical/Shutterstock, Inc.

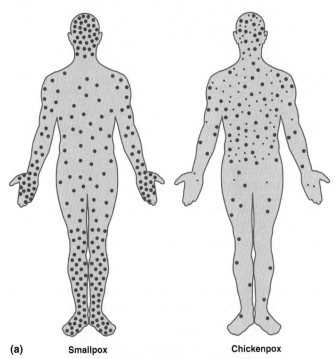

(a)   Smallpox   Chickenpox

Information from WHO, Diagnosis of Smallpox Slide Set.

Courtesy of the Library of Congress.

FIGURE 13.27 **(a)** Portrait by Edward Savage showing smallpox survivor George Washington with a smooth complexion, free of pockmarks on his face. **(b)** This rare photograph shows Abraham Lincoln (circled on the photograph) suffering from early signs of smallpox while giving a speech at Gettysburg. He complained of weakness and dizziness the day before he arrived at Gettysburg. Unfortunately, there was no close-up photograph of Lincoln just before, during, or immediately after his speech.

Courtesy of James Hicks/CDC.

FIGURE 13.28 **(a)** The centrifugal rash of smallpox compared to chickenpox in which the rash is concentrated on the truck and face. Rash distribution distinguishes the two viral diseases, aiding diagnosis. **(b)** The usual centrifugal rash pattern of smallpox (fewer lesions on the trunk and more on the face and extremities) is revealed by this photograph of a boy in Bangladesh (1973).

chickenpox appears in waves and is more concentrated on the face and trunk of the body.) The crusts dry up and fall off, leaving deeply pitted scars, particularly on the face. The variola virus has two strains, **variola major**, which has a case fatality rate of about 30%, and **variola minor**, which causes a milder disease with a 1% to 2% fatality rate.

Variola virus is transmitted directly from person to person by saliva droplets expelled from an infected individual onto a mucosal surface of another. Variola virus contaminating clothing and bedding (i.e., crusty lesions falling off onto the bedding) can also spread the virus. The first week of illness is the most infectious time because of the high numbers of poxviruses in the saliva.

The eradication of smallpox stands as a public health triumph of the 20th century. Generations to come will never know the horrors of this disease. The unique characteristics of smallpox and of variola virus that led to success in smallpox eradication and its establishment as the criterion by which to evaluate other diseases as candidates for eradication are as follows:

- It is a disease only of humans; there are no natural reservoirs or biological vectors.
- The infection is easily diagnosed because of a characteristic rash.
- The duration and intensity of infectiousness are limited.
- Recovery establishes permanent immunity.
- A safe, effective, inexpensive, easily administered, stable (even in tropical climates), one-dose vaccine is available.
- Vaccination confers long-lasting immunity.
- Vaccination often results in a permanent and recognizable scar, allowing for detection of immune versus nonimmune individuals in a population.

The WHO Smallpox Eradication campaign began in 1967, as a mass vaccination effort to immunize 100% of the world population. At the time, 10 to 15 million cases occurred annually. Over time, the strategy changed to containment, or **ring vaccination**, around newly discovered cases or outbreaks. **Smallpox recognition cards** were widely used by workers searching for the last remaining cases of smallpox in remote areas of India and Africa (**FIGURE 13.29A**). By 1970, the **bifurcated needle** replaced the jet injector to administer the smallpox vaccine (**FIGURE 13.29B**). The jet injector could do more than 1,000 vaccinations in an hour, but it was not practical and too expensive for house-to-house vaccination.

Rumor registers and **containment books** were used to follow up on every case of smallpox so that ring vaccination could be carried out. Ultimately, the most effective method to ensure prompt reporting of cases was to offer a reward to both the healthcare worker who investigated the case and the person who reported it. Rewards began

Shors' Notebook

My postdoctoral research training involved determining the function of two different genes of vaccinia virus. Vaccinia virus is the poxvirus strain that was used as the live smallpox vaccine. I was doing this research in the mid-1990s at the National Institutes of Health in the Laboratory of Viral Diseases. In order to do this research, I was required to get a booster dose of the vaccine. So, I had to go to the clinic and a nurse needed to administer the vaccine with a bifurcated needle.

The needle holds a droplet of the vaccine (vaccinia virus), and the skin is pricked multiple times. Typically, a first dose would be 3 pricks and a second dose, like I needed, would be around 13 pricks. I specifically remember the nurse counting each prick out loud and she miscounted, giving me an extra prick. She also pricked slowly instead of quickly, which was quite torturous to me. When I returned for the follow-up appointment, the nurse said I had a large reaction (a crusty itchy scab). I reminded her that I got an extra prick!

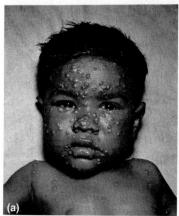

(a) Reproduced from F. Fenner et al. (1988). Smallpox and its eradication. Geneva: World Health Organization. http://www.who.int/iris/handle/10665/39485.

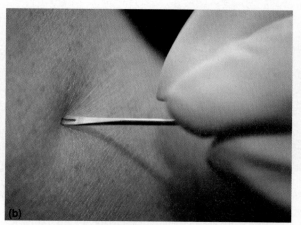

(b) Courtesy of James Gathany/CDC.

(c) Courtesy of WHO. Used with permission.

FIGURE 13.29 **(a)** Front of the WHO recognition smallpox card used during the eradication program. The cards were shown by workers to query villagers for sightings of people with a similar rash. **(b)** Bifurcated needle used to administer the smallpox vaccine. **(c)** Poster celebrating the eradication of smallpox.

in 1974 starting at 50 rupees (U.S. $6.25) for every case found in Indian states, but rewards continued to increase. In 1978, the WHO increased the reward to 8,000 rupees (U.S. $1,000) for reporting confirmed cases. In 1975, the last natural case of smallpox caused by variola major was a 3-year-old girl in Bangladesh. In 1977, the last natural case of smallpox caused by variola minor was reported in Somalia. The disease was declared eradicated on October 26, 1979 (**FIGURE 13.29C**). Routine smallpox vaccination among American civilians stopped in 1972, after the disease was eradicated in the United States.

On July 13, 2018, the FDA announced its approval of **TPOXX (tecovirimat or ST-246)** developed by SIGA Technologies as the first drug to treat smallpox as a countermeasure to address the risk of bioterrorism. Although the drug could not be tested on humans in challenge experiments in which people would be intentionally infected with variola virus and followed by treatment with TPOXX, challenge studies in which nonhuman primates were infected with monkeypox (a poxvirus related to variola virus) and other animals (e.g., rabbits challenged with rabbitpox) determined that the drug provided a 90% survival rate in treated animals infected with lethal doses of poxviruses. No severe side effects were observed when toxicity testing was performed on 359 healthy human volunteers who received doses of the drug. TPOXX inhibits the p37 protein of the variola virus. The p37 protein is involved in the formation of mature, enveloped infectious virus particles during infection of cells in culture. TPOXX blocks virus envelopment, reducing the cell-to-cell spread of variola viruses.

Besides treating smallpox, TPOXX can also be used to treat monkeypox. The FDA approved the drug for **Orphan Drug** status, which encourages the development of drugs for rare diseases that do not have treatment options. Because immunocompromised individuals cannot be vaccinated with a live attenuated poxvirus such as vaccinia, in the event of a bioterrorist attack these individuals could be treated prophylactically with TPOXX. The drug could also be given to individuals who had a severe reaction toward the smallpox vaccine.

⦿ Monkeypox

Human **monkeypox** infections are rare. Before 1970, monkeypox was known as a disease of animals in the rainforests of Central and West Africa. Monkeypox virus, a poxvirus related to variola virus, was first isolated from infected cynomolgus monkeys in 1958. Between 1970 and 1986, the first human cases of monkeypox were reported in West Africa and the Congo Basin as smallpox disappeared. *The smallpox vaccine also prevents monkeypox virus infections.* After smallpox cases dwindled and vaccination ceased, monkeypox was able to resurface, and the first outbreaks of human-to-human monkeypox transmission occurred 20 years after the eradication of smallpox in 13 villages in Zaire in 1996–1997 (**FIGURE 13.30**).

In June 2003, in a multistate outbreak in the United States, the first cases of monkeypox were reported among

Courtesy of CDC/Brian W. J. Mahy.

FIGURE 13.30 A group of epidemiologists gathered with Katako-Kombe Health Zone villagers during a 1997 investigation of a monkeypox outbreak that took place in the Democratic Republic of Congo (DRC), 1996–1997, formerly Zaire.

several people who had had contact with infected pet prairie dogs and other exotic pets. The cumulative number of cases in this outbreak was 47. Monkeypox is a milder disease than smallpox; one difference is that it causes **lymphadenopathy** (chronic abnormal enlargement of lymph nodes) during the early stages of the disease. It is fatal in 1% to 10% of cases.

From 2016 to July 2018, more than 1,000 confirmed human cases of monkeypox had been reported in the Democratic Republic of Congo, 88 cases in the Republic of Congo, 16 cases in Cameroon, 2 in Liberia, and 1 in Sierra Leone. Although there is no specific treatment for monkeypox virus infection, outbreaks can be controlled with vaccination against smallpox. Monkeypox virus has been isolated from rope squirrels, tree squirrels, Gambian rats, striped mice, dormice, and nonhuman primates. The exact natural reservoir of monkeypox is not known.

Sexually Transmitted Viral Infections (STIs)

The three major sexually transmitted viral diseases are genital herpes, genital warts, and HIV/AIDS (**TABLE 13.4**). Papillomaviruses, which cause genital warts and are involved in cervical, tongue/oral, penis, and anal cancers, are naked viruses that are highly stable in the environment. In contrast, HIV and herpesviruses, such as herpes simplex viruses 1 and 2, which cause cold sores and genital herpes, are enveloped and are not stable for very long on surfaces. Alcohol-based hand sanitizers can reduce the infectivity of HIV and herpes simplex viruses but will have no effect on papillomaviruses.

HIV/AIDS

The first AIDS cases in the United States were reported in the CDC's *Morbidity and Mortality Weekly Report* (MMWR)

TABLE 13.4 Sexually Transmitted Viral Infections (STIs)

Viral Disease	Incubation Period	Symptoms	Immunization and Comments
HIV/AIDS	A few days to a few weeks for acute symptoms; average of 10 years for AIDS symptoms	Acute infection: short duration of flulike symptoms AIDS: Weight loss, tiredness, loss of appetite, repeated infections	Always fatal, but longevity has been substantially improved; causes severe immunosuppression
Genital herpes	4–10 days	Painful and itchy sores on the penis or on the labia, vagina, or cervix; can be asymptomatic	Lifetime infection with recurrent lesions; three antivirals available for treatment: Zovirax, Famvir, and Valtrex
Human papillomavirus infections and genital warts	1–6 months	Presence of warts; can be asymptomatic	All boys and girls should be vaccinated at age 11 or 12 years: Gardasil and Gardasil-9 HPV vaccines available. High-risk types of HPV infections can lead to cervical, penile, anal, vulvar and vaginal, and throat cancers

on June 5, 1981. Doctors reported five cases of previously healthy young men in Los Angeles, California, described as "homosexuals" with opportunistic infections associated with severe immune deficiency. The men suffered from a rare skin cancer, **Kaposi's sarcoma (FIGURE 13.31A)**, later discovered to be caused by a herpesvirus; pneumonia caused by *Pneumocystis carinii* (later rename *P. jirovecii*); and *Candida albicans* infections (**FIGURE 13.31B**). This prompted other case reports from San Francisco, New York City, and other cities. In 1982, researcher Luc Montagnier at the Pasteur Institute in France determined that a retrovirus, named **human immunodeficiency virus (HIV)** was the cause of **acquired immune deficiency syndrome (AIDS)**.

The AIDS epidemic spurred medical research and development of **antiretroviral drugs** and treatment protocols, but it took time. The movie *Dallas Buyer's Club*, based on the true story of Ron Woodruff who was diagnosed in 1985 with HIV/AIDS, provides some sense of what early HIV/AIDS sufferers and healthcare providers went through. Woodruff had a T-cell count of 2 and was given a month to live. Antiretroviral drugs were not available for treatment, and there were no vaccines to prevent HIV infection. Not willing to accept his death sentence, Woodruff began to search for alternative medicines and traveled to different countries in his quest. He purchased multivitamins, immune-boosting drugs, and antioxidants. Some of the medicines he purchased were unproven therapies not available in the United States. His efforts extended his life about 7 years after his diagnosis.

In 2000, of the 34.3 million people living with HIV, fewer than 1 million were being treated with antiretroviral drugs, most of whom were living in wealthy countries. Today, of the 36.7 million people living with HIV, 20.9 million are receiving treatment, most of them living in **developing countries** where the HIV infection is most prevalent. The release of generic antiretroviral drugs, which lowered drug costs, and the development of the infrastructure

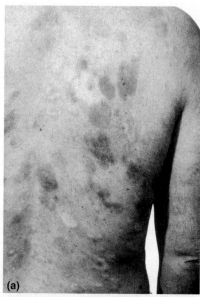

(a)

Courtesy of the National Cancer Institute.

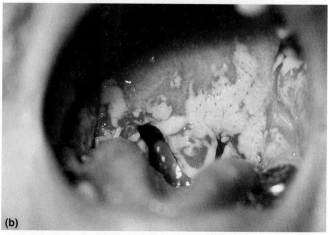

(b)

Courtesy of the CDC.

FIGURE 13.31 (a) Kaposi's sarcoma in a young man with AIDS. **(b)** This AIDS patient has chronic opportunistic oral candidiasis (thrush).

to distribute them, was unprecedented in fighting the HIV/AIDS battle once it started. It is a public health success story. Today the challenge is *preventing HIV infections*.

○ Complacency and Preventing HIV Infections

"HIV can and is raising its head again in places where risks align."

—Thomas Stopka, professor,
Tufts University School of Medicine

Today's college students are living in the shadow of an AIDS pandemic, a dreaded disease caused by HIV, which surfaced in 1981, and has since spread havoc and desperation throughout the world, particularly in Africa. A distinction needs to be made: HIV infection is not AIDS but rather signifies the presence of HIV. AIDS is a syndrome of many **opportunistic infections**.

The stories of Mark Gardner Hoyle and Ryan White remain symbolic. Mark, a student of Case Junior High School in Swansea, Massachusetts, died of AIDS on October 26, 1986, at the age of 14. His name is inscribed on home plate at the school's playing field. Ryan was an Indiana teenager who died of AIDS on April 8, 1989, at the age of 19. The Ryan White CARE (Comprehensive AIDS Resources Emergency) Act, designed to assist the growing number of Americans living with AIDS but without adequate healthcare insurance, was founded in his remembrance. At the time both boys and their families struggled with an unsympathetic, hostile, and uninformed public intent on ostracizing them. Mark and Ryan contracted and died from this so-called gay disease when, in fact, both suffered from a bleeding disorder known as **hemophilia** and required blood products in their treatment. These products were contaminated with HIV, the **retrovirus** that causes AIDS.

Today, the memories of Mark Gardner and Ryan White have been forgotten or are not known by many. Gloomy reports have emerged of the resurgence of HIV epidemics in the United States. Those at highest risk of infection are the same groups that were at risk at the start of the epidemic in the 1980s: people who inject drugs, gay and bisexual men, transgender people, and prostitutes and their sexual partners. These are the same groups that avoid AIDS services that are available to them. As a result, many do not know they are infected with HIV, and the disease spreads. Epidemics among injection drug users who share needles are making news headlines. In 2015, HIV spread through Austin, Indiana, a small town in rural America. At least 170 individuals tested positive for HIV. Of these, 96% injected Opana, heroin, or methamphetamine an average of 4 to 15 times a day. Sharing needles for injection was common, and 86% were coinfected with hepatitis C virus. The median age of those confirmed with a positive HIV diagnosis was 32 years. All individuals were Caucasian/white. One in five lived in poverty, and one in five lacked a high school degree.

In 2017, in Milwaukee, Wisconsin, a large cluster of people, involving more than 100 young adults, as well as some teens, tested positive for HIV and syphilis. Milwaukee has a reputation for being a "mecca for sex trafficking." Three children in Milwaukee were born with syphilis in 2017. The number of new HIV infections increased by 13% from 2016 to 2017. The city also saw a 29% increase in syphilis, a 12% increase in gonorrhea, and a 0.5% increase in chlamydia over the same period. Other cities across America are also witnessing a spike in STIs.

Opioid use sparked an HIV epidemic in Massachusetts in 2018. Massachusetts has near-universal healthcare insurance for everyone in the state. That being said, one would not expect an HIV outbreak, but between 2015 and 2018 there were 129 new HIV infections in two cities located 9 miles apart from each other: Lowell (population 110,000) and Lawrence (population 80,000). The major contributing factor to the outbreak was the injection use of **opioids** such as fentanyl, a pain killer similar to morphine (**FIGURE 13.32**). The majority of cases were white men between the ages of 20 and 39, and 90% were coinfected with hepatitis C virus.

So why is this happening today? Although it is a multifactorial issue, the main contributor to increasing HIV infections is that prevention efforts have started to erode. **Needle-exchange programs** are highly successful in preventing the spread of infection through intravenous drug use. Austin, Lowell, and Lawrence did not have any of these programs available. Funding has been cut for STI prevention and biomedical research. These budgetary actions prompted news headlines such as "*Make Gonorrhea Great Again,*" targeted at President Trump's request to cut $186 million dollars in funding for the CDC's center on HIV/AIDS, viral hepatitis, STIs, and tuberculosis prevention. The erosion of prevention efforts is not just happening in the United States, it is happening around the world. Now is not a time to be complacent about HIV/AIDS. *It is a time to reinvigorate the fight against HIV/AIDS.*

○ HIV and Transmission

HIV, the cause of AIDS, is an enveloped retrovirus. **FIGURE 13.33** shows an electron micrograph of HIV "budding" out of host cells. The biology of HIV and other retroviruses is unique in that the flow of genetic information starts with RNA, not with DNA as the usual starting point.

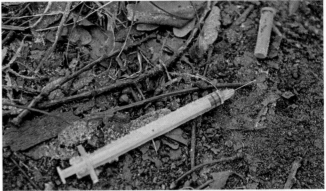
© AdamBoor/Shutterstock.

FIGURE 13.32 Injection syringe discarded on the ground in public places are a common site in Lawrence and Lowell, Massachusetts.

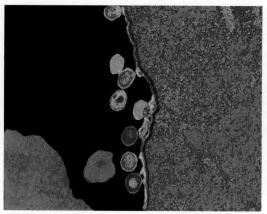

© Phanie/Alamy Stock Photo.

FIGURE 13.33 Digitally colorized transmission electron micrograph of HIV "budding" out of host cells.

To date, two types of HIV have been identified: type 1 (HIV-1) and type 2 (HIV-2). The best known is HIV-1, the most common cause of AIDS worldwide. In West Africa, HIV-2 is most common. Within each group are a number of subcategories. Diagnostic testing distinguishes both types. Because air travel is so common, in the United States HIV testing is performed to determine if a person is infected with either HIV-1 or HIV-2 (or possibly both).

It is important to understand the mode of transmission for HIV infection in order to control epidemics and to dispel myths, particularly the "I can't get it" myth that persists in many persons. HIV can be transmitted in five ways (**FIGURE 13.34**):

1. Sexual contact with an infected partner, whether it be male to male, male to female, female to male, or female to

Sexual contact

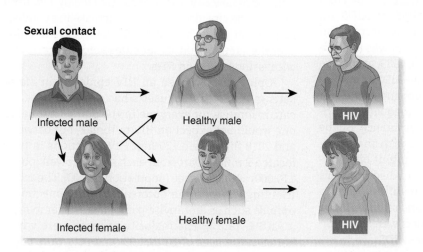

Contact with HIV-contaminated blood or blood products

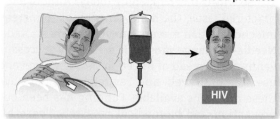

Sharing needles containing HIV-contaminated blood

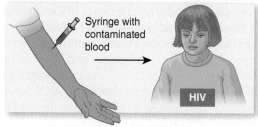

Transmission from HIV infected mother to unborn child

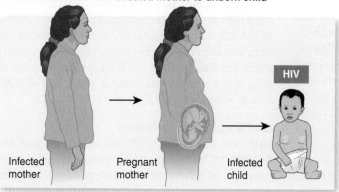

Pre-mastication

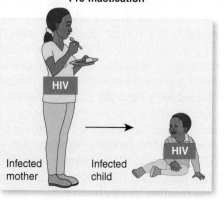

FIGURE 13.34 HIV modes of transmission.

female. HIV can penetrate the lining of the vagina, vulva, penis, rectum, or mouth during sexual activity. Some sexual behaviors are considered more risky and unsafe than others. Anal sex is the most dangerous because the lining of the anus is more subject to tears and injury than is the lining of the vagina, allowing HIV (and other microbes) easier passage into the bloodstream.

2. Contact with blood or blood products contaminated with HIV. Before identification of the cause of AIDS and the development of an HIV diagnostic test in April 1985, persons requiring transfusions of blood or blood products, including those with hemophilia, were hard hit. Many recipients of blood or blood products died as a result of AIDS, including tennis star Arthur Ashe, who died in 1995 as a result of having received contaminated blood 10 years earlier during heart surgery.

3. Sharing blood-contaminated needles and syringes, as is sometimes practiced by drug users when shooting up, has a high risk of HIV transmission. All it takes is a minute amount of contaminated blood on a needle. Accidental needle sticks and glove tears are a nightmare to all healthcare workers, although the estimated risk of transmission depends upon the viral load present in contaminated blood.

4. Transmission from mother to unborn child through the passage of HIV across the placenta carries about a 25% risk if the mother does not receive antiretroviral therapy or breastfeeds. The risk is reduced to 2% if the mother receives antiretroviral treatment. HIV can also be transmitted at the time of birth by infected vaginal secretions or, after birth, through infected breast milk.

5. **Premastication**, or prechewing food for infants, can transmit HIV from an HIV-positive caregiver chewing the food to an uninfected child. The caregiver's gums may bleed while chewing, contaminating the prechewed food. Premastication is a common cultural feeding practice in China and Africa.

○ AIDS: The Disease

A diagnosis of AIDS is terrifying because the best that an infected person can hope for is a life without too much misery. HIV does not directly cause death, as other microbes do, by their secretions of toxins or by tissue damage. Rather, HIV depletes the number of **T helper cells**, resulting in the crippling of the **adaptive immune system**. Therefore, the HIV-infected individual becomes immunocompromised

and is vulnerable to life-threatening opportunistic infections caused by an array of pathogens. Examples of such infections are listed in **TABLE 13.5**.

Infection with HIV does not constitute AIDS but does, with rare exceptions, progress to clinical AIDS over an incubation period that can vary from a few years to as many as 15 years—maybe even more. Those few people living with HIV who continue to survive after 10 years live in a state of uncertainty, not knowing when full-blown AIDS will develop, but their health can be managed well with regular monitoring of the disease through viral load testing and measuring T helper cell numbers (discussed later). It should be emphasized that no two cases are alike, and there might be considerable variation in the time and course of infection. The three stages of HIV infection are (1) acute infection, (2) clinical latency, and (3) AIDS. Treatment helps people living with HIV at all stages of disease or prevents progression from one stage to another.

Acute Infection

Acute HIV infection is the earliest stage of disease. It occurs within 2 to 4 weeks of HIV infection. During this period, most people experience flulike symptoms called **acute retroviral syndrome (ARS)**. Not everyone develops ARS. The signs and symptoms of ARS are listed in **TABLE 13.6**.

Clinical Latency

The second stage is the **clinical latency period**; neither signs nor symptoms of HIV disease are present. This phase is sometimes referred to as asymptomatic HIV infection or chronic HIV infection. At this stage, the HIV infection becomes established. People living with HIV whose health is managed by **antiretroviral therapy (ART)** may be in the clinical latency stage for several decades. For people living with HIV without ART, this period can last up to a decade, but some may progress faster. Toward the middle and end of this stage, the **viral load** begins to rise and the T helper counts begin to drop.

AIDS

The third and final stage is full-blown AIDS, the last stage of HIV infection. AIDS is defined as a T helper cell count of fewer than 200 T helper cells per microliter (µL) of blood and/or one or more opportunistic infections. Opportunistic infections rarely occur in people with competent immune systems but are deadly to those in stage 3 of HIV disease. In addition to the examples of opportunistic infections listed in Table 13.5, patients with AIDS often suffer from cancers such as Kaposi's sarcoma and non-Hodgkin's lymphoma. **Wasting syndrome**, a loss of more than 10% body weight due to fever or diarrhea for more than 30 days, is a common problem in many AIDS patients.

AIDS patients who are severely immunocompromised are managed with **prophylactic antibiotic therapy** as a prevention for certain opportunistic infections such as those from *Mycobacterium tuberculosis*, *Mycobacterium avium*, *Streptococcus pneumoniae*, cytomegalovirus, *Toxoplasma gondii*, and *Pneumocystis jirovecii*.

Krasner's Notebook

How comforting is the 0.5% statistic to someone who has accidentally stuck a finger? Not very, as my experience tells me; it happened to me 30 years ago when drawing blood from a patient at Hadassah Hospital in Jerusalem. Fortunately, I did not pick up any infections from the needle. In 1990 a dentist with AIDS was responsible for infecting six patients. Improperly sterilized needles used to deliver local anesthetic were found to be contaminated with the dentist's blood.

TABLE 13.5 Opportunistic Infections Associated with HIV/AIDS

Type of Pathogen	Opportunistic Infection/Disease
Bacteria	
Mycobacterium tuberculosis	Tuberculosis
Mycobacterium avium and *M. intracellulare*	*Mycobacterium avium* complex (MAC); extrapulmonary tuberculosis
Salmonella spp., *Shigella* spp., *Campylobacter* spp.	Gastroenterititis (salmonellosis, shigellosis, campylobacterosis)
Legionella pneumophila	Legionnaires' disease
Viruses	
Papillomaviruses	Genital warts; cervical, oral, and anal cancers
Herpes simplex viruses 1 and 2	Genital herpes, herpetic whitlow, encephalitis
Cytomegalovirus	Retinitis, enterocolitis, pneumonitis
Varicella zoster virus	Shingles
Hepatitis B and C viruses	Hepatitis, chronic cirrhosis, liver cancer
Molluscum contagiosum virus	Molluscum contagiosum
Protozoans	
Pneumocystis jirovecii	Pneumonia
Cryptosporidium parvum	Cryptosporidiosis
Toxoplasma gondii	Toxoplasmosis, retinitis, pneumonitis
Isospora belli	Isosporiasis
Fungi	
Candida albicans	Candidiasis (thrush, yeast infections)
Cryptococcus neoformans	Cryptococcosis
Histoplasma capsulatum	Histoplasmosis
Aspergillis spp.	Aspergillosis
Pneumocystis jirovecii	Pneumonia (PCP)

○ HIV Laboratory Diagnosis

The first diagnostic test to determine if someone is HIV positive involves screening for HIV-1 and HIV-2 antibodies and for HIV p24 antigens (a protein that is part of the internal structure of the virus particle) in a patient blood sample (**FIGURE 13.35**). If a patient tests positive, the results must be confirmed with a **nucleic acid amplification test (NAAT)** that will specifically detect the HIV-1 and HIV-2 RNA genomes if present in the blood sample. If both the antibody and NAAT results are positive, the individual definitively is infected with HIV.

TABLE 13.6 Signs and Symptoms of Acute Retroviral Syndrome (ARS)

Signs and Symptoms	Frequency
Fever, fatigue, rash, muscle pain, sore throat, swelling of lymph nodes	Experienced by more than 50% of HIV-positive individuals
Headache, diarrhea, nausea and vomiting, night sweats, weight loss, thrush (oral candidiasis), neurological symptoms (depression, dizziness, photophobia), oral and genital ulcers, coughs, enlargement of spleen and liver	Experienced by 5–32% of infected individuals

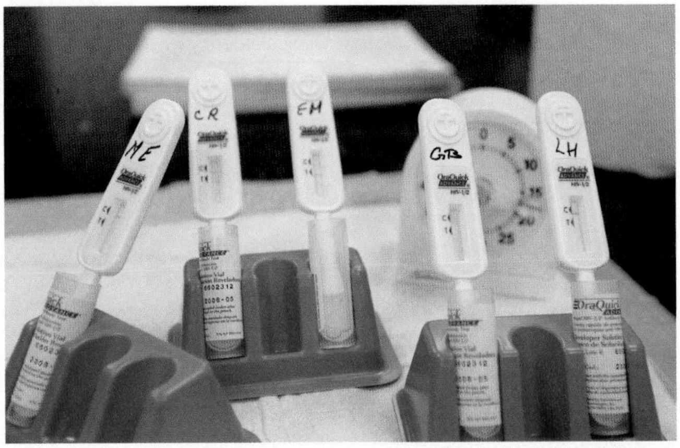

© Stephen Chernin/Stringer/Getty Images.

FIGURE 13.36 Free HIV screenings in Harlem for World AIDS Day. Rapid HIV test swabs rest in their holders incubating at the Iris House, November 30, 2007, in New York City.

The FDA has approved two home test kits (**FIGURE 13.36**). If someone tests positive using a home test, the results must be confirmed in a doctor's office. HIV testing has never been quicker or easier than it is today.

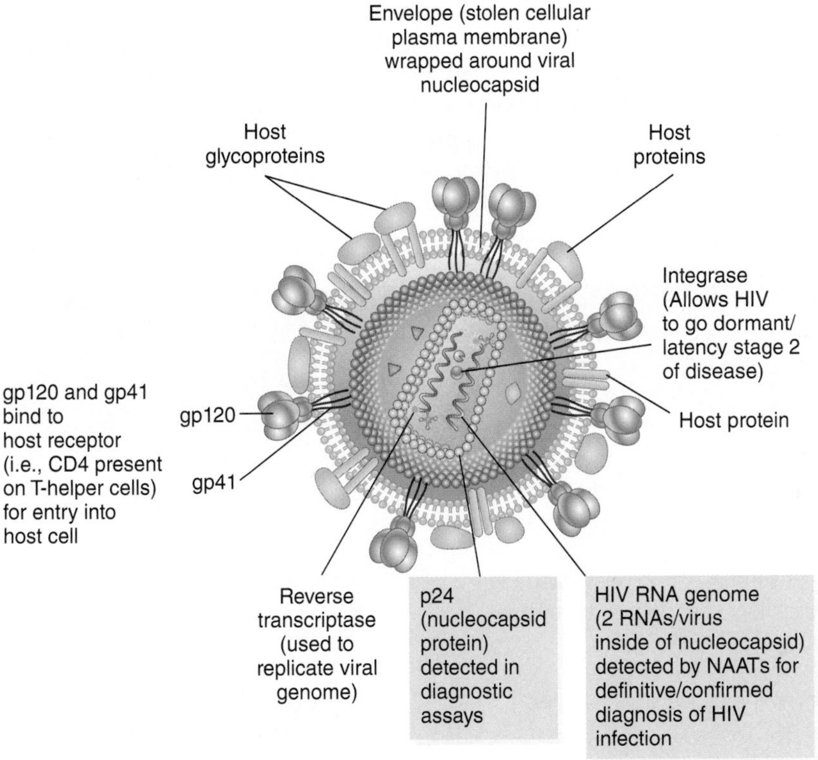

FIGURE 13.35 Structure of HIV. Antibodies present in a patient blood sample would bind to the outside of the virus. Some of these antibodies would be detected in diagnostic tests. HIV is enveloped. The viral envelope is stolen from the host cell's plasma membrane; hence, it will also contain host/cellular proteins associated with it. The p24 protein (antigen), which is a protein that makes up part of the nucleocapsid that protects the HIV RNA genome, is located inside of the virus particle. Detection of the HIV p24 antigen is used in screening patients for HIV infection. Each virus has two copies of viral RNA that are replicated into DNA by HIV reverse transcriptase (note that reverse transcriptase is packaged with the viral RNA inside of the viral nucleocapsid). The HIV integrase is used to integrate the reverse-transcribed HIV DNA into the host cell chromosome, creating a dormant state or latency (stage 2 of HIV disease).

Managing HIV Patients: Antiretroviral Therapy (ART)

The bottom line is that there is no cure for AIDS, and there is no preventive vaccine. From time to time over the past 10 years, "breakthroughs" have been announced, but they always fall short of meeting expectations. Nevertheless, considerable progress has been made: HIV has been isolated, diagnostic laboratory tests have been developed and optimized, the dynamics of HIV are better understood, treatment of opportunistic infections has improved, and mother-to-child transmission has been drastically reduced thanks to the first antiviral drug, **zidovudine**, which is also called **azidothymidine (AZT)**. Further, society is more tolerant toward individuals with HIV/AIDS, be they homosexual or heterosexual.

In order to manage treatment of a person living with HIV, viral load assays are used to quantitate the number of viral RNA genomes present in a milliliter (mL) of a patient's plasma. Viral load is a good predictor of the disease progression to AIDS. If a person living with HIV has a viral load of more than 100,000 copies of viral RNA genomes (or 50,000 retroviruses, because each virus particle contains two RNA genomes) present in their plasma sample, the person has AIDS.

A patient blood sample is also used to monitor that status of the T helper cells. A normal T-cell count is about 800 to 1,000 T cells present in 1 microliter (μL) of blood. If a person living with HIV has 200 or fewer T helper cells per microliter of blood, they are defined as having clinical AIDS (**FIGURE 13.37**).

The **replication cycles** of all viruses are complex, and this is particularly true

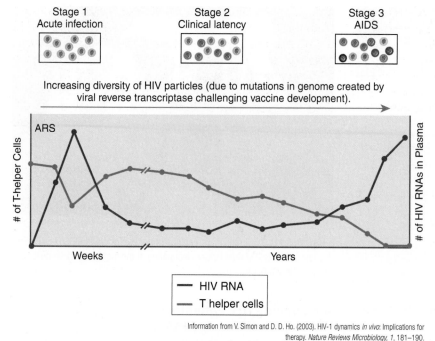

Increasing diversity of HIV particles (due to mutations in genome created by viral reverse transcriptase challenging vaccine development).

Information from V. Simon and D. D. Ho. (2003). HIV-1 dynamics *in vivo*: Implications for therapy. *Nature Reviews Microbiology, 1,* 181–190.

FIGURE 13.37 The progression of HIV disease based on T helper cell numbers and viral load (HIV RNA in plasma). As T helper cells are depleted, HIV multiplies. As the HIV disease progresses to AIDS, millions of HIV variants are present within the body of a person living with HIV. This makes treatment and vaccine development challenging.

of antiretroviral drugs to inhibit HIV infection. The cocktail contains at least three different antiviral compounds to jointly block HIV replication.

AZT (Retrovir) was the first clinically safe and effective drug used to treat HIV infection. AZT and other anti-HIV drugs are inhibitors of the reverse transcriptase enzyme and act at an early stage by preventing synthesis of DNA from RNA. The AZT molecule resembles an important building block in viral DNA synthesis, **nucleosides**, and when it is mistakenly used by the virus DNA synthesis is brought to a halt. AZT and related drugs slow the spread of HIV in the body and temporarily forestall opportunistic infections. They are not cures.

A second class of drugs, called **protease inhibitors**, has been developed and approved for treating HIV infection. These drugs act by interfering with the final assembly and maturation of the virus; whereas AZT slows viral replication, protease inhibitors act to shut down the assembly line that makes viral particles. Because HIV develops resistance to AZT and to protease inhibitors, combination treatment is now considered to be the best approach. More classes of drugs have been developed. Today, seven classes of HIV ART are currently available (**TABLE 13.7**).

of HIV, which can remain in a state of clinical latency (in hiding) for more than 10 years before returning to an active replicative cycle. ART must somehow interfere with HIV replication. ART involves the use of a cocktail

TABLE 13.7 Classes of ART Used to Slow the Progression of HIV Disease

Class of HIV Antiviral	Mechanism of Action	Examples of Antivirals (Trade Name)
Nucleoside reverse transcriptase inhibitors	Block HIV reverse transcriptase from converting RNA into DNA during replication; structure of the drug looks similar to a nucleoside but plugs up reverse transcriptase	Retrovir, Videx EC, Zerit, Epivir, Ziagen, Viread, Emtriva
Non-nucleoside reverse transcriptase inhibitors	Block HIV reverse transcriptase from converting RNA into DNA during replication; bind directly to reverse transcriptase, blocking it from functioning	Viramune, Rescriptor, Sustiva, Intelence, Edurant
Protease inhibitors	Inhibit HIV from forming infectious virus particles	Invirase, Norvir, Crixivan, Viracept, Lexiva, Reyataz, Aptivus, Prezista
Fusion inhibitors	Block HIV entry into host T helper cells	Fuzeon
CCR5 inhibitors	Block HIV entry into host T helper cells	Selzentry
Integrase inhibitors	Block HIV integrase, preventing the integration of HIV DNA provirus into the host cell genome (silent reservoirs of HIV in the body)	Isentress, Tivicay, Vitekta
Pharmacokinetic enhancers	Increase the effectiveness of ART drugs	Tybost

Even though it is not a cure, ART can keep people living with HIV healthy for decades and reduce transmission to partner(s) if taken consistently and correctly. ART reduces the viral load or numbers of HIV particles in the blood and body fluids. It is recommended for people living with HIV regardless of how long they have been infected or how healthy they are. The CDC recommends that ART begins soon after diagnosis.

Genital Herpes

There are a variety of herpesviruses, and they are all "bad actors." Herpes is a highly infectious disease caused by **herpes simplex virus (HSV)**, of which there are two closely related types: HSV type 1 and HSV type 2. HSV type 1 (HSV-1) is usually associated with painful sores on the upper body, particularly around the mouth and lips, called cold sores or fever blisters, and, occasionally, on the throat and the tongue. HSV type 2 (HSV-2) is the major cause of **genital herpes** and manifests as painful and itchy sores on the lower body such as on the penis in males (**FIGURE 13.38**) or on the labia, vagina, and cervix in females. HSV-1 and HSV-2 have an incubation period of 4 to 10 days. The disease is usually mild, but initial infection with HSV-1 and HSV-2 can be accompanied by high fever and large numbers of painful sores. Prompt treatment with antiviral drugs is somewhat effective. The disease can be severe in individuals with AIDS or other immunosuppressive conditions. Once acquired, infection is lifelong, and episodes of recurrent, painful genital ulcers occur. **Latency**, a period of "taking up residence and hiding," is characteristic of herpesviruses in general, including varicella zoster virus (the cause of chickenpox and shingles). The herpesvirus' DNA (genetic material) "hides out" in the host's chromosomes and is replicated along with the host's genetic material. The viral genetic material can be triggered into an active replicating state by a variety of factors, including stress, fever, sunlight, colds, and menstruation. In many respects latency is a state of peaceful coexistence like that described for bacterial prophages. Generally, within the first year of infection four or five symptomatic recurrences take place.

Transmission of HSV-2 occurs through direct contact with herpes sores from one person to another or from one part of the body to another. The herpes lesions are highly infectious until the sores heal within about 2 to 4 weeks. Frequently, HSV-2 is transmitted by asymptomatic individuals who have no knowledge of their disease; it is important to realize that these individuals, as well as those who are aware that they are harboring the virus in a latent state, can transmit the disease.

Pregnant HSV-2–infected women are at risk for miscarriage during the course of their pregnancy. In those carrying full-term infants, a cesarean section is performed to minimize the risk of infection to the newborn during passage through an HSV-2–infected birth canal. Finally, HSV-2–infected females are at a greater risk of developing cervical cancer than uninfected females.

Genital herpes is common in the United States. According to CDC statistics, more than 1 out of every 6 people aged 14 to 49 years has genital herpes. The disease is diagnosed by visual inspection or by culturing tissue from herpes sores to determine the presence of the virus. There is no cure, but three antivirals are available that can reduce the duration and severity of occurrences: **Zovirax (acyclovir)**, Famvir (famciclovir), and Valtrex (valacyclovir). These are all available in pill form. Intravenous Zovirax can be used to treat severe cases. There is no vaccine, and prevention is based on safe sex practices.

Genital Warts

Human papillomaviruses (HPVs) are responsible for common warts, which are benign, painless elevated growths that occur most frequently on the fingers; **plantar warts**,

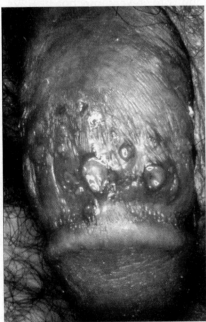

Courtesy of the CDC/Drs. N. J. Flumara and Gavin Hart.

FIGURE 13.38 Genital herpes rash. The ulcers are tender and take about 2 to 4 weeks to heal after they appear.

Shors' Notebook

For whatever reason, students in my courses find warts fascinating. I remind students to wear flip flops in communal showers in the dorms to avoid picking up papillomaviruses that cause plantar warts on the bottoms of their feet. Papillomaviruses are highly stable and can get into the crack of the skin at the bottom of a foot, causing plantar warts. The warts can spread to other locations on the souls of the feet. I ask students if they have ever had a plantar wart and how they got rid of them. This is when the enthusiasm kicks in. According to students, putting duct tape over a plantar wart smothers them and they disappear. I haven't found any research papers to support that procedure in the primary literature. Apparently, it is common advice if you search this topic via the Internet.

which are deep and painful warts on the soles of the feet (**FIGURE 13.39A**); and **genital warts**, which are fleshy growths in the genital areas of men and women. Papillomaviruses are naked viruses that are very stable in the environment. Hand sanitizer does not inhibit their infectivity.

Genital warts are the most common sexually transmitted infections in the world; in the United States an estimated 80 million people are infected with HPV. The disease is on the increase, with as many as 14 million new cases, including teens, in the United States each year. It is so common that most people will get at least one type of HPV at some point during their lives. More than 100 types of HPV have been identified and grouped as low, intermediate, and high risk. **Low-risk types of HPV**, such as

types 6 and 11, cause benign warts; in comparison, **high-risk types of HPV**, such as types 16 and 18, have been associated with cancers that invade the throat and genital tissue. Genital warts are highly contagious, and about two-thirds of people whose sexual partners have genital warts develop warts within 3 months.

In females, warts can occur on the outside and inside of the vagina, on the cervix (the opening to the uterus), and around the anus. In males, warts may occur on the shaft or head of the penis, on the scrotum, and around the anus. Oral sex with a person with genital warts can result in warts in the mouth or throat. The warts may disappear without treatment, but they can develop into large fleshy growths that resemble pieces of cauliflower (**FIGURE 13.39B**).

A major concern is the link between some strains of HPV and cancer, including cervical, vulvar, throat, and anal cancer and, more rarely, cancer of the penis. The best preventative is to avoid direct contact with the papillomavirus. Recent data indicate that 70% of HPV-related cervical cancers are caused by HPV high-risk types 16 and 18. Types 6 and 11 can also cause cancer but are more associated as being responsible for 90% of genital warts. In 2015, 12,845 American women were diagnosed with cervical cancer and 4,175 died. Cancer is the second leading cause of death in the United States. About 400 men each year are diagnosed with HPV-associated penile cancer.

Two vaccines are approved for use to prevent genital warts and cancers associated with specific high-risk types of HPV in the United States. **Gardasil** is a **noninfectious recombinant vaccine** prepared from purified **virus-like particles (VLPs)** of the major **capsid** (L1) protein of HPV types 6, 11, 16, and 18. VLPs do not contain any viral genetic material; they are basically the outside shell of the virus (**FIGURE 13.40**). Therefore, VLPs are noninfectious but can induce immunity to prevent infection by HPVs. **Gardasil-9**

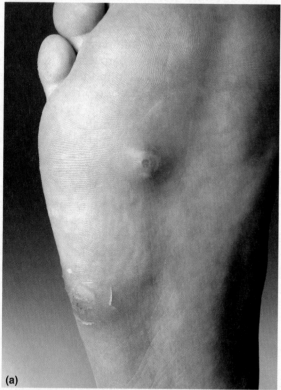

(a)

© Nau Nau/Shutterstock.

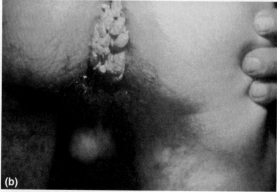

(b)

Courtesy of Dr. Weisner/CDC.

FIGURE 13.39 **(a)** Plantar wart. **(b)** Genital warts located in the anal region.

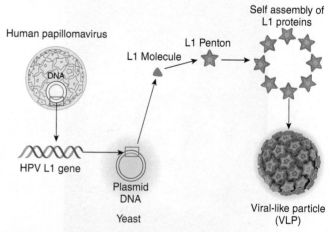

FIGURE 13.40 HPV noninfectious recombinant vaccines prepared from virus-like particles (VLPs). The gene encoding the major capsid (L1) protein of HPV is inserted into a plasmid. The plasmid is transformed into yeast cells, and large amounts of L1 proteins are produced and purified. The purified proteins self-assemble into empty or hollow VLPs that are devoid of genetic material and other HPV proteins. The purified VLPs are used as the vaccine to immunize people.

is a noninfectious recombinant vaccine prepared from purified VLPs of the major capsid (L1) protein of HPV types 6, 11, 16, 18, 31, 33, 45, 52, and 58. The additional 31, 33, 45, 52, and 58 are high-risk HPVs. The Gardasil vaccines are safe and are given as shots of three required doses. Vaccination is expected to greatly decrease the incidence of HPV-related cervical cancer.

Implementation of the use of the vaccine was highly controversial because of the ethical questions raised. For example, in 2007, the governor of Texas issued an executive order mandating that Gardasil be given to all girls entering the sixth grade as of September 2008; the Texas legislature overruled the order several months later, barring mandatory vaccination until at least 2011. In 2006, legislators in at least 41 states and Washington, D.C., introduced legislation to require the HPV vaccine, fund research for the vaccine, or educate the public about the vaccine. To date, only two bills have passed that mandate the vaccination of girls: Virginia and Washington, D.C. Major issues are the age or grade of the girls, with most states favoring entrance into sixth grade, and the role of health insurance companies in covering the cost of immunization; most pending legislation allows for parents to opt their daughters out.

The CDC recommends that girls and boys age 11 or 12 be routinely vaccinated. A catch-up vaccination for females ages 13 to 26 is also recommended. It is not mandated, but it is a public choice; for public health reasons, the vaccine is encouraged. The recommendations come at a time when a study correlating HPV and heart disease in women was published in the *Journal of the American College of Cardiology*. Could the vaccine prevent cancer and heart disease, too? The answer is that more rigorous studies are needed to determine this.

Contact Diseases (Other Than STIs) and Bloodborne Viral Diseases

This section presents several viral contact diseases (**TABLE 13.8**) currently of considerable significance in the United States and around the world.

TABLE 13.8 Contact Viral Diseases Other Than STIs and Bloodborne Viral Diseases

Viral Disease	Incubation Period	Symptoms	Immunization and Comments
Infectious mononucleosis	4–6 weeks	Mononucleosis, headache, loss of appetite, fever, sore throat, swollen glands; enlarged spleen in severe cases	"Kissing disease" (caused by Epstein-Barr virus); those ages 15 to 30 years most affected
Cytomegalovirus infections	3–8 weeks after blood transfusion; 4 weeks to 4 months after organ transplant; incubation period of sporadic cases is difficult to determine	Mononucleosis, congenital CMV syndrome in infants younger than 4 weeks, pneumonia in transplant patients, retinitis in AIDS patients	Common opportunistic infection of immunocompromised patients; most important cause of congenital viral infection during pregnancy; 40–70% of U.S. adults are infected
Hepatitis B	45–180 days	May be asymptomatic; nausea, vomiting, appetite loss, jaundice	Blood, blood-associated products, sexual practices
Hepatitis C	2–22 weeks	Asymptomatic but can cause a chronic infection and/or jaundice, leading to cirrhosis and liver cancer	Spread through blood, shared needles of injection drug users; highly expensive antivirals available to cure disease
Rabies	5 days to several years (average is 2–3 months)	Encephalitis, anxiety, hypersalivation or drooling, agitation, difficulty swallowing, hydrophobia	Almost always fatal if not given postexposure prophylaxis; immunoglobulin and vaccine on day 0 followed by additional vaccine doses on days 3, 7, and 14; zoonotic with many wild animal reservoirs
Ebola virus disease	2–21 days	Dry and wet phases of symptoms	WHO permission to use experimental vaccine in 2018 Democratic Republic of Congo Ebola outbreak; Zaire strain is the most common and deadliest

Infectious Mononucleosis and Cytomegalovirus Infections

Epstein-Barr virus (EBV) and **cytomegalovirus (CMV)** are both herpesviruses. Herpesviruses are enveloped viruses that enter a latency phase that allows them to establish a lifelong infection. These herpesviruses are most likely to be reactivated in people with weakened immune systems. It is estimated that EBV causes 79% of infectious mononucleosis cases and that CMV causes the remaining 21%.

○ Epstein-Barr Virus (EBV) Infections

Mononucleosis, often called mono or kissing disease, is a frequent and unwelcome guest on college campuses. The term **mononucleosis** refers to **monocytes**, one of the five categories of white blood cells. By adulthood, antibodies against EBV are present in most people. As many as 50 of every 100,000 Americans have symptoms of infectious mononucleosis, primarily those in the 15- to 30-year-old age group.

EBV infects and replicates in salivary glands and is transmitted by saliva and mucus during kissing, coughing, and sneezing. The incubation period is approximately 4 to 6 weeks after exposure. An infected person poses little risk to household members or college roommates, assuming there is no direct contact with the infected person's saliva, such as by kissing, sharing beverages, or eating utensils (**FIGURE 13.41A**).

Symptoms are vague, making diagnosis difficult. In fact, in many people the disease is asymptomatic or so mild that they are not even aware they are infected. The early symptoms include "not feeling great," headache, fatigue, and loss of appetite, followed by the later triad of fever, sore throat, and swollen glands, particularly in the neck. Enlargement of the lymph nodes and spleen is a more serious complication.

Diagnosis can be definitively established only by laboratory tests based on the detection of antibodies and abnormal white blood cells in a blood sample. There is no specific treatment, and symptoms generally disappear in 4 to 6 weeks. EBV, like other herpesviruses, remains latent in the body for life and is held in check by the immune system, although recurrences can occur.

○ Cytomegalovirus (CMV) Infections

CMV infections are common in all human populations, affecting 40% to 70% of adults in the United States and nearly 40% to 80% in developing countries. CMV is transmitted in utero during the first 6 months of life from exposure to the mother's genital secretions in childbirth and through breastfeeding and among preschoolers through oral and respiratory secretions. Healthy adults who become infected with CMV after birth experience few or no symptoms of the disease and no long-term health consequences. CMV is transmitted from person to person through close, intimate contact with a person shedding CMV in his or her saliva, urine, or other bodily fluids. It can be transmitted through breastfeeding, through semen (sexual contact), and from transplanted organs from a seropositive donor, and rarely through blood transfusions.

Shors' Notebook

Having taught college students for 20 years, I am familiar with a number of students who experienced mono in college. I remember one student in particular who had a severe case. Her spleen became enlarged and she slept through alarms set off by four different alarm clocks. She was a very good and conscientious student. "Heather" was a member of the Reserve Officers Training Corps (ROTC), which required her to run several miles every morning before classes. Her commander exempted her from this activity because she became so ill. This experience helped to build her character. Today she is an active Lieutenant Colonel in the United States Army and has a busy family life. Heather and her spouse (who is also in the military) are raising three children under the age of 3.

© Antonio Guillem/Shutterstock.

Courtesy of the CDC/Lauren Bishop.

FIGURE 13.41 **(a)** Infectious mononucleosis can be spread if an infected person has close contact and shares a beverage with a susceptible person. **(b)** CMV is the most important cause of congenital viral infections in the United States. About one in five newborns with CMV infection will develop complications of infection such as hearing loss, visual impairment, varying degrees of intellectual and developmental disabilities, and motor problems.

For the majority of people, CMV is not a serious disease. *CMV is the most common opportunistic pathogen in immunocompromised patients, such as AIDS patients and organ transplant recipients.* CMV infections are the most common and single most important viral infection in all solid organ transplant recipients. About 20% to 60% of transplant recipients develop symptomatic CMV infections during the first year following a transplant.

CMV is the most important cause of **congenital viral infections** (infections during pregnancy through the time of delivery/childbirth) in the United States (**FIGURE 13.41B**). *CMV-infected infants have about a one in five risk for developing complications of CMV infections. Complications of CMV infection in newborns include hearing loss, visual impairment, varying degrees of intellectual and developmental disabilities, and motor problems.* Therefore, if a woman becomes infected with CMV during pregnancy, there is the potential risk that after birth the infected infant may develop CMV-related complications. Signs and symptoms of babies suffering from congenital CMV infections at birth include:

- Rash
- Jaundice
- Microcephaly (small head)
- Low birth weight
- Enlarged liver and spleen
- Seizures
- Retinitis (damaged eye retina)
- Vision loss
- Hearing loss
- Developmental and motor delay

Congenital CMV infections can be confirmed through the diagnostic testing of newborn saliva, urine, or blood. Infants with signs of congenital CMV infection at birth are treated with the antiviral **Valcite** (valganciclovir), which can improve hearing and developmental outcomes. Infants and children who become infected with CMV *after birth*, experience few, if any, symptoms or complications. No vaccine is available to date to prevent CMV infection.

Hepatitis B and C

As previously mentioned, there are other hepatitis viruses in addition to hepatitis A and E viruses, and these are transmitted primarily by blood and blood products and by sexual practices. Infected pregnant women can transmit hepatitis B and C viruses to their infants. Hepatitis B and hepatitis C, along with other diseases, are frequently found in persons with HIV/AIDS. **World Hepatitis Day** takes place every year on July 28th (http://www.worldhepatitisday.org). The mission is to raise awareness about viral hepatitis because it is one of the largest global health threats of today.

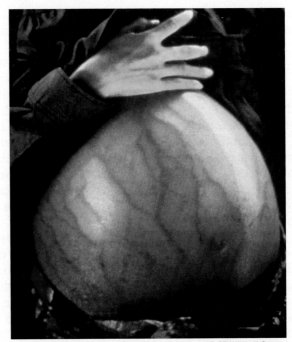

Courtesy of CDC. Used with permission of Patricia F. Walkers, MD, DTM&H, Health Partners, Center for International Health & International Travel Clinic, St. Paul, Minnesota.

FIGURE 13.42 Hepatitis B. This woman has a hepatoma (liver cancer) resulting from chronic hepatitis B infection.

Hepatitis B is 50 to 100 more times infectious than HIV; an individual pricked with a needle from an individual who has both hepatitis B and HIV has a 40% chance of acquiring hepatitis B but only a 0.5% chance of acquiring HIV. Both hepatitis B and hepatitis C are transmissible by blood and blood products, can be asymptomatic or be manifested as chronic diseases, and result in serious life-long liver problems (**FIGURE 13.42**), possibly necessitating a liver transplant.

The term hepatitis refers to inflammation of the liver. Although the clinical symptoms caused by all the hepatitis viruses are somewhat similar, their biology, circumstances of transmission and infection, and consequences of infection are different. Laboratory tests are required for a definitive diagnosis of viral hepatitis. Laboratory tests are based on serology (antibodies present in a patient's blood sample) and tests for elevated liver enzymes present in blood samples: **aspartate aminotransferase (AST)** and **alanine aminotransferase (ALT)**. Liver enzymes are normally found in the liver but spill into the bloodstream if the liver is damaged, which occurs when a patient is chronically infected. NAATs also are available.

Hepatitis B virus (HBV), originally known as **serum hepatitis**, has a long incubation time, ranging from 45 to 180 days, with an average of about 80 days. HBV is a common cause of liver cancer, second only to tobacco. HBV has been an occupational hazard for health professionals, including physicians, medical and dental hygienists, dentists, and others who have contact with blood. Individuals infected while working in service fields before a vaccine was available to them may serve as sources of HBV to the population at large. Infection can also be acquired by sharing needles, tattooing or body art, acupuncture, and ear piercing.

About 5% to 10% of people living with hepatitis B suffer from chronic infections. An algorithm that compares the disease progression for people living with hepatitis B and hepatitis C is presented in **FIGURE 13.43**. According to the WHO, 257 million people around the world are living with chronic HBV infection. In 2015, 887,000 deaths were associated with chronic hepatitis B infection that contributed to cirrhosis and **hepatocellular carcinoma** (liver cancer). HBV is most prevalent in the Western Pacific region (6.2% of population affected) and Africa (6.1% of population affected).

In the United States, 3,218 cases of hepatitis B were reported to the CDC in 2016. Because many people do not know they are infected, they are not diagnosed. Hence, the CDC estimates that the actual number of hepatitis B cases was almost 20,900 in 2016. A marked decrease in hepatitis B infection was reported from the 1990s to 2012, likely due to the widespread introduction of the hepatitis B vaccine. Since 2012, the trend in the number of cases has not been consistent, with an average of 3,000 cases per year. Overall, an estimated 850,000 people are living with chronic hepatitis B infection, but it could be as high as 2.2 million.

Although a vaccine was introduced in the early 1980s, the incidence of disease reduction was disappointing. An improved vaccine made from a component of the virus proved to be a more effective and safer vaccine, resulting in the 1992 recommendation by the American Academy of Pediatrics of universal immunization for newborns. In the same year, the WHO set a goal for all countries to implement HBV into national routine infant immunization programs by 2007, and by 2006, 84% of the 193 countries had conformed. Older children should also be vaccinated, particularly adolescents if they are sexually active. The vaccine has reduced the incidence of acute hepatitis B in children and adolescents by over 95% and by 75% in all age groups. One of the problems is that the three-shot vaccination strategy is expensive and beyond the reach of poor countries.

Hepatitis C virus (HCV) was first identified in 1988 and was responsible for almost all cases of blood transfusion–transmitted (non-A, non-B) hepatitis. Hepatitis C is a chronic bloodborne infection with an incubation period of 2 to 22 weeks; it may be subclinical or mild, but approximately 75% to 85% of the cases progress to chronic hepatitis and 10% to 20% will go on to develop cirrhosis over a period of 20 to 30 years (Figure 13.43).

The first reliable screening tests for HCV were not implemented until 1992; therefore, many people may be infected with HCV but unaware that they have the disease, particularly considering that it can take 20 years for the appearance of symptoms. An estimated 3.5 million persons in the United States are infected with hepatitis C virus. It is estimated that about 41,200 new cases occurred in 2016. Globally, 71 million people have chronic hepatitis C and approximately 399,000 people die every year from chronic hepatitis C associated with cirrhosis and liver cancer.

Intravenous drug users constitute the vast majority of hepatitis C victims, and as many as 300,000 may have contracted it from blood transfusions before screening tests were available. As a result, in 1988, the U.S. Department of Health and Human Services announced that people who received transfusions before 1992 from donors who later proved to be positive for hepatitis C would be notified.

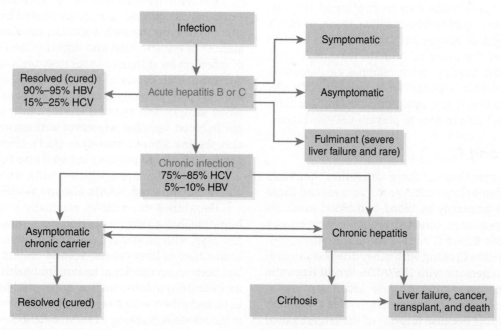

FIGURE 13.43 Comparison of courses of infection of hepatitis B and hepatitis C. Hepatitis C infection becomes chronic in 75% to 85% of those infected.

TABLE 13.9 The Cost to Cure Hepatitis C

Direct-Acting Antiviral (Trade Name)	Year of FDA Approval	Cost for Cure: 12-Week Therapy	Cost for Cure: 8-Week Therapy
Harvoni	2014	$94,800	—
Viekira Pak	2014	$83,300	—
Technivie	2015	$78,100	—
Epclusa	2016	$75,000	—
Zepatier	2016	$55,700	—
Vosevi	2017	$75,600	—
Mavyret	2017	—	$26,400

Average costs from https://www.goodrx.com

Hepatitis C is the major reason for liver transplants in the United States.

There is no recommended treatment for acute hepatitis C. Highly effective HIV protease inhibitors became available to treat chronic hepatitis C in 2011. Since that time, new drugs have become available. At the time of this writing, 13 different FDA-approved drugs are available for HCV treatment. More than 90% of HCV-infected people can be cured of HCV within 8 to 12 weeks. A downside is the hefty cost of these drugs. The average cost of treatment for the combination of **direct-acting antivirals (DAAs)** are listed in **TABLE 13.9**.

The DAAs specifically target HCV (instead of causing a general effect, such as boosting the immune system, or inhibitors that affect many different viruses such as Ribavirin, which inhibits RNA synthesis) and cure 99% of HCV patients, usually within 12 weeks (Mavyret takes 8 weeks). A few more DAAs have become available that are not on the list. No **generic antivirals** are currently available, which gives pharmaceutical companies a lot of freedom in establishing prices (which are high to recoup the money it took to develop the antiviral drug through rigorous clinical testing and obtain a license by the FDA).

Many people living with chronic HCV need financial assistance from insurance companies. Although some insurance companies pay for treatment, the criteria that many of these insurance companies use are based on whether the patient has severe liver disease; whether the patient uses alcohol or drugs; whether the doctor who prescribed the medication is a specialist in liver diseases; life expectancy, if treated; whether less expensive drugs can be used first; and whether the patient has other preexisting conditions that contribute to the liver damage. In other words, the person needs to be sick enough to require treatment or the insurer will not pay for it. It would be in society's best interest to follow the HIV/AIDS generic drug development movement and work toward making the cure for hepatitis C available to everyone who needs it.

Rabies

World Rabies Day is September 28th (https://www.cdc.gov/features/rabies/index.html). It was created in 2007 to observe awareness about the burden of rabies and establish partnerships to control rabies in the world. Rabies is not a human disease per se. It is a **zoonotic disease** in which humans are incidentally infected. Death is almost a certainty in untreated cases of **rabies**, a disease formerly known as **hydrophobia**; few survive the **encephalitis** that results from infection with the rabies virus. No treatment is available once symptoms appear, but thanks to Louis Pasteur, a French chemist, a rabies immune globulin and a vaccine is available for both prevention and prophylactic treatment (before the development of symptoms) of rabies.

Two types of rabies are described, namely **furious rabies** and **dumb rabies**, both of which cause death. About 80% of rabies cases are of the furious type and characterized by brain dysfunction; a rabid animal displays aggressiveness, excitability, and (not always) foaming at the mouth—the "mad dog" image. Dumb, or paralytic, rabies primarily involves the spinal cord; the animals display weak limbs and are unable to raise their heads and/or make sounds because of paralysis of the neck and throat muscles.

Rabies has a worldwide distribution in diverse wild mammal reservoirs, including coyotes, skunks, cats, bats, foxes, mongooses, and raccoons, as well as domestic animals. Although control measures and immunization have substantially reduced the incidence of domestic animals as vectors in the United States and in other countries, dogs remain the major source of rabies in Asia (especially China), Africa, and Latin America, causing thousands of deaths each year. So many human rabies cases associated with rabid dog bites occurred in Beijing that the country introduced a "one dog per family" policy and a size restriction: dogs taller than 35 centimeters (1.1 feet) are illegal.

The annual worldwide human death toll is approximately 55,000, of which 34,000 occur in Asia and 24,000 in Africa. At least 150 people die of rabies every day. India has the highest rate in the world, attributed to the large number of stray dogs. These statistics are underestimates because underreporting of all infectious diseases happens in developing countries.

In the United States a dramatic shift has occurred over the past 40 years in the principal wildlife carrier. In 1966, foxes were the major reservoirs, followed several years later by skunks; since about 1993 raccoons have emerged as the main reservoir on the East Coast (**FIGURE 13.44**). The raccoons frequent neighborhoods in search of food and may bite dogs and cats. Seventy-nine human cases occurred in 2006, and deaths due to rabies have been reduced to one to two per year.

The bat is a rabies vector, and most human rabies cases in the United States are caused by bat bites. Bats have acquired a bad but unearned reputation, perhaps, at least in part, due to the novel *Dracula* by Bram Stoker, published in 1897, in which bats were depicted as evil, bloodsucking creatures capable of turning their victims into horrible life forms. However, the novel contained no reference to a connection between bats and rabies. Bats have been dramatized by the escapades of Batman as he wheels around in his Batmobile. Expressions such as "bats in the belfry," "batty," "blind as a bat," and "crazy as a bat" are part of our jargon and indicate our fear of bats. But let's clear up a few misconceptions: Bats are not blind, most species do not suck blood (an exception being vampire bats of South America), and they do not particularly like to "get in

your hair." Actually, they are remarkable creatures (some say "beautiful," but that is too much) and play an important role in ecosystems around the world by eating insects (especially mosquitoes), including agricultural pests.

Despite rabies vaccination in domestic pets, wild animals that carry rabies and unvaccinated pets remain a constant threat. A rabies scare occurred in June 2008 when a dog imported from Iraq under Operation Baghdad Pups, an affiliate of the International Society for the Prevention of Cruelty to Animals, was confirmed to have rabies. Twenty-three other dogs in the shipment were exposed to the rabid dog and shipped to 16 states, but none became ill. Thirteen of the 28 persons who had contact with the dogs were administered **postexposure prophylaxis (PEP)**. Other similar cases have occurred in conjunction with other programs. The CDC is responsible for implementation of federal regulations governing the importation of dogs (and other animals).

The downside of organ transplantation is that it introduces the opportunity for the transmission of disease. Two examples make this clear. In June 2004, three organ recipients died from rabies transmitted from the organs of an infected donor; the donor had been bitten by bats. In February 2005, three persons contracted rabies as the result of organ and corneal transplants.

In 2004, Wisconsin teenager Jeanna Giese made history after contracting rabies at age 15. Jeanna had picked up a bat and suffered a minor bite on her finger (**FIGURE 13.45**). She received no medical attention and 21 days later developed the symptoms of rabies. It was too late for PEP because she had already developed an antibody response to the virus. Her doctors, aware that death from rabies is associated with brain dysfunction, decided on an experimental treatment known as the **Milwaukee protocol**. Giese was put into a drug-induced coma using **antiexcitatory drugs** and treated with antivirals (ribavirin and amantadine).

The theory was that the coma would allow her brain to recover and give her immune system time to produce antibodies to clear rabies virus from her body. After 7 days the drugs were tapered off because there were signs that her immune system was responding; she was brought out of the coma. Jeanna is the first person known to have recovered from symptomatic rabies without receiving the rabies vaccine after exposure.

Since then, the Milwaukee protocol or a modified version of it has been used to treat rabies victims in multiple countries (e.g., United States, India, Germany, Columbia, United Kingdom, South Africa, Canada, Taiwan, Brazil, Saudi Arabia, Thailand, Ireland, and The Netherlands). From 2005 through 2014, 29 cases of rabies were reported in the literature; however, only 2 patients were reported as surviving. Despite initial hope and enthusiasm for the Milwaukee protocol to treat rabies, evidence is lacking to support its use to manage rabies patients.

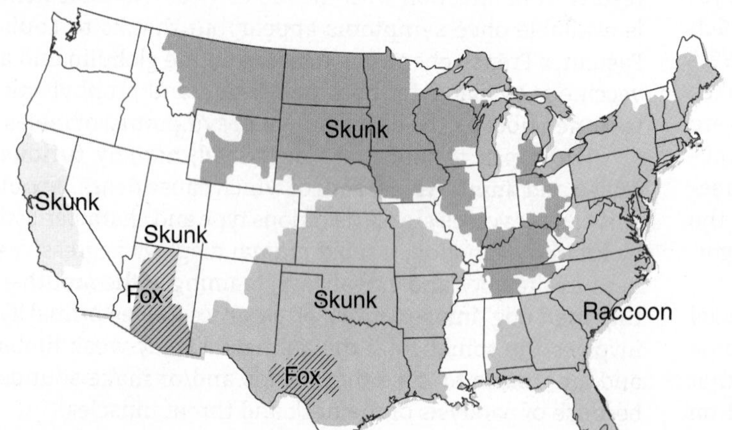

Information from J. L. Dyer, P. Yager, L. Orciari, L. Greenberg, R. Wallace, C. A. Hanlon, & J. D. Blanton. (2014). Rabies surveillance in the United States during 2013. *Public Veterinary Medicine: Public Health, 245*, 1111–1123.

FIGURE 13.44 The distribution of the major reservoirs of rabies in the United States in 2013.

© Morry Gash/AP Photo.

FIGURE 13.45 Jeanna Geise, the first rabies survivor who did not receive postexposure vaccination.

The most common mode of rabies transmission is through the bite of a rabid animal. In addition to a bite, the rabies virus can also be transmitted, although rarely, into the eyes, nose, and respiratory tract. The **prodromal period** (early symptoms) of rabies are nonspecific and flulike, and there may be pain and tingling at the site of the bite. The secondary (illness) stage is characterized by symptoms of anxiety, confusion, agitation, delirium, abnormal behavior, hallucination, drooling, and hydrophobia. Persons with rabies are tortured by thirst and, ironically, are revolted by water. The mere sight of water results in uncontrollable spasms in the muscles of the mouth and pharynx, leading to spitting and choking.

Before the work of Pasteur the public's fear of hydrophobia and the prevailing attitude that the disease could be transmitted through the saliva or breath of rabies victims was so overwhelming that violent modes of death, including suffocation, were inflicted upon those who had suffered bites of a rabid animal. Such events must have been quite frequent, because in 1810 a bill in France was written in the following terms:

It is forbidden under pain of death, to strangle, suffocate, bleed to death, or in any other way, murder individuals suffering from rabies, hydrophobia, or any disease causing fits, convulsions, furious, and dangerous madness.

Early treatments were terrible. **Cauterization** was the most frequently used method; if the wounds were somewhat deep, it was recommended to use long, sharp needles and to push them well in, even if the wound was on the face. Another practice involved the sprinkling of gunpowder over the wound and setting a match to it.

The incubation period before the onset of symptoms is related to the extent of the wound and its closeness to the brain. The incubation period for wounds on the hands is about 8 weeks, whereas for wounds on the face it is about 5 weeks. The average incubation period is 1 to 2 months, with extremes of 5 days to years. Initially,

rabies virus multiplies in the muscle and connective tissue at the site of the bite, then it migrates along nerves to the **central nervous system (CNS)** (**FIGURE 13.46**). This takes time. The rabies viruses travel within the axons of the motor nerves of the CNS at a rate of 1.3 to 38 centimeters (0.5 to 15 inches) per day. This window of time gives PEP a chance to block rabies virus before it can cause pathogenic effects. Hence, it is better to be bit on the toe than the head!

The rabies vaccine currently in use consists of inactivated virus cultured in tissue culture. Treatment consists of one dose of immune globulin (blood serum from individuals containing high levels of antibodies against rabies) and a first dose of the vaccine given as soon as possible after exposure, with the remaining four doses given on days 3, 7, 14, and 28, respectively. The injections are relatively painless and are given in the arm in the same way that other immunizations are given. Earlier versions of the vaccine were administered into the abdomen and were painful.

Whether the individual needs to be vaccinated is carefully evaluated and depends on the circumstances. Laboratory tests are available to determine whether the animal does, in fact, have rabies. If the bite is from a domestic animal, the animal can be quarantined and observed for symptoms. Vaccinations are advised if a wild animal is involved and the animal cannot be captured. The development of the immunofluorescent antibody test in 1958 allows for immediate determination of whether an animal suspected of being rabid is actually rabid; based on the test outcome, a decision as to whether the exposed person needs to be given the vaccine can be made. It may have occurred to you that, for most diseases, immunization is usually given as a preventive measure and not as a treatment; if it has, you are correct (and very astute). However, the situation with rabies is somewhat unusual in that PEP, as a treatment, is effective because of the long incubation time generally associated with rabies. However, one case of human rabies developed in only 10 days after the person had been bitten. After an animal bite, the wound should be thoroughly cleansed. Pre-exposure vaccination is recommended for those in high-risk groups, including veterinarians, animal handlers, and laboratory personnel engaged in rabies research.

A study carried out by researchers at the CDC collaborating with the Peruvian Ministry of Health determined that people living in two remote Amazon communities (Santa Marta and Truenococha) who were repeatedly exposed to rabies virus survived without vaccination. Blood samples were collected from 63 people, and 11% of them had neutralizing antibodies against rabies virus. Researchers were unable to determine if any of the people in the study experienced any symptoms of rabies. Over the past 20 years, outbreaks of fatal human rabies caused by vampire bat bites had occurred regularly. The results open the door to the idea that a small percentage of remote Peruvians encountered enough exposure

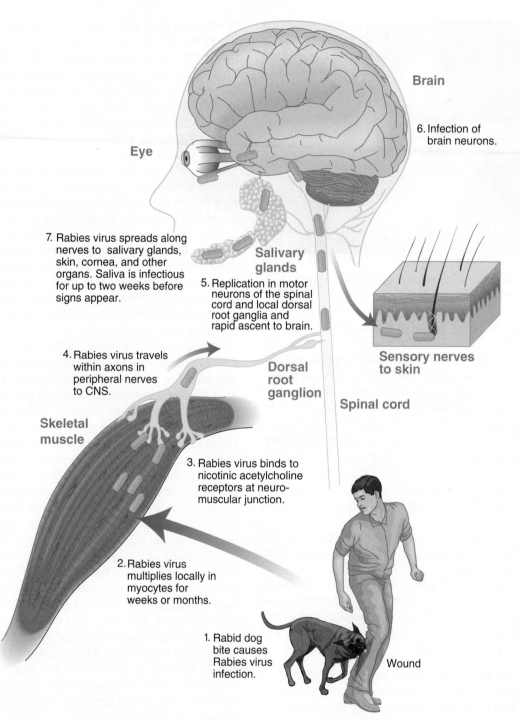

Brain

6. Infection of brain neurons.

Eye

7. Rabies virus spreads along nerves to salivary glands, skin, cornea, and other organs. Saliva is infectious for up to two weeks before signs appear.

Salivary glands

5. Replication in motor neurons of the spinal cord and local dorsal root ganglia and rapid ascent to brain.

Sensory nerves to skin

4. Rabies virus travels within axons in peripheral nerves to CNS.

Dorsal root ganglion

Spinal cord

Skeletal muscle

3. Rabies virus binds to nicotinic acetylcholine receptors at neuromuscular junction.

2. Rabies virus multiplies locally in myocytes for weeks or months.

1. Rabid dog bite causes Rabies virus infection.

Wound

FIGURE 13.46 Steps of rabies transmission and rabies virus replication after a dog bite to the lower limb.

to the vampire bat rabies virus to evoke natural immunity but not enough to kill them. This counters the traditional belief that 100% of individuals exposed to rabies virus died unless they sought PEP.

Ebola Virus Disease

Following are vivid excerpts about Ebola hemorrhagic fever from *The Coming Plague* by Laurie Garrett (1994, The Penguin Group):

And he was bleeding. His nose bled, his gums bled, and there was blood in his diarrhea and vomit. . . . They pumped Antoine full of antibiotics, chloroquine, vitamins, and intravenous fluid to offset his dehydration. Nothing worked. . . . The horror was magnified by the behavior of many patients whose minds seemed to snap. Some tore off their clothing and ran out of the hospital, screaming incoherently. . . . Some, the huts of the infected, were burned by hysterical neighbors.

The following is an excerpt from an article in *Newsweek* published on May 22, 1995:

> Then, as the virus starts replicating in earnest, the victim's capillaries clog with dead blood cells, causing the skin to bruise, blister, and eventually dissolve like wet paper. By the sixth day, blood flows freely through the eyes, ears, and nose, and the sufferer starts vomiting the black sludge of his disintegrating internal tissues. Death usually follows by day nine.

These descriptions are enough to make you break out in a sweat. It is no wonder that near panic resulted in the United States in 1989, 1990, and again in 1996, when Ebola virus was introduced into primate quarantine facilities in Pennsylvania, Texas, and Virginia from monkeys imported from the Philippines. Several individuals developed antibodies, but no human cases of disease were identified. In 2008, another scare—the first known infection of Ebola-Reston virus in pigs on a farm in the Philippines—occurred. Six workers from a pig farm and slaughterhouse developed antibodies but did not get sick.

Ebola virus, the cause of Ebola hemorrhagic fever, which was renamed Ebola virus disease (EVD) during the 2014 West Africa outbreak, is named after the Ebola River in the African nation of Zaire (now the Democratic Republic of the Congo), where it was first detected in 1976 (**FIGURE 13.47**). Ebola virus causes sporadic outbreaks of severe infection with a high fatality rate in humans and in monkeys and chimpanzees. The disease is considered to be zoonotic; the reservoir of the Ebola virus is believed to be bats. Transmission requires contact with the bodily fluids of an individual ill with the

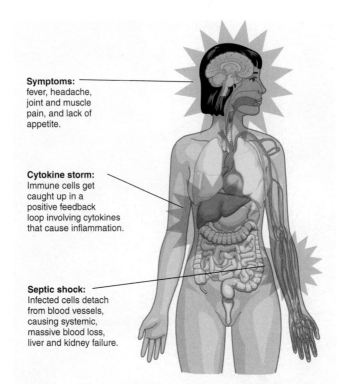

Symptoms:
fever, headache, joint and muscle pain, and lack of appetite.

Cytokine storm:
Immune cells get caught up in a positive feedback loop involving cytokines that cause inflammation.

Septic shock:
Infected cells detach from blood vessels, causing systemic, massive blood loss, liver and kidney failure.

FIGURE 13.48 Pathogenesis of Ebola virus. The viral infection begins with dry symptoms of EVD. As the infection proceeds, the body's immune cells produce cytokines that cause inflammation and a **cytokine storm** response that ultimately leads to septic shock.

disease. Sexual contact may also transmit Ebola virus, because Ebola viruses remain infective in semen and genital secretions for as long as 9 months. Similar to rabies infections, humans are "incidental" hosts—they do not "carry" Ebola virus.

The dry symptoms, including fever, chills, muscle aches, headache, stomach pain, sore throat, and abdominal pain, appear after an incubation period of 2 to 21 days. Wet symptoms include diarrhea and vomiting. The blood fails to clot, resulting in massive hemorrhage throughout the body, both externally and internally (see **FIGURE 13.48** and Figure 1 in the opening case study).

Control measures in a community or hospital environment center on isolation of those with the disease and avoidance of direct contact with infected persons and their blood or secretions and with the bodies of deceased patients. **Barrier technique** is used in caring for patients, including the use of gowns, goggles, masks, and gloves by physicians, nurses, and other healthcare workers, the restriction of visitors, and proper disposal of wastes and corpses. Treatment involves using the **best available supportive care (BASC)**. BASC is palliative care that focuses on providing the patient with relief from EVD symptoms that are life threatening. It involves balancing the patients' fluids and electrolytes, maintaining their oxygen status and blood pressure, and treating any complicating infections that emerge.

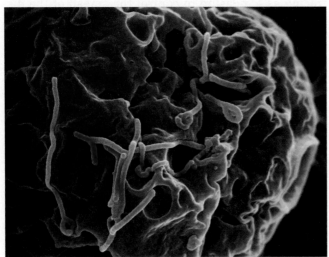

Courtesy of the National Institute of Allergy and Infectious Diseases.

FIGURE 13.47 Colorized scanning electron micrograph showing viruses budding from the surface of African green monkey cells grown in the laboratory. Ebola viruses are helical, enveloped, spaghetti-shaped viruses.

Courtesy of Dr. Scott Smith/CDC.

FIGURE 13.49 BSL-4 laboratory research. Highly virulent microbes and viruses must be handled using biosafety cabinets in BSL-4 laboratories to protect laboratory personnel and the environment. This special pathogens laboratory worker is counting viral plaques within a fixed monolayer of cells in a BSL-4 laboratory at the CDC.

Researchers and other laboratory workers handling Ebola virus and other viruses that cause hemorrhagic fevers and highly fatal viruses must work in **biosafety level four (BSL-4)** laboratories (**FIGURE 13.49**) as a matter of protection. The CDC has established biosafety levels one through four with reference to work practices, safety equipment, and facilities to minimize the consequences of escape of infectious microbes to laboratory personnel and to the environment. BSL-4 is reserved for the "hot" microbes—extremely dangerous and exotic microbes with a potential to cause life-threatening disease and for which there is no vaccine or treatment.

O Ebola Virus Disease Outbreak, West Africa: 2014–2016

"Ebola doesn't just disappear. It just goes into hiding."

—David Quammen, American science and nature author

The 2-year-battle with EVD that started in West Africa was the worst Ebola epidemic in human history. **Patient zero** was an 18-month-old toddler who died in December 2013 in the Meliandou village of Gueckedou, Guinea. The epidemic was a spillover event. It is likely that the boy had contact with the reservoir host. A large hollow tree with a colony of bats was just 45.7 meters (150 feet) from where the boy lived. He could have been infected while playing in the tree through contact with bat saliva or droppings contaminated with Ebola viruses. The child's family hunted, grilled, and ate bats (**FIGURE 13.50**).

The toddler's sister, mother, grandmother, a village midwife, and nurse were infected and also died. The nurse died in a hospital in Gueckedou before there was a definitive EVD diagnosis. Healthcare workers, not knowing what was causing the fatal disease, did not use enough precautions in preventing its spread. EVD had never been in West Africa before. The disease was unfamiliar to the people in this region. The disease also spread through **unsafe burial practices** in which touching and washing the body of deceased relatives was a cultural practice. The body fluids of Ebola victims were teeming with Ebola viruses. The infective dose of Ebola virus may be as low as 1. The largest number of secondary cases of EVD were linked to funerals of traditional healers and midwives. During the week of December 15, 2015, an investigation determined that 85 new Ebola virus infections were linked to one traditional funeral ceremony for an assistant midwife in Guinea.

Five months into the outbreak, EVD spread to Conakry, the capital of Guinea, a city with a population of about 2 million people (**FIGURE 13.51**). It was the first time in human history that Ebola virus infections occurred in such a densely populated region that had an international airport (giving Ebola virus the opportunity to quickly cross borders). Previous outbreaks in Africa were in remote areas and small villages. It made **contact tracing** as a measure to control the spread of the disease impossible. Contact tracing is a process of finding all individuals who had direct contact with an infected person, in this case an Ebola patient. Contacts are monitored for signs and symptoms of illness for 21 days (1 incubation period) from the last day they came in contact with the Ebola patient (**FIGURE 13.52**). If symptoms begin, the contact is quickly isolated, tested, and provided care. This control strategy works well when used in smaller, localized outbreaks.

Besides unsafe burial practices, there were many other challenges at the onset of the outbreak. Poor public health infrastructure resulted in a shortage of doctors, healthcare workers, protective equipment, and disinfectant. No antivirals or vaccines were available to prevent EVD. However, *the most difficult barrier to overcome was fear.* When West Africans became ill, they were taken to traditional healers, hidden from the surveillance system. Fear fed hostility, threatening the safety of response teams. Fear affected the airlines. Some airlines refused to transport personal protective equipment for response workers. Courier services refused

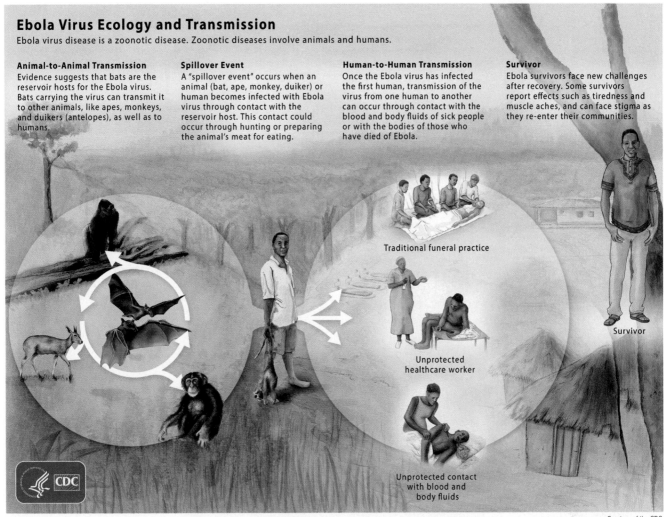

Ebola Virus Ecology and Transmission

Ebola virus disease is a zoonotic disease. Zoonotic diseases involve animals and humans.

Animal-to-Animal Transmission
Evidence suggests that bats are the reservoir hosts for the Ebola virus. Bats carrying the virus can transmit it to other animals, like apes, monkeys, and duikers (antelopes), as well as to humans.

Spillover Event
A "spillover event" occurs when an animal (bat, ape, monkey, duiker) or human becomes infected with Ebola virus through contact with the reservoir host. This contact could occur through hunting or preparing the animal's meat for eating.

Human-to-Human Transmission
Once the Ebola virus has infected the first human, transmission of the virus from one human to another can occur through contact with the blood and body fluids of sick people or with the bodies of those who have died of Ebola.

Survivor
Ebola survivors face new challenges after recovery. Some survivors report effects such as tiredness and muscle aches, and can face stigma as they re-enter their communities.

Courtesy of the CDC.

FIGURE 13.50 Ecology of Ebola viruses in Africa. It is believed that Ebola viruses circulate in the wild among bat, antelope (duiker), rodent, and nonhuman primate populations. During epizootic events (epidemics in animal populations), humans may have contact with fluid from infected bats or other infected animals, triggering incidental transmission followed by human-to-human transmission.

Courtesy of the CDC/Sally Ezra.

Courtesy of the CDC/Heidi Soeters.

FIGURE 13.51 **(a)** Apartment complexes in Conakry, Guinea, photographed during the Ebola epidemic. **(b)** Red and green gloves propped up on sticks to dry after being washed in a bleach solution to inactivate Ebola viruses. The gloves somewhat resembled a garden outside the grounds of Donka Hospital in Conakry.

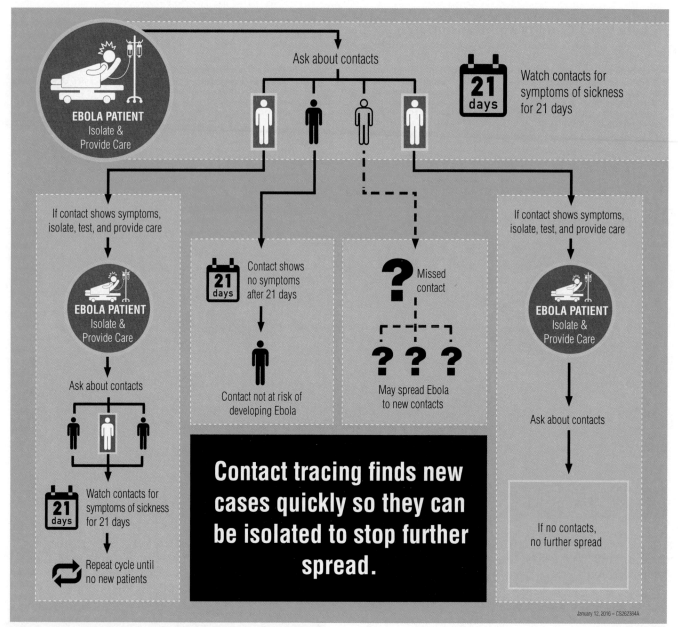

January 12, 2016 – CS262364A

Reproduced from the CDC.

FIGURE 13.52 Contact tracing. All it takes is one missed contact to keep the outbreak going. Once the epidemic spread to Conakry, it was impossible to use contact tracing as a measure to control the epidemic.

to securely package and transport patient samples to WHO-approved laboratories for testing. There were fears that imported cases would reach wealthy nations through air travel, spreading the disease into the general population (**FIGURE 13.53**). More than 339,000 people were screened flying out of Guinea, Liberia, and Sierra Leone to prevent the spread of Ebola.

In order to stop the spread of the epidemic, the WHO and the United Nations Mission for the Ebola Response established a **70-70-60 plan**, with the goals of:

- Isolating and treating 70% of cases;
- Burying 70% of victims safely (**FIGURE 13.54**);
- Within 60 days from October 1 until December 1, 2014.

Dr. Bruce Aylward, Assistant Director-General for Emergencies at the WHO, stated that "*leadership, adaptation, and innovation played important roles in the handling of the Ebola epidemic crisis in West Africa*" (http://www.who.int/features /ebola/then-and-now/en/).

© Teri Shors.   Courtesy of the CDC.

FIGURE 13.53 **(a)** Procedural changes occurred in the U.S. healthcare system. These signs that address EVD were used in the Koeller Street Clinic located in Oshkosh, Wisconsin. The display contained child- and adult-sized masks, facial tissue, and a collection basket for used tissues. **(b)** Health advisory used in the Atlanta, Georgia, airport.

Courtesy of the CDC/LCDR Medical Officer E. Kainne Dokubo.

FIGURE 13.54 The scene in Sierra Leone inside the Marfoki Cemetery during the Ebola epidemic in 2014. A burial team is performing a safe burial of a deceased Ebola victim. There are freshly dug graves in the backdrop.

It was an unprecedented epidemic. **TABLE 13.10** compares the statistics of the recent West Africa epidemic to all previously known EVD epidemics combined. *In fact, the epidemic was 11 times larger than all previous outbreaks combined.* EVD hit healthcare workers hard; they were at the highest risk of contracting the disease. From the start of the epidemic through November 2015, a total of 882 confirmed healthcare worker infections were documented and 513 deaths (58% mortality rate) in Guinea, Liberia, and Sierra Leone. Liberia lost 8% of its doctors, nurses, and midwives to EVD. Sierra Leone and Guinea lost 7% and 1% of their healthcare workers, respectively.

According to the head of the WHO Emergency Assessment Team, Dr. Rick Brennan, delivering a baby in Liberia was *"one of the most dangerous jobs in the world."* Some of the medical giants died of EVD. Additional costs of the Ebola epidemic are listed in **TABLE 13.11**. Due to the size of the epidemic, there are more survivors of EVD than ever before. This large number has allowed medical researchers to better understand how Ebola virus affects

TABLE 13.10 Burden of Ebola Virus Disease: Vital Statistics, 2014–2016

EVD Epidemic(s)	Cumulative Ebola Cases	Cumulative Ebola Deaths	Mortality Rate
2014–2016 West Africa EVD epidemic	28,616	11,310	39.5%
All previous EVD epidemics *combined*	2,427	1,597	65.8%

Information from Centers for Disease Control and Prevention. (2017). Ebola (Ebola Virus Disease) case counts. Retrieved from https://www.cdc.gov/vhf/ebola/history/2014-2016 -outbreak/case-counts.html

TABLE 13.11 Costs of the Ebola Virus Disease: 2014–2016

Impact Factor	Impact/Costs
Global economy	$2.2 billion in gross domestic product (GDP) lost in Guinea, Liberia, and Sierra Leone in 2015, threatening agriculture and food security and capital development and growth. Sierra Leone lost 50% of its private-sector workforce.
Healthcare system	881 doctors, nurses, and midwives were infected with Ebola virus in West Africa; 513 died. 8% reduction in healthcare workforce in Liberia. 23% reduction in health services in Sierra Leone.
Expansion of healthcare system	24,655 healthcare workers were trained in infection prevention and control in West Africa. 24 labs in Guinea, Liberia, and Sierra Leone were established that could test for Ebola virus.
Children	17,300 children lost one or both parents to Ebola. More than 33 weeks of education lost due to school closures. 30% decline in childhood vaccination coverage.
Response	U.S. government: $2.4 billion United Kingdom: $364 million Germany: $165 million World Bank $140 million Three operations centers set up in Guinea, Liberia, and Sierra Leone. $3.6 billion spent to fight the epidemic by the end of 2015.

Data from Centers for Disease Control and Prevention. (2017). Ebola (Ebola Virus Disease) case counts. Retrieved from https://www.cdc.gov/vhf/ebola/history/2014-2016-outbreak/case-counts.html

people who have recovered and to advise survivors how to take care of themselves and their communities. The survivors of EVD face stigma and complications of the disease (**TABLE 13.12**).

TABLE 13.12 Complications of Ebola Virus Disease Survivors: 2014–2016

	Complications
Common complications	Muscle and joint pain, headache, tiredness, weight gain, stomach pain or loss of appetite, eye and vision problems (blurry vision, pain, redness, and light sensitivity)
Other complications	Memory loss, neck swelling, dry mouth, tightness of the chest, alopecia (hair loss), tinnitus (ringing in the ears), pain or tingling in the hands and feet, inflammation of one or both testicles, inflammation of tissue around the heart, changes in menstruation, impotence, insomnia, depression, anxiety, posttraumatic stress disorder (PTSD)

Information from Centers for Disease Control and Prevention. (2017). Ebola (Ebola Virus Disease) survivors. Retrieved from https://www.cdc.gov/vhf/ebola/treatment/survivors.html

Arthropodborne Diseases

A variety of viral diseases are transmitted by **arthropod** vectors (**TABLE 13.13**). Arthropodborne diseases account for more than 17% of all infectious diseases, causing more than 700,000 deaths annually. These disease are emerging and reemerging globally. Many of these diseases are preventable through informed protective measures. People should educate themselves and take adequate measures to protect themselves from blood-feeding arthropods such as mosquitoes and ticks that carry viruses (**FIGURE 13.55**). (Arthropods can also carry other types of microbial pathogens, such as bacteria, protozoans, and helminths.)

Mosquitoes that carry pathogens may be day biting (e.g., *Aedes* spp.) or night biting (e.g., *Culex* spp.). Experts are in widespread agreement that **climate change** is one of several forces that "drives" the emergence of **infectious diseases** caused by arthropods because they spend part of their life cycle outside of its host, exposing it to the environment. Diseases transmitted by insect vectors such as ticks and mosquitoes will spread to new areas with climate change. Extreme weather events also are a factor in spreading arthropodborne diseases. In 2018, it was announced that the Asian **longhorned tick** is now present in at least eight U.S. states. It is the first new tick species in the United States in half a century. It comes from the opposite side of the globe. It has not been shown to carry

TABLE 13.13 Arthropodborne Viral Diseases

Viral Disease	Incubation Period	Vector	Symptoms	Immunizations/Comments
Zika virus disease	3–12 days	Mosquitoes (*Aedes* spp.)	1 in 5 cases experiences mild symptoms: fever, rash, headache, joint pain, conjunctivitis (red eyes), muscle pain	Emerging viral disease; Zika virus can be passed from pregnant mother to her fetus
Chikungunya virus disease	1–12 days (usually 3–7 days)	Mosquitoes (*Aedes* spp.)	Most people develop some symptoms: headache, muscle pain, joint swelling, rash; some people experience joint pain that persists for months	Emerging viral disease
Yellow fever	3–6 days	Mosquitoes (*Aedes* spp. or *Haemogogus* spp.)	Most people asymptomatic; some people experience fever, chills, headache, back pain, body aches, nausea, vomiting, fatigue, weakness; some people experience severe symptoms such as high fever, jaundice, bleeding, shock, organ failure	Reemerging viral disease; live attenuated vaccine available
West Nile encephalitis	3–12 days	Mosquitoes (*Culex* spp.)	No symptoms in most people; 1 in 5 has mild symptoms such as fever, body aches, joint aches, diarrhea, vomiting, rash, severe illness in a few people affecting central nervous system causing encephalitis or meningitis	Emerging viral disease in the United States (late 1990s); West Nile virus also infects over 300 species of wild birds; no vaccine, no antiviral treatments available
Dengue	4–7 days	Mosquitoes (*Aedes* spp.)	First infection: high fever, headache, pain behind eyes, muscle and bone pain, rash, mild bleeding; second infection increases risk of severe symptoms/dengue hemorrhagic fever	Four viral serotypes (DEN 1, 2, 3, 4); one serotype cannot protect against infection by a second serotype; if infected a second time, severe illness/ patient will likely develop dengue hemorrhagic fever; emerging viral disease
Colorado tick fever	1–14 days	Rocky Mountain wood tick	Fever, headache, chills, body aches, tiredness, weakness; some experience rash, sore throat, vomiting, abdominal pain; rare cases develop severe illness affecting the central nervous system	Spread by infected ticks in western U.S. or western Canada, 4,000 to 10,000 feet above sea level
Powassan encephalitis	Not known	Deer (blacklegged) tick and groundhog or woodchuck tick	Fever, headache, vomiting, weakness, confusion, seizures, memory loss; long-term neurological problems can occur	Emerging viral disease in U.S. Midwest/Great Lakes region
Heartland virus disease	Not known	Lone star tick	Fever, headaches, fatigue, diarrhea, muscles aches, thrombocytopenia (low platelet counts)	Emerging viral disease in U.S. Midwest (first cases in Indiana)
Bourbon virus disease	Not known	Deer (blacklegged) tick?	Fever, tiredness, headache, nausea, vomiting, rash, low white blood cell count	Emerging viral disease in U.S. Midwest (first case in Bourbon, Kansas)

FIGURE 13.55 Arthropod vectors that carry viruses that infect humans. **(a)** *Aedes* spp. day-biting mosquitoes carry Zika, Chikungunya, dengue, and yellow fever viruses. **(b)** *Culex* spp. night-biting mosquitoes carry West Nile virus. **(c)** The blacklegged, or deer, tick, carries Powassan virus. **(d)** The lone star tick carries the Heartland virus. **(e)** The longhorned tick, not known to carry human pathogens in the United States, to date (2018).

diseases in the United States so far, but in China it may carry a virus that causes **thrombocytopenia syndrome** (which results in platelet deficiency, fever, vomiting, hemorrhaging, and organ failure, with a mortality rate of 10% to 30%).

Mosquitoborne Viruses

Mosquitoes are the best known disease vector. Mosquitoes can transmit infectious diseases between humans or from animals to humans. Mosquitoes ingest viruses (or other types of microbial pathogens that can infect humans) during a **blood meal** of an infected host (human or animal) and later inject the pathogens into a new host during the next blood meal. The female mosquitoes require blood meals (the male mosquitoes do not) and transmit pathogens that cause disease in its host.

○ Zika Virus Disease

Zika virus gets its name from the **Zika Forest** (**FIGURE 13.56**). The Zika Forest is located 25 kilometers (15.5 miles) from Kampala City, Uganda. It has a virus research field station that is part of the **Uganda Virus Research Institute (UVRI)**, which is involved in studying arthropodborne diseases. Besides its use by researchers, nature lovers are allowed to hike in the forest and observe wildlife in its natural surroundings. The Zika Forest is highly suited for bird-watching activities.

In the late 1940s researchers interested in isolating viruses harbored in mosquitoes native to the forest routinely trapped mosquitoes in forest treetops, and **sentinel rhesus monkeys** were placed in cages on treetop platforms. In 1947 one of the sentinel rhesus monkeys became sick. Serum from the sick rhesus monkey was injected into healthy mice. All of the mice became sick with similar symptoms, and Zika virus was isolated from *Aedes africanus* mosquitoes trapped in the Zika Forest.

FIGURE 13.56 The Zika Forest is both a research station for arthropodborne diseases as well as a wildlife sanctuary for the public.

The first human case of **Zika virus disease (ZVD)** was reported in Nigeria in 1954. Sporadic human cases of ZVD occurred in Africa, the Pacific, and Southeast Asia during the 1960s. In Micronesia, Yap Island, 185 human cases of ZVD were reported in 2007 and, in French Polynesia, 8,510 laboratory confirmed cases occurred during 2013–2014. The later outbreak was a cause for concern. Experts estimated there were closer to 29,000 cases in the French Polynesia outbreak because many infected individuals were asymptomatic. The Zika virus has two different genetic lineages: African and Asian.

Zika virus emerged from the northeastern part of Brazil at the end of 2014. Genetically similar mosquitoborne viruses that are transmitted by the same mosquito vector (A. aegypti), such as Chikungunya virus and dengue virus, are endemic to this region and were cocirculating during this epidemic. Within months, the number of Brazilians suffering from a rash, low-grade fever, headache, joint pain, and **conjunctivitis** (pink eye) showed a sharp increase. Most of these symptoms are similar to those of Chikungunya virus disease and dengue fever.

By the end of September 2015, an increase in the number of newborns with **microcephaly** was reported (**FIGURE 13.57**). Microcephaly is a medical condition in which the circumference of the head is smaller than normal because the brain is not developed properly or has stopped growing. Depending on the severity of this **congenital defect**, children may have impaired cognitive development, delayed motor development and speech, facial distortions, seizures, and difficulties with coordination and balance. Zika virus

infection was correlated with this increase in the unique birth defects, referred to as **congenital Zika syndrome**, found among fetuses and babies infected with Zika virus. There are varying degrees of congenital Zika syndrome. Congenital Zika syndrome is described as having the following five features or characteristics:

- Severe microcephaly in which the newborn's skull is partially collapsed.
- The child has decreased brain tissue, resulting in brain damage.
- The newborn's eyes are damaged (e.g., scarring behind the eye), resulting in vision impairment.
- The joints of the newborn have limited range of motion (e.g., clubfoot).
- Too much muscle tone restricts body movement after birth, challenging the baby's ability to sit independently, sleep, or be fed properly.

Zika virus infection was also associated with **Guillain-Barré syndrome** in adults. Guillain-Barré syndrome is a rare disorder in which a person's immune system damages the nerve cells of the body, resulting in muscle weakness and sometimes **paralysis**. Symptoms can last for a few weeks to several months.

Laboratory diagnosis of ZVD was challenging because there was no **gold standard** diagnostic tool. Serology assays were not accurate because Zika virus antibodies in patient blood samples cross reacted with antibodies in patient blood samples against Chikungunya, dengue, and other similar viruses that were cocirculating in the area.

Only 1 in 5 people suffering from Zika virus infection become symptomatic; thus, diagnosis based on symptoms was difficult. Many of the symptoms of ZVD overlap with other recent mosquitoborne diseases listed in **TABLE 13.14**. However, conjunctivitis (pink eye) was common in ZVD patients and not in the other viral mosquitoborne diseases cocirculating in the same location.

At the beginning of the epidemic in Brazil, no commercial diagnostic tests were available. Clinical specimens were sent to specialized labs (e.g., the CDC or WHO). The most accurate method to screen patients and the blood supply for ZVD is through NAATs. In 2017, commercially available NAATs were approved by the FDA; however, many labs associated with hospitals still do not perform NAATs. Patient blood samples are sent to special reference labs for diagnostics.

On November 12, 2015, the Ministry of Health of Brazil declared a Public Health Emergency of National Importance. It was estimated that there were between 440,000 and 1,300,000 ZVD cases in Brazil and 5,280 suspected cases of microcephaly, most of them in northeast Brazil. Women were encouraged to not get pregnant. It was also determined that males suffering from ZVD shed Zika virus in semen; thus, Zika virus could be transmitted

Baby with microcephaly

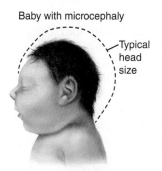

Baby with typical head size

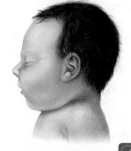

Courtesy of the CDC, National Center on Birth Defects and Developmental Disabilities.

FIGURE 13.57 Comparison of a newborn with a normal size head versus a baby with microcephaly.

TABLE 13.14 Disease Symptom Comparison of Emerging Mosquitoborne Diseases

Symptom	Zika Virus Disease*	Dengue Fever	Chikungunya Virus Disease	West Nile Encephalitis*
Fever	+++	++++	+++	++
Headache	+++	++++	+++	++
Joint pain	+++	+++	++++	++
Rash	+++	++	++	++
Conjunctivitis (pink eye)	+++		+	
Retro-orbital pain (pain behind eyes)	++	++	+	
Lymphadenopathies (inflammation of lymph nodes)	+	++	++	
Leukopenia and thrombocytopenia (decrease in white blood cells and platelets)		+++	+++	
Edema	++	+	+	
Hepatomegaly (enlarged liver)			+++	
Hemorrhage		+		

*About 1 in 5 people infected experiences disease symptoms.

Note: Plus (+) sign indicates a high percentage of cases exhibiting a specific symptom.

sexually. Some of the first ZVD cases during 2016 in the United States were indeed transmitted this way.

ZVD spread to the United States in 2015. The first cases were infected travelers returning from affected areas in South America. A few new infections were caused by sexual transmission from returning travelers with ZVD. The statistics for symptomatic ZVD cases in the United States and associated territories reported to the CDC from 2015 to August 1, 2018, are shown in **TABLE 13.15**. From 2015 to March 31, 2018, a total of 2,374 completed pregnancies in the United States were associated with possible Zika virus infections.

Of these newborns, 116 had congenital Zika syndrome; 9 miscarriages were linked to severe Zika virus–associated birth defects.

No antiviral therapies are available to treat ZVD nor vaccines to prevent it; vector control remains the only method to control and prevent the spread of ZVD (**FIGURE 13.58**). The ZVD epidemic has ended in Brazil for the time being. Several Zika virus vaccines are undergoing clinical trials.

Chikungunya Virus Disease

Before 2013, Chikungunya virus infections only occurred in Africa, Asia, and islands in the Indian and Pacific Oceans. Chikungunya virus is transmitted through the bite of an infected *Aedes* spp. mosquito. The first detailed description of **Chikungunya virus disease**, or **CHIK**, was based on a 1952 outbreak in Tanzania. This outbreak was followed by more outbreaks in Southeast Asia and a major epidemic that was reported in 1963 in Kolkata, India. After 1973, there was period of 32 years of dormancy in which no cases were reported. In 2005, Chikungunya virus reappeared, causing a large-scale epidemic in which 1.25 million CHIK cases occurred along the coastal region of Andhra Pradesh.

Krasner's Notebook

In the spring of 1999, during a field trip to the Amazon area of Brazil, our motor boat docked in a particularly remote area where there was a young native girl, obviously there to greet the tourists, with a monkey perched on one shoulder and a second monkey cradled in her arms. Within only several feet from her was a tree with, I would guess, about 15 monkeys. This girl was feeding the monkeys and playing with them as if they were her pets. I could not help but think that this could be the scenario in which yellow fever virus, Zika virus, HIV, Ebola virus, or other viruses made the species leap from monkey to human.

TABLE 13.15 ZVD Cases in the United States and Associated Territories

Time Period	Locations	Number of Symptomatic ZVD Cases Reported to the CDC	Number of Imported Cases (Infected Traveler Returning to the United States)	Vector/Mosquito Transmission (Natural Cases Acquired Locally)	Sexual Transmission
2015	U.S.	62	62	0	0
	U.S. Territories	10	1	9	0
2016	U.S.	5,168	4,897	224 Florida: 218 Texas: 6	47
	U.S. Territories	36,512	145	36,367	0
2017	U.S.	452	437	7 Florida: 2 Texas: 5	8
	U.S. Territories	666	1	665	0
2018 (January 1– August 1, 2018)	U.S.	34	34	0	0
	U.S. Territories	74	0	74	0

Information from Centers for Disease Control and Prevention. (2018, August 2). Zika cases in the U.S. reported to ArboNET. Retrieved from https://www.cdc.gov/zika/reporting/case-counts.html

In January 2014, U.S. authorities reported cases among travelers returning from the Caribbean (**FIGURE 13.59**). *In July 2014, the first acquired CHIK in a nontraveler was reported in Florida.* The number of CHIK cases reported to ArboNET in the United States since 2014 are listed in **TABLE 13.16**. No vaccine exists to prevent CHIK. The most effective way to avoid infection by Chikungunya viruses is to prevent mosquito bites. The *Aedes* spp. that transmits Chikungunya virus are day- and night-biting mosquitoes. More information on prevention strategies can be found at: https://www.cdc.gov/chikungunya/prevention/index.html.

Chikungunya virus gets its name from the Swahili word *kungunyala*, which refers to the stooped posture of patients suffering from severe joint pain. In the Bantu language Chikungunya means '*the one which bends up*,' which also refers to the posture of the patient who is experiencing excruciating pain in the joints.

© LuizSouza/Shutterstock.

© Joa Souza/Shutterstock.

© Sruilk/Shutterstock.

FIGURE 13.58 A glimpse into the Zika virus epidemic that began in Brazil in 2015. **(a)** Rio de Janeiro, Brazil, February 13, 2016. Man spraying insecticide at a construction site to kill mosquitoes that transmit Zika virus. **(b)** City of Salvador, Brazil, April 27, 2017. Child affected by congenital Zika syndrome being cared for. **(c)** Conjunctivitis (pink eye) is one of the common symptoms of ZVD that distinguishes it from other viral diseases such as Chikungunya virus disease and dengue fever that are transmitted by the same *Aedes* spp. mosquito vector.

Courtesy of the CDC.

FIGURE 13.59 During the 2014 outbreak of Chikungunya among Caribbean nontravelers, CDC entomologists collected mosquitoes for Chikungunya virus screening on the island of St. Croix, U.S. Virgin Islands.

TABLE 13.16 Chikungunya Virus Disease in the United States

Year	Travel-Associated Cases	Locally Acquired Cases
2014	2,799	12 (Florida)
2015	895	1 (Texas)
2016	248	0
2017	156	0
2018*	34	0

*As of August 7, 2018.
Information from Centers for Disease Control and Prevention. (2018, August 3). Chikungunya virus in the U.S. Retrieved from https://www.cdc.gov/chikungunya/geo/united-states.html

The natural reservoir host of Chikungunya virus is likely nonhuman primates and other vertebrates such as cattle and rodents. Symptoms of CHIK are high fever, rigors, headache, photophobia, and rash; most individuals complain of severe joint pain that is incapacitating. Joint damage affects mainly the extremities (hands, ankles, knuckles). The mortality rate is low (0.4%). In about 10% to 60% of patients, crippling and lingering **arthritis** symptoms may persist for up to 3 to 5 years. Patient management consists of managing the symptoms with medications to relieve pain and inflammation.

Dengue Fever

Dengue fever, also called **breakbone fever**, can result from infection caused by one of four different dengue viruses. The vectors are the *Aedes aegypti* and, rarely, *Aedes albopictus*. The recent appearance of *A. albopictus* in the United States is of great concern because this species is aggressive, and its biting habits could spread dengue fever, ZVD, and CHIK. Climate change will impact the range of *A. albopictus*, allowing these mosquitoes to flourish and spread upward into the Midwest and northern U.S. states.

Dengue fever is caused by dengue virus and is usually self-limiting; although the disease can be debilitating, recovery occurs in about 10 days. **Dengue hemorrhagic fever** is a serious and potentially fatal infection caused by a dengue virus strain different from the one causing the initial infection. The condition is manifested by hemorrhages occurring in the skin, gums, and other areas within the body. Shock may develop, requiring immediate treatment to prevent death, and even with intensive supportive measures as many as 40% of those infected may die. There is no immunization against any of the four dengue virus strains. Because the early characteristics of the disease are flulike, a definitive diagnosis is difficult and requires laboratory confirmation, either through virus isolation or through detection of virus-specific antibodies in the blood.

Dengue emerged as a worldwide problem in the 1950s. More than one-third of the world's population is living in areas at risk for infection by dengue viruses. As many as 400 million people are infected yearly. Dengue is endemic in many popular tourist destinations in Latin America, Southeast Asia, and the Pacific islands.

Travel-associated dengue fever and small outbreaks do occur in the continental United States; most dengue cases in U.S. citizens occur as endemic transmission among residents in some of the U.S. territories and Hawaii. Dengue fever was placed on the CDC *Morbidity and Mortality Weekly Report* (MMWR) list of "reportable diseases" on January 22, 2010, requiring laboratories and healthcare providers to report all U.S. cases of dengue fever to the CDC. There is no vaccine against dengue fever. As in all arthropodborne diseases, insect control is the best prevention.

Yellow Fever

Yellow fever, also called **yellow jack**, once had a widespread distribution, but mosquito-control measures resulted in elimination in many countries, including the United States. In 1793, a yellow fever epidemic devastated Philadelphia, America's early capital city, causing 2,000 deaths. The largest epidemic in the United States was in New Orleans in 1853, resulting in a death toll of 7,849; New Orleans was also the site of the last epidemic in 1905. The disease is still present in areas of South America, Central America, and Africa. In jungle areas monkeys serve as reservoirs, and the incidence of disease is highest in these areas because mosquitoes bite both monkeys and humans. The vector that is infected by yellow fever virus is *Aedes aegypti*; it bites by day and breeds in standing water,

Courtesy of Edwin P. Ewing, Jr./CDC.

FIGURE 13.60 The Panama Canal. Construction of the Panama Canal was possible only when yellow fever and malaria were conquered.

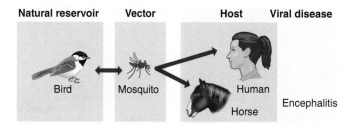

FIGURE 13.61 Cycle of transmission of West Nile virus.

such as in old tire casings. Yellow fever is manifested by fever, bloody nose, headache, nausea, muscle pain, (black) vomiting, and jaundice. In about a week the infected individual is either dead or in the process of recovery.

Immunization against yellow fever has been available since 1950 with live attenuated viruses. Protection is relatively long-lasting, and immunization is required for travelers to certain destinations.

The history of the Panama Canal is linked with yellow fever and malaria (**FIGURE 13.60**). The construction of the canal was an engineering triumph thwarted for years by the yellow fever virus and by a protozoan, the malaria parasite. Hence, its construction was also a triumph over microbes.

O West Nile Encephalitis

It appears that mosquitoes have a vengeance against New York. In late August 1999, on the heels of a St. Louis encephalitis outbreak, New Yorkers were plagued by yet another mosquitoborne virus, West Nile virus, first described in the 1930s in Africa. West Nile virus is carried by the same mosquito (*Culex* spp.) that transmits St. Louis encephalitis. In a period of only several weeks, 56 cases, including 7 deaths, had been reported in the New York City area. Mosquito-control measures, including aerial spraying, were quickly implemented.

West Nile encephalitis is a new and emerging disease in the United States. The virus's origins are in Egypt and Israel, and it is named after the Nile River. It is not certain how West Nile virus arrived in New York, but experts believe that it probably came in a bird or mosquito imported on a jet aircraft—another example of the pitfalls of technology.

Outbreaks of illness and deaths occurred in horses and in hundreds of crows in the area by 2000. At the Bronx Zoo several rare birds died and others became ill, including "a trumpeter swan doing the backstroke." A variety of mammals, including skunks, chipmunks, cats, dogs, and horses, also are susceptible. Mosquito

spraying abatement measures, along with the onset of cooler temperatures, were effective in breaking the transmission cycle (**FIGURE 13.61**).

In a matter of only several years the disease spread across the country and into Canada, Mexico, and the Caribbean. A CDC surveillance report summarizing West Nile virus covering January 1 to October 16, 2007, cited 42 states with a total of 3,022 cases of human West Nile encephalitis; 76 cases were fatal. Additionally, 1,924 birds (primarily crows, jays, and magpies) died as a result of West Nile virus infection in 34 states and in New York City.

A state of emergency was declared in Dallas, Texas, in 2012, during a West Nile encephalitis epidemic. The entire city was sprayed with mosquito insecticide by airplanes. It was the first time in 50 years that aerial spraying occurred in Dallas. Most cases occur in older adults, usually in the summer, when humans, mosquitoes, and migratory birds are in close proximity outdoors. Older patients are affected most severely. Worldwide, most cases are children and young adults.

West Nile virus has been detected in over 300 bird species, especially crows and jays, which can get sick and die from infection. *Mosquitoes become infected by biting infected birds.* Reporting of dead birds is a way to check for the presence of West Nile virus in the environment. Some predatory birds, such as hawks and owls, or scavengers, such as crows, may become infected after eating sick or dead birds that were already infected with West Nile virus (**FIGURE 13.62**). A list of all dead birds in which West Nile has been detected in the United States from 1999 to 2016 can be found at: https://www.cdc.gov/westnile/resources/pdfs/BirdSpecies1999-2016.pdf.

Human cases of West Nile encephalitis have been reported in all the continental United States. Cases occur during the summer and continue through fall. About 1 in 5 people infected with West Nile virus develops a fever and other symptoms. About 1 in 150 people develops a serious, sometimes fatal, illness. In 2017, 2,002 cases and 121 deaths were reported in the continental United States. California (509 cases), Texas (133 cases), and Arizona (109 cases) had the highest number of cases.

O La Crosse Encephalitis

La Crosse encephalitis is named after La Crosse, Wisconsin, where the first case occurred in 1963. Historically, most cases of La Crosse encephalitis were reported from

FIGURE 13.62 Reporting dead birds is a way to check for the presence of West Nile virus in the environment. Scavengers such as crows **(a)** and turkey vultures **(b)** or predators such as **(c)** hawks (e.g., red-tailed hawk) and **(d)** owls (e.g., long-eared owl) can become infected with West Nile virus through the consumption of sick or dead birds that were already infected with West Nile virus.

the upper Midwestern states of Wisconsin, Minnesota, Iowa, Illinois, Ohio, and Indiana. More recently cases have been reported in the mid-Atlantic and Southeast. Most individuals infected by La Crosse encephalitis virus are asymptomatic. About 80 cases are reported each year in the United States. Most cases of severe La Crosse encephalitis occur in children younger than age 16 who are bitten by eastern tree hole mosquitoes (*Aedes triseriatus*) in outdoor recreational areas. *A. triseriatus* lay their eggs in tree holes or manmade containers and typically bite during the day.

Mosquito Bite Prevention
The most effective way to prevent infection spread by mosquitoborne viruses is to prevent mosquito bites (https://www.cdc.gov/chikungunya/prevention/index.html). The following insect repellents contain at least one of the following active ingredients. When used according to manufacturer's directions, these are safe and effective, even for pregnant women and breastfeeding women:

- DEET
- Picardin (also known as KBR 3023; called icaridin outside the United States)
- IR3535
- Oil of lemon eucalyptus (OLE) or para-menthane-diol (PMD)
- 2-undecanone

The EPA provides insect repellent product information at: https://www.epa.gov/insect-repellents/find-repellent-right-you

Tickborne Diseases
Tickborne infections may be asymptomatic or cause mild to life-threatening illness. Viral tickborne diseases cause similar symptoms such as fever/chills, fatigue, lethargy, vomiting, headaches/migraines, myalgia (muscle pain), and distinctive rashes. Severe infections may progress to encephalitis or hemorrhagic fever, requiring hospitalization. No tests for viral tickborne diseases are commercially available.

In the United States, diagnostic testing using serology tests measuring virus-specific IgM antibodies (indicating a recent infection) in serum or cerebrospinal fluid (CSF) or NAATs to detect viral genomes in CSF or tissues or virus isolation in cell cultures must be performed at the CDC in Atlanta or selected state hygiene laboratories. Diagnosis can be difficult. It is usually based on clinical signs and symptoms and patient history (tick bites and tick exposure). Currently, no specific antiviral therapies are available to treat severe infections and no licensed vaccines are available to prevent any tickborne virus infections. Treatment is supportive care as appropriate.

Powassan Encephalitis
Between 2007 and 2017 only 75 cases of Powassan encephalitis were reported in the United States, but experts predict the number of annual cases will rise. Powassan virus is an emerging pathogen, named after Powassan, Ontario, where a young boy died of severe encephalitis in 1958. More recently, cases have been reported in the Great Lakes region (Minnesota and Wisconsin), New York State, and New England.

Powassan virus is transmitted by deer ticks (**FIGURE 13.63**). An unknown proportion of the human population infected with Powassan virus are asymptomatic or exhibit mild symptoms of Powassan disease. However, those who do manifest symptoms require hospitalization for encephalitis. Their symptoms include fever, vomiting, headache, confusion, rash, loss of coordination, speech difficulties, and confusion. *The mortality rate for Powassan encephalitis is 10% to 15% and about half of survivors suffer from permanent neurological symptoms, including memory loss, reoccurring headaches, memory problems, and muscle wasting and weakness.*

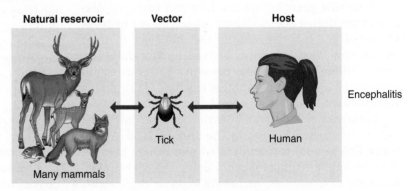

FIGURE 13.63 Cycle of transmission of tickborne diseases.

Prion Diseases

Prions are infectious proteins that cause a number of neurodegenerative diseases in animals and humans called **transmissible spongiform encephalopathies (TSEs)**. The first human TSE discovered was **kuru**. TSEs are characterized by long incubation periods of several years, but once symptoms begin the condition progresses rapidly and is fatal. There is no cure. Treatment is supportive. The prions do not evoke any host-specific immune response and are unusually resistant to disinfection. The prion disease that has created the most awareness is **bovine spongiform encephalopathy (BSE)**.

Mad Cow Disease

It all began in 1986, when an outbreak of BSE, better known as **mad cow disease**, in the United Kingdom frightened meat eaters around the world and raised the questions, "Can it happen here?" "Is it transmissible to humans?" In the United States, the mad cow disease scare was fueled by an Oprah Winfrey show about the disease and Oprah's statement, "*I will never eat another hamburger*." Television news shows wasted no time in showing pastures populated by mad cows drooling, stumbling, and unable to rise to their feet.

BSE is characterized by spongy degeneration of the brain accompanied by severe and fatal neurological damage. Cows suffering from BSE are referred to as **downer cows**. A downer cow is unable or unwilling to stand for longer than 12 hours. Other symptoms include:

○ Changes in temperament, such as nervousness and aggression toward other cattle or humans
○ Kicking when being milked
○ Incoordination
○ Difficulty walking
○ Head shyness (hanging head low)
○ High stepping hind leg gait
○ Skin tremors
○ Weight loss despite having a good appetite

Since the initial reports of BSE in 1986, according to the CDC over 180,000 cases have been confirmed in the United Kingdom. The disease spread to cattle in other European countries, primarily France, Switzerland, Portugal, and Ireland. More than 5 million cattle were slaughtered in the United Kingdom to halt the epidemic. Twenty-five infected cows have been reported in North America since 1993. The only classic case of BSE in the United States was imported from Canada.

Other forms of TSEs in addition to BSE are also characterized by spongy deterioration of the brain. **Scrapie** is a TSE of sheep and is manifested by animals scraping against trees, fencing, and whatever else might be available. TSEs that resemble BSE are also found in household cats, deer, elk, and mink, as well as in other ruminants (**BOX 13.2**).

Shors' Notebook

It was Thanksgiving week during my first year as an assistant professor at the University of Wisconsin–Oshkosh. I noticed that there were hardly any males in my classes that week. I was confused as I didn't notice a drop in female attendance as I would expect before a holiday. Naively, I asked colleagues about my class attendance observations, and the response was: "*Students are deer hunting during the week of Thanksgiving*." Deer hunting is a tradition that is deeply ingrained in the state of Wisconsin. Faculty also warned me that there are drops in attendance at Monday night classes because of Monday Night Football. Packer fans are strong and loyal through thick and thin; no matter if they are winning or losing. I was relieved to learn that I would never have to teach a Monday night class.

BOX 13.2 Where the Deer and the Antelope Play

Deer hunting is an age-old tradition, dating back to tens of thousands of years ago. According to the 2001 *National Survey of Fishing, Hunting, and Wildlife-Associated Recreation* (FHWAR), deer hunting is the most popular form of hunting in the United States. There were 10.3 million deer hunters in 2001. Nearly 1 in every 20 Americans and 8 in 10 hunters hunted deer in 2001, and their hunting-related expenditures while seeking deer totaled nearly $10.7 billion. At least 50% of hunters in all but a few states hunt deer.

The deer population is plagued by a new disease. Symptoms of the disease were first observed in mule deer grazing on northern Colorado wildlife research land in 1967. The sick animals had poor coats and appeared emaciated, starving, and "wasting away." Other symptoms of the sick mule deer were blank facial expressions, excessive drooling and thirst,

frequent urination, teeth grinding, nervousness, holding head in a lowered position, and sluggish behavior, and sick deer isolated themselves from the herd. In 1978, researchers named it **chronic wasting disease (CWD)** and classified it as a TSE. To date, CWD has been identified in 16 states: Colorado, Illinois, Kansas, Missouri, Wisconsin, Minnesota, New York, North Dakota, Utah, West Virginia, Nebraska, South Dakota, Oklahoma, New Mexico, Michigan, and Wyoming. It has also been identified in deer and farmed elk in the provinces of Saskatchewan and Alberta, Canada.

While human prion diseases are very rare, the incidence of CWD can be over 30% in wild deer populations and as high as 100% in captive deer herds. CWD prions are found in saliva, feces, blood, and urine of the infected animal. The stability of the prions makes it difficult to remove them from the environment

(continues)

BOX 13.2 Where the Deer and the Antelope Play *(continued)*

(e.g., soil, watering holes, and birthing sites). Transmission likely occurs through direct animal-to-animal contact or indirectly through contaminated water, carcasses, and food sources.

The catchphrase for returning home with deer and elk is *"no skull, no backbone."* This is based on the idea that the prions are concentrated in the nervous tissues, spinal cord, and antlers of the animal. Removing them would reduce exposure to prions. Hunters should take precautions by not eating brain, spinal cord, eyes, spleen, tonsils, or lymph nodes of deer. Proper field dressing removes most if not all of these body parts (**Figure 1**).

Because eating BSE-contaminated meat can cause Creutzfeldt-Jakob disease (CJD) in humans, what about eating CWD-contaminated meat? Between 1993 and 1999, three men who participated in wild game feasts in northern Wisconsin died of unknown degenerative neurologic illnesses. A CDC investigation confirmed one death to be caused by CJD. To date, there is no proof of an association between CWD and CJD. Research is ongoing to determine if there is a species barrier that protects humans from CWD. CWD is worrisome

because it is spreading through wild deer populations and could possibly make the species leap to other animals in the wild or in pastures, including livestock, squirrels, chipmunks, or even humans, as has happened with other diseases.

© Linda Freshwaters Arndt/Alamy Stock Photo.

Figure 1 Hunters field-dressing a deer.

Human TSEs

Creutzfeldt-Jakob disease (CJD), the model for the human TSEs, is transmitted in three ways. Sporadic cases occur throughout the world at a rate about one case per 1 million people, accounting for 85% to 90% of CJD cases. Another 5% to 10% of CJD cases are due to hereditary predisposition associated with gene mutation. Fewer than 5% of CJD cases are **iatrogenic**, meaning they are transmitted via contaminated surgical equipment, corneal transplants, or natural human growth hormone. The first human TSE was kuru, which was epidemic among Fore natives of New Guinea in the 1950s and 1960s. Kuru was spread by an ancient cannibalistic practice in which, as a sign of respect, Fore people, particularly females, ate the brains of deceased relatives, thereby infecting themselves with prions. This ritual has been abolished, leading to the disappearance of kuru.

In March 1996, a new form of CJD appeared. CJD develops at an average age of 65 years and has a mean

duration of illness of about 4.5 months, whereas the new form, **variant Creutzfeldt-Jakob disease (vCJD)**, affects individuals at an average age of 29 years and has a duration of illness of about 14 months (**TABLE 13.17**). Further, vCJD is transmissible through prion-contaminated beef. By November 2001, 106 cases had been identified in the United Kingdom, 4 cases had occurred in France, and 1 case had occurred in Ireland.

Disturbingly, on April 24, 2012, for the first time in 6 years, a United States Department of Agriculture (USDA)–confirmed case of BSE was reported in a cow in central California; tissue samples revealed "atypical BSE." This marked the fourth discovery of BSE in United States cattle, the first occurring in December 2002, followed by cases in 2005 and 2006, respectively. On a positive note, the finding indicates not only that the safeguards in place are effective, but also that there is the need for constant vigilance.

TABLE 13.17 Comparison of CJD and vCJD			
Disease	Average Age at Infection (years)	Duration of Illness (months)	Transmission
CJD	65	4.5	Sporadic (cause unknown), 85–90%; genetic mutation, 5–10%; iatrogenic, < 5%
vCJD	29	14	Prion-contaminated beef, 100%

BSE–vCJD Link

The most plausible explanation for vCJD is the contamination of beef products by CNS (brain and spinal cord) tissue. The cluster of vCJD cases is most likely due to the same agent that causes BSE. Support for this hypothesis is based on several lines of evidence, including an association between these two TSEs in time and place, a resemblance of pathological features in the brains of monkeys inoculated with either vCJD or BSE, and the nearly identical and distinctive distribution and pathology of the infectious agent in the brains of mice injected with infected cow BSE tissues or with human vCJD tissue.

Assuming that BSE and vCJD are caused by the same prion strain, how did all of this come about? The events that brought us to this stage are illustrated in (**FIGURE 13.64**). The widely held view is that the outbreak of BSE in cattle in 1986 in the United Kingdom was the result of the species leap of prions from sheep to cattle when prion-infected animal proteins were used as cattle feed to "beef up" milk production.

The term "animal protein" sounds innocuous, but animal protein is a concoction of ground-up carcasses of sheep and cattle, including their brains, spleens, thymus glands, tonsils, and intestines, a mixture called **offal**. In essence, cattle were eating the remains of other cattle, including downers (sick and dead animals found in the fields). Upton Sinclair's 1906 novel *The Jungle* describes conditions in "Packingtown" and the inclusion of downers in meats packaged for human consumption, a practice that continued for almost another 100 years. In 1989, the use of offal from cattle was banned in the United Kingdom. Bovine milk and milk products are not considered to pose any risk for transmission of TSEs.

But why did this outbreak of BSE suddenly occur in 1986? What changed on the farm or en route from farm to table? The most plausible explanation has to do with a change that occurred in the rendering process used in the production of food products for cattle. Rendering is a process somewhat like the making of stew, in which all ingredients are put into a pot and boiled. In the late 1970s the rendering process was altered by the elimination of the solvent-steam treatment phase. Solvents and steam reduce the infectivity of prions and, hence, the transmissibility of TSEs. The upshot, allegedly, is that the scrapie prion, in the absence of the solvent-steam treatment, survived the rendering process and contaminated the dietary supplement, which was then distributed throughout the United Kingdom, resulting in infected herds.

Experiments to study the rendering process and its effect on the inactivation of prions confuse the situation, because none of the different rendering processes completely inactivates prion infectivity. Paul Brown, a career investigator of TSEs at the National Institute of Neurological Disorders and Stroke, suggests that

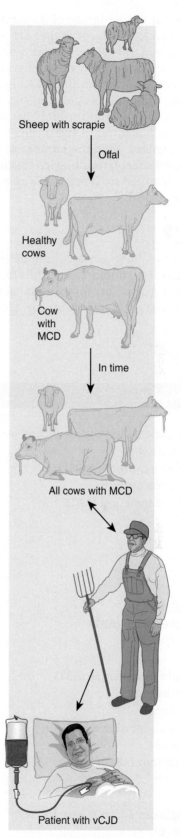

FIGURE 13.64 Transmission of spongiform encephalopathies.

even though the number of infectious scrapie particles may have been only modestly reduced by the solvent-steam treatment, the reduction may have been sufficient to make the difference if, in fact, the level of infectivity was at the borderline of the number of prions necessary to cause BSE in cattle that are fed offal. This might explain why the United States, and other countries that similarly modified the rendering process at about the same time as the United Kingdom, did not experience an outbreak of BSE. But why did the outbreak occur in the United Kingdom? Because the ratio of sheep to cattle is lower in the United States than in the United Kingdom, so is the number of prions (from scrapie-infected sheep), resulting in less than the number of prions required for infectivity. The explanation is confusing but logical.

Summary

Humans serve as hosts for a variety of viral diseases. Zika virus disease, Middle East respiratory syndrome, and Ebola virus disease are regarded as "new" infectious diseases, whereas rabies and plague are diseases of antiquity. The viral diseases discussed in this chapter are categorized on the basis of their primary mode of transmission: foodborne and waterborne, airborne, sexually transmitted infections, contact, and arthropodborne. Each of these diseases is described with attention to factors involved in the microbial cycle and pathogenicity.

Some viruses cause asymptomatic or mild infection, whereas others are almost always fatal; some are zoonoses, and some exhibit latency. Fever, muscle aches, and respiratory distress are common symptoms and are sometimes too vague to make a definitive diagnosis. Skin rashes accompany some viral infections and are significant in establishing a clinical diagnosis. Several viruses cross the placenta or are acquired by newborns during delivery and cause serious damage to the newborn.

Prions are infectious proteins that cause TSEs in humans and animals. TSEs are characterized by long incubation periods (years) but once symptoms start, the disease progresses quickly and is always fatal.

KEY TERMS

acquired immune deficiency syndrome (AIDS)
acute HIV infection
acute retroviral syndrome (ARS)
adaptive immune system
alanine aminotransferase (ALT)
antibodies
antiexcitatory drugs
antigenic drift
antigenic shift
antigens
antiretroviral drugs
antiretroviral therapy (ART)
arthritis
arthropod
aspartate aminotransferase (AST)
azidothymidine (AZT)
barrier technique
best available supportive care (BASC)
bifurcated needle
biosafety level four (BSL-4)
blender
blood meal
bovine spongiform encephalopathy (BSE)
breakbone fever
bronchiolitis
bulbar poliomyelitis
capsid
carrier

cauterization
central nervous system (CNS)
centrifugal rash
chickenpox
Chikungunya virus disease (CHIK)
chronic wasting disease (CWD)
climate change
clinical latency period
congenital defect
congenital viral infections
congenital Zika syndrome
conjunctivitis
contact tracing
containment books
convalescent blood
Creutzfeldt-Jakob disease (CJD)
cytokine storm
cytomegalovirus (CMV)
dengue fever
dengue hemorrhagic fever
developing countries
direct-acting antivirals (DAAs)
downer cows
dry phase
dumb rabies
dwell time
Ebola virus disease (EVD)
encephalitis

epicenter

epidemic

Epstein-Barr virus (EBV)

fecal–oral route

flyway

fomites

furious rabies

GamaSTAN S/D

Gardasil

Gardasil-9

gastroenteritis

generic antivirals

genital herpes

genital warts

gold standard

Guillain-Barré syndrome

H1N1 swine influenza A virus

H3N2 swine influenza A virus

H5N1 avian influenza A virus

H7N9 avian influenza A virus

hantavirus pulmonary syndrome (HPS)

healthcare-associated infections (HAIs)

hemagglutinin (H) spikes

hemophilia

hepatitis

hepatocellular carcinoma

herd immunity

herpes simplex virus (HSV)

herpesvirus

high-risk types of HPV

human immunodeficiency virus (HIV)

hydrophobia

iatrogenic

immune globulin

infantile paralysis

infectious diseases

infectious dose

infectious hepatitis

infective dose

influenza

intermediate host

intubation

iron lung

jaundice

Kaposi's sarcoma

Koplik's spots

kuru

La Crosse encephalitis

latency

latent state

life expectancy

live animal markets

longhorned tick

low-risk types of HPV

lymphadenopathy

maculopapular rash

mad cow disease

measles

melana

MERS-CoV

microbiota

microcephaly

Milwaukee protocol

measles, mumps, rubella vaccine (MMR)

MMRV (measles, mumps, rubella, varicella) vaccine

mode of transmission

monkeypox

monoclonal antibody

monocytes

mononucleosis

mumps

myalgia

naked viruses

needle-exchange programs

neuraminidase (N) spikes

noninfectious recombinant vaccine

nonmedical exemptions

nucleic acid amplification test (NAAT)

nucleosides

offal

opioids

opportunistic infections

Orphan Drug

Oxivir TB

Pacific flyway

pandemic

paralysis

paralytic poliomyelitis

patient zero

personal protective equipment (PPE)

petechial rash

Peyer's patches

plantar warts

pneumonia

Polio Global Eradication Initiative

poliomyelitis

postexposure prophylaxis (PEP)

postherpetic neuralgia (PHN)

premastication

prions

prodromal period

prophylactic antibiotic therapy

protease inhibitors

rabies

Rapivab
receptors
Relenza
replication cycles
reservoir
retrovirus
ring vaccination
route of entry
rubella
rumor registers
Sabin vaccine
Salk vaccine
Sin Nombre strain
SARS-CoV
scrapie
secondary bacterial infections
sentinel rhesus monkeys
Serious Communicable Disease Unit
serum hepatitis
70-70-60 plan
shingles
Shingrix
sialic acid
smallpox
smallpox recognition cards
species barrier
syncytium
T helper cells
Tamiflu
temperate zones
thrombocytopenia syndrome

TPOXX (tecovirimat or ST-246)
transmissible spongiform encephalopathies (TSEs)
Uganda Virus Research Institute (UVRI)
unsafe burial practices
U-shaped mortality curve
Valcite
variant Creutzfeldt-Jakob disease (vCJD)
varicella zoster virus
variola major
variola minor
Vessel Sanitation Program
viral envelope
viral load
virions
virulence factors
virus-like particles (VLPs)
wasting syndrome
West Nile encephalitis
wet phase
World Hepatitis Day
World Rabies Day
W-shaped mortality curve
yellow jack
zidovudine
Zika Forest
Zika virus disease (ZVD)
ZMapp
zoonotic disease
Zostovax
Zovirax (acyclovir)

SELF-EVALUATION

O PART I: Choose the single best answer.

1. Foodborne hepatitis is most commonly the result of infection by _____.
    a. noroviruses
    b. hepatitis virus A and E viruses
    c. hepatitis B virus
    d. hepatitis C virus

2. HSV-1 causes _____, and HSV-2 causes _____.
    a. cold sores; genital herpes
    b. fever blisters; cold sores
    c. fever blisters; shingles
    d. chickenpox; genital herpes

3. Zika virus disease can be transmitted via _____.
    a. contaminated hands
    b. other babies
    c. placental transfer
    d. intoxication

4. Chickenpox and shingles are caused by _____.

    a. varicella zoster virus

    b. papillomaviruses

    c. Epstein-Barr virus

    d. cytomegaloviruses

5. Which of the following pathogens has the longest incubation period?

    a. Ebola virus

    b. Rabies virus

    c. Norovirus

    d. MERS-CoV

6. Which of the following pathogens is the easiest to kill using disinfectants?

    a. Herpes simplex virus

    b. Norovirus

    c. BSE prions

    d. Human papillomavirus type 16

7. What is the natural reservoir for Ebola viruses?

    a. Skunks

    b. Mosquitoes

    c. Bats

    d. Camels

8. Which of the following in an HPV vaccine that can be used to prevent genital warts in boys and girls?

    a. Zostavax

    b. MMRV

    c. Gardasil

    d. Shingrix

9. Who was the lead author on the journal article (that was retracted) correlating autism and the MMR vaccine?

    a. Walter Reed

    b. Darin Dingdong

    c. Andrew Wakefield

    d. Louis Pasteur

10. Which virus caused a pandemic in 1918?

    a. Norovirus

    b. Hepatitis C virus

    c. SARS-CoV

    d. Influenza A virus

○ PART II: Match the statement on the left with the disease, vector, reservoir, or virus on the right. Not all letters are used.

1. Kissing disease

2. TSE disease of deer

3. Zika virus disease

4. Bloodborne pathogen

5. Common in infants younger than 6 months of age

6. Powassan encephalitis

7. Transmitted by rodents

8. Bats are reservoir

9. Cold sores

10. MERS-CoV

a. SARS-CoV

b. Rabies

c. CWD

d. RSV

e. Camels

f. Herpes simplex virus-1

g. Epstein-Barr virus

h. Hepatitis C virus

i. Deer ticks

j. mosquitoborne

k. Hantavirus pulmonary syndrome

○ PART III: Answer the following.

1. Discuss the reasons why the WHO program was able to eradicate smallpox.

2. Discuss why it would be difficult, if not impossible, to eradicate rabies virus.

3. "One bad apple spoils the barrel" is a popular expression. Discuss this expression and how it applies to vCJD.

4. Explain why postexposure rabies vaccination is able to prevent rabies.

5. Discuss the inherent problems associated with "chickenpox parties."

6. Why has there been an increase in measles cases in the United Kingdom and the United States?

7. Explain why contact tracing could not be used to prevent infections during the 2014–2016 Ebola epidemic.

8. Discuss reasons why experts believe that influenza A viruses have the potential to cause a catastrophic pandemic.

9. What is a zoonotic disease? List at least five viral zoonotic diseases.

10. How can viral diseases transmitted by mosquitoes be prevented? List at least five viral diseases transmitted by mosquitoes.

PROTOZOAN, HELMINTHIC, AND FUNGAL DISEASES

> "*In low-income countries, the main problems you have is infectious diseases.*"
>
> —Bill Gates
>
> *Founder of Microsoft and philanthropist*

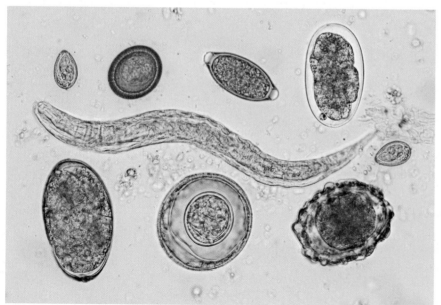

© Jarun Ontakrai/Shutterstock.

Helminth (worm) and eggs in human stool. The eggs of helminths are observed microscopically in order to diagnose helminthic diseases.

© ShutterStock, Inc. / happykanppy.

LEARNING OBJECTIVES

1. Evaluate the modes of transmission of five different diseases caused by helminths.

2. List three parasitic diseases that occur in both the United States and developing countries, and explain why these diseases have not been eliminated in the United States.

3. Compare and contrast protozoans and fungi.

4. Summarize how parasitic diseases transmitted by arthropod vectors can be prevented.

5. Explain why it is difficult to develop vaccines that prevent parasitic diseases.

6. Identify four lung infections caused by fungi that can afflict people with healthy immune systems and cause severe illness in people with weakened immune systems.

7. Generalize why CDC experts are concerned about *Candida auris* infections.

8. State the importance of cooking fish and meats thoroughly.

Case Study: Killer Bagpipes

"Neil" was a 61-year-old man who lived in England. He had a 7-year history of a dry cough and shortness of breath that was diagnosed as **hypersensitivity pneumonitis**, or **HP**. His condition was progressively worsening, so he was referred by his primary care doctor to a lung disease specialist. HP is an inflammation of the lungs caused by the body's immune response toward inhaled **antigens** (e.g., microbes). HP is associated with certain occupations and environmental exposures. It is most often seen in patients after the inhalation of bird proteins, causing **bird fancier's lung (BFL)**, or fungal spores (*Saccharopolyspora rectivirgula* and *Thermactinomycetes vulgaris*), causing **farmer's lung**. Quite often, doctors have a difficult time defining what microbes are causing HP. Treatment is based on managing symptoms of HP, such as the use of **steroids** to reduce lung **inflammation** and ease breathing.

As doctors tried to discern what was causing Neil's HP, they were able to rule out fungal infections related to exposure to birds or pigeons because he had no exposure to birds. The residence where he lived never had any water damage, which can result in mold contamination. He was not a smoker. Neil was treated with **prednisone**, a steroid used to reduce lung inflammation, which was restricting his breathing. During the 7-year period of illness, he had one 3-month period in which his symptoms abated. During this time, he was living in Australia. While living in Australia, his lungs improved immensely. He was able to walk 10 kilometers (6.2 miles) on a beach without stopping.

Upon Neil's return to England, once again his breathing rapidly deteriorated. Six months after his referral to a lung specialist, who was not able to determine the cause of the HP, Neil was admitted to the hospital, complaining of severe shortness of breath. Upon observation, he was **afebrile** (no fever), **hypoxic** (a dangerous condition in which the body tissues are deprived of oxygen), and **tachypneic** (shallow and rapid breathing due to an imbalance of carbon dioxide and oxygen in the blood). His sputum sample was sent to the clinical laboratory for testing. After microscopy analysis and culturing, sputum samples were negative for acid-fast bacteria (such as *Mycobacterium* spp., which can cause TB or other serious respiratory ailments, and *Nocardia* spp., which can also cause pulmonary disease). Culturing revealed normal microbiota and *Candida albicans*. A chest x-ray showed some fibrosis changes in his lungs from a prior medical examination (lung tissue damage and scarring).

He was treated with intravenous antibiotics and antifungal drugs because doctors suspected he had an underlying fungal pneumonia. They wanted to make sure he was treated for **atypical pneumonia** (pneumonia caused by an unidentified microbe) and *Pneumocystis jirovecii*. Neil's condition quickly worsened, and he died a month after being admitted to the hospital. His cause of death was attributed to a chronic history of HP and acute **interstitial lung disease** (sudden onset of labored breathing and respiratory failure).

It was not until after Neil's death that light was shed upon the mysterious origin of his lung disease. During

(continues)

Case Study: Killer Bagpipes (continued)

© Teri Shors.

Figure 1 Musicians who play bagpipes and other wind instruments, such as saxophones, trumpets, bassoons, or trombones, need to be aware of their risk for the development of hypersensitivity pneumonitis. Wind instruments need to be cleaned. Traditional bagpipes are made of hide and require regular "seasoning" to seal the pores of the skin. The products used to season the bags have antimicrobial properties. Bagpipes made from manmade or synthetic materials do not need to be seasoned, but they do require regular cleaning.

Neil's final admission to the hospital, his bagpipes were swabbed and cultured in the clinical laboratory. Neil's hobby was playing the bagpipes (**Figure 1**). During the time when he was in Australia, he did not play the bagpipes, which explains the reduction of lung problems or

a partial remission during that time period. When Neil lived in England, he played the bagpipes daily.

Several fungi present in Neil's bagpipes were cultured and identified. An air sample from the bag of the instrument contained *Rhodotorula mucilaginosa*, *Penicillium* species, and *Fusarium oxysporum*. The stock, or neck, of the bagpipe contained a mixture of *R. mucilaginosa* and *Trichosporon mucoides*; the chanter reed protector contained *Penicillium* species, *F. oxysporum*, and *R. mucilaginosa*.

This was the first case report of fungal exposure of a bagpipe player as a potential trigger for the development of HP. The moist environment of bagpipes provides conditions that can promote the growth of yeasts and molds, the microbial antigens that Neil was exposed to on a daily basis when playing the bagpipes. These offending antigens likely triggered the development of HP.

Cases of HP have been reported in saxophone, bassoon, and trombone players, as well as in pigeon breeders. **Table 1** identifies microbes cultured from wind instruments discussed in previous reports in the medical literature. Some of the owners of the instruments listed in Table 1 suffered from lower respiratory symptoms such as prolonged cough, including HP; asthma; and dyspnea.

Neil's doctors reported their findings and remind clinicians of the importance of patient history (e.g., Neil's hobby of playing the bagpipes) and that musicians who play wind instruments need a greater awareness of the importance

TABLE 1 Microbes Cultured from Wind Instruments

Instrument	Acid-Fast Bacteria (*Mycobacterium* spp.)	Non-Acid-Fast Bacteria	Fungi or Yeast
Saxophone	None	None	*Phoma* spp., *Ulocladium botrytis*, *Penicillium* spp., *Cladosporium* spp.
Trumpet	*Mycobacterium chelonae*, *Mycobacterium abscessus*, *Mycobacterium terrae*	Nonidentified Gram (–) rods	*Candida albicans*, *Fusarium* spp., *Penicillium* spp.
Trombone	*Mycobacterium chelonae*, *Mycobacterium abscessus*, *Mycobacterium intracellulare*	*Stenotrophomonas maltophilia*, *Escherichia col*, *Myroides* spp., *Elizabethkingia* spp., *Flavobacterium* spp., *Agrobacterium* spp.	*Fusarium* spp., *Candida albicans*, *Paecilomyces ilacinus*
Bassoon	*Mycobacterium chelonae*	None	*Phoma* spp., *Rhodoturola* spp.

Information from M. L. Metersky, S. B. Bean, J. D. Meyer, M. Mutambudzi, B. A. Brown-Elliott, M. E. Wechsler, & R. J. Wallace Jr. (2010). Trombone player's lung. *Chest, 138*, 754–756; F. Metzger. (2010). Hypersensitivity pneumonitis due to molds in a saxophone player. *Chest, 138*, 724–726; J. Moller, C. Hyldgaard, S. B. Kronborg-White, F. Rasmussen, & E. Bendstrup. (2017). Hypersensitivity pneumonitis among wind musicians—an overlooked disease? *European Clinical Respiratory Journal, 4*, 1351268.

of regularly cleaning their instruments to minimize their risk associated with the development of HP.

Questions and Activities

1. What is HP? What was determined to be the likely cause of the development of HP in Neil?

2. List the identified fungi that likely triggered Neil's lung hypersensitivity. Where are these fungi typically found?

3. What bacteria, fungi, and yeast have been cultured from wind instruments? List these microbes, and summarize where they are typically found in the environment.

4. Explain why HP is an occupational hazard of pigeon breeders and chicken farmers.

5. What is the typical treatment for HP? What are the negative side effects of this treatment?

6. How can musicians who play wind instruments decrease their risk of for developing HP?

Information from J. King, M. Richardson, A. M. Quinn, J. Holme, & Nazia Chaudhuri. (2016). Bagpipe lung: A new type of interstitial lung disease? *Thorax*, *72*, 380–382.

Preview

Malaria, leishmaniasis, African sleeping sickness, and giardiasis are examples of protozoan diseases. Ascariasis, river blindness, and tapeworm infection are examples of diseases caused by helminths (worms). Histoplasmosis, ringworm, and candidiasis are fungal diseases. Why are these diseases lumped together in this chapter? Protozoans are microscopic and unicellular microbes, whereas helminths are macroscopic and multicellular. Some fungi are microscopic and unicellular, whereas others are multicellular and macroscopic. However, protozoans, helminths, and fungi are all composed of eucaryotic cells, and this is the characteristic that links them together. *The helminths are not microbes but are included because some cause infection (sometimes referred to as infestations) with accompanying symptoms that challenge the immune system.* Many parasitic protozoans, helminths, and fungi exhibit complicated life cycles involving more than one host.

Neglected Parasitic Infections (NPIs)

Even though parasitic infections are typically associated with poor sanitation in low-income, developing countries, these infections also occur in the United States and other developed countries. **Neglected parasitic infections (NPIs)** are a group of five parasitic infections targeted by the Centers for Disease Control and Prevention (CDC) as priorities for public health action based on the number of people infected, the severity of the infection, and the ability to prevent and treat these NPIs. They are considered neglected because relatively little attention has been devoted to their surveillance, prevention, and/or treatment. They include Chagas disease, cysticercosis, toxocariasis, toxoplasmosis, and trichomoniasis. Here are a few fast facts:

- More than 300,000 people in the United States are infected with *Trypanosoma cruzi*, the cause of Chagas disease.

- At least 1,000 people a year are hospitalized with neurocysticercosis every year.

- Each year at least 70 people, mostly children, are blinded by toxocariasis.

- More than 60 million people are chronically infected with *Toxoplasma gondii*.

- Each year 1.1 million people are newly infected with *Trichomonas*.

Biology of Protozoans

Protozoans are eucaryotic microorganisms. The term "protozoan" is derived from the Greek *protos* meaning "first," and *zoon*, meaning "animal." Experts estimate that there are at least 36,400 species of protozoans, and their distribution is worldwide. They are found in freshwater, marine habitats, mud, drainage ditches, water-filled tires, in the guts of termites, and in soil.

Because they are unicellular, protozoans require that each cell bear the total burden of staying alive on its own; there is no cellular specialization and differentiation allowing for sharing of functions, as is the case in multicellular organisms. Like all eucaryotes (and as distinct from procaryotes), protozoans have a cell membrane, a membrane-bound nucleus, and other membrane-bound organelles within the cytoplasm. Some species have a protective rigid cover, the **pellicle**, outside the cell membrane; freshwater species continually take in water and eliminate it by contractile vacuoles that push water out. Most protozoans are heterotrophic and aerobic. Many ingest food particles by phagocytosis; the particles are then enclosed within food vacuoles in which digestion takes place.

Many protozoan pathogens have complicated life cycles involving more than one **host** and, in some cases, survival in nature outside of a host. **Encystation** allows for outside survival and is a process resulting in a dormant resting stage **cyst** surrounded by a thick capsule (**FIGURE 14.1**). In some helminthic life cycles, transmission between hosts takes place by ingestion of cysts. **Excystation** to the trophozoite form (the reproductive and feeding stage) occurs after ingestion of the cyst by a host.

Asexual reproduction by **binary fission** is the most common type of multiplication in protozoans.

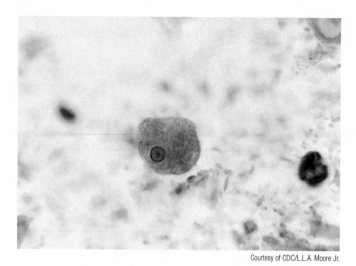

Courtesy of CDC/L.L.A. Moore Jr.

FIGURE 14.1 A photomicrograph showing ultrastructural details of an *Entamoeba histolytica* cyst.

A primitive form of sexual reproduction, called **conjugation**, occurs in some protozoans and allows for the direct exchange of genetic material during physical contact of mating pairs.

Relatively few protozoans are human pathogens; some are pathogens for other animals. Most protozoans are **free-living** and constitute a sizable portion of **plankton**, a primary food source for many aquatic organisms and the base of food chains. Although protozoan (and helminthic) diseases are more common in the tropics, they are also significant in temperate zones, particularly in areas with increased immigration and tourism.

Classification of the protozoans is complicated and controversial. Agreement between protozoologists on a universally acceptable classification is limited, but from a practical and medical point of view four groups are defined and are classified on the basis of their mechanism of **locomotion** (movement): sarcodina, mastigophora, ciliata, and sporozoa (**TABLE 14.1** and **FIGURE 14.2**).

TABLE 14.1 Classification of the Protozoans

Group	Locomotion	Disease(s)
Sarcodina	Pseudopodial or ameboid-like	Amebic dysentery
Mastigophora	Flagellate	Chagas disease, African sleeping sickness, giardiasis, trichomoniasis, leishmaniasis
Ciliata	Ciliary	Balantidiasis
Sporozoa	No locomotion	Malaria, babesiosis

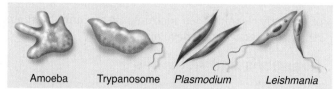

FIGURE 14.2 Illustrated examples of protozoans.

Protozoan Diseases

Protozoan diseases are organized by their mode of transmission, following the pattern for bacterial diseases and viral diseases (**TABLE 14.2**). **Climate change** is likely to increase the prevalence of waterborne infections caused by *Giardia spp.* and *Cryptosporidium spp.* in all countries. Increased rainfall and storm frequency, flooding, landslides, average temperatures, and extreme heat episodes are the environmental factors that will impact these two protozoan infectious diseases. Increased average temperatures will have the most impact on the malaria in the highlands in Africa and parts of South America and southeastern Asia. Malaria is a mosquitoborne infectious disease caused by protozoans of the genus *Plasmodium*.

Foodborne and Waterborne Protozoan Diseases

○ Giardiasis

Consider the following hypothetical scenario. A group of Boy Scouts returns from a hiking trip in the Rocky Mountains tired, hungry, and dirty but in good health and in good humor. Over the next 10 to 12 days disaster strikes, with several of the scouts falling ill with cramps, nausea, and diarrhea. The diagnosis is likely to be **giardiasis**, a common protozoan intestinal infection in the United States acquired by drinking contaminated water. If you are a hiker or a camper, beware of drinking from "pristine" mountain streams and other bodies of water. The days are long gone when you could drink directly from clear, cool, refreshing lakes and streams without running the risk of acquiring giardiasis, an infection caused by the flagellated protozoan *Giardia lamblia* (**FIGURE 14.3**). The intestinal disease caused by this parasite has been dubbed **beaver fever** after one of the protozoan's animal hosts.

Over the past 15 years giardiasis has been recognized as one of the most common waterborne human diseases in the United States, and it is the most commonly diagnosed intestinal parasite in public health laboratories. The parasite has been isolated from humans and a variety of animals, including dogs, coyotes, cats, cattle, horses, birds, and the eponymous beavers—all of which serve as reservoirs and pass cysts into water.

After excystation of the cysts to the vegetative and multiplying **trophozoites**, attachment to the lining of the intestine occurs by means of sucking disks. The parasites feed on mucous secretions and undergo reproduction, resulting in large populations. Typical symptoms include diarrhea,

TABLE 14.2 Major Protozoan Diseases in Humans

Protozoan	Protozoan Disease	Mode of Transmission	Body Site(s)
Foodborne and waterborne			
Giardia	Giardiasis	Water, direct contact	Intestinal tract
Entamoeba	Amebiasis	Water, food, direct contact	Intestinal tract
Cryptosporidium	Cryptosporidiosis	Water	Intestinal tract
Cyclospora	Cyclosporiasis	Water and food	Intestinal tract
Toxoplasma	Toxoplasmosis	Food, contact with cat feces resulting in fecal–oral transmission	Brain, heart, lungs; fetal transmission possible
Arthropodborne			
Trypanosoma brucei	African sleeping sickness	Tsetse fly	Blood
Trypanosoma cruzi	Chagas disease (South American trypanosomiasis)	Kissing bug	Heart
Leishmania	Leishmaniasis	Sand fly	Liver, spleen, mucocutaneous membranes, skin
Plasmodium	Malaria	Mosquito	Blood
Babesia	Babesiosis	Black-legged (deer) tick	Blood
Sexually transmitted infection			
Trichomonas	Trichomoniasis	Sexual contact	Urogenital tract

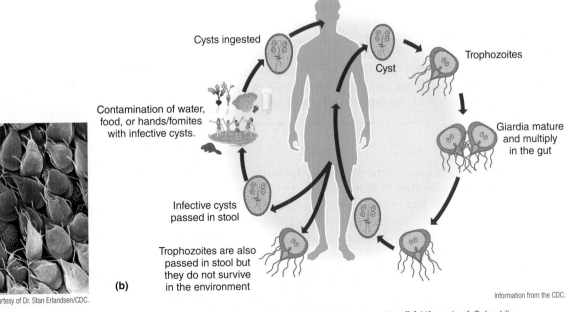

Courtesy of Dr. Stan Erlandsen/CDC.

Information from the CDC.

FIGURE 14.3 (a) Scanning electron micrograph of the small intestinal surface covered by *Giardia lamblia* trophozoites. **(b)** Life cycle of *G. lamblia*.

abdominal pain, and large amounts of gas. Although the diarrhea may be extensive, it is not bloody because the protozoans do not usually invade deeply into the lining of the intestine. Symptoms usually appear about 2 weeks after infection and may last 6 weeks or longer. Transmission occurs by swallowing water from contaminated swimming pools, water theme parks, lakes, rivers, shallow wells, springs, and even municipal water supply systems. *Giardia* is resistant to the chlorine concentration used in municipal systems, making control of giardiasis a major problem. Adding to the problem is that *Giardia* has a very low **infective dose** (or **infectious dose**)—as few as 10 cysts are enough to establish giardiasis. The protozoan parasite can be transmitted from contaminated environmental surfaces and person-to-person contact by the fecal–oral route, as might occur in day care centers. *Giardia* epidemics have broken out in day care centers, with as many as 70% of the children infected with parasites. The parasites may be accidentally acquired from contaminated toys and diaper-changing tables. An infected person may shed 1 to 10 billion cysts daily in their feces for several months. Good handwashing practices are essential. Fecally contaminated food is another source. It is imperative that uncooked vegetables and fruits be thoroughly washed in uncontaminated water. *Giardia* can survive for weeks or months outside of the body (Figure 14.3b depicts the life cycle of *Giardia*).

The finding of cysts or trophozoites is diagnostic for giardiasis, but this is problematic because the organisms may be shed intermittently, necessitating the examination of stool specimens over several days. Prior to the development of tests based on molecular biology, a person exhibiting the symptoms of giardiasis despite negative multiple stool examinations may undergo the Entero-Test, which is an alternative diagnostic procedure. The person swallows a lead sinker in a gelatin capsule attached to a string, and after 4 hours the string is withdrawn and the capsule is examined for the presence of trophozoites. New tests using probes for the detection of cysts in stool samples are available. A number of drugs are effective in treating this disease.

Giardiasis is a worldwide disease. Nearly 33% of people in developing countries have had giardiasis. *Giardia* can be found in every region of the United States and around the world.

Cryptosporidiosis

The largest outbreak of waterborne disease in the history of the United States occurred in Milwaukee in 1993. The source of the outbreak was the city's water supply system, which had become contaminated with human fecal material containing *Cryptosporidium parvum*. The episode resulted in about 100 deaths and 400,000 illnesses.

C. parvum is found in a variety of mammals, birds, and reptiles. Infectious **oocysts** are discharged into the water in fecal material, and their ingestion initiates infection. The oocysts undergo excystation to sporozoites that

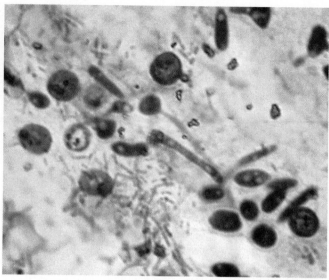
© Photo Researchers/Science Source/Getty Images.

FIGURE 14.4 *Cryptosporidium* oocysts (red), yeast stained blue-green. Three-step stool examination for cryptosporidiosis in 10 homosexual men with protracted watery diarrhea.

penetrate the intestinal cells where multiplication results in a new batch of oocysts (**FIGURE 14.4**). The release of these oocysts into the environment completes the life cycle. In patients with HIV/AIDS and in other immunocompromised hosts, potentially life-threatening diarrhea often occurs. Up to 25 bowel movements occur per day, resulting in a loss of over 10 liters (2.6 gallons) of fluid and severe dehydration. Other symptoms include weakness, fever, nausea, and abdominal pain. The incubation period ranges from several days to a few weeks. Outbreaks of cryptosporidiosis are difficult to prevent because oocysts are not inactivated by the usual doses of chlorination and can survive for days in treated water. As few as 10 oocysts are sufficient to cause disease. Further, field tests are not reliable; in tests of water samples intentionally seeded with oocysts, only about 10% were detected.

Cryptosporidium can be found throughout the world and in every region of the United States. The incidence of **cryptosporidiosis** in the United States has increased from 1 case per 100,000 people in 1999 to more than 3 cases per 100,000 people in 2008. The reason for this increase is unknown. Eighteen outbreaks occurred in 2006, and all involved public recreational water use. An outbreak of cryptosporidiosis occurred in 2009 in a summer camp in North Carolina; 40 cases were identified. The appearance of clean water does not, in itself, guarantee its safety, and the continued occurrence of outbreaks indicates the need for improved strategies of disinfection and constant vigilance.

Amebiasis

You might be fascinated by the movements of amoebas in a drop of pond water examined under the microscope. Unfortunately, these "cute" amoebas have disease-producing cousins. **Amebiasis** is primarily caused by the protozoan *Entamoeba histolytica* and occurs worldwide, especially in regions with poor sanitation. In the United States amebiasis is most prevalent in immigrants from developing countries, in people who have traveled to developing countries, in people who live in institutions where it is difficult to maintain good sanitary conditions, and in those who practice oral–anal sex.

Krasner's Notebook

In the spring of 1999, I was in Salvador, Brazil, studying tropical diseases. Despite the usual precautions of drinking only bottled water and avoiding salads, I became quite ill with diarrhea. I was determined not to give in because the last week of my trip was to be spent in the Amazon, and this was the chance of a lifetime. I completed my plans with some limitations. Upon returning home I was dehydrated to the point that hospitalization, and intravenous and oral rehydration were necessary. Ultimately, the diagnosis was giardiasis, a protozoan. I was disappointed because I could have acquired this much closer to home without going all the way to the Amazon, and further, I would rather have suffered from something much more exotic, like schistosomiasis!

The life cycle is initiated by the ingestion of cysts and excystation in the intestinal tract (**FIGURE 14.5**). Four trophozoites emerge from each cyst, move into the large intestine, and attach to the wall where they mature, multiply, and feed. About 90% of infected individuals remain asymptomatic or suffer from only mild disease characterized by diarrhea and stomach pain. Severe disease can result, however, accompanied by fever, bloody stools, and stomach pain due to destruction of the lining of the large intestine. Symptoms usually occur about 1 to 4 weeks after infection and include dysentery (bloody, mucus-filled stools), weight loss, fever, and fatigue. The dysentery may be severe enough to cause the rectum to extrude through the anus (prolapsed rectum). Sometimes amoebas invade the kidneys, skin, brain, spleen, and liver (**FIGURE 14.6**). Invasion of the liver can result in amebic hepatitis and liver abscesses. Several drugs are available for the treatment of amebiasis.

Amebiasis is one of the most common parasitic diseases worldwide and has a 10% fatality rate. An estimated 50 million cases occur each year, and more than 100,000

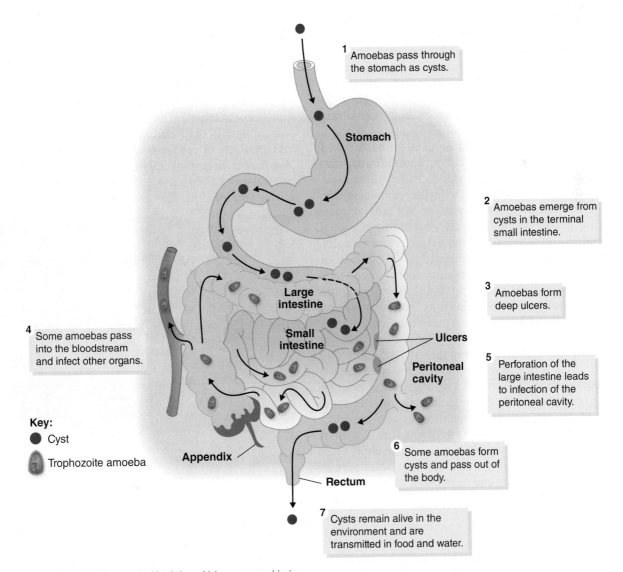

1 Amoebas pass through the stomach as cysts.

Stomach

2 Amoebas emerge from cysts in the terminal small intestine.

3 Amoebas form deep ulcers.

4 Some amoebas pass into the bloodstream and infect other organs.

Large intestine

Small intestine

Ulcers

Peritoneal cavity

5 Perforation of the large intestine leads to infection of the peritoneal cavity.

Key:
● Cyst
◯ Trophozoite amoeba

Appendix

6 Some amoebas form cysts and pass out of the body.

Rectum

7 Cysts remain alive in the environment and are transmitted in food and water.

FIGURE 14.5 Life cycle of *Entamoeba histolytica*, which causes amebiasis.

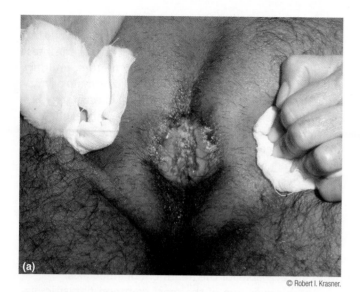

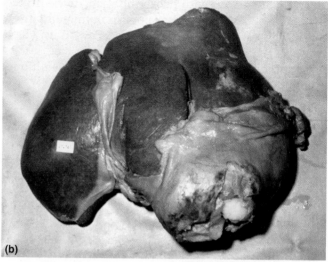

© Robert I. Krasner.

© Robert I. Krasner.

FIGURE 14.6 **(a)** Perianal cutaneous amebiasis. The amoebas have caused erosion of the tissues around the anus. **(b)** Amebic abscesses in the liver. Amoebas may invade from the intestine and produce lesions outside the intestine. Hepatic (liver) amebiasis, a serious disease, results from invasion of the liver by trophozoites.

deaths. In the United States amebiasis is relatively rare and frequently remains undiagnosed, indicating that there may be more cases of amebiasis than are reported. Fecal–oral amebic infections are more common in countries where sanitation is poor, and as many as 10% to 50% of the population in tropical countries may be infected with E. histolytica.

Diagnosis of amebiasis is based on finding trophozoites or cysts in an examination of fecal smears. It can be difficult to distinguish nonpathogenic intestinal amoebas from those that are pathogenic. Other, more definitive diagnostic laboratory tests are available to help in diagnosis.

As in the case of giardiasis and cryptosporidiosis, chlorination of water supplies may fail to kill amebic cysts. Practicing good personal hygiene, including thorough handwashing after using the toilet and before handling food, minimizes the risk. Individuals traveling to countries in which poor sanitary conditions prevail need to be particularly careful about their source of water, fruits, and uncooked vegetables; these items may have been washed with contaminated water.

Brain-eating amoebas refer to *Naegleria fowleri*, an amoeba that has caused several recent deaths (**FIGURE 14.7**). A few have been associated with the use of neti pots, devices that are filled with water that is forced into the nose and used to irrigate the sinuses. In these fatal cases tap water containing amoebas was used. The amoeba travel up the olfactory nerve and into the brain, where they then feed on brain cells. The condition is almost always fatal. Diving, water skiing, or other scenarios that propel water into the nose can also result in this form of amebiasis. However, *Naegleria* is of no consequence if swallowed in drinking water.

○ Toxoplasmosis

Toxoplasmosis is caused by the parasite *Toxoplasma gondii* and has a worldwide distribution. The disease is

a **zoonosis** and occurs in over 200 species of birds and mammals; members of the feline family (**FIGURE 14.8**), both domestic and wild, serve as the primary reservoir and host. Toxoplasmosis can be acquired by humans after the accidental ingestion of oocysts present in cat feces or by eating meat that contains cysts. Exposure to infective oocysts from cat feces usually occurs when people are careless in their handwashing habits and ingest minute amounts of cat feces containing oocysts in cat litter, sandboxes, or garden soil in which cats have defecated. In moist surroundings the oocysts are capable of remaining viable and infective for several months.

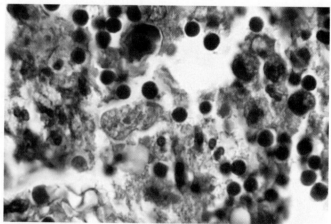

Courtesy of the CDC/Dr. George R. Healy.

FIGURE 14.7 The presence of brain-eating *Naegleria* in human brain tissue specimens. Infection with Naegleria occurs when the amoeba enters the body through the nose while the person is swimming underwater or diving. The amoeba then travels to the brain and spinal cord. Initial signs and symptoms include headache, fever, nausea, vomiting, and stiff neck. As the amoeba causes more extensive destruction of brain tissue, confusion, lack of attention to people and surroundings, loss of balance and bodily control, seizures, and hallucinations result. The disease progresses rapidly, and infection usually results in death within 3 to 7 days.

© Teri Shors.

FIGURE 14.8 Cats play an important role in the spread of toxoplasmosis. They become infected by eating infected rodents, birds, or other small animals. The parasite is then passed in the cat's feces. Kittens and young cats can shed millions of parasites in their feces for as long as 3 weeks after infection. Pictured is a Russian blue cat named Sasha, playing with her favorite string toy.

Additionally, humans may acquire toxoplasmosis through the ingestion of cysts in pigs, sheep, cattle, and poultry that have picked up oocysts in the soil. The oocysts develop into cysts in the tissue of these animals and may then be consumed by humans. Studies have indicated that pork and lamb are frequently contaminated with cysts. Eating raw or undercooked meats is a common cause of toxoplasmosis; cysts are killed by heating above 60°C (140°F).

The risk of transmission of *T. gondii* during pregnancy to the fetus is about 30%, and can result in stillbirth, neonatal death, or serious fetal defects, such as brain damage, convulsions, and retinal damage leading to blindness. *Pregnant women should not change a cat's litter box and should minimize the touching of cats because the possibility of picking up oocysts on the hands and transferring them into the mouth exists.* The complicated life cycle of *T. gondii* is illustrated in **FIGURE 14.9**.

Most cases of toxoplasmosis are asymptomatic or produce only mild symptoms, including sore throat,

1 Birds and rodents acquire the parasites from the soil.

2 The cat is infected when it consumes an infected bird or rodent.

3 The child is infected by contact with the cat or by the sand in the sandbox.

4 A woman is infected by contact with contaminated cat litter.

5 The fetus is infected by passage across the placenta.

6 Consumers are infected by eating contaminated and undercooked pork, chicken, or beef.

FIGURE 14.9 Life cycle of *Toxoplasma gondii.*

low-grade fever, and lymph node enlargement, making it difficult to assess the burden of the disease. In those individuals with HIV/AIDS or with immune systems weakened from other causes, however, toxoplasmosis is frequently a rapidly fatal disease resulting from massive invasion of the parasites into the brain. Diagnosis of toxoplasmosis is difficult because the disease resembles infectious mononucleosis.

Definitive diagnosis can be accomplished by isolating the parasites, by identifying them in samples of infected tissues, or by detecting the presence of antibodies in a blood sample. Drugs are available to treat the disease and may need to be taken for as long as 1 year to prevent recurrent infection. Toxoplasmosis is considered to be a leading cause of death attributed to foodborne illness in the United States. More than 30 million men, women, and children in the United States carry *T. gondii*, but very few have symptoms because the immune system usually keeps the parasite from causing disease. Infection is usually highest in low-lying areas in hot, humid, climates.

Arthropodborne Protozoan Diseases

○ Trypanosomiasis

African trypanosomiasis, also known as **African sleeping sickness**, is an ancient disease and one of many that have plagued Africa for millennia. The two types of African trypanosomiasis are named for the area of Africa in which they are present. West African trypanosomiasis is caused by *Trypanosoma brucei gambiense*, found in the rainforests of western and central Africa. The East African variety, caused by *Trypanosoma brucei rhodesiense*, is found in the upland savanna of East Africa. Both varieties are transmitted by the bite of infected tsetse flies, which are large and aggressive flies that inflict painful bites. The geographical distribution of the two varieties of African trypanosomiasis is a reflection of the different ecological niches occupied by the tsetse fly vectors.

The cycle of trypanosomiasis starts when a **tsetse fly** takes a **blood meal** from an infected reservoir host: wild animals (e.g., hyenas, lions, and wild pigs), domestic animals (e.g., goats, cows, and dogs), and humans are **reservoirs**. The **trypanosomes** multiply and migrate from the gut of the fly to the salivary glands, where further development takes place.

As many as 50,000 parasites can be injected by a single fly bite, far in excess of the approximately 500 that are required to establish infection. After the bite a red chancre (sore) develops at the site, and from there the parasites move into the blood, spinal fluid, lymph nodes, and brain. Early symptoms include fever, fatigue, swollen lymph nodes, and aching muscles and joints. As the disease progresses over a few months to a few years, the trypanosomes invade the brain and cause personality changes, progressive confusion, difficulty in walking, and altered sleep patterns, which become worse with time. Sleeping for long periods of the day (hence, the name "sleeping sickness"), accompanied by insomnia at night, is common. If the disease is untreated, death occurs within a few months to several years after infection.

Trypanosomes (like the influenza A and B viruses) exhibit **antigenic variation**—they change their "coats" as a defensive virulence strategy, resulting in evasion of the host's immune defense mechanisms. This property interferes with vaccine development. Estimates are that a few hundred new cases of East African trypanosomiasis are reported to the World Health Organization (WHO) each year. Over 95% of the cases of human infection occur in Tanzania, Uganda, Malawi, and Zambia. In the United States, there is one case on average per year, usually a traveler returning from a safari in East Africa.

West African trypanosomiasis is on the rise in areas of Sudan. In 1997, a team of epidemiologists examined almost 1,400 persons in 16 villages and reported the presence of the disease in every village surveyed. It is estimated that 10,000 new cases occur each year. Over 95% of the cases of human infection are found in Democratic Republic of Congo, Angola, Sudan, Central African Republic, Chad, and northern Uganda.

Trypanosoma cruzi is the causative agent of American trypanosomiasis, also called **Chagas disease**. The life cycle is similar to that of the trypanosomes that cause sleeping sickness. Wild animals, including rodents, opossums, and armadillos, serve as reservoirs for *T. cruzi*. The arthropod **vector** is a **triatomid insect**, commonly referred to as a **kissing bug**, that lives in thatched roofs, cracks in the mud walls, and dark places of adobe huts, close to humans (**FIGURE 14.10**).

The vector harbors the trypanosomes in its hindgut and has the nasty habit of defecating as it bites, causing itching and scratching, resulting in inoculation of the trypanosomes into the skin. The insect bites at night. *Trypanosoma* transmission is not limited to an arthropod vector. It can also be spread by blood transfusion and by organ transplantation. Trypanosomiasis should become less of a problem as in 2006 the U.S. Food and Drug Administration (FDA) approved a new test for the detection of antibodies against *T. cruzi*. The CDC estimates that 300,000 people living in the United States are infected with *T. cruzi*. Most people with Chagas disease in the United States acquired their infections in endemic countries.

Chagas disease occurs in Mexico and Central and South America, where an estimated 8 to 11 million people are infected, many of whom are unaware of their infection. The parasite causes slow, widespread tissue damage,

Krasner's Notebook

The photo in Figure 14.10b is of a row of traditional huts in Botswana, Africa, which has a high incidence of Chagas disease and kissing bugs. In a similar village, more than 50 bugs were isolated from a wooden bed frame, the only piece of "furniture" in the one-room hut. The bugs were examined for the presence of trypanosomes, and most of the insects were positive.

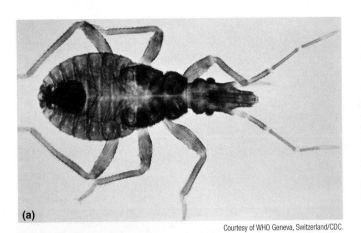

FIGURE 14.10 **(a)** A triatomid kissing bug, the vector of *Trypanosoma cruzi.* **(b)** Thatched-roof huts. This type of structure is an ideal habitat for triatomids.

particularly to the heart, causing it to enlarge and impairing its function, leading to heart failure. In about 1% to 2% of the cases the eyelid and the area around the eye become swollen, resulting in Romaña's sign (**FIGURE 14.11**). Death occurs within 30 years of infection if the disease is untreated. In addition to the usual diagnostic methods, **xenodiagnosis** is sometimes used; in this technique kissing bugs known to be free of the trypanosomes are allowed to feed on the individual suspected of having Chagas disease. After a few weeks the insects are examined for the presence of the parasites.

The following case study was described in *Kiss of Death: Chagas' Disease in the Americas* by Joseph William Bastien:

> Bertha (pseudonym) lives in La Paz, Bolivia, and her medical history provides insight into the effects of Chagas. She suffers from chronic heart ailments from Chagas disease.
>
> As a child living in the 1930s, she was bitten by vinchucas ("kissing bugs") and infected with T. cruzi when she lived in Tupiza, a small rural village in Bolivia. She later married and bore four daughters. In 1960, she moved to La Paz after her husband abandoned the family. She made a living sewing for wealthy people, but in 1974 she was diagnosed with Chagas disease.
>
> She tells the story of her life coping with Chagas. Until she was forty-four she was healthy, going up and down the hills of La Paz to do her sewing. In 1974, she felt fatigue. She began to get a swollen throat and spit blood. She didn't know what it was, she had no idea it had to do with the vinchucas bites years before. She would get tired, fatigued, and experience dizzy fainting spells. She continued to do her sewing though she sometimes would faint while she was working. The fainting spells continued for a year; the next year her fainting got more severe and she eventually suffered a stroke. Her children took her to a doctor, Dr. Jauregui, who hospitalized her. She underwent testing, xenodiagnosis, that indicated she had Chagas disease. X-rays showed that she didn't suffer from cardiomegaly (an enlarged heart), but that she probably had lesions in her heart's electrical system. These were caused by T. cruzi amastigotes being encysted in her cardiac tissue. This condition can be fatal.
>
> Dr. Jauregui implanted a pacemaker in 1980, when Bertha's heart rhythm worsened. The pacemaker keeps the heart rhythm constant and Bertha's condition improved. She was able to resume her seamstress work, although she suffered minor fatigue as she climbed the streets of La Paz at 12,000 feet. (Courtesy of the University of Utah Press.)

Chagas disease was endemic in Uruguay because blood was not screened before transfusion and kissing bugs were present in 80% of households. Through an intensive program of vector (kissing bug) reduction by both indoor and outdoor spraying and by replacing thatched roofs with metal, infection rates fell to below 0.1%, except in a single area. This was a decline of 80% in only 16 years. Uruguay was declared free of Chagas disease in 1997, a great triumph of public health measures and an inspiration to other countries that the burden of Chagas disease could be lifted.

FIGURE 14.11 Romaña's sign.

○ Leishmaniasis

Leishmaniasis is caused by several species of *Leishmania* and is endemic in many tropical and subtropical areas. Transmission occurs by the bite of female **sand flies** that become infected while taking a blood meal from wild and domestic animal reservoir hosts (**FIGURE 14.12**). In the Mediterranean region, including North Africa, jackals and dogs are important reservoirs, and dogs are significant reservoirs in Brazil. In India there are no animal reservoirs, and sand flies transmit the disease from human to human. Donor blood is not universally screened for the presence of *Leishmania*, thereby contributing to the problem.

When an infected sand fly bites a human, the **promastigotes** (extracellular form of *Leshmania* protozoans) enter the skin and are engulfed by immune cells called **macrophages** (**FIGURE 14.13a**). Macrophages are **phagocytes** and are of prime importance to the host immune system. Macrophages survey the body and engulf foreign invaders, in this case the promastigotes. Within the macrophages, the promastigotes are transformed into a nonmotile form of *Leshmania* protozoans called **amastigotes**. The amastigotes multiply and are released when a macrophage bursts. A sand fly taking a blood meal on a person suffering from leishmaniasis ingests the amastigotes, which develop into promastigotes within the sand fly (**FIGURE 14.13b**). The life cycle of *Leishmania* is illustrated in **FIGURE 14.13c**.

Three human manifestations of leishmaniasis are possible, based on the particular parasite species, the geographical location, and the host immune response. **Visceral leishmaniasis**, also called **kala-azar**, is the most severe form of the disease, with close to a 100% mortality rate within 2 to 3 years if not treated. About 90% of visceral

© Robert I. Krasner.

© Robert I. Krasner.

FIGURE 14.12 *Leishmania* reservoirs. **(a)** This photo was taken on a field trip into the Sinai Desert to search for rodent reservoirs. **(b)** *Psammomys obesus*, a rodent that can be a *Leishmania* reservoir.

leishmaniasis cases occur in parts of India, Bangladesh, Nepal, Sudan, Ethiopia, and Brazil. The parasites invade the liver and spleen and cause characteristic symptoms, including irregular bouts of fever, weakness, weight loss, anemia, and protrusion of the abdomen due to the swelling of the spleen and liver.

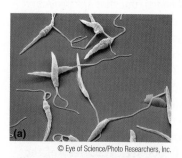

© Eye of Science/Photo Researchers, Inc.

Courtesy of James Gathany/CDC.

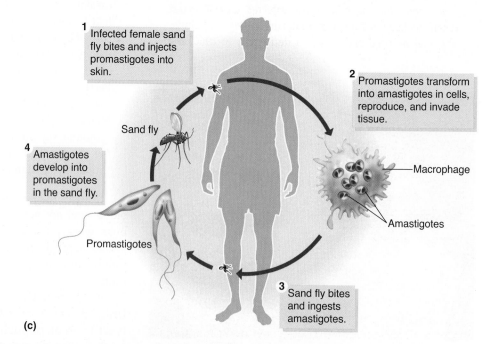

1 Infected female sand fly bites and injects promastigotes into skin.

2 Promastigotes transform into amastigotes in cells, reproduce, and invade tissue.

3 Sand fly bites and ingests amastigotes.

4 Amastigotes develop into promastigotes in the sand fly.

Sand fly

Macrophage

Amastigotes

Promastigotes

FIGURE 14.13 **(a)** A colorized scanning electron micrograph of *Leishmania* promastigotes. **(b)** *Phlebotomus dubosci*, a sand fly vector of *Leishmania*, ingesting amastigotes while taking a blood meal through human skin. **(c)** Life cycle of *Leishmania*.

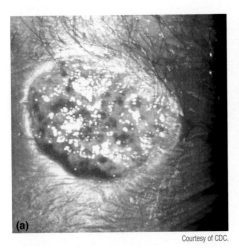
(a)
Courtesy of CDC.

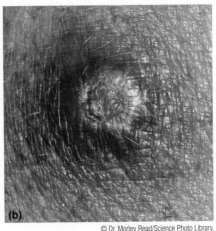
(b)
© Dr. Morley Read/Science Photo Library.

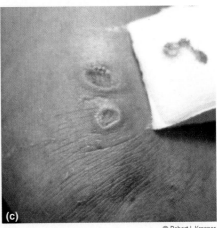
(c)
© Robert I. Krasner.

FIGURE 14.14 Cutaneous leishmaniasis. A variety of skin lesions are caused by infection with certain species of *Leishmania*. They are described as dry **(a)** or moist and diffuse, **(b)** raised, or **(c)** ulcerated.

The other two varieties of leishmaniasis are more localized and are seldom fatal. In mucocutaneous leishmaniasis, also called **espundia**, the parasites invade the skin and mucous membranes, causing destruction of the nose, mouth, and throat. **Cutaneous leishmaniasis** (**FIGURE 14.14**; also called **Oriental sore**, **Baghdad boil**, and **tropical boil**), results in mild to disfiguring skin lesions, primarily on exposed parts of the body, particularly the face, arms, and legs. Typically, only a few *Leishmania* lesions appear on an individual, but there may well be over 100. About 90% of cutaneous leishmaniasis cases occur in parts of Afghanistan, Algeria, Iran, Saudi Arabia, Syria, Brazil, Colombia, Peru, and Bolivia.

People with Oriental sores rarely get kala-azar, encouraging the practice of self-vaccination on inconspicuous areas of the body. Some parents purposely infect their children with *Leishmania* strains that cause Oriental sores as prevention against getting kala-azar. Although this practice may prevent cutaneous lesions, it does not always prevent kala-azar.

Estimates are that approximately 12 million people in the tropics and subtropics suffer from leishmaniasis. Further, 1.5 million new cases of the cutaneous form and half a million new cases of the visceral form are recognized each year. This disease is one of the most common infectious diseases among soldiers on duty in Iraq and Afghanistan. Military personnel serving in Afghanistan or Kuwait where the disease is also endemic are at risk as well.

○ Malaria
Ask almost anyone to name a tropical disease, and the chances are they will name malaria. The symptoms of malaria were described in Chinese medical writings dating back to 2700 B.C. Despite many years and many dollars spent on trying to control this disease, it remains a scourge on humankind. It is the world's biggest burden of tropical disease and, with the exception of tuberculosis, kills more people than any other **infectious disease**. The term **malaria** is derived from the Italian words *mala*

("bad") and *aria* ("air"). In the 1940s, the spraying of the insecticide **dichlorodiphenyltrichloroethane (DDT)** almost eradicated malaria, but it has since returned with a vengeance as a result of the emergence of DDT-resistant, other insecticide-resistant, and antimalarial drug-resistant strains.

Malaria is not limited to Africa, although the burden of malaria falls heavily on that continent (**FIGURE 14.15**). Gro Harlem Brundtland, a previous director general of WHO, stated, "*Malaria is hurting the living standards of Africans today and is also preventing the improvement of living standards for future generations. This is an unnecessary and preventable handicap on the continent's economic development.*" Studies have indicated that Africa's gross domestic product would now be substantially higher if malaria had been eliminated years ago. **World Mosquito Day**, August 20th, was established in 1897, when the association between mosquitoes and malaria transmission was discovered by Sir Ronald Ross (https://www.malarianomore.org.uk/world-mosquito-day). The day aims to create awareness about the causes of malaria and how it can be prevented.

Nearly 7 million lives have been saved since 2001 thanks to the intervention efforts of public health and political leaders where malaria remains a serious threat. **World Malaria Day** is April 25th (http://www.who.int/campaigns/malaria-day/2018/en/). The 2018 theme was "Ready to beat malaria." The CDC is leading a group of partners that aim to eliminate malaria transmission from Haiti and the Dominican Republic by 2020. Initially, support was provided by a $29.9 million grant from the Bill & Melinda Gates Foundation (http://www.malariazeroalliance.org/). Here are some facts about malaria:

- An estimated 3.2 billion people (almost half the world's population) across 91 countries or territories are still at risk of malaria. Except for India, all of these countries are in sub-Saharan Africa.

- Malaria killed 445,000 people in 2016.

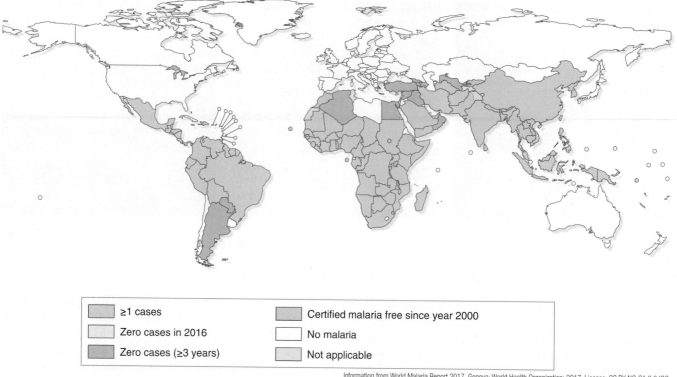

	≥1 cases		Certified malaria free since year 2000
	Zero cases in 2016		No malaria
	Zero cases (≥3 years)		Not applicable

FIGURE 14.15 Estimated global malaria risk: 2016.

- In 2016, 216 million people became ill with malaria (an increase of 5 million over the previous year). Malaria remains a major killer of children younger than 5 years, taking the life of a child every 2 minutes.

- The human and economic costs of malaria are estimated to be at least $12 billion per year.

- Malaria continues to be a threat to U.S. travelers, members of the U.S. military, and U.S. citizens living abroad, with more than 1,700 imported diagnosed each year in the United States.

- Two species of mosquitoes known to transmit malaria are still prevalent in the United States, posing a risk that malaria could be reintroduced.

- Many species of *Plasmodium* (which causes malaria) are resistant to antimalarial drugs.

- Some *Plasmodium*-carrying mosquitoes are insecticide resistant.

- To date, no licensed vaccine is available to prevent malaria.

Malaria is transmitted by the bite of female *Anopheles* mosquitoes infected with the *Plasmodium* protozoan parasite (**FIGURE 14.16**). The five most common species are *P. falciparum, P. malariae, P. ovale, P. knowlesi,* and *P. vivax.* Malaria can also be acquired by shared needles, blood transfusions, and infected mother-to-fetus transmission.

Many protozoans exhibit complicated life cycles, but the malaria life cycle is one of the most complex,

Courtesy of the CDC/James Gathany.

FIGURE 14.16 Adult *Anopheles* spp. mosquitoes are the known vector for malaria.

consisting of an asexual and a sexual stage. The sexual stage occurs in female *Anopheles* mosquitoes, and the asexual stage occurs in the liver and red blood cells of the infected individual.

When a *Plasmodium*-infected mosquito bites an individual, the infective forms of the protozoan, called **sporozoites**, pass from the salivary glands of the mosquito into the person's bloodstream. Within about an hour the sporozoites travel to the person's liver, where asexual multiplication takes place, resulting in forms known as merozoites that eventually leave the liver. A single sporozoite can produce thousands of merozoites. The merozoites enter **erythrocytes** (red blood cells), where further asexual multiplication occurs. The blood cells lyse, or burst, releasing more merozoites into the bloodstream, where they may infect other erythrocytes. If the mosquito bites while the *Plasmodium* merozoites are in the blood, the protozoa are ingested and develop into male and female **gametocytes** (sex cells), initiating the sexual cycle. This results in sporozoites that migrate into the mosquitoes' salivary glands. The life cycle is now complete (**FIGURE 14.17**).

After a *Plasmodium*-infected mosquito bites a person, symptoms usually appear in 2 to 4 weeks but may not appear for 1 year. While the merozoites are in the liver, the infected individual usually experiences no symptoms. In the symptomatic stage, the infected person experiences the characteristic shaking chills and fevers associated with malaria. The symptoms are caused by the destruction of large numbers of red blood cells accompanying the release of merozoites and their toxins into the bloodstream. This event is frequently synchronized during an attack, and many new merozoites are released at one time. It is as though the protozoa can tell time!

In the case of *P. malariae*, the chills and fever occur about every 72 hours, whereas in the other species these symptoms occur about every 48 hours. Several episodes of chills and fever constitute an attack (**paroxysm**); between attacks the individual feels normal. Muscle aches, fatigue,

diarrhea, and nausea may accompany the chills and fever. *P. falciparum* is the most dangerous of the five species, because it can cause cerebral (brain) malaria. Pregnant women have an increased susceptibility, resulting in lower birth weights, decreasing the chance of the baby's survival. This form of malaria is fatal for 1% to 2% of those infected.

Malaria can cause **anemia** (severe blood loss) if large numbers of red blood cells are lysed. The anemia, if severe, contributes to maternal deaths. Further, pregnant women who are infected with malaria and are positive for HIV are more likely to infect their unborn child with the virus.

Before the early 1900s, malaria was **endemic** in the United States, primarily in the Southeast. Massive **mosquito abatement programs** and a concomitant reduction in the number of human **carriers** were effective; the disease is no longer endemic. *Plasmodium* parasites and infected mosquitoes have been known to survive flights from countries where malaria is present and can transmit the disease in the vicinity of the airport, a phenomenon known as **airport malaria**. In 2005, 6 out of 30 U.S. students returning from East Africa as part of an educational program came down with malaria symptoms in a UK airport. They had been on a flight from Nairobi to London before their flight to the United States. This is a reminder to travelers to take **malaria prophylaxis**.

The diagnosis of malaria is often based on the detection of *Plasmodium* merozoites in red blood cells during

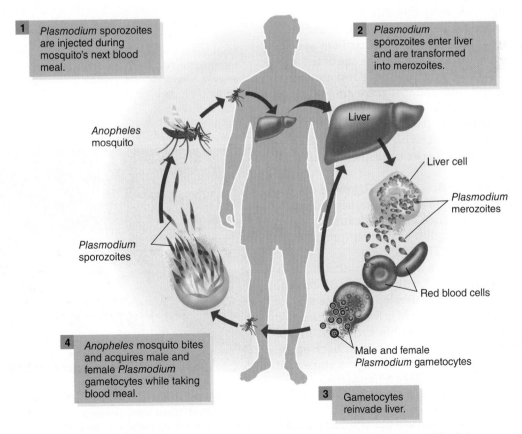

FIGURE 14.17 Simplified life cycle of *Plasmodium* sp. that causes malaria in humans, focusing on its human liver and blood stages.

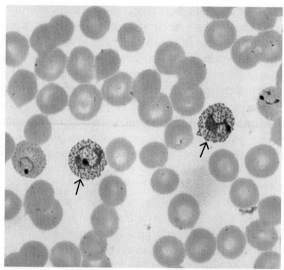

Courtesy of Dr. Mae Melvin/CDC.

FIGURE 14.18 Giemsa-stained blood smear showing *Plasmodium* merozoites inside of red blood cells. Arrows indicate the red blood cells in the center of the photo that are infected with the malarial parasite.

TABLE 14.3 Drugs Used to Treat or Prevent (Prophylaxis) Malaria	
Drug	***Plasmodium* species**
Atovaquone–proguanil combination (Malarone)	*P. vivax*
Artemether–lumefantrine combination (Coartem)	*P. falciparum*
Mefloquine (Lariam)	*P. falciparium*
Chloroquine	*P. malariae, P. knowlesi, P. falciparium, P. ovale*
Quinine	*P. falciparium*
Doxycycline (used in combination with quinine)	*P. falciparium*
Clindamycin (used in combination with quinine)	*P. falciparium*
Artesunate[a]	Severe malaria; patient unable to take oral medication
Primaquine[b]	*P. vivax, P. ovale*

[a] Not licensed for use in the United States but available through the CDC malaria hotline.
[b] Should not be taken by pregnant women or individuals with G6PD deficiency.
Information from Centers for Disease Control and Prevention. (2017). Malaria treatment. Retrieved from https://www.cdc.gov/malaria/diagnosis_treatment/treatment.html

microscopic examination of a drop of blood smeared onto a slide (**FIGURE 14.18**). Rapid test kits that provide results within 2 to 15 minutes detect antigens derived from the malaria parasites. The rapid kits are a useful alternative in situations where microscopy is not available. Malaria is a curable disease if promptly diagnosed and treated with appropriate drugs.

Several drugs are available for treatment and prevention; the particular drug prescribed depends on the circumstances, including the *Plasmodium* species identified, the severity of disease, and the age of the individual (**TABLE 14.3**). Antimalarial drug resistance is a serious problem, and international travelers and travel clinics need to be aware of this and appropriately advised. From January to March 2000, two U.S. citizens died from malaria as the result of receiving malaria-preventive drugs to which *Plasmodium* species in the area of their travels were resistant.

In several areas of Africa, such as Uganda, the use of **bed nets** regularly treated with recommended insecticides reduced the incidence of child deaths by about 30% (**FIGURE 14.19**). Mosquito abatement programs in malaria-endemic countries are vital.

The world eagerly awaits a vaccine against malaria. Efforts are ongoing, and there have been many contenders over the years. It is not easy to develop a vaccine against *Plasmodium* spp.; one of the problems is that the *Plasmodium* parasite changes forms during its life cycle as it moves along from the blood to the liver and then back again, exposing different protein surfaces as antigens, somewhat analogous to the way trypanosomes "change their coats."

Babesiosis

Babesiosis, sometimes described as a malaria-like parasite, is caused by the protozoans *Babesia microti* and *Babesia*

Courtesy of the CDC/BK Kapella; M.D.; (CDR; USPHS).

FIGURE 14.19 Bed net in Uganda. In 2013, the President's Malaria Initiative helped Uganda achieve universal coverage with 21 million, long-lasting insecticide-treated bed nets that can help save the lives of an estimated 53,000 Ugandan children.

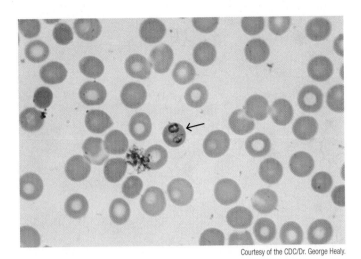
Courtesy of the CDC/Dr. George Healy.

FIGURE 14.20 Photomicrograph of a Giemsa stained blood smear. An arrow points to the presence of *Babesia* sp. inside of a red blood cell.

divergens. **Babesiosis**, or "the Babe" as some biologists call it, is transmitted by black-legged (deer) ticks. The ticks become infected by feeding on infected vertebrate hosts (rodents, cattle, and wild animals). The protozoa multiply and develop in the ticks and are transmitted to the next vertebrate host, where they invade the red blood cells. As in malaria, the protozoan undergoes cycles of multiplication and reinvasion of red blood cells (**FIGURE 14.20**).

Based on its similarity to malaria, babesiosis can be easily misdiagnosed as malaria in areas where malaria is endemic. Cases may be asymptomatic, but when symptoms occur they include severe headache, high fever, and muscle pain. Because red blood cells are destroyed, anemia and jaundice can also appear. The symptoms may last for several weeks, and in some cases a prolonged carrier state may develop. Almost all untreated cases of *B. divergens* babesiosis are fatal, whereas cases of *B. microti* babesiosis are seldom fatal in healthy persons. Avoiding tick bites is the best means of protection. Babesiosis is distributed worldwide; in the United States it has been most frequently identified in the Northeast and Midwest.

This section has presented several arthropodborne protozoan diseases. Although they are most common in tropical and semitropical areas, they remain a constant threat to other areas of the world, including the United States. Measures of control center around programs designed to control populations of the arthropod vectors involved and include spraying and avoidance of standing water in which mosquitoes breed.

Sexually Transmitted Protozoan Infections

Trichomonas vaginalis causes **trichomoniasis**. It is estimated that as many as 3.7 million people have the infection in the United States and close to 180 million cases occur annually worldwide, making this disease one of the most common sexually transmitted infections. Rarely, it is also transmitted by contaminated towels and articles of clothing. The human urogenital tract is the reservoir, and many infected persons, both female and male, are asymptomatic carriers. *Trichomonas* has no cyst stage and therefore cannot survive for long periods outside the host. It frequently accompanies infection with the bacterial diseases chlamydia and gonorrhea.

In females, symptoms include intense itching, urinary frequency, pain during urination, and vaginal discharge. In males, the disease may be characterized by pain during urination, inflammation of the urethra, and a thin milky discharge, but most infections are asymptomatic. Diagnosis is accomplished by direct microscopic examination of the discharge and searching for the presence of actively swimming **trichomonads**. The infection is successfully treated by the oral administration of appropriate antimicrobial drugs, and it is recommended that both sex partners be treated simultaneously to prevent "ping-pong" reinfection.

Biology of Helminths

The worms crawl in,
the worms crawl out,
in your belly,
and out your snout.

—Anonymous

This may sound like a cute little nursery rhyme, but in reality it describes a gruesome picture "and opens up a can of worms." Live worms of several species have been known to crawl out the mouth, nose, and umbilicus (belly button) and to be discharged into toilet bowls or onto the ground from the anuses of infected individuals, particularly in impoverished countries where poor sanitation is frequent. As is the case with bacteria and viruses, it should be emphasized that most **helminths** (worms) in nature are not parasites, but there are a number of worms that cause disease in humans, in other animals, and in plants.

The idea of a worm, a eucaryotic, multicellular animal, slithering around within your body is repugnant (**FIGURE 14.21**). All the bacteria, viruses, and protozoans discussed thus far are subcellular, procaryotic, or eucaryotic single-celled microscopic biological agents. They may, in fact, be more serious but are not necessarily viewed as loathsome *because* they are not visible. In other words, what you cannot see may not be so bad, but looking at worms is another story!

The helminths are divided into two morphological categories: **roundworms** (**nematodes**) and **flatworms**. Worms are at the systems level of biological organization and possess circulatory, nervous, reproductive, excretory, and digestive systems. Depending on the species, some systems may be lacking or may be rudimentary. The parasitic mode of existence of worms eliminates, or at least greatly minimizes, the need for a sophisticated digestive system, nervous system, and method of locomotion. They are surrounded by absorbable nutrients already digested, have little to react to, have no place to go in the quest for

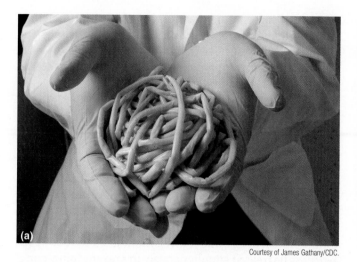

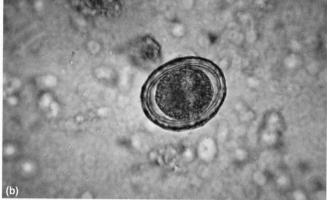

Courtesy of James Gathany/CDC.

Courtesy of the CDC/Dr. Mae Melvim.

FIGURE 14.21 **(a)** A handful of *Ascaris lumbricoides* worms, which had been passed by a child in Kenya, Africa. **(b)** Fertilized *Ascaris* roundworm egg. They are structurally round with a thick shell. Magnification 400×.

food, and are passively transferred from host to host. However, worms have complex reproductive systems, ensuring the production of very large quantities of fertilized eggs and maximizing the continuance of the species. In some species males and females are **dioecious** (the two sexes have different reproductive organs), whereas in others single animals are **hermaphroditic** and produce both sperm and eggs. Cross-fertilization or self-fertilization may occur, depending on the species. In some cases, their life cycles are very complex and include both larval and adult forms as well as multiple **intermediate hosts**.

Helminthic Diseases

Helminthic diseases are major problems in tropical countries. Infection with worms may result in death, but more frequently worm diseases are chronic and debilitating and may lower the quality of life in an entire community. In some areas, primarily in developing countries, as many as 90% of the population have at least one type of worm; they are almost like normal microbiota. Their presence in the United States and in other temperate areas is becoming of increasing concern because of immigration, travel, and HIV/AIDS.

Helminthic diseases are organized by their mode of transmission (**TABLE 14.4**). The mode of transmission varies with the particular species and may include food and water, arthropods, and direct contact. Climate change will play a significant role in the emergence or decline of helminthic diseases that affect humans.

Foodborne and Waterborne Helminthic Diseases

○ Ascariasis

Ascariasis is caused by *Ascaris lumbricoides*, large roundworms that resemble earthworms and set up housekeeping in the small intestine. Female *Ascaris* roundworms can grow to over 1 foot (30.5 centimeters) in length; adult

males are 8 to 10 inches (20 to 25 centimeters) long. An estimated 807 million to 1.2 billion people in the world are infected with *A. lumbricoides*. Ascariasis is now uncommon in the United States.

Infective *Ascaris* eggs are shed onto the soil in the feces of infected individuals. People become infected when they ingest food or water contaminated with the eggs. The eggs are very hardy; they resist drying out and thrive in warm, moist soils. Once in the intestine, the eggs hatch into **larvae** (immature worms), which penetrate the intestinal wall and begin their journey through the body. They enter the blood vessels and are carried along by the blood flow into the heart and then into the capillaries of the lungs, migrate up the respiratory tract, and enter the back of the throat. They are swallowed and returned to the small intestine, where they mature into male and female worms over a period of 2 to 3 months. Female *Ascaris* roundworms produce 200,000 or more eggs per day, and adults live approximately 2 years.

Many people experience no symptoms when infected with *Ascaris*, but in some people damage to the lungs may occur as the roundworms journey through the lung tissue, causing a cough and the threat of secondary bacterial infection. Rarely, adult worms may be coughed up or block the pharynx, causing suffocation. Further malnutrition may result from a large number of worms feeding on intestinal contents (**FIGURE 14.22**). When the roundworm burden is high, intestinal obstruction and perforation of the intestinal tract may occur, leading to death (**FIGURE 14.23**).

Krasner's Notebook

The photo in Figure 14.23a was taken in a hospital in San Jose, Costa Rica. The cause of death was due to intestinal blockage caused by a massive ball of entangled live and writhing worms, some of which had migrated into the bile duct and liver. During my professional career I have witnessed many gross events, but that scene may have been the worst.

TABLE 14.4 Major Helminthic Diseases

Helminth	Classification	Helminthic Disease	Mode of Transmission*	Body Site(s)
Foodborne and waterborne				
Ascaris sp.	Roundworm	Ascariasis	Food, water	Intestinal tract, lungs
Trichuris sp.	Roundworm	Trichuriasis	Food, water	Intestinal tract
Trichinella sp.	Roundworm	Trichinellosis	Pork consumption	Intestinal tract
Dracunculus sp.	Roundworm	Guinea worm disease (dracunculiasis)	Water	Skin
Taenia saginata	Flatworm	Taeniosis	Consumption of undercooked beef	Intestinal tract
Taenia solium	Flatworm	Cysticercosis and neurocysticercosis	Consumption of undercooked pork	Intestinal tract, brain
Diphyllobothrium spp.	Flatworm	Diphyllobothriasis	Raw fish consumption/ salmon sashimi	Intestinal tract
Vectorborne				
Wuchereria bancrofti and *Brugia* spp.	Roundworm	Lymphatic filariasis (elephantiasis)	Mosquitoes	Lymphatic system
Onchocerca	Roundworm	Onchocerciasis (river blindness)	Black flies (*Simulium* spp.)	Skin, eyes
Contact				
Toxocara cati (cat roundworms) or *Toxocara canis* (dog roundworms)	Roundworm	Visceral or ocular toxocariasis	Zoonosis; children have contact with cat or dog feces, eating contaminated soil	Eyes, multiple body organs
Ancylostoma/Necator	Roundworm	Hookworm disease	Contact	Intestinal tract, lungs, lymph
Enterobius sp.	Roundworm	Pinworm disease, enterobiasis, or oxyuriasis	Contact	Intestinal tract
Schistosoma sp.	Trematode worms or blood flukes	Schistosomiasis	Water contact	Skin, bladder, blood vessels
Strongyloides sp.	Roundworm	Strongyloidiasis	Contact	Small intestine, skin, bladder, small blood vessels

*Transmission may occur by more than one mechanism.

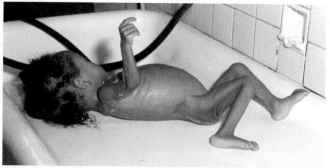

© Robert I. Krasner.

FIGURE 14.22 Malnutrition associated with ascariasis. Lack of food or a poor diet results in malnutrition, a problem particularly acute in less-developed countries. Most people with malnutrition are infected with one or more species of worms. The swollen belly of the child in the photo is characteristic of malnutrition accompanied by roundworm infection.

The presence of *Ascaris* eggs or roundworms in the feces is diagnostic (Figure 14.21b); an adult worm may be coughed up or passed in fecal material. A number of drugs are effective to rid the body of adult worms; they work by temporarily relaxing the roundworms, resulting in their passage through the anus.

Pet cats, dogs, reptiles, and birds (but not goldfish) that share our homes and sometimes beds are potential sources of *Ascaris* worms and other infections (**BOX 14.1**). Simple measures of personal hygiene, handwashing in particular, and veterinary care to maximize healthy pets minimize transmission from pets to people.

○ Whipworm Infection

Trichuris trichiura resembles a whip, hence its English name "whipworm." **Whipworms** are morphologically distinct from *Ascaris* but have a similar life cycle (**FIGURE 14.24a**). *Trichuris* eggs are passed in fecal material onto moist, warm soil and become infectious about 3 to 6 weeks after ingestion. The eggs hatch in the small intestine and release larvae that mature, migrate to the large intestine, and take up residence. After sexual maturation and fertilization, the females begin to lay eggs 60 to 70 days after infection and produce as many as 5,000 eggs per day, which can be shed onto the soil (**FIGURE 14.24b**). The life span of the adults is about 4 to 7 years.

In most cases **trichuriasis** produces no symptoms, but it can cause abdominal pain, diarrhea, and rectal prolapse, particularly in small children. Identification of whipworm eggs in feces is proof of infection. Drugs are available for treatment. Trichuriasis is the third most common roundworm disease of humans and is more frequent in the tropics and subtropics. It is estimated that 604 to 795 million people in the world are infected with *Trichuris* whipworms.

○ Trichinellosis

Beware of eating raw or undercooked pork, because it may be infected with *Trichinella spiralis* and cause **trichinellosis**. For that matter, if you like to eat wild game, including bear, boar, walrus, or seal, make sure the meat is well cooked. (Remember that the next time you order a walrus sandwich!) This parasitic nematode has a direct life cycle, meaning that there is no intermediate host.

When pork or other meat that contains *Trichinella* cysts is ingested, digestive juices dissolve the hard covering of the cysts and roundworms emerge in the small intestine (**FIGURE 14.25**). Maturation occurs in 1 to 2 days, and, after mating, adult females lay eggs that develop into larvae. Larvae migrate through the blood and lymph vessels and are transported to muscles, including the eye, tongue, diaphragm, and chewing muscles, within which they form cysts. The human host is a dead end for trichinellosis (barring that the person is consumed by a cannibal). Infected rats and other rodents are fed upon by meat-eating animals, including pigs, and thus play a major role as reservoirs in maintaining trichinellosis.

Early symptoms of trichinellosis occur within only 1 to 2 days after infection and include nausea, diarrhea, vomiting, fatigue, and fever, followed by later symptoms of headaches, chills, aching muscles, and itchy skin. If the

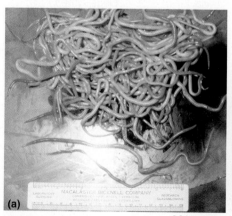

(a)

(b)

© Robert I. Krasner.

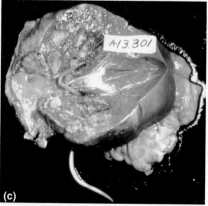

(c)

© Robert I. Krasner.

FIGURE 14.23 Intestinal obstruction due to *Ascaris* roundworms. **(a)** A tangle of *Ascaris* roundworms caused an intestinal blockage, resulting in death. **(b)** *Ascaris* roundworms invaded the gallbladder. **(c)** *Ascaris* roundworms invaded the liver.

BOX 14.1 Man's Best Friend

Most would agree that dogs are man's best friend (**Figure 1**). Dogs have been domesticated for thousands of years and have been used as sheep and cattle herders and as guard dogs. They conjure up visions of St. Bernard dogs trudging through deep snow in an effort to rescue stranded mountaineers, Dalmatians accompanying fire engines, and Collies named Lassie playing with children and alerting household members of a fire or other danger. They are known for their gentleness, loyalty, and affection. No matter how stressful a dog owner's day may be, upon returning home the owner will be relieved by the sight of an excited, happy animal anxious to bestow kisses. A pat on the head, a soothing "good dog," or a treat is all that is asked for in return. Dogs, of course, can be pests at times, such as when they demand to "go out" after having just "gone out."

More significantly, dogs (and other pets) can be a source of disease. Fortunately, most microbial diseases, including worms, that dogs acquire cannot be transmitted to humans, and, fortunately for dogs, most human microbial diseases are not suffered by dogs. Nevertheless, there are exceptions.

Most puppies are born with *Ascaris* and can be a threat to children because of the possibility of the accidental ingestion of *Ascaris* eggs; the eggs mature and subsequently migrate throughout the body. The incidence of this happening, however, is low.

A variety of bacteria, including *Salmonella, Campylobacter,* and *Escherichia coli* O157:H7 have been documented to be transmitted from pets to humans. To minimize the risk of pet-to-human transmission, practice good personal hygiene, particularly handwashing; deworm the dogs as necessary; clean up the area frequently where the dogs relieve themselves; and, most important, do not feed pets raw or undercooked fish or meats.

© Robert I. Krasner.

Figure 1 "Man's best friend." Dogs and other pets can be a source of bacterial, viral, worm, and fungal infections. This dog, Charlotte, owned by microbiology textbook author Robert I. Krasner is a Coton de Tulear, a relatively rare breed in the United States.

---

burden of worms is heavy, the individual may experience cardiac and respiratory problems; in severe cases death may occur. Mild cases may never be diagnosed and are limited to flulike symptoms.

The incidence of trichinellosis caused by the ingestion of undercooked pork is now relatively rare in the United States. Worldwide, an estimated 10,000 cases of trichinellosis occur each year. According to the CDC, in the late 1940s, an average of 400 cases and 10 to 15 deaths occurred in the United States each year; from 2008 through 2015 the number declined to 15 cases per year, with most cases associated with eating wild game.

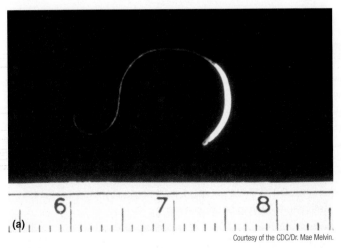

(a)

Courtesy of the CDC/Dr. Mae Melvin.

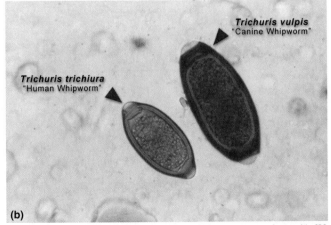

Trichuris vulpis "Canine Whipworm"

Trichuris trichiura "Human Whipworm"

(b)

Courtesy of the CDC.

FIGURE 14.24 (a) Photomicrograph of an adult *Trichuris* sp. female human whipworm that is approximately 4 centimeters (1.6 inches) in length. Note that it resembles a whip. **(b)** Photomicrograph of a human and a dog whipworm egg.

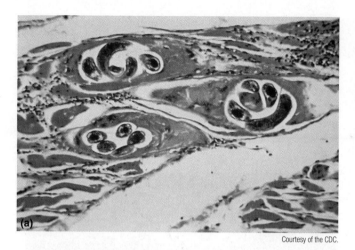

Courtesy of the CDC.

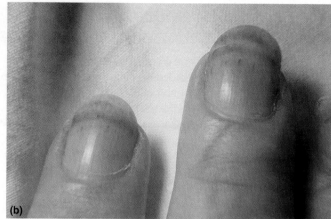

Courtesy of Dr. Thomas F. Sellers/Emory University/CDC.

FIGURE 14.25 (a) *Trichinella* cysts within human muscle tissue. **(b)** Splinter hemorrhages under the fingers of a patient suffering from trichinellosis.

The best way to prevent trichinellosis is to use a food thermometer to measure the internal temperature of cooked meat. The United States Department of Agriculture (USDA) recommendations for meat preparation are presented in **TABLE 14.5**.

Curing (salting), drying, smoking, or microwaving meat will not consistently kill infective *Trichinella* roundworms. Homemade sausage and jerky have been the cause of trichinellosis reported to the CDC in recent years. Freezing pork less than 6 inches (15 centimeters) thick for 20 days at 5°F (−15°C) will kill *Trichinella* pork roundworms. However, freezing wild game meats, unlike freezing pork, may not effectively kill the *Trichinella* roundworms, because some *Trichinella* species that infect wild game animals are resistant to freezing. Pigs or wild animals should not be allowed to eat uncooked meat, scraps, or carcasses of any animals, including rats, which may be infected with *Trichinella*.

The decline reflects a combination of factors, the primary being the passage of laws prohibiting the feeding of **offal** (the organs and trimmings of butchered animals),

raw meat, and uncooked garbage to pigs. It is therefore highly unlikely that trichinellosis will be acquired from pigs raised in the United States. Awareness on the part of the public regarding the danger of eating raw or undercooked pork or wild game has also contributed to the decline of trichinellosis. Infection occurs worldwide but is mostly found in areas where raw or undercooked pork and pork products (ham and sausage) are consumed. A muscle biopsy and a blood test can reveal the presence of trichinellosis. Several drugs with limited effectiveness are available to treat this infection.

○ Dracunculiasis (Guinea Worm Disease)

Dracunculiasis, also called **guinea worm disease**, is caused by the parasitic roundworm *Dracunculus medinensis*, which is present in poor communities in Africa that lack access to safe drinking water. The life cycle is initiated by drinking water contaminated with *Dracunculus* larvae; small **copepods** (water fleas) feed on these larvae. Copepods are small crustaceans in the phylum *Arthropoda*. When someone ingests water containing infected

TABLE 14.5 Meat Safety Temperatures and Trichinellosis Prevention

Meat(s)	Type of Meat Preparation	Temperature
All meat, excluding poultry and wild game	Whole cuts	145°F (63°C) with food thermometer placed in thickest part of the meat and allowed to rest for 3 minutes before carving or consuming.
All meat, excluding poultry and wild game	Ground meat	Cook to at least 160°F (71°C). No rest time required.
Poultry	Whole cuts and ground meat	Cook to at least 165°F (74°C), and for whole poultry allow to rest for 3 minutes before carving or consuming.
Wild game	Whole cuts and ground meat	Cook to at least 160°F (71°C).

Information from Centers for Disease Control and Prevention. (2012). Trichinellosis FAQs. Retrieved from https://www.cdc.gov/parasites/trichinellosis/gen_info/faqs.html

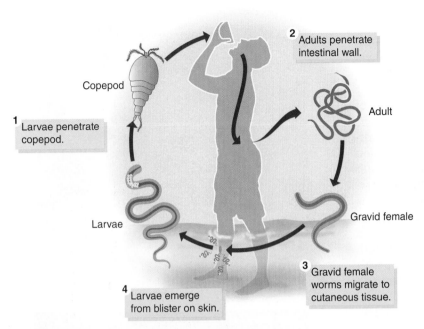

FIGURE 14.26 Life cycle of guinea worms in humans.

Copepod

1 Larvae penetrate copepod.

2 Adults penetrate intestinal wall.

Adult

Gravid female

3 Gravid female worms migrate to cutaneous tissue.

4 Larvae emerge from blister on skin.

Larvae

copepods, the copepods are digested but not the worm larvae, which penetrate the stomach and intestinal wall and enter the abdominal cavity (**FIGURE 14.26**).

Over the next year, the larvae mature into adult worms, copulation takes place, the male worms die, and the females migrate toward the skin surface, most commonly the foot or the leg. A painful blister that eventually ruptures occurs at the site. In many cases, persons immerse the blistered area in water in an attempt to gain some relief from the pain. The temperature change can induce the blister to open, exposing the adult worm (**FIGURE 14.27a**). The gravid (i.e., "pregnant" with internal eggs or larvae) female releases a milky white fluid that contains millions of larvae into the water, a process that continues each time the worm comes in contact with

water. The larvae are ingested by copepods and mature in 2 weeks (and after two molts) into infective larvae, completing the cycle.

Symptoms of infection occur about 1 year after the ingestion of water contaminated with infected copepods. A few days to several hours before rupture of the blister and emergence of the worm, fever, pain, and swelling in the area occur. After the rupture the meter-long worm begins to enzymatically bore its way out through the skin, cutting the flesh during its migration and causing excruciating and debilitating pain (**FIGURE 14.27b**). The worm is the diameter of a paper clip or a piece of thin spaghetti so it cannot be pulled out. The risk is too great that part of it will be left inside the leg, leading to a potentially life-threatening bacterial infection. The most you can do is coil the worm around a small stick as it is expelled from the blister and hope the agony will end in 2 to 3 weeks and not drag on for as long as 3 months.

Diagnosis is made clinically and does not require laboratory confirmation. Difficulty in walking, loss of appetite, and exhaustion are common symptoms. No antihelminthic medication is available to prevent or end the infection. Surgical removal before the formation of a blister is possible, assuming medical care is available. Healing at the blister site takes place in about 8 weeks, but **secondary bacterial infections** and permanent crippling and disability may result.

In addition to the human misery associated with dracunculiasis, the disease poses an economic and financial burden for families. While the worm is emerging, an infected person is often unable to work or go about daily activities for a few months. The inability to farm and to harvest results in heavy crop losses, and children are often kept out of school to work in the fields.

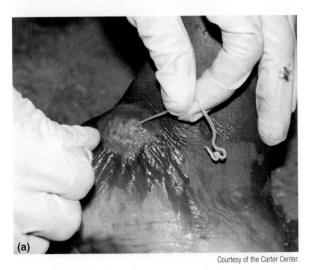

Courtesy of the Carter Center.

Courtesy of the Carter Center.

FIGURE 14.27 (a) After a female guinea worm matures, it induces formation of a blister at the site from which it will emerge. **(b)** A worm is removed from a young girl's foot, a painful process that can take several weeks.

Once the larvae are ingested, guinea worm disease cannot be cured. *Prevention is based on teaching villagers in affected communities, many of whom are illiterate, to filter their water through cloth or nylon filters to rid the drinking water of copepods.* Additionally, people need to avoid seeking relief from the burning sensation of emerging worms by immersing their feet or other body areas in reservoirs of drinking water.

Guinea worm disease is on a path toward eradication. In 1986, the dracunculiasis afflicted 3.5 million people a year in 21 countries in Asia and Africa. Former President Jimmy Carter took the initiative in 1986 in marshaling global resources against dracunculiasis and was knighted in Mali, Africa, in April 1998 in recognition of his efforts. Since then, a coalition of organizations, initiated by the Carter Center (https://www.cartercenter.org/health/guinea_worm/index.html) and the CDC, in concert with the WHO and UNICEF, the World Bank, the Bill & Melinda Gates Foundation, and numerous bilateral and multilateral agencies, private corporations, and foundations, is working toward the eradication.

Guinea worm disease is set to become the second human disease eradicated (smallpox, a viral disease, was the first human disease eradicated from Earth). As of April 2018, a total of 187 countries are certified free of transmission, and only three cases have been reported to the WHO. Chad, Ethiopia, and Mail are the only endemic countries. It will be the first parasitic disease eradicated, and the first disease to be eradicated without a vaccine or medication. *Dog infections with guinea worm pose a challenge to the program.* In 2017, more than 800 dogs in Chad, 11 dogs in Ethiopia, and 9 dogs in Mali were reported with guinea worm emergence.

Education about filtering drinking water and avoidance of entering water when worms are emerging, treating water sources with compounds to kill copepods, and the provision of clean water are the necessary steps to eradicate guinea worm. Further, the provision of clean drinking water is essential.

○ Tapeworms: Cysticercosis, Taeniosis, and Diphyllobothriasis

The two most common species of tapeworm are *Taenia saginata* (the beef tapeworm) and *Taenia solium* (the pork tapeworm). Their life cycles are similar and are initiated by the ingestion of undercooked beef or pork containing encysted larvae (**FIGURES 14.28** and **14.29**) or **cysticerci** (the larvae of tapeworms with a scolex head retracted into a bladderlike structure; a bladder worm). The beef tapeworm is the larger of the two and may reach 20 to 25 feet (6 to 7 meters) in length, whereas the pork tapeworm reaches approximately 15 to 20 feet (4.5 to 6 meters) in length. An individual with a tapeworm might have a 20-foot animal dangling into their intestinal tract! The renowned parasitologist Asa Chandler had a tapeworm he named Homer.

Infection is acquired by eating meat (muscle) containing larvae that have been encysted into forms known as

© Regien Paassen/Shutterstock, Inc.

FIGURE 14.28 A butcher shop stall in a less-developed country. The unrefrigerated pieces of meat are open to the environment and contamination.

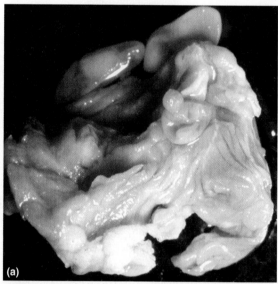

(a)

© Robert I. Krasner.

(b)

© Robert I. Krasner.

FIGURE 14.29 **(a)** A tapeworm larvae–contaminated piece of beef in a butcher shop in Costa Rica. **(b)** Tapeworm cysticerci resemble small bladders.

cysticerci (Figure 14.29b). Upon ingestion, these larvae are digested, except for the **scolex** (the head), which attaches to the intestinal wall by suckers. The scolex produces compartment-like segments known as **proglottids**, each of which is a reproductive bag containing both testes and ovaries. The tapeworm body continues to lengthen as new proglottids are produced, and the worms reach adult size. Each worm has anywhere from 1,000 to 2,000 proglottids, which mature and contain 80,000 to 100,000 eggs per proglottid. Proglottids detach from the worm and migrate to the anus, and approximately six per day are passed in the stool. The eggs can survive for long periods in the environment; cattle and other herbivorous (plant-eating) animals become infected by grazing on contaminated vegetation. Larvae hatch from the eggs, penetrate through the intestinal wall, and become encysted as cysticerci in the muscle of the animal. When humans (or other animals) ingest these cysticerci, the cycle is completed.

Both *T. saginata* and *T. solium* produce only mild abdominal symptoms. Laboratory diagnosis is made to identify eggs and proglottids in the feces. It takes about 3 months after infection for adult tapeworms to develop; therefore, laboratory confirmation is not initially possible. Eggs of *T. saginata* and *T. solium* cannot be differentiated microscopically, but examination of proglottids and, rarely, the scolex allows species identification.

Both species are worldwide in distribution. *T. solium* is more common in poor communities where people may live in close contact with pigs and eat undercooked pork. Medication is available for the treatment of tapeworms. In the United States the USDA is responsible for the inspection of beef that is intended for human consumption.

A serious disease called **cysticercosis** can result from the ingestion of pork tapeworm eggs (as opposed to ingesting cysticerci). These eggs are passed in the feces of an infected person and may contaminate food, water, or surfaces. Those infected with pork tapeworm can reinfect themselves by poor sanitary habits. The tapeworm eggs hatch inside the stomach, penetrate the intestine, and travel through the bloodstream; ultimately, they may cause cysticerci in the muscles, brain, or eyes. Lumps may be present under the skin, and infection in the eyes may cause swelling or detachment of the retina.

Neurocysticercosis (cysticerci in the brain or spinal cord) is serious and may cause death. It is considered to be the most frequent preventable cause of epilepsy in developing countries. More than 80% of people in the world who are affected by epilepsy live in developing countries. Seizures and headaches are the most common symptoms and can occur months, or even years, after infection. Antiparasitic drugs are available to treat cysticercosis, and their use depends on the number of brain lesions judged to be present and the individual's symptoms.

Other species of tapeworms can be acquired by humans. Fish tapeworm infections are common in Scandinavia, Russia, Japan, and other areas of the world, including the Great Lakes region of the United States. The disease is acquired by humans through the ingestion of raw or undercooked contaminated fish. Sushi eaters beware (**BOX 14.2**)!

BOX 14.2 Sushi Eaters Beware!

Sushi lovers all over the world enthusiastically devour the tasty and exquisitely presented little cakes of rice topped with a fillet of raw fish. Sushi bars can be found in just about all large cities, as well as many small ones. Platters of these delicacies are found on buffet tables at weddings, anniversary parties, and other festive occasions.

The making of sushi dates back to at least the 1600s; it was introduced in Southeast Asia as a method of preservation of fish. The fish were packed with rice, which fermented and produced lactic acid, resulting in their being pickled. Why is sushi so popular? By no means is it inexpensive; a single little tidbit will cost you anywhere from $4.50 to $8.50—even more if you crave the exotic. The vinegar in sushi rice is claimed to have antibacterial properties, reduce fatigue, and lower blood pressure. Nutritionists recommend two to three servings of fish per week, because fish is an excellent source of omega-3 fatty acids thought to be effective in the prevention of heart disease, rheumatoid arthritis, cancer, and maybe even depression. In short, sushi, according to sushi lovers, is a health food (**Figure 1a**).

What about the dangers of acquiring worms from eating raw fish? About 20 million people worldwide are believed to harbor fish tapeworms. In 1906, the first description of a human infected with *Diphyllobothrium latum* fish tapeworms came from Minnesota. The majority of fish tapeworm cases since that time have been caused by the ingestion of raw or undercooked freshwater fish such as perch or pike from the Great Lakes region.

Statistics indicate that from 2001 to 2016, 825 cases of diphyllobothriasis caused by *D. nihonkaiense* fish tapeworms present in Asian Pacific coast salmon were reported in Japan; statistics are not kept in the United States. The fish tapeworm can reach 40 feet (12 meters) in length in a person's intestinal tract. The presence of these worms causes abdominal pain that is frequently misdiagnosed as appendicitis, resulting in numerous unnecessary appendectomies. Adding to its miserable reputation, the worm excretes waste from its face and has a borelike tooth that enables it to drill holes into the intestinal wall. Fortunately, the human is a dead-end host, and the worm dies after several days.

Recently, Japanese *D. nihonkaiense* fish tapeworms were found in wild-caught Pacific coast Alaskan salmon (**Figure 1b**). Carnivores such as bears that consume the wild Alaskan salmon were confirmed to be infected with the fish tapeworms, and

(continues)

BOX 14.2 Sushi Eaters Beware! *(continued)*

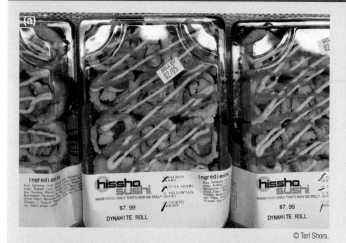

(a)

© Teri Shors.

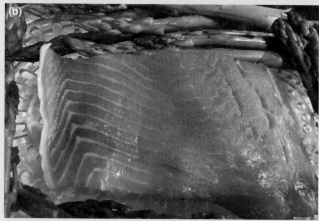

(b)

© Teri Shors.

(c)

© Teri Shors.

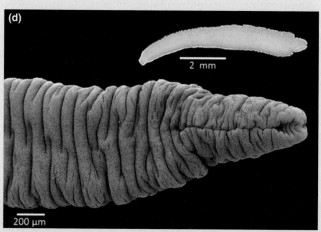

(d)

Courtesy of CDC.

Figure 1 **(a)** Sushi at a supermarket. **(b)** Alaskan wild-caught salmon with asparagus in a steamer. **(c)** Alaskan wild-caught salmon for sale in a supermarket. The Japanese *Diphyllobothrium nihonkaiense* fish tapeworms have been found in Pacific coast Alaskan wild-caught chinook, Coho, pink, and sockeye salmon and linked to human cases of diphyllobothriasis in the United States. **(d)** Scanning electron micrograph of *D. nihonkaiense* showing the scolex, which has a long slitlike bothrium (groove).

human cases have been reported in which individuals ingested salmon sushi or sashimi. Pacific salmon are frequently exported unfrozen, on ice; the tapeworms may survive the active transport and cause human infections in locations where the fish tapeworm is not endemic, such as China, Europe, New Zealand, and the United States.

In January 2018, a human infected with the Japanese *D. nihonkaiense* fish tapeworm went viral in the news. A man showed up in an emergency room at the Community Regional Medical Center in Fresno, California, complaining of bloody diarrhea. The man asked the attending physician, Dr. Kenny Banh, if he could be treated for worms. Then the unidentified man handed the doctor a bag that contained a toilet paper roll which had a tapeworm wrapped around it. He had expelled the tapeworm before his arrival to the emergency room. The fish tapeworm was 5 feet, 6 inches long (1.6 meters; **Figure 2**)—not a

Courtesy of Dr. Kenny Banh.

Figure 2 Worm expelled by emergency room patient. He wrapped the fish tapeworm around a toilet paper roll.

record-breaker, but still enough to make one uncomfortable. The man admitted that he ate salmon sashimi almost every day. Treating the parasite was simple—a single pill of deworming medicine, **praziquantel** or **niclosamide** (the same pill given to dogs), will kill all the worms.

Infections by fish tapeworms are uncommon. The bottom line is this: the more likely risk is in handling the fish. *The best sushi restaurants routinely freeze their fish before its appearance on cakes of rice, which kills the tapeworms, affording a measure of protection to potential sushi eaters.* Sushi chefs are meticulous and frequently wash their hands, counters, and utensils, but choose your sushi restaurant carefully (as you should any restaurant), and then go ahead and order whatever strikes your fish fancy. You might choose the **ahi** (raw tuna), **hamachi toro** (raw yellowtail belly), **uni** (raw sea urchin), or salmon **sashimi** (raw salmon without rice). The biggest danger is that you might ingest too much wasabi and feel like you are choking to death! For the uninitiated, a minute amount of wasabi is used on the sushi. It is even more potent than curry and will cause you to gasp, bring tears to your eyes, and burn your lips. Enjoy!

Arthropodborne Helminthic Diseases

Lymphatic Filariasis (Elephantiasis)

FIGURE 14.30 shows a dramatic example of **elephantiasis**, also known as **lymphatic filariasis**. Note that the term "elephantitis" is not correct, but, as illustrated, the grotesque swelling in the legs is elephant-like, and the texture and the appearance of the skin resemble the hide of an elephant.

What causes elephantiasis and brings about this deformity? The disease is caused by tiny adult thread-like roundworms, called filarial worms, which block the lymphatic vessels and cause the accumulation of large amounts of lymphatic fluid in the limbs. The lymphatic vessels function in the body's fluid balance and return **lymph** back into the circulation. Hence, blockage of the **lymphatic system** results in a "backup" of lymph, causing swelling, particularly in the legs. The affected areas may swell to several times their normal size. In males, the scrotum and penis may be involved, and scrotal swelling can be so severe that a wheelbarrow is used to support the scrotum to allow limited walking. In females, the breasts may become considerably enlarged. Individuals with filariasis are prone to bacterial infections.

Wuchereria bancrofti, *Brugia malayi*, and *Brugia timori* are the three major species responsible for lymphatic filariasis. *W. bancrofti* causes 90% of cases. *W. bancrofti* infections are initiated by the bite of mosquitoes carrying filarial (slender, filamentous) larvae. The larvae migrate into the lymphatic vessels where, within about a year, they grow into adult worms with a life span of up to several years. The adult **filarial** roundworms mate, and the females release into the blood millions of **microfilariae** that may be ingested by mosquitoes taking a blood meal. The mosquitoes transmit the microfilariae to other humans, completing the parasite's life cycle.

About half of the people with lymphatic filariasis have overt clinical disease. The early symptoms are chills, fever, and inflammation of lymph nodes and lymphatic vessels; some individuals remain asymptomatic and may never develop elephantiasis. The remaining people harbor infections with millions of microfilariae and dozens of adult worms in their bodies but with internal damage undetected and untreated. Despite the elephant-like deformities, the disease is not usually fatal. The WHO ranks filariasis as the second leading cause of permanent and long-term disability worldwide. The disease carries a social stigma because of its extreme disfigurement, and individuals are frequently shunned to the point their families neglect them. Their disability may cause them to be unable to work.

Examination of blood samples allows the identification of microfilariae. In patients with elephantiasis, swollen limbs are wrapped in compressive bandages in an effort to force out lymph, thereby reducing discomfort and swelling. Lymphatic filariasis is most common in tropical and subtropical areas and is fostered by poor sanitation and population growth, which create increased breeding

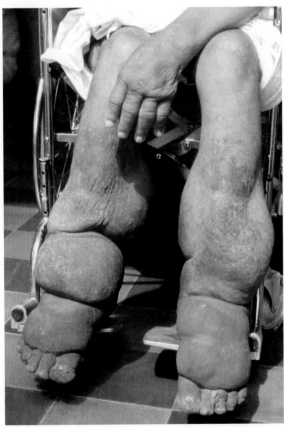

© Robert I. Krasner.

FIGURE 14.30 Elephantiasis. Infection with filarial worms can block lymphatic vessels, causing severe edema (swelling).

areas for mosquitoes. (Population growth is a factor in emerging and reemerging infections.) Filariasis affects 856 million people in 52 countries where it is endemic, including numerous countries in Africa, Southeast Asia, and the western Pacific and at least seven countries in the Americas. About 65% of cases occur in Southeast Asia and 30% in Africa. Approximately one-third occur in India.

A strategy to eradicate lymphatic filariasis involves treating at least 65% of the at-risk population through *preventive* chemotherapy called **massive drug administration** (MDA) repeated annually for at least 5 years. Drug regimens involve treatment combinations of diethylcarbamazine citrate (DEC) plus albendazole, ivermectin plus albendazole, or albendazole alone. More than 6.7 billion treatments have been delivered to stop the spread of infection since 2000. Approximately 499 million people no longer require preventive chemotherapy due to successful implementation of WHO strategies. Lymphatic filariasis is one of only six diseases considered potentially eradicable. As a result of an effective single-dose treatment, global elimination of lymphatic filariasis is now possible. A number of agencies in the public and private sector are collaborating to reduce the burden of filariasis.

Onchocerciasis (River Blindness)

Onchocerciasis, also called **river blindness**, is one of the world's leading causes of blindness; it exists in many parts of Africa and Central America and is caused by *Onchocerca volvulus*, a filarial roundworm. Onchocerciasis can devastate entire communities. In many small villages, nearly all villagers over the age of 40 are blind, presenting a serious obstacle to socioeconomic development (**FIGURE 14.31**).

The disease arthropod vector is a black fly (*Simulium* spp.), which abounds in fertile riverbanks and breeds in fast-flowing rivers and streams. When the black fly bites, larval worms are deposited under the skin. The larvae mature into adult roundworms about a year later and live for as long as 14 years. They produce millions of microfilariae that migrate throughout the body, causing rashes, intense itching, depigmentation of the skin, and nodules that form around the adult worms. Some migrate into the eyes and invade the cornea, causing blindness. These manifestations begin to occur 1 to 3 years after injection of the larvae by the black flies. The parasite's life cycle is perpetuated when a biting black fly ingests microfilariae from the blood of an infected person while taking a blood meal and then injects larvae into a new host (**FIGURE 14.32**).

Diagnosis is difficult, and there is no blood test to detect the presence of microfilariae. The presence of nodules under the skin is diagnostic, and their removal by minor surgery reduces the number of microfilariae produced

© Robert I. Krasner.

FIGURE 14.31 A statue at the WHO headquarters in Switzerland, *A Little Child Shall Lead*, commemorates the battle against onchocerciasis. In some African villages a child leading a blind person is a common sight.

(**FIGURE 14.33**). Several anthelmintic drugs are available; **ivermectin** is particularly effective because it kills migrating microfilariae. In practice, ivermectin is the only drug suitable for treatment, because other drugs cause severe reactions.

Krasner's Notebook

The photos in Figure 14.33 were taken on a field trip in Costa Rica. Figure 14.33b demonstrates nodules under the skin, which were easily removed by minor surgery. I took my turn with the scalpel.

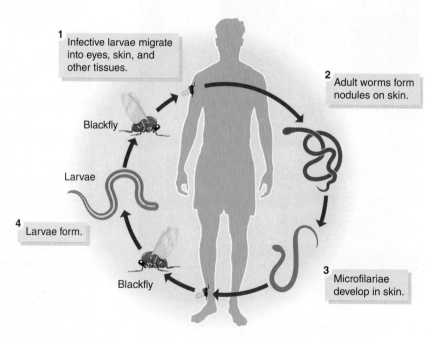

FIGURE 14.32 Life cycle of *Onchocerca volvulus,* the helminth that causes river blindness.

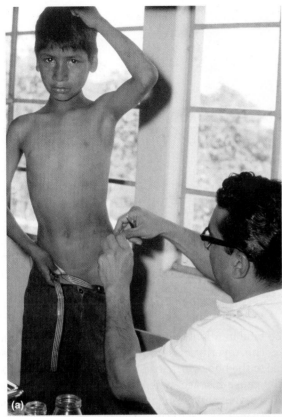

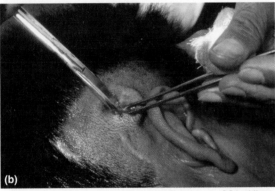

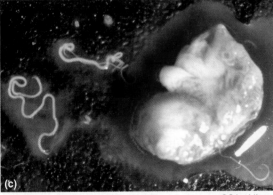

FIGURE 14.33 Onchocerciasis intervention. A bite by an infected black fly causes the formation of a nodule containing the larval form of the worm. **(a)** A healthcare worker examining nodules. **(b)** Surgical removal of worms from nodules. **(c)** Appearance of worms.

The WHO estimates that 25 million people are infected worldwide with *O. volvulus*. Of those infected, 300,000 are blind and 800,000 have some kind of visual impairment. Some 123 million people worldwide (those who live near streams or rivers where there are *Simulium* black flies), of whom 90% are in Africa, are at risk for onchocerciasis. In some western African communities, 50% of the men over the age of 40 have been blinded by the disease. River blindness is also found in six countries in Latin America, as well as Yemen in the Arabian Peninsula. More than 6.5 million suffer from severe itching and skin disease; about half a million are blind or severely visually impaired due to the disease. The disease is endemic in 37 countries, of which 30 are in sub-Saharan Africa and 6 are in the Americas.

Until the 1980s, control focused on the use of chemicals to kill immature black flies in rivers. A renewed and major effort to eliminate river blindness was initiated in 1987, when the drug ivermectin was licensed by Merck & Co. and provided free of charge for treating the disease. The Carter Center, the Lions Club Sight First Project, the CDC, and Merck are collaborating in an effort to curb river blindness.

Direct Contact Helminthic Diseases

Hookworm Disease

In the United States, rural areas of the South were, at one time, burdened with **hookworm disease**, a disease that insidiously saps the strength of those parasitized. Thanks to the pioneering work of the Rockefeller Sanitary Commission in the early 1900s, the incidence of this major parasitic disease has markedly declined as the result of improved sanitation, particularly the installation of indoor plumbing, and a better way of life, including the wearing of shoes.

Hookworm disease is caused by two roundworm species: *Ancylostoma duodenale* (**Old World hookworm**) and *Necator americanus* (**New World hookworm**). *A. duodenale* is found in the Mediterranean countries, Iran, India, Pakistan, and the Far East. *N. americanus* is the most common hookworm worldwide and is present in North America (southeastern U.S.), South America, Central Africa, Indonesia, islands of the South Pacific, and parts of India. Over 500 million people globally are burdened by hookworms, particularly in tropical and subtropical climates.

Infection occurs by direct contact with soil contaminated with hookworm larvae while walking barefoot, touching soil with bare hands, or accidentally swallowing bits of contaminated soil. Eggs require warm, moist, shaded soil and hatch in 1 to 2 days into barely visible larvae. After 4 to 10 days the larvae reach the infective stage and penetrate the skin; they are carried to the lungs, where they are coughed up and swallowed, reaching the intestinal tract in about a week. In the small intestine the larvae develop into half-inch-long adult worms. The mature adult worms live and mate attached to the wall of the small intestine and produce thousands of eggs,

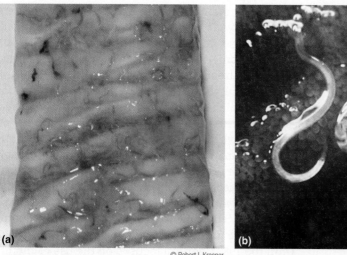

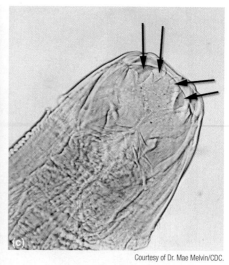

© Robert I. Krasner. Courtesy of CDC. Courtesy of Dr. Mae Melvin/CDC.

FIGURE 14.34 **(a)** Hookworm in wall of intestine. **(b)** Hookworm attached to intestinal mucosa. **(c)** Mouth parts of a hookworm. The hookworm uses its teeth (indicated by the arrows) to grasp the intestinal wall.

which are passed in the feces (**FIGURE 14.34**). The eggs ultimately reach the soil, completing the life cycle. The cycle takes about 5 to 6 weeks from larval invasion to egg deposition by female worms (**FIGURE 14.35**). Person-to-person contact is not possible, because development of the larvae must be in soil as part of the life cycle.

The adult worms have mouth parts, called **biting plates**, by which they attach and suck blood, causing anemia, protein deficiency, and fatigue; in children, continual infection can cause stunted growth and developmental learning disabilities as the direct result of blood loss. Serious infection can be fatal in infants and causes children to be malnourished. The severity of symptoms appears to be directly related to the hookworm load. Fewer than 25 worms do not usually cause disease, whereas 500 to 1,000 hookworms result in severe and often fatal damage. Diagnosis of hookworm is confirmed by the presence of eggs in a fecal sample. Medication is available for treatment.

Strongyloidiasis

Strongyloidiasis is caused by the nematode *Strongyloides stercoralis*, sometimes called the "threadworm" because of its minute size. Around 30 to 100 million people globally are infected with this parasitic helminth. *S. stercoralis* is unique in that the females produce eggs by **parthenogenesis**, a process that does not require fertilization by males (tough luck, boys). Infective larvae in the soil penetrate the skin and are sequentially transported by blood to the heart, lungs, and bronchial tree and into the pharynx, where they are coughed up and swallowed, finally reaching the small intestine. There, they mature into adult female worms and produce eggs. Some eggs develop into larvae that penetrate the small intestine or the skin around the anus and repeat the life cycle through the heart, lungs, bronchial tree, and pharynx and to the small intestine where they mature into adult worms (**FIGURE 14.36**). This repeated cycle results in internal **autoinfection**, a process unique among worms in humans. Autoinfection accounts for persistent infections for over 20 years in infected persons who have moved to areas where the parasite is not present. Further, autoinfection can result in **hyperinfection**, characterized by the dissemination of the parasite within the body.

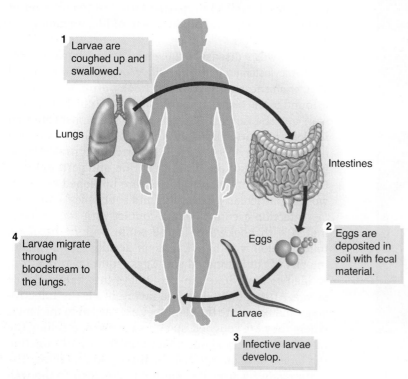

1 Larvae are coughed up and swallowed.

Lungs

Intestines

4 Larvae migrate through bloodstream to the lungs.

Eggs

2 Eggs are deposited in soil with fecal material.

Larvae

3 Infective larvae develop.

FIGURE 14.35 Life cycle of hookworms.

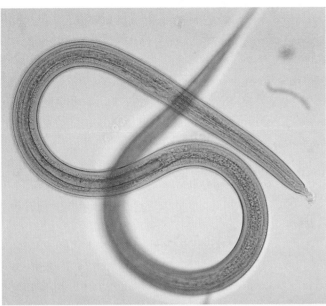

Courtesy of Dr. Mae Melvin/CDC.

FIGURE 14.36 A *Strongyloides* larva.

The free-living life cycle is an alternative strategy. It results from the passage of eggs in the feces that develop into free-living roundworms that produce large numbers of infected larvae. The free-living life cycle is an important reservoir of human infection.

Strongyloides infection may be mild and go unnoticed, but it can be severe, resulting in nausea, vomiting, anemia, weight loss, and chronic bloody diarrhea. In 1958, a Japanese biologist infected himself by the intradermal injection of larvae to follow the course of the disease. The severity of infection appears to be related to the number of worms, the site of involvement, and the immune status of the host. In individuals who are immunocompromised, large numbers of invasive larvae can cause extensive damage, leading to death.

As with many helminthic infections, the presence of larvae in the feces and duodenal contents establishes a definitive diagnosis. A reliable blood test to detect antibodies to *Strongyloides* is available, as are drugs to treat strongyloidiasis.

Enterobiasis (Pinworm Disease)

Pinworm disease, also known as **enterobiasis** or **oxyuriasis** is the most common helminthic disease in the United States, with an estimated 40 million people (most commonly school- and preschool-aged children) infected, according to the CDC. *Enterobius vermicularis*, a white roundworm that is about the length of a staple, is the culprit.

The life cycle is a simple one. The adult male and female worms live and mate in the infected person's intestine. At night, while the person sleeps, egg-bearing females exit through the anus, deposit as many as 15,000 eggs around the skin in the anal area, and then die. Egg laying is triggered by the lower temperature outside the body. Infective larvae develop within the eggs in about 4 hours.

The eggs are swallowed and hatch in the intestine, releasing larvae that mature into adults and become established in the intestine. The interval from ingestion of larvae to egg-producing females is about 1 month; the life span of the adults is about 2 months.

Self-infection is a common mode of transmission, particularly in children. Pinworms cause an intolerable itch, resulting in intense scratching. Eggs cling to the fingers, are lodged under the fingernails, and are then swallowed, reinitiating infection. Further, an infected person can transfer eggs through direct contact, on bedclothes, and even by inhalation of airborne eggs.

Anal itching, restlessness, and loss of appetite are common and relatively mild symptoms; the disease is rarely debilitating. School-age children have the highest rate of infection, and in some day care centers and other institutional settings more than 50% of children are infected. If one member of a family has pinworms, there is a good chance that other family members are infected.

Diagnosis is based on the **Scotch tape test**; a piece of tape is applied to the anal region during the night or early morning hours. If eggs are present, they stick to the tape and are identified under a microscope (**FIGURE 14.37**). Treatment involves a two-dose course of an anthelmintic drug; the second dose should be given 2 weeks after the first. In 2009, more than 300 million children were dewormed. It is usually wise to treat all members of a family to break the cycle. According to the CDC, cleaning and vacuuming the entire house and washing sheets every day are probably not necessary or effective.

Schistosomiasis

In 2016, more than 206.4 million people required preventive treatment for **schistosomiasis**, and of those more than 89 million were reported to have been treated for an active infection. Shistosomiasis has been reported in 78 countries in Africa (accounting for 91.4% of reported infections), South America, the Caribbean, the Middle East, and the Far East. However, preventive chemotherapy treatment is only needed in 52 endemic countries that have moderate to high transmission rates. It is caused by a number of species of *Schistosoma*, including *S. mansoni*, *S. mekongi*, *S. guineensis*, *S. japonicum*, *S. intercalatum*, or *S. haematobium*, based on geographical distribution of the parasite; these parasites are trematode worms, which are frequently called **blood flukes**. Certain species of *freshwater snails serve as intermediate hosts* in these geographical regions; they release immature, free-swimming microscopic forms of blood flukes, called **cercariae**, resembling miniature tadpoles that can survive in the water for about 48 hours. Exposure to infested water containing cercariae allows their penetration into the skin (**FIGURE 14.38**).

Penetration is followed by migration into blood vessels and, in about a week, entrance into the liver, where the liver flukes achieve sexual maturity. The adults have evolved a clever strategy of coating themselves with antigens of the host, thereby escaping detection by the host's immune system. Female blood flukes reside in a groove in

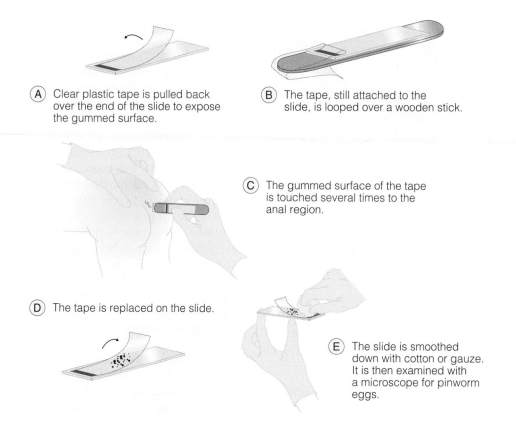

Ⓐ Clear plastic tape is pulled back over the end of the slide to expose the gummed surface.

Ⓑ The tape, still attached to the slide, is looped over a wooden stick.

Ⓒ The gummed surface of the tape is touched several times to the anal region.

Ⓓ The tape is replaced on the slide.

Ⓔ The slide is smoothed down with cotton or gauze. It is then examined with a microscope for pinworm eggs.

FIGURE 14.37 Diagnosing pinworm disease. The transparent tape technique is used in the diagnosis of pinworm disease.

the body of the larger male, and the two remain entwined as a (happy) mating pair for up to 10 years. The pair migrate into small **venules** (veins) of the intestinal tract or of the bladder, depending on the particular *Schistosoma* species. Ancient Egyptian writings describe males with blood in their urine as "menstruating"; possibly, this was the result of blood flukes in their urinary tract. The trematode worms feed on blood, and the females produce eggs within several weeks; the eggs are discharged by infected persons defecating or urinating into the water. The eggs hatch into actively motile larvae that penetrate the snails and develop into cercariae, completing the life cycle of *Schistosoma* spp.

Itchiness develops at the penetration site within several days and may be followed by fever, chills, cough, and muscle aches within the next several weeks, but some

Courtesy of the CDC.

Courtesy of the CDC.

FIGURE 14.38 Larval forms called cercariae are present in water and penetrate the skin, initiating an infection that causes schistosomiasis. **(a)** Two boys wading in water despite a warning sign. **(b)** A boy with a swollen belly resulting from schistosomiasis.

people experience no symptoms during this early stage. The major consequence of schistosomiasis is the damage caused to the liver, intestine, or bladder. The eggs cause allergic reactions. More rarely, *Schistosoma* eggs may migrate into the brain or spinal cord, causing seizures, paralysis, or inflammation of the spinal cord. The adult *Schistosoma* blood flukes can also damage the spleen, liver, lungs, intestine, and bladder, causing the abdomen to become distended (Figure 14.38b). The adult worms escape detection by the immune system and can survive for years.

Diagnosis is confirmed by the presence of *Schistosoma* eggs in the urine or stool. Safe and effective drugs are available for treatment, but they are relatively expensive. Prevention through improved sanitation and control of the snail vectors is a more efficient and cost-effective approach. In some areas of Africa, the introduction of a species of small fish that feeds on snails (biological control) is reducing the cercarial population.

Technological development has often been cited as a causative factor of new and emerging infections. An unfortunate and unpredictable consequence of the building of the Aswan Dam in Egypt in the 1960s was the significant increase in schistosomiasis because of the favorable conditions created for the intermediate snail hosts. *In the United States the required intermediate snail host is not present so schistosomiasis does not occur.*

A condition known as swimmer's itch, or cercarial dermatitis, occurs sporadically and unpredictably in the United States, particularly in the northern lakes and in Mexico and Canada. Swimmer's itch is due to penetration of the skin by cercariae from eggs of water birds infected with the avian species of schistosomes; humans are **incidental hosts**. The body's immune system destroys the cercariae, resulting in an allergic reaction that causes a rash and itching.

Biology of Fungi

Fungi are a category of eucaryotic microbes; some are unicellular, and others are multicellular. They have distinctive characteristics as compared to other microbes. Fungi are a ubiquitous and varied group, as evidenced by the many kinds of mushrooms that seemingly spring onto your lawn overnight or the fuzzy moldy growth on a pair of old sneakers thrown into your closet or an aging piece of vegetable or fruit (**FIGURE 14.39**).

Basically, fungi have two distinct forms; namely, the unicellular microscopic yeasts and the branching filamentous molds characterized by their fuzzy appearance (**FIGURE 14.40**). Fungi are a difficult group to place in a taxonomical (classification) sense. More than a million species of fungi have been identified but surprisingly only about 300 cause disease in humans. Many fungi have complex life cycles exhibiting both asexual and sexual stages. Fungal spore formation is the mode of multiplication (unlike bacterial endospores). Fungal spores are

© Teri Shors.

FIGURE 14.39 Moldy orange.

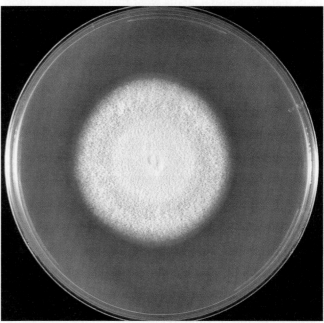
Courtesy of Dr. Libero Ajello/CDC.

FIGURE 14.40 Fungal colony.

easily inhaled or land on the skin, as manifested by the frequency of cutaneous and respiratory tract infections.

This chapter deals with fungi that can be a serious nuisance and others that cause disease in humans. However, some fungi are beneficial in the production of certain foods. The soil mold *Penicillium notatum* is the original source of the antibiotic **penicillin**; the mold *Streptomycin griseus* is the original source of **streptomycin**, the first effective antibiotic used to treat tuberculosis (caused by *Mycobacterium tuberculosis*). Numerous other fungi are harvested as a source of antibiotics. *Penicillium roqueforti* is used to produce blue cheese.

The authors have not attempted to cover fungi to the extent given to other microbes and worms; only a handful are presented here. Hence, the usual organization by mode of transmission does not apply.

Fungi in the Environment

Mushrooms and other fungi are common inhabitants in the environment, and some mushrooms are quite beautiful, but don't be fooled. Other fungi, like *Stachybotrys chartarum*, may take your house down. "*Stachy*" (as it is affectionately called, but I don't know by whom in their right mind) is truly a bad actor and is responsible for sick house and **sick building syndrome**. It is a black and rapidly producing mold capable of growing on the surface of many building materials, including dry wall and surface tile. Once it gets into a building it is very difficult to get rid of and usually requires professional remediation; in some cases, as a last resort, homes have been burned down to get rid of *Stachybotrys*.

Monster Category 4 Hurricane Charley hit land at Caya Costa, Florida, near Captiva Island in southeastern Florida on August 14, 2004, with winds of 140 miles per hour (225 kilometers per hour) and water surges as high as 15 feet (4.5 meters). Roofs went flying off, and in short order the interior of houses became soaked. The hurricane ended but the misery continued for months. After a few weeks, it was obvious that *Stachybotrys* and other molds had invaded. As one homeowner stated, "*After losing my roof, I wasn't prepared for mold taking over my house.*" Hurricane Katrina hit Louisiana and parts of neighboring states in Mississippi and Alabama in 2005. Again, once the hurricane was over, *Stachybotrys* and other molds became a major problem, resulting in the need for entire blocks of houses to be torn down. Remember that these fungi grow well on dry wall, a major building material (**FIGURE 14.41**).

However, it does not take a hurricane for *Stachybotrys* and other molds to move into a house or into a building; leaky and poorly placed gutters, drains, and poor landscaping may allow seepage. In buildings constructed in the last 20 years or so great effort was made to build them "airtight"; unknowingly, this impaired the ability of the building to breathe in the sense of outdoor–indoor air exchange. Further, some of the materials used were found to exude noxious odors that induced the usual

© Julie Dermansky/Photo Researchers, Inc.

FIGURE 14.41 A room covered in mold after Hurricane Katrina and the flooding that occurred after the levees broke.

allergy-like symptoms. Hopefully, new green regulations, developed to staunch the risk of the development of mold, will minimize construction-based problems associated with sick homes and sick buildings. Despite these negative scenarios, fungi play very significant roles based on their fermentative capability in the food and alcoholic beverage industries. It's hard to beat a generous slice of Roquefort cheese on a slice of bread with a glass of beer to wash it down!

Many examples of interactions between humans and other animals and plants and numerous species of fungi can be cited (**BOX 14.3**). In the wake of the tornado that hit Joplin, Missouri, on May 22, 2011, at least 13 people developed a laboratory-confirmed rare necrotizing soft tissue infection caused by **mucormycetes**, resulting in five deaths. Mycormycetes is an old name for Zygomycetes, a class of fungi. Genera in this class include *Mucor*, *Rhizopus*, *Absidia*, *Rhizomucor*, and others. In nature these fungi are found on vegetation; however, if the vegetation gets driven into the skin via wind or water, they can cause disease. Mucormycetes are opportunists, causing problems for individuals with untreated diabetes and those who have compromised immune systems (as in HIV/AIDS, causing pneumonia). Mucormycetes have been reported following a number of natural disasters.

Fungal Diseases of Animal Species

○ "A Froggy Would A-Wooing Go" or "Where Have All the Froggies Gone?"

Walk the streets of Chinatown in San Francisco and you will note many street vendors selling bullfrogs from their tanks for use in laboratories, as pets, or as food. Perhaps you dissected a frog in an early course in biology. Care is taken by the vendor to prevent these amphibians from escaping from their tank or that during a sale drops of tank water do not fall to the pavement. Why the concern? It turns out that many of the frogs are infected with the fungus *Batrachochytrium dendrobatidis* (the worst tongue-twister yet cited), causing a thickening of their skin that even **Botox** wouldn't help.

The thickening of the skin interferes with the frogs' gas exchange and water absorption abilities but does not usually cause death. Most of these frogs have been imported from Taiwan. "*Batra*" can spread to native frog populations by an infected frog or by water from the holding tank. Efforts to ban the importation of frogs from Taiwan has met with strong resistance from the large Asian American population in the area where many make their living selling the frogs. Further, the bullfrogs are a part of their cuisine. As stated by one Chinese resident of the community, "*For over 5,000 years, it has been the practice of both the Chinese and Asian Americans to consume these particular animals. They are a part of our staple. They are a part of our culture. They are a part of our heritage.*"

The other side of the battle is presented by the nonprofit organization Save the Frogs!, whose membership

BOX 14.3 "Oh, Christmas Tree, Oh Christmas Tree, How I Love My Christmas Tree"

Yes, you may love your beautifully decorated blinking tree with its fresh pine-scented aroma, but it may not love you (**Figure 1a**). The tree may well bear invisible mold spores of *Phytophthora cinnamoni* or other species that can bring you misery in the form of itchy and teary eyes, stuffy nose, headache, sore throat, sneezing, and coughing, all of which are the usual symptoms of a mold allergy that occurs in about 15% of the population. A "Grinch," in the form of a mold, might steal your holiday causing you to, instead of loving the tree, end up approaching it with a chain saw!

The Fraser fir is one of approximately 10 species of fir trees; it has graced the Blue Room of the White House as a Christmas tree more than any other tree based on its fragrance, long limbs, and delayed loss of needles. In its natural environment it has its own problems as a result of infection by flagellated spores produced by *Phytophthora* that cause root rot (**Figure 1b**).

© casenbina/iStockphoto.

Courtesy of Stacy Clark, USDA Forest Service.

Figure 1 **(a)** Fraser fir, one of the most common Christmas trees, is especially vulnerable to fungal infection. Airborne mold counts increase after live Christmas trees are placed in a room. **(b)** Root rot caused by *Phytophthora cinnamomi.*

petitioned the governor to take action banning the importation and sale of bullfrogs in California fearing that a *Batra*-infected bullfrog would escape into the environment and threaten the native species. The solution may be that the importation of frogs will be prohibited unless there is appropriate documentation certifying that they are free of *Batra*.

Bats

There may not be as many bats in the belfries around Washington, D.C., these days because millions of the "little brown bats," without even batting an eyelash, have been destroyed by the fungus *Geomyces destructans*. The bats develop **white-nose syndrome**, which is characterized by a powdery white substance that develops on their muzzle, ears, and wings. The interactions between fungal species and plant and animal species is endless.

Fungal Diseases of Humans

Fungi are frequently secondary invaders in a number of viral and bacterial diseases in humans. The leading cause of death in AIDS patients is a pneumonia caused by the fungus *Pneumocystis jirovecii*; other **opportunistic infections** caused by fungi are associated with HIV/AIDS because of the patients' compromised immune system (**TABLE 14.6**). Fungi are the primary cause of a variety of human diseases known as **mycoses** (**TABLE 14.7**); a few are further detailed in this chapter. Other fungi exist "in the wild" (**TABLE 14.8**). Most of the diseases are not fatal but, once contracted, could be a constant source of irritation and can lead to permanent scarring. Fungal diseases do not occur as epidemics, as seen with bacterial disease outbreaks.

Fungal Lung Infections Endemic in the United States

Fungi are everywhere. Four different fungal diseases that can affect the lungs are endemic in the soil of different regions of the United States (**FIGURE 14.42**): histoplasmosis, blastomycosis, cryptococcosis, and coccidioidomycosis (Valley fever). For people who are healthy, exposure to the fungi causes mild or no symptoms. However, fungal lung infections can be fatal unless treated and occur primarily in those with HIV/AIDS, organ recipients

TABLE 14.6 Fungal Infections in Humans with Compromised Immune Systems

Fungal Pathogen	Disease
Candida albicans	Candidiasis (thrush, yeast infections)
Candida auris	Healthcare-associated infection/serious invasive infections
Histoplasma capsulatum	Histoplasmosis
Blastomyces dermatitidis	Blastomycosis
Coccidioides immitis	Coccidioidomycosis (Valley fever)
Cryptococcus gattii or Cryptococcus neoformans	Cryptococcosis
Penicillium marneffei	Penicilliosis
Aspergillus sp.	Aspergillosis
Pneumocystis jirovecii	Pneumonia
Mucormycetes	Mucomycosis or zygomycosis

TABLE 14.7 Fungal Infections in Humans with Normal Immune Systems

Fungal Pathogen	Disease
Candida albicans	Candidiasis (yeast infections)
Coccidioides immitis*	Coccidioidomycosis (Valley fever; lung infection)
Histoplasma capsulatum*	Histoplasmosis cave disease (Ohio Valley disease; lung infection)
Blastomyces dermatitidis*	Blastomycosis (lung infection)
Trichophyton*	Athlete's foot (tinea pedis; ringworm of the foot)
Trichophyton*	Dermatomycosis (ringworm of trunk, arms, or legs; also called tinea corpis)
Trichophyton*	Jock itch (tinea cruris; ringworm of the groin)

*Also infects animals (e.g., dogs and cats).

taking immune suppressive drugs to prevent organ rejection, cancer patients on chemotherapy, or people taking immune suppressive medication for autoimmune disorders.

Histoplasma capsulatum, the cause of **histoplasmosis**, grows in soil and in materials contaminated with bird and bat droppings (guano). It is most common in the central and eastern United States, especially in states around the Ohio and Mississippi River valleys (note that it can also be found in parts of Central and South

TABLE 14.8 Fungal Infections in the Wild

Fungal Pathogen	Host	Disease
Batrachochytrium dendrobatidis	Amphibians (e.g., wild and pet frogs)	Chytridiomycosis
Phytophthora cinnamomi	Fraser fir (Christmas) trees; infects many plant and tree species (firs, oaks, avocado)	Severe root rot
Geomyces destructans	Hibernating bat populations in the United States	White-nose syndrome
Stachybotrys chartarum	Grows on building materials after water damage; can make humans sick	Sick building syndrome
Apophysomyces trapeziformis	Humans wounded during tornado	Cutaneous mucormycosis or zygomycosis (rare infection following Joplin tornado, May 2011)

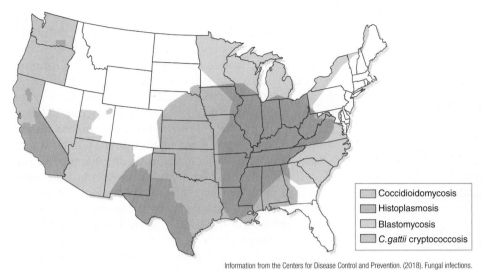

Coccidioidomycosis
Histoplasmosis
Blastomycosis
C. gattii cryptococcosis

Information from the Centers for Disease Control and Prevention. (2018). Fungal infections.
Retrieved from https://www.cdc.gov/features/fungalinfections/index.html

FIGURE 14.42 Fungi present in soil can cause lung infections in people with healthy or compromised immune systems. The four different lung infections found in soils in the United States are histoplasmosis, blastomycosis, cryptococcosis, and coccidioidomycosis (Valley fever).

America, Africa, Asia, and Australia). Approximately 60% to 90% of the people who live in these areas have been exposed to the fungus during their lifetime. Whenever the ground is disturbed, be it by a shovel or bulldozer, spores of all sorts are released, become airborne, and may be breathed in. (Remember this every time you see a bulldozer preparing land for another useless mall when the one down the street is half empty.) *Histoplasma* fungal spores have been found in litter used in poultry houses, bird roosts, and areas harboring bats. The infection is also called **cave disease** or **Ohio Valley disease**.

Histoplasma has an interesting lifestyle. In the environment it grows as a **mycelial phase** (brown patches), but at body temperature it morphs into a yeastlike phase. Antifungal medication is available and is a lifesaver for those with the disseminated form, assuming treatment is started early.

Blastomycosis is caused by *Blastomyces dermatitidis*, which lives in moist soil and in decomposing organic matter such as wood and leaves. It is found mainly in the Mississippi River valley and the Great Lakes region. People can get blastomycosis by inhaling the fungal spores from the air. Wisconsin has the highest incidence of infection, ranging from 10 to 40 cases per 100,000 persons. Some blastomycosis outbreaks have occurred when people were exposed to large amounts of fungal spores through the disruption of soil such as during construction, or excavation, or during recreational activities near lakes or rivers (e.g., the Wolf River, Wisconsin, blastomycosis outbreak during the summer of 2015). Like *Histoplasma*, *Blastomcyes* has a similar mycelial and yeastlike phase.

Valley fever, also called **coccidioidomycosis** is a fungal lung infection that occurs when dust containing *Coccidioides immitis* spores is inhaled. This fungus lives in the soil of the southwestern United States and parts of Mexico and South America. Usually, those who get symptoms of Valley fever improve within a week, with the exception of people with weakened immune systems who will need antifungal medication. Arizona, California, Nevada, New Mexico, and Utah collectively report an average of 40 cases per 100,000 people each year; rates of infection are highest among people aged 60 to 79 years of age. Phoenix and Tucson are highly endemic metropolitan areas for Valley fever cases.

Cryptococcus gattii is present in British Columbia and the Pacific Northwest as well as in many tropical and subtropical areas of the world. *C. gattii* infections are rare lung infections, referred to as **cryptococcosis**. Besides the lungs, the central nervous system may also be affected. *Cryptococcus neoformans* lives in the environment throughout the world. Like *C. gattii* infections, these infections are very rare but can cause serious infections in people with HIV/AIDS or people with weakened immune systems caused by other conditions.

❍ Ringworm

Ringworm, or the tinea group of mycoses, is caused by **dermatophytic fungi** (Table 14.7). These fungi thrive on **keratin**, the dead protein layer of the surface of skin. The condition was once thought to be due to a worm, but, although this is not the case, the name persists. Heat and moisture favor the growth of these fungi, resulting in their being found primarily in skin folds such as in the groin and between the toes. The word *tinea* (Latin) refers to the overall wormlike appearance, but the full name takes on the location of the outbreak on the skin. For example, *tinea capitis* identifies ringworm of the scalp, *tinea cruris* refers to ringworm of the groin area (**jock itch**), and *tinea pedis* signifies ringworm between the toes (**athlete's foot**).

The symptoms are manifested by itchiness; scaly appearance; and, if on the scalp or in the beard, patches of baldness. Athlete's foot (and you don't have to be an athlete to get it) is quite common and easily acquired by walking on wet surfaces, such as around a pool or on a shower floor. It won't kill you but it is a big nuisance, with sometimes almost extreme burning and itchiness (**FIGURE 14.43**). It can be transmitted by person-to-person contact and by sharing combs, brushes, clothing, and other personal articles. Dog and cat lovers beware—these pets can carry and transmit the fungus (a personal account of the disease is presented by one of the authors; **BOX 14.4**); you are well advised to stick to goldfish!

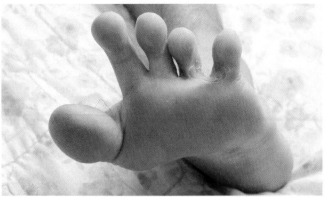

© Kanjanee Chaisin/ShutterStock, Inc.

FIGURE 14.43 A person with athlete's foot (tinea pedis). The fungi grow between the toes. It is very itchy and causes a burning sensation.

● *Candida alibicans* and *Candida auris* infections

Candidiasis is caused by yeast cells; recall that yeasts are unicellular fungi. *Candida albicans* is the most common species, but numerous other species of *Candida* can cause

mycoses. A common yeast infection is **thrush**, in which the yeast is in the oral cavity and the throat. It can affect both males and females. In females, *Candida* grows in the vagina and is influenced by vaginal acidity and hormonal changes, resulting in intense itching, burning, and discharge. Yeast infection is about the second most common cause of females seeing their physician. *Candida* is a component of the normal body microbiota and is found on the skin and mucous membranes. In both oral thrush (**FIGURE 14.44**) and **vaginitis**, the infection is due to an imbalance of the **normal microbiota** of the area as caused by overgrowth of yeast. Consider, for example, that some individuals on an antibiotic may develop thrush due to the fact that the antibiotic kills off not only the pathogen for which the medication was given, but also members of the normal microbiota. The consequence is somewhat Darwinian-like in the sense that the medication does not kill the yeasts and, thereby reduces the competition, giving the yeast a "growth kick."

One of the authors (Robert I. Krasner) developed thrush after surgery and a course of antibiotics and can personally tell you it "isn't fun," but it is not at all life

---

BOX 14.4 Ringworm: An Author's Experience (Teri Shors)

I have poignant memories of a ringworm infection when I was about 5 years old growing up on a dairy farm in central Minnesota. Many litters of kittens were running around in the barn, and I used to love holding them to my chest and playing with them. The barn kittens had very sharp claws and could crawl up the old brick walls of the barn; frequently, they would scratch me. My memories are those of ringworm patches or lesions all over my tummy or chest. Every day for about 2 weeks my mother would put iodine on the ringworm patches. Eventually the ringworm went away. The iodine probably did nothing, but this was in the 1960s, a time before there were antifungal over-the-counter medicines.

The culprit of ringworm is not a worm but instead common types of soil fungi known as **dermatophytes**. *Trichophyton* is the cause of ringworm in companion animals (dogs, cats, rabbits, rats, guinea pigs, horses, etc.). Infected animals may show no signs or symptoms. Ringworm, or **dermatomycosis**, affecting the trunk of the body is called tinea corporis. Anyone at any age can get ringworm. It is diagnosed by the classic rash (**Figure 1**). Ringworm responds well to topical antifungal creams such as **miconazole**, **clotrimazole**, **ketoconazole**, and **terbinafine**, which must be applied twice a day to the lesions for at least 3 weeks.

Ringworm is highly contagious and is spread by direct contact, which is common between pets and kids, especially pets that often get smothered with attention like the barn kittens in my case. Good hygiene practices, particularly attention to handwashing, can certainly help, but some degree of risk will remain. *Trichophyton* spores can survive for more than a year. Therefore, once a pet has ringworm it is suggested that the

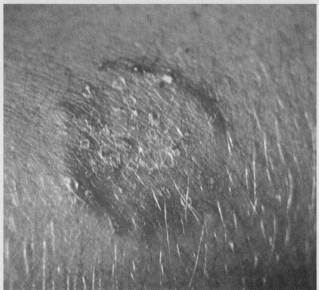

Courtesy of Dr. Lucille K. Georg/CDC.

Figure 1 Ringworm, or tinea corporis, present on the arm of a patient.

pet's bedding be thrown away. Any grooming tools should be treated with bleach diluted with water at 1:10, or a 0.2% enilconazole solution. Carpets should be vacuumed frequently to remove hair and spores. Any areas that the pet frequents should be washed and sterilized. Hands should be thoroughly washed after touching an infected pet. Controlling ringworm outbreaks can take time and be frustrating, but it's not a serious disease and is controllable.

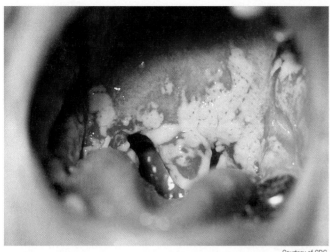

Courtesy of CDC.

FIGURE 14.44 Oral thrush (candidiasis) caused by *Candida albicans*.

threatening. Your mouth and throat are painfully dry and covered with white patches of yeast and it feels like your tongue is sticking to the roof of your mouth. Antifungal mouthwashes solve the problem usually within a few days.

Invasive candidiasis, unlike thrush and vaginitis, which are considered superficial infections, is a serious and potentially life-threatening systemic condition. It can be caused by *C. albicans* or other species of *Candida*. It is more likely to occur in those with an immunocompromised immune system as in HIV/AIDS or those with a compromised immune system due to another cause. The yeast enters the blood causing bloodstream infection and spreads throughout the body. Drug therapy is available.

Candida auris is an emerging fungus that presents a serious global health threat (**FIGURE 14.45**). It causes

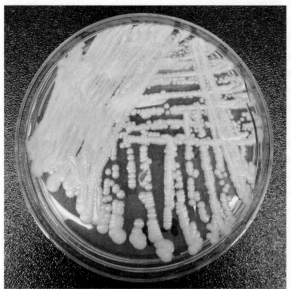

Courtesy of the CDC./Shawn Lockhart.

FIGURE 14.45 *Candida auris* cultured on a solid media.

serious invasive infections in hospitalized patients. *C. auris* is resistant to commonly used antifungal drugs. About 30% to 60% of patients infected with *C. auris* die. Infections are spread through contact with contaminated environmental surfaces or equipment and from person-to-person contact. It is difficult to identify using standard laboratory procedures. The rise of this multidrug-resistant pathogen is troubling and unexpected. CDC experts are quite concerned about this emerging pathogen.

Summary

Protozoans are microscopic, eucaryotic, unicellular organisms; helminths are macroscopic, eucaryotic, multicellular organisms. Some fungi are microscopic, eucaryotic, unicellular organisms; others are macroscopic, eucaryotic, and multicellular. All of these groups contain species that are pathogens of humans. Depending on the specific organism, transmission is by food or water, vectors, sexual contact, and direct (other than sexual) contact.

The diagnosis of protozoan and worm diseases is based largely on microscopic examination of body discharges, blood, and secretions for the presence of parasite eggs, larvae, or mature forms. In some cases, examination of skin biopsies, bone marrow, and spinal fluid may be necessary. Symptoms and history, particularly travel history, are important parts of the diagnostic process.

Many of the protozoan and helminthic diseases are found in rural regions of tropical and subtropical areas that suffer from poverty and poor sanitation. Strategies of prevention and control are targeted at vector abatement, minimizing contact between humans and parasites, and implementation of sanitary measures. Sanitation is extremely important and includes the use of latrines, provision of safe drinking water, avoidance of nightsoil (human feces as fertilizer), proper cooking and preparation of foods, and the wearing of shoes.

Drugs are available for the treatment and, in some cases, prevention of these diseases, but the treatment may be worse than the disease. There are no vaccines. Protozoans are unicellular eucaryotic organisms, as are humans, meaning that drugs toxic to them are also toxic to humans. The same is even more true for multicellular eucaryotic worms.

The fungi are a diverse group with extremely complex life cycles. Some fungi are beneficial, others are a nuisance value, and others cause disease in humans. Hence, it is a juggling act to choose the drugs that are more toxic to protozoans, worms, and fungi that cause diseases in humans than they are to the human host.

KEY TERMS

afebrile
African sleeping sickness
African trypanosomiasis
ahi
airport malaria
amastigotes
amebiasis
anemia
antigenic variation
antigens
ascariasis
asexual reproduction
athlete's foot
atypical pneumonia
autoinfection
babesiosis
Baghdad boil
beaver fever
bed nets
binary fission
bird fancier's lung (BFL)
biting plates
blastomycosis
blood flukes
blood meal
brain-eating amoebas
Botox
Candida auris
candidiasis
carriers
cave disease
cercariae
Chagas disease
climate change
clotrimazole
coccidioidomycosis
conjugation
copepods
cryptococcosis
cryptosporidiosis
cutaneous leishmaniasis
cyst
cysticerci
cysticercosis
dermatomycosis
dermatophytes
dermatophytic fungi
dichlorodiphenyltrichloroethane (DDT)
dioecious
dracunculiasis
elephantiasis

encystation
endemic
enterobiasis
erythrocytes
espundia
excystation
farmer's lung
filarial
flatworms
free-living
gamatocytes
giardiasis
guinea worm disease
hamachi toro
helminths
hermaphroditic
histoplasmosis
hookworm disease
host
hyperinfection
hypersensitivity pneumonitis (HP)
hypoxic
incidental hosts
infectious disease
infectious dose
infective dose
inflammation
intermediate hosts
interstitial lung disease
invasive candidiasis
ivermectin
jock itch
kala-azar
keratin
ketoconazole
kissing bug
larvae
leishmaniasis
locomotion
lymph
lymphatic filariasis
lymphatic system
macrophages
malaria
malaria prophylaxis
massive drug administration
merozoites
microfilariae
micronazole
mosquito abatement programs
mucormycetes

mycelial phase
mycoses
neglected parasitic infections (NPIs)
nematodes
neurocysticercosis
New World hookworm
niclosamide
normal microbiota
offal
Ohio Valley disease
Old World hookworm
onchocerciasis
oocysts
opportunistic infections
Oriental sore
oxyuriasis
paroxysm
parthenogenesis
pellicle
penicillin
phagocytes
pinworm disease
plankton
praziquantel
prednisone
proglottids
promastigotes
protozoans
reservoirs
ringworm
river blindness
roundworms
sand flies
sashimi
schistosomiasis

scolex
Scotch tape test
secondary bacterial infections
self-infection
sick building syndrome
sporozoites
steroids
streptomycin
strongylodiasis
sushi
tachypnea
terbinafine
thrush
toxoplasmosis
trichinellosis
trichomonads
trichomoniasis
trichuriasis
triotomid insect
trophozoites
tropical boil
trypanosomes
tsetse fly
uni
vaginitis
Valley fever
vector
venules
visceral leishmaniasis
whipworms
white-nose syndrome
World Malaria Day
World Mosquito Day
xenodiagnoses
zoonosis

SELF-EVALUATION

○ **PART I: Choose the single best answer.**

1. Mosquitoes are vectors for _____.
   - a. leishmaniasis
   - b. giardiasis
   - c. malaria
   - d. Chagas disease

2. Intestinal blockage is most likely to occur after infection with _____.
   - a. filariasis
   - b. tapeworms
   - c. earthworms
   - d. *Candida albicans*

3. Which emerging fungal pathogen is a multidrug-resistant global health threat today?
   - a. *Candida auris*
   - b. Ringworm
   - c. *Cryptosporidium*
   - d. *Plasmodium* spp.

4. Drinking what appears to be clean water from a mountain stream may put you at risk for infection with _____.

    a. guinea worms
    b. *Giardia*

    c. *malaria*
    d. tapeworms

5. Which parasite can be acquired from ingestion of eggs?

    a. *Trypanosoma*
    b. *Entamoeba*

    c. *Ascaris*
    d. *Blastomyces dermatitidis*

6. Which disease can be acquired from eating raw Alaskan salmon?

    a. Hookworm
    b. Diphyllobothriasis

    c. Guinea worm disease
    d. Schistosomiasis

7. Ringworm is caused by a _____.

    a. dermatophyte
    b. neophyte

    c. saprophyte
    d. parasite

8. Which of the following is *not* a fungal infection?

    a. Tinea pedis
    b. Candidiasis

    c. Jock itch
    d. River blindness

9. Which of the following causes thrush?

    a. *Candida albicans*
    b. *Trichophyton*

    c. *Toxoplasma*
    d. *Plasmodium*

10. Which of the following causes sick building syndrome?

    a. *Stachybotrys*
    b. *Histoplasma*

    c. *Batrachochytrium*
    d. *Penicillium*

⬤ PART II: Match the statement on the left with the disease (or microbe) on the right. A letter can be used more than once or not at all.

1. Worm migrates out through skin
2. Valley fever
3. Invasive candidiasis
4. Ingested cysts
5. Pneumonia
6. Malaria
7. Blastomycosis
8. River blindness
9. Chagas disease
10. Cats are carriers
11. *Ascaris*
12. Whipworm disease
13. Tapeworm
14. Ringworm
15. Ohio Valley disease
16. Filariasis
17. Hookworm disease
18. Thrush
19. Sick building syndrome
20. Pinworm disease
21. Hypersensitivity pneumonitis
22. Salmon sashimi

a. *Onchocerca*
b. *Pneumocystis jirovecii*
c. *Giardiasis*
d. Dermatophyte
e. *Plasmodium* spp.
f. *Blastomyces*
g. Roundworm
h. *Diphyllobothrium nihonkaiense*
i. *Candida auris*
j. *Necator americanus*
k. Inflammation of the lungs
l. *Trypanosoma*
m. *Toxoplasma*
n. *Wuchereria*
o. *Coccidioides immitis*
p. *Histoplasma*
q. *Candida albicans*
r. *Enterobiasis*
s. *Dracunculus*
t. *Stachybotrys*
u. *Trichuris*
v. *Taenia*

Fill in the vector for each disease.

Disease   Vector (Transmission)

1. Leishmaniasis

2. Chagas disease

3. Sleeping sickness

4. Malaria

5. Onchocerciasis

6. Dracunculiasis

7. Filariasis

○ **PART III: Answer the following.**

1. Choose a protozoan disease that is transmitted by food or water and describe its life cycle. A diagram would be helpful.

2. You are a health worker involved in the eradication of guinea worm disease. Prepare an article to be handed out to the inhabitants of a village telling them about the disease and how to avoid becoming infected.

3. You are the director of public health services in an African country that is plagued with a variety of protozoan and helminthic diseases. Prepare a document outlining your plans to reduce the number of people who will become infected over the next years. (A budget will be developed based on your agenda.)

4. Create a short list of fungi that produce antibiotics. Discuss why soil fungi produce antibiotics.

5. Explain why is it difficult to remediate buildings that are contaminated with *Stachybotrys*.

6. You will be traveling outside the United States to an area where malaria is endemic. What precautions will you take to prevent coming down with malaria?

7. Explain why a pregnant woman should not maintain a pet cat's litter box.

8. Why is trichinellosis relatively rare in the United States today?

9. Summarize why there are more cases of helminthic diseases in the tropics compared to the United States.

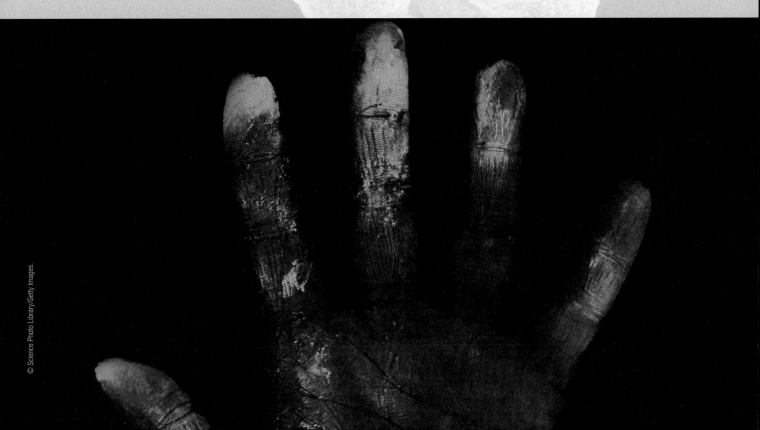

PART 4

MEETING THE MICROBIAL CHALLENGE

CHAPTER 15

THE IMMUNE RESPONSE

Terri S. Hamrick, PhD
Campbell University

> "*The immune system is complex; if it was simple, it would be practically useless.*"
>
> —Unknown

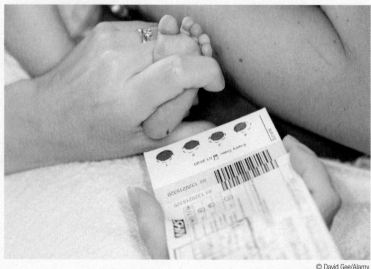

NIH. Test reliably detects inherited immune deficiency in newborns. Retrieved from https://www.nih.gov/news-events/news-releases/test-reliably-detects-inherited-immune-deficiency-newborns.

© David Gee/Alamy.

The immune system is critical for fighting the many antigens we encounter in our lives. A newborn who does not have an immune system at birth is subject to a multitude of serious infections. A screening test for severe combined immunodeficiency (SCID) is available and has been adopted as part of routine newborn screening in many, but not all, states. The Recommended Universal Screening Panel (RUSP) collects a small amount of blood from the newborn's heel as shown in the picture. The collected sample is used to screen for a number of disorders.

LEARNING OBJECTIVES

1. List the cells and organs of the immune system.

2. Identify the principal components of innate and adaptive immunity.

3. Explain the mechanisms utilized by innate and adaptive immunity to identify self and nonself (foreignness).

4. Describe the roles of lymphocytes (T cells and B cells) in the adaptive immune response, including their activation and effector functions.

5. Identify hypersensitivity and autoimmune disorders.

6. Classify the different types of vaccines commonly used in the United States, including the advantages and disadvantages of each.

7. Explain passive and active immunity and provide examples of natural and artificial immunity of each.

Case Study: Baby Jacob

The day Jacob was born was a very happy day for his parents. He was their first child, and his mother, Elizabeth, was looking forward to being a stay-at-home mom for his first 6 months. She had a normal pregnancy and uneventful delivery. He was born near his due date and weighed a healthy 7 pounds, 3 ounces (3.2 kilograms) at birth. Before he left the hospital, he received his first hepatitis B vaccine. At his 2-month well-baby check, he received the following vaccines: rotavirus (RV5); diphtheria, tetanus, and acellular pertussis (DTaP); *Haemophilius influenzae* type b (Hib); pneumococcal conjugate (PCV13); and inactivated poliovirus (IPV). Six days later, he returned to the pediatrician with severe diarrhea and difficulty breathing. He had lost weight since his last visit and was now small for his age on the growth chart. The pediatrician sent him straight to the hospital. Bloodwork showed severe lymphopenia (very low lymphocyte count) at 135 cells per cubic millimeter (normal is 2,500 to 16,500). Rotavirus was detected in stool samples. A bronchoscopy was performed, which included collection of an aspirate sample. Tests of the aspirate revealed *Pneumocystis jirovecii*. Testing of a nasal-wash specimen confirmed respiratory syncytial virus (RSV) as well. Jacob was very sick.

Several clues pointed to a problem with Jacob's immune system, including the diagnosis of an infection with *P. jirovecii*. Rotavirus and RSV were not uncommon in infants, but *P. jirovecii* is a fungal infection that is typically only associated with people who have a medical condition that weakens the immune system. His physicians tested him for a number of conditions that might be the cause of an immune system problem.

Genetic analysis revealed homozygous mutations in a gene known as recombination activating gene 1 (RAG1). Homozygous means that Jacob inherited a mutant (nonfunctional) gene from each of his parents. RAG1 is a gene that encodes for one of the enzymes that is involved in making the genes that encode the receptors for the surface of lymphocytes. This means that Jacob could not make the genes for the T-cell receptor of T cells nor the immunoglobulins for B cells. Without these genes, his lymphocytes never completed their development. Without lymphocytes, Jacob's immune system was very weak, a condition known as severe combined immunodeficiency (SCID). Several different mutations can lead to SCID, all of which involve the loss of functional T cells and often B cells, too.

Jacob received excellent care in the hospital, including hydration for diarrhea and medicine for the fungal lung infection. At 8 months of age, he received transplantation of bone marrow from a relative. He needed medications to help the transplanted bone marrow stem cells settle into his body without excessive immune reactivity, but by 14 months of age he had functional lymphocytes. His stool samples remained positive for rotavirus for many months, but by 14 months they were negative.

Testing of the rotavirus isolated from him revealed that the virus in his system was the live attenuated strain used in his vaccine. Unfortunately for Jacob and his family, he lived in a state that did not routinely screen for SCID as part of routine newborn screening. Many states now include SCID screening. The Advisory Committee on Heritable Disorders in Newborns

(continues)

Case Study: Baby Jacob (continued)

and Children advises that all newborns be tested using a Recommended Universal Screening Panel (RUSP), which, as of November 2016, included 34 core and 26 secondary conditions. SCID is one of the newest additions to the RUSP, and not all states screen for it yet.

Questions and Activities

1. What role do lymphocytes play in responding to infections?

2. Why did Jacob appear normal at birth?

3. What immune defenses were still working for Jacob even though he did not have lymphocytes?

4. Why was a bone marrow transplant from a healthy donor able to cure his condition?

5. Find out more about *Pneumocystis jirovecii*. What other diseases are commonly associated with it?

6. The rotavirus vaccine is typically very safe for infants to receive, even at 2 months. Look up the recommendations for rotavirus vaccine administration. List the

contraindications for rotavirus vaccine administration. Which one applied to Jacob?

7. Jacob received several vaccines at his 2-month visit. Look up the formulations of the other vaccines. Why was the rotavirus vaccine the only one to cause problems for him?

8. Look on the Internet to identify conditions for which your state screens newborns. Does your state screen for SCID? (Screening recommendations are made by a federal advisory committee, but the Department of Health for your state probably has the current list.)

9. If a newborn is found to have SCID, what precautions do you think will need to be taken?

Information from N. C. Patel, P. M. Hertel, M. K. Estes, M. de la Morena, A. M. Petru, L. M. Noroski, . . . S. L. Abramson. (2010). Vaccine-acquired rotavirus in infants with severe combined immunodeficiency. *New England Journal of Medicine, 362*(4), 314–319; D. Bogaert, K. Van Schil, T. Taghon, V. Bordon, C. Bonroy, M. Dullaers, . . . F. Haerynck. (2015, August 6). Persistent rotavirus diarrhea post-transplant in a novel *JAK3*-SCID patient after vaccination (Letter to the editor). *Pediatric Allergy and Immunology, 27*, 92–106.

Preview

Based on your own experiences or even just watching the news, you probably have a good idea of the many different types of microbes in the world and some of the infectious diseases that can come about because of them. You may even be scared to leave your house. In this chapter, you will learn that your immune system is ready for many of these germs. You will learn that your immune system is poised to act on the microbes you encounter today, and is ready for those that you have not yet encountered, too.

The word **immunity**, in its broadest sense, means "freedom from a burden," be it legal action, taxes, or, in the present context, disease. In a legal connotation, the term means to be immune from civil liability that can be granted to a charitable or nonprofit organization; a hospital is a good example. Witnesses who may have been involved in criminal activity can be granted immunity from prosecution so they may testify against those involved in more serious crimes. In the early years of the colonization of the United States, the battle cry of the colonists was "immunity from taxes."

The ultimate outcome of infection is the result of the dynamic interplay between the microbe and its virulence factors and the function of the host immune system. The interplay determines whether no infection, mild infection, or severe and, possibly, fatal infection will result. The early hours of encounter between the microbe and the immune

system are particularly crucial in determining the eventual outcome.

The immune system functions to recognize and destroy foreign molecules as embodied in invading microbes and their products and in mutant, damaged, and worn out cells. This is accomplished by both **innate immunity** and **adaptive immunity**. As their names imply, innate immunity is preexisting, whereas adaptive immunity develops more slowly and provides defenses that are more specialized and effective because they have adapted to the specific infection. Innate immunity is always present in healthy people and serves to prevent entry of pathogens or to act quickly to eliminate them if they enter. Adaptive immunity, also referred to as specific or acquired immunity, requires selection and specialization of immune cells in response to the presence of foreign components (molecules). Impairment of the immune system as occurs in acquired immune deficiency syndrome (AIDS), leukemia, use of immunosuppressive drugs, or congenital disorders renders the immunocompromised individual subject to repeated life-threatening infections.

Basic Concepts

Immunology is a challenge to learn, in part because all of the components and functions are interconnected. It is hard to decide where to start. You will understand it the best when you are able to picture all of the parts working

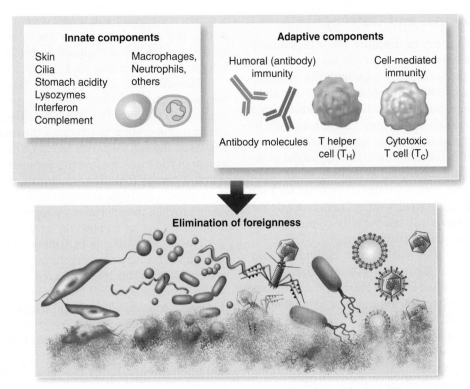

FIGURE 15.1 Innate and adaptive components of the immune system protect against foreignness.

together. This very basic overview will give you the framework upon which to put more details from the subsequent sections. Try to keep reminding yourself of the connections as you read the rest of this chapter.

The immune system functions to recognize and destroy molecules that are foreign to the body. Foreign includes molecules on invading microbes and their secretions, toxins, and enzymes, but foreignness can also exist on cells of the body that have become "non-normal" self cells if they become cancerous or are infected by a virus. Functionally, the immune system consists of innate immunity, which includes the defensive components of the skin and mucous membranes, as well as quick cellular and noncellular responses, and adaptive immunity, which takes time to develop initially but then provides lasting protection that is highly specific for the molecules called **antigens** that elicited the response. (**FIGURE 15.1**).

Think about innate and adaptive immunity this way. Say that a man is going to a formal event for which he will need a tuxedo. He could go to a rental store, where he will be able to get one pretty quickly. It might not fit exactly right, but it will be mostly appropriate for his event. He will need to return it when the event is over, so if he wants to use this tuxedo again he will need to rent it again. The fit and style will be the same each time—mostly ok, but probably not exactly right. This is like the innate immune system. It is a pretty good system for the infection, and it is available pretty quickly, but it will not be an exact match for the pathogen. If the man wants to be ready for important events, it might be better for him to buy a custom-made tuxedo. He will be fitted for it using a few samples that the tailor has in the store, but his final suit will not be ready for a week or more. In the meantime, he can wear his rental tuxedo. When his new tuxedo comes in, it will be exactly what he wants for his event, and he will have it forever. Anytime that event comes up again, he will have his custom-made tuxedo ready to go. This is what adaptive immunity is like. Adaptive immunity is essentially custom-made from a small number of immune cells in response to a specific infection. While the adaptive immune response is developing, innate immunity can be working on the infection. After the adaptive immune response (tuxedo) is made, it will be always ready, but unlike a tuxedo that could be worn to any formal event, an adaptive immune response is highly specific for only one infection.

Krasner's Notebook

In writing this text the original author was initially in a dilemma. Should the discussion of the microbes and their virulence mechanisms precede explanation of the body's immune defense, or the other way around? It is the "what came first, the chicken or the egg?" puzzle. Because a strong focus in this text is on disease prevention, it seemed to him more logical to first present what it is that the immune system is combating. It makes sense to me, and I hope you agree!

Efficacy, Variations, and Detriments

There is no one good way to measure the effectiveness of the immune system because of its many complex interactions. In fact, the ability to cope with the constant barrage of microorganisms is not constant, nor is it measurable. Age, sex, race, nutrition, and health status, along with physical and mental stress, play a significant role in immune function. Think about final exam time. Lights are on late in dormitories and in study halls around the campus, and fast foods are consumed at an even more rapid rate than usual as you review a semester's work for four or five courses in about a week's time. There is no doubt that your immune system is temporarily compromised, making you more vulnerable to infection.

Immune status varies within the individual and between individuals. Consider, for example, that 100 people may be crowded into a lecture hall, movie theater, or restaurant, and that all are exposed to circulating cold viruses. Some will "catch the cold," and others will not. Although they were all exposed to the same pathogen with the same virulence factors, whether infection results is largely a function of the immune status at that time. A recent news article reported the death of an 84-year-old man because of a West Nile virus infection, which was acquired from the bite of an infected mosquito. In this case, and for most people who develop severe West Nile virus encephalitis, the patient's advanced age was a factor. Immune system function wanes in the elderly. Genetics, too, plays a key role; it is well established that there are ethnic and racial differences in susceptibility to microbial infection.

Although the primary role of the immune system is as a mechanism of defense against microbial infection, it is also a system of internal surveillance leading to elimination of tumor cells, repair of damaged tissue, and destruction of old, worn out, and damaged cells.

The immune system is beneficial for the most part, but there is another side to the coin. Sometimes the immune system goes into overdrive, leading to harm. This is known as a **hypersensitivity reaction**, which you may know as an allergic reaction. Many people suffer from allergies, which are adverse immune responses to protein molecules associated with pollens, dust, foods, mites, antibiotics, and bee stings, to name a few common **allergens**. These allergies run the gamut from being a nuisance (sneezing, watery eyes, and runny nose) to life-threatening anaphylactic shock. **Anaphylactic shock** occurs when the body releases an overwhelming amount of **histamine** in response to an allergen. The histamine causes the blood vessels to dilate, which lowers blood pressure, and severely constricts the bronchioles in the lungs, making breathing very difficult. The truth of the matter is that an **allergy** can develop at any time. A dramatic example is the case of a student who died of an allergic reaction after eating shrimp in the dining hall at his college. Unfortunately, many similar examples can be offered. Before prescribing an antibiotic, a physician should ask whether you are allergic to any antibiotics and not prescribe one to which you are allergic. Otherwise, the consequences could be worse than the infection for which you are being treated.

Autoimmune disease is the failure to distinguish between nonforeign (self) and foreign (nonself) antigens and is also technically a hypersensitivity. In other words, the immune system is attacking the body's own cells and tissues as though they were foreign, a clearly nonbeneficial aspect of the immune system. Some examples of autoimmune diseases include rheumatoid arthritis, lupus erythematosus, multiple sclerosis, and a variety of anemias. Paul Ehrlich, a pioneer immunologist, referred to this dysfunction as "*horror autotoxicus*," or a fear of self-poisoning.

The major obstacle to transplantation of organs is rejection of transplanted organs by the recipient's immune system. Should this be considered a nonbeneficial aspect of the immune system? Most immunologists would not think so, because, after all, the recipient is responding to foreign molecules of the donor cells in a manner nature evolved to thwart microbial invasion and to conduct internal surveillance. Transplantation technology is not a natural phenomenon but is the result of advanced medical science.

Foreignness

"Who am I?" is a simple question but is a critical one for effective functioning of the immune response. When the immune system encounters cells or molecules, it must determine whether they are self, meaning they belong in the body, or nonself (foreign). The innate and adaptive immune systems use different strategies to differentiate self from nonself.

The cells and immune components of the innate immune system rely on **pathogen-associated molecular patterns (PAMPs)** to identify and respond to an infection. PAMPs are components that microbes have but human cells do not, and are thus a direct way for innate immune components to identify a microbe as foreign and respond accordingly. Examples of PAMPs include cell wall components, flagella, and certain carbohydrate structures. They do not all look exactly the same for all microbes, but they have similar structure. Cells of the innate immune system, most notably many of the white blood cells, or **leukocytes**, have **pattern recognition receptors (PRRs)** that recognize PAMPs. The leukocytes important in innate immunity include **phagocytic cells** such as macrophages, neutrophils, and dendritic cells that can engulf microbes. **Natural killer (NK) cells** are cells of innate immunity that can identify stressed or infected cells and destroy them. Innate immunity plays an important role in initiating the adaptive immune response by alerting the cells involved in adaptive immunity, which can then examine the microbial remains as they prepare to respond (**FIGURE 15.2**).

Adaptive immunity utilizes a different strategy for identifying foreignness. The leukocytes important for adaptive responses are **T lymphocytes (T cells)** and

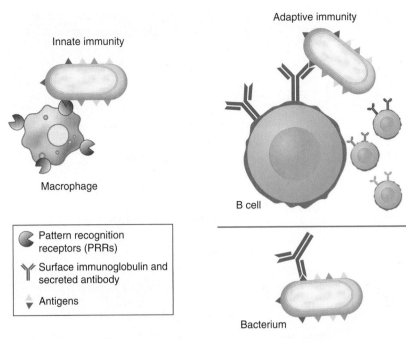

FIGURE 15.2 Recognition of foreignness by macrophages of innate immunity and B cells of adaptive immunity. Phagocytic cells of innate immunity have pattern recognition receptors (PRR) that recognize molecular patterns common to many pathogens (pathogen-associated molecular patterns, or PAMPs). The PRR–PAMP binding might not be an exact fit but can be helpful for the innate response. Immunoglobulin on B cells and its corresponding secreted antibodies can be an exact fit for microbial antigens. Notice that not all of the B cells "fit" for this particular infection. T cells have a similar cell surface receptor that can also have a nearly exact match to a microbial component.

B lymphocytes (B cells). These lymphocytes have receptors able to identify and bind to a far greater assortment of molecular components than the PRRs of innate immunity. Each lymphocyte builds the genes that encode for the proteins that will assemble as a receptor that can detect, with extraordinary precision, a highly specific molecular component. The structural component recognized by a lymphocyte receptor is called an antigen. A person's repertoire of lymphocytes contains a very small number of lymphocytes with any given antigen specificity. When an antigen enters the body, lymphocytes able to respond to that specific antigen will proliferate and differentiate, leading to the important immune response, which, for a new antigen, takes about a week or two for full immune activity.

It is theoretically possible to develop B cells and T cells that can recognize virtually every substance that you may encounter. Due to the vast array of potential lymphocyte receptors, during early development lymphocytes go through a selection process to make sure they are not reactive to self-antigens. During this selection process, any self-reactive lymphocytes are eliminated from the repertoire as they are being made. However, sometimes self-reactive lymphocytes slip through, which can lead to autoimmunity.

Anatomy and Physiology of the Body's Defenses

Having now established a few basic concepts involving the immune system, attention needs to be focused on anatomical and physiological considerations. Perhaps in a precollege course in biology you dissected a frog; recall that the digestive system is a series of defined anatomical structures starting with the oral cavity, leading sequentially to the esophagus, the stomach, and the intestinal tract. The same organization can be said for all the systems of the body, in that there is a hierarchy of cells, tissues, and organs in biological systems. The immune system is, however, unique in that the anatomical structures are often shared with more defined systems. The spleen, in most anatomy manuals, is described with the circulatory system, and the lymph nodes are considered components of the lymphatic system, but these structures are also parts of the immune system.

Leukocytes

The three categories of blood cells are red blood cells (**erythrocytes**), white blood cells (leukocytes; **FIGURE 15.3**), and **platelets** (thrombocytes, which are differentiated cell fragments, important in clotting). Leukocytes or their products are responsible for most immune functions.

White blood cells found in the circulation are divided into five categories based on morphology (**TABLE 15.1**). **Neutrophils** (also known as polymorphonuclear cells or PMNs), basophils, and eosinophils are referred to as **granulocytes** because microscopic examination reveals the presence of granules in their cytoplasm. **Lymphocytes** and **monocytes** are referred to as **agranular cells**, because they do not contain visible cytoplasmic granules. The lymphocytes include the B lymphocytes (B cells) and the T lymphocytes (T cells), which are the key cells of adaptive immunity,

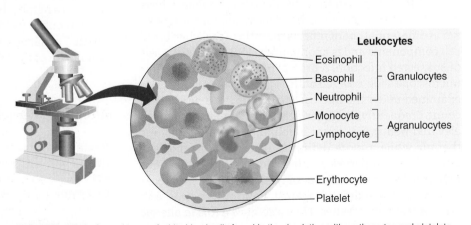

FIGURE 15.3 Several types of white blood cells found in the circulation with erythrocytes and platelets.

TABLE 15.1 Categories of White Blood Cells (Leukocytes)

Cell Type	Percentage of Total Blood Leukocytes	Function(s)
Agranulocytes		
Lymphocytes	25–35%	Antibody formation, cell-mediated immunity
B lymphocytes (B cells)		Produce antibody response
T lymphocytes (T cells)		Necessary for adaptive immunity
Natural killer (NK) cells		Innate response to viral infection
Monocytes	3–8%	Phagocytosis
Granulocytes		
Neutrophils	60–70%	Phagocytosis
Eosinophils	2–4%	Inflammatory response, limited phagocytosis
Basophils	0.5–1%	Not clear; contain histamine

and the natural killer (NK) cells, which are involved in innate immunity. Under the microscope, B cells and T cells look similar, but they have very different functions. Upon further maturation and help from other immune cells, the B cells become antibody-secreting cells known as **plasma cells**, which become stationed in various sites in the body. During an adaptive immune response, T cells also mature into cells with specific functions (discussed in the section on adaptive immunity).

Neutrophils and monocytes are phagocytic cells in the blood and circulation. **Phagocytosis**, or "cell eating," is an important defense mechanism by which microbes are engulfed and destroyed; phagocytosis is described in more detail under the section on innate immunity. Monocytes that enter the tissues of the body are known as **macrophages**. **Dendritic cells** are another phagocytic cell found in the tissues of the body. Dendritic cells and macrophages have important roles in initiating the adaptive immune response.

A **differential count** reflects the ratio of the white blood cell components and is an important tool in the diagnosis of suspected infection. A drop of blood is smeared onto a microscope slide, covered with a specific stain, rinsed, and examined under the microscope to perform a differential count. Frequently, in bacterial (versus viral) infections an elevated neutrophil count is found, justifying antibiotic therapy while a more definitive laboratory diagnosis is awaited. By comparison, with the viral disease infectious mononucleosis, the number of lymphocytes is elevated.

Eosinophils defend against helminth worms and other intestinal parasites. They also have a role in allergic responses and asthma. The role of basophils is not entirely clear. **Basophils** are rich in granules of histamine that are released during allergic reactions and are responsible for watery eyes, itchy throat, sneezing, and runny nose. **Mast cells**, which reside in all connective tissues, have granules like those of the basophil, but mast cells are developmentally distinct from basophils. The activation and degranulation of mast cells contributes to the inflammation seen during infections and allergic responses. Antihistamine medications, of which there are many over-the-counter choices available, can provide symptomatic relief for allergies. You may be one of the unlucky sufferers. Don't feel too bad; approximately 50% of the population has allergies.

Blood Plasma

The noncellular part of blood is referred to as **plasma**. This fluid is straw colored and somewhat viscous. It consists of water and a number of proteins (**TABLE 15.2**). Some of these proteins contribute to innate immunity, adaptive immunity, or both. The globulin fraction is particularly important to the function of the immune system; a subfraction, **gamma**

Author's Notebook

The term *antigen* was applied to components (molecules) that when injected were *antibody generators*—thus *antigen*. We now know that antibodies are not the only aspect of adaptive immunity that has specificity. T cells have specificity similar to that of antibodies, so the term antigen is now broadened to mean the components that lead to an adaptive immune response, whether it is binding the T cell receptor, the B cell immunoglobulin, or antibodies.

TABLE 15.2 Composition of Blood

Component	Function
Plasma	
Water (80–90%)	Solvent
Proteins	
Albumin	Blood volume
Globulins	Antibody molecules
Fibrinogen	Blood clotting
Complement	Immune amplification
Interferon	Antiviral properties
*Blood cells**	
Red blood cells (erythrocytes)	Oxygen and carbon dioxide transport
White blood cells (leukocytes)	Antibody formation, cell-mediated immunity, phagocytosis
Thrombocytes (platelets)	Blood clotting

* Each microliter of blood has about 4.5 to 5.5 million erythrocytes, 5,000 to 10,000 leukocytes, and 250,000 to 400,000 thrombocytes.

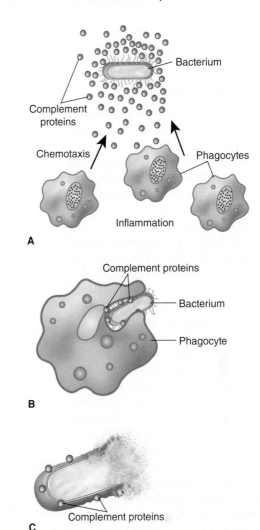

FIGURE 15.4 Functions of complement. Complement components are initially made in an inactive form. **(a)** When complement becomes activated near a microbe, some of the complement pieces float away to set up a concentration gradient for phagocytes to follow as they migrate to the site of the infection. These pieces also promote inflammation. **(b)** Complement attached to the microbe acts as a handle for phagocytes to use. **(c)** Further activation of complement and assembly at the bacterial surface forms a pore in the microbe.

globulin, contains most of the antibodies. Antibodies have a number of jobs within an effective immune response, including neutralization of antigens and recruiting other immune components to act on bound antigens.

Complement, a series of blood proteins, is a significant defense mechanism against potential disease-causing microbes. Complement proteins are made in an inactive form, and can be activated in a cascadelike manner as part of innate immunity or an antibody can trigger the cascade as part of adaptive immunity. Complement is a system of biological amplification. Consider the role of the amplifier in a music system as an analogy. It does not actually produce the music; if the amplifier were unplugged the music would continue, but the quality would suffer. Complement activation amplifies the immune response in three ways (**FIGURE 15.4**):

1. Complement activation leads to **chemotaxis** of phagocytes and initiates inflammation.

2. Complement enhances phagocytosis, a defense mechanism by which bacteria are engulfed and destroyed.

3. Complement can bring about lysis (destruction by building a pore) of the target cell, such as a bacterium, causing it to "spill its guts" through the now-leaky membrane.

Interferon, another component of blood, was discovered in 1957, and was so named because of its ability to interfere with viral replication. It is not a single component but a group of related compounds. Interferon is released from virus-infected cells and acts to signal nearby cells to be on the defense against viral infections. When interferon was discovered, the medical community had lofty expectations that it would prove to be a powerful antiviral drug that would be as effective against viruses as antibiotics are against bacteria. Unfortunately, interferon has not lived up to expectations. It has been only partially successful in the treatment of certain forms of hepatitis.

The Lymphatic System

Good immune function requires effective communication between cells. This is especially true in the early stages of the adaptive immune response, when the important lymphocytes may be in very small numbers. Rather than rely on the appropriate cells to find each other anywhere in the body, the vast majority of lymphocytes are located in specialized tissues known as **lymphoid tissues** or **lymphoid organs**. The major lymphoid organs are shown in **FIGURE 15.5**.

The **lymphatic system** is anatomically connected to the blood circulatory system. **Lymph**, a tissue fluid derived from blood but without blood cells, is collected by the lymphatic vessels, which then open into the blood circulatory system, allowing return of lymphatic fluid to the circulatory system. In elephantiasis, worms block the lymphatic vessels, leading to the grotesque elephant-like limbs. Hence, the two systems are interdependent: lymph is derived from blood and drains back into blood (**FIGURE 15.6**). Whereas the blood is driven by a pump (the heart) and is a true circulatory system, the lymph vessels move lymph fluid without the assistance of a pump. One-way valves in the lymph vessels only allow the lymph fluid to go one direction. When you move or use your muscles, you are helping to move your lymph. Situated along the path of the lymphatic vessels are the lymph nodes.

Primary Immune Structures

Primary, or central, lymphoid tissues are where lymphocytes develop and mature to the point where they can respond to a specific pathogen. The primary lymphoid tissues are functionally divided into the bone marrow and the thymus.

○ Bone Marrow

In adults, the **bone marrow** is the major site of **hematopoiesis**, which is the process by which all blood cells are formed (i.e., red blood cells, white blood cells, and

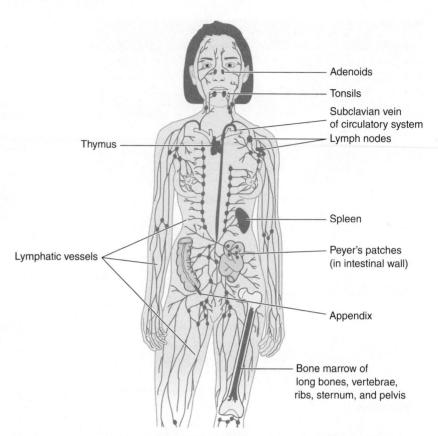

Adenoids

Tonsils

Subclavian vein of circulatory system

Lymph nodes

Thymus

Spleen

Peyer's patches (in intestinal wall)

Lymphatic vessels

Appendix

Bone marrow of long bones, vertebrae, ribs, sternum, and pelvis

FIGURE 15.5 Anatomy of the immune system. Lymphocytes develop in the primary lymphoid organs shown in red: the bone marrow and thymus. Immune stimulation, activation, and communication for adaptive immunity occur in the peripheral lymph tissue also known as the secondary lymph tissues, shown in purple. The lymph vessels open into the subclavian veins of the circulatory system to return the lymph fluid to the blood volume.

platelets). Because blood cells are short-lived, they have to be continuously generated from **hematopoietic stem cells** located in the bone marrow of the skull, ribs, sternum (breastbone), vertebrae, pelvis, and femurs. There are two pathways of blood cell maturation (**FIGURE 15.7**): the myeloid path and the lymphoid path. The **lymphoid lineage** leads to lymphocytes—B cells, T cells, and NK cells—whereas the **myeloid lineage** leads to essentially all of the other blood cells. Most cells leave the bone marrow as mature cells, ready to perform their functions. When the B cells leave the marrow, they enter the circulation and then populate the secondary structures of the immune system. When T cells leave the bone marrow, they are not quite ready to function in an immune response.

○ Thymus

T cells leave the bone marrow in an immature state and migrate to the **thymus** (not thyroid), an organ located behind the sternum and just above the heart (Figure 15.5). In the thymus the T cells build their genes that encode for their antigen-specific receptors and learn what is self. When T cells leave the thymus, they go to the secondary, or peripheral, lymphoid tissues, ready to respond to an infection.

At birth, the thymus is a relatively large organ and attains its maximum size by puberty. The early years are the ones in which numerous childhood diseases and

---

Author's Notebook

If you have been sitting still reading for a long time, it might be a good idea to stand up and move your lymph around (stretch). Hopefully, you haven't been reading so long that you have swollen legs, but when patients are immobile in the hospital, especially in the intensive care unit, their mattresses are programed to move and they receive regular physical therapy to move their arms and legs, in part so that lymph fluid won't collect in their extremities.

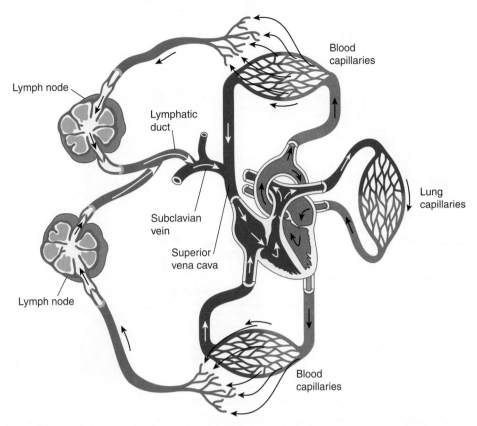

FIGURE 15.6 Interrelationship of the lymphatic and circulatory systems. The lymph vessels are like a drainage system to collect fluid, known as lymph, from the peripheral tissues and return it to the blood circulation. Along the way, the lymph passes through lymph nodes, which will survey the fluid for foreign antigens.

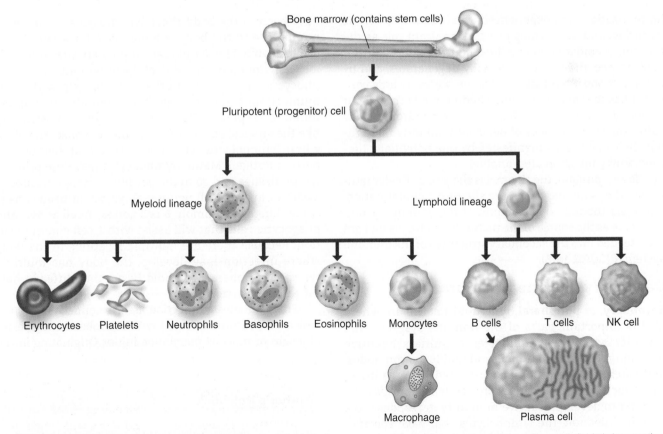

FIGURE 15.7 Blood cell lineages. All blood cells begin their maturation in the bone marrow and then mature along two pathways. Some cells finish their maturation in other locations throughout the body.

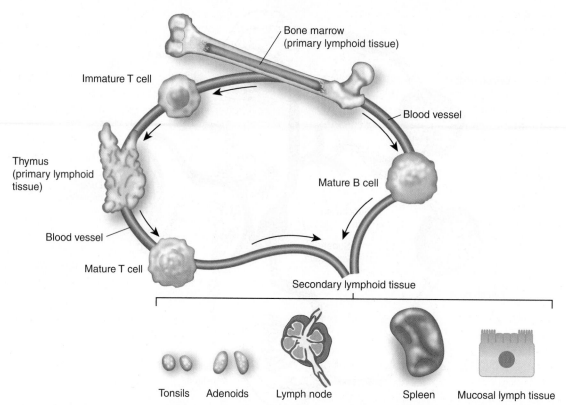

FIGURE 15.8 Maturation and "seeding" of B and T lymphocytes. Lymphocytes develop in the primary lymphoid tissues. Mature lymphocytes enter the secondary lymphoid tissues, where they will respond to foreign antigens. Some lymphocytes may circulate through the different secondary lymph tissues.

immunizations are experienced; the immune system is maturing, and the thymus plays a significant role as the training ground for building the repertoire of T cells. After adolescence, the thymus gradually degenerates, and by late adulthood it is mostly fatty tissue, with very little lymphoid tissue remaining. Some children are born without a thymus, a condition called **DiGeorge syndrome**. They suffer the consequences of deficient immunity throughout their lifetime, as manifested by one potentially life-threatening infection after another.

To recapitulate, the thymus is the site of T cell maturation; the bone marrow is the site of B cell maturation. These are the primary structures of the immune system and serve as the source of the mature T and B cells that are then seeded into the secondary structures of the immune system (**FIGURE 15.8**).

Secondary Immune Structures

Secondary, or **peripheral**, **lymphoid tissues** are where all the important parts of the immune system come together with the lymphocytes to initiate adaptive immunity. The **spleen**, **tonsils**, **adenoids**, lymph nodes, and patches of lymph tissue associated with the intestinal tract constitute the secondary lymphoid tissues. In order to develop adaptive immunity in response to a specific infection, the lymphocytes that have specificity for the antigens of that microbe need to encounter those antigens. If this is the first time these antigens

have been in the body, those lymphocytes are in small numbers, and the body is a big place for a very small cell to search. The peripheral lymphoid tissues provide a local environment where all of the important parts (lymphocytes, antigens, and other cells to help with communication) can come together to initiate an adaptive immune response. The secondary lymph tissues are like the singles' bars of the immune response: the place where lymphocytes can come to try to find their perfect partner antigen. Mature lymphocytes leave the primary lymph tissues and go to the peripheral lymph tissues to await an antigen. All peripheral lymphoid tissues have some things in common: **B cell zones**, **T cell zones**, and phagocytic cells that will assist with T cell communication. Because there are multiple locations in the body where infection may develop, the body has multiple types of peripheral lymphoid tissue that differ in how the antigens arrive.

Frequent sites of infection are the connective tissues, which can become infected with microbes as the result of a skin or mucous membrane injury. Originating in the

Author's Notebook

The human body contains about 450 lymph nodes strategically placed throughout the body.

connective tissue is the network of open-ended lymphatic vessels that collect the extracellular lymph fluid. Remember that lymph is plasma that continually leaks out of the blood vessels and into the tissues. The lymphatic vessels in the periphery connect to larger lymph vessels, ultimately returning the fluid to the circulation by emptying into the venous circulation. Lymph nodes are small kidney bean–shaped organs that are stationed along the lymphatic vessel network. The lymph fluid filters through the lymph node, allowing the lymphocytes in the lymph node the opportunity to examine any antigens that may have been present in the connective tissue and collected along with the lymph. Residing in the tissues of the body are dendritic cells, which are specialized cells for presenting antigens. They travel in the lymph to the nearest lymph node to deliver antigens from the tissues to the T cells in the lymph node.

If the infection is in the blood rather than the connective tissue, it would be important to act fast to initiate an adaptive immune response. Blood infections are dangerous! The spleen is a lymphoid organ that is organized much like a lymph node in that it has mature B and T cells as well as phagocytic cells that can help with communication. The difference is that the immune cells in the spleen survey the blood for antigens. The spleen is a spongy, fist-sized organ located in the upper-left portion of the abdominal cavity. Because of its location and its relatively thin connective tissue covering, it is subject to injury as might be sustained in an automobile accident. Severe trauma to this organ may result in its rupture, with severe hemorrhaging necessitating its removal (**splenectomy**). As you might expect, splenectomy patients may not be able to respond as quickly to infections in the blood. As such, a patient who has had a splenectomy needs to be very aware of any signs of an infection and should always be up-to-date on vaccinations.

Other common locations for microbial infections are the mucosal surfaces of the respiratory or gastrointestinal tracts, so we have multiple lymphoid aggregates in those sites. The tonsils and adenoids are located at the back of the throat, just in front of the pharynx. Their lymphocytes play a role in protection against microbes entering through the nose and throat. A generation or so back it was routine for young children to undergo removal of their tonsils

(**tonsillectomy**) because the swelling of these tissues was thought to contribute to the frequency of throat and ear infections; this is no longer a common procedure because of the availability of antibiotics. Doctors in the early 1900s often performed these operations in the home, perhaps on the kitchen table. Occasionally, tonsillectomy is performed on individuals with repeated infections that do not respond well to antibiotics. (As a benefit, you are indulged with a lot of ice cream and popsicles to help alleviate a severe sore throat!) Patches of lymphoid tissues are distributed in the lining of the gastrointestinal and respiratory tracts as well. In the small intestine, these collections are known as **Peyer's patches**. The source of antigens for the Peyer's patch is the lumen of the gut. Specialized cells sample the contents of the gut, and deliver those samples to the B and T cells in their respective zones within the Peyer's patch.

Via the different routes of antigen delivery to peripheral lymphoid tissues, essentially all antigens entering the body (usually as microbial components) can encounter cells of the immune system.

Immune Function: Innate and Adaptive Immunity

Recall that the immune system consists of two components: innate immunity and adaptive immunity (**FIGURE 15.9**). Adaptive immunity, in turn, has two arms: humoral, or antibody-mediated, immunity and cell-mediated immunity. Key to an effective immune response is identifying the threat and then determining which cells and immune components should act.

Innate Immunity

Innate immunity was originally thought to be nonspecific, mostly defensive, tactics to keep infection out of the body. However, we have since learned that some of innate immunity involves defensive strategies that function in a very general way to prevent infection, but other innate immune defenses are induced by the presence of an infection. These innate defenses need a few hours or a few days to be fully functional and are dependent on the type of pathogen. Innate immunity is still not as precise as adaptive immunity, but it does take into account the type of infection so cannot be thought of as nonspecific. For example, innate defenses recognize and respond to a viral infection differently than to a bacterial infection. Innate immune defenses are present since birth and do not require any previous exposure to the microbe. This means that innate immune defenses respond exactly the same each and every time the microbe is encountered, whereas adaptive immunity is more exact, but takes time to develop.

Innate immunity is characterized by physiological defenses that operate either to prevent microbes in the external environment from gaining access into the body or to eliminate them quickly if they are inside. **TABLE 15.3** summarizes innate immunity.

Author's Notebook

You may have heard the term "swollen glands" in reference to a throat infection such as strep throat. These are actually enlarged lymph nodes, which are not glands at all! Glands are structures that are important for secreting substances, most commonly, hormones. The purpose of the lymph node is for immune cell communication and activation. Increased fluid and activation of cells in the nearest lymph node during an infection results in enlarged lymph nodes. So the next time you have a sore throat and you can feel your enlarged lymph nodes, you can say, "My innate and adaptive immune response is getting revved up" (and then go to the doctor, especially if you also have a fever).

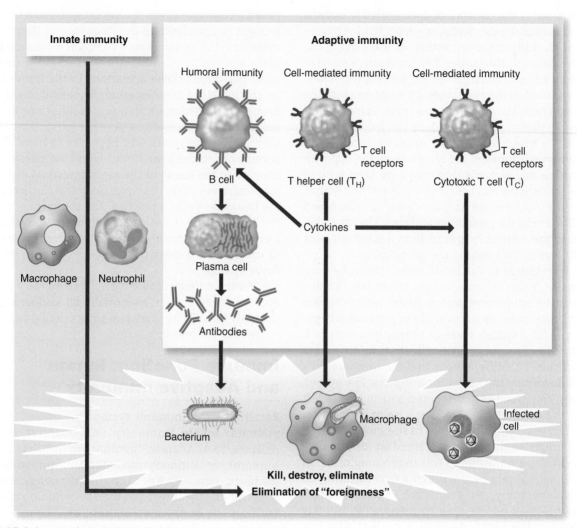

FIGURE 15.9 Innate and adaptive immunity. Innate immunity responds quickly to an infection. Adaptive immunity involves lymphocytes: B cells and T cells. B cells will ultimately make plasma cells to secrete specific antibody. Cytotoxic T cells target infected cells for destruction. As the command center of the immune response, T helper (T_H) cells assist with each of these functions by secreting cytokines, which will help with antibody production, cytotoxic T cell development, and killing by phagocytes.

TABLE 15.3 Innate Immunity

Mechanism	Function
Skin and mucous membranes	Mechanical barriers
Cilia	Found along respiratory tract and have "upward" motion, pushing microbes up to pharynx where they are swallowed
Phagocytosis	A system of phagocytic cells in the blood and scattered throughout the body
Lysozyme	Found in tears; breaks down bacterial cell walls
Interferon	Found in blood; has antiviral properties
Acid pH of stomach	Many microbes are killed by strong acid environment

These physiological defense mechanisms are present at the time of birth and target microbes in a general manner. A number of innate internal defense mechanisms, including complement and interferon, as previously described, are inhibitory to those microbes that breach these mechanisms.

Mechanical and Chemical Barriers

The skin is the first line of defense against microbial infection. Based on its structural composition, the intact skin is an excellent physical barrier that blocks the entry of microorganisms into the body. The external surface of your skin is not very hospitable for most microbes, and very few microbes can penetrate intact skin. Obviously, there are circumstances in which the skin is broken, allowing possible penetration by microbes. Cuts and abrasions, insect bites, and injections with hypodermic needles are familiar examples. Staphylococci, normal residents on the surface of the skin, are particularly prone to enter the body as opportunists through these breaks and establish infection. Many diseases, including malaria, yellow fever, babesiosis, and sleeping sickness (other than in the classroom), are transmitted by insects whose bites penetrate the skin.

The acidity of the stomach is detrimental to most microbes. Certainly, food and eating utensils are not sterile, and hence large numbers of microbes gain access to the stomach. In addition, mucous membranes line the internal body cavities and act as internal barriers. For example, the digestive tract running from the mouth to the anus is lined by a mucous membrane that protects the body from invasion by microbes that reside within the digestive tube. Penetration of this mucous membrane by intestinal microbes, even those that normally reside within the digestive tract, is extremely serious and leads to a life-threatening infection known as peritonitis. Surgeons are very much aware of this when performing procedures on the digestive system; nicking the lining of the intestinal tract allows microorganisms to spill out into body cavities where they do not belong.

The airways in certain areas of the respiratory system have cells with hairlike projections, called **cilia** that propel the microbes into the pharynx, where they are swallowed and killed by stomach acid. This system is called the **mucociliary escalator**. Smokers are more prone to respiratory infections than nonsmokers, because smoke damages this system.

The enzyme **lysozyme** destroys the cell walls of some bacteria. Lysozyme is found in saliva, tears, and sweat (but don't sweat it!).

Normal Microbiota

In addition to the chemical and physical barriers provided by the skin and mucous membranes, large microbial communities known as commensal microorganisms

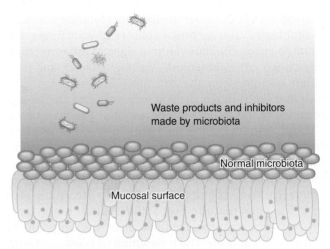

FIGURE 15.10 Normal microbiota provide protection from infection. The skin and most of the gastrointestinal tract have normal microbiota, also called commensal bacteria. These bacteria, shown in blue, protect from infection by pathogenic organisms, shown in red; by physically interfering with access to the mucosal surface or by making products that discourage the growth of other organisms; and by providing some immune stimulation.

or **normal microbiota** colonize the skin and many mucosal surfaces of healthy people. A pathogen trying to cause an infection through body sites colonized by normal microbiota must compete with the resident microbes for space and nutrients. The mucosal surfaces of the gut are a particularly good habitat for normal microbiota, and more than a thousand different species of microorganisms call the human gut home. This is a **symbiotic** relationship in that both the human and the microbes benefit (**FIGURE 15.10**). Before birth, we don't have any commensal microbes and the immune system is also relatively underdeveloped. After birth, acquisition of the normal microbiota influences our health and the development of the immune system by providing immune stimuli, protection from pathogens, nutritional support, and many other benefits that we are only just now starting to understand. Indeed, we are learning that aspects of our overall health are affected by the composition of the normal microbiota, which is influenced by diet, lifestyle, stage of life, and antibiotic use.

Phagocytosis

Phagocytosis is a highly significant innate defense mechanism by which monocytes, macrophages, and neutrophils engulf and destroy foreign substances, including microbes

Author's Notebook

You actually have more procaryotic cells in your body than human cells thanks to the microbiota in your gut. Does that make you eucaryotic or procaryotic? If asked on an exam, you should probably answer that you are eucaryotic.

(**FIGURE 15.11**). Consider the following familiar scenario: Several hours after getting a splinter in your finger, the wound displays the four cardinal signs of **inflammation**: redness (rubor), heat (calor), swelling (edema), and pain (dolor). Microbes from the skin, most likely staphylococci, have invaded through the site of injury. What is the significance of the inflammatory reaction? Resident tissue macrophages are in the tissue. They have PRRs that allow them to be on the lookout for signs of infection. In addition to killing bacteria they find, they also send chemical signals known as **cytokines** to initiate inflammation and recruit neutrophils. Complement activation also creates signals that promote inflammation in the local area. Inflammation can continue if the warning signs of an infection continue. The inflammation associated with the incision and stitches in **FIGURE 15.12** was roughly 10 days after the surgery.

The four classic signs of inflammation are due to activity in the local small blood vessels. In response to cytokines and complement the local capillaries become engorged with blood (vasodilation), accounting for the redness and the heat, and some of the fluid from the blood leaks from the vessels into the surrounding tissue spaces, causing swelling. Pain at the site is a result of increased pressure and the effect of products released by the injured tissue on nerve endings. It hardly seems like a defense mechanism. Within these blood-engorged capillary beds, however, phagocytic cells, mostly neutrophils, pile up and stick to the vessel walls surrounding the site of the injury. These cells, attracted in large numbers to the inflamed tissue (chemotaxis), migrate out of the capillaries and kill the bacteria, preventing or minimizing infection (**FIGURE 15.13**). Dendritic cells will also engulf microbial components, and then get the adaptive

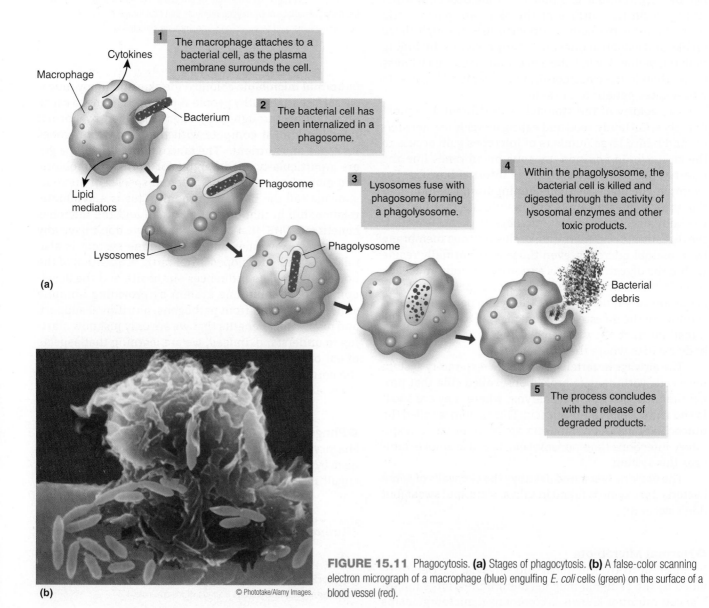

1 The macrophage attaches to a bacterial cell, as the plasma membrane surrounds the cell.

Cytokines

Macrophage

Bacterium

2 The bacterial cell has been internalized in a phagosome.

Lipid mediators

Phagosome

Lysosomes

3 Lysosomes fuse with phagosome forming a phagolysosome.

4 Within the phagolysosome, the bacterial cell is killed and digested through the activity of lysosomal enzymes and other toxic products.

Phagolysosome

Bacterial debris

(a)

5 The process concludes with the release of degraded products.

(b) © Phototake/Alamy Images.

FIGURE 15.11 Phagocytosis. **(a)** Stages of phagocytosis. **(b)** A false-color scanning electron micrograph of a macrophage (blue) engulfing *E. coli* cells (green) on the surface of a blood vessel (red).

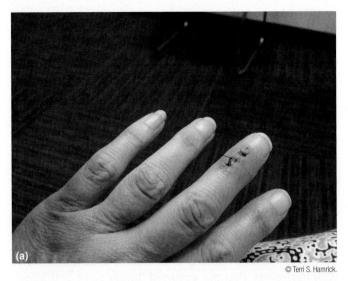

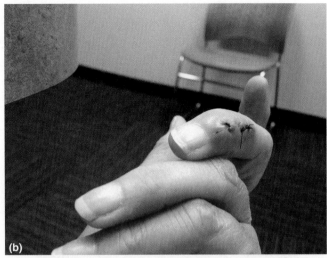

© Terri S. Hamrick.   © Terri S. Hamrick.

FIGURE 15.12 Inflammation associated with a surgical incision and stitches. Roughly 10 days after surgery, this surgical site and stitches showed marked inflammation **(a)** and swelling **(b)**. Initial inflammation after the surgery was minimal; however, after bacteria entered the wound about a week later, innate immune mechanisms resulted in marked inflammation. After removing the stiches and completing a course of antibiotics the site healed without incident.

immune response started. The discovery of phagocytosis by Elie Metchnikoff in 1884 was a landmark in immunology for which Metchnikoff received a Nobel Prize in 1908 (**FIGURE 15.14**).

Phagocytosis is an extremely efficient host defense mechanism in its own right but is rendered even more efficient in the presence of antibodies and complement. Antibodies and complement bound to the surface of the microbe become handles for the phagocyte, which has receptors that can bind tightly to the antibodies and complement. Monocytes, neutrophils, and macrophages are very efficient at phagocytosis and play a key role in the destruction of microbes. Macrophages are located strategically in nearly all tissues throughout the body; monocytes and neutrophils are located in the blood, but can be quickly recruited to wherever infections are found.

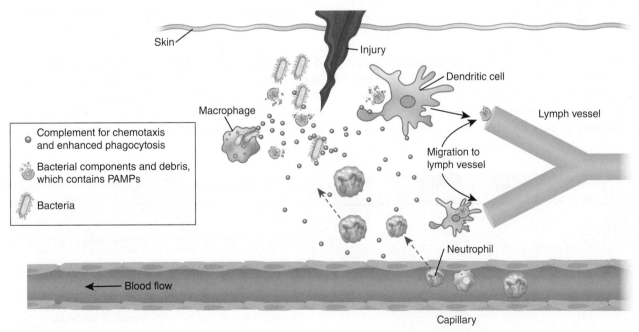

FIGURE 15.13 Inflammation in the tissue. Complement activation and cytokines produced by macrophages at the site of the infection initiate inflammation at the local capillaries. Leukocytes in the capillaries, mostly neutrophils, migrate out of the blood vessels (capillaries). Using chemotaxis (indicated by dashed red arrows), they arrive at the site of the infection and work to destroy the bacteria. Dendritic cells will also engulf microbial components to initiate the adaptive immune response.

© North Wind Pictures Archives/Alamy Images.

FIGURE 15.14 Elie Metchnikoff (1845–1916).

Adaptive Immunity

The adaptive immune system responds to the presence of foreign microbes that have breached the external and internal innate defense mechanisms. At this point an all-out war between microbes and host is in effect. A key to understanding immunity is the concept of specificity. Think of antigen–antibody specificity as the complementarity that exists between a lock and a key or between two pieces of a puzzle that fit together.

As previously mentioned, the adaptive immune system is divided into two categories: **humoral** (antibody-mediated) **immunity** and **cell-mediated immunity (CMI)**. These two categories are the "big guns." Although these systems operate in very different ways, the mission is a common one: to eliminate foreign antigens—the hallmark of adaptive immunity. Most antigens are large protein molecules associated with microbes, tumor cells, damaged cells, allergens, or foods. They engage and "sensitize" lymphocytes to generate an effective immune response that may produce antibodies specific for that antigen (humoral immunity) or T cells directed against that antigen (CMI).

⊙ T Lymphocytes

T lymphocytes have a role in both humoral immunity and CMI, so it is important to first understand these cells. T cells have been broken down into multiple categories, collectively referred to as the **T-cell subset** (**TABLE 15.4**). Each category of T lymphocyte has a specific role and is identifiable by molecules referred to as **cluster of differentiation (CD)** molecules that are acquired during the T-cell maturation process.

Some T cells, called **cytotoxic T (T_C) lymphocytes (CTLs)**, are the effectors of cell-mediated immunity. They are identified by **CD8 molecules**. The T_C cells are capable of becoming sensitized to ("angry at") the foreign molecules (antigens) carried by the invading microbes, especially viruses; CTLs require **T helper (T_H) cells** in order to become sensitized.

The T_H cells are identified by **CD4 molecules**. CD4 T cells are the command cells of the immune response. The T_H cells are crucial in that nearly all B cell and T_C cell responses depend on help from T_H cells. T_H cells also help with the functioning of other immune cells, such as macrophages. T_H cells exert their helper function through direct contact with the other cell and by secreting chemical messengers known as cytokines, which activate the target cell(s).

⊙ Antibody-Mediated (Humoral) Immunity

The concept of using an infectious agent to prevent and recover from certain diseases has been realized for centuries. Before the 1700s, medical practitioners had some success preventing what later was termed smallpox by

TABLE 15.4	T-Cell Subsets		
Cell Type	Abbreviation	Cluster of Differentiation Cell Marker	Function(s)
T-helper cell	T_H	CD4	Activates B cells to produce antibodies; activates cytotoxic T cells; activates inflammatory cells
Cytotoxic T cell	T_C or CTL	CD8	Killer cell that targets cells with "foreign" intracellular antigens, including viruses, bacteria, and cancer cells
T regulatory cell*	T_R	CD4	Regulatory T cell that turns down immune responses

*Not discussed in text.

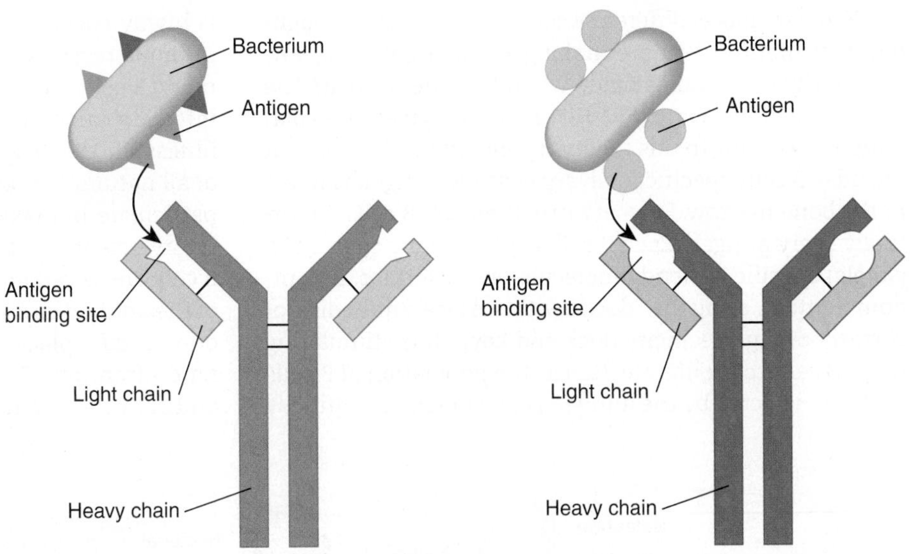

picking scabs from pox-infected individuals. The scabs were dried, pulverized, and introduced into the nose of those susceptible. Many of those treated contracted a mild case of smallpox, from which they recovered and were resistant to subsequent infections. In 1796, Edward Jenner introduced a vaccination against smallpox by inoculation of the cowpox virus, resulting in a benign infection in humans. What these early pioneers in immunology did not know was the biological mechanisms behind the resistance—the production of adaptive immunity and antibodies.

Humoral immunity is mediated by antibodies, products made by B cells (with the aid of T_H cells) in response to antigens. The antibody molecule is a four-chained structure (**FIGURE 15.15**) that binds with antigens on bacterial cells, viral particles, toxins, and internal cells, including tumor cells and dead cells.

The binding of antibodies to antigens has several outcomes (**FIGURE 15.16**), each of which facilitates destruction of the foreign antigen–bearing microbe. The immunoglobulin, initially located on the surface of the B cell, binds to its corresponding antigen, which is necessary to activate the B cell. T_H lymphocytes send cytokine signals to the activated B cell so that the activated B cell proliferates and differentiates into a plasma cell, which makes and secretes antibody. Estimates are that a single

FIGURE 15.15 Structure of antibody molecules. The basic antibody structure consists of two identical heavy chain proteins and two identical light chain proteins covalently bonded together. The ends of the light chain and the heavy chain create the antigen binding site.

plasma cell can produce thousands of specific antibodies per second. These antibodies are produced with identical specificity. Aside from the antigen specificity, antibodies can be one of five different classes, or isotypes, of antibodies or immunoglobulins (Ig), termed IgG, IgA, IgD, IgE, and IgM. Each isotype has unique properties. IgG accounts for approximately 80% of the antibody molecules in the circulation and is the best characterized. The antibody molecule binds to antigens on bacterial cells, viral particles, toxins, and internal cells, including tumor cells and dead cells, resulting in a "kill, destroy, or eliminate" outcome.

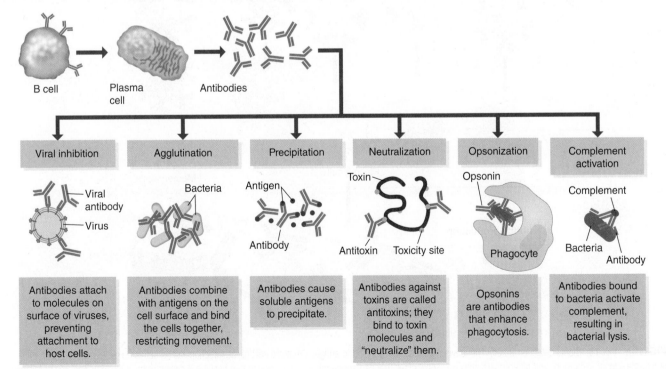

FIGURE 15.16 Protective effects of antibodies binding to antigens (humoral immunity).

You have lots of different B cells that are made throughout your lifetime. You are making B cells right now. How are the right antibodies made when they are needed? The clonal selection theory (**FIGURE 15.17**) provides an explanation. According to this theory, a population of B cells that includes B cells specific for every possible antigen is made in the bone marrow. Even at birth, there are B cells for virtually every antigen. Each B cell displays many copies of a (single) specific antibody molecule on its surface. By random contact, antigens "dock" with surface antibodies of corresponding specificity (lock and key), thus stimulating those B cells to proliferate. Hence, the population of B cells that was selected by the antigen is expanded. An antibody is highly specific for its corresponding antigen, but in an immune response, you will make antibodies derived from many B cells such that the antibodies produced during an infection can bind all over the microbe that is causing the illness. In this way, **polyclonal antibodies** are the result of all natural antibody responses, because multiple B cells participate in making the pool of antibodies. In the laboratory, scientists can create cell lines originating from one B cell. These cell lines can produce lots of antibody, but all of the antibodies are exactly the same and will bind to only one specific place on one specific antigen. Having originated from one B cell, these are **monoclonal antibodies** and are used in a number of therapies (**BOX 15.1**).

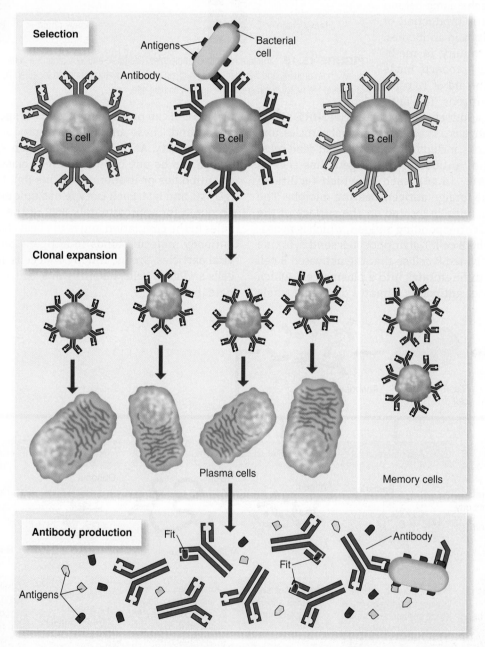

FIGURE 15.17 Clonal selection theory. Those lymphocytes that can bind to antigens that are part of the infection will be activated (selected); the cells will proliferate (clonal expansion) and differentiate, making lymphocytes that can act on the antigens. Notice that only the B cells that have the lock-and-key fit for the antigens will proliferate and become plasma cells. The plasma cells will make and secrete antibodies with the same antigen specificity as the B cell.

BOX 15.1 Send in the Monoclonal Search Team

Wouldn't it be great if we could make a little search team that could go into the body to find and grab the tiny things that cause disease? That is essentially what antibodies are. Antibodies are very exact in the target to which they will bind. The exact binding of antibodies is used today in a number of medical therapies. Most of the antibody therapies used today are what we call monoclonal antibodies (MAbs). Originally, MAbs were made by fusing a B cell that contained the antibody genes of interest to a myeloma cell from a mouse to make a **hybridoma**. A **myeloma cell** is a plasma cell that can be cultured indefinitely in the lab. Through screening, the lab can then find the hybridoma that is making the antibody of interest. With today's molecular techniques, the human antibody genes can be put into a human plasma cell line in the lab to make MAbs for therapeutic use. It is usually easy to spot drugs that are MAbs. The generic name, which is included in parentheses after the brand name, usually has "mab" at the end of the name. For example, the preventive treatment given

to high-risk infants to protect them from respiratory syncytial virus is called paliviumab.

The way these therapies work is that the MAb is injected into the patient. The MAb then finds the target and binds to it so that the target cannot do what it would normally do. For example, remember that autoimmune disorders are due to the immune system in overdrive. For some autoimmune disorders, the immune system is making too much of a cytokine known as **TNF-alpha** (TNF-α), which leads to excess inflammation. A number of MAb therapies to treat rheumatoid arthritis, Crohn disease, and inflammatory bowel disease specifically bind the TNF-α. The TNF-α does not then bind to receptors that would lead to inflammation. **Figure 1** illustrates how MAbs that act on TNF-α work. Patients taking these medications have to be very careful about infections, though, because all inflammation is turned down, even inflammation that might be helpful.

We don't yet have a little army of designer antibodies to seek out each malady, but we are headed that direction.

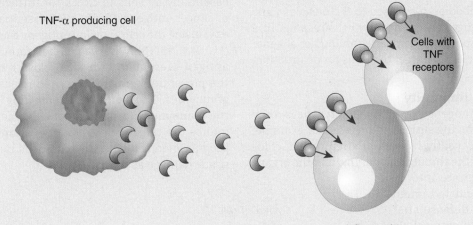

(a) Inflammatory state. TNF-α binding to receptors leading to excess inflammation

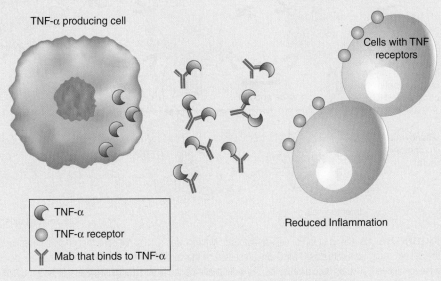

(b) MAb therapy. MAb binding to excess TNF-α, reducing inflammation.

Figure 1 Use of monoclonal antibodies (MAbs) in the treatment of some autoimmune disorders. Inflammation associated with some autoimmune disorders is reduced by the use of MAb therapy that specifically binds to TNF-α. Normally, TNF-α is made in small amounts and leads to an appropriate amount of inflammation in response to infection or injury. For some autoimmune disorders, there is too much inflammation. The MAb binds to the excess TNF-α, reducing inflammation.

After antigen encounter and appropriate signals, usually from a T_H cell, the B cell differentiates into a plasma cell, which will secrete antibody with essentially the same antigen specificity as the surface immunoglobulin. After a lag time of approximately 10 to 12 days, antibodies are present in the blood in amounts large enough to be detected. After an initial burst of IgM, the T_H cell will also direct which type of antibody is made (IgG, IgA, or IgE), depending on what kind of antibody response is needed.

The T_H cells also plan for the future by directing some of these B cells to become memory cells that are retained for years. This accounts for the fact that, except in rare circumstances, certain so-called childhood diseases (e.g., measles, rubella, mumps, and chickenpox) are acquired only once in a lifetime despite repeated exposure. Assume, for example, that you had measles as a child. In later years your child has measles and, despite close contact with the child, you do not acquire the disease. This is because a part of the B-cell population making antibodies against the measles virus was reserved as memory cells, including some long-lived plasma cells that will continue to trickle out protective antibodies. If exposed to the measles virus again, these preprogrammed memory cells respond in a matter of hours by pouring out measles antibodies at a rate and quantity sufficient to target the virus for destruction.

Cell-Mediated Immunity

Most bacterial pathogens are extracellular; that is, they take up residence on the surface of the cells. For example, streptococci, the causative agents of strep throat, colonize the surface of the throat and pharynx, and the organisms do not penetrate into the cells. Antibody-mediated immunity is the body's defense strategy for coping with extracellular microbes, but what about intracellular bacteria and viruses (all of which are intracellular)? What about mutant and damaged cells in the body? Viruses that have just entered the body are extracellular, which is why generating antibodies through vaccination works to prevent viral infections. However, antibodies play little role in protection against microbes that have already found their way into the intracellular environment. So what is the defense? Cell-mediated immunity (CMI) is the major defensive strategy against intracellular bacteria, viruses, and tumor cells. CMI is mediated by T_C cells that have become sensitized (angry) to a specific antigen

and react against that antigen. (Think of CMI in this way: when you are angry at an individual, your anger is vented against that person.)

Unlike B cells, which will bind to antigen directly with surface immunoglobulin, T cells are more precise about how they view antigens. The T cell receptor has antigen specificity, but it can only "see" its corresponding antigen when it is presented by another cell. The other cell, known as an antigen-presenting cell, has chewed up the antigen, complexed the pieces with a molecule known as **major histocompatibility complex (MHC)**, and presented the antigen–MHC complex on its surface for the T cell to view. MHC is like a serving platter for antigen, and T cells will only attempt to recognize antigen if it is presented on a serving platter (MHC). What fussy cells! Two different kinds of MHC assist T cells with antigen recognition. **Class I MHC** is present on all nucleated cells, whereas **class II MHC** is expressed only on certain cells.

Activated T_C cells, also known as cytotoxic lymphocytes (CTL), need to recognize and kill infected or cancerous cells; therefore, T_C cells need to be able to look at essentially all cells to check for intracellular pathogens or cancer antigens. Class I MHC essentially acts as a window inside the cell, showing antigens that are found inside (**FIGURE 15.18**). When a cell has foreign (or cancerous) antigen presented with MHC class I, the CTL binds to the presenting antigen, and then releases perforins and granzymes. The **perforins** create tiny holes that allow the **granzymes** into the target cell. The granzymes then induce the target cell to undergo **apoptosis**, also called programmed cell death, which does not cause as much tissue damage as if the cells lysed.

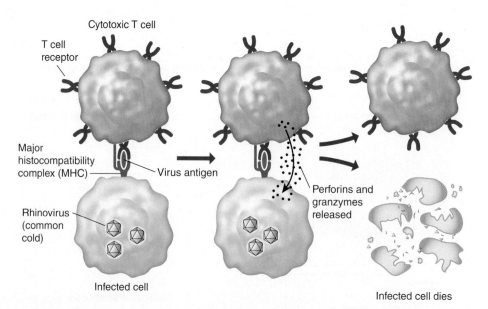

FIGURE 15.18 Killing of target cells by cytotoxic lymphocytes (CTLs). Antigens that originate from inside the cell are complexed with class I MHC and presented on the surface for CTLs to examine. If the CTL identifies a foreign antigen, such as from a virus, the CTL will release perforins and granzymes that will induce apoptosis in the target cell. The CTL moves on to find another infected cell to kill.

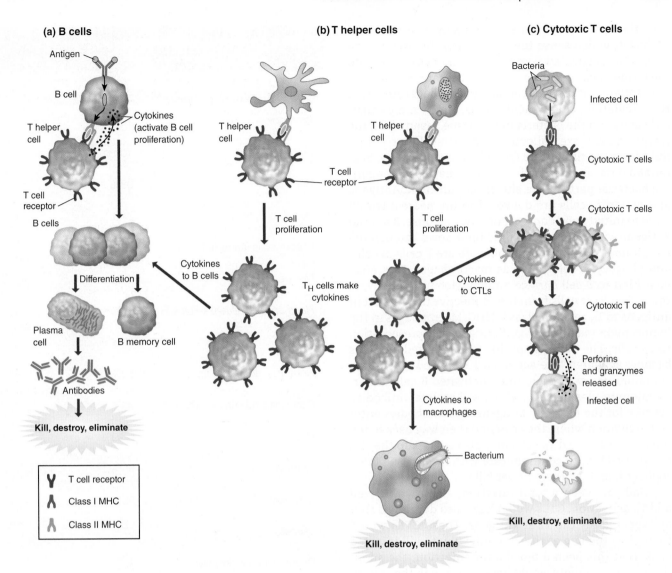

FIGURE 15.19 T$_H$ cell function and the importance of major histocompatibility complex (MHC) for T cell communication and activity. B cells, macrophages, and dendritic cells are professional antigen presenting cells, which utilize class II MHC to present antigens. **(a) B cells** able to bind to antigen associated with the infection will present the antigen with class II MHC to T$_H$ cells as a way to ask for help. The T$_H$ cells will produce cytokines that will direct the B cells to proliferate and differentiate into plasma cells and memory cells. The plasma cells will secrete antibody. **(b) T helper cells:** Dendritic cells and macrophages are able to engulf antigen to then present to T$_H$ cells with class II MHC. This will lead to more activated T$_H$ cells (and more help!). The cytokines made by these activated T$_H$ cells can make macrophages better killers, direct the kind of antibody response, or assist with CTL killing. **(c) Cytotoxic T cells (CTLs)** recognize intracellular antigens presented with class I MHC on infected cells, which are then killed when the CTLs release perforins and granzymes that induce apoptosis.

The T$_H$ cells help with all of the above functions and more. They are the command cells of the immune response. It would not be a very efficient system if all of the cells of the body tried to communicate with the command cells. It is important to follow the chain of command. Cells known as **professional antigen-presenting cells** express class II MHC in order to present antigens to the T$_H$ cells (**FIGURE 15.19**). The professional antigen-presenting cells are dendritic cells, macrophages, and B cells. These cells are designed to pick up antigens, show them to T$_H$ cells, and elicit their help. The T$_H$ cells can then help form the appropriate immune response by producing cytokines. For the B cells, T$_H$ cells

produce cytokines to elicit the correct type of antibody response. For a cell-mediated immune response, the T$_H$ cells produce cytokines that encourage CTLs to kill target cells. T$_H$ cells can also release cytokines that encourage macrophages and other phagocytic cells to kill better.

Immune Response Summary

You now know all of the parts of the immune response. To put it all together, let's look back at the skin infection we talked about briefly in an earlier section as an example. Bacteria penetrate the skin. Parts of the innate immune system that are present and able to act on the bacteria include complement and resident tissue macrophages, which sense the presence of the bacteria via their PRRs. Activated complement and cytokines from the tissue macrophages act at the local capillaries to initiate inflammation. This recruits more phagocytes from the circulation, chiefly neutrophils, which are particularly active on the bacteria in the tissue. Meanwhile, the lymph vessels are collecting the extracellular lymph fluid from the tissue. The lymph contains bacteria, bacterial debris, and dendritic cells that have been engulfing bacterial parts at the site of the infection. Antigens and dendritic cells are delivered to the nearest lymph node to initiate an adaptive immune response. Bacterial antigens within the lymph fluid filter down through the lymph node. Within the lymph node are T cells, B cells, and phagocytic cells, such as macrophages. Free antigens bind to B cell surface immunoglobulin to activate them. Dendritic cells from the connective tissue present antigens to T_H cells via class II MHC. Macrophages in the lymph node may also engulf antigens for presentation to T_H cells. The T_H cells specific for presented antigens become activated. The activated T cells and B cells can communicate with each other. Activated B cells differentiate into plasma cells, which will secrete antibodies specific for the bacterial antigens. The antibodies enter the circulation where they can find their way back to the infection. Activated T_H cells may also go back to the site of the infection to communicate with local tissue macrophages, making them better killers.

Had this been a viral infection, along with B cell and T_H activation, CTLs would have also developed that are specific for the viral antigens. These CTLs would then go to the site of the viral infection to kill infected cells. Had this been a bloodborne infection, the adaptive immune activity would have occurred in the spleen. Had this been a gastrointestinal infection, the adaptive immune activity would have occurred in the lymphoid tissue there. Regardless of which pathway of adaptive immunity was involved, long-lived memory cells of the effective lymphocytes (T and B cells) will ensure that if this pathogen reenters the body the response will not take as long.

Immunization

The dramatic decrease in infectious diseases since the middle of the 19th century is largely due to two factors: improved sanitation and the widespread use of vaccines. In the United States alone, the lives of 3 million children are saved each year because of routine immunization. For some diseases the decline over the past century in the United States has been 100%, or close to 100% (**TABLE 15.5**). As an example, since 1988, polio has decreased by almost 100%, and by 1997, it was eliminated from the United States. Since 2014, polio cases have been identified in only three countries (Afghanistan, Nigeria, and Pakistan), down from more than 125 in 1988.

TABLE 15.5 Impact of Vaccines in the 20th and 21st Centuries

Infectious Disease	Decline
Smallpox	100%
Diphtheria	100%
Hepatitis A	98%
Hepatitis B, acute	71%
Poliomyelitis (paralytic)	100%
Measles	> 99%
Haemophilus influenzae type B	> 99%
Mumps	97%
Rotavirus, hospitalizations	82%
Congenital rubella syndrome	99%
Rubella	> 99%
Tetanus	95%
Pertussis	92%
Varicella	97%
Pneumococcus (invasive)	
All ages	54%
< 5 years	89%

Information from the Centers for Disease Control and Prevention. (2015). *Epidemiology and prevention of vaccine-preventable diseases.* (13th ed.). J. Hamborsky, A. Kroger, & S. Wolfe (Eds). Washington D.C. Public Health Foundation.

Smallpox was the first disease to be eradicated thanks to Edward Jenner's pioneering work with vaccination and to the fact that the disease has no animal reservoirs. In 1881, almost a century after Jenner's work, Louis Pasteur developed a vaccine against anthrax in animals and, only a few years later (1885) developed a vaccine against human rabies. Before 1900, vaccines against three additional diseases were available. Sixteen diseases can be prevented with vaccines that are commonly used in the United States. In the United States, the Food and Drug Administration (FDA) is the agency responsible for issuing a license to vaccine manufacturers, allowing the vaccine to be widely distributed. Data on efficacy and disease trends helps to shape subsequent vaccine recommendations and formulations. Some vaccines are available in different forms, each of which requires approval before marketing. **TABLE 15.6** is a list of common and special-use vaccines and the types of vaccines available for each.

Some of these diseases, like whooping cough and chickenpox, are still somewhat common. Some vaccine-preventable diseases, such as diphtheria and polio, are uncommon and have faded from our collective memory.

TABLE 15.6 Recommended Immunizations and Vaccine Formulations for Adults and Children in the United States, 2018

Vaccine	Type of Vaccine	Typical Use	Special Use
Hepatitis B (HepB)	Subunit of the virus	Routine childhood	Healthcare providers, if not vaccinated previously
Rotavirus (RV) RV1 (two doses), RV5 (three doses)	Live attenuated	Routine childhood	
Diphtheria, tetanus, acellular pertussis (DTaP)	Toxoids (D and T), several subunit components (aP)	Routine childhood	
Tetanus, diphtheria, acellular pertussis (Tdap)	Toxoids (T and d), several subunit components (ap)	Routine during adolescence	One dose instead of Td for all adults; one dose during last trimester of each pregnancy
Tetanus, diphtheria (Td)	Toxoids	Every 10 years	If tetanus-suspicious wound, boost if > 5 year since last dose
Haemophilus influenzae type b, (Hib)	Capsular polysaccharide attached to protein	Routine childhood	
Pneumococcal conjugate (PCV13)	Capsular polysaccharide attached to protein	Routine childhood	One dose, adult (≥ 65 years of age)
Pneumococcal polysaccharide (PPSV23)	Pure polysaccharide	Adult (≥ 65 years of age)	
Influenza (IIV)	Inactivated virus	Annually for children and adults	High dose available for the elderly
Measles, mumps, rubella (MMR)	Live attenuated viruses	Routine childhood	Adults: One or two doses if not demonstrated immunity
Varicella (VAR)	Live attenuated virus	Routine childhood	One or two doses for adults if not vaccinated before
Zoster	Live attenuated virus or recombinant subunit	One or two doses at 50 or 60 years of age (depending on formulation)	
Hepatitis A (HepA)	Inactivated virus	Routine childhood	
Meningococcal (MenACWY-D)	Capsular polysaccharide	Adolescent and at 16 years	
Meningococcal B	Subunit protein		Adolescent or young adult if in high-risk area
Human papillomavirus (HPV)	Capsid proteins	Adolescent or young adult	

Information from Advisory Committee on Immunization Practices (ACIP). (2018). Recommended immunization schedule for children and adolescents aged 18 years or younger, and recommended immunizations schedule for adults aged 19 years or older, United States, 2018. Retrieved from https://www.cdc.gov/vaccines/hcp/acip-recs/index.html

However, it is important to remember that they became rare because we have been vaccinating against them. Before there was a vaccine, nearly everyone in the United States got measles and hundreds died from it each year. Today, you have probably never seen a real case of measles. In 1921, more than 15,000 people in the United States died from diphtheria. In the past decade there have been only a couple of reported cases. In 1964–1965, an epidemic of rubella (German measles) infected more than 12 million Americans, killed 2,000 infants, and caused 11,000 miscarriages. In 2015, the Americas became the first region in the world to be declared free of in-country transmission of rubella. Whereas this news is encouraging, as long as the disease still exists in the world, it is important to continue immunizing. Even if there are only a few cases of disease, if we take away the protection provided through vaccination, more and more people can get sick and spread the disease to others. Consider that in 2000, the Center for Disease Control and Prevention (CDC) declared measles "eliminated" in the United States, yet in 2014, there were 667 cases of measles in 23 different outbreaks.

The availability of vaccination is a story of "good news and bad news." The good news is that much of the world is immunized against a variety of microbial diseases, whereas the bad news is that underdeveloped countries have not shared in this victory. Tragically, whooping cough, measles, and tetanus take the lives of more than 3 million children each year, despite the fact that immunization is available. The Bill & Melinda Gates Foundation, the United Nations International Children's Emergency Fund (UNICEF), and the Global Alliance for Vaccines and Immunization (GAVI) are but a few examples of organizations focused on delivering vaccines to the world's poorest countries. Vaccine development and improvement of existing vaccines is also a top priority. Malaria, HIV/AIDS, and tuberculosis are responsible for suffering and deaths by the millions throughout the world; many children in developing countries do not grow up to become adults.

Active Immunization

TABLE 15.7 outlines the categories of immunization. **Active immunization** is the result of stimulating a person's immune system to produce antibodies and memory cells and generally confers immunity over a relatively long time. The two categories of active immunization are natural and artificial. Natural active immunity is achieved by the process of recovering from a particular disease. Analysis of an individual's blood serum may reveal the presence of antibodies against which there is no clinical history of disease, indicating that the disease had occurred at a subclinical level.

The use of vaccines, however, is artificial in the sense that vaccines are administered into the body to provoke an immune response as a future protective measure. As

TABLE 15.7 Outline of Immunization

Active Immunization (individual makes own antibodies)	Passive Immunization (individual receives preformed antibodies)
Natural (subclinical or clinical disease and recovery)	Natural (in utero mother-to-infant passage; breast milk)
Artificial (vaccines for immunization "shot")	Artificial (use of immune globulin)
Live attenuated microbes	
Killed microbes	
Toxoids and other purified microbial components	
New and experimental vaccines	

indicated in Table 15.7, artificial active immunization can be accomplished in more than one way; the method chosen reflects the best potential for the particular disease. All strategies must meet the three basic requirements of safety, effectiveness, and stability (**TABLE 15.8**). Additionally, an ideal vaccine needs to be affordable to developing countries.

TABLE 15.8 An Ideal Vaccine

General Vaccine Requirements	Ideal Vaccine Requirements
Safety	Safety
Effectiveness	Effectiveness
Stability	Stability
	Affordability
	Administration as a nasal spray or edible vaccine (no needle)
	No need for refrigeration; stability at ordinary "tropical" temperatures
	One dose or one shot
	Long shelf life

Safety issues are of prime importance and are further addressed in the section on vaccine safety. For a vaccine to be effective it must stimulate an immune response affording protection to vaccine recipients. Vaccine preparations need to be stable over time to make them cost-effective. Some vaccines can be stored at room temperature, whereas others need to be refrigerated, presenting a problem in distribution of these vaccines to the developing world. On too many occasions vaccines need to be destroyed because refrigeration requirements are not observed. Strict attention needs to be paid to expiration dates and to conditions of storage.

Live Attenuated Microbes

Vaccines made from live attenuated microbes, most of which are viruses, confer long-lasting, frequently lifetime immunity. (Poetic license is taken with the term "live" to describe viruses.) The word "attenuated" means weakened virulence. **Live attenuated vaccines** tend to produce the longest-lasting immunity, because as "live" microbes they replicate in the body and engage the immune system in much the same way as the pathogen. However, as live organisms, these vaccines may produce mild and limited symptoms but not overt clinical disease, as manifested by fever, headache, fatigue, and soreness at the site of injection for about 24 hours; frequently, there are no symptoms.

Some live attenuated viral vaccines have been achieved by serial (repeated) transfer in tissue culture, allowing the production of random and unpredictable mutants. These mutants are tested in laboratory animals, and those strains producing no symptoms are selected for further trial. The **bacillus Calmette-Guérin (BCG) vaccine** against tuberculosis continues to use a mycobacterium strain (*Mycobacterium bovis* BCG) attenuated by repeated subculturing between 1908 and 1918 on laboratory media with no reversion to virulence for more than 80 years. Pasteur's vaccine against rabies was the result of gradually drying virally infected spinal cords of dogs and rabbits, a procedure that rendered the live virulent virus nonvirulent but continued to provoke protective antibodies against the virus. The major concern of the use of some older live attenuated vaccines is the possibility of reversion to virulent forms. With any live attenuated vaccine, individuals with impaired immune systems may have difficulty fighting the vaccine strain, even though it is weakened. Examples of vaccines using the live attenuated microbes are the Sabin (oral) polio, measles, mumps, rubella, rotavirus, and yellow fever vaccines.

Killed (Inactivated) Microbe Vaccines

Killed microbe vaccines are used when attenuation has not been accomplished or when reversion to the virulent type is considered to be too risky. Virulent microbes are heat killed or killed with particular chemical reagents. They present no risk but are not as effective at stimulating lasting immunity as those containing live microbes. Some require multiple doses to achieve and maintain protective antibody levels. The Salk (inactivated) polio vaccine and the hepatitis A vaccine are examples.

Toxoids and Other Subunit Vaccines

Some of the most serious bacterial diseases (diphtheria, tetanus, and botulism) result from the production of very potent microbial **exotoxins**. These toxins can be inactivated by heat or formaldehyde, resulting in a loss of toxicity, but they retain their antigenic shape to stimulate specific antibody production; inactivated toxins are referred to as **toxoids**. The **diphtheria tetanus acellular-pertussis (DTaP) vaccine**, commonly administered to children at about the age of 2 months, contains diphtheria and tetanus toxoids (inactivated toxins). Some vaccines contain other purified microbial components, such as virus proteins, bacterial adherence proteins, or bacterial capsule carbohydrates. The pertussis part of the DTaP vaccine contains several bacterial components of the whooping cough bacterium. The pneumococcal vaccines are examples of capsule carbohydrate vaccines.

Some subunit vaccines are made using recombinant DNA technology. In the laboratory, the gene for the antigen of interest is put into another nonvirulent host microbe, such as yeast. That gene is then replicated and expressed in the new host. The protein antigen can then be safely purified in large quantities to be used in a vaccine. Hepatitis B and human papilloma virus (HPV) vaccines are examples of subunit vaccines made through recombinant DNA technology.

Experimental Vaccines

DNA vaccines are a new and promising approach to immunization. In this strategy, microbial DNA is inserted into plasmids that are then injected directly into the host, which will then make the protein antigens. Subsequently, these proteins are recognized as foreign by the host's immune system and stimulate an immune response. DNA vaccine development for a variety of microbial diseases, including influenza, tuberculosis, malaria, Lyme disease, and hepatitis C, is underway but has not yet been shown to be effective.

Passive Immunization

Active immunization is based on stimulating a recipient's immune system to produce antibodies and memory cells (**FIGURE 15.20**). In **passive immunization**, by contrast, the recipient receives ready-made preformed antibodies (immune serum) from human or animal sources by injection. For diseases caused by the activity of a bacterial toxin, having antibody, known as an antitoxin, can neutralize the toxin activity. To make some antitoxins, pharmaceutical companies inject toxoids into horses (or sheep or goats in some cases). After a brief time the animal's immune system produces antibodies to the toxin (antitoxins); the blood is taken and processed to recover the antitoxin component as an effective immunization

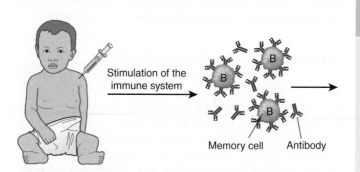

Stimulation of the immune system

Memory cell   Antibody

"Down the line" exposure to microbe

Antigen-antibody defense mechanisms
• Agglutination
• Precipitation
• Opsonization
• Complement activation
• Neutralization

FIGURE 15.20 How immunization works. Antigens are introduced to the body, most often via an injection, although some live attenuated vaccines are oral. The antigens in the vaccine stimulate an immune response that elicits the production of antibody, memory B cells, and, depending on the type of vaccine, some subsets of T cells. Antibodies provide lasting protection.

strategy. Botulism and tetanus antitoxins are examples. A big advantage to passive immunization is that antibodies are delivered immediately upon injection and can be of lifesaving value, as in cases where exposure or symptoms have already occurred. The immunity gained, however, is relatively short-lived and limited to the duration of the administered antibodies in the recipients; there is no immunological memory. **TABLE 15.9** summarizes the distinctions between active and passive immunization.

Before the 1940s and the advent of antibiotics, immunotherapy, the use of immune serum, was a common practice, particularly for diphtheria, tetanus, and pneumococcal pneumonia. The first two diseases are toxemias, meaning that they are a manifestation of the production of a lethal toxin. As described earlier, antitoxins (antibodies against toxoids) are produced by the injection of toxoids into horses to produce immune serum. In the treatment of diphtheria, about all that could be done in the days of the "horse-and-buggy" doctors was the administration of the antitoxin-containing immune serum in a desperate attempt to save lives.

Immunotherapy can be hazardous because of possible complications arising from the fact that the antibodies, along with other (blood) serum components, are often "seen" as foreign protein by the recipient's immune system, resulting in antibody production against the foreign protein in the serum, causing a condition known as **serum sickness**. Serum sickness is characterized by the formation of antigen–antibody complexes that are deposited in the skin, kidney, and other body sites. Nevertheless, under

certain circumstances immunotherapy is still used when immediate antibody protection is required.

The use of human immune serum, taken from individuals after vaccination or during their convalescence period from a specific disease, minimizes but does not eliminate the risk of serum sickness in the recipient. This human immune serum (also called immune globulin therapy referring to the globulin fraction of the blood) contains high levels of specific antibodies. For example, tetanus immune globulin is rich in antibodies against tetanus, and varicella-zoster immune globulin has a high concentration of antibody against the virus that causes chickenpox and shingles.

Consider a case in which an individual reports to an emergency room having sustained a puncture and is at risk of tetanus. Should that person receive active immunization by a booster shot with tetanus toxoid or passive immunization with tetanus immune globulin, or both? The answer depends on the person's immune history. If the individual has, within the last 5 to 10 years, received tetanus toxoid as a booster vaccination, all that is necessary is another booster to effectively stimulate those memory cells preprogrammed to produce tetanus antibodies almost immediately. However, if the individual has not received (or is not certain of) past immunizations against tetanus, immediate protection against the tetanus toxin is necessary. Because there is not sufficient time to make antibodies from scratch, tetanus immune globulin should be administered for immediate protection, along with tetanus toxoid for future protection.

Antitoxins against the deadly toxins injected into the body by certain species of snakes and arachnids (e.g., spiders and scorpions) are examples of lifesaving passive immunization. If bitten by a poisonous snake or scorpion, immediate antibody protection is vital.

Vaccine Safety

Vaccines have greatly reduced the burden of infectious diseases around the world, but as these diseases have declined attention has focused on the risks associated

TABLE 15.9 Properties of Active Versus Passive Immunization

	Protection	Duration	Adverse Reactions
Active	Waiting period	Extended memory	Possible
Passive	Immediate	Limited (no memory)	Possible

with vaccines: How safe are the vaccines on the market? No vaccine (or other medication) is 100% safe and without risk. The better question to be asked is, "Do the benefits of the vaccine outweigh the risks?"

For example, oral polio vaccine, a live vaccine preparation, carries a risk of vaccine-associated polio of about one case of polio in 2 to 3 million doses of vaccine administered. While this risk is tiny when compared to the high transmission rate of the natural virus and the 1 in 200 risk of paralysis from natural infection, now that polio is nearly eradicated it starts to become significant. For this reason, it makes sense to use the Salk inactivated virus vaccine, even though it costs more and does not protect from the poliovirus growing in the gastrointestinal tract. The current strategy for polio immunization is that all four doses consist of the Salk inactivated virus vaccine, which provides protection from paralytic disease, although the Sabin live attenuated oral vaccine may still be used in other parts of the world where polio is still prevalent.

Vaccines are constantly monitored and modified or withdrawn as circumstances dictate. For example, in 1976, Fort Dix, New Jersey, was threatened by the appearance of a new and deadly strain of swine flu virus; in response a vaccine was quickly developed, and 45 million people were vaccinated. Unexpectedly, in some cases the vaccine was associated with an increase in the risk of a neuromuscular disorder called **Guillain-Barré syndrome (GBS)**. GBS is a rare disorder in which the person's immune response damages nerve cells causing muscle weakness and sometimes paralysis. Most people fully recover, but some have permanent nerve damage. GBS is not fully understood, and more typically follows an infection when the immune system is active. Studies suggest that it is more likely that a person would get GBS after getting the flu than after the vaccine, but the CDC always looks at the incidence as part of the close monitoring of all vaccines.

Rotavirus infection is a potentially fatal disease in children, and the development of the RotaShield vaccine in 1998 was heralded as a preventive measure against this disease. However, 1 year later RotaShield was withdrawn because of an association with the occurrence of intussusception (twisting and obstruction of the bowel) in 23 infants 1 to 2 weeks after vaccination. As disappointing as this was, the rapid response remains as a tribute to the FDA because of their swift reaction to this unexpected circumstance. Subsequently, two rotavirus vaccines are currently in use for infants in the United States: RotaTeq (RV5) approved in 2006, and Rotarix (RV1) approved in 2008. Both vaccines are live attenuated (weakened) virus vaccines that are given by putting drops in the infant's mouth. Both demonstrate excellent protection against severe rotavirus infections.

The pertussis (whooping cough) component of the combined diphtheria, tetanus, and whooping cough vaccine (DTP), formerly made from whole cells, has, since 1991, been derived from components of the microbe, resulting in fewer adverse effects. The new DTP vaccine is called the DTaP vaccine because of the use of acellular pertussis (aP). Fever and other mild systemic events were common with the old DTP as were concerns about vaccine safety. DTP is no longer used in the United States.

Although it is true that some adverse reactions have occurred after vaccination, it is also true that vaccines may be falsely blamed because unrelated events may coincidentally occur shortly after vaccine administration. Although adverse reaction examples are worrisome, there is some comfort in the realization that the FDA attempts to stay on top of the situation.

In response to the vaccine safety concerns, Congress passed the National Childhood Vaccine Injury Act in 1986, mandating that healthcare providers furnish a vaccine information sheet to recipients describing the risks and benefits of the vaccine. Providers are also required to report certain side effects after vaccination to the **Vaccine Adverse Event Reporting System (VAERS)** jointly administered by the FDA and CDC. Reporting to VAERS is encouraged even if the vaccine provider is not certain that the incident is vaccine related. A number of other CDC initiatives also focus on improved vaccine safety, including the Vaccine Safety Datalink, which is a database to link multiple sources that may identify vaccine safety issues and the Clinical Immunization Safety Assessment Project, which seeks to improve understanding of adverse effects following immunization at the individual patient level. Further, the Vaccine Injury Compensation Program (VICP) provides "no fault" vaccine compensation to those who experience certain health issues following vaccination.

The FDA, the licensing agent, does not approve a vaccine unless initial trials indicate the benefits clearly outweigh the risks. Licensure is a rigorous process involving three phases of clinical trials and may take 10 or more years. Vaccines are subject to particularly high safety standards, because, unlike other health treatments, they are given as preventives to healthy people.

Vaccines are manufactured by pharmaceutical companies; each batch of vaccine must be approved by the FDA before it can be released for use by health providers. Issues of safety, effectiveness, sterility, and purity are all evaluated by laboratory procedures, and postmarketing surveillance is conducted to identify undesirable side effects that might occur in large groups of people over long periods.

Childhood Immunization

The burden of infectious disease has been reduced throughout the world, most notably in the United States and in other industrialized countries, through the routine practice of childhood immunization. Immunization recommendations and schedules are updated almost every year. **FIGURE 15.21** shows the 2018 recommended childhood immunization schedule (aged newborn to 6 years) in the United States, supported by the CDC Advisory Committee on Immunization Practices, the American Academy of Pediatrics, and the American Academy of Family Physicians. Included are routine immunizations against 14 diseases.

Vaccine	Birth	1 mo	2 mos	4 mos	6 mos	9 mos	12 mos	15 mos	18 mos	4–6 yrs
HepatitisB (HepB)	1st dose	2nd dose			3rd dose					
Rotavirus (RV) RV1 (2-doses), RV5 (3-doses)			1st dose	2nd dose	(3rd RV5)					
Diphtheria, tetanus, a cellular pertussis (DTaP)			1st dose	2nd dose	3rd dose			4th dose		5th dose
Haemophilus influenzae type b, (Hib)			1st dose	2nd dose	3rd dose, some versions		3rd or 4th dose			
Pneumococcal conjugate (PCV13)			1st dose	2nd dose	3rd dose		4th dose			
Inactivated poliovirus (IPV)			1st dose	2nd dose	3rd dose					4th dose
Influenza (IIV)					Annual vaccination, 1 or 2 doses					
Measles, mumps, rubella (MMR)							1st dose			2nd dose
Varicella (VAR)							1st dose			2nd dose
Hepatitis A (HepA)							2-dose series			

Data from CDC and American Academy of Pediatrics.

FIGURE 15.21 Childhood immunization recommendation schedule in effect as of February 1, 2018. The CDC publishes a schedule for recommended immunizations that is approved by the Advisory Committee on Immunization Practices, the American Academy of Pediatrics, and the American Academy of Family Physicians.

Some of the immunizations are against bacterial diseases, but most are against viral diseases; attenuated, killed, and subunit vaccines are all represented. Two additional vaccinations are recommended for adolescents: human papillomavirus and meningococcal disease. Adults need protection from infections as well, so the adult schedule includes some of the same vaccinations as for children, such as the tetanus booster, as well as some different vaccines (shingles vaccine and some formulations of the pneumococcal vaccine). Some populations receive vaccines based on their potential exposures, often related to occupation, including the rabies vaccine for veterinarians.

Despite the low cost and effectiveness of immunization, thousands of children and adults have never had basic immunizations or are not up-to-date. Multiple factors affect vaccine compliance, including misinformation about vaccines (**BOX 15.2**). Almost 100,000 adults die every year from influenza, pneumonia, and other vaccine-preventable diseases.

Clinical Correlates

Despite these elaborate mechanisms of defense against potential pathogens, people become ill and sometimes die because of microbial diseases. It may be that the immune system is defeated in the dynamic interplay between microbe and immunity in a virtual tug-of-war by highly infectious agents such as Ebola, HIV, and rabies viruses and bacteria that cause meningococcemia or pneumonia. The person's own immune system may be impaired, as is the case with leukemia, or the person may have an immunodeficiency. Treatment with immunosuppressive drugs also increases a person's susceptibility to infection.

Human Immunodeficiency Virus

Human immunodeficiency virus (HIV), the virus responsible for AIDS, destroys T_H cells, resulting in the failure of an infected person to mount an appropriate immune response either by antibody formation or by T_C cells. The HIV-infected individual becomes severely **immunocompromised** (i.e., has a weakened immune system), which results in one infection after another or several infections at one time, eventually causing death from infection. Blood from HIV-infected individuals is routinely monitored to determine the level of T_H lymphocytes and the amount of circulating virus. The prognosis is grave when the number of T_H cells drops significantly, signaling that the individual no longer has the capacity to combat invading microbes via the adaptive immune system. When

BOX 15.2 Vaccination and Autism

When discussing vaccines, one question that always seems to come up is whether vaccines cause autism. This concern was initially based on a paper published in the late 1990s that included a small number of individual cases, selected with bias, suggesting that there was an association between the MMR vaccine and autism. It is not a bad question to ask. How autism happens is a mystery. However, the way to really answer that question is to look at large numbers of children who either did or did not get the MMR vaccine to see if there were more autism diagnoses in the group that got the vaccine. Almost immediately after the 1999 publication, scientists the world over did those well-designed studies. Multiple studies later, collectively looking at hundreds of thousands of children, they all found the same thing: whether or not a child was vaccinated, the risk of autism was the same. Even when looking at younger siblings of a child with autism, whether that younger child was vaccinated or not, the risk of developing autism was the same.

In total, at least 17 studies have shown that the MMR vaccine doesn't cause autism. Yet, in 2014, the United States experienced 23 measles outbreaks resulting in 667 cases, including a large outbreak that accounted for more than half of the cases that year as measles spread through communities in Ohio with high numbers of unvaccinated individuals. In 2015, declining vaccination contributed to a measles outbreak that started at the Disneyland theme park and spread throughout Southern California and then to 25 states. Ultimately, 188 people, mostly children, were infected before the virus spread into Canada, where a few hundred more people were infected.

The bottom line is that the cause of autism is still unknown, but what we do know is that it is not vaccines.

you hear that an individual with AIDS has a low CD4 count, this refers to a low level of T_H cells.

Leukemia

Leukemia, unfortunately, is an all too familiar term. It is a form of cancer characterized by uncontrolled reproduction of white blood cells. The two major categories of leukemia reflect the two pathways of blood cell maturation (Figure 15.7). Myeloid leukemia is the result of overproduction of monocytes and/or granular leukocytes (neutrophils, basophils, and eosinophils). Consider that immature neutrophils may be produced in excess and are unable to carry out phagocytosis, resulting in an immunocompromised individual. Lymphocytic leukemia is the result of overproduction of lymphocytes, leading to abnormally large numbers of immature and nonfunctional lymphocytes and their spread by metastasis into tissues throughout the body, crowding out normally functioning cells. Leukemia is classified as acute or chronic. In acute leukemia the symptoms appear suddenly and progress rapidly; death occurs in a few months unless the condition is successfully treated. Chronic leukemia can remain undetected for many months, and life expectancy without treatment is somewhere around 3 years. Acute lymphocytic leukemia is the most common form of leukemia in children, but fortunately treatment of this condition has a high success rate. The Dana-Farber/Boston Children's Cancer and Blood Disorders Center is world renowned for its success in treating children with leukemia and continues to be the world leader in leukemia clinical trials of new treatments. Fans of baseball history will be delighted to know that baseball legend Ted Williams of the Boston Red Sox was a frequent visitor and contributor to this institution.

Immunodeficiencies

A variety of immunodeficiencies may be present at birth as a result of inheritance, including deficiencies in complement production, phagocytes, B cells, and T cells. A properly functioning immune system requires the interplay of both innate and adaptive mechanisms; impairment results in an immunocompromised individual subject to repeated life-threatening infections. **TABLE 15.10** gives a sampling of immunodeficiency disorders, three of which are described here.

Severe combined immunodeficiency disorder (SCID) is a devastating congenital disease in which individuals lack functional T and B cells and therefore cannot mount either an antibody- or cell-mediated immune response. Death is a certainty, resulting from repeated infections. This disorder is commonly known as "bubble boy disease," because infants born with this condition must be kept in a germ-free environment, such as a plastic bubble. [One of John Travolta's early movies, *The Boy in the Plastic Bubble* (1976), dramatized the plight of a boy who spent his first 12 years in a germ-free bubble because he was born with SCID.] Those afflicted cannot even have contact with their parents, and all items introduced into the bubble, including air, food, and water, must be sterilized. There are several different genetic causes for SCID. In recent years, gene therapy and early bone marrow transplants (within 3 months of birth) have proved to be effective in some cases.

DiGeorge syndrome is a disorder resulting from the absence or incomplete development of the thymus. T-cell maturation is abnormal, resulting in impairment of both humoral immunity and CMI. In incomplete DiGeorge, the small amount of thymus that the child has may eventually function enough to produce T cells. With complete DiGeorge, the child will need a thymus transplant.

Chronic granulomatous disease is an inherited disorder of phagocytes. Because of the inability of phagocytes to kill, serious, life-threatening, and persistent infections result.

TABLE 15.10 Sampling of Immunodeficiency Disorders

Deficiencies in Complement	Deficiencies in Phagocytosis	Deficiencies in B Lymphocytes	Deficiencies in T Lymphocytes	Deficiencies in Both B and T Lymphocytes
Complement factor 3 deficiency	Chronic granulomatous disease	X-linked agammaglobulinemia (Bruton's agammaglobulinemia)	DiGeorge syndrome	Severe combined immunodeficiency disorder (SCID)
Complement factor 5 deficiency	Chediak-Higashi syndrome	IgA deficiency	Bare lymphocyte syndrome (MHC deficiency)	

Summary

The immune system is a defense mechanism against invading microbes and, additionally, an internal surveillance system. The ultimate outcome of exposure to pathogenic microbes is the result of the dynamic interplay between the disease-producing (virulence) mechanisms of the microbe and the immune system. The recognition of foreignness is the key element in triggering mechanisms of immune defense. Both innate and adaptive immune systems can identify those things that belong in your body and those things that don't. The immune system may incorrectly target self as foreign and mount an immune response against self, resulting in a variety of autoimmune diseases.

Anatomically, the immune system is unique in that it shares cells, tissues, and organs with other functional systems of the body. Blood and blood cells, thymus, bone marrow, tonsils, lymph nodes, and spleen are key components of the immune system.

The innate immune system includes the skin, phagocytic cells, components of blood, and ciliated cells that line part of the respiratory tract. Adaptive immunity responds to the presence of foreign and potentially harmful microbes through the activity of antibody-producing (B) cells, cytotoxic (T_C) cells that kill target cells, and T helper (T_H) cells that can direct the activity of many cells. Innate immunity and adaptive immune mechanisms work together to target microbial invaders for destruction.

Disorders of the immune system may be the result of infection, cancer, or immunosuppressive drugs or may be congenitally inherited. Allergic reactions and autoimmune diseases are adverse reactions of the immune system. Allergic reactions are hypersensitivity (exaggerated responses) reactions to external antigens. Autoimmune diseases are characterized by the misrecognition of self as nonself (foreign), resulting in the immune system mounting an attack against self. Immunization developed from its obscure and nonscientific beginnings to result in the eradication of smallpox and the elimination, or near elimination, of many infectious diseases. New and improved vaccines are sought and continue to become available.

KEY TERMS

active immunization
adaptive immunity
adenoids
agranular cells
allergens
allergy
anaphylactic shock
antigens
apoptosis
autoimmune disease
B cell zones
B lymphocytes (B cells)
bacillus Calmett-Guérin (BCG) vaccine
basophils
bone marrow
CD4 molecules
CD8 molecules

cell-mediated immunity (CMI)
chemotaxis
chronic granulomatous disease
cilia
Class I MHC
Class II MHC
cluster of differentiation (CD)
complement
cytokines
cytotoxic T (T_C) lymphocytes (CTLs)
dendritic cells
differential count
DiGeorge syndrome
diphtheria tetanus acellular-pertussis (DTaP) vaccine
eosinophils
erythrocytes
exotoxins

gamma globulin
granulocytes
granzymes
Guillain-Barré syndrome (GBS)
hematopeoisis
hematopoietic stem cells
histamine
humoral immunity
hybridoma
hypersensitivity reaction
immunity
immunocompromised
inflammation
innate immunity
interferon
isotypes
leukemia
leukocytes
live attenuated vaccines
lymph
lymphatic system
lymphocytes
lymphoid lineage
lymphoid organs
lymphoid tissues
lysozyme
macrophages
major histocompatibility complex (MHC)
mast cells
monoclonal antibody
monocytes
mucociliary escalator
myeloid lineage

myeloma cell
natural killer (NK) cells
Neutrophil
normal microbiota
passive immunization
pathogen-associated molecular patterns (PAMPs)
pattern recognition receptors (PRRs)
perforins
peripheral lymphoid tissues
Peyer's patches
phagocytic cells
phagocytosis
plasma
plasma cells
platelets
polyclonal antibodies
professional antigen-presenting cells
secondary lymphoid tissues
serum sickness
severe combined immunodeficiency (SCID)
spleen
splenectomy
symbiotic
T cell subset
T cell zones
T helper (T$_H$) cells
T lymphocytes (T cells)
TNF-alpha
thymus
tonsillectomy
tonsils
toxoids
Vaccine Adverse Event Reporting System (VAERS)

SELF-EVALUATION

○ **PART I: Choose the single best answer.**

1. Which of the following white blood cells are primarily involved in phagocytosis?
   a. Neutrophils
   b. Eosinophils
   c. Basophils
   d. Lymphocytes

2. Which of the following is a primary (central) immune structure?
   a. Tonsils
   b. Lymph nodes
   c. Thymus
   d. Blood cells

3. The lymphocytes of the adaptive immune response generate numerous receptors with which of the following features?
   a. Detect self-components, only
   b. Are expressed by phagocytic cells
   c. Detect specific foreign components
   d. Recognize virally infected cells only

4. Human immunodeficiency virus (HIV) specifically infects and destroys which of the following cells?
    a. T_H cells
    b. Red blood cells
    c. T_C cells
    d. Neutrophils

5. Myeloid stem cells give rise to all the following *except*:
    a. Lymphocytes
    b. Neutrophils
    c. Platelets
    d. Basophils

6. Antibodies are secreted from which of the following cells?
    a. T cells
    b. Plasma cells
    c. NK cells
    d. Macrophages

7. Which of the following is *not* considered a secondary immune tissue?
    a. Spleen
    b. Tonsils
    c. Bone marrow
    d. Lymph nodes

8. Serum sickness can result from which of the following?
    a. Attenuated vaccines
    b. Live vaccines
    c. Toxoid administration
    d. Passive immunization

9. A 6-month-old infant received the Hib vaccine, which contained capsular polysaccharide from *Haemophilus influenzae* type b attached to a protein component. What type of vaccine is this?
    a. Live attenuated vaccine
    b. Killed whole bacteria vaccine
    c. Subunit vaccine
    d. Recombinant DNA vaccine

○ **PART II: Fill in the blank.**

1. B cells mature in the _____.

2. _____ diseases fail to distinguish between nonforeign and foreign cells.

3. The immune system functions to recognize and destroy that which is _____.

4. _____ is the lymph organ that is important for responding to infections in the blood.

5. CMI stands for _____.

6. Children born with the condition known as DiGeorge syndrome are born without a(n) _____.

7. _____ are live or inactivated viruses or microbes administered to prevent infectious diseases.

8. _____ are lymphocytes that are involved in innate immunity.

9. _____ is a form of cancer characterized by uncontrolled reproduction of white blood cells.

○ **PART III: Answer the following.**

1. What are autoimmune diseases? Discuss the immunological basis of these diseases.

2. Explain the differences between innate and adaptive immunity.

3. An individual with AIDS has a low T_H count. What does this mean? What is the significance of this low count?

4. Explain why there aren't vaccines available for every infectious microbe or virus.

5. Why are immunocompromised individuals more at risk for microbial infections?

6. List the different subsets of T cells and describe how each subset functions within the immune system.

7. List the different protective effects of antibodies.

8. Why are memory cells important?

9. Distinguish between active and passive immunization.

10. List some requirements or properties of an ideal vaccine.

CONTROL OF MICROBIAL DISEASES

Terri S. Hamrick, PhD
Campbell University

> "*It is a disturbing fact that Western Civilization, which claims to have achieved the highest standards of health in history, finds itself compelled to spend ever increasing sums for the control of disease.*"
>
> —Rene Dubos
>
> *microbiologist and Pulitzer Prize winner, 1987*

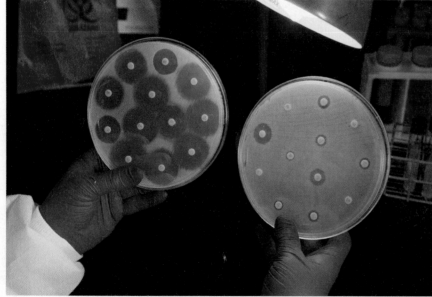

Courtesy of the CDC/ Melissa Dankel.

An antibiotic sensitivity test using disc diffusion. The two different petri dish cultures are two different bacterial strains or species growing in the presence of discs containing various antibiotics. The isolate on the left plate appears to be susceptible to the antibiotics on the discs and is unable to grow near the discs. The plate on the right contains a bacterial isolate that is resistant to the antibiotics tested and is able to grow nearer to the discs. The plate on the right was inoculated with a carbapenem-resistant *Enterobacteriaceae* (CRE) bacterium.

1. Identify appropriate uses for disinfectants and antiseptics.

2. List different types of antibiotics and identify the specific microbial target of each.

3. Define *selective toxicity*.

4. Apply the principle of selective toxicity to antibiotics.

5. Explain mechanisms by which microbes become resistant to antibiotics.

6. Discuss challenges to antibiotic therapy in the clinical setting.

7. Apply strategies to reduce the prevalence of antibiotic-resistant organisms.

8. List antiviral targets for antiviral medications.

9. Identify challenges to selective toxicity specific to viral infections.

10. Provide examples of antiviral resistance.

Case Study: A Pain in the Back

Mabel Anderson, a 55-year-old woman, presented to her physician's office to talk about her back pain. She thought that the source of her back pain had been fixed when she had back surgery 8 months ago, but here she was again at her doctor for lower back pain and a fever that she just couldn't shake. She told her doctor that the pain and fever had been going on for about a month, and that it felt different than the pain from before her surgery. Her previous surgery had been a lumbar spine fusion that involved implantation of screws, a bar, and a cagelike device to provide needed stability to her lower back. She was frustrated that she was hurting again. Other than a few urinary tract infections since her surgery, she had been in good health, and had been doing the exercises her physical therapist had recommended to recover from the back surgery.

A physical exam revealed a fever of 102°F (38.9°C) and some numbness near her surgery site. The numbness was normal after extensive surgery, but the fever was not. Bloodwork showed an elevated white blood cell count and suggested to the physician that she had an infection. Finding the infection proved quite elusive. The fluid from her spinal column didn't contain any microbes, even though the lab used very sensitive molecular techniques to look for pathogens that might have been the source of her infection. Blood and urine were also free of microbes by culture techniques. A magnetic resonance imaging (MRI) scan of the lumbar spine showed evidence of inflammation consistent with an infection of the lumbar spine, especially around the cage and screws from her previous surgery. Seeing evidence of an infection, but not knowing what it might be, her physician administered intravenous antibiotics:

cefuroxime (a strong antibiotic that affects bacterial cell walls) and vancomycin (an antibiotic typically reserved for dangerous infections with microbes resistant to other commonly used antibiotics).

Eighteen days later, she again had a fever of 103°F (38.9°C). A culture of her urine revealed *Enterococcus faecium*. *E. faecium* is common in the human gastrointestinal tract, and doesn't usually cause harm in the body unless it finds its way to other sites, such as the urinary tract and bone. Antibiotic sensitivity testing of the strain isolated from Mrs. Anderson revealed that it was resistant to ampicillin (a cell wall antibiotic) and vancomycin. The isolate was sensitive to several antibiotics, two of which, rifampicin and daptomycin, were administered to her at this time. However, her fever persisted. The surgical team decided to remove the implanted hardware from her lumbar spine, clean out the area, and obtain a direct sample of the infected area. Culture of the isolated disc from her back also showed *E. faecium* with the same antibiotic-susceptibility profile as the strain isolated from the urine. She left the hospital taking three antibiotics prescribed for 6 months. Over the next 6 months, the inflammation in her back resolved, and the fever did not return. The neurosurgery team suggested a new surgery to stabilize her lumbar spine. She decided that for now, she was OK.

Questions and Activities

1. Consider the normal location of *E. faecium*. Propose a likely **mode of transmission** by which the bacteria infected the lumbar spine and screws.

2. Enterococci occur naturally in the gastrointestinal tract of humans. Why then is it particularly concerning that this strain was resistant to vancomycin?

(continues)

Case Study: A Pain in the Back (*continued*)

3. Following culture of her urine, the physicians were able to administer antibiotics to which her infecting strain was sensitive, yet her fevers persisted. Why? How did the surgery to remove the hardware in her back help to resolve the infection?

4. Why did her physician administer antibiotics for so long?

5. Find images of the human skeleton that include the cerebrospinal column. Locate the lumbar spine. What other areas of the body could be infected when bacteria gain access to the lumbar spine?

Information from: C. Guli, G. Iacopino, P. Di Carlo, C. Colomba, A., . . . R. Maugeri. (2017). Vancomycin-resistant *Enterococcus faecium* (VRE) vertebral osteomyelitis after uneventful spinal surgery: A case report and literature review. *Interdisciplinary Neurosurgery, 7*, 12–16.

Preview

Advances in public health during the 20th century have decreased the burden of microbial disease on a worldwide basis, particularly in the United States and in other industrialized nations. Sanitation and clean water, food safety, immunization, and antibiotics are major factors in the control of infectious diseases. This chapter will focus on antibiotics and other means of controlling microbial diseases.

Is the general health of your generation better than that of your parents' or grandparents' generations? Most decidedly, your response should be "yes," despite the current problem of new and reemerging infections. Consider that a person born in the United States in the early 1900s could anticipate an average life span of 45 years and that the death rate at birth was slightly higher than 10%. Today, life expectancy in the United States, as estimated by the

Central Intelligence Agency (CIA) *Worldfact Book*, is 80 years (77.7 for males and 82.2 for females), an increase of 35 years from the early 20th century. The overall worldwide life expectancy is 69 years, ranging from a high of 89.4 years in Monaco to a low of 50.6 years in Chad. Surprisingly, the United States ranks 43rd in a list of 224 countries despite the fact that it spends the most on health care. Wealthy Americans are among the world's healthiest people, whereas Americans on the bottom rungs have a life expectancy characteristic of sub-Saharan Africa.

Some of the remarkable gains in life expectancy over the decades can be explained by a combination of technological improvements (e.g., safer cars, safety in the workplace, and better medical technologies) and social factors (e.g., antipoverty programs, better nutrition, and reduced cigarette smoking). The 20th century has also seen advances in infection control, prevention, and treatment (**FIGURE 16.1**). In 1900,

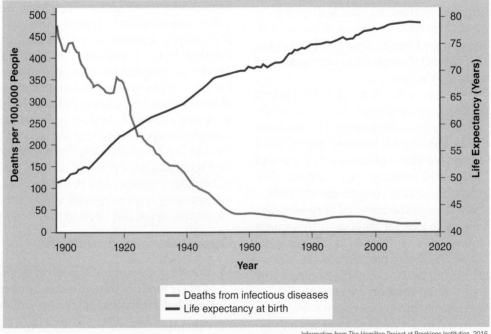

Information from The Hamilton Project at Brookings Institution, 2016.

FIGURE 16.1 Major gains in life expectancy and crude death rates, 1900–2014. **(a)** Infectious diseases were a major cause of morbidity and mortality in the early 1900s.

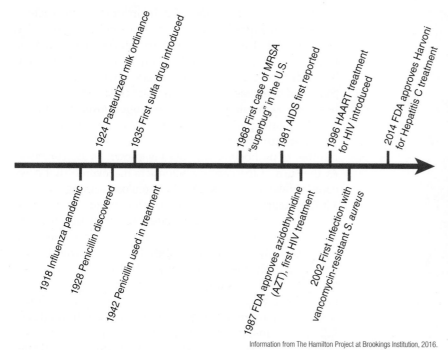

1918 Influenza pandemic
1924 Pasteurized milk ordinance
1928 Penicillin discovered
1935 First sulfa drug introduced
1942 Penicillin used in treatment
1968 First case of MRSA "superbug" in the U.S.
1981 AIDS first reported
1987 FDA approves azidothymidine (AZT), first HIV treatment
1996 HAART treatment for HIV introduced
2002 First infection with vancomycin-resistant S. aureus
2014 FDA approves Harvoni for Hepatitis C treatment

Information from The Hamilton Project at Brookings Institution, 2016.

FIGURE 16.1 (b) Timeline of selected significant events in the modern history of infectious diseases.

Disinfection Methods

TABLE 16.1 lists physical and chemical methods used to control microbes. Temperature is one of the most common physical methods of microbial control. Refrigeration and freezing can halt or slow microbial growth. High temperatures to the point of incineration is one method of **sterilization**, which involves the destruction of all live microbes, spores, and viruses present on an object. Boiling can kill many microbes but may not kill bacterial spores. **Autoclaves** use steam under pressure to inactivate biohazardous materials collected in hospitals or healthcare facilities or to sterilize instruments used in surgical procedures or in a dentist office (**FIGURE 16.2**). Some microbes are more difficult to inactivate than others (**TABLE 16.2**). For example, **prions** are the most difficult infectious agents to inactivate. Prion-contaminated

the three leading causes of death were pneumonia, tuberculosis, and diarrhea/enteritis. These infections, together with diphtheria, caused more than one-third of all deaths for that year. Contrast that to 2015, when the only infections in the top 10 leading causes of death were pneumonia and influenza, coming in at number eight. In the battle against infectious diseases, prevention strategies and treatment both play a role in keeping us healthy.

Disinfection and Disease Control

The connection between germs and disease was validated in the late 19th century and has led to important innovations such as antibiotics, new antibacterial products, and hygiene practices. Disinfection is one of the cornerstones of microbial infection prevention and control. Our world is becoming sterilized, sanitized, and "germ-free." How many times have you washed your hands today? Global Handwashing Day (October 15th) was created for children and schools in 2008 to raise awareness about the benefits of handwashing and to foster handwashing with soap in every country. Hand sanitizer dispensers have become commonplace. There is a growing sense of germ phobia.

Disinfection is defined as killing or inactivating microbes that cause disease. Disinfection applications date to the 18th century, a time when people were ignorant of pathogenic microbes yet employed methods that controlled diseases of animals and humans. Methods of disinfection were classified into three categories: chemical (by derivatives of sulfur, mercury, copper, alkalis, and acids), physical (**fumigation**, heating, and **filtration**), and biological (burial). Today, physical and chemical methods remain important means of controlling and preventing infectious diseases.

TABLE 16.1 Disinfection and Sterilization Methods

Physical Methods	Chemical Methods
Heat (sterilization): incineration, autoclaving (steam under pressure)	Heavy metals: mercury (e.g., topical mercurochrome), silver (e.g., silver sulfadiazine dressings on wounds), copper (e.g., copper bedrails in hospitals)
Heat: pasteurization, dry heat (oven), boiling, dehydration	70% alcohol (e.g., hand sanitizer or skin antiseptic, instrument disinfectant)
Low temperature control: refrigeration and freezing (retards growth)	Aldehydes (preservation of dead animals)
Radiation (sterilization): x-rays, gamma rays	**Halogens**: chlorine (e.g., swimming pools), bromine, iodine (e.g., topical on wounds), fluorine
Radiation: ultraviolet radiation, microwaves	Quaternary ammonium compounds (QUATS; disinfectant in hospital and healthcare settings)
Physical removal: filtration of air and liquids	Salt (preservation of foods)
	Hydrogen peroxide (wound treatment)
	Ethylene oxide gas (sterilization of plastic ware)

© Terri S. Hamrick.

FIGURE 16.2 An autoclave uses steam under pressure to sterilize equipment, liquids, and other items.

samples must be placed in sodium hydroxide or undiluted fresh household bleach and autoclaved at 130°C (266°F) for 4.5 hours, whereas materials contaminated with all other infectious agents can be sterilized for 15 minutes at 121°C (250°F). Ethylene oxide gas or ionizing radiation are alternative methods of sterilization for items that cannot be exposed to the high temperatures of an autoclave. Without a reliable means of sterilization, many of the major advances in medicine, such as surgery, medication therapy, and wound management, would not be possible.

Pasteurization, named after its inventor, Louis Pasteur, was originally developed to keep beer from going bad. Yes, beer! In 1856, Pasteur, a chemistry professor researching fermentation, studied spoiled and fresh beer from local breweries for over a year. He observed with a microscope that good beer was full of round yeast cells and that spoiled beer was swimming with long microbes. After heating the beer for a brief period, the long microbes died but the yeast cells were still thriving. Today, heating liquids to reduce the number of pathogenic microbes is known as *pasteurization*. At the turn of the century, scientists discovered that cows could spread diseases, such as tuberculosis, through milk. By 1907, pasteurization of milk was implemented and the number of disease outbreaks associated with milk plummeted.

Heavy metals are very reactive and have an **oligodynamic effect** on microbes. In other words, it takes very few molecules of a heavy metal to kill or be toxic to microbes. For example, a few drops of copper sulfate kill algae in a fish aquarium. Recent studies have shown that copper room surfaces in intensive care units (ICUs) kill 97% of hospital bacteria, reducing infections by 40%. Don't be surprised if hospitals are eventually refitted with copper bed rails, sinks, and toilets (**FIGURE 16.3**)! The mechanisms of the killing action of chemicals used in disinfection methods vary—from binding, denaturing, or oxidizing proteins; to dissolving membranes, causing them to leak; or reacting with nucleic acids.

Killing of microbes does not generally happen instantaneously, but rather a fixed proportion of microorganisms are killed over a given amount of time. The temperature, type of microbe, concentration of the agent, environment,

TABLE 16.2 List of Microbes and Their Resistance to Disinfection

Infectious Agent	Examples (disease or specific infectious agents)	**Most Resistant**
Prions	Creutzfeldt-Jakob disease (CJD), chronic wasting disease (CWD)	
Bacterial spores	*Bacillus, Clostridium*	
Helminth eggs	*Ascaris*	
Mycobacteria	*Mycobacterium tuberculosis, M. leprae*	
Small, nonenveloped viruses	Poliovirus, norovirus, papilloma viruses	
Protozoan cysts	*Giardia*	
Fungal spores	*Aspergillus, Penicillium*	
Gram-negative bacteria	*Escherichia, Salmonella, Pseudomonas*	
Yeast	*Candida*	
Gram-positive bacteria	*Staphylococcus, Streptococcus*	
Enveloped viruses	HIV, influenza, herpes simplex virus	**Least Resistant**

Information from: G. McDonnell & P. Burke. (2011). Disinfection: Is it time to reconsider Spaulding? *Journal of Hospital Infections, 78*, 163–170.

FIGURE 16.3 Water running into a copper sink. The copper will inhibit bacterial growth on the sink surfaces.

© Robert I. Krasner.

FIGURE 16.4 Cleaning products on shelves in a store.

and other conditions influence the death rate. In general, more microbes die as the temperature rises and as the concentration of the chemical agent increases. You may have noticed that disinfectant labels in the store report efficacy as a percentage: "Kills 99.9% of viruses and bacteria" is a common claim. This means that if you started with 1,000,000 bacteria, this particular disinfectant will leave behind approximately 1,000 microorganisms after the measured amount of time. Because each disinfectant has a specific killing rate, it is important to think through the microbial burden when considering how long it may take to reduce the number of disease-causing microbes to a safe level. For example, swimming pools have a baseline concentration of chlorine that will kill the small number of bacteria coming from swimmers while they enjoy the pool. However, if a child has a diaper accident, and lots of fecal bacteria enter the water, then the pool needs to be treated with a higher concentration of chlorine for a longer time to kill the microorganisms and make the water safe again.

Cleaning Products, Soap, and Handwashing

A vast array of disinfectants and antiseptics are available for hospitals and other users to choose from (**FIGURE 16.4**); they differ by odor, color, and mechanism of dispersal (foam, spray, direct from bottle, diluted, not diluted). Disinfectants and antiseptics are defined by their properties. Disinfectants can be used on inanimate objects and surfaces, but they are too toxic for use on body tissues. **Antiseptics** are disinfectants that can be used on body tissues (e.g., a wound). The presence of dirt, blood, and damaged tissue can make it more difficult for antiseptics to kill bacteria in a wound. Your parent was correct for washing your scraped knee to clean out the dirt before using an antiseptic ointment and bandage.

We pay high prices for disinfectants and antiseptics, but only a few different chemical formulations are involved. The labels of the products will advertise their effectiveness

and scents. For example, the label of a product containing *n*-alkyl and dimethyl benzyl ammonium chlorides emphasizes that it "Kills 99.9% of germs in 60 seconds and powers through tough grease and grime!" Some products advertise that they have an original fresh lemon or citrus scent. The newest trends are natural cleaning products that are "chlorine free" but contain active ingredients that are at least 95% naturally derived, such as citric acid, enzymes, boric acid, lactic acid, ethanol, fragrance with essential oils, calcium chloride, alkyl polyglucosides from plants, sodium lauryl sulfate (detergent), sodium gluconate, and others.

Humans have been living with bacteria for centuries. In fact, the human body has 10 times more bacterial cells living in or on it than the number of human cells. Bacteria live inside us, on us, and all around us. Microbes strengthen our immune systems. If we are not adequately exposed to germs, our immune systems will atrophy such that the next time a pathogen is encountered we may not have an adequate defense. A small percentage of bacteria can make us sick. Part of the challenge is to know when to seek the elimination of microbes, when to merely reduce the microbial burden, and when to just leave them alone.

Handwashing is one of the most important ways to protect yourself and others from infection. Soap and water is an effective way to remove contaminants, including pathogenic bacteria and viruses from your hands. Handwashing is easy and effective if you take enough time to wash thoroughly.

For a long time the popular view was that if handwashing is good, then handwashing with antibacterial soap is better. However, that does not appear to be the case. Many studies have found no advantage to antibacterial soap over

Krasner's Notebook

When washing your hands, scrub for at least 20 seconds, or about as long as it will take you to hum the "Happy Birthday" song, twice.

plain soap and water. One reason for this is that antibacterial agents need time to work. Industries were testing antimicrobial soaps as if they were testing an antibiotic drug or a disinfectant—adding the test agent to bacterial cultures, and then looking at growth over a few days. However, that isn't at all the way we use soap. In one study, the bacteria had to be exposed to the antibacterial agent for 9 hours before there was a difference in viability as compared to plain soap. That is a long time to leave soap on your hands! Additionally, some of these antibacterial agents had unwanted consequences in the body, including allergies, hormone disruption, toxicity, accumulation in the body, and microbial resistance.

The bottom line is that when used with good handwashing practices, soap and water work just as well as antiseptic soaps, without the additional health concerns. The Food and Drug Administration (FDA) agreed, and in 2016 issued a statement declaring that the risks posed by the use of certain antibacterial agents in soap outweigh the benefits. As such, most soaps will no longer be marketed with added antibacterial agents.

Handwashing does not remove all microbes from your hands. This is one of those instances where we need to remember that having bacteria on our skin is OK. However, certain activities may put us at risk for picking up pathogens that we will want to wash away before we ingest them or pass them on to others. The following activities require handwashing: before, during, and after preparing food; before eating; before and after caring for someone who is sick; before and after treating a cut or wound; after using the restroom; after handling anything that might be dirty or contaminated, such as dirty diapers, garbage, animals, or animal bedding; or after blowing your nose, coughing, or sneezing (**FIGURE 16.5**).

© Terri S. Hamrick.

FIGURE 16.5 Employees are reminded to wash their hands after going to the bathroom at work.

© Terri S. Hamrick.

FIGURE 16.6 Technology gadgets may harbor pathogenic bacteria, including bacteria from human feces.

Washing hands with soap and water is the best way to get rid of germs. If soap and water are not available, use an alcohol-based hand sanitizer that contains at least 60% alcohol, which is the concentration necessary to denature microbial proteins and membranes. Hand sanitizers are effective at reducing the number of viable pathogenic microbes on the skin in most circumstances, but they don't remove dirt, grease, or blood, all of which can be helpful for microorganisms. Alcohol-based hand sanitizers are also not effective against bacterial spores and some viruses. As a general rule, soap and water work the best.

Tech Gadgets and Germs

This is the age of technology gadgets. Most of us own a smartphone (**FIGURE 16.6**), and nearly all televisions use a remote control. A number of studies have looked at the degree of microbial contamination on these devices. Researchers have found that smartphones, tablets/e-readers, game controllers, keyboards, and remote controls each house more bacteria per square inch than that found on a typical home toilet seat. So, what do we do about it? The following are a few good suggestions for gadget hygiene:

- The bathroom is no place for a phone, unless you want it covered in fecal bacteria.

- Wash your hands before handling your devices or at least use a hand sanitizer. You have more bacteria per square inch than a toilet seat. Wash your hands to keep your bacteria off your devices.

- Use gadget-friendly wipes to keep your technology clean.

Antibiotics

Killing microbes is not hard at all. Heat, radiation, and bleach are examples of relatively good ways to kill microbes. In the case of an infection, the tricky part is to kill the microbes while not harming the human. In the treatment of infections, we use medications that target the specific microbes while sparing the human cells and tissues. **Antibiotics** can rightly be considered the single most important discovery for the treatment of bacterial infections in the history of medicine. Ironically, the microbes themselves provided the groundwork for the first antibiotic medications and serve as testimony that "nature knows best." The term *antibiotic* literally means "against life." Some microorganisms produce antibiotics that are released into their local habitat to affect nearby microbes so that they don't have to share the neighborhood.

Many of the antibiotics used today are derived from the natural products made by microorganisms as a strategy to eliminate the competition. Technically, antibiotics are natural products; however, the term antibiotic is often used for natural as well as synthetic antimicrobials. The secretion of metabolic products by soil bacteria and fungi that inhibit the growth of other microbes is an example of ecological antagonism at the microbial level.

Types of Antibiotics

There is no such thing as a universal antibiotic. Bacteria vary in their antibiotic susceptibility, and each antibiotic has a spectrum of activity against certain bacteria. For example, some antibiotics are more effective against Gram-positive organisms, whereas others exhibit greater activity against Gram-negative bacteria (**FIGURE 16.7**).

A **broad-spectrum antibiotic** is inhibitory to a large variety of Gram-positive and Gram-negative bacteria, whereas a **narrow-spectrum antibiotic** is inhibitory to a limited range of bacteria. Some antibiotics are extremely

effective but, unfortunately, exhibit marked toxicity, rendering them not useful. In addition, some antibiotics are better at getting to some parts of the body than others. Some antibiotics stay in the body longer than others do. Adjustments to the basic chemical structure of an antibiotic can also change the spectrum of activity and the availability of the drug in different parts of the body.

Some antibiotics are very expensive, whereas others are not, and some are more prone to result in antibiotic-resistant strains than others. In prescribing an antibiotic from among the many that are available, cost and antibiotic resistance are considered, but effectiveness, location of the infection, and lack of toxicity are the central factors. Broad-spectrum antibiotics are generally prescribed when an individual is seriously ill and the causative bacteria have not been identified, so as to target a broad range of suspects. The downside of a broad-spectrum antibiotic is that a large number of bacterial species of the normal **microbiota** are killed, causing ecological disruption and allowing organisms that are not susceptible to the antibiotic to flourish. Some people receiving antibiotics develop oral thrush, a painful yeast infection, resulting from disruption of the normal bacterial communities and allowing yeasts to overgrow the normal microbiota. The tongue has a whitish appearance as a result of the colonies of yeast that have colonized it; fortunately, thrush responds well to mouth rinses with antiyeast drugs. In females receiving a course of antibiotics, vaginal thrush may develop and cause pain, burning, and itching, treatable by vaginal creams. Narrow-spectrum antibiotics cause less disruption in the ecological balance of microbes and also minimize the likelihood of antibiotic resistance.

Mechanisms of Antimicrobial Activity

Antibiotics act by interfering with or disrupting vital structures and metabolic pathways of the bacterial cell (**FIGURE 16.8**). Antibiotic drugs target components of a microbe that human cells do not have. This concept is known as **selective toxicity**, which essentially means that the drug is toxic to the microbe, but not to human cells. Based on what you now know about bacteria, you can probably identify a number of structures that bacteria have that your cells do not, the most obvious of which is the peptidoglycan cell wall of Gram-negative and Gram-positive bacteria. Penicillin exhibits selective toxicity because this drug interferes with cell wall production, and human cells do not have cell walls. Most antimicrobials target an important process in the microbe rather than destroying an already existing structure. For antibiotics that are cell wall specific, this means that the drug interferes with the building of the cell wall rather than affecting the existing cell wall. Some antibiotics are **bactericidal** (they kill directly), whereas others are **bacteriostatic** (they keep the population from growing, thus allowing the body's defense mechanisms to get rid of the invaders). Five mechanisms of antibiotic action are described in the following sections.

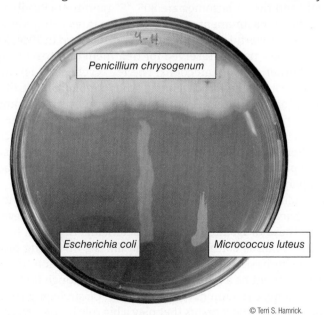

© Terri S. Hamrick.

FIGURE 16.7 Penicillin activity against a Gram-positive bacterium and a Gram-negative bacterium. The fungus *Penicillium chrysogenum* was streaked across the top of the plate. The antibiotic penicillin, which is made by the fungus, diffused into the agar of the plate. *Escherichia coli*, a Gram-negative bacterium, is in the streak on the left side of the plate. The *E. coli* is resistant to the effects of the penicillin, because the penicillin cannot get through the porins of the *E. coli* outer membrane. Therefore, the *E. coli* can grow very close to the fungus, even though the antibiotic concentration is high. *Micrococcus luteus*, a Gram-positive bacteria, is in the streak on the right side of the plate. *M. luteus* is very sensitive to the activity of penicillin and can only grow very far away from the fungus.

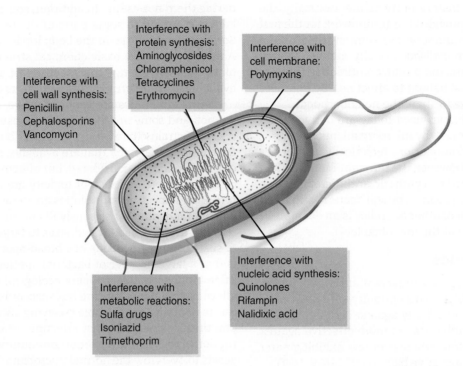

Interference with
protein synthesis:
Aminoglycosides
Chloramphenicol
Tetracyclines
Erythromycin

Interference with
cell membrane:
Polymyxins

Interference with
cell wall synthesis:
Penicillin
Cephalosporins
Vancomycin

Interference with
metabolic reactions:
Sulfa drugs
Isoniazid
Trimethoprim

Interference with
nucleic acid synthesis:
Quinolones
Rifampin
Nalidixic acid

FIGURE 16.8 Mechanisms of antimicrobial activity.

◉ Interference with Cell Wall Synthesis

Bacterial cells have rigid cell walls (**peptidoglycan**) that afford protection against lysis when bacteria are exposed to the low osmotic pressure of body fluids. Think about the peptidoglycan cell wall like layers and layers of chain link fence wrapped around the bacterium. The crosslinks connecting the chains of wire contribute to the strength of a chain-link fence. Similarly, as bacterial peptidoglycan is assembled around the bacterium, the long chains of the cell wall are crosslinked together by the activity of an enzyme known as **transpeptidase** to provide the overall structural integrity of the cell wall.

The antibiotics known as penicillins and cephalosporins contain structures (β-lactam rings) that associate and interfere with the enzymes responsible for building the crosslinks during cell wall synthesis. Collectively known as **β-lactam antibiotics**, these drugs result in the synthesis of a weakened cell wall, so that as the bacteria grow, they lyse. β-lactam antibiotics are examples of bactericidal antibiotics. Vancomycin, sometimes considered the "last antibiotic stronghold," also interferes with the construction of the peptidoglycan, but it acts by binding to the target of the crosslinking enzymes, thus interfering with the formation of the crosslinks.

◉ Interference with Protein Synthesis

Protein synthesis is an integral part of a cell's activity and is the culmination of expression of DNA. Bacterial **ribosomes**, cytoplasmic structures on which protein synthesis takes place, are targets for some antibiotics because bacterial ribosomes differ in size and structure from human ribosomes. (Bacterial cells are procaryotic and their ribosomes are 70S, whereas human cells are eucaryotic and their ribosomes are 80S. "S" represents Svedberg units, a measurement of sedimentation rates.) Streptomycin is a powerful antibiotic that was discovered in 1944; its use is now usually reserved for treatment of tuberculosis and other serious infections. The tetracyclines, chloramphenicol, and erythromycin are other antibiotics that, like streptomycin, interfere with protein synthesis by binding with procaryotic ribosomes. Tetracycline antibiotics are used for the treatment of several tickborne diseases, such as Rocky Mountain spotted fever, although repeated use is typically avoided in young children because it can lead to permanent tooth discoloration.

Many of the protein synthesis inhibitor antibiotics are bacteriostatic. You might wonder why we would ever use these drugs rather than a bactericidal drug. If a bacterial toxin is contributing to the illness, inhibiting protein synthesis, and thus inhibiting the production of the toxin, might be a really good choice, even though it might take longer to eliminate the bacteria. *Clostridium perfringens* makes many toxins that play a big role in the severity of gas gangrene. Tetracycline and clindamycin are protein synthesis inhibitor antibiotics and are more effective in the treatment of gas gangrene than the bactericidal antibiotic penicillin.

◉ Interference with Cell Membrane Function

Cell membranes function in a vital capacity as "gatekeepers." Based on their chemical and physical structure, they control what goes into and out of the cell. Antibiotics that distort and interfere with membrane function result in increased membrane permeability and rapid cell death. Antibiotics that target cell membranes are examples of antibiotics that act on an existing structure rather than its synthesis. You might remember that Gram-negative bacteria have an outer membrane in addition to the plasma membrane, whereas Gram-positive bacteria have only a plasma membrane. This difference matters when it comes to drugs that interfere with membrane function. Polymyxin B is an antibiotic that binds to and distorts the bacterial outer cell membrane, and is therefore active on many Gram-negative bacteria, but not Gram-positive ones. Daptomycin can bind to and disrupt the plasma membrane, but only in Gram-positive bacteria, because the drug can't get through the outer membrane of Gram-negative bacteria.

Interference with Nucleic Acid Synthesis

The replication and synthesis of the nucleic acids, RNA and DNA, are steps in the expression of DNA, a long and complicated series of chemical reactions that can be targeted for antibiotic activity. Rifampin blocks RNA synthesis by binding to the bacterial enzyme **RNA polymerase** as it initiates RNA synthesis. The quinolone antibiotics act by inhibiting the action of enzymes, such as **DNA gyrase**, that bacteria use to coil and uncoil DNA during replication of the bacterial chromosome. Nalidixic acid was the first quinolone used clinically and was effective for the treatment of Gram-negative urinary tract infections. More recently developed fluoroquinolones are effective drugs for the treatment of many Gram-negative and Gram-positive infections. Mammalian cells use different enzymes for these activities, and hence are not affected by this antibiotic—another example of selective toxicity.

Interference with Metabolic Activity

Metabolism, the chemical reactions involved in the production and synthesis of cellular materials, is a key characteristic in the distinction between life and nonlife. **Antimetabolites** are drugs that are structurally similar to natural compounds involved in metabolism and that competitively bind with these enzymes. The sulfa drugs (sulfonamides) work in this fashion. Bacteria synthesize folic acid, which is then used in a number of metabolic processes to make the building blocks of the cell. Sulfonamides are structurally similar to a component that bacteria use to make folic acid. Mammalian cells do not make folic acid but obtain this compound from their diet; hence, sulfa drugs can be used in human therapy.

Clinical Challenges to Antibiotic Therapy

Even without considering resistances acquired to specific antibiotics, we find that there are a number of challenges to effective use of antibiotics in clinical medicine. In these cases, the antibiotic may be biologically effective for a particular organism, but in the patient the drug doesn't work well enough to cure the patient.

Nearly all of the antibiotics that we use in clinical medicine today target a microbial process. This means that when the bacteria are growing more slowly, the antibiotic does not have as many opportunities to act. For example, if an antibiotic acts on the cell wall, when the organism is growing slowly it will take a longer time for the antibiotic to result in a cell wall weak enough to lead to bacterial lysis. Some bacteria are naturally slow growing, which means that an antibiotic will need to be consistently taken for a long time to effectively treat that infection.

For *Mycobacterium tuberculosis*, the causative agent of tuberculosis, under ideal growing conditions in the lab, it takes about 2 weeks to grow a visible colony on an agar plate. Compare that to about 2 days for *E. coli*. An antibiotic acting on a cellular process in *M. tuberculosis* will be waiting around for a long time to act on the cellular process. Effective antibiotic therapy to treat active tuberculosis is typically about 6 to 9 months' duration. Slow growth for *M. tuberculosis* also means that antibiotic susceptibility testing in the lab takes a long time. By the time the results are available the patient may have been on the wrong antibiotic for weeks.

Another challenge in the treatment of an infection is that the antibiotic may not have easy access to the bacteria in the patient. This is especially true for infections that have formed a **biofilm**. A biofilm can form when the growing bacteria adhere to a surface and develop an organized community. The biofilm is the result of the bacteria sticking to a surface, the bacteria sticking to each other, and extracellular polysaccharides made by the bacteria that help them to stick even better.

Biofilms form on implanted medical devices such as catheters, stitches, replacement joints, and other nonbiological surfaces. As the biofilm persists, clumps of the biofilm, which contain bacteria, can break off to allow the infection to persist and spread to other sites in the body. It is bad enough that the biofilm helps the bacteria to persist, but bacteria associated with a biofilm are also more resistant to antibiotics. Bacteria near the surface of the biofilm are usually metabolically active and may be susceptible to antibiotic activity. However, bacteria deep within the biofilm may be metabolically inactive or dormant such that antibiotics that target growth processes are ineffective (**FIGURE 16.9**). Additionally, the antibiotic may not be able to penetrate through the surface layers of the biofilm to gain access to the bacteria within. Very often the medical device, and thus the site of the biofilm, must be removed before the infection can be successfully treated.

Another challenge in the clinical setting is the access of the antibiotic to the growing bacteria. Some anatomic sites have easier access to the circulation or gastrointestinal tract, where the antibiotic will be. Bacteria growing in tissue that is poorly perfused or walled off by immune cells will not have good access to drugs, which can be an advantage to the bacteria. Bone infections typically need a long treatment course. Infections of the central nervous system will need antibiotic drugs that can cross the blood–brain barrier.

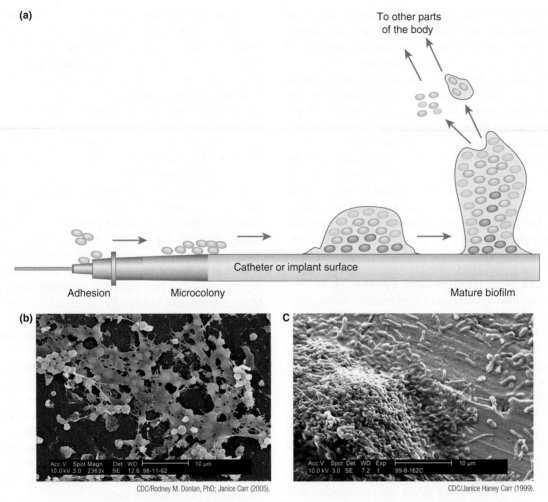

(a)

To other parts
of the body

Catheter or implant surface

Adhesion   Microcolony   Mature biofilm

(b)

Acc.V Spot Magn Det WD
10.0 kV 3.0 2363x SE 12.6 98-11-62   10 µm

CDC/Rodney M. Donlan, PhD; Janice Carr (2005).

c

Acc.V Spot Det WD Exp
10.0 kV 3.0 SE 7.2 1   99-8-162C   10 µm

CDC/Janice Haney Carr (1999).

FIGURE 16.9 Biofilm formation and bacterial community. **(a)** Bacteria, depicted as purple cocci in this figure, adhere to a nonbiologic surface, which is easier for the bacteria to adhere to than normal healthy tissues. These single bacteria may be susceptible to the action of antibiotics. As the bacteria proliferate and form a microcolony (very small, microscopic colony of bacteria), they stick to each other and to the surface. The bacteria secrete a sticky polysaccharide substance that facilitates better adherence and creation of a complex community of bacteria known as a biofilm. Bacterial cells at the surface are actively growing and may be susceptible antibiotics. Bacteria deeper within the biofilm are protected from the action of antibiotics, depicted in this image as the darker purple color. After continued growth within the biofilm, individual bacteria and clumps of the biofilm may be released into the body to further the infection or set up new sites for biofilm formation. **(b)** Biofilm containing *Staphylococcus aureus* from the inside of an indwelling catheter. An important observation from this image is the sticky-looking substance between the round cocci bacteria, which was secreted by the bacteria to create the protective biofilm. Scanning electron microscopic image, magnification 2,363×. **(c)** Biofilms may also be **polymicrobial**, meaning that they may contain more than one bacterial species, as shown in this picture of a biofilm formed in the lab on a stainless steel surface. A mixture of Gram-negative and Gram-positive bacteria are contained within this biofilm community. Scanning electron microscopic image.

Acquisition of Antibiotic Resistance

Scientists recognized that some microbes were intrinsically resistant to some antibiotics, but when bacteria previously susceptible to certain antibiotics developed resistance to their action it became clear that antibiotics might not be the "wonder drugs" that they were initially thought to be. In order to stay ahead of the bacteria, it is helpful to understand how they develop resistance and how they share those resistances with other bacteria.

The experience of the past 50 years has revealed that bacteria can be thought of as "smart" because of their development of mechanisms of resistance to the antibiotics designed to bring about their death. (You really can't blame them!) But they are not really smart; when antibiotics are used, resistance has a survival advantage for the bacterium. This is essentially a manifestation of the Darwinian process of natural selection—"survival of the fittest"— resulting from the widespread use and misuse of antibiotics.

How do cells develop resistance to antibiotics? What are the biological factors involved? When a new antibiotic-resistant strain appears, the development of that resistance is accomplished through genetic changes. The bacterial cell, like all cells, has DNA. Additionally, some bacteria possess **plasmids**, which are extra bits of DNA independent from the DNA of the bacterial chromosome. Antibiotic resistance can result from mutations in the chromosomal DNA or plasmid DNA. Mutations in DNA occur spontaneously and randomly in populations of growing cells at a rate higher than 1 in 10 million; they are not caused by selective pressure but rather are selected for survival by selective pressure. Only the survivors multiply and, in so doing, pass on their new "survival genes" along with all the other genes. The outcome is that the next generation carries these new survival genes. This is the basis for Charles Darwin's concepts of survival of the fittest and of evolution applied at the microbial level.

As an analogy, consider a hypothetical population of trees in a geographical area that has suffered a drought for several years. Gradually, most of the trees die, except for a few survivors that are "lucky" enough to have random preexisting mutations in their genes that allow them to survive with less water. In the years to come the forest will be populated by drought-resistant trees. The drought did not cause the mutations but selected those trees with spontaneous, random, and preexisting mutations that allowed the trees to survive—hence, survival of the fittest. In the same fashion, antibiotics do not cause mutations but select those preexisting mutations that confer antibiotic resistance; the genes with those mutations are passed on to successive generations during cell division.

In addition to spontaneous mutation of preexisting genes, some antibiotic resistance is the result of the acquisition of resistance genes, which may be carried on R (resistance) plasmids or transferred from one bacterium to another as a complete gene (**FIGURE 16.10**). Whereas chromosomal mutations usually confer resistance to only a single antibiotic, **R plasmids** can confer resistance to several antibiotics at one time, a phenomenon that was first reported in Japan in 1959 when lab personnel noted the emergence of *Shigella* bacteria that were resistant to several antibiotics. The origin of these R plasmids is not known.

Transposons, or segments of DNA that can "jump," are another strategy of antibiotic resistance. They may carry genes for antibiotic resistance and can integrate into chromosomes or plasmids of bacteria, allowing for rapid dissemination of antibiotic resistance.

Mechanisms of Antibiotic Resistance

The major mechanisms of bacterial resistance to antimicrobials are: (1) alteration of the target of the drug; (2) inactivation of the drug by an enzyme produced by the bacterium; and (3) keeping the drug away from the target, by either excluding it from entry or quickly pumping it out of the bacterial cell (**FIGURE 16.11**). Each of these mechanisms to develop antibiotic resistance is accomplished through genetic changes such as the accumulation of point mutations in an already existing bacterial gene or by acquiring a new gene or set of genes via some mechanism of genetic exchange, such as plasmid transfer.

After getting into the bacterial cell, antibiotics act by binding to and inactivating the target, usually an important enzyme or structure. If that target changes by either acquiring a new gene for that function or by acquiring

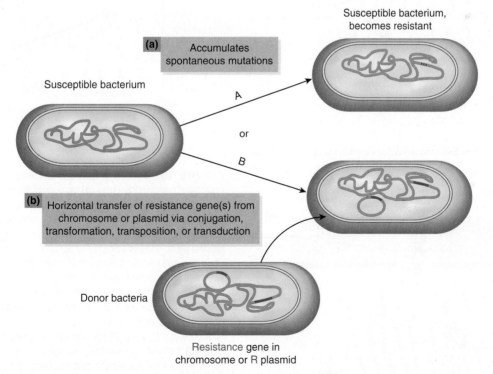

FIGURE 16.10 Acquisition of antibiotic resistance. Susceptible bacteria acquire resistance to antibiotics by spontaneous mutation or horizontal gene exchange. **(a)** Spontaneous point mutations occur that change the susceptibility, typically by changing the antibiotic target. In the absence of antibiotic selective pressures, these spontaneous mutations might not be preserved. **(b)** Horizontal gene transfer can lead to the acquisition of resistance gene(s) from the chromosome or R plasmids of donor bacteria through genetic exchange mechanisms such as transformation, conjugation, transduction, or transposition.

(a)

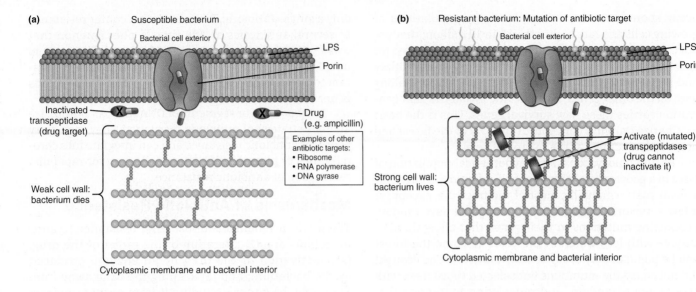

(b) Resistant bacterium: Mutation of antibiotic target

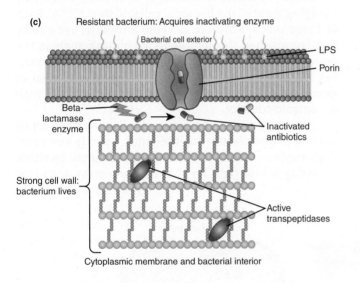

(c) Resistant bacterium: Acquires inactivating enzyme

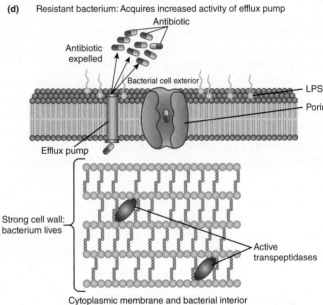

(d) Resistant bacterium: Acquires increased activity of efflux pump

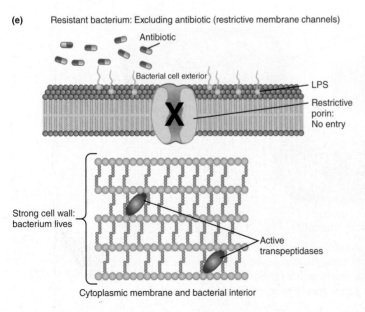

(e) Resistant bacterium: Excluding antibiotic (restrictive membrane channels)

FIGURE 16.11 Bacteria fight back. Some bacteria have developed resistance to antibiotics. Panel **(a)** represents the activity of an antibiotic that inhibits cell wall synthesis: the antibiotic is the red and white pill and acts on the oval transpeptidase target. The antibiotic gains entry into the cell wall area of a Gram-negative bacterium by passing through a channel called a porin. Some antibiotics that act on cell walls are useless against Gram-negative bacteria because their porins don't allow the drugs to pass through them in order to gain entry into the cell. Ampicillin and amoxicillin can gain entry through many different types of Gram-negative porins. In panel **(b)** the bacterium has acquired resistance to the antibiotic via acquired mutations within the transpeptidase, such that the transpeptidase target doesn't "fit" the antibiotic anymore (indicated by shape change). In panel **(c)** the bacterium has acquired resistance via a gene that encodes for an enzyme, beta-lactamase, that inactivates the antibiotic. Some bacteria are resistant because of a very active efflux pump which forces out the antibiotics that entered the bacterial cell. **(d)** or they exclude the antibiotic completely, in this example by changing the porin so that the antibiotic cannot enter, indicated by porin color change **(e)**.

mutations in the gene encoding it, then the bacterium may be resistant to the drug (Figure 16.11B). Penicillin is an antibiotic that interferes with cell wall synthesis by acting on the bacterial enzyme transpeptidase. Resistance to penicillin can occur through mutation of the gene that encodes the transpeptidase, such that the antibiotic cannot act on the new transpeptidase. Modification of the ribosomal target is a mechanism by which bacteria become resistant to macrolide antibiotics such as erythromycin. Resistance to quinolone antibiotics is mainly the result of mutations to the genes encoding the target enzyme, gyrase. These mutations reduce the binding of the drug to the gyrase.

Some bacteria are resistant to an antibiotic because they make an enzyme that can break down and inactivate the antibiotic (Figure 16.11C). This mechanism of resistance is not the result of point mutations, but rather the bacteria acquire a complete gene that encodes the new enzyme. A principal mechanism of resistance to antibiotics that contain a β-lactam ring is via acquisition of a bacterial enzyme known as a β-lactamase. A β-lactamase cleaves the β-lactam ring of the antibiotic, which permanently inactivates the antibiotic. So far, more than 300 β-lactamase enzymes have been identified. The gene that encodes the β-lactamase is often on an R-plasmid or transposon, making the resistance transferable. In the early 1970s, penicillin was universally used to treat gonorrhea. In 1976, highly resistant strains of gonorrhea containing a transposon with the gene for a β-lactamase made penicillin ineffective.

Most antibiotic targets are active inside the cell. Some antibiotic-resistance strategies involve pumping out the drug as quickly as it enters the cell or excluding the antibiotic from entering the cell (Figure 16.11D and E). Resistance to the antiribosomal drugs known as tetracyclines involves a bacterial pump that actively excretes the drug out of the cell. Active efflux pumps are also the basis for some resistance to fluoroquinolones.

Antibiotic resistance was not caused by the misuse of antibiotics; the genes for antibiotic resistance were present long before the development of antibiotics. Pathogenic (and other) bacteria can acquire these preexisting genes from other bacteria through **horizontal gene transfer**, a mechanism by which genes are passed from one mature bacterial cell to another. (**Vertical gene transfer** is the passage of genes from parent to offspring.) The widespread use and misuse of antibiotics has fostered the emergence of antibiotic-resistant strains.

It appears that evolutionary forces are constantly at play in humans' attempts to rein in bacteria; we fight bacteria with new drugs and they fight back by adaptation. Darwin was right. Nobel Laureate Joshua Lederberg stated, "Pitted against microbial genes we have mainly our wits."

Antibiotic Misuse

Antibiotic misuse is a global problem, but the meaning of antibiotic use and "misuse" can take many forms. In developing countries antibiotics are either unavailable or unaffordable to the majority of the population. For those fortunate enough to procure antibiotics, it is common for people to take only a few pills and to save the rest for a later time. In developed countries, where antibiotics are readily available, "misuse" is due to antibiotics being too readily available. Further, in some countries antibiotics can be purchased over the counter in markets and pharmacies without a prescription. Millions of prescriptions for antibiotics are written around the world each year, of which many are unnecessary. Prescriptions are written for colds and flu, despite the fact that antibiotics are not effective against these viral illnesses. The justification is "just in case." Too frequently, patients demand and receive antibiotics from their physicians even though antibiotics are not indicated; the particular disease is running its normal course, or the infection is caused by a virus. Studies have indicated that patients who walk out of their physician's office without a prescription for an antibiotic will often complain that their physician "billed them for nothing."

A patient's lack of knowledge about antibiotics adds to the problem. The patient begins to feel better after a few days and, because he or she is not well informed, does not take the remaining doses. These practices favor the selection of antibiotic resistance, because only the more susceptible strains will be wiped out by only a few doses. The consequence of this misuse of antibiotics, many biologists and health professionals warn, is that we may be forced back to the pre-antibiotic era. Use of antibiotics, especially if they aren't necessary, create the potential for resistant organisms (**FIGURE 16.12**).

Aside from the potential to develop superbugs, antibiotic use has an effect on the microbiota and function of the immune system. Antibiotic treatments intended to target a pathogen have collateral damage on the other microbes within the body (Figure 16.12). Your body actually has more bacterial cells in it than human cells. These bacterial cells, collective known as the **normal microbiota**, are organized into highly complex microbial communities that function to keep us healthy in ways that we don't completely understand yet. However, we do know that one of the major functions of the microbiota is to prevent colonization by other organisms. The microbiota can physically interfere with pathogens should they try to move into the body. Another important function of the human microbiota is through instruction of the immune system. There is growing concern that factors that disrupt the composition of the microbiota, particularly in the early years, may contribute to the development of autoimmunity, allergies, and other inflammatory diseases.

Krasner's Notebook

You frequently hear the expression, "I'm resistant to antibiotics"; it is not you but your bacteria that display antibiotic resistance. You may be allergic, but that is a whole different story.

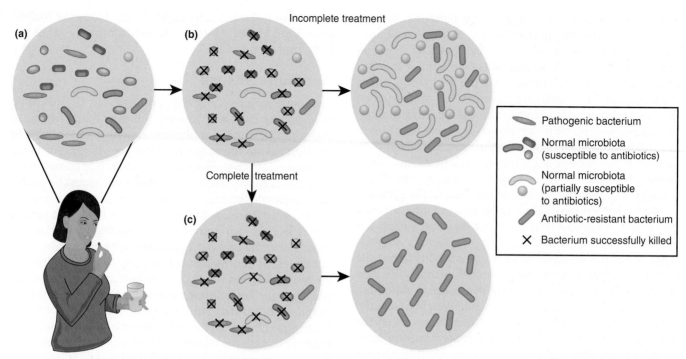

Incomplete treatment

Complete treatment

Pathogenic bacterium

Normal microbiota
(susceptible to antibiotics)

Normal microbiota
(partially susceptible
to antibiotics)

Antibiotic-resistant bacterium

X Bacterium successfully killed

Information from CDC.

FIGURE 16.12 Selection of resistance through antibiotic use and misuse. **(a)** The human body has many bacteria. A few may be antibiotic resistant (shown in red), but in the absence of antibiotics they are in the vast minority as compared to the normal microbiota (shown in blue). **(b)** During an infection, antibiotics kill the bacteria causing the illness (shown in green), as well as other bacteria that are sensitive to the drug (blue). Antibiotics will kill many of the bacteria early in the treatment. Some bacteria may not die immediately (pink); some bacteria may be resistant to the drug (red). If antibiotics are not taken as directed or for the full treatment, the resistant bacteria and those that are moderately sensitive may remain alive and grow. These remaining bacteria can develop additional resistances over time. **(c)** Even completing the full course of treatment may lead to proliferation of resistant organisms (shown in red). Resistant organisms that were in the vast minority before treatment can become the dominant organism after treatment.

Here are a few alarming facts resulting from the misuse of antibiotics:

- *Clostridium difficile* causes life-threatening diarrhea in people who have been in a healthcare setting and have received broad-spectrum antibiotics, which reduce the protective microbiota. In a 2015 study, the Centers for Disease Control and Prevention (CDC) found that *C. difficile* caused more than half a million infections in a single year, resulting in an estimated 15,000 deaths.

- Gonorrhea is increasingly more difficult to treat because of antibiotic resistance. Each year, roughly a third of the 820,000 cases of gonorrhea are resistant to one or more of the antibiotics previously used to treat the infection.

- Tuberculosis (TB), caused by the bacterium *Mycobacterium tuberculosis*, is the leading cause of death from an infectious disease worldwide. Treatment of tuberculosis requires a long course of antibiotics. In some cases, *M. tuberculosis* can be resistant to one or more of the drugs to treat it. Extensively resistant TB (XDR TB) is resistant to most TB drugs.

- *Streptococcus pneumoniae* (also known as the pneumococcus) is the leading cause of bacterial pneumonia and meningitis in the United States. Drug-resistant *S. pneumoniae* is responsible for approximately 1,200,000 infections per year, resulting in roughly 7,000 deaths. The increasing threat of antibiotic resistance makes adherence to pneumococcal vaccination recommendations even more important.

- Staphylococcal bacteria, including methicillin-resistant *Staphylococcus aureus* (MRSA), are a leading cause of healthcare-associated infections. MRSA is resistant to penicillin and other antibiotics. MRSA is responsible for more than 80,000 severe infections and roughly 11,285 deaths per year.

- Resistance to vancomycin, once considered the "last stronghold," has been reported to occur in staphylococci and enterococci. When these bacteria become resistant to vancomycin, there are few treatment options available.

- Carbapenem-resistant *Enterobacteriaceae* (CRE) bacteria are on the rise among patients in medical facilities. CRE have become resistant to all or nearly all of the

BOX 16.1 MRSA, VRE, CRE, and Others: A Very Dangerous Alphabet Soup

Have you heard about methicillin-resistant *Staphylococcus aureus*? Perhaps you are more familiar with its acronym, MRSA. Or maybe you have heard about its cousin, VRE, vancomycin-resistant enterococcus. MRSA, VRE, and other similar acronyms are not the names of new bacterial species, but rather refer to the antibiotic-resistance properties of microbes that we already know about. According to the Centers for Disease Control and Prevention (CDC), antibiotic-resistant bacteria cause at least 2 million illnesses and 23,000 deaths per year. The CDC has identified 18 drug-resistant threats in the United States, labeling each as "urgent," "serious," or "concerning." We might call this the CDC's "most wanted list," or perhaps "most unwanted list" would be more appropriate.

MRSA is in the CDC's "serious" threat category. Staphylococci are part of the normal microbiota of the skin and mucous membranes and, for the most part, are not typically disease producers. About 30% of the population carries *S. aureus* in their nose. MRSA refers to the resistance properties of some strains of *S. aureus*, namely resistance to antistaphylococcal drugs, including methicillin, oxacillin, penicillin, and cephalosporins. In the past, MRSA infections were found almost exclusively in the healthcare setting where the use of antibiotics enabled the selection of strains with acquired resistance. These strains were particularly dangerous to people in the hospital if the MRSA invaded their blood or internal organs. Disturbingly, many new cases of MRSA infection are showing up in people who have no known exposure to these facilities. Fortunately, these community-acquired infections tend to be skin infections, such

as a pimple or cut that turns red, swollen, and purulent (pus filled) with the presence of a yellow or white center, or "head"; draining pus; and fever.

Vancomycin-resistant *Enterococcus* (VRE) is another bacterium placed in the "serious" threat category. Enterococci are commonly found in the normal bacteria of the human intestinal tract. Vancomycin is an antimicrobial often reserved for serious infections such as MRSA. Use of this last-line-of-defense antibiotic has led to the selection of resistant organisms, such as VRE.

Carbapenem-resistant *Enterobacteriaceae* (CRE) is in the "urgent" category. You may not have heard of CRE, because it is not a very common infection, thankfully. However, CRE infections are on the rise among patients in medical facilities. Carbapenem is one of the big-gun antibiotics developed to tackle antibiotic-resistant organisms. CRE have become resistant to all or nearly all the antibiotics we have today.

What can you do to prevent the spread of antibiotic-resistant organisms? Practice good hygiene. Infection control strategies, such as good handwashing, prevent infections—all infections. If we prevent infections, we prevent the spread of resistance. Another important strategy is good antibiotic stewardship: use antibiotics only when necessary, choose the right antibiotic, and use it as directed. Although infections with antibiotic-resistant organisms are still a major public health threat, life-threatening infections with resistant organisms such as MRSA are declining. Further vigilance is important, but progress is encouraging. Keep up the good work!

---

antibiotics we have today. Almost half of hospital patients who get bloodstream infections from CRE die from the infection, resulting in about 600 deaths a year. *Enterobacteriaceae* can spread resistances to other bacteria (**BOX 16.1**).

Working Toward the Solution

The seriousness of antibiotic resistance as an impending global crisis has been established. What is the solution? The answer lies in the hands of everyone who uses

antibiotics: physicians, patients, and those who manage food animals, all of whom share the responsibility for the misuse and overuse of antibiotics resulting in the emergence of **superbugs**, and therefore all are obligated to work toward the solution. Perhaps you unintentionally misuse antibiotics and demand that your physician prescribe an antibiotic when you are ill with what seems to be a cold. You may be guilty of not following instructions to take the full dose, because after a few days you feel better. In so doing you contribute to the emergence of antibiotic-resistant bacteria. Another too common scenario is that you do not feel well and pull some leftover antibiotic out of the medicine cabinet or ask your roommate if he or she has any antibiotics on hand without knowing the identity of the microbe. Physicians share in the responsibility for the antibiotic crisis and in its control; too often, they fail to spend the time explaining to patients why they do not need an antibiotic and succumb to patient pressure. Also, physicians fear being sued by a patient, claiming that his or her illness is a result of not being "put on an antibiotic."

Krasner's Notebook

Antibiotic resistance can move from the farm to our tables. For a long time antibiotics were included in animal feed for growth promotion or increased feed efficiency. This is an inappropriate antibiotic use. The CDC estimates that more than 400,000 people in the United States become ill with infections caused by antibiotic-resistant foodborne bacteria every year.

Supermarket shelves are loaded with a tremendous variety of sprays, mists, and bubble-producing solutions, all designed to kill bacteria, viruses, molds, and mildew. Use them appropriately. You can fight back by frequently washing your hands. Plain soap and water are usually best to prevent infection. Preventing infection is much easier than treating an infection. The judicious use of antibiotics can stave off and minimize an already impending antibiotic-resistance crisis, allowing society to continue to enjoy the benefit of Fleming's serendipitous observation of the antagonism between a mold and a bacterium.

Antiviral Drugs

Compared with the number of drugs available to treat bacterial infections, relatively few **antiviral drugs** are available. Developing drugs that specifically target viral processes is particularly challenging because viruses use human cells for many of their viral processes. In other words, many substances that may inhibit viral replication may also be toxic to the human cells because the viral processes are using the human host machinery. An effective antiviral drug will interfere with successful viral replication while sparing the host cells (**FIGURE 16.13**).

Krasner's Notebook

Now that you have learned that viruses are subcellular and do not have the target sites for antibacterial activity, you know that you should not expect or ask for an antibiotic when you have a viral infection. Frequently, students complain to me about the health services on campus and complain that "the doctor didn't even give me an antibiotic!" You now know better.

Antiviral drugs in current use accomplish this either by targeting specific viral processes or by enhancing the antiviral processes of the human immune response. Developing antiviral drugs requires that we understand the intricate details of viral replication within human cells to find those targets that are unique to the virus. Note that antibiotics are not effective against viruses because viruses lack the target components against which the antibiotics are directed.

Types of Antiviral Drugs

In order to replicate, a **virus** needs to find a susceptible **host cell**, get inside the cell, use the viral nucleic acid to make new copies of the virus genome and make viral proteins, and assemble new viruses. The virus then leaves the cell to find a new host cell. Antivirals can target each of these steps, if the viral processes are different enough from the host cell processes. For example, some viruses, such as human immunodeficiency virus (HIV) and herpesviruses, use a viral enzyme for replication of the viral genome. That viral polymerase enzyme is often a good target for antiviral drugs. Many antiviral drugs are structurally similar to the nucleosides, which are the building blocks for nucleic acids. The viral polymerase uses the drug instead of the real nucleosides, interrupting the replication of the viral nucleic acid. A number of antiviral agents are available, and research is ongoing to develop new ones (**TABLE 16.3**).

Several antiviral drugs work by enhancing the immune response to the virus. Imiquimod is an immune response modifier that does not have any direct antiviral activity, yet is a topical agent used for the treatment of warts caused by human papillomavirus. Imiquimod stimulates immune cells to release chemical signals known as **interferons**.

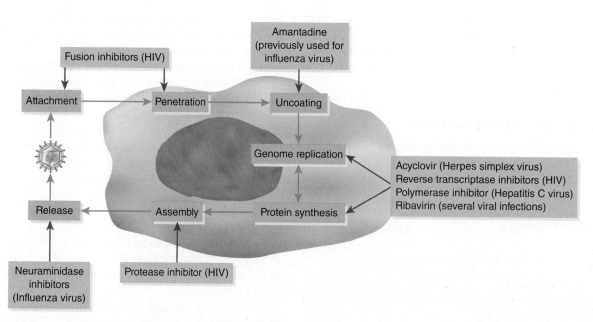

FIGURE 16.13 Viral life cycle and antiviral targets. Blue arrows indicate the steps that are involved in a typical virus infection of a host cell. The arrows and boxes red and pink indicate target and antiviral drugs, respectively.

TABLE 16.3 Sampling of Antiviral Agents and Their Activity by Generic Name (Brand Name)

Activity	Mechanism of Activity
Viral entry (to synthesis)	
Enfuvirtide (Fuzeon), maraviroc (Selzentry), ibalizumab (Trogarzo)	Interferes with binding of HIV to host cell receptors
Viral synthesis and replication	
Acyclovir (Zovirax), penciclovir (Denavir), valacyclovir (Valtrex)	Interferes with herpesvirus DNA replication
Cidofovir (Vistide)	Inhibits viral DNA synthesis of cytomegaloviruses
Interferons (multiple forms, including: Intron A, Betaseron, Extavia, and Alferon N)	Interferes with viral replication by enhancing the immune system
Ledipasvir/sofosbuvir (Harvoni)	Suppresses viral replication by interfering with hepatitis C viral polymerase
Nevirapine (Viramine)	Interferes with binding site of HIV reverse transcriptase
Ribavirin (Rebetol)	Inhibits polymerase activity of hepatitis C and respiratory syncytial viruses
Zidovudine (Retrovir, formerly AZT)	Interferes with DNA replication in HIV by targeting reverse transcriptase
Viral assembly, release	
Oseltamivir (Tamiflu), zanamivir (Relenza), peramivir (Rapivab)	Interferes with influenza virus neuraminidase enzyme necessary for release of virus from host cell
Saquinavir (Invirase), fosamprenavir (Lexiva)	Interferes with action of HIV protease, resulting in noninfectious viruses
Simeprevir (Olysio)	Interferes with action of hepatitis C virus protease

Interferons essentially signal cells to be less permissive for viral replication and stimulate immune cell activity. Some antiviral drugs are synthetic versions of interferons. During a natural viral infection, interferons are responsible for some of the achy symptoms associated with some viral infections; synthesized interferon drugs tend to have similar side effects.

Over the last several decades, many advances in antiviral drug discovery have come from research directed toward chronic viral infections such as HIV, hepatitis B (HBV), and hepatitis C (HCV). Whereas HIV treatment does not result in a cure, combination antiretroviral therapy targeting several steps in the HIV life cycle have made long-term management of HIV infection possible. An estimated 3.5 million people are living with chronic hepatitis C virus infections. Hepatitis C is a major cause of liver cirrhosis and liver failure.

In 2014, the FDA approved Harvoni (ledipasvir/sofosbuvir) for the treatment of some patients with chronic hepatitis C infections. Drug development is typically a very costly venture for a pharmaceutical company (**BOX 16.2**). As such, when a new drug is released to the market, the cost can be particularly high for a new drug. Harvoni reports a greater than 90% cure rate; however, the cost of a 12-week course is very high. Balancing the cost of drug development for the company with the lifetime cost of treating a patient with chronic infection is not an easy task.

Also on the horizon are new treatments for hepatitis B infections. A multifaceted approach to global hepatitis B eradication includes already approved drugs that inhibit the viral enzymes and enhance the immune response. In development are drugs that utilize molecular techniques to target the persistent hepatitis B viral DNA. Together with a robust hepatitis B vaccination program, a global path to eradiation may be in sight.

BOX 16.2 Drug Development

When you go to the pharmacy with a new prescription, have you ever wondered what went into making sure it is safe? Just how did that drug get approved for use? In the United States, a drug must show safety and efficacy for its intended use. The Food and Drug Administration (FDA) is the agency that determines if a drug has met these criteria.

It is a long and difficult process to bring a potential drug candidate to the point of where a physician can prescribe the drug for a patient (Figure 1). Many potential new drugs are eliminated at each step along the way.

Initial discovery of a potential new drug involves the collective knowledge of the scientists and the research team. They need to know a great deal about drugs, biochemistry, disease states, and many other aspects of medicine to conceptualize and screen for the new drug compound. Thousands of compounds may be considered. **Preclinical testing** of a drug candidate includes laboratory and animal testing to provide very basic information about the drug's safety and efficacy. If the drug compound shows potential, the research team will submit an investigational new drug application to the FDA for approval to conduct clinical trial drug testing in people. Human subject testing can proceed if the FDA decides that the preclinical data show potential and the clinical trials do not place human subjects at unreasonable risk of harm.

Clinical trial drug testing occurs in four phases:

- Phase 1 involves a small number of healthy volunteers to study the safety of the drug in humans.

- Phase 2 involves patients with the disease or condition, but the number of patient volunteers is small. This phase will emphasize effectiveness in addition to safety. At the end of phase 2, the FDA will examine the data before allowing large-scale phase 3 studies.

- Phase 3 enrolls thousands of patients. These studies will collect more information about safety and effectiveness in different patient populations, including those with pre-existing conditions, and at different dosages. At the end of phase 3, the FDA will examine all of the data to decide if the drug can be marketed for use.

- Phase 4 trials are carried out after the drug is approved by the FDA. Even though clinical trials provide a lot of data about safety and efficacy, it isn't always possible to know how the drug will behave in all people and all situations. The FDA requires safety updates as part of postmarket safety monitoring. This safety monitoring can detect unexpected serious adverse events, which may lead to additional drug label warnings or, in rare cases, withdrawal from use.

A drug developer invests a great deal of time and money into the development of a new drug. In 2017, just 12 new drugs for the treatment of infections and infectious diseases received final approval, yet thousands of potential new drug candidates would have started the process years before. When a new drug comes to market, it is patent protected, which means that the developer has exclusive marketing rights and only its

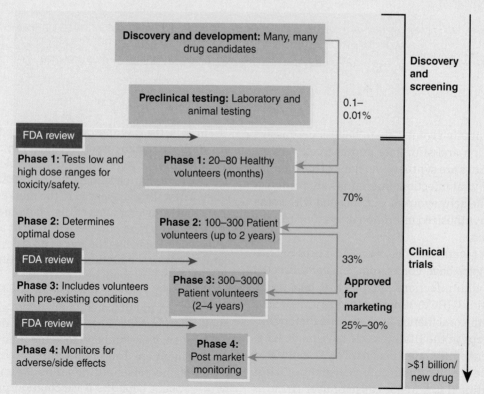

Figure 1 The path of drug discovery and FDA approval.

brand-name drug is available. When the patent expires, other drug manufacturers can make the drug, which will be known as a generic version of the drug. These companies will not need to do safety and efficacy studies. They will only need to show that their generic version of the drug is comparable to the original drug. This is much easier and far less expensive, which is part of the reason generic drugs are much less expensive than their brand-name counterparts.

Antiviral Resistance

In order to become resistant to a specific antiviral drug, the virus must change the viral component that the drug targets. This will involve mutation of the gene encoding the target of the drug. Viruses do not acquire genes to inactivate drugs like bacteria often do. Some viruses are more error prone than others when replicating their nucleic acids, which may lead to drug resistance over time. The high error rate of HIV during replication of its genome creates mutations in HIV genes and promotes the development of drug-resistant strains. Using a combination of several antiviral drugs together, even in patients newly diagnosed with HIV, is one strategy to address resistance, because it is more difficult for HIV to develop resistance to multiple drugs at the same time. Herpesvirus resistance to acyclovir is typically due to mutations in one of two herpesvirus genes. Fortunately, these mutations appear infrequently.

Summary

The 20th century witnessed an increase in life expectancy in many nations of the world. Today, U.S. residents live 35 years longer, on average, than they did in 1900. At least 25 of those gained years are attributable to public health achievements in disinfection and antibiotics—the topics of this chapter.

The connection between germs and disease was validated in the late 19th century and has led to important innovations such as antibiotics. Despite major advances in the treatment of infectious diseases, these infections continue to present substantial challenges to health and healthcare resources worldwide. Antimicrobial resistance has become a global health crisis that requires cooperation from each individual and collaboration with the healthcare community, too.

Antimicrobial resistance is closely associated with the use of antimicrobial drugs, which allows for the selection of resistant organisms. Appropriate use of antibiotics is important for preventing the further spread of antibiotic-resistant microbes. Disease prevention through handwashing, appropriate use of disinfectants, and knowing when to be careful about the potential for infection will go a long way toward keeping us healthy, thus avoiding antibiotic drugs. Failure to use good infection control measures can result in the spread of infections, including antibiotic-resistant organisms. Finding a healthy balance between good hygiene practices and germ phobia is key.

KEY TERMS

antibiotic
antimetabolite
antiseptic
antiviral drugs
autoclave
bactericidal
bacteriostatic
ß-lactam antibiotics
biofilm
broad-spectrum antibiotic
disinfection
DNA gyrase
filtration
fumigation
halogen
horizontal gene transfer
host cell
interferons
microbiota
microcolony

mode of transmission
narrow-spectrum antibiotic
normal microbiota
oligodynamic effect
pasteurization
peptidoglycan
plasmid
polymicrobial
preclinical testing
prion
R plasmid
ribosome
RNA polymerase
selective toxicity
sterilization
superbugs
transpeptidase
transposon
vertical gene transfer
virus

○ **PART I: Choose the single best answer.**

1. Which of the following should be used to sterilize a solution of saline for use in the hospital?
   - a. Autoclave
   - b. Boiling
   - c. Pasteurization
   - d. Microwaves

2. Which of the following is generally the most resistant to disinfection?
   - a. Giardia
   - b. HIV
   - c. Mycobacteria
   - d. Yeast

3. Which of the following is *not* an antibiotic?
   - a. Chloramphenicol
   - b. Erythromycin
   - c. Interferon
   - d. Penicillin

4. Penicillin acts on the synthesis of which of the following bacterial components?
   - a. Cytoplasmic membrane
   - b. DNA polymerase
   - c. Peptidoglycan
   - d. Ribosome

5. Which of the following is a bacterial enzyme that can inactivate penicillin?
   - a. ß-lactamase
   - b. Gyrase
   - c. Polymerase
   - d. Transpeptidase

6. Which of the following can be used on skin?
   - a. Antiseptic wipe
   - b. Chlorine bleach
   - c. Ethylene gas
   - d. Formaldehyde

7. Which of the following interferes with influenza virus release from host cells?
   - a. Acyclovir
   - b. Fuzeon
   - c. Oseltamivir
   - d. Ribavirn

8. Which of the following agencies has oversight for approval of new drugs in the United States?
   - a. CDC
   - b. FDA
   - c. USDA
   - d. NCAA

9. Which of the following might provide an advantage for bacteria growing on an intravenous catheter?
   - a. Being Gram-negative
   - b. Biofilm formation
   - c. Fever
   - d. High blood pressure

10. Which of the following mechanisms is most likely to be involved in the spread of antibiotic resistance from one bacterium to another?
    - a. Apoptosis
    - b. Conjugation
    - c. Reverse transcription
    - d. Translation

○ PART II: Fill in the blank.

1. The antibiotic tetracycline interferes with the synthesis of _____.

2. Acyclovir is used in the treatment of _____.

3. Penicillin is not toxic for humans because it interrupts bacterial _____.

4. _____ can be used on skin before drawing blood.

5. An antibiotic that halts growth, but doesn't outright kill a bacterium is referred to as being
   _____.

6. _____ are the hardest infectious agents to destroy or inactivate.

7. _____ are circular extrachromosomal DNA elements that carry antibiotic resistance
   genes.

8. Streptomycin acts on bacterial _____.

9. Heating milk to reduce the number of pathogenic microbes is known as _____.

10. _____ is/are an example of a drug that has bactericidal activity.

○ PART III: Answer the following.

1. Give an example of when you would use each of the following for microbial control: isopropyl alcohol, chlorine
   bleach, autoclave.

2. Explain the concept of selective toxicity and give three examples of drugs that exhibit selective toxicity.

3. From a Darwinian point of view, describe the emergence of antibiotic-resistant strains.

4. Discuss why antibacterial agents are no longer routinely added to hand soap.

5. Identify six cleaning products and list their active ingredients.

6. List three innovations in medicine since the 1900s that may have played a role in the increase in life expectancy.

7. List three or four microbes or infectious agents that are difficult to inactivate with disinfectants. What characteristics make them resistant to disinfection?

8. Describe a clinical situation when it might be best to use a bacteriostatic antibiotic.

9. List the features of *Mycobacteria tuberculosis* that make it difficult to treat promptly.

10. List three different mechanisms by which bacteria become resistant to antibiotics.

PART 5

CURRENT MICROBIAL CHALLENGES

HARNESSING THE POWER OF MICROBES: PERIL AND PROMISE

Nancy Boury, PhD
Iowa State University

OUTLINE

"The impression sometimes created among the public is that scientists are working away in their labs, and maybe they're not always thinking about the implications of their work. But we are."

—Jennifer Doudna

American biochemist at UC Berkley and codiscoverer of CRISPR/Cas9, New Yorker article, November 16, 2015

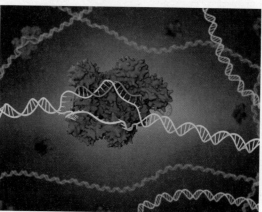

© Meletios Verras / ShutterStock, Inc.

The CRISPR-Cas genome editing system allows for precise DNA manipulation that has revolutionized genetic engineering. The blue Cas9 enzyme is guided to a target sequence by an RNA guide (purple sequence).

© ShutterStock, Inc. / happykanppy.

LEARNING OBJECTIVES

1. Explain why diseases are more common during war.

2. Define *biological warfare* and *biological terrorism*, giving examples of each.

3. Distinguish between category A, Category B, and Category C biological agents.

4. Explain the advantages and disadvantages of using biological weapons.

5. Describe what makes both anthrax and smallpox potential biological weapons and the appropriate countermeasures to these agents.

6. Summarize the advantages and disadvantages of gain-of-function research on deadly diseases and form an informed opinion on the issue.

7. Define genetically modified organisms, give examples, and describe the controversies surrounding their use.

8. Diagram the process of CRISPR/Cas genome editing.

9. Compare and contrast the 1975 Asilomar Conference and the 2015 Napa Conference in terms of each meeting's purpose and conclusions.

10. Explain how and why genetic engineering is regulated in the United States.

Case Study: Fighting the World's Deadliest Animal

When you think of the world's deadliest animal, you likely envision a Bengal tiger, a great white shark, or more cynically, humans. You'd be wrong. While humans commit approximately 400,000 homicides each year, the deadliest animal kills almost twice than many people. Mosquitoes and, more important, the diseases they carry are responsible for three-quarters of a million human deaths around the globe each year. This number reflects the **mortality**, or death rate, not **morbidity**, or rate of illness. For example, Zika virus, which is carried by mosquitoes, alters fetal development, causing thousands of babies to be born with small heads and brains, a condition known as microcephaly. Zika virus aside, the worldwide **prevalence**, or number, of illnesses caused by mosquito-borne diseases is staggering. Each year 3.2 million people become infected with dengue fever, and more than 200 million cases of malaria are identified. Add to that an additional 150,000 severe cases and approximately 50,000 deaths caused by yellow fever, and it's easy to see how the mosquito is truly humans' deadliest foe.

These diseases pose a serious threat to human health, and insect-control methods, such as bed nets and insecticides, are expensive, often not fully adopted, and are only partially successful. One company, Oxitec, was started at Oxford University in in the early 2000s with the ambitious goal of controlling the *Aedes aegypti* mosquito population without the use of pesticides like DDT.

Working with the Bill & Melinda Gates Foundation, Oxitec genetically engineered a strain of mosquitoes that contain a gene, **tTAV (tetracycline-controlled transactivator)**, which upregulates production of the tTAV protein by a positive-feedback loop. The more tTAV protein produced, the more the cell increases production of tTAV. The overproduction of tTAV overburdens the mosquito's genetic machinery, so the mosquitoes expressing this gene cannot make proteins needed for life and die.

In the Oxitec laboratories, these genetically altered mosquitoes, called Friendly mosquitoes, are fed the antibiotic tetracycline during larval development and as they mature (**Figure 1**). Tetracycline binds to the tTAV protein and inactivates it, sparing the life of the mosquito as long as it has a diet that includes tetracycline. For these mosquitoes tetracycline is an antidote to a deadly condition they are born with.

How can these Friendly mosquitoes help combat killers like dengue, Zika, and malaria in the wild? The Friendly mosquitoes that are released are male, which means they do not bite humans and do not carry disease. They do, however, mate with the potentially disease-carrying females. Because the Friendly males carry two copies of this deadly tTAV gene, they pass the trait on to all of their offspring. Tetracycline is not normally found in nature, so the males that are released have a very short life expectancy but do survive long enough to breed and produce offspring that then die very early

(continues)

Case Study: Fighting the World's Deadliest Animal (continued)

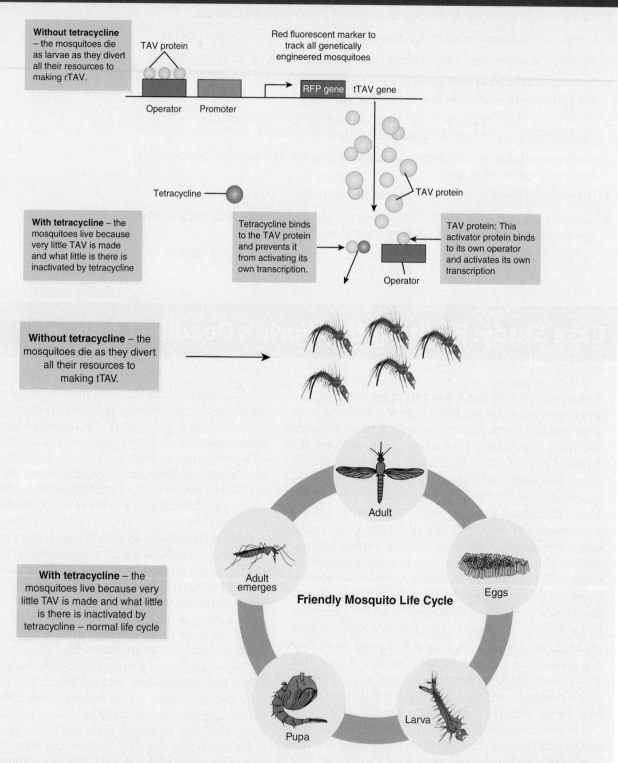

Without tetracycline – the mosquitoes die as larvae as they divert all their resources to making rTAV.

TAV protein

Red fluorescent marker to track all genetically engineered mosquitoes

Operator   Promoter   RFP gene   tTAV gene

Tetracycline

TAV protein

With tetracycline – the mosquitoes live because very little TAV is made and what little is there is inactivated by tetracycline

Tetracycline binds to the TAV protein and prevents it from activating its own transcription.

TAV protein: This activator protein binds to its own operator and activates its own transcription

Operator

Without tetracycline – the mosquitoes die as they divert all their resources to making tTAV.

With tetracycline – the mosquitoes live because very little TAV is made and what little is there is inactivated by tetracycline – normal life cycle

Adult

Adult emerges

Friendly Mosquito Life Cycle

Eggs

Pupa

Larva

Figure 1 Genetic modifications found in Friendly mosquitoes. A gene cassette with two features is inserted into these mosquitoes. The first feature (represented by the red box) is that these mosquitoes will have a red glow that can be detected with ultraviolet light. The second feature is the *tTAV* gene that promotes production of tTAV protein through a positive-feedback mechanism. This *tTAV* gene encodes a protein that, when expressed, instructs the cell to make even more tTAV protein. This causes the cells to die as they waste energy and resources on the tTAV protein. Tetracycline binds to and inactivates the tTAV protein, which prevents the massive and deadly overexpression seen when tetracycline is absent. The Friendly mosquitoes are bred and maintained on a tetracycline-laced diet. Once released, the male Friendly mosquitoes will mate and then die. Their offspring will have the deadly *tTAV* gene and die as well. (Note: The **operator** and **promoter** regions are DNA sequences involved in the expression or mRNA transcription of the gene cassette shown in this illustration.)

in development. The genetic engineers at Oxitec put a **red fluorescent protein (RFP)** next to the *tTAV* gene so they can track the presence of this gene construct in the dead larvae. This provides the researchers a way to track the range and impact of this construct on local mosquito populations.

In the fall of 2009, Oxitec released thousands of male mosquitoes in Grand Cayman and reported an 80% drop in *A. aegypti* numbers in 2010. This success paved the way for the release of the Friendly mosquito in Brazil and Malaysia and application for approval in Sweden and the United States. In a later study in Brazil, introduction of Friendly males resulted in an 81% drop in the mosquito population and an impressive 91% drop in dengue cases during the test period. Encouraged by these results, Oxitec has built a factory in Piracicaba, Brazil, that can produce tens of millions of Friendly mosquitoes every year, and as of 2018 plans to expand by setting up smaller laboratories across the country.

Although the results in the Cayman Islands and Brazil are promising, the Oxitec project has hit multiple roadblocks in its proposal to release Friendly mosquitoes in the Florida Keys. The first hurdle has been identifying the appropriate government agency responsible for regulating the release of these mosquitoes. Because these mosquitoes are a genetically modified animal, Oxitec applied to the U.S. **Food and Drug Administration (FDA)** for permission to release these Friendly mosquitoes in a small efficacy trial. In 2016 the FDA determined that the release of the genetically engineered mosquitoes would not have a significant impact on the environment. Part of the FDA approval process is the gathering of public comments about the proposed experimental trial. The thousands of comments submitted to the FDA were polarized. Individuals against the release were either opposed to genetically modified organisms, wanted more research to be done on potential long-term consequences, were concerned about the natural balance of animals in the area, and often simply opposed the idea of adding more mosquitoes to the area. Those in support of the efficacy trial mentioned the severity of mosquito-borne illness and the positive preliminary results seen in Brazil, the fact that *A. aegypti* is not a native species in Florida, the cost and environmental impact of pesticides that are currently being used, and finally, voiced concerns about the spread of Zika virus to Florida and its effect on tourism. The issue has been divisive, with strong emotions on both sides. As of 2018, the FDA has passed regulatory control of the Friendly mosquito project to the **Environmental Protection Agency (EPA)**, as these genetically engineered mosquitoes are designed to control the native mosquito population, which effectively makes them pesticides and part of the EPA's regulatory responsibility. It remains to be seen if the United States will join Brazil, the Cayman Islands, and Malaysia in using these Friendly mosquitoes to control the unfriendly and dangerous A. *aegypti* mosquito.

Questions and Activities

1. How does morbidity differ from mortality and prevalence? If you were trying to determine if a disease has spread into new areas, which statistic would be most informative?

2. What is the native range of A. *aegypti* mosquitoes? Where are they invasive species?

3. Why do Oxitec researchers only release male mosquitoes?

4. Because these mosquitoes are homozygous for a lethal mutation, what percentage of their offspring will carry this deadly mutation?

5. What is the function of tetracycline, the *tTAV* gene, and the red fluorescent protein genes? If a new technician forgot to add tetracycline to the growth media for the Friendly mosquitoes, what would happen to the study using these mosquitoes?

6. If you were part of the Oxitec communications team, how would you propose the release of the Friendly mosquito in the Florida Keys? Write a paragraph to convince the residents to approve of the trial.

7. If you were a resident of the Florida Keys, would you vote to approve or reject the proposal to release Friendly mosquitoes in an effort to control A. *aegypti*? Write a paragraph to state and support your opinion.

Information from C. De Koning, A. Sica De Campos, S. Hartley, J. Lezaun, & L. Velho. (2017). Responsible innovation and political accountability: genetically modified mosquitoes in Brazil. *Journal of Responsible Innovation, 4*(1), 5–23; A. M. Nading. (2015). The lively ethics of global health GMOs: The case of the Oxitec mosquito. *Biosocieties, 10*(1), 24–47; E. Waltz. (2016). GM mosquitoes fire first salvo against Zika virus. *Nature Biotechnology, 34*, 221–222.

Preview

Microbes are both useful and harmful. Like a knife that can be either a weapon or a chef's tool, bacteria and viruses can be chosen, designed, and honed for a particular purpose. At times, that purpose is a positive one, such as designing bacteria to produce human insulin that is used to treat diabetes. Other times, however, bacteria and viruses can be used as weapons, to cause damage. This chapter describes the perilous use of biological agents as weapons and infectious diseases in war, the promising use of microbial processes in genetic engineering, and the ethical dilemmas scientists and policy makers must consider as new discoveries bring both new opportunities and new challenges.

War and disease have a long history. Soldiers at war are challenged by poor hygiene, crowded living conditions, and often limited nutrition on the battlefield. These battles and the threat of invasion often cause civilians to flee and congregate in overcrowded refugee camps. Because of this, it is no surprise that diseases such as cholera and typhus have been part of several pivotal battles in history and have added to the misery of war-torn countries such as Yemen in 2017 and 2018.

While managing their own inevitable outbreaks, armies learned that the outbreak of disease is deadly to the opposition. Biological weapons used for warfare or for acts of bioterrorism are described as the "poor man's weapons of mass destruction" because of the relative ease and low cost of their deployment. In all wars, armies have had to defend against microbes within their own troops as a natural consequence of battlefield conditions, as well as against their opposition. In stark contrast, however, the deliberate use of microbes to kill is a heinous crime and challenges the rules of warfare. Because of this, international treaties strictly prohibit the use of biological weapons in war. The U.S. Department of Health and Human Services (DHHS), of which the **Centers for Disease Control and Prevention (CDC)** is an agency, is charged by presidential directive with the responsibility of preparing a national response to emergencies resulting from the use of **biological weapons**.

In the modern era, genetic engineering has provided opportunities for great advances, but it also serves as a toolbox to increase the potency of biologic weapons. The advent of new genetic engineering technologies has enabled researchers to create crops that resist insect predation and herbicides. These technologies increase crop yields. The same technologies that can put beneficial genes in crops can be used to add toxin genes to pathogens or otherwise increase the damage that biological weapons can cause. The U.S. Food and Drug Administration, the U.S. Department of Agriculture, and the Environmental Protection Agency each regulate different aspects of biotechnology.

Biological Agents as Weapons

Biological weapons typically employ microbes deployed to grow in (or on) their target host or microbial products that are deliberately used to incapacitate or kill individuals. They are considered weapons of mass destruction because they can affect large numbers of people, soldiers and civilians alike. **Biological warfare** is the use of biological weapons by nations in the conduct of war. **Biological terrorism**, on the other hand, is the use of biological weapons by groups, such as terrorist organizations, or individuals.

No countries and no communities are safe from this type of weaponry. In 1998, then Secretary of State Madeleine Albright remarked that "*biological weapons know no boundaries.*" In 1999, President Bill Clinton stated that the possibility of germ attack kept him awake at night and, further, "*a chemical attack would be horrible, but it would be finite, but a biological attack could spread . . . kind of like the gift that keeps giving.*" Civilians and soldiers, enemies and allies, all could easily fall victim to a weaponized pathogen.

War and Disease

Historically, microbes have always played a role in battles and wars; in many cases microbes were perhaps even more influential than the conventional weaponry forces brought to bear by the opposition. Battlefield conditions are conducive to the spread of infectious diseases because of unsanitary conditions, particularly resulting from the problem of disposal of large amounts of human fecal material, the unavailability of clean drinking water, poor personal hygiene, crowding, inadequate supplies of safe foods, and overwhelmed and inadequate medical facilities. These poor living conditions contributed to massive typhus outbreaks in Napoleon's army as he invaded Russia. History books often describe how Napoleon was defeated in the Battle of Waterloo, which is true. These same books regularly fail to mention that almost a third of the soldiers killed and buried nearby had "spotted fever," or typhus. In the American Civil War, diseases such as malaria, typhoid fever, and dysentery claimed the lives of more U.S. soldiers than were lost in World War I (**FIGURE 17.1**).

The circumstances of war force people to move into overcrowded and unsanitary conditions (e.g., refugees, prisoners of war, and inhabitants of concentration camps). These conditions often lead to outbreaks of **infectious diseases**, particularly those diseases that are transmitted through contaminated food and water. Tragically, this played out in the 2017–2018 cholera outbreak in war-torn Yemen. The World Health Organization (WHO) estimates that more than a million people were infected between April 2017 and March 2018, with several thousand deaths. International relief agencies, including the Doctors Without Borders, the World Health Organization (WHO), the International Red Cross, and the United Nations, play a vital role in attempting to minimize human misery by providing people with nutrition and sanitary latrines.

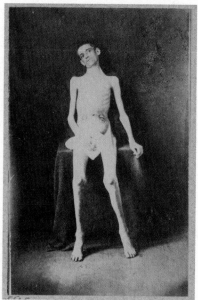

Courtesy of the Library of Congress.

FIGURE 17.1 Emaciated soldier as he was released from a POW camp during the American Civil War.

History of Biological Weaponry

Documentation of the use of biological weapons is difficult because they have been prohibited by international law. This makes confirmation of allegations of use difficult. Reliable epidemiological data are often lacking, and propaganda surrounds any accusation of biological weapons use. This makes differentiating naturally caused diseases from those caused by biological weapons difficult. Nevertheless, there is ample evidence that biological weapons have been used since ancient times. This historical overview of the use of biological weapons examines three time periods: early history to World War II, World War II to the 1972 Biological Weapons Convention, and 1972 to September 11, 2001.

Early History to World War II

Ancient warriors recognized the devastating impact of infection and poor sanitation on armies. One particularly devastating tactic was to hurl the bodies of plague victims at the enemy. During the siege of Kaffa (now Feodossia, Ukraine) in 1346, Tatar military leaders recognized that plague-ridden corpses could be used as weapons and catapulted their own dead soldiers into the midst of the enemy. According to one account:

> The Tatars, fatigued by such a plague and pestiferous disease, stupefied and amazed, observing themselves dying without hope of health, ordered cadavers placed on their hurling machines and thrown into the city of Kaffa, so that by means of these intolerable passengers, the defenders died widely.

During a plague outbreak, bodies of dead soldiers, along with large quantities of excrement, were hurled into the ranks of the enemy at Karlstein, Bohemia, in 1422. In 1710, Russian troops battling Swedish forces in Reval (now Tallinn, Estonia) threw their plague victims over the walls into the midst of the opposition.

Oftentimes the biological attacks were subtler than flinging bodies and human excrement at target populations. In the 16th century, Spanish conquistadors may have given smallpox-contaminated clothing to South American natives. Although there is some debate as to whether the South American natives were intentionally infected, there are documented reports of British colonial soldiers handing out blankets from a smallpox hospital to the local Delaware tribe in North America.

The end of the 19th century and the beginning of the 20th century marked the emergence of modern microbiology, a period known as the "Golden Era of Microbiology." The pioneering work of Louis Pasteur, Robert Koch, and other microbe hunters established that infectious diseases are caused by microbes and not by miasmas or "bad air." Ironically, because of the development of the **germ theory of disease**, specific pathogens could be identified and then used for biological warfare. During World War I, Germany was accused of conducting biological warfare by shipping infected horses and cattle to the United States and to other countries. It was also alleged, but never proven, that in

1915, German forces attempted to spread cholera into Italy and Russia and air-dropped bacterially contaminated fruit, chocolate, and children's toys into Romania. Germany was exonerated by the League of Nations because of the lack of hard evidence.

After the war, the Protocol for the Prohibition of the Use in War of Asphyxiating, Poisonous, or other Gases and of Bacteriological Methods of Warfare, often referred to as the **1925 Geneva Protocol**, was drafted. It has been signed by 140 nations, including Iraq (in 1931), as of June 2018. The term "bacteriological" has subsequently been reinterpreted to include viruses and fungi. The protocol does not contain provisions prohibiting the use of microbes for basic research or mandating inspection for purposes of monitoring compliance. These limitations were at the heart of Iraq's defiance of UN inspections for the alleged production of bioweapons, prompting escalation of the Persian Gulf War in 2002.

World War II to 1972

The occasions on which biological weapons were or may have been used during the next period of approximately 50 years (to the 1972 Biological Weapons Convention) are numerous, and the evidence is confounded with allegations, secrecy, intrigue, charges, and countercharges. Japan used biological warfare agents in Manchuria, China, from 1932 until the end of World War II. The infamous Japanese Army Units 731 and 100 were biological warfare units that conducted torturous experiments on prisoners in China. The prisoners were injected with pathogens, including *Bacillus anthracis*, *Neisseria meningitidis*, *Vibrio cholerae*, *Shigella* spp., and *Yersinia pestis*. At least 3,000 prisoners died. Reports suggest that in at least 11 Chinese cities the Japanese contaminated water and food supplies with a variety of pathogens, threw live cultures into homes, or sprayed infective aerosols from aircrafts. Japan's attack on Changteh in 1941 using *V. cholerae* as a weapon backfired, resulting in 1,700 deaths among Japanese troops. This illustrates the danger of using biological weapons. Although some of these incidents have not been conclusively substantiated, the Japanese government later stated that its conduct was "*most regrettable from the viewpoint of humanity.*"

In the events leading up to World War II and during the war, there is no documentation that Adolf Hitler's Nazi Germany conducted biological warfare. Winston Churchill, prime minister of the United Kingdom from 1940 to 1945, is said to have considered anthrax as a retaliatory measure if Germany used biological agents against Britain. During the years after World War II, allegations regarding the use of biological agents were carelessly thrown about. Often the accusations of biological weapon use were purposely used as blatant propaganda and scare techniques during the Cold War.

In response to the international threat of the use of biological weapons, the United States embarked on an offensive biological weapons program in 1942. The Soviet Union also began such a program during World War II. The U.S. initiative was headquartered at Camp Detrick (later renamed Fort Detrick) in western Maryland. At its peak

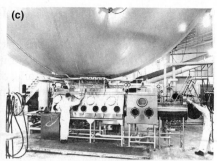

All photos courtesy of Fort Detrick Public Affairs Office/U.S. Army.

FIGURE 17.2 Fort Detrick, the location of the U.S. Army's main facility for research on biological warfare defense. **(a)** An old photo of the main gate, probably taken in the 1950s. **(b)** Technicians working with highly virulent microbes in biological containment hoods. **(c)** Huge vat used to culture microbes on a large scale.

in 1945, almost 2,000 scientific and military personnel worked at this maximum-security campus, which had the dubious distinction of being the world's most advanced biological warfare unit (**FIGURE 17.2**).

Weapons systems disseminating a variety of pathogens were tested at Fort Detrick, major cities in the United States, and remote desert and Pacific sites. New York and San Francisco were unknowingly used as models to test aerosolization and dispersal methods using "harmless bacteria." In 1966, *Bacillus atrophaeus,* a nonpathogenic **endospore** formerly related to *B. anthracis,* was released into the subway system of New York City to determine the effectiveness of aerosolization and the city's vulnerability. The results indicated that the entire subway tunnel system can be infected by release of endospores into a single station because of the air currents generated by the trains. Alarmingly, the researchers determined that the bacteria could be carried on the trains and then released onto station platforms at distant locations when the doors open, which was, and still is, a bigger threat.

During the 1960s, international concern peaked regarding the risk, unpredictability, lack of control measures, and the failure of the 1925 Geneva Protocol for preventing biological weapons proliferation. Nevertheless, in a surprise move in November 1969, President Richard Nixon terminated the U.S. offensive biological weapons program, stating, "*We'll never use the damn germs . . . If someone uses germs on us we'll nuke 'em.*" The 1972 Convention on the Prohibition of the Development, Production, and Stockpiling of Bacteriological (Biological) and Toxin Weapons and on their Destruction, commonly known as the **Biological Weapons Convention of 1972**, was convened, and a treaty was signed by 103 nations. One point of the treaty prohibited the development of delivery systems intended to dispense biological agents and, further, required signatory nations to destroy their stocks of biological agents, delivery systems, and equipment within nine months of signing. The treaty was ratified in April 1972, and went into effect in March 1975.

1972 to September 11, 2001

As a result of the Biological Weapons Convention of 1972 and the termination of the offensive biological warfare program, the biological arsenal at Fort Detrick was destroyed. Agents that were destroyed included the causative agents of the potentially lethal diseases anthrax, tularemia, and Venezuelan equine encephalitis and of botulinum toxin, staphylococcal enterotoxin B, and several biological agents targeting crop plants. The focus at Fort Detrick shifted from offensive to defensive strategies and was placed under the auspices of the U.S. Army Medical Research Institute of Infectious Diseases. The availability of high-level containment facilities at the Fort Detrick facility still allows for the study of diseases caused by highly virulent pathogens, such as Ebola virus.

As was the case following the 1925 Geneva Protocol, the 1972 Biological Weapons Convention has proved to be ineffective. The Soviet Union continued an offensive biological weapons program through the 1980s. As an example, on April 2, 1979, inhabitants of the community of Sverdlovsk (now Ekaterinburg), a city 2.5 miles (4 kilometers) downwind of a Soviet military microbiology facility, reported to hospitals with symptoms of anthrax. At least 94 cases and 64 autopsy-confirmed deaths resulted, the largest epidemic of inhalational anthrax in history (**FIGURE 17.3**). The outbreak also caused the death of livestock. Western intelligence suspected the Sverdlovsk facility was a biological warfare research center and that the epidemic was due to the accidental airborne release of *B. anthracis* endospores. The Soviets denied the allegation and insisted that the

Krasner's Notebook

I spent my first sabbatical leave from Providence College at Fort Detrick in 1965; security was tight and enforced. Before my appointment at Fort Detrick, I was extensively investigated to get a security clearance. On occasion, reports of hospitalization of Fort Detrick laboratory workers surfaced. For approximately 10 years after leaving, I periodically received a questionnaire inquiring as to my health and whether I had suffered from any unusual illnesses. Fortunately, all I got was athlete's foot—a fungal disease.

© Lynn Johnson/National Geographic Creative.

FIGURE 17.3 Russian woman visits the grave of her son who died in the 1979 Sverdlovsk incident in the Soviet Union.

incident was due to ingestion of contaminated meat. However, in 1992, almost 15 years later, Russian President Boris Yeltsin admitted that the facility was part of an offensive biological weapons program and that a failure to activate air filters resulted in the accidental release of *B. anthracis* endospores. Yeltsin then proclaimed he would terminate biological warfare initiatives.

Iraq was once considered to be a major threat in terms of biological warfare. Between 1986 and the end of the first Gulf War in April 1991, Iraqi scientists investigated the potential of a large number of pathogens to serve as biological weapons, including fungi, bacteria, viruses, and botulinum toxin. In 1990, the United Nations imposed sanctions on Iraq because of its invasion of Kuwait and the threatened use of biological weapons. Iraq's failure to allow inspection of its biological weapons by the UN Special Commission led to air strikes by U.S. and British forces in December 1998. The alleged possession of weapons of mass destruction in Iraq was a major factor in that country being invaded by a multinational force in March 2003, which began the second Gulf War.

Emergence of Biological Terrorism

In the past few decades the threat of biological weaponry has taken on a new face: that of biological terrorism. Nongovernmental agents, including religious cults, terrorists, and individuals, can use biological weapons to further their own personal or political agendas. In some respects, biological terrorism is an even greater threat than biological warfare; it is more difficult to detect and control, and terrorists can strike without warning. In 1981, members of a religious commune, followers of Bhagwan Shree Rajneesh, intentionally contaminated 10 restaurant salad bars in an Oregon community with *Salmonella enterica* serovar Typhimurium. More than 750 people fell ill with *Salmonella* gastroenteritis in a community in which fewer than 5 cases per year are

usually reported. This was intended as a "rehearsal" for a plan to cause an epidemic on Election Day to influence the outcome of the county elections. The terrorist threat posed by radical groups was again demonstrated in 1995 by the Aum Shinrikyo ("Supreme Truth") cult, possibly the world's most infamous apocalyptic sect, when they released sarin gas into a Tokyo subway, killing at least 11 people and making thousands of others ill. Perhaps even more frightening is the fact that the cult previously attempted at least nine biological attacks with the intent of causing mass murder. Their lack of sophistication, not their lack of determination, resulted in the failure of these activities.

In still another incident, on May 5, 1995, Larry Harris, a septic tank and well water inspector in Ohio, received three vials of *Y. pestis*, the causative agent of plague, from the American Type Culture Collection (ATCC). His intent was not clear, but the fact that he needed only a credit card and false letterhead is disturbing to say the least. This is only one example of numerous others indicating the necessity of establishing tighter controls on microorganisms to minimize their getting into the hands of terrorists. To this end, three major statutes have been passed by the U.S. Congress: (1) the Biological Weapons Act of 1989, (2) the Chemical and Biological Weapons Control and Warfare Elimination Act of 1991, and (3) the Anti-Terrorism and Effective Death Penalty Act of 1996.

Assessment of the Threat of Biological Weaponry

National security is a primary concern to the people of all countries, large or small, developed or developing. **Epidemics** and **pandemics** of infectious diseases negatively affect national security because they subject populations to the threat of disease and create the potential for transmission beyond a country's borders. The nature of outbreaks of infectious disease is devastating, not only to the citizenry but also to the economic development of a country. The ongoing pandemic of HIV/AIDS on the African continent is witness to this point.

The use of biological weapons, whether by nations or by terrorists, is focused on promoting outbreaks of disease. This focus adds to any nation's public health burden, because it must protect not only against naturally occurring outbreaks but also those intentionally caused. An article, "Security and Public Health: How and Why Do Public Health Emergencies Affect the Security of a Country?," from the Monterey Institute for International Studies Center for Nonproliferation Studies (2007) cites four ways in which a public health emergency, including those caused by biological weapons, threatens a nation's security: (1) it puts pressure on a country's economy at both a micro and macro level, (2) it causes social disruption, (3) it leads to political destabilization, and (4) it affects national defense.

It is within the capability and resources of many countries to use microbes as tools of mass murder.

Biological agents are accessible because of their use in legitimate research activities throughout the world. The equipment and facilities for their mass production are commonly found in the pharmaceutical, agricultural, and food industries. This makes them more accessible than agents of nuclear or chemical warfare, which require special facilities that would make their detection relatively easy. The weapons used to deliver biological agents are proliferating and are cost-effective. A group of technical experts estimated that "*for a large-scale operation against a civilian population, casualties might cost about $2,000 per square kilometer with conventional weapons, $800 with nuclear weapons, $600 with nerve gas weapons, and $1.00 with biological weapons.*"

The use of biological weapons, whether by governments, groups, or individuals, has inherent advantages and disadvantages (**TABLE 17.1**). The advantages of biological weapons include deadly or incapacitating effects on the target population, low cost, continued microbial proliferation, difficulty of immediate detection, and lack of physical damage to the area. The disadvantages include danger to the health of the aggressors, the effects of physical factors such as weather conditions on the success of an attack, public backlash, and the environmental persistence of some agents.

Considering the numerous factors necessary for effective deployment, relatively few biological weapons meet the criteria; anthrax, smallpox, plague, and botulinum toxin are the front-runners.

The psychosocial responses after even the threat of a biological attack add another dimension to the horror of biological weaponry. The fear of "germs" and infection is terrifying; the resulting panic would produce staggering numbers of psychiatric casualties that could overwhelm healthcare facilities. The threat, in itself, of the release of biological agents in a particular community, transportation system, school, or shopping mall causes a crippling effect. On December 26, 1998, 750 people were quarantined after police received a call claiming that the anthrax-causing bacteria had been released into a popular nightclub in Pomona, California. This incident was reported to be the sixth anthrax hoax in the area in 2 weeks.

TABLE 17.1 Advantages and Disadvantages of Biological Weapons

Advantages	Disadvantages
The potentially deadly or incapacitating effects on a susceptible population	The danger that biological agents can also affect the health of the aggressor forces
The self-replicating capacity of some biological agents to continue proliferating in the affected individual and, potentially, in the local population and surroundings	The dependence of effective dispersion on prevailing winds and other weather conditions
The relatively low cost of producing many biological weapons	The effects of temperature, sunlight, and desiccation on the survivability of some infectious organisms
The insidious symptoms that can mimic endemic diseases	The environmental persistence of some agents, such as endospore-forming anthrax bacteria, which can make an area uninhabitable for long periods
The difficulty of immediately detecting the use of a biological agent, due to the current limitations in fielding a multiagent sensor system on the battlefield as well as to the prolonged incubation period preceding onset of illness (or the slow onset of symptoms) with some biological agents	The possibility that secondary aerosols of the agent will be generated as the aggressor moves through an area already attacked
The sparing of property and physical surroundings (compared with conventional or nuclear weapons)	The unpredictability of morbidity secondary to a biological attack, since casualties (including civilians) will be related to the quantity and the manner of exposure
	The relatively long incubation period for many agents, a factor that may limit their tactical usefulness
	The public's aversion to the use of biological warfare agents

Information from the *Textbook of Military Medicine, Part I, Medical Aspects of Chemical and Biological Warfare.* Office of the Surgeon General, United States Army, 1997.

Category A Biological Threats

Biological agents and the diseases they cause have been categorized into three groups based on their risk to national security (**BOX 17.1**). Anthrax, botulism, plague, smallpox, tularemia, and viral hemorrhagic fevers make up Category A. We will take a closer look at two agents that have a long history in biological warfare: anthrax and smallpox.

Anthrax

Anthrax, or *Bacillus anthracis*, is the most likely biological weapon to be used and is thought to be in the arsenal or in the process of being developed in at least 10 countries.

Anthrax is a rapidly progressing acute infectious disease caused by the organism *B. anthracis* (**FIGURE 17.4**), a rod-shaped, endospore-forming bacterium. During the Middle Ages it was known as the "**Black Bane**" and was responsible for nearly destroying the cattle herds of Europe. The ability of *B. anthracis* to form endospores makes anthrax a choice weapon of mass destruction, because these endospores can be disseminated by bombs, artillery shells, aerial sprayers, and other methods of dispersal. Endospores are the bacterial genetic material, surrounded by multiple protective layers that makes them resistant to most heat and chemical disinfection protocols. They are hard to kill and reside in soils worldwide for many years.

BOX 17.1 Categories of Biological Diseases and Agents

Category A Agents

The U.S. public health system and primary healthcare providers must be prepared to address varied biological agents, including pathogens that are rarely seen in the United States. High-priority agents, or **Category A agents**, include organisms that pose a risk to national security because they can be easily transmitted from person to person, cause high rates of mortality with the potential for major public health impact, might cause social disruption due to public panic, and require special action for public health preparedness. The CDC and **U.S. Department of Agriculture (USDA)** have listed several viruses, bacterial species, and types of toxin as "select agents" with the potential to cause harm by physically causing disease in humans or economically by causing disease in crops and farm animals (**Table 1**).

Category A agents include variola virus (smallpox), *B. anthracis* (anthrax), *Y. pestis* (plague), *C. botulinum* toxin (botulism), *Francisella tularensis* (tularemia), filoviruses (Ebola and Marburg viruses), and arenaviruses (Lassa fever) (**Table 2**).

TABLE 1 Partial List of CDC and USDA Regulated Biological Agents

Viruses	Bacteria	Toxins
Eastern equine encephalitis	*Bacillus anthracis*	Botulinum neurotoxins
Marburg viru	Botulinum neurotoxin producing species of *Clostridium*	Ricin toxin
Ebola virus	*Rickettsia prowazekii* (typhus)	*Clostridium perfringens* epsilon toxin
Monkeypox virus	*Yersinia pestis* (bubonic plague)	Staphylococcal enterotoxins
Smallpox virus	*Francisella tularensis* (rabbit fever)	
Reconstructed 1918 influenza virus	*Brucella abortus* (brucellosis)	
Nipah virus		
Foot-and-mouth disease virus		
Avian influenza virus		

Information from CDC-USDA Federal Select Agent Program (https://www.selectagents.gov/selectagentsandtoxinslist.html).

(continues)

BOX 17.1 Categories of Biological Diseases and Agents (continued)

TABLE 2 Category A Agents and Diseases They Cause

Disease	Agent(s)	Method of Transmission	Incubation Time*	Symptoms	Treatment
Anthrax	*Bacillus anthracis*	Inhalation	1–7 days	Flulike symptoms, fatigue, breathing difficulty, possible death	Antibiotics (early administration), prevention by vaccination
Botulism	*Clostridium botulinum* toxoid	Foodborne	12–36 hours	Double vision, blurred vision, slurred speech, difficulty swallowing, dry mouth, muscle weakness, paralysis of breathing muscles leading to death	Antitoxin
Plague (pneumonic)	*Yersinia pestis*	Inhalation	1–6 days	Fever, headache, cough, weakness	Antibiotics
Smallpox	Variola virus	Inhalation	7–17 days	Characteristic rash, fever, fatigue	Prevention by vaccination (within 4 days after exposure)
Tularemia	*Francisella tularensis*	Inhalation	1–14 days	Fever, ulcerated skin lesions, chills and shaking, headache, weakness; can progress to respiratory fever and shock	Antibiotics
Viral hemorrhagic fevers	Ebola virus, Marburg virus, Lassa virus	Close contact with infected person or his or her body fluids	7–14 days (Ebola fever), 5–10 days (Marburg fever), 6–21 days (Lassa fever)	Fever, fatigue, dizziness, weakness, internal bleeding, bleeding from body orifices, shock, delirium	Supportive therapy, ribavirin (depending on circumstances)

*Considerable variation.
Information from CDC Emergency Preparedness & Response Information.
A. S. Khan & A. M. Levitt. (2000). Biological and chemical terrorism: Strategic plan for preparedness and response. *Morbidity and Mortality Weekly Report, 49*(RR 4), 1–14.

Category B Agents

The second highest priority agents include those that are moderately easy to disseminate, cause moderate morbidity and low mortality, and require specific enhancements of the CDC's diagnostic capacity and enhanced disease surveillance. **Category B agents** include *Coxiella burnetii* (Q fever), *Brucella* species (brucellosis), *Burkholderia mallei* (glanders), ricin toxin, epsilon toxin from *Clostridium perfringens*, and *Staphylococcus* enterotoxin B.

Category C Agents

The third highest priority agents include emerging pathogens that could be engineered for mass distribution in the future because of availability, ease of production and dissemination, and potential for high morbidity and mortality and major health impact. **Category C agents** include Nipah virus, hantaviruses, tickborne hemorrhagic fever viruses, tickborne encephalitis viruses, yellow fever virus, and multidrug-resistant *Mycobacterium tuberculosis*.

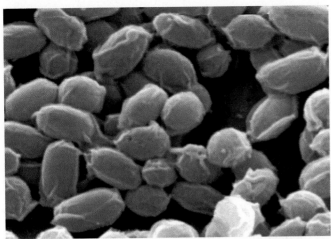

Courtesy of Janice Haney Carr/CDC.

FIGURE 17.4 Colorized scanning electron micrograph of *B. anthracis*.

In 1942, the British used Gruinard Island, off the northwest coast of Scotland, as a site for anthrax experiments. The soil remained contaminated with B. *anthracis* endospores for more than 50 years, making the area unsafe for human entry. Anthrax endospores are like seeds in that under appropriate conditions they germinate into actively multiplying and toxin-producing bacteria. Human and animal lungs and skin provide these appropriate conditions. The reasons responsible for anthrax as a "top contender biological weapon" are summarized in **TABLE 17.2**.

Three varieties of anthrax disease affect humans. First, **cutaneous (skin) anthrax** is a hazard to those employed in occupations requiring frequent close contact with the infected animals. About 20% of untreated cases result in death, but death is rare with appropriate antibiotic therapy. Second, **gastrointestinal anthrax** is rare but can be acquired by the ingestion of inadequately cooked contaminated meat. Death occurs in 25% to 60% of untreated cases. Third is **inhalation anthrax**, which poses the greatest threat. Initially, symptoms resemble a common cold, which progress to severe breathing problems. The incubation period is 1 to 7 days; death usually results in 1 to 2 days after the occurrence of acute symptoms. It takes an incredibly small dose (8,000 to 10,000 endospores) of B. *anthracis* to kill a human. Antibiotic therapy is only effective if it is administered within the first 24 to 48 hours after exposure. If untreated, inhalation anthrax has a case-fatality rate of over 90%.

In 1995, Iraq admitted to the United Nations that it had loaded B. *anthracis* spores into warheads during the Gulf War. A few years later a series of bioterrorist threats of exposure to endospores were reported by the CDC that turned out to be hoaxes. After a 3-year study, on May 18, 1998, then Secretary of Defense William Cohen approved the Pentagon's plan to vaccinate all U.S. military service members for anthrax, a decision causing continuing controversy. As a show of confidence Secretary Cohen was immunized. Full immunization required six injections administered over an 18-month period. At the time it had a price tag of about $60 to $80 per person for the full six-dose regimen and a total cost in excess of $200 million for the estimated 2.4 million personnel.

Criticism of the vaccine is based on the following major points:

1. It has a difficult immunization schedule (five or six shots are required over a period of 18 months).

2. Full immunity is not conferred until after 18 months.

3. An annual booster is required.

4. The vaccine was approved by the FDA in 1970 for prevention of cutaneous anthrax; opponents claim that the vaccine has not been proven to be effective against inhalation anthrax, which would be seen in aerosolized endospores.

In December 2003, the U.S. District Court ruled that the Department of Defense could not make anthrax vaccination in the military mandatory (unless through presidential order). In early 2007, the Department of Defense resumed mandatory anthrax vaccination for all service personnel and civilian employees deployed to Afghanistan, Iraq, and the Korean Peninsula (South and North Korea) because these areas were considered "high threat." However, in 2008, a federal judge ruled that involuntary immunization was a violation of federal law, as the vaccine was not approved for use against inhalation anthrax. The federal government has since stockpiled millions of doses of anthrax vaccine. With the 18-month vaccination schedule, however, this may not be useful in the event of a real threat.

Smallpox

Smallpox, the first infectious disease to be eradicated, is now a top biological weapon candidate. Smallpox has unique properties that made its eradication possible. Routine immunization against smallpox ended in 1972 in the United States and has since ended around the world.

TABLE 17.2 Anthrax as a Biological Weapon
The bacterium (*Bacillus anthracis*) is relatively easy to grow.
It has a high fatality rate if untreated.
The treatment is only partially successful and only if administered within 24 to 48 hours of exposure.
B. anthracis produces endospores that remain viable for years.
The endospores can be easily dispersed by missiles, rockets, aerial bombs, crop dusters, and sprayers.
There is no easy indication of exposure.

Ironically, the science that led to the eradication of smallpox has resulted in a world population under the age of 50 that is susceptible to smallpox (including over half the population of the United States). Those individuals that were immunized have a waning immunity. If smallpox is ever reintroduced into the world population, it has the potential to spread like wildfire through the naïve population because of a lack of **herd immunity**.

If smallpox has been eradicated from the face of the Earth, why is it still a threat? The answer is that the disease has been eradicated but not the virus; it exists in freezers under lock and key at a designated facility in Russia and at the CDC in Atlanta, Georgia. The problem is that covert stocks of the virus are thought to be in the hands of potential terrorists and unreliable foreign governments.

Smallpox is transmitted directly from person to person by variola virus-infected droplets of saliva. Infected clothing and bedding can also spread the virus. The first week of illness is the most infectious time because of the high numbers of viral particles in the saliva. The early symptoms of smallpox are nonspecific, which means people may continue to go out and spread the virus. Later stages of smallpox include an extensive pustular rash over the body that then crusts over. Eventually, crusts dry up and fall off the pustules, leaving deeply pitted scars, particularly on the face. The case fatality rate for the more **virulent**, or deadly, strains is about 30%.

Some have voiced concerns that there is not enough smallpox vaccine in the United States to immunize and treat the public, but that is not the case. The CDC's plan advocates that the vaccine should not be used until after an attack occurs and then should be rapidly deployed. Others call for mass vaccinations, fearing that an epidemic may develop too rapidly to be contained, particularly if an attack were to occur simultaneously in several cities. Unfortunately, the smallpox vaccine does have serious side effects. If all Americans were vaccinated tomorrow, 180 to 400 people could die just from the vaccine's side effects. In July 2018, the FDA approved **TPOXX**, a drug that has been shown to be effective in laboratory testing and nontoxic to humans. The U.S. government has stockpiled 2 million doses, to be used in the event of a terror attack.

Countermeasures to Biological Weaponry

Perhaps you find the account of biological warfare and terrorist activities unsettling and fearsome. If you do, it is with good reason. Compared with the use of conventional weapons or even chemical or nuclear warfare, the use of biological weapons presents a unique set of challenges to any country's public health infrastructure:

- An attack might not be immediately discernible and would remain unnoticed until individuals became seriously ill; incubation and infectious periods vary with the biological agent involved and also from person to person.

- If an attack were to occur in a subway station, a train terminal, or an airport, victims would be widely dispersed and, within 24 hours, could spread the disease literally across the globe before they were aware of their own illness.

- If the agent is communicable from person to person, each infected individual would seed the agent into an ever-increasing circle of disease in successive waves.

- Studies have shown that bioterrorist events affect the general public to a greater extent than other forms of terrorism, triggering chaos, panic, and general civil disorder.

- Bioterrorist attacks are directly targeted at human populations.

- Dispersal by aerosolization, drinking water, or food supply can be inexpensively accomplished with crop dusters, spray cans, and other simple devices (**FIGURE 17.5**).

- The pathogen selected would be one that is not routinely seen in a population. This would mean that the population would be particularly susceptible due to a lack of immunity (e.g., in the case of smallpox) because vaccination is no longer carried out.

Courtesy of U.S. Environmental Protection Agency.

FIGURE 17.5 Public flyers created by the EPA to educate and alert communities about water security.

○ Diagnosis may be delayed because the disease would most likely be one with which medical personnel have little or no familiarity or one in which the symptoms are vague and may resemble the symptoms of a variety of microbial infections. For example, smallpox has not been seen for over 35 years, and anthrax initially has coldlike symptoms.

○ Healthcare personnel would be the most exposed, and therefore most likely to succumb to infection, thwarting further efforts of later responses.

Initiatives and strategic plans for coping with potential bioterrorism are as follows:

○ Deterrence by controlling access to and handling of dangerous pathogens, including monitoring of the facilities and procedures currently in use with these agents. The CDC is charged with this responsibility.

○ Surveillance and rapid detection, based on the number of individuals becoming ill and their symptoms, of whether a biological agent has been released.

○ Medical and public health response aimed at strengthening the response at the local level, including the capability of mass immunization or prophylactic management, safe disposal of the deceased, infection control, and assessment of extent of the problem.

○ Creation of a stockpile of pharmaceuticals for treatment and prevention of illness from potential agents of bioterrorism as a national resource. It would be beyond the resource of local governments to develop and maintain such a facility. Appropriate pharmaceuticals would be deployed to reach victims within 24 hours. The CDC has the responsibility for developing the stockpile. President George W. Bush passed legislation to create Project BioShield in July 2004, to develop and stockpile vaccines and drugs.

○ Expansion of support for research and development related to pathogens that might be used as biological weapons, particularly the development of rapid diagnostic methods because "time is of the essence." New vaccines and new or improved antiviral and antibiological agents are critical.

More recently the United States has placed emphasis on the coordination of clinical, public health, and law enforcement activities to deal with suspected bioterrorism. Hospitals, fire and police departments, and other agencies charged with protecting the public have developed strategic plans to cope with disaster (**FIGURE 17.6**). President George W. Bush created a new government agency, the Department of Homeland Security, in 2003, to develop, coordinate, and implement measures to protect the nation. A plan of readiness is vital at all levels, from local to national, and is continually updated by the CDC and other agencies as the nation recognizes the use

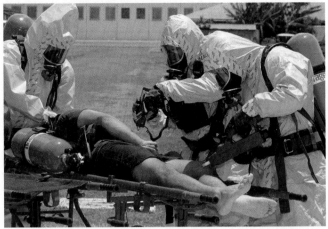

Courtesy of Photographer's Mate 1st Class William R. Goodwin/U.S. Navy.

FIGURE 17.6 Emergency workers in biosafety suits.

of biological weapons as a real threat. Being prepared to intelligently meet a threat is the best rational approach. The use of biological weapons is a clear and dramatic example of the interfaces of biology, public health, international law, and ethics.

In the Aftermath of September 11, 2001

In the days following Al Qaeda's September 11, 2001, attack on the World Trade Center and the Pentagon, Americans were gravely concerned and anxiety stricken about what terror might follow. In a *Time*/CNN poll, 53% of those surveyed feared a chemical or biological attack, whereas 23% feared a nuclear attack would follow.

The reality of using *B. anthracis* endospores as biological weapons surfaced only 2 weeks later when, on September 25, a 38-year-old assistant to NBC's Tom Brokaw developed a case of cutaneous anthrax, presumably as the result of opening a letter, postmarked September 18, containing a powder. Over the next few weeks more cases of inhalational anthrax and cutaneous anthrax targeting primarily those in the media emerged; New York City was particularly hard hit, with seven cases of cutaneous anthrax (two at NBC, one at CBS, three at the *New York Post*, and one at ABC) and one case of inhalational anthrax. The buildings housing these offices tested positive for anthrax endospores. The offices of Senate Majority Leader Tom Daschle and Senator Patrick Leahy and dozens of personnel were contaminated on October 14 by a finely milled version of "anthrax powder" sent in letters postmarked from Trenton, New Jersey. Mail facilities in the Washington, D.C. area became centers for cross-contamination of mail delivered into the city by mail trucks serving as vectors of disease. Two postal workers died of inhalation anthrax, and more than 2,000 postal workers in the nation's capital were placed on a 10-day course of the antibiotic ciprofloxacin (Cipro).

In the aftermath, government bodies, including the U.S. Senate, the Central Intelligence Agency, and the

U.S. Supreme Court, closed for various times; Americans were asked not to hoard ciprofloxacin following a buying frenzy of the antibiotic. The U.S. Postal Service instituted new security measures and distributed information about how to recognize and report suspicious mail. The nation's airlines were threatened with bankruptcy, and security measures in and around airports were heightened to protect the flying public. Northwest Airlines even went as far as to ban the sweetener Sweet'N Low because of its resemblance to the powdery substance associated with *B. anthracis* endospores. Fire and police departments and ambulance services were bombarded with calls to check out "suspicious powders," and emergency room facilities were overwhelmed with people fearful that they had been exposed to anthrax endospores.

The country continues to be on alert in the post 9/11 world and worries, perhaps excessively, about the threat of bioterrorism. Reports of suspicious bioterrorist type activity that turn out to be of no consequence continue to close buildings, streets, parks, schools, and other gathering places at the cost of anxiety-stricken people, lost wages, and millions of dollars spent in response. Suspicious package reports and letters, all which need to be investigated, continue to stretch law enforcement agencies around the country. New York and Washington, D.C., are particularly affected. New York City police have investigated over 83,700 calls since the terrorists attack in 2001. Washington, too, has been besieged by calls, partially due to the Department of Homeland Security's "If You See Something, Say Something" program.

Past and Future Threats: Influenza and Digital Attacks

Influenza

Influenza remains a cause of concern, particularly now that a team of virologists at Erasmus Medical Center in Rotterdam and an independent team at the University of Wisconsin–Madison have created in their laboratories a strain of airborne avian influenza A virus (**H5N1**), some details of which were presented at a conference in Europe in September 2011. The animal model was the ferret, the usual model for influenza studies, and the experiment resulted in influenza spreading by air within the colony. The original avian influenza A virus (H5N1) is not airborne and is not easily transmitted from human to human; it is spread only by direct contact between persons and sick or deceased birds.

Since 2013, a new strain of "bird flu" has caused concern. Avian influenza strain H7N9 has infected more than 1,500 people in China, killing over 600. Like H5N1, this strain is not directly transmissible from one human to another. The high mortality rate, however, has officials concerned. The CDC has developed a vaccine for the **H7N9** strain, and the Chinese government started clinical trials of the vaccine in 2017.

The fear is that the modified "superstrain" of either form of avian influenza could easily spread from human to human by a cough, sneeze, talking, and any other means of sending droplets into the air, potentially causing a pandemic rivaling the 1918 Spanish flu pandemic that killed over 40 million people in only 2 years. The fear is justified and focuses on the use of this information by terrorist groups to produce biological weapons as well as the possibility of accidental release of this altered virus (see **BOX 17.2**). The genetic modification of microbes is dangerous territory as is the closely related topic of synthetic biology. In fact, the prospects are so frightening that the federal advisory board has delayed publishing details.

Digital Attacks

In the modern era, millions of people live with implanted devices that are meant to improve their quality of life. These devices include pacemakers, which regulate a person's heart rate, and insulin pumps, which deliver a specific dose of insulin based on a person's blood sugar. In recent years, upgrades allow patients and physicians to adjust them via wireless communication between an external remote control and the implanted device. In 2017, half a million pacemakers were recalled due to a potential security risk, in that the wireless inputs could be hijacked

BOX 17.2 Creating the Armageddon Virus?

Virologists made an alarming announcement to scientists attending the European Scientific Working Group on Influenza Conference held September 11–14, 2011. They genetically manipulated an H5N1 influenza A virus into a version of virus that could easily pass from ferret to ferret by airborne transmission (the best animal model for influenza research; **Figure 1**). The collaborative research was led by Dr. Ron A. M. Fouchier from the Erasmus Medical Center in the Netherlands and Dr. Yoshihiro Kawaoka from the University of Wisconsin–Madison and the University of Tokyo.

While experts at the meeting acknowledged that this research helps influenza experts "prepare for the unpredictable," the National Science Advisory Board for Biosecurity (NSABB) began reviewing their findings and placed an embargo on the manuscripts for publication describing their research. The lab-made avian influenza (H5N1) strain viruses are kept under high security but there were concerns that publishing the recipe for these viruses in scientific journals could fuel the efforts of terrorists who would try to create a biological weapon or "Armageddon virus." Or, worse yet, that

© Jagodka/ShutterStock, Inc.

Figure 1 The ferret model has been used to evaluate H5N1 viruses that might infect and cause illness in humans.

the Fouchier and Kawaoka viruses could escape the laboratory, triggering an influenza pandemic with millions of deaths! Their findings ignited intense public debates and criticism in the media on the benefits and potential harm of this type of research. Concern was so great that on December 7, 2011, U.S. Secretary of State Hillary Clinton attended a summit on biological weapons held in Geneva. No American official of her ranking had attended the summit in decades. Just before Christmas, the NSABB advised that the manuscripts be published as long as the methods to create the viruses would be excised or too vague for would-be terrorists. This decision put the burden of ethics on the shoulders of the editors of *Nature* and *Science*, two leading scientific journals.

Fouchier and Kawaoka braced themselves for the media storm. On January 20, a letter signed by over 30 senior influenza researchers was published in *Nature* and *Science* declaring a 2-month moratorium on research involving the highly pathogenic strains of H5N1 that focused on what makes the H5N1 bird flu strain more transmissible between mammals.

The intention of the researchers was to obtain knowledge that would benefit public health, allowing scientists to prepare aggressive measures if a mutated H5N1 virus showed up in the wild. The lab-made virus allowed them to perform studies testing whether current H5N1 vaccines and antivirals would be effective against a new H5N1 influenza strain.

In February 2012, at a meeting assembled by the UN agency, experts agreed that full disclosure of the information needed to create these lethal mutant H5N1 lab strains was preferable over a redacted version of the research of mutant flu viruses. Editors of journals have been open to more discussions by influenza researchers, policy experts, and scientists outside of the immediate field before the manuscripts are published. Details of the secret experiments will inevitably leak out. Electronic information could be leaked or hacked. The peril of altered avian influenza viruses getting into the wrong hands is balanced with the promise of free information exchange leading to the development of better vaccines and treatments.

Journal editors have faced biosecurity issues before. Another serious example of research affected by publishing considerations occurred as early as 2001, when scientists genetically modified a mousepox virus to be 100% fatal in mice by the addition of a single gene introduced into the mousepox virus. Mousepox is not known to hurt or infect humans. The work was published purely as a scientific paper at a time when there was no form of biosecurity review.

We are at a crossroads in which well-intentioned scientific research has the potential to be misused for nefarious purposes, a dual-use dilemma. When physicists observed atomic fission, they thought their research may have beneficial applications in medicine and energy production, but on the flip side, that it could also lead to the creation of a devastating atomic bomb. Some of the same discoveries that lead to advancements in medicine or public health can be adapted to the development of weapons of mass destruction (**Figure 2**).

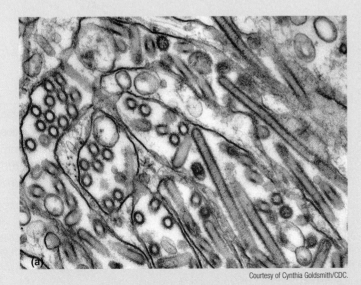

Courtesy of Cynthia Goldsmith/CDC.

Courtesy Joe Kosstatscher, United States Navy.
Harry S. Truman Library & Museum.

Figure 2 Examples of research with a dual-use dilemma. **(a)** H5N1 influenza viruses. **(b)** Atomic bomb blast over Nagasaki, Japan in 1945 (WWII).

by hackers intercepting or overriding the wireless signal. The FDA cited concerns that such tampering could alter a person's heartbeat, with potentially fatal consequences. Rather than remove and reimplant all of the potentially defective devices, patients were instructed to visit their cardiology clinic to receive the latest software update. It is important to note, however, that while these devices were vulnerable to attack, no actual attacks were reported.

Engineering Genetic Modifications: Customizable Organisms

As demonstrated by the tinkering with the H5N1 influenza A virus strains, researchers can engineer pathogens to be even more deadly, and they can engineer viruses to create vaccines. They can also engineer organisms to contain a gene of choice in their whole bodies, making **transgenic** microbes, plants, or animals. The Oxitech Friendly mosquitoes are transgenic mosquitoes that are designed to die unless given tetracycline. Researchers can add vitamins, such as vitamin A, to plants in order to supplement the diet of individuals living in nutrient-poor areas. Golden rice, as it is called, can prevent blindness by providing vitamin A to people that use rice as a staple.

Bt Corn and Roundup Ready Crops

Corn and soybeans can also become transgenic with the addition of a toxin from *Bacillus thuringiensis* (**Bt toxin**). This toxin is fatal to insects, but extensive research has demonstrated it is nontoxic to mammals. The cells lining the mammalian gut simply lack the receptors for this toxin, and so do not absorb it and are therefore unaffected by it. The scientists chose the toxin from *B. thuringiensis* because the bacteria itself had been used as a natural pesticide since the 1920s. In fact, you can order large bottles of *B. thuringiensis* bacteria from organic farm suppliers. In 2009, approximately 65% of the corn and cotton planted in the United States were Bt varieties, with the *B. thuringiensis* toxin gene added to protect them from insect predation. By 2017, that number had jumped to 80% for corn and 85% for cotton, according to the U.S. Department of Agriculture (**FIGURE 17.7**).

Another commonly added gene confers resistance to a **glyphosate**, which is the active ingredient in the herbicide Roundup. Glyphosate kills plants by preventing their ability to synthesize several amino acids. The gene that is added, called *EPSP*, is from a bacterial variant that glyphosate does not recognize. Other plants, with their normal *EPSP*, are sensitive to glyphosate, and will die when they cannot make all the amino acids they need to make their cellular

proteins. Crops with the herbicide-resistance gene added are **herbicide tolerant (HT)**. These HT strains enable farmers to reduce weeds in their fields and increase their crop yields. There are crops that are both HT and have Bt toxin as **stacked traits**. These are both important considerations as the human population reaches 8 billion people, which is projected to happen by 2025.

Controversies Surrounding Genetically Modified Organisms

These innovations may seem fairly straightforward, but they have not been universally accepted. In fact, several groups see potential peril in genetic engineering rather than the promise of greater crop yields and protection from mosquitoborne diseases. These groups have held protests, set up websites, and lobbied extensively against the use of **genetically modified organisms (GMOs)**. Some concerns are valid, including the impact of Bt crops on nontarget insect populations, such as the monarch butterfly. One early corn strain, Bt 176, did harm monarch caterpillars. This strain was removed from the market in 2003. Others have expressed concerns about accidental ingestion of the *B. thuringiensis* toxin, potential food allergies, and even blaming the decline of bee populations on Bt crops. Because mammals lack the Bt toxin receptors, they are unaffected by it. Mice given as much as 5 milligrams per kilogram (mg/kg) of purified toxin had no ill effects in multiple safety studies. Exposure to either Bt or HT varieties of corn and soybeans has not been found to produce an increase in allergy symptoms or IgE production, as measured by skin-prick tests of over 100 test subjects.

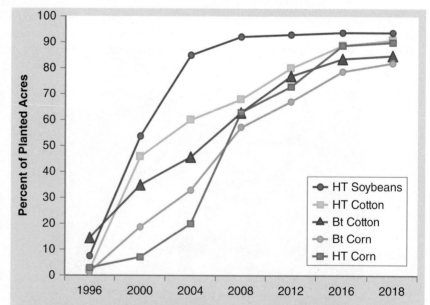

Note: HT indicates herbicide-tolerant varieties; Bt indicates insect-resistant varieties (containing genes from the soil bacterium bacillus thuringiensis). Data for each crop category include varieties with both HT and Bt traits (called stacked traits).

Source: Information from USDA Economic research station.

FIGURE 17.7 Timeline of the use of GMO corn, cotton, and soybeans in the United States.

© Protasov AN/ShutterStock, Inc.

FIGURE 17.8 Foods labeled as non-GMO (genetically modified organism) or GMO free, are free from organisms produced in a laboratory.

Although the science behind both Bt and HT crops has been demonstrated to be safe, public perception of these products remains a problem. The perception of potential harm has caused large companies and small shops alike to market products as GMO-free (**FIGURE 17.8**). For example, in 2014, McDonald's specifically stated it would not use genetically modified potatoes in its French fries. The United States passed a federal mandatory labeling law in 2016, but it had not implemented it as of summer of 2018.

Current Concerns: The Power and Peril of CRISPR/Cas Technology

What Is CRISPR?

You may have heard of the terms *gene editing* or *CRISPR/Cas* in the past few years. Clustered regularly interspaced short palindromic repeats, or **CRISPR**, work as a bacterial cell's immune system to defend against viral attack. This CRISPR region produces a short RNA molecule that binds to and guides **Cas9 proteins** to sequences that match the RNA guide. The Cas9 protein then cuts the region of DNA that is recognized by the guide (**FIGURE 17.9**). This gives researchers the ability to make precise changes to the DNA in the cells of an organism, at any stage of its development. They can edit the genome of cells in an adult or in the entire organism if they make the changes in a developing embryo.

This seems fairly complex, so why is there such hype about **genome editing** with **CRISPR/Cas**? In 2012, Jennifer Doudna and Emmanuelle Charpentier made an impressive breakthrough. The Cas9 protein will cut DNA specifically according to the guide RNA, making it possible to design guide RNA segments. This opened the door to a wide range of possibilities. One of the first experiments done was to try to **knockout** expression of mouse genes that cause disease. Since then, researchers have shown that this technology can be used to target and destroy antibiotic-resistance

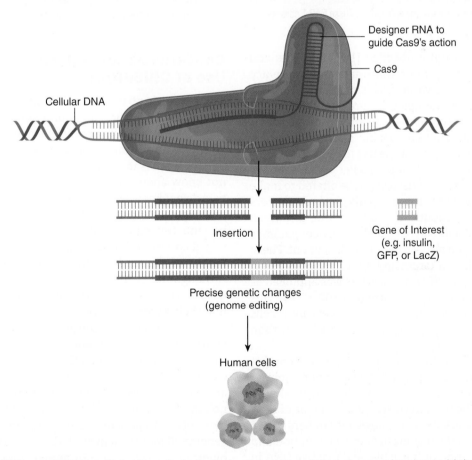

FIGURE 17.9 CRISPR/Cas cuts DNA at specific locations, allowing for precise genome manipulation such as gene insertions or deletions.

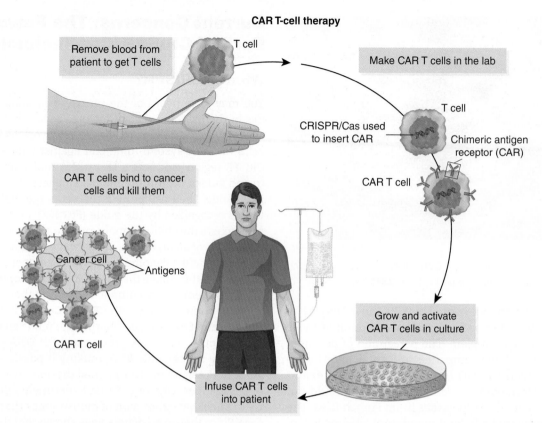

FIGURE 17.10 CRISPR/Cas–activated CAR T cells. The CAR T-cell therapy was the first CRISPR/Cas-mediated gene therapy to be approved by the FDA. In this treatment, the patient's T cells are removed from a blood sample and CRISPR/Cas is used to add the gene for a chimeric antigen receptor (CAR) that is a fusion of a T cell receptor and an antibody that recognizes cancer cells. The use of an antibody segment as an antigen receptor allows the lab to grow a large number of these cells in a short time, without antigen presenting cells. These CAR T cells are given back to the patient, where they attack cancer cells.

genes in MRSA, edit out embedded HIV, as well as add **wildtype** (functional) genetic information in the place of a dysfunctional gene variant. This means CRISPR/Cas could be used in gene therapies designed to treat genetic disorders such as cystic fibrosis or muscular dystrophy. Researchers can also make specific alterations in the genes encoding the enzymes algae use to produce oil. This has the potential to make biofuels easier and cheaper to produce. Plants and animal tissues can be altered to make them a model for genetic diseases. This gives researchers better, more reliable models to study genetic disorders. Genes to activate T cells can be added to a cancer patient's white blood cells, then infused back into the patient. These super-active T cells can target and kill tumor cells better than unaltered cells. In 2017, CRISPR/Cas was approved as a human gene therapy to treat cancer (**FIGURE 17.10**). This therapy improves the ability of T cells to target cancer cells and keeps them actively killing their targets longer than regular T cells would do.

Once the details of the CRISPR/Cas system were discovered, researchers began altering the Cas9 protein to do more than simply cut DNA at a specific target. Altered Cas9 proteins can be used to label specific regions of DNA with **fluorescent markers**. This allows researchers to see proteins bind to specific regions to start transcription and to map out the location of the active and inactive DNA in the nucleus of eukaryotic cells in real time.

Concerns About Widespread Use of CRISPR/Cas

While there is no doubt that CRISPR/Cas is an immensely powerful tool in the hands of a skilled geneticist, some are concerned that its use may be a bit *too* easy, outpacing both safety and ethical concerns. As a relatively new technology, there is much that the scientific community does not know about how CRISPR works. For example, scientists do not know whether there are undetected changes to nontarget DNA sequences that could lead to problems in patients treated with CRISPR/Cas–edited cells. Some have raised concerns that we could have designer babies born within a few decades, with key traits altered on demand. Although this seems far-fetched, in 2018, CRISPR/Cas was used to edit the fur color of mouse embryos, with white mice being born with two true-breeding gray parents.

While the previous concerns focus on the safety of individuals undergoing therapy or the ethics of a society that can customize embryos, a third concern centers on biosecurity, or preventing intentional harm. Failures of **biosafety** and **biosecurity** demonstrate the potential perils of biotechnology. Harm may be caused accidentally due to failure of biosafety, with patients developing cancer from nontarget genome alterations. It may also be caused intentionally, with **bioterrorists** using CRISPR/Cas and other genetic engineering techniques to alter the genomes of known pathogens,

TABLE 17.3 The Use of Genetic Engineering to Increase the Impact of a Viral Disease

Viral Trait	Examples
Transmissibility	Avian Influenza viruses (H5N1 and H7N9) do not transfer easily from human to human. They can, however, be mutated to bind to human cells and be transmitted from one ferret to another through the air[1].*
Viral Replication in Cells	The viral genome can be tweaked to increase or decrease the amount of an RNA virus (echovirus7) from infected cells[2].*
Host immunity to a repeat infections	Adding an immune gene to mousepox enabled it to infect previously vaccinated mice[3].*
Detection by Physicians	Changing the viral genome corresponding to the primers used to detect the virus would make the virus undetectable by PCR, forcing clinical labs to do expensive and time-consuming culture-based test for diseases like adenoviruses or HIV.
Stability outside a host	Designing viruses to withstand drying out, temperature extremes, and ultraviolet radiation.

*These experiments have been done and published.

1. De Vries, R. P., Peng, W., Grant, O. C., Thompson, A. J., Zhu, X., Bouwman, K. M., ... & Yu, W. (2017). Three mutations switch H7N9 influenza to human-type receptor specificity. *PLoS pathogens, 13*(6), e1006390.
2. Atkinson, N. J., Witteveldt, J., Evans, D. J., & Simmonds, P. (2014). The influence of CpG and UpA dinucleotide frequencies on RNA virus replication and characterization of the innate cellular pathways underlying virus attenuation and enhanced replication. *Nucleic acids research, 42*(7), 4527–4545.
3. Jackson, R. J., Ramsay, A. J., Christensen, C. D., Beaton, S., Hall, D. F., & Ramshaw, I. A. (2001). Expression of mouse interleukin-4 by a recombinant ectromelia virus suppresses cytolytic lymphocyte responses and overcomes genetic resistance to mousepox. *Journal of virology, 75*(3), 1205–1210.

Information from National Academies of Sciences, Engineering, and Medicine. (2018). Biodefense in the Age of Synthetic Biology.

making them more dangerous in a variety of ways (**TABLE 17.3**). For example, increasing the speed at which viruses replicate or lengthening the amount of time they can stay active in the environment would potentially make these pathogens spread faster and over a wider geographic area. The longer a pathogen is active and people are exposed to it, the greater the chance it will be transmitted to a new host. There is also concern that small alterations could make pathogens harder to detect, which increases the time people could be spreading a given disease before being treated.

Bioethics Catching Up with Technology

Regulating Genetic Engineering: The Asilomar Conference

In the early 1970s, advances were being made in technologies that allowed scientists to sequence and manipulate DNA. They were able to transplant DNA from one species and express it in another for the first time. The DNA of these organisms was called **recombinant DNA** because it was a new combination of genetic information from two different species. At the time very little was known about how recombinant organisms would behave in the lab, and nothing was known about how they would interact with nonrecombinant organisms in the environment.

In response to concerns about the safety of recombinant technology and the threat of restrictions being placed on this type of research, more than a hundred prominent scientists, including Nobel Laurates Paul Berg and David Baltimore; physicians; and lawyers met at the Asilomar conference center in 1975. This historic meeting, called the **Asilomar Conference on Recombinant DNA** created the core guidelines regulating DNA technology. The group recommended that safe experimentation with recombinant DNA needed to include containment protocols, and that the level of containment needed to be increased based on the potential for harm. In short, working with riskier organisms meant tighter containment protocols. The original Asilomar recommendations included a ban on manipulating toxin genes or working with highly pathogenic organisms. This was in part due to inadequate containment and disinfection technology available in the 1970s.

In 1986, the U.S. government established the **Coordinated Framework for the Regulation of Biotechnology**, which is under the purview of the FDA, the USDA, and the EPA. The three agencies have overlapping jurisdiction, which can at times be confusing. For example, most genetically modified animals are regulated by the FDA's Center for Veterinary Medicine (CVM). They may also be regulated by the USDA if these modified animals will be used as food for humans or other animals. If, however, the genetically modified animals are used to control populations of insect pests, it is considered a pesticide, and regulated by the EPA.

Need for Regulation with CRISPR/Cas Genome Editing

In October 2017, self-proclaimed **biohacker** (someone who manipulates DNA outside of laboratory channels and regulations) Aaron Traywick, the CEO of a biotechnology company, went on stage, dropped his pants, and injected himself with an untested CRISPR/Cas–based therapy for herpes. A few weeks earlier an employee at the same company publicly injected himself with an experimental HIV treatment. The FDA promptly condemned these demonstrations as stunts and reminded the public that sale of "do it yourself" gene-editing kits is illegal and that none of these products have been tested for safety.

What makes CRISPR/Cas revolutionary is its ease of use. Experiments that were once done by graduate students are now available as kits to be used by middle school children during summer camp. It is this same ease of use that has caused a rise in "biohacker"-type experimentation, which is for the most part unregulated, as the individuals are often doing this work in spaces that have not been approved for laboratory work. While the FDA has a long history of protecting the population from careless or callous distribution of harmful substances (**BOX 17.3**),

BOX 17.3 Sulfanilamide and the Birth of the Modern FDA

In the early 1930s, German bacteriologist Gerhard Domagk worked with a team of organic chemists to develop and screen hundreds of different formulations of chemical dyes for their ability to limit the growth of bacteria. One such formula was a red dye called prontosil rubrum. This dye could kill *Streptococcus* spp. and cure infections in mice. According to some, Domagk used an early version of prontosil to treat his 6-year-old daughter's infected arm, sparing her from an amputation and quite possibly saving her life.

Soon after prontosil's discovery, French scientists determined that it wasn't the dye portion of the molecule that had activity, it was the sulfanilamide portion. This opened doors for other companies to produce and sell sulfanilamide, because the chemical had first been synthesized in 1908, and thus was no longer patentable. Because this was the first known antibacterial compound, and infectious diseases were often deadly, many different drug companies made their own formulations and sold them to the public.

The S. E. Massengill Company realized that all of the sulfanilamide sold in the United States was either in the form of tablets or powder, not as a liquid, because there was no liquid formulation for the antibiotic. *Streptococcus pyogenes* throat infections were common in children, who could not easily swallow pills. The chief chemist at Massengill knew that sulfanilamide could not be dissolved in water, so he made a solution of 10% sulfanilamide, 16% water, raspberry extract, and 72% diethylene glycol. The diethylene glycol was used because it tasted sweet. This liquid form of sulfanilamide was intended for children's use. The Massengill labs tested small amounts of the newly formed "Elixir Sulfanilamide" and reported that its taste and appearance were acceptable.

Hundreds of gallons of Elixir Sulfanilamide were produced. The only ingredients listed on the label were sulfanilamide, flavorings, and inert or inactive ingredients. More than 1,300 bottles of the drug were shipped across the country in September of 1937 (**Figure 1**). By the middle of October that same year, a number of patients were seeing their doctor for kidney failure after taking Elixir Sulfanilamide. Several of these patients died. Once these deaths were reported to the FDA, the agency launched the first large-scale investigation of a drug in U.S. history.

The FDA originally thought that the Elixir Sulfanilamide had been contaminated with lead or arsenic during its manufacture.

(a)

Courtesy of FDA Histroy Office.

(b)

Courtesy of FDA Histroy Office.

Figure 1 More than 700 bottles of Elixir Sulfanilamide were distributed to unsuspecting patients. The FDA had to locate and seize them to prevent further poisonings.

The FDA tested several bottles and found no such contaminant. At this point, sulfanilamide had been used for years in powder or tablet form, without reports of kidney damage in patients using it.

It was unlikely the antibiotic itself was causing the problem. The FDA investigative team then tested the diethylene glycol that was revealed to be one of the inert ingredients. Toxicity tests showed that animals given diethylene glycol died after a few days. The cause of death? Rapid onset kidney failure, the same as doctors were seeing in patients that had been given Elixir Sulfanilamide. Little was known about diethylene glycol at the time; however, it is still used today—as brake fluid and antifreeze.

At the time, the FDA had very little authority over drug distribution and safety. The only reason it was able to recall thousands of bottles of the deadly drug was a technicality. In 1937, safety testing was not required prior to releasing a drug. There was, however, a regulation that required drugs marketed as "elixirs" to contain alcohol. Elixir Sulfanilamide contained diethylene glycol, not alcohol, which gave the FDA the power to seize all bottles of this poisonous medicine that killed 105 people in 4 weeks.

Within a few months of this disaster, the U.S. Congress passed the Food, Drug, and Cosmetics Act of 1938. This law required that all drugs be tested for safety, and that FDA approval needed to be given before they could be sold to the public. It is this law that gives the FDA the authority to ban the sale of the "treatments" that biohackers have been publicly using for self-experimentation.

it is poorly equipped to establish policy recommendations concerning CRISPR/Cas technology.

Regulating Genetic Engineering: The Napa Conference

In early 2015, approximately 40 years after the historic Asilomar Conference, researchers met in Napa, California, to discuss the use and potential for misuse of CRISPR/Cas technologies. This group included two of the most influential researchers at the 1974 Asilomar Conference, David Baltimore and Paul Berg. It also included physicians, lawyers, and Jennifer Doudna, one of the codiscoverers of CRISPR/Cas as a genome editing tool. Shortly after this **2015 Napa Conference**, the group published guidelines for the use of this new technology. These recommendations include the following:

- Discourage **germline modifications** in humans. These are modifications that affect a treated individual and can be passed down to any sons or daughters that the individual may have. In the United States, government funds cannot be used to perform germline modifications in humans.

- Encourage discussions between scientists and bioethics professionals. Both groups would benefit by discussing the potential advantages of genome editing as well as its effects on society as a whole.

- Encourage further research to determine the specificity of CRISPR/Cas–based genetic alterations. This is a new technology and more research needs to be done.

- Form an international group of scientists, policy makers, ethicists, lawyers, and the public to discuss how, when, and where these technologies are appropriate and to recommend policies to be followed at the international level.

Summary

Biological warfare has been deployed since ancient times as armies realized the devastating effects of causing disease among the opposition. Acts of biological terrorism emerged in the 20th century and pose a threat more difficult to detect than outright warfare. The horrors of biological warfare sparked the signing of the 1925 Geneva Protocol, followed soon afterward by other treaties to prohibit their use.

Advances in technology have revolutionized how we study microbes, grow plants and animals, and even how we communicate. Genetic engineering techniques have given us the ability to manipulate the genomes of microbes, plants, and animals that have great potential to help feed the world by reducing insect and weed populations on farm fields. These same genetic engineering techniques, however, can be used to make deadly pathogens, such as influenza, easier to spread and make existing pathogens deadlier. Advances in wireless communication enable doctors to adjust pacemaker settings and diabetics to more efficiently regulate their insulin dosing. This wireless technology may also open the door for a new form of bioterrorism. Malicious signals can be sent to wireless devices, with potentially fatal consequences.

In 2012, the world of genetic engineering changed dramatically with the discovery of CRISPR/Cas as a genome-editing tool. This system revolutionized how researchers manipulate genes, making it much easier and cheaper to make specific alterations. As a new technology, there has been much discussion on what it can do and how this genome-editing process should be regulated. As in the early days of genetic engineering, researchers in 2015 gathered to discuss and recommend how CRISPR/Cas and similar genome editing techniques should move forward. The main concern with all of these technologies is both biosafety, protecting against accidental catastrophe, and biosecurity, protecting the country from intentional, malicious acts. Researchers at both the 1975 Asilomar conference and the 2015 Napa conference paved the way forward for biotechnology, enabling researchers to explore and expand the promise held in these technologies while taking steps to avoid perilous missteps in both biosafety and biosecurity.

KEY TERMS

anthrax
Asilomar Conference on Recombinant DNA
biohacker
biological terrorism
biological warfare
biological weapons
Biological Weapons Convention of 1972
biosafety
biosecurity
bioterrorists
Black Bane
Bt toxin
Cas9 proteins
Category A agents
Category B agents
Category C agents
Centers for Disease Control and Prevention (CDC)
Coordinated Framework for the Regulation of Biotechnology
CRISPR
CRISPR/Cas
cutaneous (skin) anthrax
endospore
Environmental Protection Agency (EPA)
epidemic
fluorescent markers
Food and Drug Administration (FDA)
gastrointestinal anthrax
1925 Geneva Protocol
genetically modified organisms (GMOs)

genome editing
germ theory of disease
germline modifications
glyphosate
H5N1
H7N9
herbicide tolerant (HT)
herd immunity
infectious diseases
influenza
inhalation anthrax
knockout
morbidity
mortality
2015 Napa Conference
operator
pandemic
prevalence
promoter
recombinant DNA
red fluorescent protein (RFP)
smallpox
stacked traits
TPOXX
transgenic
tTAV (tetracycline-controlled transactivator)
U.S. Department of Agriculture (USDA)
virulent
wildtype

SELF-EVALUATION

○ PART I: Choose the single best answer.

1. The expression of a gene from one species in another species is called _____.
   a. germline modifications
   b. biological terrorism
   c. biosafety violations
   d. recombinant DNA technology

2. Failures in _____ will often result in the accidental exposure of people to pathogens.
   a. biosecurity
   b. biosafety
   c. 1925 Geneva Protocol
   d. biohacking

3. Researchers and regulatory agencies are concerned with _____ mutation work being done on deadly pathogens because they are concerned that these altered pathogens could fall into the wrong hands.
   a. germline
   b. knockout
   c. gain of function
   d. morbidity attenuated

4. Which disease has been eliminated from the human population, but still exists as frozen samples and may be used as a bioweapon?

   a. Rabies

   b. Anthrax

   c. Bubonic plague

   d. Smallpox

5. Which form of anthrax poses the greatest threat?

   a. Cutaneous anthrax

   b. Inhalation anthrax

   c. Gastrointestinal anthrax

   d. Waterborne anthrax

6. Which protein cuts DNA at a specific region, based on an RNA guide?

   a. Insulin

   b. CRISPR

   c. ribosomes

   d. tAV

7. Which army facility has been involved in biological warfare defense?

   a. CDC

   b. Plum Island

   c. Fort Detrick

   d. Walter Reed Naval Hospital

8. Which virus was recently altered to make it more dangerous?

   a. Smallpox

   b. Yellow fever

   c. H5N1 influenza A

   d. Rabies

9. Which of the following is a Category A biological agent?

   a. *Brucella* spp.

   b. Variola virus

   c. *Mycobacterium tuberculosis*

   d. Yellow fever virus

10. Which agency is not part of the coordinated framework for the regulation of biotechnology?

    a. FDA

    b. NIH

    c. CDC

    d. USDA

⊙ PART II: Fill in the blank.

1. The _____ of a pathogenic organism is its ability to cause harm in the form of disease symptoms, not necessarily death.

2. A new form of biological weaponry, called _____, is associated with small groups of people, not governments, intentionally releasing pathogens with the intent of making people sick and causing civil unrest.

3. An epidemic of _____ occurred in Sverdlovsk, Soviet Union, in 1979.

4. A relatively long _____ for many infectious agents plays a factor in their tactical usefulness, as people may spread a pathogen for a long time before becoming symptomatic with it.

5. Biological agents and the diseases they cause have been categorized into three groups based on their risk to the _____.

6. _____ is a Category A bacterium that produces a toxin that causes double vision, blurred vision, slurred speech, difficulty swallowing, dry mouth, muscle weakness, and paralysis of breathing muscles, leading to death.

7. A pathogen selected as a bioterrorist agent would be one that is not routinely seen in the population. This would mean that the population would be particularly susceptible due to a lack of _____.

8. A bacterial toxin, called _____, which kills insects that ingest it has been added to crop plants to protect them from insect predation.

9. Failures in _____ may result in a population being targeted by pathogens spread by terrorists or foreign governments.

10. The 2015 Napa Conference recommended a ban on _____, or changes to an organism's genome that can be passed on to offspring.

● PART III: Answer the following.

1. Cite three landmark treaties associated with biological warfare. Describe each, and give an approximate date for each.

2. Briefly describe the three types of anthrax and how they are acquired. Why is anthrax on the list of biological weapons?

3. How do you feel about the Pentagon's decision to make the anthrax vaccine mandatory for those in the military? Defend your position.

4. Anthrax is a top choice as a biological weapon. Name two properties that make anthrax an ideal weapon.

5. How is the U.S. government preparing for acts of bioterrorism?

6. List at least three advantages and three disadvantages of biological weapons.

7. What is the controversy surrounding genetically modified crop plants, such as Bt corn? State your opinion on the use of genetically modified crops.

8. How does the Coordinated Framework for the Regulation of Biotechnology regulate how genetic engineering is done in the United States?

9. Diagram the process of CRISPR/Cas action on DNA. Why is this system revolutionary?

10. Compare and contrast the 1975 Asilomar Conference and the 2015 Napa Conference in terms of each meeting's purpose and conclusions.

PARTNERSHIPS IN THE CONTROL OF INFECTIOUS DISEASES: THE FUTURE OF PUBLIC HEALTH

OUTLINE

"*I am my brother, and my brother is me.*"

—Ralph Waldo Emerson

American poet, essayist, and philosopher

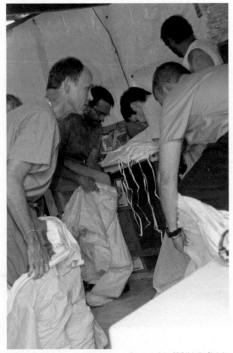

Courtesy of the CDC/Athalia Christie.

Centers for Disease Control and Prevention (CDC) staffer Dr. Jordan Tappero; Dr. Armand Sprecher of *Médecins Sans Frontières* (MSF), also known as Doctors Without Borders; former CDC director Dr. Tom Frieden, and CDC director of Global Health Dr. Joel Montgomery donned **personal protective equipment (PPE)** before entering the Ebola treatment unit (ETU) known as ELWA 3, which was opened and operated by MSF on August 17, 2014, in Monrovia, Liberia's capital city.

1. Explain how partnerships achieve public health goals and advance medical science.

2. Define *Disease X*.

3. Summarize the importance of real-time surveillance of infectious disease outbreaks.

4. Explain the association between emerging pathogens and the species barrier.

5. Describe the effects of poverty on children's health.

6. Explain the importance of research in microbiology and public health.

7. Define the CDC's One-Health Initiative.

8. List at least five organizations and define their mission to improve all health.

Case Study: The Threat of a Pandemic Keeps Virus Hunters Awake at Night

"People are beginning to understand there is nothing in the world so remote that it can't impact you as a person. It's not just diseases. Economists are now beginning to say if we are going to have good markets in Africa, we're going to have to have healthy people in Africa."

—Dr. William Foege (1936–), American epidemiologist who devised the strategy that led to the global eradication of smallpox in the 1970s

Sami Sayer and Teri Gross listened intently as Professor Cronn lectured about **emerging viruses**. Sami raised the question of how many different viruses could cause a pandemic today. Dr. Cronn smiled, his eaglelike eyes quickly scanning the room to determine if students were paying attention as he began to address Sami's thought-provoking question. He explained that the **Global Virome Project (GVP)** had been launched in 2018 in an attempt to answer Sami's question. Scientists planning the GVP's 10-year plan of partnerships are concerned about the threat of a future **pandemic** that could potentially be caused by an emerging virus.

Prior to the initiation of GVP, the first phase of PREDICT, a project of the U.S. Agency for International Development's (USAID) Emerging Pandemic Threats program led by the UC Davis One Health Institute (part of the UC Davis School of Veterinary Medicine), was conducted from 2009–2014. One of the goals of the program was to compile the most comprehensive database on the risk of unknown viruses in wildlife (mammals) with the potential to **spillover** or cross the **species barrier** from animals into human populations. The species barrier is a simplistic concept in which there is a natural mechanism that prevents a pathogen from crossing from one species to another. PREDICT was focused on strengthening the infrastructure in over 30 countries in areas of the world at highest risk for zoonotic disease emergence (**Figure 1**). The program trained more than 2,500 people in biosafety, laboratory diagnostics, field epidemiology, disease surveillance, data analysis, and modeling to produce disease hotspot maps.

Professor Cronn explained that nearly all recent pandemics or threatening epidemics occurred because a virus crossed the species barrier from animals or birds into humans. Crossing the species barrier has been recognized for many **zoonotic diseases** (**Figure 2**). For example, bats are the natural **reservoir** (the host where a pathogen naturally resides and replicates) for severe acute respiratory syndrome-coronavirus (SARS-CoV) and Ebola viruses; it is dromedary camels for Middle East respiratory syndrome-coronavirus (MERS-CoV); deer mice for hantavirus pulmonary syndrome (HPS); and pigs or ducks for influenza.

Teri became impatient and asked why GVP is a priority given that humans have repeatedly survived pandemics and epidemics. Dr. Cronn continued that with all past and current epidemics or pandemics, healthcare workers and **infectious disease control experts**, including **epidemiologists**, had reduced the impact of outbreaks using a *reactive or defensive approach* but were ill-prepared to handle challenges at the get-go (e.g., the Ebola outbreak in Guinea, West Africa). The goal of the GVP is to move forward with a *proactive approach* and to prepare before a pandemic occurs. The GVP utilizes many **partnerships** to identify every virus in the world

(continues)

Case Study: The Threat of a Pandemic Keeps Virus Hunters Awake at Night (continued)

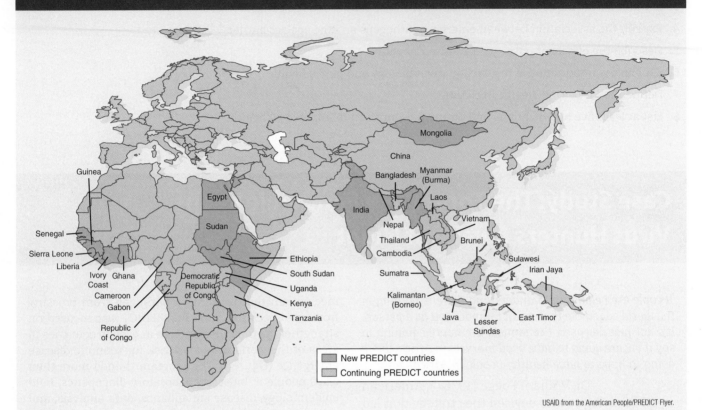

USAID from the American People/PREDICT Flyer.

Figure 1 New and continuing PREDICT countries.

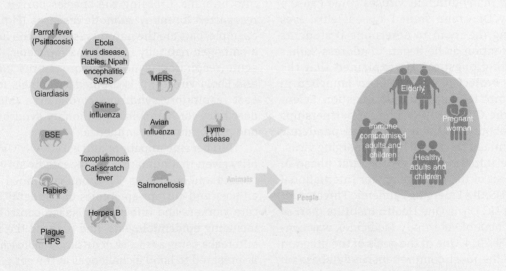

Courtesy of the CDC.

Figure 2 Zoonotic diseases spread between animals or birds (parrots, bats, beavers, pigs, dromedary camels, ducks, cats, cows, mice, macaques, skunks, turtles, deer) and people with diverse immunities (e.g., healthy vs. immune compromised). Abbreviations: HPS: hantavirus pulmonary syndrome; MERS: Middle East respiratory syndrome; SARS: severe acute respiratory syndrome; BSE: bovine spongiform encephalopathy.

that has the potential to jump from animals to humans. *It differs from PREDICT in that it is a large-scale biodiversity approach aimed to predict future viral outbreaks.* The GVP is set up to:

○ Operate in more than 25 developing countries representing **emerging disease hotspots**.

○ Place field teams in each country.

○ Capture and sample over 80,000 wildlife (especially rodents and bats, which are known to harbor emerging zoonotic viruses) and domestic animals.

○ Develop and apply diagnostics and discovery testing.

Sami chimed in to ask where the emerging disease hotspots are. Dr. Cronn continued. Viral diversity occurs in those areas where the most mammalian and bird diversity exists. Researchers have used geographical range and habitat suitability models to estimate the overlapping locations with the highest mammalian species diversity (**Figure 3**). Migrating waterfowl diversity hotspots (which have harbored pandemic influenza viruses) are based on data collected by two bird conservation and management groups, BirdLife International (http://www.birdlife.org/) and NatureServe (http://www.natureserve.org/) and are shown in **Figure 4**. Therefore, *emerging disease hotspots are places where humans are at critically close contact to animals or birds carrying zoonotic viruses.*

Going back to Sami's original question, Professor Cronn stated that based on current databases, researchers estimate that 263 viruses infect humans. Of these, 188, or 71%, are zoonotic (viruses detected in humans and at least one other mammalian species). They also predict that there are about 1.67 million unknown viruses to be discovered in mammals and birds. *Of the 1.67 million, researchers estimate that about 631,000 to 827,000 unknown viruses have the potential to cross the species barrier into humans.* Note that this is not 71%. This is because other variables were considered in the estimation, such as what is known about previous patterns of emerging viruses and the history of zoonoses.

Teri asked if there were other approaches being considered, given that the GVP was estimated to cost at least $1.2 billion, with funding from sources all over the world. Professor Cronn continued by stating that not all experts agree that the GVP is the best approach. He reiterated that GVP advocates believe that along with computer systems and **machine learning** (allowing computers to automatically use algorithms to analyze the GVP data, learn from it without any human intervention or supervision, and adjust the algorithms to new data) the task to accurately predict disease outbreaks will become similar to predicting the weather.

Some experts caution that the public's support and trust will erode if GVP supporters cannot hold true to the promise of preventing disease outbreaks. If predictions

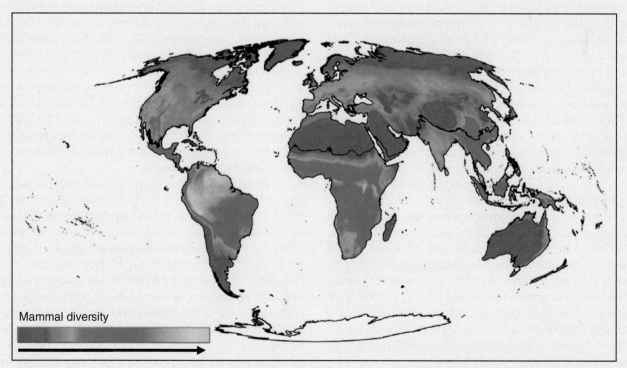

Mammal diversity

Information from C. Rondinini, M. Di Marco, F. Chiozza, G. Santulli, D. Baisero, P. Visconti, . . . L. Boitani. (2011). Global habitat suitability models of terrestrial mammals. *Philosophical Transactions of the Royal Society B, 366,* 2633–2641.

Figure 3 Emerging hotspots for viral diversity are correlated with the geographic ranges of mammalian species diversity.

(continues)

Case Study: The Threat of a Pandemic Keeps Virus Hunters Awake at Night (continued)

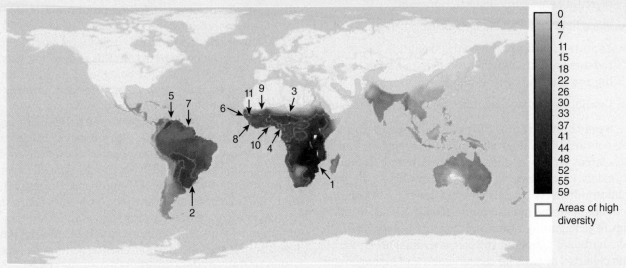

Information from L. Williamson, M. Hudson, M. O'Connell, N. Davidson, R. Young, T. Amano, & T. Szekley. (2013). Areas of high diversity for the world's inland breeding waterbirds. *Biodiversity and Conservation, 22*(6–7), 1501–1512).

Figure 4 Emerging hotspots for viral diversity are correlated with the geographic ranges of waterfowl species diversity. The world map shows the top 10 geographical locations that contain the most species diversity of inland breeding wild waterfowl.

are not accurate, stockpiling new vaccines and antiviral drugs developed based on predictions would be unnecessary, further crippling trust. Experts also argue that not enough data are available for intelligent systems. They cannot predict if a virus will cross the species barrier based on the sequence of its **genome**. Not enough data are available that will indicate how a virus will interact with host cells and how the human body will respond to the infection. There is not enough data to understand how, if, and when a virus will spillover into the human population. Therefore, non-GVP supporters argue that it is impossible to accurately make predictions.

Those experts who do not support the GVP approach prefer simpler and more cost-effective efforts such as real-time, proactive **syndrome/symptom surveillance** of *human* populations in the countries and specific locations where repeated outbreaks of severe flulike illnesses, atypical fevers, "abortion or congenital birth defect storms," or other mysterious illnesses that cannot be easily diagnosed occur. Monitoring can be done by local physicians; Doctors Without Borders; Human, Heredity and Health in Africa (H3Africa; https://h3africa .org); and global organizations such as the World Health Organization (WHO).

Blood and other clinical samples collected from patients with mysterious illnesses will undergo **metagenomics** and **serology** analysis. Portable sequencing

devices are being developed for use in remote jungles, at sea, and even on Mars. Large-scale screening of blood collected from human populations can be used to track current and past outbreaks. The **real-time surveillance** data must be accessible on servers (e.g., http://virological .org) so that making the connection between genomic and epidemiological data occurs within days of an outbreak, improving the response time. As soon as an emerging virus is identified, the virus must be studied to determine its mechanism of crossing the species barrier, how it spreads through the human population, and how it affects those infected.

Professor Cronn digressed to remind students that the virus hunters in this case study are chasing "**Disease X,**" which is listed on the WHO's **Blueprint Priority Diseases** list. In February 2018, during their second annual review, WHO experts ranked diseases that should be prioritized for research and development of vaccines and antivirals. All of the diseases on the list are *caused by viruses and have the potential to cause a major public health crisis* (**Table 1**). As Dr. Cronn finished the lecture, he assigned students to compose a 500-word essay by the next lecture period that discusses which approach they preferred to prepare for a pandemic and why. Or, they had the option to come up with a new approach and outline how it would be implemented. He ended by saying, "*new viruses will emerge unexpectedly*."

TABLE 1 WHO Blueprint Priority Disease List, 2018

Disease	Mode of Transmission
Crimean-Congo hemorrhagic fever	Transmission through tick bites or contact with infected livestock or humans
Ebola virus disease and Marburg virus diseases	Transmission through contact with contaminated body fluids and sexual intercourse with an infected individual
Lassa fever	Transmission to humans through contact with food or household items contaminated with urine or feces from infected rodents
Middle East respiratory syndrome and severe acute respiratory syndrome	Transmission through contact with infected animals; does not pass easily person-to-person
Nipah and henipaviral encephalitis	Transmission through direct contact with infected bats, pigs, or humans
Rift Valley fever	Transmission through contact with blood, body fluids, or tissues from infected animals, mainly livestock; less commonly through mosquito bites
Zika virus disease	Transmission through mosquito bites, transplacental transmission from pregnant mother to fetus, and sexual transmission
Disease X	**Unknown**

Information from World Health Organization. (2018, February). Second Annual Review: List of Blueprint Priority Diseases. Retrieved from http://www.who.int/blueprint/priority-diseases/en/

Questions and Activities

1. Former CDC director Dr. Richard Besser once said, *"You can't put a price tag on preparation for a pandemic."* State three reasons why you agree or disagree with Dr. Besser.

2. Approximately 263 viruses are known to cause disease in humans, but only a small subset of viruses have caused major epidemics in the 20th and 21st centuries. List three reasons why you believe so few viruses cause large outbreaks.

3. Will the results of the GVP make it possible to predict viral outbreaks similar to predicting weather patterns? State three reasons to support your answer.

4. Briefly summarize the two different approaches to preparing for the threat of a pandemic discussed in this case study.

5. Define *species barrier*. List three viruses that have undergone spillover into the human population. For each virus, list what species it originated from.

6. Define *emerging disease hotspot*, and list at least five countries located in an emerging disease hotspot.

7. Are the emerging disease hotspots of mammals and waterfowl in the wild exactly the same? Using examples, explain your answer.

8. Table 1 lists the WHO Blueprint Priority Diseases. Perform Internet research through the CDC or WHO. Choose three diseases and create a world map that highlights where those diseases affect human populations.

9. Currently, no PREDICT countries are located in South America, yet the world maps in Figures 3 and 4 show that both mammalian and waterfowl species diversity occur there. Why might South America not contain any emerging disease hotspots?

10. Viruses can evolve more rapidly in an immune-compromised host. Is it possible for an HIV/AIDS patient or an organ recipient (taking immune-suppressive drugs) to be the reservoir for a viral pandemic? Explain.

11. List five wild animal or bird reservoirs and the zoonotic pathogens they are known to harbor.

12. List three domesticated animals or birds and the zoonotic pathogens they harbor.

13. Based on expert estimates, how many unknown zoonotic viruses with the potential to spillover into

(continues)

Case Study: The Threat of a Pandemic Keeps Virus Hunters Awake at Night (continued)

humans have yet to be discovered? Explain why this number is so high.

14. Summarize what types of partnerships, technologies, and other resources are needed to combat a pandemic in progress.

15. What human populations are most vulnerable to a disease outbreak? Explain.

16. Define what component(s) present in human blood samples can help epidemiologists track a viral pathogen's movement through humans around the world.

17. In 1918, the Spanish influenza pandemic killed 50 to 100 million people globally. If the same 1918 influenza A virus strain reemerged today, would it kill as many people? Why or why not? How fast could

the 1918 influenza A virus strain spread around the world? Explain.

18. Are all pandemics created equal? In other words, do all pandemics result in deaths? Cause disease complications? Affect all human populations? Caused by pathogens that cross the species barrier? Cite examples in your answer.

19. Define *vaccine* and *antiviral*. If you were given the resources to manufacture a vaccine or an antiviral for combatting a viral disease, which would you choose to produce and why?

Sources: Information from D. Carroll, P. Daszak, N. D. Wolfe, G. F. Gao, C. M. Morel, S. Morzaria, . . . J. A. K. Mazet. (2018). The Global Virome Project. *Science, 359*, 872–874. Retrieved from http://www.globalviromeproject.org; E. C. Holmes, A. Rambaut, K. G. Andersen. (2018). Pandemics: Spend on surveillance, not prediction. *Nature, 558*, 180–182.

Preview

Enormous strides made in the 20th century decreased the burden of infectious diseases worldwide, particularly in developed countries. These successes were the result of collaborative partnerships in pooling funds and targeting talents and resources toward common goals. Continued progress depends on partnerships that range from local to national and international levels and requires cooperation between the public and private sectors. The disparity in health care between more developed and less-developed countries requires the implementation of strategies to ensure that the poorer nations of the world receive their fair share of the benefits of progress.

Surgeon General William H. Stewart stated in 1967, with good reason, that it was "*time to close the book on infectious diseases.*" Sanitation, vaccines, and antibiotics had made enormous strides against pathogenic microbes, and there was no reason to suggest that this wave of optimism would not continue.

But the forces of natural selection and the ability of microbes to adapt, along with urbanization, complacency, technological advances, ecological disruption, and social evolution reflected in human behavior, have proved otherwise. One scientist has remarked, "*The war has been won, but, by the other side*"; this facetious statement reflects the theme of this book, namely, the challenge of emerging and reemerging infections. Outbreaks of infectious diseases are frequent subjects of current news and indicate the need for continued surveillance and avoidance of complacency.

Background

"*The attainment for all people of the world by the year 2000 of a level of health that will permit them to lead a socially and economically productive life*" was the stated goal that emerged from the International Conference on Primary Health Care held in Alma-Ata, USSR (now Almaty, Kazakhstan) that was cosponsored by the World Health Organization (WHO) from September 6–12, 1978. Further it was declared:

> *The Conference strongly reaffirms that health, which is a state of complete physical, mental, and social wellbeing, and not merely the absence of disease and infirmity, is a fundamental human right and that the attainment of the highest possible level of health is a most important worldwide social goal whose realization requires the action of many other social and economic sectors in addition to the health sector.*

Four decades have passed since the Alma-Ata conference. Where do we stand? Has there been progress toward the attainment of these ambitious goals? Life expectancy has dramatically increased since the beginning of the 20th century, and Alma-Ata and other international alliances have contributed to these gains. The World Health Assembly adopted the slogan "Health for all by the year 2000." That year has come and gone, and the goal has yet to be realized. Nevertheless, progress has been made, partially as a result of Alma-Ata and its emphasis on primary health care. Health for all is an elusive and moving target and may not be realistic, but striving toward it can only have positive consequences. The achievements of **public health** and advances in medical science have not been equally distributed across the board, presenting an ongoing challenge.

The attainment of health in all its dimensions, including reducing the burden of infectious diseases, is a matter of public health concern at several levels ranging from the individual to the community, to state departments of health, to national agencies, and to international organizations. Barry Bloom, former dean of the Harvard University School of Public Health, stated, "*One of the myths of the modern world is that health is determined largely by individual choice and is therefore a matter of individual responsibility.*" Realistically, individuals are limited in their efforts to achieve and maintain good health, necessitating public health strategies aimed at populations to reduce the burden of infectious and other diseases.

The eradication of **infectious diseases** has been a goal since the establishment of **Koch's postulates** and the **germ theory of disease**. Smallpox was the first infectious disease in humans to be eradicated. Rinderpest, also known as "cattle plague," was the second disease to be wiped from the world, as declared by the United Nations in June 2011. Thomas Jefferson, a few years after the introduction of smallpox vaccination in 1796, commented, "*One evil more [smallpox] is withdrawn from the condition of man.*" In 1892, a contagious pleuropneumonia of cattle (which was imported into the United States in 1847) was declared eradicated from the country as the result of a 5-year, $2 million campaign to identify and slaughter infected animals. The Rockefeller Foundation ambitiously campaigned to eradicate yellow fever and hookworm disease in the early years of the 20th century but failed because of the complexity of eradication programs. Malaria eradication seemed plausible in the period from 1955 to 1965 but was unsuccessful, primarily because of the emergence of drug-resistant malaria parasites and mosquitoes.

The International Task Force for Disease Eradication (ITFDE) convened for the 18th time since 1989, at the Carter Center in April 2011. The task force focused its assessment on the elimination strategies of two worm diseases in Africa—onchocerciasis and lymphatic filariasis. The ITFDE defines an *elimination strategy* as the "*reduction of infection and transmission to the extent that interventions can be stopped but post-surveillance is still necessary.*" Further, elimination strategies generally refer to a limited geographical area (a single country or continent), whereas *eradication* is used in a global sense.

The eradication of smallpox in 1980 stands as a public health triumph of the 20th century. This generation, and those that follow, will never know the horrors of this disease, other than through photographs. Visually, smallpox is characterized by a **centrifugal rash** concentrated on the extremities of the bodies of those infected (**FIGURE 18.1**). Several disease features of smallpox and its causative agent, variola virus, were unique in leading

Courtesy of Jean Roy/CDC.

FIGURE 18.1 Smallpox: A disease of the past. Smallpox is characterized by the appearance of pustules concentrated on the head and ends of the extremities body (hands and feet), referred to as a *centrifugal rash*. The use of smallpox by terrorists is a potential worldwide threat.

to its eradication and its establishment as the criterion by which to evaluate other diseases as targets for eradication:

- It is a disease only of humans; there are no natural reservoirs or biological **vectors**.

- The infection is easily diagnosed because of its characteristic rash.

- The duration and intensity of infectiousness is limited.

- Recovery establishes permanent immunity.

- A safe, effective, inexpensive, easily administered, stable (even in tropical climates), one-dose vaccine is available.

- Vaccination usually results in a permanent and recognizable scar, allowing for detection of immune versus nonimmune individuals in a population.

The degree to which other diseases mimic smallpox reflects their potential for eradication, but these are not absolute criteria (**TABLE 18.1**). For example, a biological vector is a part of the guinea worm life cycle, polio immunization requires four doses, and neither disease produces visible early manifestations. Despite these considerations, both diseases are on the "hot list" for eradication, and considerable progress has been made toward that achievement. The last case of wild polio in the Americas occurred in Peru in 1991. Not all diseases reviewed by the ITFDE are considered candidates for eradication, highlighting the

TABLE 18.1 Criteria for Assessing Eradicability of Diseases and Conditions

Scientific feasibility	Epidemiologic susceptibility (e.g., no nonhuman reservoir, ease of spread, natural cyclical decline in prevalence, naturally induced immunity, ease of diagnosis)
	Effective, practical intervention available (e.g., a vaccine, a curative treatment)
	Demonstrated feasibility of elimination (e.g., documented elimination from an island or other geographic unit)
Political will and popular support	Perceived burden of the disease (e.g., extent, deaths, or other effects; relevance to rich and poor countries)
	Expected cost of eradication o Synergy of eradication efforts with other interventions (e.g., potential for added benefits or savings) o Need for eradication rather than control

Information from Centers for Disease Control and Prevention. (1993). Recommendations of the International Task Force for Disease Eradication. *Morbidity and Mortality Weekly Report, 42*(RR-16), 1–25; D. R. Hopkins. (2013). Disease eradication *New England Journal of Medicine, 368*, 54–63.

complexity of eradication programs. Of the 56.9 million deaths worldwide in 2016, more than half (54%) were due to the top 10 causes of deaths, which included the infectious diseases listed in **TABLE 18.2**. The other top causes of worldwide deaths were noncommunicable diseases: heart disease; stroke; chronic obstructive pulmonary disease; Alzheimer's disease and other dementias; trachea, bronchus, and lung cancers; diabetes; and road injuries.

TABLE 18.2 Infectious Disease Global Causes of Death, 2016

Cause of Death	Deaths in 2016
Lower respiratory infections	3.0 million
Diarrheal diseases	1.4 million
HIV/AIDS	1.0 million
Tuberculosis	1.3 million

Data from World Health Organization. (2018, May 24). The top 10 causes of death. Retrieved from http://www.who.int/news-room/fact-sheets/detail/the-top-10-causes-of-death

Partnerships in Infectious Disease Control

Reducing the incidence of infectious diseases is a tremendous challenge. The successes to date are largely the result of collaborative partnerships involving the sharing of funds, talents, and resources in the work toward common goals. Continued successes in public health will depend on partnerships within and between the public and private sectors. Numerous agencies have been cited for their leadership; this chapter describes some of them.

Today's crowded societies and sharing of resources are far removed from hunter-gatherer societies, where family units were relatively isolated and depended only on their own efforts and ingenuity to stay alive. As populations grew and urbanization developed, a sharing of community responsibilities emerged in all aspects of life, including health. A negative aspect of all this "togetherness" was that the sharing of pathogens also increased. Individuals were limited in measures that could be taken to minimize exposure to pathogens. In a collective effort, the early 1900s saw the establishment of community and state departments of health that evolved into a complex network from the local (state) level to the national and international levels and to partnerships in the public and private sectors.

At the Local Level

o Community, City, and State Health Departments

Every state has a department of health, together with subordinate health departments at the community and city levels. Although the organizational charts vary from state to state, they are to a large extent a reflection of the size of the population covered. Health departments focus on the prevention of disease and the promotion of health and safety of the people within their jurisdiction. They watch for the hazards of foodborne and waterborne diseases, including bacterial, viral, protozoan, and helminthic (worm) diseases. A major responsibility of health departments is to establish and implement safety regulations pertaining to food and water sanitation. For example, there is rigorous control of the milk industry in each state involving the health of dairy cows and conditions "down" at the dairy farm. Every dairy must submit milk and milk products on a strict schedule to state departments of public health laboratories to ensure the safety of the product before delivery to markets.

Public health restaurant inspectors make spot visits to eating establishments to ensure adherence to proper temperatures for the cooking and storing of foods; the absence of mice, rats, roaches, and other vermin; and the practice of appropriate measures of sanitation and hygiene by food workers. Salad bars must have shields to protect the food from coughs and sneezes. Increasingly, food workers must use disposable gloves (**FIGURE 18.2**).

Some communities, in addition to imposing fines and closing noncompliant establishments, make restaurant inspection reports available to the public on a website and have implemented other methods to alert the public.

© DenisProduction.com/ShutterStock, Inc.

FIGURE 18.2 Protection of the public. Local departments of health require food handlers to wear disposable gloves.

Los Angeles requires restaurants to post inspection grades in the window, and the local newspapers in central Florida and other areas publish the detailed results of restaurant inspections and the fines imposed on those failing to achieve a clean bill of health. In some cases, the results are hard to swallow (pun intended). Violations include potentially hazardous, uncooked food held at unsafe temperatures; food handlers preparing foods with their bare hands; raw chicken stored over other raw meat in the cooler; foods like soft cheeses such as brie, mayonnaise, greens, and milk kept at improper temperatures; and roaches crawling over counters and food. Additionally, some communities require that food workers, in those restaurants failing to pass inspection, attend a hospitality education program.

It is not uncommon for foodborne infections to hit college campuses. Students suffer from the usual and unpleasant symptoms of gastroenteritis. Noroviruses, *Escherichia coli*, and *Salmonella* spp. are frequently identified as the causative pathogens. Each incident and its containment illustrate the epidemiological detective work necessary for tracking down the source and implementing preventive measures for the future. Investigation of such outbreaks requires partnerships at the college, community, state, and, in some cases, the national and international levels. State health departments are required to notify the CDC of specific diseases so that a national network of surveillance and communication can be maintained (**FIGURE 18.3**).

Surveillance and control of infectious diseases are major functions of local health departments to which they respond with control strategies, including vaccination. Vaccine campaigns and implementation of regulations requiring immunization against infectious diseases are another function of state health departments.

Local departments of health are involved in numerous other endeavors to foster the health and welfare of citizens, including toxicology, vital statistics, public awareness health programs, water quality testing, prevention and treatment of drug abuse, cancer surveillance, and lead paint screening.

Courtesy of Massachusetts Department of Public Health.

FIGURE 18.3 A reporting form that is filled out in the event of a foodborne illness and sent to a state health department.

At the National Level (United States)

○ Centers for Disease Control and Prevention (CDC)

The Centers for Disease Control and Prevention (CDC; https://www.cdc.gov) based in Atlanta, Georgia, is the nation's premier public health facility, and its impact is global (**FIGURE 18.4**). Its functions are to:

- Detect and investigate health problems.
- Conduct research to enhance prevention.
- Develop and advocate sound public health policies.
- Implement prevention strategies.
- Promote healthy behaviors.
- Foster safe and healthful environments.
- Provide leadership and training.

The agency was founded in 1946 and employs nearly 15,000 people, including approximately 840 commissioned corps officers, in 170 occupations. The CDC is a key member of partnerships with local health departments and with national and international organizations (**FIGURE 18.5**).

The CDC is the nation's main line of defense against threatening epidemics and plague. The CDC's research labs are in some ways like a large microbial zoo; within its locked freezers are every known pathogen on Earth (including variola virus) contained in small vials under the watchful eye of a microbe keeper or in the live bodies of rabbits, mice, cotton rats, and monkeys. Locked corridors

FIGURE 18.4 The CDC headquarters in Atlanta, Georgia. The CDC conducts infectious disease surveillance and works in conjunction with state and local health departments and with the WHO and other agencies.

Courtesy of the CDC/James Gathany.

house the "deadliest of the deadly" viruses, including those that cause HIV, rabies, Ebola virus disease, hantavirus pulmonary syndrome, avian influenza, Zika virus disease, Powassan encephalitis, bubonic and pneumonic plague, and Middle East respiratory syndrome (MERS).

The CDC has come to the rescue on numerous occasions around the world helping to control or investigate incidents, including the multistate *E. coli* outbreak associated with contaminated romaine lettuce in 2018, the Ebola virus disease outbreak in Guinea in 2014, the *Elizabethkingia* outbreak in Wisconsin hospitals in 2016, the MERS epidemic in the Middle East since 2012, the Zika virus disease outbreak in Brazil in 2015, and the emergence of various diseases in the aftermath of hurricanes Harvey, Irma, Maria, and Nate in the United States in 2017, and numerous norovirus outbreaks in the United States. The CDC fields over 1,000 calls for help each year. As deemed necessary, its "SWAT" teams of epidemiologists head into the field, frequently

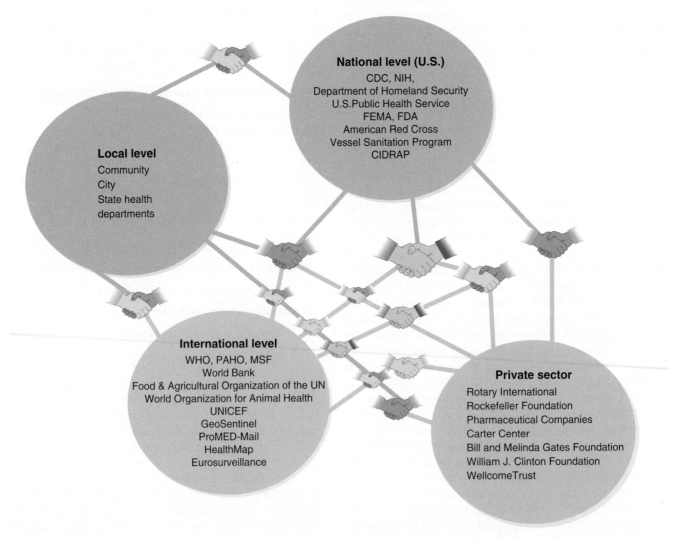

FIGURE 18.5 Partnerships to achieve health for all people have evolved into a complex network from the local (state) level to the national and international levels and to partnerships in the public and private sectors.

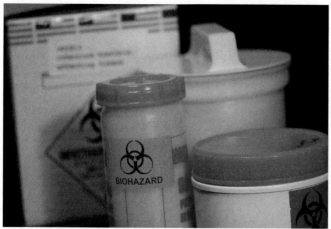

Courtesy of the CDC/Debora Cartagena.

FIGURE 18.6 The biohazard symbol. All infectious materials are placed in containers with this symbol, including the shipping containers in this photograph. Potentially infectious materials shipped for analysis from hospitals, laboratories, and other facilities must be contained in clearly marked biohazard containers.

in collaboration with partnership agencies, equipped with ready-to-go containers stocked with syringes, needles, vaccines, intravenous fluids, examination gloves, refrigerators to store samples of blood and other tissues, generators, stacks of questionnaires, and other items to conduct guerrilla warfare against the microbes.

Frequently, these "disease detectives" must ship tissue samples from sick and dead victims to the CDC labs in Atlanta for identification. The labs are designed to work with deadly microbes, and their locked doors bear large biohazard signs (**FIGURE 18.6**). The CDC's Building 15 is the "hot zone"—a biosafety level 4 (BSL-4) facility—prepared to handle the deadliest of microbes, including Ebola and hantaviruses, for which there is neither cure nor vaccine (**FIGURE 18.7**). Air leaving Building 15 is passed through a series of filters, water is boiled before entering sewer lines, and the fortress-like building has a camera trained on its one entrance. Researchers strip naked, shower, and don biohazard suits, known as orange suits, and as many as three pairs of gloves before entering the lab; air is supplied to them through a tube attached to the orange suit. Their protection against the deadliest of pathogens is limited to a layer of fabric, which can be penetrated by the jagged glass of a broken test tube or by a syringe needle. As much as possible, researchers work in pairs and monitor each other for fatigue and for tears in their gloves and suits.

In its role as the nation's watchdog, the CDC has developed a four-point strategy for the 21st century to counter the threat of new, emerging, and reemerging infections. These goals are elucidated in **TABLE 18.3**.

Public Health on the Seas: CDC Vessel Sanitation Program

Do you want to go on a cruise to a faraway destination? If so, you would be well advised to consult the inspection score referred to as the "Green Sheet," the CDC's

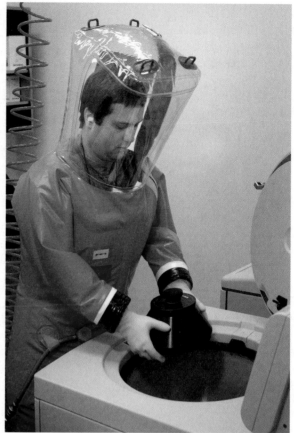

Courtesy of the CDC/James Gathany.

FIGURE 18.7 CDC investigator wearing airtight PPE while working in the biosafety level 4 (BSL-4) laboratory. The researcher is removing a centrifuge rotor that contains tubes of pelleted viruses that will be used in experiments.

Summary of Sanitation Inspections of International Cruise Ships (https://wwwn.cdc.gov/InspectionQueryTool /InspectionSearch.aspx). The Vessel Sanitation Program (VSP) was established in the early 1970s to minimize the risk of diarrheal diseases among passengers. All vessels with a foreign itinerary that carry more than 13 passengers and call on U.S. ports are subject to unannounced twice-yearly inspections by VSP staff and to reinspections when necessary (**FIGURE 18.8**). To pass, the ship must score a minimum of 86 points on a 100-point scale. Further, the general cleanliness (inspecting the medical center, potable water systems, galleys and dining rooms, swimming pools and whirlpools, housekeeping, pest and insect management, child activity centers, heating, ventilation, and air conditioning), personal hygiene and physical condition of the crew, along with training programs in environmental and public health practices, are evaluated.

The cruise ship industry has been subject to waves (pun intended) of criticism because of a few outbreaks caused by noroviruses among passengers and staff. Some of the cruises and their ships have been dubbed "ships from hell" by passengers as they experienced their dream vacations explode into nightmares. In reality though, the number of norovirus cases on cruise ships is extremely low. The reported rates of diarrhea reported among passengers and

TABLE 18.3 CDC Framework for Preventing Infectious Diseases: Sustaining the Essentials and Innovating for the Future

The CDC's roadmap for improving our ability to prevent known infectious diseases and to recognize and control rare, highly dangerous, and newly emerging threats has three elements:

Element I	Strengthen public health fundamentals, including infectious disease surveillance, laboratory detection, and epidemiological investigation.	Modernize infectious disease surveillance to drive public health action.
		Expand the role of public health and clinical laboratories in disease control and prevention.
		Advance workforce development and training to sustain and strengthen public health practice.
Element II	Identify and implement high-impact public interventions to reduce infectious diseases.	Identify and validate high-impact tools for disease reduction, including new vaccines, strategies and tools for infection control and treatment, and interventions to reduce disease transmitted by animals or insects.
		Use proven tools and interventions to reduce high-burden infectious diseases, including vaccine-preventable diseases, healthcare-associated infections, HIV/AIDS, foodborne infections, and chronic viral hepatitis.
Element III	Develop and advance policies to prevent, detect, and control infectious diseases.	Ensure the availability of sound scientific data to support the development of evidence-based and cost-effective policies.
		Advance policies to improve prevention, detection, and control of infectious diseases to help integrate clinical infectious disease practices into U.S. health care: (1) increase community and individual engagement in disease prevention efforts; (2) strengthen global capacity to detect and respond to outbreaks with potential to cross borders; (3) address microbial drug resistance; and (4) promote "One Health" approaches to prevent emergence and spread of zoonotic diseases.

Information from Centers for Disease Control and Prevention. (2011, October). *A CDC framework for preventing infectious diseases: Sustaining the essentials and innovating for the future.* Retrieved from http://www.cdc.gov/oid/docs/ID-Framework.pdf

crew members on cruise ships in the United States from 2008–2014 are shown in **TABLE 18.4**. About 73.5 million passengers sailed on voyages that required a VSP Green Sheet during 2008–2014. During that time, 0.18% of the

passengers and 0.15% of the crew members reported gastroenteritis/diarrhea. Of the patient samples tested, 92% of the cases were caused by noroviruses. Hand hygiene was vital to preventing outbreaks.

Which shots do you need before traveling to the Amazon, Mozambique, or Tahiti? These exotic foreign destinations are potential sources of disease that can make you very ill and even kill you. Before embarking, you can consult the CDC's website (https://wwwnc.cdc.gov/travel /notices/) for up-to-date traveler's health information, including vaccinations, availability of safe food and water, and disease outbreaks in the land of your dreams.

○ National Institutes of Health (NIH)

The National Institutes of Health (NIH; https://www.nih .gov) is a component of the U.S. Department of Health and Human Services (HHS). The NIH is one of the largest and most distinguished medical research centers in the world. Promising scientists from all over the world can gain experience in clinical research training and medical education. In fact, 153 Nobel Prize winners have received support from NIH.

Vintagepix ShutterStock, Inc.

FIGURE 18.8 Overview of the pool and deck area of a large, crowded cruise ship on the Caribbean Sea, January 2017.

TABLE 18.4 Diarrheal Illness on Cruise Ships, 2008–2014

Outbreak Features	2008	2009	2010	2011	2012	2013	2014
Number of voyages analyzed	4098	3964	4155	4189	4168	4146	4387
Number of outbreaks among passengers per 1,000 voyages	4.4	4.0	3.8	3.3	6.5	4.2	3.0
Number of outbreaks among crew members per 1,000 voyages	0.3	0.8	0.5	0.0	0.3	0.3	0.7

Information from A. L. Freeland, G. H. Vaughan, & S. N. Banerjee. (2016). Acute gastroenteritis on cruise ships—United States, 2008–2014. *Morbidity and Mortality Weekly Report, 65*(1), 1–5.

It has come a long way since its founding in 1887 as a one-room laboratory created within the Marine Hospital Service. Today, it comprises 75 buildings spread over a 300-acre campus in Bethesda, Maryland (**FIGURE 18.9**). Composed of 27 institutes and centers, one of which is the National Institute of Allergy and Infectious Diseases, a primary function of the NIH is to administer and support biomedical research at over 3,000 sites in the United States and abroad. Many of the research findings have direct applicability to public health measures. The NIH also houses the National Library of Medicine (NLM; https://www.nlm.nih.gov/), the world's largest medical library.

Department of Homeland Security (DHS)

The Department of Homeland Security (DHS; https://www.dhs.gov) was proposed by President George W. Bush in June 2002 in response to the September 11, 2001, terrorist attacks on the United States. The DHS replaced the earlier Office of Homeland Security and entailed a major reorganization of government agencies. The mission of the DHS, *as applicable to the use of biological weapons* (public health), is to:

○ Reduce the vulnerability of the United States to terrorism.

Courtesy of National Institutes of Health.

FIGURE 18.9 Building 1 at the NIH. This grant-awarding and research-oriented organization is located in Bethesda, Maryland.

○ Minimize the damage and assist in the recovery from terrorist attacks that do occur within the United States.

The DHS also responds to natural disasters or other large-scale emergencies. It works with federal, state, and local agencies and the private sector to ensure a rapid and effective recovery effort. Over 240,000 people, skilled and talented in a wide range of jobs, from aviation to cybersecurity, are employed by the DHS.

U.S. Public Health Service (PHS)

The U.S. Public Health Service (USPHS; https://www.usphs.gov/), headed by the U.S. Surgeon General, consists of a 6,000-member Commissioned Corps and support staff and is a component of the U.S. Department of Health and Human Services. It originated as the Marine Hospital Service in 1889; in 1912, the name was changed to the USPHS. Initially, the USPHS focused on sailors and their medical care in an attempt to alleviate the burden on public hospitals in caring for merchant seamen. In 1891, the service was charged with being the nation's medical gatekeeper by providing medical inspection of arriving immigrants to weed out "*idiots, insane persons, persons likely to become a public charge, and persons suffering from a loathsome or a dangerous contagious disease.*"

The commissioner general of immigration clearly stated in 1902 that America should not become "*the hospital of the nations of the Earth.*" Immigrants were screened by USPHS officers (referred to as line doctors) at the Ellis Island Federal Immigration Station for "germ diseases," including cholera, tuberculosis, fungal skin ailments (**favus**), typhus, plague, smallpox, yellow fever, and **trachoma** (blindness). Ellis Island was dubbed "The Island of Hope" and also "The Island of Tears"—"hope" for the new and promising way of life for those who made it through the inspection line, but "tears" for those who were separated from their families and returned to their place of origin.

The USPHS has taken a leading role, particularly during wars, in campaigns against **sexually transmitted infections (STIs)** (formerly called venereal diseases), in immunization campaigns, in vector control programs, and in public health awareness programs (**FIGURE 18.10**).

DONT VISIT HOUSES OF ILL-FAME

If in doubt call at the nearest Hospital
for free and confidential advice

SAVE YOURSELF & YOUR FAMILY FROM V. D.

(a)   Courtesy of the CDC.

A MESSAGE OF IMPORTANCE TO PARENTS...

There is no reason for your child to suffer from measles. Vaccinated once, there is every indication that he is protected for life from this dangerous disease. And measles is dangerous. Even if a child has only an ordinary case he may be very sick with a high fever, harsh cough, puffy and light-sensitive eyes, and an itchy rash.

Many cases are followed by complications such as pneumonia, deafness or blindness. Some children develop inflammation of the brain which can leave them mentally retarded. Every year hundreds of children die as a result of measles.

The United States Public Health Service recommends that every infant be vaccinated when he is about one year old. Children over this age who have not been vaccinated and who have not had measles, should also be immunized.

It is especially important to vaccinate children in nurseries, primary grades of schools, and other groups where children may be exposed. If one child in a classroom has measles, all the other children who are unprotected usually will become infected. They then carry the infection home to brothers and sisters.

So please, take your child to your family physician or your local health department and get him vaccinated.

William H. Stewart
SURGEON GENERAL,
U.S. PUBLIC HEALTH SERVICE

(c)   Courtesy of the CDC.

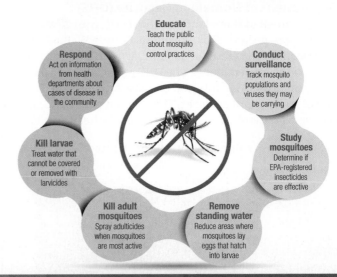

Mosquito Control:
What state and local mosquito control programs do

Why is local mosquito control important?
Some mosquitoes can spread viruses like Zika, West Nile, and dengue. Other mosquitoes bother people, but don't spread viruses. Mosquito control activities reduce all types of mosquitoes.

Who conducts mosquito control?
Mosquito control districts or state and local government departments work to control mosquitoes.

What do local mosquito control programs do?

Educate
Teach the public about mosquito control practices

Conduct surveillance
Track mosquito populations and viruses they may be carrying

Respond
Act on information from health departments about cases of disease in the community

Study mosquitoes
Determine if EPA-registered insecticides are effective

Kill larvae
Treat water that cannot be covered or removed with larvicides

Kill adult mosquitoes
Spray adulticides when mosquitoes are most active

Remove standing water
Reduce areas where mosquitoes lay eggs that hatch into larvae

Mosquito Control: You Have Options.
Learn more: http://www.cdc.gov/zika/prevention/controlling-mosquitoes-at-home.html

CDC   U.S. Department of Health and Human Services
Centers for Disease Control and Prevention

(d)   Courtesy of the CDC.

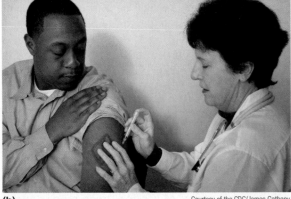

(b)   Courtesy of the CDC/James Gathany.

FIGURE 18.10 (a) The wartime fight against sexually transmitted infections. These diseases remain a very significant public health problem in the United States and around the world. **(b)** A person being immunized in a USPHS immunization campaign. The USPHS plays a major role in immunization campaigns around the world to halt epidemics by immunizing the population. **(c)** A USPHS announcement during the 1960s from the Office of the Surgeon General, Dr. William H. Stewart, helped to educate parents about the complications associated with measles and the importance of vaccination to prevent it. Even though there is an effective vaccine to prevent measles, there continue to be outbreaks of measles around the world. This message is still needed today. **(d)** Poster to educate the public on state and local mosquito-control programs in 2016 after the unprecedented Zika virus disease outbreak in Brazil. Mosquitoes carrying Zika virus spread to Florida, and the USPHS played a role in vector-control programs in the United States.

The mission of the USPHS is to improve the health of every individual by conducting research, engineering systems for safe delivery of water and disposal of waste, overseeing food and drugs, studying and developing means to contain or eliminate disease, and promoting a safe and healthful environment at work or home. As America's uniformed service of public health professionals, the Commissioned Corps achieves this mission through:

- Rapid and effective response to public health needs
- Leadership and excellence in public health practices
- Advancement of public health science

○ Federal Emergency Management Agency (FEMA)

The Federal Emergency Management Agency (FEMA; https://www.fema.gov) was created in 1979 by President Carter as an independent agency of the federal government that reports directly to the president. Its slogan is

"*Helping people, before, during, and after disaster.*" Natural disasters such as hurricanes, floods, earthquakes, and tornadoes can destroy the public health infrastructure, leading to polluted waters and compromised sanitation followed by disease. It is common to hear in such instances that the president has declared a community to be eligible for FEMA funds and assistance in coping with the disaster. FEMA works in partnership with local, national, and international agencies in performing its role. FEMA responded to a crippling snowstorm in West Virginia in the winter of 2012, Maryland in 2014, and Boston, Massachusetts in 2015; to wildfires in California in 2007 and 2018; to a catastrophic EF5 multiple-vortex tornado that struck Joplin, Missouri, in May 2011; to hurricanes Harvey, Irma, and Maria in 2017; and to volcanic activity in Hawaii in 2018 (**FIGURE 18.11**).

Hurricanes Katrina (2005) and Maria (2017) presented major challenges to FEMA; Katrina is considered to have been the largest natural disaster in the United States. The

(a)
Courtesy of Norman Lenburg/FEMA.

(b)
Courtesy of Dominick Del Vecchio/FEMA.

(c)
Courtesy of Grace Simoneau/FEMA.

(d)
Courtesy of Eduardo Martinez/FEMA.

FIGURE 18.11 FEMA responses to catastrophic events in the United States and Puerto Rico. **(a)** Parsons, West Virginia, November 5, 2012. Snow removal to clear the roads so that technicians can restore services. **(b)** Wildfires completely destroyed this Santa Rosa, California, neighborhood in October 2017. **(c)** Volcanic lava flowing toward the Kapoho Bay area, bursting homes into flames and altering the landscape of the Big Island. FEMA staff were on the ground to support local officials with life-saving emergency measures such as debris removal and the repair or replacement of publicly owned facilities. Photo taken at Leilani Estates, Hawaii, May 2018. **(d)** On January 18, 2018, over 15,000 utility poles were delivered to restore electricity in Ponce, Puerto Rico, after Hurricane Maria knocked out electricity to the power grid. FEMA partnered with the U.S. Army Corps of Engineers and private companies to deliver the poles and restore power.

agency came under severe criticism because of a delayed response time, but, to its credit, implemented major policy changes that led the way in the nation's emergency response strategies. However, when Hurricane Maria tore through Puerto Rico during the 2017 hurricane season, FEMA was criticized again for responding slowly. Eventually, 19,000 FEMA employees were deployed to Puerto Rico. The hurricane winds damaged or destroyed more than a third of the homes and crippled the island's power grid. Seven months after the hurricane hit landfall, more than 100,000 Americans were still without power on the island.

O U.S. Food and Drug Administration (FDA)

The U.S. Food and Drug Administration (FDA) is part of the U.S. Department of Health and Human Services (HHS). It focuses on the safety of foods, vaccines, antibiotics, other medicinal (biological) products, and medical devices, all of which play a role in the prevention and control of infectious disease. This agency has the last word before these products are approved for release. The safety of the nation's blood supply system is under the umbrella of the FDA; inspectors routinely test blood and blood products for contamination. The agency has the authority to direct withdrawal of products, either voluntarily or legally, as in the withdrawal of the RotaShield vaccine against gastroenteritis shortly after its approval or the banning of triclosan and other chemicals in antibacterial hand and bar soaps. Products not directly related to infectious diseases also fall under the scrutiny of the FDA. The agency is charged with protecting the consumer by enforcing the Federal Food, Drug, and Cosmetic Act and related public health laws, including the truth in labeling laws.

O American Red Cross (ARC)

Clara Barton (1821–1912) founded the American Red Cross (ARC; http://www.redcross.org) in 1881. Barton risked her life to bring food, clothing, and other supplies to soldiers in the battlefield during the American Civil War. She had extraordinary devotion to serving others. She founded the ARC at the age of 60 and led it for the next 23 years. With the outbreak of World War I, the organization's membership grew from 3,864 in 1918, to more than 20 million adults and 11 million junior Red Cross members during World War II. The ARC staffed hospitals and recruited 20,000 registered nurses to serve in the military (**FIGURE 18.12**).

The ARC is a humanitarian nonprofit organization that relies on volunteers dedicated to helping people in need throughout the United States and, in association with other Red Cross networks, throughout the world. For example, the ARC supported response efforts toward the Ebola outbreak in West Africa in 2014. It supplies about 40% of the nation's blood supply. ARC volunteers and employees provide support today in five critical areas:

1. People affected by disasters in the United States
2. Support for military and their families
3. Blood collection, processing, and distribution

Courtesy of the Library of Congress.

FIGURE 18.12 Poster created during World War I that was used as part of an ARC membership drive. It shows a Red Cross Nurse holding a large red cross.

4. Health and safety education and training
5. International relief and development

Each year, the ARC responds to nearly 64,000 disasters, the vast majority are home fires. In 2014, the ARC launched its Home Fire Campaign, which installed 1,367,300 smoke alarms in homes, ultimately saving 444 lives.

O Center for Infectious Disease Research and Policy (CIDRAP)

The Center for Infectious Disease Research and Policy (CIDRAP; http://www.cidrap.umn.edu) is committed to advancing the practice of public health through research, communication, education and training, policy-making, and forming interdisciplinary partnerships. They are "*the super planners, the worst-case scenario strategists, the global leaders in the discipline of preparedness.*" Dr. Michael Osterholm, director of CIDRAP, was one of five U.S. Science Envoys selected in 2018. He is working with priority countries on infectious disease preparedness and antimicrobial stewardship.

At the International Level

O World Health Organization (WHO)

The days are long gone of when the public health problems of a continent, or a country for that matter, were unique and limited to that geographical area. The planet's microbes know no boundaries and travel freely in or on their hosts without passport from hemisphere to hemisphere in a matter of hours. The WHO (http://www.who.int/), with its headquarters in Geneva, Switzerland, functions as a command post in its extensive partnerships and uses sophisticated surveillance and communication systems to track microbial diseases on a global level. However, its activities are multifold, and not limited to microbial disease surveillance and control.

WHO partnerships date back to the early years of the organization; its influenza surveillance network is responsible for identifying each year the strains of influenza A and B viruses to be used in the seasonal influenza vaccine, thereby serving as a global watchdog for surveillance of influenza. The WHO has entered into many partnerships in

an effort to combat diseases, including smallpox, tuberculosis, poliomyelitis, HIV/AIDS, rabies, malaria, leprosy, cholera, sleeping sickness, filariasis, and guinea worm disease.

The growth of information technology provides increased opportunities for disease surveillance and response, requiring rapid assessment to initiate control efforts with minimal delay and to screen out unsubstantiated reports. The WHO has an innovative approach to global disease surveillance aimed at improving epidemic disease control by rapid verification of diseases and quick response to potentially significant outbreaks by health professionals (**FIGURE 18.13**). The WHO, in its alliance

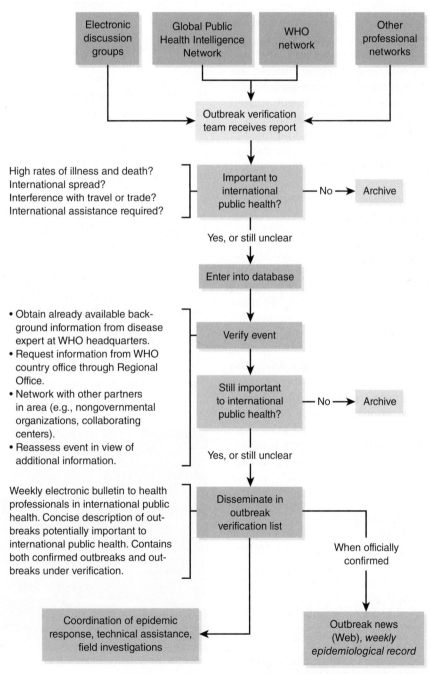

Information from T W. Grein, K.-B. O. Kamara, G. Rodier, A. J. Plant, P. Bovier, M. J. Ryan, . . . D. L. Heymann. (2000). Rumors of disease in the global village: Outbreak verification. *Emerging Infectious Diseases, 6*(2), 97–102.

FIGURE 18.13 Outbreak verification program. This organization has teams ready to go in the event of verification of an impending outbreak of disease.

with 194 member countries, across 6 regions, and from more than 150 offices, is in an ideal position to monitor infectious disease surveillance and control.

Pan American Health Organization (PAHO)

The Pan American Health Organization (PAHO; https://www.paho.org/hq/) has "*100 years of experience in working to improve the health and living standards of the countries of the Americas.*" It is a component of the United Nations and serves as the Regional Office for the Americas of the WHO. Its member states include all 35 countries in the Americas. Its mission is achieved in association with other governmental and nongovernmental agencies, universities, and community groups.

PAHO promotes primary health strategies and assists countries in combating cholera, dengue, tuberculosis, and AIDS and is committed to ensuring that blood for transfusion is safe and not a vehicle of disease. Further, the organization aims to eliminate all vaccine-preventable diseases. The agency's polio eradication efforts, started in 1985, ultimately paid off, resulting in a declaration of polio-free Americas in 1994; the last case was identified in August 1991. Improvement of drinking water, sanitation, and health care remains a top priority for PAHO, with a focus on equity.

United Nations (UN) Foundation

The UN Foundation (http://www.unfoundation.org/) was created in 1988, with entrepreneur and philanthropist Ted Turner's historic $1 billion gift to support UN causes and activities. The UN Foundation builds and implements public–private partnerships to address the world's most pressing problems and also works to broaden support for the United Nations through advocacy and public outreach; the UN Foundation is a public charity.

The UN Foundation and its partners, including the ARC and the CDC, have raised large sums of money for the Measles & Rubella Initiative (formerly the Measles Initiative). Its global health priorities include preventing malaria deaths, eradicating polio, reducing measles mortalities, developing **mobile health** technologies, innovating health finance, and improving health for every woman, every child.

Food and Agriculture Organization of the United Nations (FAO)

The Food and Agriculture Organization of the United Nations (FAO; http://www.fao.org/home/en/) is a specialized branch of the United Nations. Its mission is to defeat global hunger. Even though the world has the capacity to produce enough food to feed everyone adequately, 815 million people suffer from chronic hunger today. It is estimated that 155 million children younger than 5 years are chronically malnourished (stunted in growth) and more than 52 million are acutely malnourished (wasted). Most of the world's poor live in rural areas. The world is predicted

to increase to 9 billion people by 2050. People living in areas where populations are increasing rapidly and that are dependent on crops, livestock, forestry, and fisheries face high rates of food insecurity for sustainability. In meeting these challenges, the FAO has five strategic objectives:

1. Help eliminate hunger, food security, and malnutrition.
2. Make agriculture, forestry, and fisheries more productive and sustainable.
3. Reduce rural poverty.
4. Enable inclusive and efficient agriculture and food systems.
5. Increase the resilience of livelihoods to threats and crises.

The FAO is forging partnerships to fulfill its objectives. It has partnered with WorldFish (International Center for Living Aquatic Resources Management) to boost the efforts of fishers and small-scale fish farmers and to promote sustainable aquaculture. Continued progress is needed to strengthen fisheries management, reduce waste, and find solutions to tackle problems associated with illegal fishing, contamination of aquatic environments, and climate change.

United Nation's International Children's Emergency Fund (UNICEF)

United Nation's International Children's Emergency Fund (UNICEF; https://www.unicefusa.org/) is a humanitarian organization that puts children first all over the world. Its mission is to provide children with health care and immunizations, safe water and sanitation, nutrition, education, emergency relief, and more (**FIGURE 18.14**). It is a leader in responding to disasters affecting children. Some of UNICEF's points of pride include the following facts:

- UNICEF works in more than 190 countries and territories to save and protect children.

© Djohan Shahrin/Shutterstock.

FIGURE 18.14 Rohingya refugee children in Cox's Bazar, Bangladesh, in 2017, receive new school bags from UNICEF.

- UNICEF works in more than 100 countries to improve access to safe drinking water and sanitation facilities.

- UNICEF has implemented over 12 million diagnostic tests to prevent the deaths of children suffering from malaria.

- UNICEF has immunized 45% of the world's children.

- UNICEF's lifesaving interventions have reduced child deaths by more than 90% since 1990.

For the first time in history, an HIV/AIDS-free generation is within reach. It is estimated that UNICEF USA's efforts since 2000 have prevented 30 million new infections and have averted nearly 8 million deaths. In addition, 15 million people living with HIV are now receiving treatment, which is preventing mother-to-child transmission of HIV. More progress is still needed to prevent infected adolescents, especially girls between the ages of 15 and 19. HIV/AIDS is a leading cause of death for teens globally.

Médecins Sans Frontières (MSF)

Médecins Sans Frontières (MSF), or Doctors Without Borders (https://www.doctorswithoutborders.org), is a volunteer organization founded in France in 1971, by physicians and journalists to assist those "*whose survival is threatened by violence, neglect, or catastrophe primarily due to armed conflict, epidemics, malnutrition, exclusion from health care, or natural disasters*" (**FIGURE 18.15**). The organization cofounded the Drugs for Neglected Diseases project in 1999 and was awarded the Nobel Peace Prize the same year for its humanitarian work. MSF has treated tens of millions of people since 1971.

MSF has intervened in almost all recent epidemics of Ebola virus disease. At the peak of the Ebola outbreak that started in Guinea, West Africa, in 2014, MSF employed nearly 4,000 national staff and more than 325 international staff to combat the epidemic in Guinea, Liberia, and Sierra Leone. It set up Ebola treatment units that admitted 10,376 patients, of which 5,226 were confirmed cases of Ebola virus disease. MSF spent over 96 million euros responding to the outbreak. The organization continues to invest in Ebola survivor clinics to provide comprehensive care packages, including medical and psychosocial care and protection against stigma.

World Bank

The World Bank (https://www.worldbank.org) is an international financial institution founded in 1944 that provides loans to developing countries; it has a history of project support to low- and middle-income countries to assist them in meeting their developmental needs (**FIGURE 18.16**). The World Bank has evolved to become the world's largest external funder of health. In 1947, it made 4 loans totaling $497 million, as compared to 302 loans totaling $60 billion in 2015. The World Bank has supported public health projects associated with onchocerciasis, tuberculosis, HIV/AIDS, malaria, cholera, meningitis, and other diseases associated with microbes and fostering immunization. A part of its function is "*to overcome poverty, enhance growth with care of the environment, and to create individual opportunities and hope.*" Toward this end, the World Bank has set two goals for the world to achieve by 2030:

1. End extreme poverty by decreasing the percentage of people living on less than $1.90 a day to no more than 3%.

2. Promote shared prosperity by fostering the income growth of the bottom 40% for every country.

World Organization for Animal Health (OIE)

Rinderpest, also known as "cattle plague," causes high mortalities in cattle but can also affect domesticated buffalo, antelope, deer, giraffes, and a few other species. In 1920, an unexpected rinderpest outbreak occurred in Belgium at a time when it was thought to have been eradicated in

Courtesy of Brian Bird, Veterinary Medical Officer, Viral Special Pathogens Branch, CDC.

FIGURE 18.15 Water and sanitation staff and clinical care nurses with *Médecins Sans Frontières* preparing for a typical day in an Ebola isolation ward, Democratic Republic of Congo, 2012.

© iStockphoto/qingwa.

FIGURE 18.16 The World Bank employs more than 10,000 people in more than 120 offices worldwide. Its headquarters are in Washington, D.C.

FIGURE 18.17 Educational poster created by OIE with Global Affairs Canada and additional partners in 2017 to promote rinderpest vigilance.

Europe. A herd of infected zebus from India, en route to Brazil, were being held at a quarantine facility in the port of Antwerp. The infected zebus had contact with American cattle. The infected American cattle were shipped to markets in Brussels and Ghent, Belgium. These infected cattle spread the disease to cattle in Germany, causing multiple outbreaks. An international meeting was held in 1920 to control the spread of contagious animal diseases. The meeting resulted in the creation of the Office International des Epizooties (OIE), now known as the Organization for Animal Health (the OIE acronym remains the same). In 2011, rinderpest became the second disease in the world to be eradicated globally (**FIGURE 18.17**).

The OIE (http://www.oie.int/) is an international organization with 181 member countries and territories with a mission of improving animal health. More than 1.3 million people depend on livestock for their livelihoods. In 2017, the OIE partnered with 75 organizations to control animal health risks, to ensure that the animal disease situation in the world is transparent in communication (including diseases that are transmissible to humans), and to strengthen safety associated with the world trade of animals and animal products.

The Private Sector

O Rotary

Rotary (https://www.rotary.org/en) has been addressing challenges around the world for more than 110 years. Paul Harris, a Chicago lawyer, formed the Rotary Club in 1905 so that professionals with diverse backgrounds could network, exchange ideas, and find ways to give back to their communities. The name "Rotary" came from the group's early practice of "*rotating meetings*" among the offices of its

members. Within 16 years, it had become an international club on six continents.

Rotary has been a key player in the polio eradication program since 1979, when it launched a program to vaccinate 6 million children in the Philippines. Since then, 2.5 billion children have been vaccinated against polio. Rotary's causes are focused on promoting peace; fighting disease; providing clean water; saving mothers and children through medical care, education, and economic opportunities; fighting illiteracy and poverty; supporting education; and growing local economies.

O The Rockefeller Foundation

The Rockefeller Foundation (https://www.rockefeller foundation.org) was formed in 1913 by Standard Oil owner John D. Rockefeller Senior and his son to promote the well-being of humanity throughout the world. They sought to accomplish this together with partners and grantees. The private foundation is based in New York City. The Rockefeller Foundation's very first grant was awarded to the ARC. The ARC received $100,000 to purchase property for its headquarters in Washington, D.C. The Rockefeller Foundation has a long history with health initiatives, from eradicating hookworm in the American South to developing the yellow fever vaccine. It supports universal health coverage and disease surveillance networks.

O Wellcome Trust

The Wellcome Trust (https://wellcome.ac.uk) is a UK organization founded by Sir Henry Wellcome (1853–1936). Wellcome was a leader in the pharmaceutical industry. He established two laboratories that developed antitoxins for tetanus, diphtheria, and gas gangrene. When Wellcome died in 1936, his will stipulated that a charity organization be established, which became known as the Wellcome Trust. It was dedicated to the advancement of medical research to improve health. The Wellcome Trust continues to thrive today, supporting scientists and researchers to improve health for everyone.

O The Bill & Melinda Gates Foundation, Pharmaceutical Companies, the Carter Center, and the Clinton Foundation

The role of government agencies as partners in the control of infectious diseases has been outlined, but the success achieved over the past century is, in no small measure, also attributable to the role of the private sector and nongovernmental organizations as equal partners in the ongoing battle to lessen the world's burden of infectious diseases. In 2001, the Washington Office of the Bill & Melinda Gates Foundation (https://www.gatesfoundation.org) opened. The foundation has pledged over $100 million in pursuit of a vaccine to prevent HIV infection and has challenged others to pitch in.

Pharmaceutical companies, too, have been significant partners. Ivermectin, an important drug in the onchocerciasis control program, was donated free of charge by Merck & Co. to countries where this disease is prevalent. SmithKline Beecham (now part of GlaxoSmith-Kline) supplied the drug albendazole, an orally administered broad-spectrum anthelmintic drug, to countries as needed. Other drug companies have also made generous contributions.

The Carter Center (https://www.cartercenter.org) has committed to a variety of projects, including partnerships with both governmental and nongovernmental organizations, for the eradication of guinea worm disease, onchocerciasis, and filariasis. The Clinton Foundation (https://www.clintonfoundation.org/) was established in 1997, by former President Bill Clinton. The Clinton Foundation HIV/AIDS Initiative, created in 2002 and extended to malaria in 2007, aims to bring affordable, high-quality treatment to infected persons in developing areas.

Partnerships: The Way to Go

The enormous public health strides of the 20th century resulted in an increased life span for people in many countries and decreased the burden of infectious diseases around the world. These achievements were largely the result of partnerships from the community level to the international level, as described in this chapter. The potential threat of microbes is too enormous to be handled without teamwork, particularly given the ease of transmission resulting from globalization and the increased threat of emerging, and reemerging, infections.

Partnerships are the key to preventing and coping with epidemics and pandemics and in responding with minimal delay to populations endangered by the ravages of infectious disease as a consequence of floods, hurricanes, wildfires, earthquakes, and other natural disasters. History and current events provide many examples of the misery and deaths that occur in the aftermath of catastrophic events because of the collapse of the public health infrastructure. The lack of clean drinking water and safe foods, sanitation, and personal hygiene, coupled with overcrowded and makeshift living quarters, create an environment that is conducive to outbreaks of diarrheal, respiratory, and a multitude of other diseases. Whereas the primary motive of these partnerships is humanitarianism, there is also a selfish motivation of self-protection.

Examples of partnerships in action in response to catastrophic events are abundant. A massive earthquake, the worst in 50 years, rumbled through western India on January 28, 2001, causing thousands of houses to collapse in Bhuj, a city of 150,000 close to the quake's epicenter, and spreading damage in its path. The earthquake was quickly followed by the arrival of international disaster teams and pledges of financial aid from Britain's International Rescue Corps, the Swiss Red Cross, the International Federation of the Red Cross, the United Nations, the European Union, Germany, Turkey, Norway, and the Netherlands. Pakistan, India's archrival, was among the first to offer condolences

and help and demonstrated that human tragedy transcends politics.

The earthquake in India (2001); the SARS epidemic in China (2003); the tsunami in Indonesia (2004); Ebola outbreaks in central Congo (2005), Uganda (2012), and West Africa (2014); and the devastating hurricanes of Katrina (2006), Harvey, Irma, and Maria (2017) in the United States and Puerto Rico; cholera outbreaks in Zimbabwe (2008) and Haiti (2010); and the Zika outbreak in Brazil (2016) are excellent examples of partnerships in action at the local, national, and international levels, involving both public and private sectors, working together to prevent, monitor, and control outbreaks. The world is threatened by possible pandemics of avian and swine influenza, Nipah encephalitis, or a coronavirus that causes a severe respiratory disease, and if it were to happen partnerships at all levels would be called on to develop and implement containment strategies. There is no doubt that everything works better when everyone works together. *Partnerships are the way to go.*

An Ongoing Battle

Microbes have challenged their hosts throughout history and have influenced the course of civilization. Despite major advances in sanitation and the availability of clean water, antibiotics, and immunization, infectious diseases remain a significant cause of death, disability, and socioeconomic burden worldwide. Are microbes deliberately challenging humans? Is it in a microbe's best interest to annihilate its human (or other) host, and is it in humans' best interest to annihilate microbes? The answer to each of these questions is a definite "no." The biotic component of ecosystems is complex and varied and characterized by constant adaptation in the community. *Adaptation*, not annihilation, is the key.

Host–microbe interactions are dynamic. There are no scientific equations that can predict the outcome of specific interactions; the outcome depends on a variety of circumstances reflecting the time and place when microbes and humans are on a collision course. History reveals that epidemics come and go as a result of shifts in the seesaw of host–microbe relationships. On the surface it might seem that microbes have the upper hand. Microbes can reproduce quickly without being hampered by requiring suitable partners (sexual reproduction) with which to share their genetic material. Bacteria, in particular, can replicate quickly by **binary fission**. When large numbers of the bacterial population are killed, generation times can, in many cases, be expressed in minutes, leading to a rapid recovery of the population. This rapid multiplication results in enormous genetic variability that, in turn, allows for *genetic adaptation*.

Antibiotic resistance remains prevalent as a global challenge in the microbial community. Outbreaks of measles, chickenpox, and whooping cough still occur, along with a variety of other infectious diseases, indicating the need for strict measures of surveillance and reassessment of immunization schedules. An outbreak of cholera began raging in Tanzania in August 2016, continuing into 2017 and 2018. Currently, there is strong concern about the potential for a pandemic of avian (bird) flu or Nipah encephalitis. Many public health officials believe it will happen, and that the question is "when" and not "if." In April 2009, an outbreak of H1N1 swine flu started in Mexico and rapidly spread to many areas of the world. It resembled the 1918 Spanish flu pandemic and was a cause of great concern as a potential pandemic during the "regular" flu season starting in the fall months of the year.

Foodborne diseases continue with a high profile in the United States and around the world. In 2018, 36-state outbreak of *Escherichia coli* O157:H7 caused by contaminated romaine lettuce resulted in 210 confirmed cases, 5 deaths, and 96 hospitalizations. Malaria continues and, as a result of international travel, has increased in the United States. **Ecotourism**, "living with the natives," and "going on safari" is a growing trend for those seeking *Tarzan of the Apes* adventures. These travelers may get more than they bargained for, including a belly full of worms and malaria. And what about the effect on animal populations as human diseases are introduced into their ecosystems? A group of mountain gorillas in Rwanda's Volcanoes National Park became ill with severe respiratory problems and at least one succumbed. Monkeys in Silver Springs, Florida, tested positive for herpes B virus. Herpes B is highly fatal in humans. In 2012, bats were found to carry an H17 influenza A virus. A new virus spread by midges (*Culicoides* spp.) that causes birth defects and stillborn births in cows, sheep, and goats was reported in Schmallenberg, Germany, in 2011; it quickly spread to the Netherlands, Belgium, France, Luxembourg, Italy, and the United Kingdom. Will it "jump species" into humans, as happened with mad cow disease and the Nipah virus (**BOX 18.1**)?

Advancements in technology have dramatically improved the life span and health of humans. However, some of the biggest medical breakthroughs have not come without prior unanticipated complications. Progress in immunosuppressive treatment has resulted in a dramatic increase in the number of living **immunocompromised patients**, many of whom are organ recipients. Over the past 20 years organ transplantation has become the treatment of choice for many diseases that in the past led to patient death due to the failure of one or more vital organs. Each year, 40,000 organ transplants are performed worldwide with high success rates (90% of recipients survive at least 1 year).

BOX 18.1 A-Pork-Alypse Now

In September of 1998, a small Malaysian village, Nipah, was turned upside down when pigs began suffering from a respiratory illness and paralysis. Farmers who came in close contact through respiratory droplets of the sick pigs developed fevers, headaches, drowsiness, and confusion, lapsed into a coma 2 days after the onset of symptoms, and then died from what appeared to be viral encephalitis. Cats and dogs in close contact to pigs also became infected.

Initially, government authorities thought this "unknown ailment" in pigs was classic swine fever, or "hog cholera." The human deaths were speculated to be caused by the Japanese encephalitis virus, an endemic mosquitoborne illness. Various insecticide-control measures were implemented, but nevertheless the disease in both pigs and humans spread to the biggest pig-farming region in Malaysia and then to Singapore. Transmission was rapid and the fatality rate was high (40% in Malaysia), causing more than 265 human encephalitis cases and 105 deaths in Malaysia and 11 cases of encephalitis and 1 death in Singapore over the next 18 months.

Blood sera and cerebral spinal fluid samples from sick patients sent to the CDC revealed a new paramyxovirus, named Nipah virus after the village in which the first cases were identified. More than 1 million pigs of the 2.4 million pig population were disposed of during outbreaks in Malaysia. This was economically devastating to the pork industry; some referred to it as an "a-pork-alypse."

The government suffered losses of approximately $450 million; about 618 homes and 111 shops, schools, and banks were evacuated; and 36,000 people lost employment due to the closure of farms.

Why did this happen? Where did this emerging zoonotic Nipah virus come from? Why was the outbreak hotspot a tiny village in Malaysia? There are no clear answers, but whatever the answers, the outbreak is a glaring example of how a change in the environment led to the outbreak. The environment (favorable for disease development and transmission), a susceptible host, and a pathogen are the "big three" components of disease causation. Conversely, the disease may be prevented or halted by elimination of any one of these three components, referred to as the disease triangle model (**Figure 1**).

Prior environmental changes led to the Nipah virus outbreak. Approximately 5 million hectares (12,355,269 acres!) of tropical rainforest were cleared to allow room for piggeries (**Figure 2**). Use of a "slash-and-burn" strategy resulted in a blanket of haze over the region. Consequently, a number of flowering and fruit trees were destroyed. The bats (flying foxes), which are the natural reservoir of the Nipah virus, rely on the fruit and nectar from these trees as their primary foods (**Figure 3**). Ultimately, Nipah virus crossed the species barrier from bats into human hosts, primarily through infected pigs.

Nipah virus is present in the saliva and urine of the wild bats and in partially eaten fruit. Scientists suspect that fruit trees

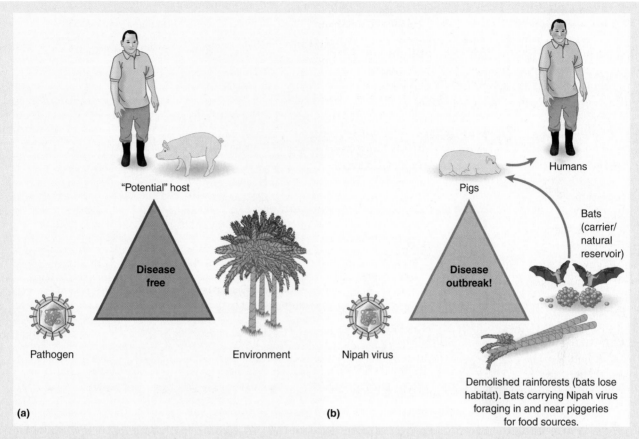

Figure 1 **(a)** Disease triangle model. **(b)** Disease triangle model and the Malaysian Nipah virus outbreak.

close to the pig pens were foraged by the bats and the virus was spread by this close proximity of pigs with bat saliva or urine, contaminated water, or dead bats. Humans can also be infected by consuming contaminated fruit or juice.

In 2012, a more virulent Nipah virus outbreak with 100% mortality in northern Bangladesh was traced to individuals drinking raw date (fruit tree) sap or juice, causing widespread panic throughout the country. Nipah virus is a recurring threat to human health in Southeast Asia. Continued education of healthcare workers, veterinarians, and other responders

about infection-control procedures, surveillance, and rapid diagnostic tests is essential in high-risk areas. In May 2018, the Nipah virus returned, causing an outbreak in the southern state of Kerala, India. It killed 17 out of the 19 people confirmed to be infected.

Courtesy of James Roth, Iowa State University.

Figure 2 Several hog confinement barns were affected during the Malaysian Nipah virus outbreak. The fruit bats that carry the Nipah virus live in caves and feed on the fruit trees that are in close proximity to the hog confinement barns.

© iStockphoto/Thinkstock.

Figure 3 Flying fox (*Pteropus* bat), natural reservoir of the Nipah virus.

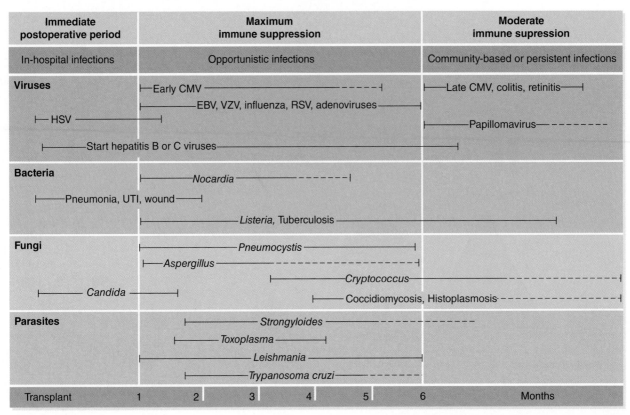

Immediate postoperative period	Maximum immune suppression		Moderate immune supression
In-hospital infections	Opportunistic infections		Community-based or persistent infections

Viruses
- Early CMV
- EBV, VZV, influenza, RSV, adenoviruses
- HSV
- Start hepatitis B or C viruses
- Late CMV, colitis, retinitis
- Papillomavirus

Bacteria
- *Nocardia*
- Pneumonia, UTI, wound
- *Listeria*, Tuberculosis

Fungi
- *Pneumocystis*
- *Aspergillus*
- *Cryptococcus*
- Candida
- Coccidiomycosis, Histoplasmosis

Parasites
- *Strongyloides*
- *Toxoplasma*
- *Leishmania*
- *Trypanosoma cruzi*

| Transplant | 1 | 2 | 3 | 4 | 5 | 6 | Months |

Information from M. Salavert, R. Granada, A. Diaz, & R. Zaragoza. (2011). Role of viral infections in immunosuppressed patients. *Medicina Intensiva* 35(2), 117–125.

FIGURE 18.18 Increased susceptibility of immunocompromised persons to infection. CMV: cytomegalovirus; HSV: herpes simplex virus; VZV: varicella zoster virus (causes chickenpox and shingles); RSV: respiratory syncytial virus; UTI: urinary tract infection.

Microbial infections are the most frequent complications in transplant recipients and are the main cause of death during their first year after transplantation (**FIGURE 18.18**). Immunosuppressive drugs are necessary to prevent organ rejection, but the downside is that they weaken host immune responses. The source of infection may come from the donor, the reactivation of latent viruses of the donor or recipient, community-acquired infections, or travel-associated diseases. Organ recipients encounter potentially drug-resistant pathogens during hospital treatments and are frequently given antimicrobials prophylactically to prevent common healthcare-associated infections.

Emerging pathogens pose a unique challenge for physicians to recognize, diagnose, and treat. Immunocompromised patients are high-risk individuals that, in some ways, are like the canary in the coal mines that miners once used to warn of a level of dangerously low oxygen. High-risk individuals are the first ones to become infected, making it imperative to monitor these "human canaries" for the emergence of new pathogens that challenge the world's population.

Natural catastrophic events will continue to occur, and in their aftermath will be an array of microbial-related public health problems (**FIGURE 18.19**). These events may be nature's way of "cleaning house," but they demand attention to minimize human suffering.

FIGURE 18.19 Catastrophic events will continue to occur and so will microbial diseases in their wake.

Just as there are developments that are worrisome and threatening, numerous examples are available of progress that has been decidedly beneficial. For the first time, plant-produced HIV antibodies are undergoing clinical trials in the United Kingdom. They were produced in the leaves of tobacco plants grown in a greenhouse in Germany and are anywhere from 10 to 100 times less expensive to produce than by using conventional methods. A vaccine for

BOX 18.2 Good Personal Health Practices

- Wash hands frequently (but not obsessively).
- Cook foods properly and refrigerate them promptly.
- Stay up to date on immunizations.
- Use antibiotics prudently.
- Practice abstinence or safe sex.

- Check on immunization requirements and recommended medications before traveling, particularly if traveling to a developing country.
- Avoid complacency.
- Use common sense in health matters.

BOX 18.3 Important Public Health Strategies to Contain Infectious Disease

- Develop new partnerships between the public and private sectors.
- Continue the war on poverty in the developing and developed world.
- Maintain and improve vector-control programs.
- Continue and increase monitoring and surveillance on a worldwide basis.
- Reduce disparity in standards of living between developed and developing countries.

- Increase research in public health as related to epidemiology.
- Educate the public and promote behavioral changes related to public health.
- Develop and maintain the public health infrastructure.
- Be prepared to handle outbreaks.
- Avoid complacency.

humans may also be on the horizon against the Ebola virus. Ebola virus is a potential candidate to be used as a **biological weapon**, which has prompted the U.S. government to allocate increasing funds for Ebola vaccine research.

So why have humans (and other hosts) been able to survive the periodic epidemics and pandemics? Nobel Laureate Joshua Lederberg stated, in reference to the 1918 Spanish influenza epidemic, "*If the mortality rate had been another order of magnitude higher, our species might not have survived.*" This statement makes it clear that the battle between microbes and humans has been a close call at times.

A notable point to be made is the ability of microbes, based on their genetic variability, to be transmitted across the species barrier such as from animal hosts to human hosts. Many biologists believe that tuberculosis and plague jumped from animal reservoirs to human hosts centuries ago. More recently, the emergence of HIV/AIDS, Nipah encephalitis, Zika virus disease, Middle East respiratory syndrome, and mad cow disease and their human variants support this fact.

The ongoing saga of infectious diseases is a reflection of the ongoing evolutionary dance. The practices of personal and public health can be brought to bear on the prevention and control of outbreaks (**BOXES 18.2** and **18.3**). Humans and microbes coexist; thus individuals and public health officials need to continue surveillance on

a worldwide basis to nip new threats in the bud as they emerge (**BOX 18.4**). Real-time surveillance is being used to track disease outbreaks and the natural disasters that are often associated with them. Examples of real-time surveillance networks are listed in **TABLE 18.5**.

Research

Ongoing scientific research leads to continued advancement in society's understanding of its environment and establishes direction for future investigation. The advancements in human welfare that have been made over the centuries, particularly in the 20th century, are the result of rigorous research. Microbiological research is directed toward an understanding of the microbial world and its interactions with humans and other species.

The scientific and medical community once found it difficult to accept that stomach and intestinal ulcers are caused by the bacterium *Helicobacter pylori*, a bacterium present in the acidic environment of the stomach. Scientists are still finding links indicating that microbes may play a role in chronic diseases in which no one suspected their involvement. Research is necessary to uncover the causal relationships.

Microbes may act as causative agents, as cofactors, or as triggers that cause damage through autoimmune reactions. The list of diseases is formidable and includes

BOX 18.4 Are You Ready for a Zombie Apocalypse?

On May 16, 2011, Dr. Ali S. Khan, director of the CDC's Office of Public Health Preparedness and Response, posted an entry on the Public Health Matters blog titled "Preparedness 101: Zombie Apocalypse," containing information on preparation for a zombie apocalypse. This may seem far-fetched, but according to a CDC spokesman the idea for the post came after the word "zombie" caused Internet traffic to spike. Zombies are fictional, undead mindless creatures that must eat humans (often brains) or animal flesh to survive. They are now a part of pop culture in Hollywood blockbuster movies (e.g., *Contagion*, *Resident Evil*, *I Am Legend*, *Night of the Living Dead*, *World War Z*), in popular American TV series (e.g., *The Walking Dead*, *Helix*, and *The Last Ship*), and in gaming (e.g., *Pandemic*, *Plants vs. Zombies*, and many others).

The purpose of the blog is to prepare the public for real emergencies, including tsunamis, hurricanes, tornadoes, and pandemics. Emergency supply kits, plans for families to regroup following zombie invasion, evacuation routes, and isolation and quarantine of those who become infected are a part of the blog. The blog has links to downloadable posters, badges (**Figure 1**) and a graphic novel, teaching moments from *The Walking Dead* TV series, and other disaster preparedness information.

Courtesy of the CDC.

Figure 1 CDC badge promoting emergency preparedness using a zombie as a hook to engage individuals to prepare before a disaster strikes.

Is there an infectious agent that could turn humans into zombies? Could a "zombie microbe" be engineered in the laboratory? Probably the closest known infectious agents to cause "crazed" symptoms are the rabies virus, prions that cause kuru and variant Creutzfeldt-Jakob disease (vCJD), the parasite *Toxoplasma gondii*, and brain-eating amoebas. A "zombie microbe" is just fiction, but the symbolism of zombies with the word "preparedness" is a catchy way to educate the public on how to respond to natural disasters and pandemics.

TABLE 18.5 Real-Time Disease Surveillance Systems

Real-Time Surveillance Network	URL
GeoSentinel	http://www.istm.org/geosentinel
ProMED-Mail	https://www.promedmail.org
HealthMap	http://www.healthmap.org/en/
Emergency and Disaster Information Service	http://hisz.rsoe.hu/alertmap/index2.php

a multitude of bacteria and viruses. In fact, some biologists believe that many chronic illnesses are microbial in nature. These diseases might then be amenable to treatment with antimicrobials or prevented by vaccines. More research suggests that the disturbance, or **dysbiosis**, of healthy gut bacteria may play a role in certain diseases. Examples of chronic diseases possibly caused by microbes are:

- Diabetes
- Atherosclerosis (heart disease)
- Thrombosis or strokes
- Allergies
- Arthritis
- Obesity
- Crohn's disease
- Inflammatory bowel disease (IBD)
- Ulcerative colitis
- Colon cancer
- Parkinson's disease
- Alzheimer's disease
- Anxiety disorders
- Autoimmune diseases
- Multiple sclerosis
- Asthma

In a sense, microbes are our partners in life. Compared to how many microbes live on and inside of us, a very low number of different microbes or viruses can cause disease. Disease may also occur if the microbial population on or within us are out of balance, allowing the pathogens to cause disease. For this reason, medical researchers study pathogenic microbes and viruses in order to develop treatments or preventions (e.g., vaccines). In doing so, we are opening Pandora's Box as more and deeper knowledge into life processes and "life" itself garners an increasing number of ethical issues. Are we opening a cauldron of trouble? Should we be conducting research to make bigger and better biological weapons, tinkering with ways to make avian influenza viruses highly transmissible, or delving into **synthetic biology** to create new microbes that may cross the species barrier? Is research for the sake of research without concern for its significance and implementation justifiable?

A system of risk-benefit analysis needs to be in place. Are scientists immune to the consequences of their research, or do they share a responsibility along with nonscientific groups? Well-intended and seemingly justifiable research may fall under a dark cloud as the products may turn out to be unpredictable—that is what research is about, to explore the unknown. Accidents (including unintended release of microorganisms), implementation of knowledge to create bioweapons, and failure by countries and individuals to adhere to established guidelines are ethical considerations that become more pronounced as "life" continues to be explored at deeper and deeper levels. Research is vital for continued survival, but the knowledge/products gained need to be handled with wisdom.

Microbes are essential for the maintenance of life on the planet, and others have been harnessed for a better quality of life. The following fields represent key research areas in microbiology and public health:

- Microbial diversity and versatility
- Microbial ecology and physiology
- Pathogenic microbiology and immune responses
- Microbial control of environmental pollution and recycling
- Biotechnology
- Microbial genomics

Genomics, the study of the functions and interactions of all genes in a genome, and microbial genome sequencing projects have led to the sequencing of numerous microbes (and other organisms), an accomplishment with a strong positive impact on an understanding of virulence factors, vaccine design, countermeasures to antibiotic resistance, and microbial diversity.

Health for All

Over half a century has passed since the proclamation of the Universal Declaration of Human Rights, and nearly half a century has transpired since the 1978 Alma-Ata Conference. Article 25 of the Declaration speaks of the highest attainment of health for all people as a fundamental right, and Alma-Ata defined its goal as "*the attainment for all people of the world by the year 2000 of a level of health that will permit them to lead a socially and economically productive life.*"

Substantial progress has been made toward these goals, as witnessed by impressive gains in healthy life expectancy for citizens fortunate enough to live in the industrialized world. But an estimated 85% of the world's population live in the developing countries and will not reap the full benefits of modern health care. Their lives are spent in abject poverty with little access to decent housing, clean water, sanitary waste disposal systems, appropriate nutrition, and education. Their future is bleak, and they live life on the edge (**FIGURE 18.20**).

The reality is that the Declaration of Human Rights and the goal of Alma-Ata remain as empty words with little promise. The disparity in life expectancy between industrialized countries and Swaziland is approximately 30 years. It is true that the successes achieved in the 20th century were remarkable, but developing countries still bear the brunt of the burden of disease (**FIGURE 18.21**). Balancing the scale and reducing the disparity between the developed industrialized nations and developing nations are priorities in international public health. New strategies and updated public health infrastructures must be implemented (**FIGURE 18.22**). This is the single most significant item of an unfinished agenda.

Poverty is considered to be the root of all evils; poverty lies at the heart of the burden of inequality and needs to be the target of social, economic, and public health interventions. Poverty, not microbes, is the fundamental cause

© K. Jensen/ShutterStock, Inc.

FIGURE 18.20 Life on the edge.

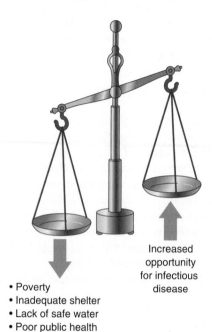

FIGURE 18.21 The scales of (in)justice.

- Poverty
- Inadequate shelter
- Lack of safe water
- Poor public health
- Infrastructure
- Illiteracy

Increased opportunity for infectious disease

of disadvantaged populations suffering more than their fair share of infectious (and other) diseases. Microbes are agents of disease that flourish in the midst of poverty. *The war against malaria and tuberculosis, as well as numerous other diseases, is a war against poverty and encompasses social and economic issues as well as public health issues.*

The poor exist in pockets of poverty in developed and developing countries. In response, the WHO launched new initiatives for health that could save millions of lives in the 21st century. A former director-general of the WHO stated the following:

> *The world could end the first decade of the twenty-first century with notable accomplishments. Most of the world's poor people would no longer suffer today's*

© Stanislav Komogorov/ShutterStock, Inc.

FIGURE 18.22 The old must give way to the new to improve the public health infrastructure in less-developed countries.

> *burden of premature death and excessive disability, and poverty itself would thereby be much reduced. Healthy life expectancy would increase for all. . . . The financial burdens of medical needs would be more fairly shared, leaving no household without access to care or exposed to economic ruin as a result of health expenditure. And health systems would respond with greater compassion, quality, and efficiency to the increasingly diverse demands they face.*

But the responsibility of addressing poverty does not rest only with the WHO; partnerships at the international level are vital. It is obvious that new initiatives targeted at addressing the disparity that exists in healthcare delivery are necessary for us to realize the intent of the Universal Declaration of Human Rights and the Alma-Ata Conference. The cycle of disease is complex and decidedly multifaceted (**FIGURE 18.23**).

One Health Initiative

One Health (https://www.cdc.gov/onehealth/index.html) is a worldwide transdisciplinary approach in which communications and partnerships are formed at the local, national, international, and private-sector levels with the goal of achieving optimal health. It recognizes that people, animals, plants, and their shared environment are interconnected. Most emerging pathogens have crossed species from animals to humans. Hence, it is important to focus on improving the health of animals and people in their shared environment.

One Health will advance the health of all life in the 21st century and beyond. Prioritizing the research of zoonotic diseases, continuing advancements in technologies, and improving disease surveillance and prevention measures will be necessary to attain health for all. In turn, the scientific knowledge base, education, and public and environmental health will advance. One Health has the potential to protect and save millions of lives today and for future generations. **FIGURE 18.24** illustrates the collaborative, transdisciplinary, and synergistic nature of One Health.

Summary

The 20th century witnessed tremendous accomplishments in public health that led to a worldwide decrease in the burden of infectious diseases, particularly in developed countries. The attainment of "*health for all*" is a goal yet to be realized. Further successes require teamwork at the local, national, and international levels and between the public and private health sectors. The players on the team are described in this chapter.

Surveillance and control of microbial disease at the community level are the responsibility of local and state departments of health. They are responsible for alerting the CDC to the occurrence of specified communicable diseases in order that a national network

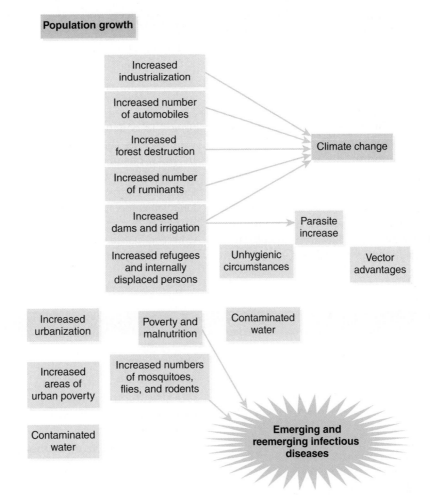

Reproduced from D. B. Louria, in *Emerging Infections I*, W. M. Scheld, D. Armstrong, and J. M. Hughes (eds.), 1998, ASM Press, Washington, D.C.

FIGURE 18.23 Relationships among determinants of emerging infections.

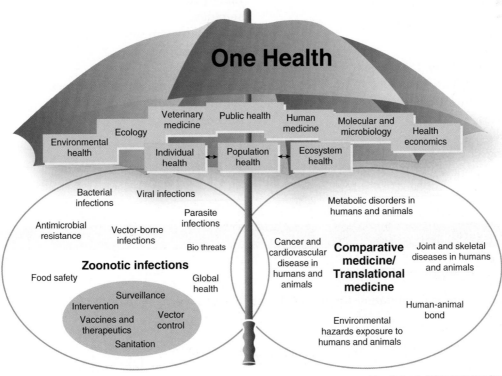

Information from the CDC One Health Home Page.

FIGURE 18.24 The One Health concept showing the interconnections between, people, animals, and their shared environments.

of surveillance and communication be maintained. At the national level a number of agencies, most notably the CDC, cooperate to bring about the containment of infectious diseases. Microbes spread readily and quickly from continent to continent, creating the need for international agencies like the WHO, that have the capacity to respond to outbreaks of disease at the global level. The One Health concept, which uses a transdisciplinary approach that recognizes the interconnection between people, animals, and the environment, will lead to the advancements needed to improve health for all in the 21st century.

The Harvard University School of Public Health bears the following inscription, repeated in several languages, on an outside wall: "*The highest attainable standard of health is one of the fundamental rights of every human*

being." To conclude this text the **Red Queen hypothesis**, proposed by the evolutionary biologist Leigh van Valen in 1973, based on the novel *Through the Looking Glass* authored by Lewis Carroll in 1871, is appropriate. The Red Queen, a character in the novel said, "*It takes all the running you can do to stay in the same place.*" According to the principle, "*for an evolutionary system, continuing development is needed just to maintain its fitness relative to the systems it is co-evolving with.*"

KEY TERMS

binary fission
biological weapon
Blueprint Priority Diseases
centrifugal rash
Disease X
dysbiosis
ecotourism
emerging disease hotspots
emerging viruses
epidemiologists
favus
genome
genomics
germ theory of disease
Global Virome Project (GVP)
immunocompromised patients
infectious disease control experts
infectious diseases
Koch's postulates
machine learning

metagenomics
mobile health
One Health
pandemic
partnerships
personal protective equipment (PPE)
public health
real-time surveillance
Red Queen hypothesis
reservoir
rinderpest
serology
sexually transmitted infections (STIs)
species barrier
spillover
syndrome/symptom surveillance
synthetic biology
trachoma
vector
zoonotic disease

SELF-EVALUATION

○ Part I: Choose the single best answer.

1. Despite major advances in sanitation and the availability of clean water, antibiotics, and immunization, infectious diseases remain _____.

    a. low

    b. an ongoing battle

    c. a mystery

    d. unchallenging

2. _____ are the most susceptible to opportunist infections.
   a. Healthy people
   b. Healthcare workers
   c. Organ recipients
   d. Zombies

3. _____ are used in the CDC blog to educate the public about emergency preparedness.
   a. Bacteria
   b. Vampires
   c. Viruses
   d. Zombies

4. Travelers may get more than they bargained for including a belly full of worms and malaria by _____.
   a. ecotourism
   b. cruise ship vacations
   c. gardening
   d. farming

5. Which agency is the primary support of medical research in the United States?
   a. PAHO
   b. AMC
   c. NIH
   d. FDA

6. Which agency focuses on the safety of foods, vaccines, antibiotics, and other medicinal and medical products?
   a. CDC
   b. FDA
   c. MSF
   d. World Bank

7. Continued advancement in society's understanding of its environment and established direction for future investigation is accomplished by _____.
   a. evolution
   b. sanitation
   c. research
   d. quarantine

8. The fundamental cause of disadvantaged populations suffering more than their fair share of infectious (and other) diseases is _____.
   a. poverty
   b. financial security
   c. politicians
   d. global change

9. Microbes may play a greater role in _____ diseases than previously suspected.
   a. acute
   b. respiratory
   c. brain
   d. chronic

10. In a sense, our partners in life are _____.
    a. plants
    b. insects
    c. microbes
    d. animals

O **Part II: Fill in the blank.**

1. Progress in _____ has resulted in a dramatic increase in the number of living immunocompromised patients.

2. The cruise ship industry has been subject to criticism because of outbreaks caused by _____.

3. _____ are essential for the maintenance of life on the planet, and others have been harnessed for a better quality of life.

4. Sanitation, vaccines, and _____ have made enormous strides against pathogenic microbes.

5. Microbes may act as causative agents, as cofactors, or as _____ that cause damage through auto-immune reactions.

6. The practices of personal and public health can be brought to bear on the prevention and control of _____.

7. The scientific and medical community once found it difficult to accept that stomach and intestinal ulcers are caused by _____, a bacterium present in the acidic environment of the stomach.

8. _____ was the first infectious disease eradicated from the world in 1980.

9. _____ are used as effective, safe, preparations that induce long-lasting immunity and prevents infections caused by specific microbes or viruses.

10. _____ is the city that is the headquarters of the CDC.

O Part III: Answer the following.

1. Explain the following statement: "The ongoing saga of infectious diseases is a reflection of the ongoing evolutionary dance."

2. Explain the One Health concept and its relevance to optimal health in the 21st century.

3. Create a list of items that should be included in an emergency supply kit if a pandemic were to occur.

4. List at least five ways in which microbes have challenged their hosts throughout history and have influenced the course of civilization.

5. Discuss why infectious disease outbreaks occur after natural catastrophic events.

6. List at least five examples of advancements in technology that have dramatically improved the life span and health of humans.

7. Explain the importance of real-time infectious disease surveillance on a worldwide basis.

8. List common infections in transplant recipients during their first year after transplantation.

9. Research and create a list of vaccines you would need to travel to the Republic of the Marshall Islands, Jekyll Island, or Peru.

10. Explain why partnerships are needed to make enormous health strides in many countries.

APPENDIX

MAIN SOURCES FOR CASE STUDIES

Chapter	Title of Case Study	Sources
1	Wounded Civil War Soldiers Who Glowed in the Dark	Glowing wounds. (n.d.). AAAS Science NetLinks. Retrieved from http://sciencenetlinks.com/science-news/science-updates/glowing-wounds/ Inman III, F. L., Singh, S., & Holmes, L. D. (2012). Mass production of the beneficial nematode *Heterorhabditis bacteriophora* and its bacterial symbiont. *Indian Journal of Microbiology, 52*, 316–324.
2	Saved by a Syringe Full of Dodge Pond Bacteriophages	Chan, B. K., Turner, P. E., Kim, S., Mojibian, H. R., Elefteriades, J. A., & Narayan, D. (2018). Phage treatment of an aortic graft infected with *Pseudomonas aeruginosa*. *Evolution, Medicine, and Public Health*, 2018(1), 60–66. doi10.1093/emph/eoy005 Chan, B. K., Sistrom, M., Wertz, J. E., Kortright, K. E., Narayan, D., & Turner, P. E. (2016). Phage selection restores antibiotic sensitivity in MDR *Pseudomonas aeruginosa*. *Scientific Reports*, 6, 26717. doi:101038/srep26717
3	The Romans and Toilet Phobia	Koloski-Ostrow, A. O. (2018). *The archaeology of sanitation in Roman Italy: Toilets, sewers, and water systems* (Reprint edition). Chapel Hill: The University of North Carolina Press. Wald, C. (2016). The secret history of ancient toilets: By scouring the remains of early loos and sewers, archaeologists are finding clues to what life was like in the Roman world and in other civilizations. *Nature, 533*, 456–458.
4	Don't Forget to *Safely* Wash Your Hands	Halden, R. U., Lindeman, A. E., Aiello, A. E., Andrews, D., Arnold, W. A., Fair. P., . . . Blum, A. (2017). The Florence Statement on Triclosan and Triclocarban. *Environmental Health Perspectives, 125*(6), 064501.
5	Microbes in the New York City Subway System	Afshinnekoo, E., Meydan, C., Chowdhury, S., Jaroudi, D., Boyer, C., Bernstein, N., . . . Mason, C. E. (2015). Geospatial resolution of human and bacterial diversity with city-scale metagenomics. *Cell Systems, 1*, 1–15.
6	Take Two Fecal Pills and Call Me in the Morning	Jiang, Z. D., Ajami, N. J., Petrosino, J. F., Jun, G., Hanis, C. L., Shah, M., . . . Dupont, H. L. (2017). Randomised clinical trial: Faecal microbiota transplantation for recurrent *Clostridium difficile* infection—fresh, or frozen, or lyophilized microbiota from a small pool of healthy donors delivered by colonoscopy. *Alimentary Pharmacology & Therapeutics, 45*, 899–908.
7	Don't Touch That Armadillo!	Domozych, R., Kim, E., Hart, S., & Greenwalk, J. D. (2016). Increasing incidence of leprosy and transmission from armadillos in Central Florida: A case series. *JAAD Case Reports, 2*, 189–192.

(continues)

Chapter	Title of Case Study	Sources
8	Dr. Crozier's Puzzling Eye Color Change	Varkey, J. B., Shantha, J. G., Crozier, I., Kraft, C. S., Marshall Lyon, G., Mehta, A. K., . . . Yeh, S. (2015). Persistence of Ebola virus in ocular fluid during convalescence. *New England Journal of Medicine, 372*, 2423–2469.
9	Solving the Mystery of Why Vampire Bats Can Live on Blood	Lisandra Zepeda Mendoza, M., Xiong, Z., & Thomas Gilbert, M. P. (2018). Hologenomic adaptations underlying the evolution of sanguivory and the common vampire bat. *Nature Ecology & Evolution, 2*, 659–668. Schneider, M.C., Romijn, P. C., Uieda, W., Tamayo, H., da Silva, D. F., Belotto, A., . . . Leanes, L. F. (2009). Rabies transmitted by vampire bats to humans: An emerging zoonotic disease in Latin America? *Revista panamericana de salud pública* (Pan American Journal of Public Health), *25*, 260–269.
10	The Smelly Chicken Factory Worker	Mills, C. M., Llewelyn, M. B., Kelly, D. R., & Holt, P. (1996). A man who pricked his finger and smelled putrid for 5 years. *Lancet, 348*, 1282.
11	You Can Get Fatal Herpes from a Monkey!	Wisely, S. M., Sayler, K. A., Anderson, C. J., Boyce, C. L., Klegarth, A. R., & Johnson, S. A. (2018). Macacine herpesvirus 1 antibody prevalence and DNA shedding among invasive rhesus macaques, Silver Springs State Park, Florida, USA. *Emerging Infectious Diseases, 24*, 345–350.
12	From Sea to Sepsis	Baker-Austin, C., Trinanes, J., Gonzalez-Escalona, N., & Martinez-Urtaza, J. (2017). Non-cholera vibrios: The microbial barometer of climate change. *Trends in Microbiology, 25*, 76–82. Centers for Disease Control and Prevention. (1996). *Vibrio vulnificus* infections associated with eating raw oysters—Los Angeles, 1996. *MMWR, 45*, 621–624. Centers for Disease Control and Prevention. (2017). *Vibrio vulnificus* infections and disasters. Fact Sheet. Retrieved from https://www.cdc.gov/disasters/vibriovulnificus.html Hendren, N., Sukamar, S., & Glazer, C. S. (2017). *Vibrio vulnificus* septic shock due to contaminated tattoo. *BMJ Case Report.* doi:10.1136/bcr-2017-220199 Sheer, A. J., Kline, K. P., & Lo, M. C. (2017). From sea to bloodstream: *Vibrio vulnificus* sepsis. *American Journal of Medicine, 130*, 1167–1169.
13	Surviving Ebola	Lyon, G. M., Mehta, A. K., Varkey, J. B., Brantly, K., Plyler, L., McElroy, A. K., . . . Ribner, B. S. (2014). Clinical care of two patients with Ebola virus disease in the United States. *New England Journal of Medicine, 371*, 2402–2409. Saving Dr. Brantly [video]. NBC News. Retrieved from https://www.nbcnews.com/feature/saving-dr-brantly
14	Killer Bagpipes	King, J., Richardson, M., Quinn, A.-M., Holme, J., & Chaudhuri, N. (2016). Bagpipe lung: A new type of interstitial lung disease? *Thorax, 72*, 380–382.
15	Baby Jacob	Bogaert, D., Van Schil, K., Taghon, T., Bordon, V., Bonroy, C., Dullaers, M., . . . Haerynck, F. (2016). Persistent rotavirus diarrhea post-transplant in a novel JAK3-SCID patient after vaccination [Letter to the editor]. *Pediatric Allergy and Immunology, 27*, 92–106. Patel, N. C., Hertel, P. M., Estes, M. K., de la Morena, M., Petru, A. M., Noroski, L. M., . . . Abramson, S. L. (2010). Vaccine-acquired rotavirus in infants with severe combined immunodeficiency. *New England Journal of Medicine, 362*(4), 314–319.
16	A Pain in the Back	Guli, C., Iacopino, D. G., Di Carlo, P., Colomba, C., Cascio, A., Giammanco, A., . . . Maugeri, R. (2017). Vancomycin resistant *Enterococcus faecium* (VRE) vertebral osteomyelitis after uneventful spinal surgery: A case report and literature review. *Interdisciplinary Neurosurgery, 7*, 12–16.

Chapter	Title of Case Study	Sources
17	Fighting the World's Deadliest Animal	Nading, A. M. (2015). The lively ethics of global health GMOs: The case of the Oxitec mosquito. *Biosocieties, 10*(1), 24–47.
		Ostera, G. R., & Gostin, L. O. (2011). Biosafety concerns involving genetically modified mosquitoes to combat malaria and dengue in developing countries. *JAMA, 305*(9), 930–931.
		Shelly, T., & McInnis, D. (2011). Road test for genetically modified mosquitoes. *Nature Biotechnology, 29*(11), 984.
		Waltz, E. (2016). GM mosquitoes fire first salvo against Zika virus. *Nature Biotechnology, 34*, 221–222.
18	The Threat of a Pandemic Keeps Virus Hunters Awake at Night	Carroll, D., Daszak, P., Wolfe, N. D., Gao, G. F., Morel, C. M., . . . Mazet, J. A. K. (2018). The Global Virome Project. *Science, 359*, 872–874. Retrieved from http://www.globalviromeproject.org
		Holmes, E. C., Rambaut, A., & Andersen, K. G. (2018). Pandemics: Spend on surveillance, not prediction. *Nature, 558*, 180–182.

abscess   An accumulation of pus containing immune cells and microbes in the body tissue due to microbial infection.

acellular (aP) vaccine   A pertussis vaccine that contains purified components of *Bordetella pertussis*.

acid-fast bacteria   Bacteria that resist decolorization by acid alcohol during the acid-fast staining procedure due to the presence of mycolic acids in their impermeable, waxlike cell walls. *Mycobacterium* and *Nocardia* are acid-fast genera.

acid-fast staining procedure   Bacterial differential staining method developed by Franz Ziehl and Friedrich Neelsen to identify bacteria that resist decolorization by acid alcohol during staining due to the presence of mycolic acids in their impermeable, waxlike cell walls.

acquired immune deficiency syndrome (AIDS)   A disease caused by the human immunodeficiency virus in which the immune system becomes severely compromised due to the depletion of T helper lymphocytes.

active carriers   Individuals who have a microbial disease that can be transmitted to others.

acute HIV infection   The earliest stage of HIV infection that occurs within 2 to 4 weeks after infection with HIV. Characterized by flulike symptoms as HIV multiples and spreads throughout the body. Also referred to as *acute retroviral syndrome (ARS)*.

acute retroviral syndrome (ARS)   The earliest stage of HIV infection that occurs within 2 to 4 weeks after infection with HIV. Characterized by flulike symptoms as HIV multiples and spreads throughout the body. Also referred to as *acute HIV infection*.

acyclovir   An antiviral drug used to treat herpesvirus infections, including shingles and genital herpes. Note that it does not cure herpesvirus infections but reduces the frequency of outbreaks. Original trade name was Zovirax® but today it has over 148 trade names and many generic forms of the antiviral drug.

adaptive immune system   Subset of the human immune system composed of highly specialized groups of cells—lymphocytes, T cells, and B cells—that eliminate specific pathogens and act to prevent future infections by the same pathogens through the development of immune system "memory."

adaptive immunity   Activation of the subset of the human immune system composed of highly specialized groups of cells—lymphocytes, T cells, and B cells—that eliminate specific pathogens and act to prevent future infections by same pathogens through the development of immune system "memory."

adenoids   Tissues located in the back of the throat that are associated with the secondary immune structures; their lymphocytes play a role in protection from microbes entering through the nose and mouth.

adenosine triphosphate (ATP)   A high-energy molecule that drives most cellular processes.

adhesins   Molecules on the surface of some bacteria that enable them to stick to host receptor molecules in a Velcro-like manner.

adsorption   The first stage in the viral replication cycle during which viruses attach to the surface of host cells.

aerobes   Microbes that require oxygen for their metabolic activities.

aerosols   Suspensions of airborne particles ranging from 1 to 5 micrometers that are a means of transmission of microbes.

afebrile   Absence of fever.

African sleeping sickness   A protozoan disease caused by the bite of a tsetse fly carrying the parasite. Two types (West and East African trypanosomiasis) are known.

African trypanosomiasis   A protozoan disease caused by the bite of a tsetse fly carrying the parasite. Two types are known.

agar   A heat-stable gelling agent used to prepare solid bacteriological media derived from seaweed. It cannot be broken down by microbial enzymes. Its development was instrumental in the development of the field of microbiology.

agar-agar   A heat-stable gelling agent used in making jams and jellies that was derived from seaweed. It cannot be broken down by microbial enzymes. It was applied in the late 1880s to the preparation of bacteriological media in Robert Koch's laboratory due to the suggestion of laboratory assistant Fran Angelina Hess.

agranular cells   White blood cells that do not have granules. Includes monocytes and lymphocytes.

ahi   A type of tuna used in sushi. Is generally consumed raw, and thus may contain parasites such as tapeworms.

AIDS (acquired immune deficiency syndrome)   A disease caused by the human immunodeficiency virus in which the immune system becomes severely compromised due to the depletion of T helper lymphocytes.

airport malaria   Malaria that is the result of the survival of malaria-infected mosquitoes from other countries and is transmitted to people within the vicinity of an airport.

alanine aminotransferase (ALT)   A liver enzyme that is detected in the bloodstream due to liver damage or disease. Chronic hepatitis B and C, alcohol abuse, or toxicity caused by certain medications can cause elevated levels of this enzyme in the bloodstream.

allergens   Substances that initiate an allergic reaction due to activation of the adaptive immune system (IgE antibodies that trigger the release of histamines).

allergy   An adverse immune response to molecules associated with a variety of nonbiological or biological triggers such as pollen, dust, foods, mites, nickel, antibiotics or other drugs, or bee stings.

ALT (alanine aminotransferase)   A liver enzyme that is detected in the bloodstream due to liver damage or disease. Chronic hepatitis B and C, alcohol abuse, or toxicity caused by certain medications can cause elevated levels of this enzyme in the bloodstream.

alveoli   The tiny air sacs at the end of the bronchioles of the lungs where gas exchange with the blood occurs.

amastigote   A nonmotile form of a protozoan that lacks cilia or flagella. It develops inside of vertebrate host cells during the life cycle of protozoans that belong to the family *Trypanosomatidae*, such as *Leishmania* spp.

ambulatory surgical centers   Healthcare facilities that perform outpatient surgeries that do not require overnight hospital stays.

amebiasis   A disease caused by the protozoan *Entamoeba histolytica*; it is prevalent in areas with poor sanitation.

amino acid   Nitrogen-containing molecule that is the building block of protein.

anaerobes   Microbes that do not require oxygen for their metabolic activities.

anaerobic process   Metabolic processes that occur in the absence of oxygen such as fermentation.

anaphylactic shock A life-threatening allergic reaction to an antigen in the body that requires immediate medical attention. Examples of symptoms include swelling, hives, impaired breathing, and low blood pressure.

anemia Lack of iron in the blood. Some microbial or viral infections can cause anemia through a variety of mechanisms.

anesthetic A substance that reduces a person's ability to feel pain.

Angel's Glow Phenomenon observed during Battle of Shiloh during the U.S. Civil War where the injuries of soldiers on the battlefield exhibited a blue glow and healed more quickly. Eventually determined to be caused by infection of wounds by the bioluminescent bacterium *Photorhabdus luminescens*.

animalcules The name for microbes first described by Antonie van Leeuwenhoek during his examination of tooth scrapings with a primitive microscope.

anthrax A potentially fatal disease caused by *Bacillus anthracis*; the bacterium has been used in bioterrorism and is a potential agent of biological warfare.

antibiotic resistance The ability of bacteria to resist the actions of an antibiotic. Can arise in two ways. First, a bacterium that survives because it has the ability to neutralize or escape the effect of the antibiotic can then multiply and replace all the bacteria that were killed off. Second, bacteria that were at one time susceptible to an antibiotic can acquire resistance through mutation.

antibiotic sensitivity testing Microbiological techniques used to determine which antibiotics will be most effective in the treatment of a patient suffering from a specific bacterial infection (e.g., placing antibiotic disks on lawn of bacteria grown on Mueller Hinton agar).

antibiotics Metabolic products of bacteria and fungi that kill or inhibit the growth of other microbes.

antibodies Protein molecules produced by an infected host in response to antigens; they react specifically with the antigen that triggered their production and are an important defense of the adaptive immune system.

anticodon A series of three nucleotides on tRNA that relate to codons on mRNA.

antiexcitatory drugs Neuroprotective drugs that reduce neurotransmitter activity in the brain by inducing a comatose state, allowing the body to heal from a serious infection. They were used in combination with antivirals in the experimental Milwaukee protocol used to treat human rabies.

antigenic drift A minor change in the H or N spikes of influenza A and B viruses occurring over a period of years. Antigenic drift produces seasonal viral strains that cause epidemics.

antigenic shift A major and abrupt antigenic change in the H or N spikes of the influenza A virus, resulting in a novel subtype or pandemic strain of the virus.

antigenic variation When a bacterium, virus, or protozoan changes its surface proteins in order to evade recognition by the host's immune response.

antigen-presenting cells Host cells that phagocytize microbes and present antigenic components to other cells of the immune system.

antigens Components of microbes, usually protein structures, that are recognized as foreign by the immune system and are targeted for destruction.

antimetabolite Antibiotics that inhibit a bacterium's synthesis and use of a normal metabolite such as sulfanilamides that inhibit bacteria by competing with para-aminobenzoic acid (PABA) synthesis. Some can be used as chemotherapy agents to target tumors.

antiparallel Characteristic of the DNA double helix whereby the two complementary strands run in opposite directions alongside each other.

antiretroviral A drug that is effective against retroviruses, such as HIV.

antiretroviral therapy (ART) HIV/AIDS treatment that involves the use of multiple antiretroviral drugs to suppress the progression of HIV disease.

antiseptic An agent used to minimize and destroy the growth of microbes in a wound.

antitoxin therapy The use of immune globulin or antibodies produced against bacterial toxins to treat intoxication (e.g., the treatment of botulism or lockjaw/tetanus with antitoxins).

antiviral drugs Class of drugs that target different steps during the viral replication of specific viruses; used to inhibit or stop the production of new infectious viruses in patients suffering from a viral disease.

apoptosis Programmed cell death that occurs naturally or may be triggered by immune system responses towards pathogens during the course of an infection.

arboviruses Viruses that are carried by arthropods.

Archaea One of three domains in the Woese system of classification, to which the archaebacteria are assigned.

ARS (acute retroviral syndrome) The earliest stage of HIV infection that occurs within 2 to 4 weeks after infection with HIV. Characterized by flulike symptoms as HIV multiples and spreads throughout the body. Also referred to as *acute HIV infection*.

arsenic Naturally occurring element that may contaminate groundwater resources. Usually excreted from the body, but ingestion of excess amounts can result in accumulation and adverse health effects (i.e., poisoning).

ART (antiretroviral therapy) HIV/AIDS treatment that involves the use of multiple antiretroviral drugs to suppress the progression of HIV disease.

arthritis Inflammation of the joints that can be caused by certain pathogens such as *Borrelia* spp. (cause of Lyme disease) or Chikungunya virus.

arthropods Animals belonging to the phylum Arthropa. They are characterized by jointed appendages; insects, spiders, snails and lobsters. Insects are important arthropod vectors in the transmission of multiple and diverse diseases affecting humans and other animals.

ascariasis Infection with parasitic *Ascaris* spp. roundworms. Most people do not experience symptoms, but those with large numbers of worms may experience abdominal pain and nausea.

aseptic technique Use of procedures such as handwashing to prevent contamination by pathogens and thus reduce risk of infection. In the laboratory, the set of procedures used to grow pure cultures of bacteria (e.g., flaming or incinerating loop before and after inoculating bacteriological medium with bacteria).

asexual reproduction When an offspring is produced from the genetic material of a single organism. Genetic recombination does not occur and the offspring is identical to the parent. Binary fission by a bacterium is an example.

Asilomar Conference on Recombinant DNA Conference held in 1975 to discuss the regulation and potential hazards of genetic engineering. Established guidelines for experiments using recombinant DNA to limit the potential risks of such experiments to humans and the larger environment.

aspartate aminotransferase (AST) A liver enzyme that is detected in the bloodstream due to liver damage or disease. Chronic hepatitis B and C, alcohol abuse, or toxicity caused by certain medications results in elevated levels of this enzyme in the bloodstream.

assembly A stage in the replication cycle of viruses in which the viral components are assembled into a particle or virion.

asymptomatic An organism that is infected with a pathogen but displays no signs and symptoms.

asymptomatic carrier   Organism that is infected with a pathogen but displays no signs and symptoms and can transmit the pathogen to others.

athlete's foot   A fungal infection affecting the feet, most typically between the toes. Also called *tinea pedis*.

ATP (adenosine triphosphate)   A high-energy molecule that drives most cellular processes.

attenuated strain   A weakened strain of a virus or microbe that can be used in a vaccine to stimulate an immune response but not the illness.

attenuated virus   A weakened strain of a virus that can be used in a vaccine to stimulate an immune response but not the illness.

atypical pneumonia   Respiratory tract infection caused by an unknown pathogen during routine testing for known pathogens in the medical technology laboratory.

autoclave   A device that generates steam under pressure to sterilize materials.

autoimmune disease   A disease in which the immune system fails to distinguish "self" from "nonself," thereby resulting in the host's attack against its organs or tissues (e.g., rheumatoid arthritis, lupus, or celiac disease).

autoinducer   Signaling molecule produced when a population of cells reaches a certain density. Plays a role in quorum sensing by bacteria.

autoinfection   A process in which infection is perpetuated within the body; pinworm and strongyloidiasis are examples.

autotrophs   Microbes and plants that are capable of utilizing the energy of the sun or derive energy from the metabolism of inorganic compounds.

avian influenza A virus (H7N4)   A subtype of avian influenza A virus isolated from a patient for the first time in February 2018. The virus originated in wild water fowl in China.

avian influenza A virus (H7N9)   A subtype of avian influenza A virus first isolated from a patient in China for the first time in 2013. The strain originated in wild water fowl. It is highly lethal in humans. As of September 6, 2018, there were 1,567 laboratory confirmed human cases in China (615 deaths: 39% mortality). To date, it is not highly transmissible from person to person.

avirulent   A microbe or virus that is unable to cause disease.

AZT (azidothymidine)   The first clinically safe and effective drug used for the treatment of AIDS; it acts as an inhibitor of the reverse transcriptase enzyme. Also called *zidovudine*.

B cell zones   Regions within the secondary lymphoid tissue containing mostly B cells. Regions are situated such that the antigen entering the lymphoid tissue can interact with the B cells, and the T cells are nearby for good cellular communication.

B cells   White blood cells that produce antibodies; component of the adaptive immune response.

B lymphocytes   White blood cells that produce antibodies; component of the adaptive immune response.

babesiosis   A tickborne protozoan disease caused by species of *Babesia*.

bacilli   Rod-shaped bacteria.

bacillus Calmette-Guérin vaccine   The only vaccine available against tuberculosis and has limited effectiveness. It is a live, attenuated strain of *Mycobacterium bovis*.

bacitracin   Topical antibiotic that blocks bacterial cell wall synthesis. It is used topically to prevent infection caused by Gram (+) bacteria in cuts or abrasions.

bacteremia   Presence of bacteria in the bloodstream.

Bacteria   One of three domains in the Woese system of classification. Unicellular microbes with distinct properties.

bactericidal   Any substance that can be used to kill bacteria, such as antibiotics and disinfectants.

bacteriocins   Proteins produced by bacteria that kill or inhibit the growth of similar or closely related bacterial species.

bacteriophage therapy   The use of bacteriophages to treat bacterial infections. Today it is being developed in Western medicine to treat multidrug-resistant bacterial infections but it has been used in Russia for over a century to treat bacterial infections.

bacteriophages   Viruses that infect and kill bacteria. Also called *phages*.

bacteriostatic   Agent that prevents bacteria from replicating, but not necessarily killing them.

Baghdad boil   Skin ulcer resulting from leishmaniasis, which is caused by the *Leishmania* spp., a parasite spread by sandflies.

barrier technique   Infection control practices such as wearing of gloves and masks in healthcare settings that are intended to prevent cross-contamination between healthcare workers and patients.

basal bodies   The structures by which flagella are anchored to the cell wall and cell membrane.

BASC (best available supportive care)   Treatment used to care for patients suffering from Ebola virus disease because there are no known antiviral therapies available. The standard protocol is to keep patients hydrated, correct electrolyte imbalances, provide nutritional support, and to critically manage care for respiratory and kidney failure.

basophils   White blood cells that are rich in histamine granules.

beaver fever   An intestinal disease caused by the protozoan *Giardia lamblia* and acquired by drinking contaminated water. Also referred to as *giardiasis*.

bed nets   Lightweight mesh placed as a canopy over beds in order to prevent bites from mosquitoes while sleeping. Preventive measure against mosquitoborne diseases such as malaria, dengue fever, and Chikungunya virus disease.

best available supportive care (BASC)   Treatment used to care for patients suffering from Ebola virus disease because there are no known antiviral therapies available. The standard protocol is to keep patients hydrated, correct electrolyte imbalances, provide nutritional support, and to critically manage care for respiratory and kidney failure.

BFL (bird fancier's lung)   A form of hypersensitivity pneumonitis caused by exposure to bird droppings and feathers.

β-lactam antibiotics   Class of broad-spectrum antibiotics that have a β-lactam ring in their structure. Includes common antibiotics such as penicillins and cephalosporins.

bifurcated needle   A type of needle with two prongs on the end that was designed to hold a dose of smallpox vaccine (variola virus). It provided a way to easily and cost-effectively deliver millions of doses of the smallpox vaccine, playing a large role in the global eradication of smallpox.

binary fission   An asexual mode of reproduction in which a cell splits into two new cells.

binomial system of nomenclature   A system of nomenclature established by Carolus Linnaeus in 1735 in which the genus and species are identified that is also used in the taxonomy of microbes (e.g., *Escherichia coli*).

bioaugmentation   Spraying of nutrients on beaches or on other microbe-contaminated sites to foster the growth of microbes indigenous to the area in order to accelerate degradation of pollutants.

biocide   Any chemical that destroys life, such as a pesticide, herbicide, or fungicide.

biocontainment   Management practices implemented to control spread of microbes.

biofilm   A community of microbes consisting of a single species or multiple species that produces a slimy, gluelike substance that anchors the microbes to living or inert surfaces. Can cause problems when they form in places such as inside of pipes or on medical implants or devices such as catheters.

Biofresh   Trademark name of clothing that has had triclosan, a controversial antimicrobial, added to the fabric.

biogeochemical cycles   Processes such as the carbon and nitrogen cycles in which microbes play a critical role.

biohacker   A person who uses genetic techniques to manipulate DNA for entertainment or malice.

bioinsecticides   Toxins produced by certain bacteria used to control insect pests (e.g., *Bacillus thuringiensis*, or Bt toxin).

biological terrorism   Employment of biological weapons by nonstate governments, religious cults, militants, or individuals with malicious intentions.

biological terrorist   Person who uses biological weapons for terroristic purposes.

biological vectors   Arthropods, including mosquitoes, ticks, lice, and flies, that transmit microbial disease.

biological warfare   The use of microbes, such as *Bacillus anthracis* and variola virus, as weapons in the conduct of war.

biological weapons   Biological agents that are deployed with the intention of producing clinical disease in an attempt to incapacitate or kill large numbers of individuals (e.g., *Bacillus anthracis*, botulinum toxin, or variola virus).

Biological Weapons Convention of 1972   Multilateral agreement banning the development, use, or stockpiling of biological weapons.

bioluminescent   Ability by some organisms, particularly bacteria, to undergo a chemical reaction that emits light, causing them to "glow" (e.g., *Photorhabdus luminescens*).

biopsy   Tissue removed from the body to check for the presence of disease.

bioremediation   Use of microbes, particularly bacteria, to solve an environmental problem, such as degrading pollutants in water (e.g., oil spills or cyanide used in extracting gold from ore) or soil (e.g., munitions on military bases).

biosafety   Use of preventative measures to reduce the risk of exposure and possible infection of humans or animals to potentially dangerous pathogens.

biosafety level 4 (BSL-4)   The highest of the four biosafety levels. Used to contain those pathogens that are particularly contagious and frequently fatal, such as Ebola virus or variola virus (cause of smallpox).

biosecurity   Use of preventative measures to reduce the risk of exposure of humans or animals to potentially dangerous pathogens.

biosolids   The product of the conversion of waste material that may be used as fertilizer; also called sludge.

biotechnology   Area of molecular biology that uses living microbes to develop new products (e.g., bacteria that produce human insulin or growth factor for treatment).

biotic   Of or relating to life; refers to the living components of an ecosystem.

bird fancier's lung (BFL)   A form of hypersensitivity pneumonitis caused by exposure to bird droppings and feathers.

biting plates   Mouth parts found in some worms that allow them to attach to and suck blood from their host.

Black Death   A bacterial disease caused by *Yersinia pestis*, a highly virulent bacterium; also known as the *plague*.

blacklegged tick   Arthropod vector responsible for the transmission of the *Borrelia burgdorferi*, the bacterium that causes Lyme disease, or Powassan encephalitis caused by Powassan virus. Also called *deer tick*.

blastomycosis   Fungal infection of the lungs caused by inhaling spores of *Blastomyces* spp. The fungi lives in moist soils and are found primarily in the Great Lakes region of the United States and Canada.

blender   An organism that facilitates the creation of a novel strain with the potential to cause severe disease and a possible pandemic; an example is that of a pig that becomes co-infected with a human strain and an avian strain of influenza A virus, resulting in a novel recombinant influenza A virus.

blood flukes   Common name for *Schistosoma* species and some other worms.

blood meal   The blood ingested by a blood-sucking insect such as a mosquito or tick (required by the insect vector to produce eggs or undergo morphogenesis).

bloodborne pathogens   Pathogens that are present in human blood and can cause disease in humans such as HIV, hepatitis B and C viruses or *Plasmodium* spp. (cause of malaria).

Blueprint Priority Diseases   Diseases and pathogens identified by the World Health Organization as having priority with regard to research and eradication efforts.

body orifices   Openings in the body of an animal, such as the mouth, nose, ear canals, etc. Provide a portal of entry or exit for pathogens.

boils   Localized skin infections frequently caused by staphylococci.

bone marrow   Source of all blood cells.

borreliosis   More commonly known as *Lyme disease*. A disease primarily caused by bites from ticks carrying the bacteria *Borrelia burgdorferi*. The disease is often characterized by a bull's-eye rash, flulike symptoms, and joint pain.

botox   Minute doses of *Clostridium botulinum* exotoxin used to reduce wrinkles and to treat common disorders associated with muscle overactivity.

botulism   A form of intoxication caused by the ingestion of foods contaminated with *Clostridium botulinum*; if untreated, the neurotoxin produced by the bacterium causes flaccid paralysis.

bovine spongiform encephalopathy (BSE)   A neurological condition in cattle resulting from abnormally folded proteins called prions that convert normally folded prions into the defective form; also referred to as *mad cow disease*.

brain-eating amoebas   An amoeba (*Naegleria fowleri*) found in warm water, including warm lakes and ponds, untreated well water, and untreated spas and pools. When a person gets water from a contaminated source into his or her nasal passages, the amoebas can travel to the brain, which they consume as a food source.

breakbone fever   Another name for *Dengue fever*, a mosquitoborne disease associated with muscle and joint pain.

breath test   A means of diagnosing peptic ulcers based on the production of urease by *Helicobacter pylori*.

broad-spectrum antibiotic   Antibiotics that are effective against a range of Gram (+) and Gram (−) bacteria.

bronchiolitis   A respiratory tract infection affecting the bronchioles that lead to the lungs.

BSE (bovine spongiform encephalopathy)   A neurological condition in cattle resulting from abnormally folded proteins called prions that convert normally folded prions into the defective form; also referred to as *mad cow disease*.

Bt toxin   Toxin produced by *Bacillus thuringiensis* that kills insects. Manufactured for use as an organic biological insecticide.

buboes   Enlarged lymph nodes that occur when bacteria localize in the lymph nodes, such as in bubonic plague caused by *Yersinia pestis*.

bubonic plague A form of *Yersinia pestis* infection in which the bacteria localize in lymph nodes, causing them to swell to the size of hens' eggs.

budding A mechanism by which some enveloped viruses such as HIV or measles virus are released from the host cell at the end of its replication cycle without killing it; also a cell division feature of certain yeast such as *Saccharyomyces cerevisiae*.

bulbar poliomyelitis An extremely serious form of polio in which individuals have difficulty swallowing and breathing because of muscle paralysis.

bull's-eye skin rash A red target-shaped rash characteristic of Lyme disease.

bullae Air pockets filled with fluid (i.e., blisters) characteristic of bubonic plague.

burst The release of bacteriophages from a cell as a result of the splitting open (lysis) of the bacterial cell wall. As many as a few hundred new bacteriophages may be released during the lysis of one infected bacterium.

campylobacteriosis Foodborne illness resulting in diarrhea, bloating, and fever caused by *Campylobacter* spp.

CA-MRSA (community-associated methicillin-resistant *Staphylococcus aureus*) MRSA infection in patients who did not acquire the infection in a hospital or other healthcare setting (e.g., high school athletes suffering from CA-MRSA infections that occurred during contact sports such as football).

Candida auris A type of yeast that causes infection in humans. It is considered an emerging health threat because it is often resistant to multiple antifungal drugs. Outbreaks have occurred in healthcare facilities and in patients with compromised immune systems.

candidiasis A yeast infection that forms in the mouth or throat (thrush) or vagina. Can also spread to other parts of the body. It is an opportunistic infection that frequently occurs in people with suppressed immune systems, such as AIDS patients or cancer patients undergoing chemotherapy.

capsid The protein coat or shell that encloses and protects the nucleic acid genome of a virus.

capsomeres Identical structural protein units that make up the viral capsid; they confer helical, polyhedral, or complex shapes.

capsule A component of the bacterial envelope that may contribute to virulence; it is not present in all species of bacteria.

carbapenemase A β-lactamase enzyme produced by certain species of bacteria that renders the bacteria resistant to the effects of a variety of antibiotics.

carbapenem-resistant *Enterobacteriaceae* (CRE) A healthcare-acquired infection (HAI) caused by a highly antibiotic resistant bacteria that are able to break down carbapenems. This is a severe type of HAI with high mortality rates.

carbapenems A broad-spectrum antibiotic that kills both Gram (+) and Gram (–) bacteria.

carbohydrates Organic molecules that function as an energy source (e.g., glucose, lactose, and mannitol) and are found in some cellular structures including capsules; also referred to as *sugars*.

carbolic acid Antiseptic used to prevent contamination of surgical wounds by pathogens. Use was pioneered by Joseph Lister in the late 1800s. Also called *phenol*.

carbuncles Localized skin infections caused by staphylococci that are larger and deeper than boils and can reach baseball size.

carnivore An animal that feeds on the meat of other animals.

carrier An organism that is infected with a pathogen and can transmit the pathogen to others.

Carter Center Nonprofit foundation founded by former president Jimmy Carter. One of the center's missions is to eradicate guinea worm disease based on community-based interventions such as filtering drinking water and keeping people with emerging guinea worms from drinking water sources.

Cas 9 proteins Proteins that bind to DNA using a CRISPR RNA guide and make site-specific cuts to the DNA sequence.

Category A agents Biological agents and the diseases that pose the highest risk to national security. Include anthrax, botulism, plague, smallpox, tularemia, and viral hemorrhagic fevers.

Category B agents Biological agents and the diseases that pose the second highest risk to national security. They are moderately easy to disseminate, cause moderate morbidity and low mortality, and require specific enhancements of the CDC's diagnostic capacity and enhanced disease surveillance. Include *Coxiella burnetii* (Q fever), *Brucella* species (brucellosis), *Burkholderia mallei* (glanders), ricin toxin, epsilon toxin from *Clostridium perfringens*, and *Staphylococcus* enterotoxin B.

Category C agents Biological agents and diseases that pose the third highest risk to national security. They include emerging pathogens that could be engineered for mass distribution in the future because of availability, ease of production and dissemination, and potential for high morbidity and mortality and major health impact. Include Nipah virus, hantaviruses, tickborne hemorrhagic fever viruses, tickborne encephalitis viruses, yellow fever virus, and multidrug-resistant *Mycobacterium tuberculosis*.

cauterization An early method of treating wounds inflicted by rabid animals in which long, sharp, hot needles were inserted deeply into the wounds.

CAUTI (catheter-associated urinary tract infection) Healthcare-associated infection caused by introduction of pathogens into the urinary tract due to catheter use.

cave disease Another name for *histoplasmosis*. A fungal infection caused by breathing spores of *Histoplasma capsulatum*, which is found in soil and bird and bat droppings. Most common in the Mississippi and Ohio River Valleys. Can cause serious respiratory infections in infants and people with compromised immune systems.

CD (cluster of differentiation) molecules Distinctive molecules found on T cells that identify their specific role.

CD4 molecules Receptor molecules found on some T cells that identify them as T-helper cells.

CD8 molecules Receptor molecules found on some T cells that identify them as cytotoxic T cells.

CDC (Centers for Disease Control and Prevention) Federal agency tasked with protecting people from health, safety, and security threats. Focus is on disease control and prevention.

ceftazidime An antibiotic used to treat bacterial infections. Acts by inhibition of bacterial cell wall synthesis.

cell The basic unit of life.

cell culture The cultivation of mammalian cells outside of the host that supports their growth under controlled conditions (e.g., monkey kidney cells, HeLa cells, mouse L cells); cell cultures can be used as a host system to propagate viruses in the laboratory.

cell-mediated immunity A type of specific immunity that results in the production of sensitized T lymphocytes directed against a particular antigen.

cell membrane A component of bacterial cell walls that are composed of lipids and proteins that regulate the passage of molecules between the bacterium and its external environment; eucaryotic cells also contain a cellular plasma membrane that surrounds the cytoplasm and performs a similar function in the exchange of molecules into and outside of the cell. Membranes are also wrapped around the internal organelles, including the nucleus of eukaryotic cells. Enveloped viruses steal parts of cellular membranes, which are wrapped around the viral capsid.

cell theory   Theory that all living organisms are composed of one or more cells, that the cell is the basic unit of structure and organization or organisms, and that all cells arise from preexisting cells.

cell wall   An outer component of a bacterium that confers its characteristic shape and structural integrity.

cellulitis   A bacterial skin infection that appears red and hot to the touch. Often caused by *Staphylococcus* and *Streptococcus* spp.

Center for Disease Research and Policy (CIDRAP)   Center located at the University of Minnesota that provides disease surveillance in its efforts to address public health preparedness and response to emerging infectious diseases.

Centers for Disease Control and Prevention (CDC)   Federal agency tasked with protecting people from health, safety, and security threats. Focus is on disease control and prevention.

central line–associated bloodstream infection (CLABSI)   Healthcare-associated infection caused by introduction of pathogens into the bloodstream by tubes placed into veins.

central nervous system (CNS)   The part of the nervous system comprising the brain and the spinal cord.

centrifugal rash   A pattern of rash where the lesions are concentrated on the face and extremities with fewer lesions on the abdomen; characteristic of smallpox.

cephalosporin   An antibiotic containing a β-lactam ring structure that disrupts cell wall synthesis similar to penicillin.

cercariae   Free-swimming immature forms of *Schistosoma* that are responsible for "swimmer's itch."

CF (cystic fibrosis)   Genetic disease that causes buildup of sticky, thick mucus in the lungs. Results in breathing problems and susceptibility to bacterial lung infections, in particular *Pseudomonas aeruginosa* infections.

Chagas disease   A protozoan disease caused by *Trypanosoma cruzi* that leads to widespread tissue damage, particularly to the heart, causing it to enlarge and impairing its function.

Chamberland porcelain ultrafilter   An ultrafilter that retains nonfilterable bacteria but allows viruses in a liquid to pass through. Used to remove bacteria from drinking water.

chancres   Sores on the penis or on the cervix that are characteristic of the primary stage of syphilis.

cheater bacteria   Bacteria that respond to quorum-sensing signals but do not turn on genes that code for proteases. However, they do show up and use the nutrients made available through the group activity. The formation of biofilms by quorum-sensing bacterial pathogens restricts these bacteria.

chemosynthetic autotrophs   Microbes that derive energy from the metabolism of inorganic compounds.

chemotaxis   The process of moving toward or away from a chemical stimulus.

chickenpox   A disease characterized by blisterlike lesions on the body and typically occurring in childhood.

Chikungunya virus disease (CHIK)   A disease caused by the Chikungunya virus. The virus is transmitted to humans by *Aedes aegypti* and *Aedes albopictus* mosquitoes. Main symptoms are joint pain, fever, and headache. Occurs worldwide, but primarily in Africa, Asia, and India.

childbed fever   A postpartum streptococci infection characterized by fever and lower abdominal pain that begins the second or third day after delivery. Can result in death. Incidence greatly decreased after Ignaz Semmelweis promoted the practice of handwashing to reduce the transmission of bacteria by obstetricians during the late 1800s in Europe. Also called *puerperal fever*.

chlamydia   A common sexually transmitted infection caused by *Chlamydia trachomatis*. Often presents without symptoms but when symptoms do emerge they include genital pain and discharge from the penis or vagina. Curable with antibiotics.

Chlorella   A fresh water photosynthetic algae.

cholera   A disease caused by *Vibrio cholerae* that is characterized by severe diarrhea and dehydration.

chorioamnionitis   A complication of pregnancy caused by a bacterial infection of the fetal amnion and chorion membranes in which bacteria ascend from the vagina, rectum or anus into the uterus. It occurs in 2% of births in the United States and is a cause of premature delivery.

chromosome   The structure into which organismal DNA is organized.

chronic carriers   Individuals who harbor a pathogen for long periods without becoming ill with the disease but who may spread the disease to others.

chronic granulomatous disease   An inherited disorder of phagocytes characterized by their inability to kill bacteria.

chronic wasting disease (CWD)   A form of transmissible spongiform encephalopathy (TSE) that affects deer. Characterized by wasting syndrome leading to death.

CIDRAP (Center for Disease Research and Policy)   Center located at the University of Minnesota that provides disease surveillance in its efforts to address public health preparedness and response to emerging infectious diseases.

cilia   Hairlike projections present on the surface of eucaryotic cells. Ciliated eucaryotic epithelial cells that line the respiratory tract assist in the removal of mucus-trapped microbes and other inhaled particles.

ciprofloxacin   An antibiotic used to treat bacterial infections. Acts by inhibiting the ability of bacteria to replicate and repair their DNA.

cirrhosis   Irreversible scarring of the liver. Caused by a variety of conditions, including hepatitis B and C.

CJD (Creutzfeldt-Jakob disease)   A genetic and sporadic form of a human transmissible spongiform encephalopathy in which the brain becomes sponge-like, resulting in a fatal neurodegenerative disease.

CLABSI (central line–associated bloodstream infection)   Healthcare-associated infection caused by introduction of pathogens into the bloodstream by tubes placed into veins.

claw hand   Characteristic of leprosy where neurological damage impairs a person's ability to extend the fingers and open the hand.

climate change   Observable increase in global temperatures over the past century that is impacting Earth's climate system. Anticipated effects include rising sea levels, changing precipitation, and expansion of deserts in the subtropics. The largest human influence on increasing temperatures has been the emission of greenhouse gases.

clinical latency period   Second stage of HIV infection where the virus is present and replicates slowly in the body.

Clostridium difficile **(C. diff.) infections**   Difficult-to-treat infection that causes colitis, an inflammation of the colon. Often spread in healthcare facilities.

Clostridium difficile **(CDI) event**   Healthcare-associated infection caused by *Clostridium difficile*. Chronic antibiotic therapy can result in the overgrowth of *C. difficile* in the gut.

clotrimazole   A topical antifungal medication used to treat candidiasis (yeast infection).

cluster of differentiation (CD) molecules   Distinctive molecules found on T cells that identify their specific role.

CMV (cytomegalovirus)   A herpesvirus known to infect most people but that can cause blindness in those with suppressed immune systems (e.g., AIDS patients and patients taking immunosuppressants) and congenital defects in a developing fetus if a mother becomes infected with CMV during pregnancy.

CNS (central nervous system)   The part of the nervous system comprising the brain and the spinal cord.

coagulase An enzyme produced by pathogenic *Staphylococcus* species that coagulates or causes blot clots through the conversion of fibrinogen to fibrin.

cocci Spherical-shaped bacteria (e.g., *Streptococcus* or *Staphylococcus*).

coccidioidomycosis Valley fever, a fungal infection caused by *Coccidioides immitis*. The fungus is endemic to parts of the southwestern United States and northern Mexico. Primarily affects the lungs and is caused by inhalation of airborne fungal spores that are swept into the air by disruption of contaminated soils during dust storms.

codon A series of three nucleotides on mRNA that relates to anticodons.

colitis A chronic inflammatory bowel disease recently associated with dysbiosis of human gut microbiota.

colonies Visible masses of bacterial cells growing on an agar surface of solid bacteriological media, each presumably derived from a single cell or clone.

colonization Growth and multiplication of a microbe in a particular niche, resulting in large numbers of cells that can either cause infection or provide protection from invading pathogens.

colonization resistance Various mechanisms whereby gut microbiota can prevent invading pathogens from colonizing the intestinal tract.

commensal microbe A microbe in a relationship with another organism whereby the microbe obtains food or another benefit without hurting or helping the other organism. Normal microbiota contain commensal microbes.

commensalism A symbiotic relationship between two species in which one benefits and the other is neither harmed nor benefited.

common-source epidemics Outbreaks of disease arising from contact with a single contaminated source, typically associated with fecally contaminated food or water.

communicable disease Contagious disease caused by microbes that can be transmitted from one person to another.

community-associated methicillin-resistant *Staphylococcus aureus* (CA-MRSA) MRSA infection in patients who did not acquire the infection in a hospital or other healthcare setting (e.g., high school athletes suffering from CA-MRSA infections that occurred during contact sports such as football).

competent Bacterial cells that are able to take up DNA from the environment resulting in genetic transformation.

complement A series of blood proteins that constitute a significant defense mechanism against disease-causing microbes; component of the innate immune system.

complementary The two strands of double-stranded DNA contain complementary base pairs held together by hydrogen bonds. Adenine (A) always pairs with thymine (T) and guanine (G) always pairs with cytosine (C).

complex Term used to describe the shape of viruses that do not fit the typical icosahedral or helical structure shape (e.g., variola virus and herpesviruses).

compound microscope A light microscope that has two or more convex lenses. Allows for higher magnification than single-lens magnifiers, typically enabling a magnification range between 10× and 2000×.

congenital defect A medical condition that is present at birth. May be caused by exposure to pathogens in utero, chemicals, or certain medications, in addition to those caused by genetic defects (e.g., Zika and measles virus, and cytomegaloviruses are associated with congenital defects in newborns).

congenital syphilis Syphilis resulting from the passage of spirochete-shaped bacteria (*Treponema pallidum*) across the placenta from mother to baby.

congenital viral infections A viral infection of a fetus or newborn; often results in birth defects (e.g., cytomegalovirus infections are the number one cause of hearing loss in newborns).

congenital Zika syndrome Syndrome affecting infants and children who were exposed to Zika virus in utero. Symptoms may include microcephaly, decreased brain tissue, eye damage, clubfoot, and hypertonia.

conjugation A recombinational process in bacteria resulting in the transfer of DNA from donor to recipient during physical contact with a pilus.

conjunctivitis An inflammation of the conjunctiva of the eye resulting in redness of the eyeball and inner eyelid. Also called *pink eye* and can be caused by a variety of bacterial or viral pathogens. It is also a symptom of Zika virus disease that distinguishes it from other mosquitoborne viral diseases.

consumers Organisms that take in oxygen and release carbon dioxide (e.g., humans and mammals).

consumption Historical term used to refer to progressive wasting away of the body from tuberculosis.

contact tracing Measures used in the identification and diagnosis of people who may have come into contact with an infected person in an effort to stop the spread of infectious disease. It is used to prevent the spread of Ebola virus disease during outbreaks in remote locations of Africa.

contagion Historical term used to describe spread of infectious disease.

containment books Record-keeping of dates and cases of smallpox used in smallpox outbreak investigations during the WHO's smallpox eradication program.

convalescence period The time in which recovery from an illness takes place, strength is regained, repair of damaged tissue occurs, and rashes disappear.

convalescent blood Passive immunization method where blood from a person who has recovered from an infection is given to a person with an active infection in an attempt to treat the disease. Used as an experimental treatment method for Ebola virus disease (EVD).

Coordinated Framework for the Regulation of Biotechnology Formal policy established in 1986 by the federal government that outlines the responsibilities of the USDA's Animal and Plant Health Inspection Service (USDA-APHIS), the U.S. Environmental Protection Agency (EPA), and the Department of Health and Human Services' Food and Drug Administration (FDA) in ensuring the safety and regulation of genetically engineered agricultural products.

copepods Water fleas that can harbor infective guinea worm larvae.

coreceptor A second cellular receptor necessary for viral entry into a host cell (e.g., HIV binding to a chemokine coreceptor for entry).

cough etiquette Coughing or sneezing into the crook of one's elbow or hand to prevent the airborne spread of influenza A or B viruses and other respiratory pathogens.

course of a disease Generalized model of five general stages of an infectious disease: incubation, prodromal period, period of illness, period of decline, and convalescence period.

CPEs (cytopathic effects) Molecular and cellular changes that occur in eucaryotic cells as a result of viral replication.

CRE (carbapenem-resistant *Enterobacteriaceae*) A healthcare-acquired infection (HAI) caused by a highly antibiotic resistant bacteria that are able to break down carbapenems. This is a severe type of HAI with high mortality rates.

Creutzfeldt-Jakob disease (CJD) A genetic and sporadic form of a human transmissible spongiform encephalopathy in which the brain becomes sponge-like, resulting in a fatal neurodegenerative disease.

CRISPR In bacteria, snippets of viral DNA that have been incorporated into the bacterial genome during past attacks. The presence of the viral DNA enables the bacterium to respond more quickly to future infections by similar viruses.

CRISPR/Cas A complex of a Cas protein with a segment of Clustered Regularly Interspaced Short Palindromic Repeats (CRISPR) that is used by prokaryotic cells to defend against viruses. This system can be used to alter DNA at specific locations, a process known as *genome editing*.

cross-resistance The use of soaps or other products that contain antimicrobial compounds, such as triclosan, that leads to resistance to antibiotics used to treat infections of patients.

cryptococcosis Fungal disease caused by *Cryptoccocus* spp. Typically affects people with compromised immune systems, such as those with HIV/AIDS, causing pneumonia-like symptoms when it affects the lungs and encephalitis when it affects the brain.

cryptosporidiosis A protozoan disease transmitted by drinking fecally contaminated water.

crystal toxins Insecticidal protein-based crystals released by some bacteria such as *Photorhabdus luminescens* and *Bacillus thuringiensis*.

crystal violet A dye used in the Gram stain method used in bacterial classification. Also referred to as *gentian violet*.

cytotoxic T lymphocytes (CTLs) Members of the T-cell subset designated CD8.

cutaneous anthrax A form of anthrax that is acquired by contact with *Bacillus anthracis* or its spores via wool, hides, leather, or hair products.

cutaneous leishmaniasis A form of leishmaniasis that results in skin lesions.

CWD (chronic wasting disease) A form of transmissible spongiform encephalopathy (TSE) that affects deer. Characterized by wasting syndrome leading to death.

cyst Resistant structure formed during the life cycle of certain protozoans such as *Giardia* and *Cryptosporidium* that are responsible for the transmission of disease. Cysts can withstand harsh environmental conditions, including chlorination in swimming pools.

cystic fibrosis (CF) Genetic disease that causes buildup of sticky, thick mucus in the lungs. Results in breathing problems and susceptibility to bacterial lung infections, in particular *Pseudomonas aeruginosa* infections.

cysticerci Tapeworm larvae that are enclosed within a membranous sac.

cysticercosis A disease resulting from the ingestion of pork tapeworm eggs.

cytokine storm Overproduction of cytokines by immune cells that causes the lungs to accumulate fluid. Has been associated with particularly deadly influenza A virus strains, especially the virus that caused the 1918 Spanish flu pandemic.

cytokines Products that are released by lymphocytes in response to stimuli and that trigger responses in other cells.

cytomegalovirus (CMV) A herpesvirus known to infect most people but that can cause blindness in those with suppressed immune systems (e.g., AIDS patients and patients taking immunosuppresants) and congenital defects in a developing fetus if a mother becomes infected with CMV during pregnancy.

cytopathic effect (CPE) Molecular and cellular changes that occur in eucaryotic cells as a result of viral replication.

cytoplasm The fluid (which also contains associated cellular components, e.g., ribosomes and chromosomal DNA) that is enclosed by a cell wall in procaryotes and the fluid (which also contains ribosomes, membrane-bound organelles including the nucleus) contained by the plasma membrane of eucaryotic cells. The processes of DNA replication, transcription, and translation occur in the cytoplasm of procaryotes. The processes of DNA replication and RNA transcription occur in the nucleus and translation of proteins occurs in the cytoplasm of eucaryotic cells.

cytotoxic T lymphocytes (CTLs) Members of the T-cell subset designated CD8; involved in killing virally-infected cells; plays a role in the adaptive immune system response.

cytotoxins Bacterial toxins that damage or kill host cells at the site of infection.

DAAs (direct-acting antivirals) A new class of antiviral drugs that specifically target proteins involved in the replication cycle of hepatitis C virus, thereby disrupting viral replication or the production of progeny infectious viruses. They have been hailed as a cure for hepatitis C.

DDT (dichlorodiphenyltrichloroethane) An organochloride compound that was widely used as an insecticide throughout the world until it was found to have negative effects on wildlife (e.g., a decline in the bald eagle population) in the environment. It is no longer used in the United States.

death phase The final phase of the bacterial growth curve in which bacteria die as nutrients and resources are depleted.

debridement Removal of dead (necrotic), damaged, or infected tissue to improve the healing process (e.g., debriding necrotic tissues caused by flesh-eating bacteria).

decolorization Removal of dye from a stained bacterial specimen (e.g., in the Gram stain procedure or the acid-fast staining procedure).

decomposers Microbes that break down compounds into simpler constituents.

deer tick Arthropod vector responsible for the transmission of the *Borrelia burgdorferi*, the bacterium that causes Lyme disease, and Powassan virus that causes Powassan encephalitis. Also called *blacklegged tick*.

defensive strategies Adaptations that allow microbes to escape destruction by the host immune system.

deforestation Widespread clearance of forests and conversion of forestland to nonforest use such as gold and diamond mining in the Amazon rainforest. Will expand the range of infectious disease vectors (e.g., mosquitoes) and result in a loss of habitat for species that carry zoonotic pathogens (e.g., bats carrying rabies or Ebola viruses), increasing their proximity and risk of infection in human populations.

deletion mutation Removal of one (or more) nucleotides in DNA; can occur naturally during DNA replication or in genetic engineering applications.

dendritic cells Antigen-presenting cells (APCs) of the immune system that present antigens on the cell surface to the T cells of the immune system. They act as messengers between the innate and adaptive parts of the immune system.

dengue Another name for *Dengue fever*, a mosquitoborne disease associated with muscle and joint pain; also called *breakbone fever*.

dengue hemorrhagic fever A potentially fatal disease that can result from infection with a dengue virus strain different from the one causing the initial infection.

denitrifying Returning nitrogen to the atmosphere.

deoxyribonucleic acid (DNA) Biological macromolecule that stores the genetic information of all living organisms.

deoxyribose The five-carbon sugar molecule that alternates with phosphate groups to form the backbone of DNA.

dermatomycosis A fungal infection of the skin. Includes conditions such as tinea and ringworm.

dermatophytes Term used to describe pathogenic fungi that cause infections of the skin, mucous membranes, scalp, and nails.

dermotropic Term used to describe viruses that enter the skin to cause infection (usually via a puncture or wound in the skin, e.g., papillomaviruses).

developing countries   Countries that have a less developed industrial base and less infrastructure than more developed countries; populations are generally poor and reliant on agriculture.

diarrhea   Frequent passage of loose, soft stools. Can be caused by a variety of bacterial, parasitic, or viral pathogens.

diatom   A type of unicellular alga.

dichlorodiphenyltrichloroethane (DDT)   An organochloride compound that was widely used as an insecticide throughout the world until it was found to have negative effects on wildlife (e.g., a decline in the bald eagle population) in the environment. It is no longer used in the United States.

differential count   A reflection of the ratio of the white blood cell categories.

differential staining   Use of two or more dyes in order to differentiate between different types of bacteria in the same clinical specimen (e.g., the Gram and acid-fast staining).

diffusion   The passive movement of a substance or molecules from an area of high concentration to an area of low concentration.

DiGeorge syndrome   An immune disorder resulting from abnormal development of the thymus gland.

dinoflagellate   A type of unicellular algae containing two flagella found in marine and fresh water environments; dinoflagellates are the primary source of food in the oceans.

diphtheria   An upper respiratory tract infection caused by *Corynebacterium diphtheriae*.

diplococci   Groups of cocci that occur in pairs (e.g., *Neisseria* spp. are diplococci).

direct fluorescent antibody test   A diagnostic test used for the detection of antigens using fluorescent tagged antibodies.

direct observational therapy short course (DOTS)   A strategy to ensure that individuals infected with tuberculosis take their prescribed antibiotics.

direct-acting antivirals (DAAs)   A new class of antiviral drugs that specifically target proteins involved in the replication cycle of hepatitis C virus, thereby disrupting viral replication or the production of new or progeny viruses. They have been hailed as a cure for hepatitis C.

disconjugate gaze   Inability of the eyes to turn together in the same direction (e.g., a complication experienced by some Ebola virus disease survivors).

disease   A possible outcome of infection in which health is impaired in some fashion.

disease biomarkers   A measurable protein that indicates the presence of the disease biological or pathological processes, such as inflammatory proteins or signal proteins that evoke inflammation that can be targeted with therapeutic intervention (anti-inflammatory proteins that bind to and that block their activity).

disease-specific surveillance   Surveillance of a disease through numerous tracking systems. Targets a specific disease or set of symptoms in a defined population.

disease triangle model   Addresses the interactions among the host (e.g., humans), infectious agent, and environment that produce infectious disease. Whenever there is a change in the host, pathogen, or environment, an infectious disease has the potential to occur.

disease vector   Any carrier of an infectious agent capable of transmitting a pathogen into a living host. A disease vector can be living or nonliving. Examples include fomites, dust particles that are inhaled, and mosquitoes, ticks, or fleas that harbor pathogens.

Disease X   The as yet unknown infection that epidemiologists look for as the source of the next global pandemic.

disinfection   A process of killing or minimizing microbes, usually on surfaces.

distilled spirits   Alcoholic beverages resulting from microbial fermentation (e.g., yeast) and having a high alcoholic content (e.g., brandy, rum, and whiskey).

DNA   Biological macromolecule that stores the genetic information of all living organisms.

DNA gyrase   Bacterial enzyme involved in coiling of circular DNA during DNA replication.

DNA polymerase   Enzyme that synthesizes DNA molecules by assembling nucleotides, the building blocks of DNA.

DNA replication   Cellular process whereby a DNA molecule consisting of two complementary strands of double-stranded DNA is synthesized by DNA polymerase by reading and copying the original DNA template molecule.

DNA replication fork   Region during replication of DNA where the DNA double helix is open and the single-stranded DNA is exposed, acting as a new single-stranded DNA template.

DNA sequencing   Molecular technique used to determine the order of nucleotides (adenine, thymine, guanine, and cytosine) in a DNA molecule; useful in diagnostics, research, and metagenomics applications.

DNase   Enzyme that degrades DNA into fragments. Used in a variety of genetic engineering applications.

Doctors Without Borders   Humanitarian organization with French origins (*Médecins Sans Frontières*, or MSF) that provides medical care in areas affected by war and endemic diseases that are often lacking in basic healthcare services.

dormant state   Strategy where a microbe, such as a bacterium that produces endospores, minimizes metabolism to conserve energy. Often used as a strategy to survive adverse environmental conditions. When conditions improve, the microbe becomes metabolically active again. Certain bacteriophages and mammalian viruses (e.g., herpesviruses) can also exist in a dormant state (e.g., prophage or provirus) and become active through external triggers.

DOTS (direct observational therapy short course)   A strategy to ensure that individuals infected with tuberculosis take their prescribed antibiotics.

downer cow   A livestock cow that is unable to stand on its own and thus is killed. Often due to the prion disease bovine spongiform encephalopathy (BSE), also known as mad cow disease.

dracunculiasis   A disease caused by the parasitic guinea worm, *Dracunculus medinensis*.

droplet transmission   Spread of pathogens through airborne respiratory secretions.

dry phase   Beginning of Ebola virus disease (EVD) characterized by sudden onset of fever, intense weakness, muscle pain, headache, and sore throat.

DT (diphtheria, tetanus) vaccine   Vaccine used for immunization against diphtheria and tetanus.

DTaP (diphtheria, tetanus, acellular-pertussis) vaccine   Combination vaccine used to prevent diphtheria, tetanus, and acellular-pertussis.

dumb rabies   Form of rabies where the animal becomes increasingly lethargic, depressed, and uncoordinated. As the disease progresses paralysis results. Also called *paralytic rabies*.

dwell time   The required amount of time that a disinfectant must remain wet on a surface in order to inactivate or kill pathogens.

dysbiosis   Imbalance of microbiota, in particular gut microbiota, which can result in disease (e.g., colitis, obesity etc.).

Earth Day   An annual worldwide event that is celebrated on April 22nd to support environmental protection.

Earth Week   An annual worldwide event that is celebrated the week of April 22nd to support environmental protection.

Ebola virus disease (EVD)   A severe viral infection characterized by extreme hemorrhaging with a high fatality rate.

EBV (Epstein-Barr virus)   A herpesvirus that is the primary cause of infectious mononucleosis (also known as the *kissing disease*).

ecosystem   A population of organisms in a particular physical and chemical environment.

ecotourism   Sustainable tourism focused on visiting natural areas with the goal of having a low impact on the environment and supporting local peoples.

ecotoxicity   The adverse effect of a chemical agent on the environment and the organisms living in it.

edema   Accumulation of fluid in tissues of the body. Certain infections can cause edema (e.g., the 1918 Spanish influenza caused edema).

ehrlichiosis   Tickborne disease caused by *Ehrlichia* spp. It results in flulike symptoms. It may take as long as 2 weeks for symptoms to appear after infection.

electron microscope   A type of microscope that uses a beam of electrons to resolve an image of a microbe or virus (which is too small to be seen with the naked eye or light microscope).

Electronic Surveillance System for Early Notification of Community-Based Epidemics (ESSENCE I)   A symptom-based surveillance network implemented in 1999 as a collaboration between the U.S. Department of Defense and Johns Hopkins University's Applied Physics Laboratory to provide real-time tracking of data on the occurrence of syndromes or symptoms to identify potential public health threats. It was used to perform worldwide monitoring of the health status of all army personnel in all U.S. treatment facilities.

elephantiasis   A disease caused by filarial worms that results in blocked lymphatic vessels and the accumulation of large amounts of lymph fluid in the tissues; also called *lymphatic filariasis*.

Ellis Island Federal Immigration Station   Processing center for incoming immigrants into the United States from 1892–1954. Located on Ellis Island, Upper New York Bay harbor, where over 12 million immigrants were screened and isolated for showing signs or symptoms of an infectious disease, as needed.

embryonated (fertile) chicken eggs   Chicken eggs containing live embryos; they are sometimes used to cultivate viruses that do not replicate in established cell lines; can be used to manufacture influenza vaccines.

Emergency and Disaster Information Service   A well-known event surveillance system that monitors and documents all the events on Earth that may cause disaster or emergency, with a focus on events endangering the United States.

emergency treatment units (ETUs)   Temporary facilities constructed to treat people with Ebola virus disease (EVD) during the unprecedented outbreak that occurred from 2014–2016 in West Africa.

emerging disease hotspots   Regions where there is the high potential for the development of new zoonotic pathogens that can adversely affect human populations.

emerging infectious disease   New diseases that have increased in incidence in the recent past and have the potential to increase in the future.

emerging pathogens   Pathogens responsible for new or emerging infectious diseases—diseases that have increased in incidence in the recent past and have the potential to increase in the future.

emerging viruses   Newly identified viruses responsible for emerging infectious diseases—diseases that have increased in incidence in the recent past and have the potential to increase in the future.

Emory University Hospital's Serious Communicable Disease Unit   One of the few hospital units in the United States that had the necessary facilities to treat patients with Ebola

virus disease and other diseases caused by biosafety level 4 pathogens.

encapsulated   Bacteria that are enclosed within a capsule or sugar coat. These bacteria are more virulent than those lacking such capsules because they resist phagocytosis by host immune cells.

encephalitis   An inflammation of the brain caused by certain pathogens such as Powassan or rabies virus.

encephalopathy   A neurodegenerative condition involving brain pathology as in transmissible spongiform encephalopathy.

endemic infectious disease   A disease that is continually present at a steady level in a population and poses little public threat.

endocarditis   An infection of the inner lining of the heart. Usually caused by bacteria such as *Streptococcus* spp. or coxsackie viruses.

endocrine disruptor   Chemicals that interfere with thyroid function and alter estrogen and testosterone regulation. Exposure may lead to developmental and/or reproductive problems in mammals, including humans. Triclosan was banned from soaps in the United States because research suggested it was an endocrine disruptor.

endocytosis   A process of engulfment by certain viruses to facilitate their entry into host cells.

endospore   A highly resistant structure produced by certain *Bacillus* spp. and *Clostridium* spp. that enables them to enter a dormant state during adverse environmental conditions (e.g., when nutrient availability is low/starvation).

endotoxin   A toxin produced by all Gram (–) bacteria; it is usually released upon death and disintegration of the microbe as a result of antibiotic treatment.

enology   The science of wine making.

enterobiasis   A parasitic infection caused by *Enterobius vermicularis*, a type of roundworm. Common in children; symptoms include itching in the anal area. Also called *pinworm disease*.

enterotoxin   Bacterial toxins released by pathogens that affect the intestines, such as those causing gastroenteritis or foodborne illness.

enveloped virus   Virus that contains a membrane wrapped around its capsid. Such viruses are not very stable in the environment and are sensitive to inactivation by hand sanitizer (e.g., influenza A and B viruses, HIV, and herpesviruses).

Environmental Protection Agency (EPA)   Federal agency tasked with protecting the environment by maintaining and enforcing national standards under a variety of environmental laws. The agency conducts environmental assessment, research, and education.

enzyme-linked immunosorbent assay (ELISA)   A diagnostic technique used to detect antibodies or antigens in a clinical specimen collected from a patient.

enzymes   Proteins that act on specific biological substrates (e.g., proteases, DNases, RNases, lipases).

eosinophils   White blood cells that contain granules that are easily stained with eosin; sometimes associated with helminthic infections and allergies.

EPA (Environmental Protection Agency)   Federal agency tasked with protecting the environment by maintaining and enforcing national standards under a variety of environmental laws. The agency conducts environmental assessment, research, and education.

epicenter   The geographic focal point of an infectious disease outbreak.

epidemic   A disease that has a sudden increase in morbidity and mortality in a particular population.

epidemiologists   Scientists who study the origin and spread of infectious diseases with the goal of preventing or reducing such diseases.

epidemiology The study of the sources, causes, and distribution of diseases and disorders and how they affect a community.

EPS (extracellular polysaccharide polymers) Carbohydrate polymers produced by quorum-sensing bacteria that facilitate their adhesion to surfaces, thus enabling the formation of biofilms.

Epstein-Barr virus (EBV) A herpesvirus that is the primary cause of infectious mononucleosis (also called the *kissing disease*).

erythrocytes Red blood cells.

erythrogenic toxin Bacterial toxins that cause rashes produced by certain strains of *Streptococcus pyrogenes*.

espundia A form of mucocutaneous leishmaniasis characterized by invasion of the parasite into the skin and mucous membranes, causing destruction of the nose, mouth, and pharynx.

ESSENCE I (Electronic Surveillance System for Early Notification of Community-Based Epidemics) A symptom-based surveillance network implemented in 1999 as a collaboration between the U.S. Department of Defense and Johns Hopkins University's Applied Physics Laboratory to provide real-time tracking of data on the occurrence of syndromes or symptoms to identify potential public health threats. It was used to perform worldwide monitoring of the health status of all army personnel in all U.S. treatment facilities.

ETUs (emergency treatment units) Temporary facilities constructed to treat people with Ebola virus disease (EVD) during the unprecedented outbreak that occurred from 2014–2016 in West Africa.

Eucarya One of three domains in the Woese system of classification.

eucaryotic cells Complex cells containing a membrane-bound nucleus and other membrane-bound organelles.

Euro-GASP (Gonococcal Antimicrobial Surveillance Programme) A disease-specific surveillance system implemented to provide data on the antibiotic susceptibility of *Neisseria gonorrhoeae* strains circulating in Europe.

event surveillance Gathering of data from diverse Internet sources in real time or near-real time. Data are collected from news or online discussion platforms in various languages to detect potential or confirmed epidemics.

***ex vivo* gene therapy** Techniques that manipulate a person's genes in order to treat or prevent disease. The genes are manipulated in cells removed from the body and then infused back into the patient.

excystation The process by which a cyst comes out of dormancy resulting in an active stage.

exoenzymes Enyzmes secreted by pathogens that foster the spread of invading bacteria throughout the tissues by causing damage to host cells in their immediate vicinity and, in so doing, break down tissue barriers. Also called *spreading factors*.

exotoxins Protein molecules that cause toxic effects. They are released by bacteria during cell division and metabolism.

extracellular polysaccharide polymers (EPS) Carbohydrate polymers produced by quorum-sensing bacteria that facilitate their adhesion to surfaces, thus enabling the formation of biofilms.

extremely drug-resistant tuberculosis (XDR-TB) Tuberculosis cased by difficult-to-treat multi-drug antibiotic resistant strains of *Mycobacterium tubercuolsis*.

extremophiles Archeae bacteria that grow in harsh environmental conditions (e.g., extremes of temperature etc.).

F (fertility) factor A DNA F plasmid in a bacterium that allows plasmid DNA to be transferred from one bacterium to another through conjugation. Also called a *sex factor*.

F⁻ bacteria Bacteria that do not possess a fertility factor or F plasmid but can act as a recipient of the F plasmid through conjugation with an F⁺ bacterium.

F⁺ bacteria Bacteria that contain an F plasmid or fertility factor.

F plasmid The plasmid DNA or fertility factor that is present in F⁺ bacteria.

facultative anaerobes Microbes that grow best in the presence of oxygen but are capable of survival in its absence.

farmer's lung A form of hypersensitivity pneumonitis caused by exposure to hay dust or mold spores in agricultural products.

favus Historical term used for a severe form of tinea capitis caused by the dermatophyte *Trichophyton*. Immigrants entering the United States through Ellis Island were screened for favus.

FDA (Food and Drug Administration) Federal agency charged with protecting public health by assuring the safety, efficacy, and security of human and veterinary drugs, biological products, medical devices, the food supply, cosmetics, and products that emit radiation.

fecal microbiota transplantation A process whereby feces from a donor is administered into the intestinal tract of the recipient. Used to colonize the digestive tract with bacteria that may be missing from the recipient's microbiota.

fecal transplants A process whereby feces from a donor is administered into the intestinal tract of the recipient. Used to colonize the digestive tract with bacteria that may be missing from the recipient's microbiota.

fecal–oral mode of transmission Route of pathogen transmission where pathogens present in the feces of one person are passed to the mouth of another person. Primarily caused by inadequate sanitation and poor hygiene practices (not washing hands after going to the bathroom). Diseases passed in this manner also occur through the ingestion of fecally contaminated food or water. Examples of diseases passed in this manner include cholera, hepatitis A, winter vomiting disease caused by noroviruses, and salmonellosis.

fermentation A metabolic process used by some bacteria and yeast that is regulated by enzymes that break down sugars to acids and carbon dioxide. Microbial fermentations are used to produce foods such as yogurt, cheeses, and breads, and alcoholic beverages.

filtration Process of separating microbes from liquid or the air (e.g., HEPA filters) based on the size of the microbe.

fission ring During binary fission by bacteria, this structure directs the formation of a septum that divides the bacterium.

flaccid paralysis A loss of voluntary movement or paralysis because weakness or loss of muscle tone has caused damage to the nerves in muscle (e.g., as a result of poliovirus infections)

flagella Bacterial structures composed of the protein flagellin; involved in motility and chemotaxis.

flagellin The protein of which flagella are composed.

Flu Near You A web-based event surveillance system that has the objective of monitoring influenza outbreaks.

fluorescent markers Biomolecules used by researchers to label target structures inside living cell in order to study cellular processes. These molecules will glow when exposed to the appropriate wavelength of light.

flyway The path that migrating birds take as they travel each season. Experts believe that avian influenza A viruses can travel from migratory birds into human populations, potentially causing a pandemic.

fomites Inanimate objects that serve as means of transmission pathogens (e.g., used facial tissue, door knobs, contaminated toys at a daycare).

Food and Drug Administration (FDA) Federal agency charged with protecting public health by assuring the safety, efficacy, and security of human and veterinary drugs, biological products, medical devices, the food supply, cosmetics, and products that emit radiation.

food intoxication The result of ingestion of bacterial toxins (e.g., botulism).

food poisoning Also referred to as *intoxication*. The ingestion of already-produced bacterial toxins in contaminated food products; the bacteria may, in fact, no longer be present. Ingestion of toxins results in symptoms of nausea, vomiting, diarrhea, and, possibly, bloody stools and fever.

Food Safety News A web-based publication dedicated to reporting illnesses associated with food safety.

foodborne infection The ingestion of bacterial-contaminated foods and subsequent bacterial multiplication in the intestinal tract, secretion of bacterial exotoxins, and, possibly, invasion of the intestinal tract. Production of exotoxins by the bacteria results in symptoms of gastroenteritis such as nausea, vomiting, diarrhea, and, possibly, bloody stools and fever.

foreign gene Genes in an organism's genome that have been acquired from another organism.

free-living Term used to describe microbes that live independently of other organisms; that is, they are not parasites.

fried rice diarrheal syndrome Syndrome associated with the ingestion of improperly prepared rice (rice naturally contains *Bacillus cereus*). The symptoms are caused by the enterotoxins produced during the multiplication of *Bacillus cereus* in the small intestine. Symptoms include abdominal pain, watery *diarrhea*, seldom vomiting, and no fever.

fried rice emetic syndrome Syndrome associated with the ingestion of improperly prepared rice (rice naturally contains *Bacillus cereus*). Syndrome caused by enterotoxin produced during the multiplication of *Bacillus cereus* in the small intestine. Symptoms include *vomiting* and abdominal pain.

fumigation Disease prevention and control method where gaseous chemicals are applied to kill pathogens present on nonliving surfaces (e.g., U.S. library books were fumigated after the 1918 influenza pandemic ended).

fungi A category of eucaryotic microbes including mushrooms and yeasts.

furious rabies Form of rabies where the animal is confused, hyperactive, and excitable, often exhibiting strange behaviors. Symptoms also include problems swallowing, excess salivation, and fear of water (hydrophobia). Usually fatal.

fusion A mechanism host cell entry employed by enveloped viruses in which there is contact between the viral envelope and the host cell plasma membrane.

GamaStan S/D An immune globulin medication used for postexposure prophylaxis against hepatitis A virus, varicella zoster virus, measles virus or rubella virus.

gametocytes Sexual cells of *Plasmodium* spp. (the cause of malaria). They are taken up by mosquitoes during a bloodmeal and are responsible for the transmission of *Plasmodium* spp. from humans to another host.

gamma globulin A fraction of the globulin component of blood plasma in which most of the antibodies are present.

Gardasil A vaccine developed against certain types of high-risk human papillomaviruses (HPV) that are responsible for cervical, penile, anal, and oral cancers.

Gardasil-9 A vaccine developed against nine high-risk types of human papillomavirus (HPV) that are responsible for a genital warts, cervical, penile, anal, and oral cancers.

gas gangrene A condition brought about through contamination of a wound with particles of soil containing *Clostridium perfringens* or its endospores.

gastroenteritis Inflammation of the stomach and intestines resulting in diarrhea, cramps, nausea, vomiting, and fever. Caused by oral–fecal transmission of a variety of pathogens.

gastrointestinal anthrax A form of anthrax resulting from the ingestion of inadequately cooked meat contaminated with *Bacillus anthracis*.

GBS (Guillain-Barré syndrome) A rare autoimmune reaction in which the immune system mistakenly attacks the body's peripheral nervous system resulting in rapid-onset of muscle weakness ; it sometimes occurs after administration of a vaccine.

gene therapy Techniques that manipulate a person's genes in order to treat or prevent disease; can be *in vivo* or *ex vivo* therapy.

generalized transduction DNA recombination process whereby bacterial DNA from one bacterium is introduced into another bacterium by a bacteriophage. The DNA may or may not be incorporated into the recipient bacterium's chromosome.

generally regarded as safe (GRAS) Designation of the Food and Drug Administration that an ingredient (including probacteria) added to a food item is considered safe for human consumption.

generation time The length of time between rounds of binary fission (e.g., the generation time for *E. coli* is 20 minutes).

generic antivirals Copies of brand-name antivirals that can be sold more cheaply because they are manufactured by pharmaceutical companies that did not incur the expense of research, developing, and marketing of the antiviral drug. The generic drugs are equally effective in the treatment of viral infections as the nongeneric version. They are used to treat HIV/AIDS in developing countries as a cost-effective means of controlling the progression of the disease in those with HIV/AIDS.

genes Segments of the DNA that encode a specific protein.

genetic engineering A means of manipulating genes using recombinant DNA technology in order to create a genetically modified organism for use in a specific application.

genetically modified organism (GMO) An organism whose genetic material has been modified using molecular biology techniques through the introduction of foreign DNA from another organism to confer the expression of new traits that were not present in the original organism (e.g., genetically modifying mosquitoes to prevent the transmission of Zika viruses, genetically modifying bacteria to produce insulin for treatment of diabetes, or genetically modifying yeast to produce a hepatitis B vaccine).

Geneva Protocol (1925) Formally titled the Protocol for the Prohibition of the Use in War of Asphyxiating, Poisonous or other Gases, and of Bacteriological Methods of Warfare. This treaty prohibits the use of chemical and biological weapons in wars and armed conflicts.

genital herpes A sexually transmitted infection caused by herpes simplex virus 2 (HSV-2) that causes painful sores on the genitals. Can be treated with anitivirals to reduce the incidence of outbreaks but not cured.

genital warts One of the most common sexually transmitted diseases, caused by human papillomaviruses.

genome editing Use of genetic engineering techniques to make changes in the DNA sequence of a cell or organism (e.g., CRISPR-Cas applications).

genome The complete set of chromosomes in a cell.

genomics Branch of biology that focuses on the sequencing and understanding of genomes.

genus A category in the binomial system of nomenclature above the species level (e.g., *Escherichia* is a bacterial genus).

germ theory of disease Theory that infectious diseases are caused by microbes.

germline modifications A type of genetic engineering that focuses on making changes to the DNA in reproductive cells (i.e., sperm, eggs) and embryos in the earliest stages of development. The modified genes would appear in the resulting recombinant organism, as well as any progeny and future generations produced by the recombinant organism.

giardiasis An intestinal disease caused by the protozoan *Giardia lamblia* and acquired by drinking contaminated water. Also called *beaver fever*.

gigantobacteria Very large bacteria; often have multiple copies of chromosomes (e.g., *Thiomargarita namibiensis* found in ocean sediments on the continental shelf of Namibia).

giruses Very large viruses (e.g., mimivirus and mamavirus).

glioblastoma An aggressive type of cancer that occurs in the brain or spinal cord.

Global Handwashing Day Day dedicated to increasing the awareness and understanding the importance of handwashing with soap as an easy, effective, and affordable way to prevent the spread of infectious diseases and save lives. Occurs on October 15th of each year.

Global Virome Project (GVP) Global partnership among various organizations with the goal of characterizing all of the planet's threatening viruses in order to mitigate the emergence and spread of novel zoonotic viruses in the human population, thus heading off the next pandemic.

glomerulonephritis A kidney disease that is a complication of untreated *Streptococcus pyogenes* infection.

glyphosate Widely used broad-spectrum herbicide that kills many broadleaf plants and grasses. Various crop plants have been genetically engineered to be resistant to it, so that farmers can spray it on their fields to kill weeds but not the genetically modified crop plants, thus improving crop yields.

GMO (genetically modified organism) An organism whose genetic material has been modified using molecular biology techniques through the introduction of foreign DNA from another organism to confer the expression of new traits that were not present in the original organism (e.g., genetically modifying mosquitoes to prevent the transmission of Zika viruses, genetically modifying bacteria to produce insulin for treatment of diabetes, or genetically modifying yeast to produce a hepatitis B vaccine).

gold standard The best treatment or diagnostic test for a certain infectious disease or condition.

Gonococcal Antimicrobial Surveillance Programme (Euro-GASP) A disease-specific surveillance system implemented to provide data on the antibiotic susceptibility of *Neisseria gonorrhoeae* strains circulating in Europe.

gonococcal ophthalmia neonatorum A form of blindness in newborns that was later proven to be caused by *Neisseria gonorrhoeae*, the bacterium that causes gonorrhea.

gonorrhea A sexually transmitted disease caused by *Neisseria gonorrhoeae*.

Gram (–) negative bacteria A cell wall-staining property of some bacteria; they do not retain the crystal violet stain after decolorizing with ethanol and appear pink to red after being counterstained with safranin.

Gram (+) positive bacteria A cell wall-staining property of some bacteria that causes them to retain the crystal violet stain after decolorizing with ethanol; they appear purple.

Gram stain A differential staining technique used to classify bacteria based on differences in their cell wall composition developed by Hans Christian Gram. Bacterial cells are stained with crystal violet, an iodine mordant is added, and then the cells are decolorized with ethanol. A counterstain is then applied. Bacteria are classified as Gram (–) or Gram (+) based on their cell wall differences. Gram (–) cells can be decolorized and are stained red. Gram (+) cells cannot be decolorized with ethanol and remain stained purple throughout the procedure.

granulocytes White blood cells that contain enzymes such as proteases that digest or degrade pathogens. Part of the innate immune system, they include neutrophils, eosinophils, and basophils.

granuloma An accumulation of cells associated with chronic inflammation that walls off the inflammatory agent.

granzymes A type of enzyme in cytotoxic T cells and natural killer cells that breaks down cells that are infected with viruses or bacteria. Also play a role in apoptosis.

GRAS (generally regarded as safe) Designation of the Food and Drug Administration that a bacteria used to produce a food product are considered safe for human consumption.

growth curve A graphical representation of the multiplication of a bacterial culture. Liquid culture samples are taken and then plated onto solid media in petri dishes. Population counts are performed by allowing the bacteria to develop into (macroscopic) colonies that can be counted. The four phases of growth of bacterial cultures are the lag phase, log phase, stationary phase, and death phase.

Guillain-Barré syndrome (GBS) A rare autoimmune reaction in which the immune system mistakenly attacks the body's peripheral nervous system resulting in rapid-onset of muscle weakness; it sometimes occurs after administration of a vaccine.

guinea worm disease Parasitic infection caused by consumption of larvae of *Dracunculus medinesis*, a nematode roundworm, in contaminated drinking water. Results in the formation of excruciatingly painful lesions as the mature worms slowly emerge from the body. Disease is a problem in Africa and the target of numerous eradication efforts. Also known as *dracunculiasis*.

gummas Tumorlike lesions in connective tissues that are characteristic of tertiary syphilis.

GVP (Global Virome Project) Global partnership among various organizations with the goal of characterizing all of the planet's threatening viruses in order to mitigate the emergence and spread of novel zoonotic viruses in the human population, thus heading off the next pandemic.

H spike Hemagglutinin glycoproteins protruding from the envelop of influenza A and B viruses; enables influenza A and B viruses to attach to and enter host epitheleal cells lining the respiratory tract.

H1N1 swine influenza A virus A subtype of swine influenza A virus isolated from pigs that infected humans, causing an influenza pandemic in 2009. The virus was isolated from pigs located in Mexico (the epicenter of this pandemic).

H3N2 swine influenza A virus A subtype of human influenza A virus that was isolated from a human influenza patient in the Philippines in 1982.

H5N1 avian influenza A virus A subtype of avian influenza A virus isolated from a human patient in Hong Kong for the first time in 1997. The strain originated in wild water fowl and spread to chickens and subsequently humans. It is highly pathogenic in domesticated birds and humans but does not spread easily from person to person to date.

H7N9 avian influenza A virus A highly infectious avian influenza that primarily affects birds; virulent strains have emerged that can infect humans.

HAIs (healthcare-associated infections) Infections acquired while receiving medical care. Most are caused by bacteria.

halogen Periodic table group that includes fluorine, chlorine, bromine, iodine, astatine. Compounds containing these elements have been found to be effective in killing bacteria.

halointolerant Trait possessed by bacteria that are unable to grow on bacteriological media containing high salt (7.5–10%) concentrations, such as mannitol salts agar.

halophiles Archaea bacteria that live in environments containing extreme salt concentrations.

halotolerant Trait possessed by bacteria that can grow on bacteriological media containing high salt (7.5–10%) concentrations, such as mannitol salts agar.

hamachi toro Sushi made from yellowtail tuna. Consumed raw; may contain parasites such as tapeworms.

hand hygiene   Act of cleaning hands to remove dirt and microbes. Single most important practice in preventing healthcare-associated infections (HAIs), as well as infectious diseases in daily living. Was initially promoted by Ignaz Semmelweis in the late 1800s. Its incorporation into medical practice and sanitation resulted in sharp decreases in childbed fever and other infectious diseases.

handwashing   Act of cleaning hands to remove dirt and microbes. Single most important practice in preventing healthcare-associated infections (HAIs), as well as infectious diseases in daily living. Was initially promoted by Ignaz Semmelweis in the late 1800s. Its incorporation into medical practice and sanitation resulted in sharp decreases in childbed fever and other infectious diseases.

Hansen's disease   A disease caused by *Mycobacterium leprae* that causes skin inflammation or tumorlike lesions; formerly known as leprosy.

hantavirus pulmonary syndrome (HPS)   A disease caused by a hantavirus that produces severe flulike respiratory problems and can result in death.

healthcare-associated infections (HAIs)   Infections acquired while receiving medical care. Most are caused by bacteria.

HBOT (hyperbaric oxygen therapy)   Therapy where a patient is administered almost pure oxygen in a pressurized room or tube. It is thought to promote tissue healing and provide other therapeutic benefits.

HealthMap   An Internet event surveillance system that aggregates information from multiple informal sources for real-time surveillance of public health threats.

healthy carriers   Individuals who have no symptoms of a particular microbial disease but harbor the microbes and may unwittingly pass the disease on to others.

heat fixation   Application of heat to a microscopic specimen of bacteria in order to cause the microbe to adhere to a slide, kill it and denature any harmful toxins or enzymes.

helical   Term used to describe the shape of viruses in which capsomers are wrapped around the viral genome in the form of a long filament (Ebola virus is an example of a helical-shaped virus).

helminths   Parasitic worms, including tapeworms, whipworms, and hookworms, that can infect the human gastrointestinal tract.

hemagglutinin   A glycoprotein consisting of spikes that protrude the envelop of influenza A and B viruses; it aids in the attachment/fusion to and entry of cells by the virus.

hemagglutinin (H) spikes   Glycoproteins on the outer surface of influenza A and B viruses that enable them to infect cells lining the respiratory tract of susceptible hosts (e.g., humans, pigs, wild water fowl, and domesticated birds, such as chickens).

hematopoiesis   The process by which blood cells are formed and differentiated.

hematopoietic stem cells   An immature blood cell that has the potential to develop into any type of blood cell.

hemolysins   Secretions produced by some microbes that destroy red blood cells through the destruction of cell membranes.

hemolytic uremic syndrome (HUS)   Kidney damage occurring primarily in young children as a result of infections caused by microbes such as *E.coli* O157:H7.

hemophilia   Genetic disorder that results in an inability of the blood to clot properly due to a defective genes that code for clotting factors.

hepatitis A virus   A virus found in feces and transmitted through contaminated drinking water and food. It is the most common hepatitis virus and usually causes only mild disease.

hepatitis B virus   A virus transmitted in body fluids. It causes subclinical to severe and chronic disease that may lead cirrhosis and liver cancer.

hepatitis C virus   A virus transmitted mainly by intravenous drug use; it causes a chronic infection in the majority of those infected, leading to cirrhosis and liver cancer.

hepatitis E virus   A virus transmitted by the fecal–oral route or the consumption of undercooked pork. It causes a gastrointestinal disease that is usually of moderate severity.

hepatitis   Inflammation of the liver.

hepatocellular carcinoma   A common form of liver cancer that often develops in people suffering from chronic hepatitis B or C.

herbivore   Organisms that feed on plants.

herd immunity   Immunity of enough members of a population to protect against an epidemic; can be acquired through natural infections or immunization programs.

herpes simplex virus (HSV)   Type of herpesvirus. HSV-1 is the main cause of cold sores and fever blisters around the mouth, HSV-2 is the main cause of genital herpes.

herpesvirus   Family of viruses responsible for a range of diseases in humans, including oral and genital herpes, chickenpox, shingles, cytomegalovirus infections, and Epstein-Barr disease (mononucleosis or the kissing disease).

heterotrophs   Microbes that require organic compounds as an energy source.

Hfr bacteria   Bacteria that exhibit a much higher frequency of DNA recombination with F⁻ bacteria.

high-risk types of HPV   Types of human papilloma virus that have the greatest potential to cause cervical cancer, such at types 16 and 18.

high-throughput sequencing   A fast, inexpensive DNA sequencing technique that is particularly well suited to sequencing large genomes rather quickly.

histamine   Chemical released by mast cells during immune response to an allergen that causes inflammation.

histoplasmosis   Fungal infection caused by breathing spores of *Histoplasma capsulatum*, which is found in soil and bird and bat droppings. Most common in the Mississippi and Ohio River Valleys. Can cause serious respiratory infections in infants and people with compromised immune systems. Also called *cave disease*.

HIV (human immunodeficiency virus)   The virus that causes AIDS. HIV-1 is the most common cause of HIV/AIDS worldwide, and HIV-2 is the most common cause of HIV/AIDS in West Africa.

hologenome   The genome of an organism as well as the genomes of all of its associated symbiotic microbes (e.g., the genomes of humans and their microbiome).

homologous recombination   Form of genetic recombination where similar or identical nucleotide sequences are exchanged between DNA molecules during replication.

hookworm disease   A disease that is caused by the roundworms *Ancylostoma duodenale* and *Necator americanus*.

hospitalism   Term coined by Sir James Young Simpson in the late 1800s who observed that hospitals had a negative effect on patient outcomes. Simpson collected data that showed that patients had a greater chance of dying on the operating table in a hospital than those being operated on in a private or country practice.

host   A metabolically active organism that is susceptible to infection by microbes and viruses.

host cell   A cell that is infected by a virus or intracellular pathogens. Once inside the cell, the viruses or intracellular microbes replicate and produce progeny.

host range   The variety of species and types of host cells that a virus can infect and replicate in.

HP (hypersensitivity pneumonitis)   Disease of the lungs caused by an allergic reaction to repeatedly inhaled triggers, such as fungi, bacteria, bird droppings, or dust particles; HP is associated with *bagpipe lung*.

HPS (hantavirus pulmonary syndrome)   A serious disease caused by a hantavirus that produces severe respiratory problems and can result in death.

HPV (human papillomavirus)   A dermotropic virus transmitted by skin-to-skin or into punctures or abrasions of the skin. Numerous varieties, a number of which are passed by sexual contact. Can cause plantar's warts, genital warts, and is associated with cervical, penile, and oral cancers.

HSV (herpes simplex virus)   Type of herpesvirus. HSV-1 is the main cause of cold sores and fever blisters around the mouth, HSV-2 is the main cause of genital herpes.

Human Genome Project   The mapping of the genes located on the 23 pairs of human chromosomes.

human growth hormone   A hormone that is used for the treatment of dwarfism and is now produced by genetic engineering.

human immunodeficiency virus (HIV)   The virus that causes AIDS. Type 1 is the most common cause of AIDS worldwide, and type 2 is the most common cause of AIDS in West Africa.

human papillomavirus (HPV)   A virus passed by skin-to-skin contact. Numerous varieties, a number of which are passed by sexual contact. Can cause genital warts and cervical cancer.

human insulin   A hormone that regulates the amount of glucose in the bloodstream. Much of the insulin now used to treat diabetes is a product of genetic engineering.

humoral immunity   The type of adaptive immune response mediated by antibodies. This type of immunity is particularly good at defending against extracellular microbes and toxins.

HUS (hemolytic uremic syndrome)   Kidney damage occurring primarily in young children as a result of bacterial toxins caused by E. coli O157:H7 infection.

hyaluronidase   An enzyme produced by certain bacteria such as *Staphylococcus aureus* that breaks down hyaluronic acid found in connective tissue.

hydrophobia   Fear of water; an old term for rabies.

hygiene   Practices that help to maintain health and prevent the spread of infectious diseases.

hyperbaric oxygen therapy (HBOT)   Therapy where a patient is administered almost pure oxygen in a pressurized room or tube. It is thought to promote tissue healing and provide other therapeutic benefits.

hyperinfection   Repeated infection with larvae produced by helminths present in the body of persons suffering from strongyloidiasis.

hypersensitivity pneumonitis (HP)   Disease of the lungs caused by an allergic reaction to repeatedly inhaled triggers, such as fungi, bacteria, bird droppings, or dust particles (associated with "bagpipe lung").

hypersensitivity reaction   Over activation of the immune system in response to an allergen.

hyperthermophiles   Archaeae bacteria that live in extremely hot environments.

hyphae   Intertwined filaments characteristic of fungi in the environment.

hypothermia   Condition that occurs when the body rapidly loses heat, causing the body temperature to deteriorate below 35°C (95°F). Can result in death.

hypothesis   A proposed explanation for a phenomenon that can be tested experimentally using the scientific method.

hypoxic   Low blood oxygen levels in the body's tissues.

iatrogenic   Infection induced in a patient by a medical procedure such as through a contaminated needle when administering a drug or CJD acquired during cornea transplantation in which the cornea donor was contaminated with prions.

icosahedral   Term used to describe the shape of viruses containing a three-dimensional, 20-sided structures with triangular sides; icosahedral-shaped viruses are structurally very stable.

ICTV (International Committee on Taxonomy of Viruses)   Global organization that has developed a standardized naming scheme for viruses and seeks to classify every virus that infects living organisms.

ICU (intensive care unit)   Department of a hospital that provides treatment for critically ill patients who require multiple interventions and close monitoring.

ID (infectious dose)   The minimum number of microbes or viruses needed to cause an infection within a host.

IDPs (internally displaced persons)   Persons or groups of persons for whom social disruptions have forced to move but have not crossed international borders.

Ig   Classes of antibodies produced by the body (immunoglobulins A, D, E, G, and M), each possessing specific properties and functions of the immune system.

IGRA (interferon gamma release assay)   A blood test that can be used to diagnose infection with *Mycobacterium tuberculosis*.

immune globulin   Preformed antibodies that can be given intravenously as passive immunity in the treatment of specific infections (e.g., botulism antitoxin or convalescent serum that contains these antibodies is given to patients suffering from Ebola virus disease as an experimental therapy).

immunity   An organism's ability to defend against pathogens due to activation of antibodies by the adaptive immune system. Often the result of past exposure to a pathogen through disease or immunization.

immunization   Administration of a vaccine that contains biological molecules from specific pathogens that will stimulate an immune response and result in the development of adaptive immunity to a specific pathogen.

Immunization Action Coalition   Organization that works in partnership with the Centers for Disease Control and Prevention to provide the public with information on vaccine-preventable diseases and vaccinations.

immunocompromised   People with impaired immune systems due to diseases such as HIV/AIDS, elderly with weakened immune systems, patients that use medications to prevent rejection of organ transplants, or patients undergoing chemotherapy cancer treatment.

immunofluorescent staining   Microscopy technique where a fluorescence microscope is used to view clinical specimens such as tissue biopsies or blood samples that have been stained with fluorescently labeled antibodies.

immunoglobulins   Preformed antibodies that can be given intravenously as passive immunity in the treatment of specific infections (e.g., botulism antitoxin or convalescent serum that contains these antibodies is given to patients suffering from Ebola virus disease as an experimental therapy).

impetigo   A superficial infection of the skin that is typically manifested by blisters around the mouth and is usually caused by staphylococci.

imported case   A person who contracts an infectious disease when travelling abroad and then introduces the infectious disease to susceptible persons in an area when they return home.

in vivo gene therapy   Techniques that manipulate a person's genes in order to treat or prevent disease. The genes are manipulated while the cells are still inside of the body.

incidental hosts   Infection of an intermediate host by a parasite in which it harbors the parasite but it is not spread or transmitted to a final or definitive host.

inclusion bodies   In bacteria: aggregates of insoluble proteins present in the cytoplasm of some bacteria. In viral infections: "viral debris" consisting of viral parts (genomes, protein capsids, envelopes) discarded in the process of viral assembly. They are important in the diagnosis of specific viral diseases because they can be identified microscopically.

incubation period   The time between a pathogen's entry into the body and the appearance of signs and symptoms.

indirect transmission   The passage of infectious material from a reservoir to an intermediate host and then to a final host.

inducer   A chemical or protein whose presence turns on the expression (transcription) of a particular gene (e.g., TAV protein activates the expression of its own tTAV gene).

infantile paralysis   A form of poliomyelitis in infants and young children, characterized by muscle paralysis.

infection control officer   Public health officer responsible for coordinating infection monitoring and prevention in a hospital environment or community.

infectious disease   Disease caused by microbes or viruses that can be transmitted, directly or indirectly, from one person to another.

infectious dose   The minimum number of microbes or virus needed to establish an infection in a host.

infectious hepatitis   Liver infection caused by hepatitis A virus. Highly contagious by fecal–oral transmission but generally with mild symptoms.

infective dose   The minimum number of microbes or virus needed to establish an infection in a host.

inflammation   Swelling, pain, redness, and heat in an area of infected or damaged tissue.

influenza   A highly contagious respiratory infection caused by influenza A and B viruses. Usually resolves on its own, but can be deadly to young children, older adults, and people with compromised immune systems.

influenza A virus   A type of influenza virus that infects humans, pigs, birds, and some other mammals. It causes epidemics and occasionally pandemics.

influenza B virus   A type of influenza virus that infects humans and causes epidemics. It does not have an animal reservoir.

infusion   A sterilized broth used by Louis Pasteur to disprove the theory of spontaneous generation.

inhalation anthrax   Form of anthrax resulting from the inhalation of *Bacillus anthracis* endospores.

innate immunity   An immediate, nonspecific first-line defenses or immune responses against any type of pathogen. It includes white blood cells (leukocytes, phagocytes, monocytes, natural killer cells) that seek out and eliminate pathogens that could cause infection.

inorganic compound   Generally, a chemical compound that does not contain carbon (although carbon dioxide and several other carbon-containing compounds are considered inorganic).

insect vectors   Insects that carry or transmit pathogens to other organisms by biting or taking a blood meal (e.g., mosquitoes carry Zika virus biting a human host).

insecticide   A substance that kills insects, such as insecticidal toxins produced by *Bacillus thuringiensis* and *Photorhabdus luminescens*.

insects   A large class of arthropods characterized by three body segments and six legs. Insects such as mosquitoes and ticks are arthropod vectors that carry certain pathogens.

insertion mutation   Addition of one (or more) nucleotides; can occur naturally during DNA replication or are intentionally introduced through genetic engineering techniques.

intensive care unit (ICU)   Department of a hospital that provides treatment for critically ill patients who require multiple interventions and close monitoring.

interferon gamma release assay (IGRA)   A blood test that can be used to diagnose infection with *Mycobacterium tuberculosis*.

interferon   A type of protein released by a host cell when it senses the presence of a virus that heightens the cell's defenses. Also manufactured as therapies to treat hepatitis C in combination with antiviral drugs.

intermediate host   A host that supports a parasite for a short period, during which it might complete some portion of its development, before it moves on to the final host (e.g., role of snails in liver fluke life cycle).

intermediate vehicle   A source of an infectious agent that is involved in indirect transmission to a host.

internally displaced persons (IDPs)   Persons or groups of persons for whom social disruptions have forced to move but have not crossed international borders.

International Committee on Taxonomy of Viruses (ICTV)   Global organization that has developed a standardized naming scheme for viruses and seeks to classify every virus that infects living organisms.

interspecies communication   Quorum-sensing communication between different species.

interstitial lung disease   A progressive disease that results in scarring of the lungs. Can be caused by pathogens, such as the bacterium *Mycoplasma pneumoniae*, or as the result of an autoimmune condition.

Intestiphage®   A commercial bacteriophage therapy that is used to treat gastrointestinal infections.

intraspecies communication   Quorum-sensing communication within the same species.

intubation   Insertion of a plastic tube down the mouth and into the trachea to serve as an airway in emergency situations.

invasive candidiasis   A *Candida* spp. (yeast) infection that spreads throughout the body in immune compromised individuals. It is a serious infection that rarely occurs outside of healthcare environments.

invasive species   An organism that is introduced to a new environment where it is not native and spreads easily, causing environmental or economic damage.

iron lung   An artificial respirator in the shape of a cylindrical steel chamber or tank containing pumped air, causing the lungs to expand; used from the 1930s to the 1950s to help victims of polio in breathing.

isoniazid   An antibiotic primarily used in the treatment of tuberculosis.

isotypes   Classes of immunoglobulins: IgA, IgD, IgE, and IgM.

ivermectin   A drug used to treat onchocerciasis.

jaundice   Yellowing of skin or whites of the eyes due to buildup of bilirubin from the liver. Can be caused by hepatitis viruses, yellow fever virus, and certain medications.

jock itch   Fungal infection affecting the genital region, inner thighs, and buttocks.

kala-azar   A severe and usually fatal form of leishmaniasis in which the parasites invade the liver and other organs.

Kampung Improvement Program   Program in Indonesia developed to improve public sanitation in order to reduce disease caused by exposure to the pathogens present in human waste.

Kaposi's sarcoma   An aggressive type of skin cancer caused by a herpesvirus that is observed in people suffering from AIDS.

keratin   Protein found in the skin, nails, and hair. Serves as nutrient source for dermatophytes.

keratitis   Inflammation of the cornea of the eye caused by microbial contamination of a contact lens solution or by herpesviruses.

ketoconazole   An antifungal medication used to treat candidiasis and tinea.

kissing bug   An insect that is the vector for Chagas disease.

Klebsiella pneumoniae carbapenemase (KPC)   An emerging bacterial pathogen that is known to cause healthcare-associated infections. It produces carbapenemase, making the infection difficult to treat with commonly used antibiotics.

knockout   The process of genetically modifying the genome of an organism so that the function of a particular gene is inactivated. The function of the gene is "knocked out" of the organism.

Koch's postulates   A set of procedures by which a specific organism (typically bacteria) can be associated as the cause of a specific infectious disease.

Koplik's spots   Spots that appear in the mouth in the early stage of measles.

KPC (Klebsiella pneumoniae carbapenemase)   An emerging bacterial pathogen that is known to cause healthcare-associated infections. It produces carbapenemase, making the infection difficult to treat with commonly used antibiotics.

kuru   The first transmissible spongiform disease discovered n humans. It is a fatal neurodegenerative disease that occurred among members of the Fore tribe of Papua New Guinea at epidemic levels during the 1950s and 1960s. It was caused by the tribal members, especially the women and children, eating human nervous tissue infected with prions (cannibalism) as part of their funeral ceremonies.

Kyoto protocol   An international treaty whereby countries commit to reducing their greenhouse gas emissions in an effort to slow climate change. The United States has not signed the treaty.

La Crosse encephalitis   Mosquitoborne viral disease that causes swelling of the brain, resulting in seizures, coma, or paralysis. The first human case occurred in La Crosse, Wisconsin.

lag phase   In the bacterial growth curve, this is the "get ready for growth" stage that represents a time of adaptation to the new environment. There may be an initial drop in the bacterial count because of adjustment and survival of only a part of the microbial population.

larva   A stage in arthropod development that occurs after hatching and before maturity is reached.

laudable pus   During the Civil War period, doctors thought that the presence of thick, creamy pus in a wound or surgical site was involved in the healing process. The pus was most likely a *Staphylococcus* bacterial infection.

LD$_{50}$ 50% lethal dose; a laboratory measurement of virulence to determine the dose that kills 50% of the test animals in a given time.

legionellosis   A type of pneumonia caused by infection with *Legionella* bacteria spread primarily by breathing in mist containing the bacteria, such as that generated by air conditioners. Symptoms of cough, shortness of breath, fever, muscle pain, and headache being within a week after exposure; also called Legionnaires' disease.

Legionnaires' disease   A type of pneumonia caused by infection with *Legionella* bacteria spread primarily by breathing in mist containing the bacteria, such as that generated by air conditioners. Symptoms of cough, shortness of breath, fever, muscle pain, and headache being within a week after exposure; also called legionellosis.

leguminous plant   A type of plant with swellings or nodules along the root system containing *Rhizobium* and other nitrogen-fixing bacteria.

Leishmaniasis   A disease caused by *Leishmania* parasites transmitted by the bite of female sand flies.

leper colony   Places where lepers were quarantined so that leprosy would not spread to other people. Kalaupapa, located on the island of Molokai, was Hawaii's leper colony in which more than 8,000 lepers were sent into exile over the course of a century.

lepromas   Nodular skin lesions caused by *Mycobacterium leprae*, cause of leprosy.

leprosarium   A hospital dedicated to the care or quarantine of people with leprosy.

leprosy   A disease caused by *Mycobacterium leprae* that causes skin inflammation or nodularlike lesions called lepromas.

leptospirosis   A zoonotic disease caused by the bacterial spirochete *Leptospira spp.* Transmission occurs through the accidental exposure of contaminated water or soil with urine from an infected rodent. Symptoms range from mild to severe; also called swamp fever, mud fever, autumn fever, rice-field fever, and Cane cutter's fever.

leukemia   A type of cancer characterized by uncontrolled growth and division of white blood cells.

leukocidin   A cytotoxin produced by some bacteria that kill white blood cells.

leukocytes   White blood cells.

life expectancy   The expected amount a time a person is expected to live based on their year or birth. The U.S. average was 78.7 years in 2018.

light microscope   A type of microscope that uses visible light and lenses to magnify microbes 10–2000 times their normal size.

lipases   An enzyme that breaks down fats for digestion, transport, or processing.

Listerine®   Antiseptic mouthwash product named after Joseph Lister, the pioneer of antiseptic surgery.

listeriosis   Foodborne illness caused by the bacterium *Listeria monocytogenes*. Symptoms include diarrhea and fever. If it occurs during pregnancy it can result in an increased risk for serious health problems for the newborn.

live animal markets   Markets common in developing countries where live animals, including bushmeat and birds, are available for purchase and consumption. Such markets have been identified as risk factors for the transmission of emerging pathogens such as SARS-CoV and avian influenza A virus to human populations.

live attenuated vaccines   Vaccines made of live but weakened strains of a virus that can be used to stimulate an immune response to prevent infections.

live bird markets   Markets common in developing countries where live birds are available for purchase and consumption. Such markets in China and Southeast Asia have been identified as risk factors for the transmission of avian A influenza to human populations.

live vaccines   Vaccines that consist of a live, replication-competent weakened or attenuated virus.

lockjaw   A symptom of infection caused by *Clostridium tetani* characterized by contraction of the muscles in the jaw; also an alternative name for *tetanus*.

locomotion   Means by which an organism moves. In microbes, it often involves cilia or flagella.

log phase   In the bacterial growth curve, the period during which the number of bacteria increases exponentially, governed by the generation time, which is the time it takes a bacterium to undergo binary fission.

lone star tick   Arthropod found throughout the eastern United States that is a vector for a variety of bacteria, including *Ehrlichia* spp., which causes ehrlichiosis and the Heartland virus.

longhorned tick   A tick originating in Asia but increasingly found in the United States that can cause the cattle disease theileriosis.

lower respiratory tract   Portion of the respiratory system that includes the trachea, bronchi, bronchioles, and alveoli.

low-risk types of HPV   Low-risk types of human papillomavirus that cause benign warts.

Lyme disease   A disease primarily caused by bites from ticks carrying the bacterium *Borrelia burgdorferi*. The disease is characterized by a bull's-eye rash, flulike symptoms, and joint pain. Also called *borreliosis*.

lymph nodes   Structures along lymphatic vessels that act as microbe filters.

lymph   A tissue fluid derived from blood that is returned to the blood by lymphatic vessels.

lymphadenitis   Inflammation of one or more lymph nodes.

lymphadenopathy   Swelling of lymph nodes that occurs in AIDS and other infections.

lymphatic filiariasis   Parasitic disease caused by *Wuchereria* spp. of roundworms that invade the lymph nodes. Also called *elephantiasis*. Primarily occurs in the tropics and classified as a neglected parasitic infection.

lymphatic system   A system in which lymph is transported in lymphatic vessels.

lymphocytes   A category of white blood cells that play a key role in adaptive immunity.

lymphoid lineage   The development of blood cells that leads to the production of lymphocytes.

lymphoid organs   The red bone marrow and thymus gland. Organs of the lymphatic system where lymphocytes are formed and mature.

lymphoid tissues   The cells and organs of the lymphatic system. Include leukocytes, bone marrow, thymus, spleen, and lymph nodes.

lysis   Bursting of host cells during the release of viruses.

lysogenic bacteriophage   A bacteriophage that infects its bacterial host, integrates its DNA genome into a specific region of the bacterial chromosome, and then replicates as a prophage along with the host cell. No infectious bacteriophages are produced.

lysogenic cycle   A bacteriophage infection whereby the phage infects the host and its dsDNA genome is integrated into the DNA of the host's chromosome. The host cell is not destroyed by lysis. The integrated phage DNA replicates as a provirus, along with the host cell DNA. No infectious bacteriophages are produced during a lysogenic cycle.

lysogenic phage conversion   The phenotype of a bacterium is altered by genes within an prophage integrated into the bacterial chromosome.

lysogenized   A bacterium that contains a prophage within its chromosome.

lysozyme   An enzyme present in tears that contributes to the nonspecific immune system by disrupting bacterial cell walls.

lytic bacteriophage   A bacteriophage that has a lytic (virulent) replication cycle, thus destroying the infected bacterial host cell upon the release of the new bacteriophages.

lytic cycle   A bacteriophage infection that results in the lysis of the host cell membrane and destruction of an infected host cell during the release of new infectious progeny bacteriophages at the end of its replication cycle.

M protein   A protein found in the cell walls of streptococci that confers resistance to phagocytosis.

MAC (*Mycobacterium avium* complex)   A group of mycobacteria that can cause a serious, life-threatening opportunistic bacterial infections, particularly in people with suppressed immune systems, such as those with HIV/AIDS and organ transplant patients on immunosuppressive drugs.

machine learning   An automated method of data analysis where a computer uses statistical techniques to "learn" about the data.

macrophages   Phagocytic cells that function in the innate immune system.

macroscopic   Term meaning visible to the naked eye. There is no need for the aid of a microscope.

maculopapular rash   A rash characterized by both flat and raised lesions on the skin caused by allergy or infection.

mad cow disease   A transmissible spongiform encephalopathy of cattle caused by prions that results in spongy degeneration of the brain accompanied by severe and fatal neurological damage; also known as bovine spongiform encephalopathy.

magnetotactic bacteria   Bacteria that orient and migrate along the lines of Earth's magnetic field.

major histocompatibility molecules (MHC)   Molecules associated with presentation of antigenic material to receptor molecules on cytotoxic T cells.

malaise   General feeling of discomfort.

malaria   A tropical disease caused by the bite of female *Anopheles* mosquitoes infected with the *Plasmodium* protozoan parasite.

malaria prophylaxis   Use of antimalarial medications before infection that kill the *Plasmodium* parasite if the person is bitten by an infected mosquito.

Mallon, Mary   Better known as "Typhoid Mary" (1869–1938), a notorious healthy carrier of *Salmonella typhi*, the cause of typhoid fever. She was forcibly quarantined to living in a cottage on North Brother Island until her death.

Mantoux turberculin skin test   A method for diagnosis of tuberculosis; purified protein derived from *Mycobacterium tuberculosis* is injected into the arm of a patient, and the presence of an induration between 48 and 72 hours after injection indicates past or present exposure to the tubercle bacillus.

March for Science   An international series of rallies and marches held on Earth Day, April 22, 2017, in Washington, D.C., and more than 600 other cities across the world to emphasize that science upholds the common good and to call for evidence-based policy in the public's best interest.

mast cells   White blood cells with granules rich in histamine and heparin that are part of the immune system.

MDR TB (multiple drug–resistant tuberculosis)   A strain of *Mycobacterium tuberculosis* that is resistant to the two best anti-TB antibiotics, rifampicin and isoniazid.

measles   Vaccine preventable, highly infectious airborne disease caused by the measles virus. Infection results in a red blotchy rash, fever, and cough.

melana   Dark, tarlike feces that contains digested blood.

meningitis   Inflammation of the meninges that can be caused by microbes such as *Neisseria meningiditis*.

meningococcal disease   A severe and potentially life-threatening bacterial infection of the meninges surrounding the brain and spinal cord caused by *Neisseria meningitidis*.

meningococcal septicemia   A severe and potentially life-threatening bacterial infection caused by *Neisseria meningitidis* present in the bloodstream.

meningococcemia   A severe and potentially life-threatening bacterial infection of the blood caused by *Neisseria meningitidis*.

merozoites   A form of the protozoan *Plasmodium* spp. (the cause of malaria) that occurs through the asexual multiplication of sporozoites in the liver. It is one of several forms that develop during the malarial parasite's life cycle.

MERS-CoV   Middle East respiratory syndrome-coronavirus is an emerging virus associated with dromedary camels that causes severe respiratory illness in humans with a fatality rate of about 35%. The majority of cases occur in Saudi Arabia.

messenger RNA (mRNA)   RNA molecule transcribed from the DNA that is translated in protein.

metabolites   Substances formed through metabolic processes (e.g., para-aminobenzoic acid, or PABA, that bacteria use to synthesize folic acid).

metagenomics   The analysis of the genetic material obtained from environmental samples.

methicillin-resistant *Staphylococcus aureus* (MRSA)   Strains of *Staphylococcus aureus* that are resistant to the methicillin, a narrow-spectrum antibiotic.

methicillin-resistant *Staphylococcus aureus* (MRSA) bacteremia event   The development of bacteremia due to spread of methicillin-resistant *Staphylococcus aureus*.

MHC (major histocompatibility molecules)   Molecules associated with presentation of antigenic material to receptor molecules on cytotoxic T cells.

miasma   A term, used before microbes were identified as pathogens causing infectious diseases, meaning "bad air" or "swamp air."

miasma theory   Theory that contagious diseases were caused by "bad air" and thus the product of environmental factors such as contaminated water, foul smelling air, and poor hygienic conditions. Was the predominant theory of disease transmission prior to the development and acceptance of the germ theory of disease in the 19th century.

miconazole   A topical antifungal medication sold under the brand name Monistat, among others, used to treat ring worm and vaginal yeast infections.

Microban   Registered trademark of products with built-in antimicrobials to protect it against damaging microbes such as bacteria, mold, and mildew.

microbe   A microscopic organism such as a bacterium, fungus, or protozoan.

microbial diversity   The variability of all types of microbes or subcellular agents (e.g., viruses) in a specified environment (e.g., seawater, the dessert, soil, or a rainforest).

microbial umbrella   The organisms that often are included in the study of microbiology. Includes bacteria, fungi, protozoans, viruses, prions, and unicellular algae.

microbiome   The microbes that form a community in a particular environment, such as the human digestive system or skin. In humans, the gut community may play a key role in maintaining health.

microbiota   The microbes inhabiting a particular body site of living organisms.

microcephaly   Condition where a baby's head is smaller than expected, often due to abnormal brain development. Can be caused by exposure to pathogens, such as Zika virus, during pregnancy.

microfilariae   Mature filarial worms in the blood that can be taken up by mosquitoes to infect other people with elephantiasis.

micrometer   A unit of measurement equal to one-millionth of a meter.

micronazole   A topical antifungal medication used to treat ring worm and yeast infections.

microorganism   A microscopic organism such as a bacterium, fungus or a protozoan.

microscopic   Visible only with the aid of a microscope.

Milwaukee protocol   Experimental rabies treatment protocol whereby rabies victims (e.g., Wisconsin native Jeanna Giese) were put in a coma with antiexitory drugs and treated with an antiviral cocktail. It is no longer considered a valid treatment for human rabies.

mimivirus   A girus found in amoebas first isolated from a cooling tower in England.

MMR (measles, mumps, and rubella) vaccine   Combination measles, mumps, and rubella vaccine.

MMRV (measles, mumps, rubella, varicella) vaccine   Combination vaccine formulated to prevent four of the most common and dangerous infectious viral diseases of childhood.

mobile health   Provision of healthcare services outside of the standard healthcare infrastructure, such as in people's homes or in the field.

mode of transmission   The method by which a pathogen moves from a reservoir and gains entry into a susceptible host. Examples include person to person, airborne, waterborne, vectorborne, zoonotic, transplacental, and oral–fecal modes of transmission.

molecular biology   The branch of biology that is concerned with the molecular basis of biological activity such as that involving DNA, RNA, and other macromolecules.

monkeypox   Disease caused by monkeypox virus that occurs most often in Central and West Africa in primates, exotic pets (e.g., Gambian rats and prairie dogs), as well as humans. Symptoms include fever, headache, muscle pain, and swollen lymph nodes. Related to cowpox and smallpox. The vaccine used to prevent smallpox is also effective against this virus.

monoclonal antibody   Antibodies produced toward a specific antigen in the laboratory from hybridomas (the fusion of myeloma cells with B cells). Monoclonal antibodies are being developed as experimental therapies (e.g., ZMapp to treat Ebola virus disease) and to treat certain cancers.

monocytes   A category of phagocytic white blood cells.

mononucleosis   A disease caused by Epstein-Barr virus in which the salivary glands are infected by the virus; frequently referred to as "mono" or the *kissing disease*.

Monster Soup   Term used to refer to the sewage-contaminated water drawn from the Thames River in London in the 19th century.

morbidity   A measure of the rate of illness of a particular disease.

mordant   A chemical or physical process (e.g., steaming) used to fix a primary stain to a microbe (e.g., application of iodine to form a complex with the crystal violet bound to the peptidoglycan of bacterial cell walls in the Gram stain procedure).

morphology   The shape of an organism; for example, rod or coccus-shaped bacteria. It can also refer to the shape of bacterial colonies (e.g., flat, circular, raised).

mortality   A measure of the rate of death resulting from a particular disease.

mosquito abatement programs   Application of insecticides and efforts to eliminate stagnant water sources in order to decrease mosquito populations and thus the incidence of mosquitoborne diseases.

motility   Flagellar movement of bacteria involving chemotactic responses toward or away from a substance.

MRSA (methicillin-resistant *Staphylococcus aureus*)   Strains of *Staphylococcus aureus* that are resistant to the methicillin, a narrow-spectrum antibiotic.

mucormycetes   Fungi that cause mucormycosis, a fungal infection affecting the lungs or sinuses that can spread to other parts of the body.

mucosal-ciliary escalator   Process whereby debris that enters the respiratory tract is moved out of the lungs in mucus that is moved upward by cilia lining the epithelial tissue toward the pharynx, where it is then swallowed and eliminated by the digestive system.

multidrug-resistant (MDR) bacterial infections   Infections caused by bacteria that are resistant to the inhibitory effects of more than one antibiotic. Such infections are on the rise due to excessive use of antibiotics, which fosters the accumulation of bacteria with genes that confer antibiotic resistance.

mumps   Highly contagious viral disease that primarily causes severe swelling of the salivary glands.

mutagens   Physical or chemical agents that cause mutations in DNA.

mutation   A change in DNA that is transferred to subsequent generations.

mutualism   A symbiotic relationship in which both organisms benefit from the association.

myalgia   Muscle pain.

mycelial phase   A two-phase process exhibited by fungal pathogens. While in the environment, the fungi exist in the environment as mycelium or a mass of threadlike hyphae absorbing nutrients in contrast to its presence inside of the body in which it undergoes a yeast phase.

***Mycobacterium avium* complex (MAC)**   A group of mycobacteria that can cause a serious, life-threatening opportunistic bacterial infections, particularly in people with suppressed immune systems, such as those with HIV/AIDS and organ transplant patients on immunosuppressive drugs.

mycolic acids   Fatty acids found in the cell walls of certain acid-fast bacteria such as *Mycobacterium* and *Nocardia* spp. that gives them a waxlike property such that they are resistant to decolorization by acid alcohol during the acid-fast differential staining procedure.

mycoses   Diseases caused by fungi.

mycotoxin   Toxins produced by fungi that are capable of causing disease and death in other organisms.

myeloid lineage   The path of blood cell maturation that leads to the production of platelets, red blood cells, monocytes, neutrophils, eosinophils, and basophils.

N (neuraminidase) spikes   Glycoproteins on the outer surface of influenza A and B viruses that cleave sialic acid present on the surface of host cells lining the respiratory tract, enabling new progeny viruses to be released from the host cell, facilitating viral spread. Antivirals such as Tamiflu, Relenza, and Rapivab block neuraminidase activity.

NAAT (nucleic acid amplification test)   Diagnostic technique that is used to amplify the genomes of a particular pathogen in clinical sample.

naked virus   A virus that does not have a viral envelope wrapped around its capsid. Such viruses are more stable in the environment and resistant to the effects of hand sanitizer (e.g., noroviruses and papillomaviruses).

nanometer   A unit of measurement equal to one-billionth of a meter. Many human viruses are in the nanometer range in size.

nanotechnology   Technology that focuses on the use of nanoparticles in drug delivery, diagnostics, gene therapy, and other clinical applications and research.

Napa Conference (2015)   Conference that addressed the ethical use of CRISPR/Cas technology in the manipulation of the human genome.

Narrow-spectrum antibiotic   An antibiotic that is inhibitory to limited range of bacteria.

National Institutes of Health (NIH)   Largest research institution in the United States tasked with biomedical and public health research.

National Tuberculosis Surveillance System (NTSS)   A disease-specific surveillance system for tuberculosis that was created in 1953. Mandates that state health departments report when a person tests positive for *Mycobacterium tuberculosis*.

natural killer (NK) cells   Lymphocytes that kill virus-infected cells as part of innate immunity.

necrotizing enterocolitis   Death of tissue in the intestine and bowel, possibly due to a bacterial infection. It is the most common and serious intestinal disease of premature babies, with a 25% mortality rate.

necrotizing fasciitis   A rare condition caused by highly invasive streptococci and other bacteria in which the subcutaneous tissue is infected; the streptococci are sometimes referred to as "flesh-eating."

needle-exchange programs   Community-based programs that supply free, sterile syringes in exchange for used needles. Promotes the use of sterile syringes among drug abusers in order to prevent the spread of infectious diseases such as HIV and hepatitis C through the reuse of contaminated syringes.

neglected parasitic infections (NPIs)   Diseases caused by parasites that have had relatively few resources dedicated to their control and eradication. Most occur in impoverished or developing countries that lack safe water and sanitation. Examples include African trypanosomiasis and leishmaniasis.

nematodes   Roundworms that inhabit many environments and play an important role in nutrient cycling. Many are parasitic or mutualistic.

neonatal (newborn) tetanus   A manifestation of tetanus in newborn children that results from unsanitary conditions during delivery.

neuraminidase (N) spikes   Glycoproteins on the outer surface of influenza A and B viruses that cleave sialic acid present on the surface of host cells lining the respiratory tract, enabling new progeny viruses to be released from the host cell, facilitating viral spread. Antivirals such as Tamiflu, Relenza, and Rapivab block neuraminidase activity.

neurocysticercosis   The presence of cysticerci in the brain or spinal cord resulting in seizures, headaches, and possibly death.

neurotoxins   Toxins released by microbes that interfere with the transmission of neural impulses (e.g., *Clostridium tetani* and *Clostridium botulinum* neurotoxins).

New World hookworm   The *Necator americanus* species of hookworm. It is the most common roundworm helminth worldwide and is present in the southeastern United States.

next-generation DNA sequencing   Extremely fast DNA sequencing techniques that can be used to sequence entire genomes in a day.

niclosamide   A medication sold under the tradename of Niclocide, among others, used to treat tapeworm infections.

NIH (National Institutes of Health)   Largest research institution in the United States tasked with biomedical and public health research.

nitrification   A stage in the nitrogen cycle in which ammonia is converted into nitrogen.

nitrogen cycle   A cycle in nature in which atmospheric nitrogen is recycled through a pathway involving bacteria.

NK (natural killer) cells   Lymphocytes that kill virus-infected cells as part of innate immunity.

non-acid-fast bacteria   Bacteria that do not contain mycolic acids in their cell walls and thus are sensitive to decolorization by acid alcohol during the acid-fast differential staining procedure. They are counterstained with methylene blue.

nonenveloped viruses   Viruses that do not have an envelope around them; also referred to as "naked."

noninfectious recombinant vaccine   A vaccine that consists of the major outer capsid proteins of human papillomavirus types 6, 11, 16, and 18 produced through genetic engineering. The vaccine consists of the outer "shell" of the human papillomaviruses that do not contain any genetic material. Hence the "empty" papillomaviruses cannot replicate or cause an infection but can induce immunity, resulting in protection against exposure to these types of papillomaviruses known to be involved in cancers and genital warts.

nonmedical exemptions   Religious or other reasons whereby a person does not have to submit to legal immunization requirements.

nonspecific immunity   Physiological defenses that prevent microbes from gaining access into the body or eliminate those that have penetrated the body.

nontuberculous mycobacteria (NTM)   *Mycobacterium* spp. present in the environment that do not cause tuberculosis or leprosy but that may cause pulmonary infection (e.g., *Mycobacterium avium*).

normal microbiota All of the microbes that reside on (i.e., the skin) or inside of the human body (i.e., gastrointestinal tract, respiratory tract).

NPIs (neglected parasitic infections) Diseases caused by parasites that have had relatively few resources dedicated to their control and eradication. Most occur in impoverished or developing countries that lack safe water and sanitation. Examples include African trypanosomiasis and leishmaniasis.

NTM (nontuberculous mycobacteria) *Mycobacterium* spp. present in the environment that do not cause tuberculosis or leprosy but that may cause pulmonary infection (e.g., *Mycobacterium avium*).

NTSS (National Tuberculosis Surveillance System) A disease-specific surveillance system for tuberculosis that was created in 1953. Mandates that state health departments report when a person tests positive for *Mycobacterium tuberculosis*.

nucleic acid amplification test (NAAT) Diagnostic technique that is used to amplify the genomes of a particular pathogen in clinical sample.

nucleic acid Biological macromolecules composed of chains of nucleotides. DNA and RNA are nucleic acids.

nucleocapsid A viral structure consisting of the viral nucleic acid and the protein coat.

nucleoid The DNA-rich area in procaryotic cells; it is not surrounded by a membrane.

nucleosides A purine or pyrimidine linked to a ribose or deoxyribose sugar.

nucleotides Monomers composed of a nitrogenous base, a five-carbon sugar, and a phosphate group that comprise DNA and RNA. In DNA, they are adenine, thymine, guanine, and cytosine. In RNA, thymine is replaced by uracil.

nutrient broth A liquid media made of peptone and beef extract that is used to grow bacteria.

nymph A preadult stage in tick development.

offal A ground-up mixture of organs and trimmings of dead animals used to feed other animals.

Ohio Valley disease Another name for *histoplasmosis*. A fungal infection caused by breathing spores of *Histoplasma capsulatum*, which is found in soil and bird and bat droppings. Most common in the Mississippi and Ohio River Valleys. Can cause serious respiratory infections in infants and people with compromised immune systems.

Old World hookworm The *Ancylostoma duodenale* species of hookworm that is found only in southern Europe, northern Africa, northern Asia, and parts of South America.

oligodynamic effect Killing or toxic effect of heavy metals on bacteria.

OMKO1 phages Outer membrane knockout dependent phage 1, a group of phages that infect the bacterium *Pseudomonas aeruginosa*. The phages were isolated from Dodge Pond, East Lyme, Connecticut, and modified by researchers to treat multidrug-resistant *Pseudomonas aeruginosa* infections.

omnivores Animals that eat both plants and other animals as food sources.

onchocerciasis A parasitic disease caused by *Onchocerca volvulus* that frequently results in blindness.

oncolytic viruses Viruses that infect and destroy cancer cells.

One Health An integrative effort between multiple disciplines to work globally to attain optimal health for all people, animals, and the environment.

oocyst Infectious form of *Cryptosporidium parvum*.

operons A group of functionally related genes that act as "on and off" switches.

ophthalmia neonatorum A condition in which the corneas are damaged as a result of the transmission of *Neisseria* into the eyes of newborns during delivery.

opioids Class of narcotics that includes pain relievers available by prescription (e.g., oxycodone, morphine, codeine) and illegal drugs such as heroin and fentanyl.

opisthotonos A body position in which the back is severely arched; characteristic of tetanus.

opportunistic infections Infections that are caused by microbes that do not cause disease in people with healthy immune systems but will cause infections in persons with weakened immune systems, such as those with HIV/AIDS.

oral–fecal mode of transmission Mode of pathogen transmission where pathogens present in the feces of one person are passed to the mouth of another person. Primarily caused by inadequate sanitation and poor hygiene practices (not washing hands after going to the bathroom). Diseases passed in this manner also occur through the ingestion of fecally contaminated food or water. Examples of diseases passed in this manner include cholera, hepatitis A, winter vomiting disease caused by noroviruses, and salmonellosis.

Orangi Pilot Project Program in Pakistan developed to improve public sanitation in order to reduce disease caused by exposure to the pathogens present in human waste.

organ systems A group of organs that work together to perform one or more functions; for example, the digestive system includes all of the organs that facilitate the digestion of food.

organelles Membrane-bound structures present in eucaryotic cells that serve specialized functions.

organic compound Generally, a chemical compound that contains carbon (although carbon dioxide and several other carbon-containing compounds are considered inorganic).

organisms Living entities composed of cells.

organohalides A novel group of chlorinated phenols with antimicrobial properties synthesized by chemists in the laboratory by substituting hydrogen atoms present on the aromatic rings of phenol with chlorine atoms.

organs Structures composed of more than one tissue type.

Oriental sore A form of leishmaniasis that results in skin lesions. Also known as *cutaneous leishmaniasis*.

origin of replication A sequence of DNA that denotes where replication should be started.

osteomyelitis An infection of the bone.

Otitis media An infection of the ear canal.

outbreak The sudden appearance of multiple cases of an infectious disease.

outer membrane porin M protein Cell wall surface protein of *Pseudomonas aeruginosa* that is used as a receptor by OMKO1 bacteriophages for infection.

Oxivir TB A surface disinfectant based on proprietary hydrogen peroxide technology that contains a low dwell time (1 minute) that is fast and effective against viruses (including noroviruses), bacteria (including *Mycobacterium* spp. and MRSA), and fungi.

oxyuriasis A parasitic infection caused by *Enterobius vermicularis*, a type of roundworm. Common in children; symptoms include itching in the anal area. Also called *pinworm disease*.

Pacific flyway The major north–south migratory pathway for American birds that extends from Alaska to the tip of South America. Experts are concerned that wild waterfowl that may carry highly pathogenic avian influenza A viruses could shed viruses along this flyway, causing a potential influenza pandemic.

palindrome A sequence of letters or numbers that reads the same backward and forward, such as "madam" or "GAATTC" (the restriction enzyme site for *EcoR1*). Restriction enzymes cleave DNA at these sequences to generate "sticky ends" allowing one to create recombinant DNA molecules.

PAMPs (pathogen-associated molecular patterns)   Molecules not found in or on human cells, but are associated with groups of organisms and are recognized by the innate immune system. The presence of these molecules indicates an infection. Examples include peptidoglycan, LPS, mannose sugar, etc.

pandemic   A worldwide outbreak of an infectious disease.

paralysis   The loss of the ability to move or have sensation in part or all of the body due to nerve injury or disease. Some bacterial infections or viral infections can cause paralysis.

paralytic poliomyelitis   A form of polio caused by replication of poliovirus in nerve cells, sometimes resulting in severely deformed limbs and paralysis.

parasitism   A form of symbiosis in which one biological agent lives at the expense of another.

paresis   A type of partial paralysis caused by nerve damage or disease.

paroxysm   A period when someone infected with *Plasmodium* spp. experiences symptoms of an attack of malaria that consists of episodes of chills and fevers, in contrast to periods in between in which the person has no signs and symptoms of malaria and feels normal.

partnerships   Collaborative efforts between different organizations to further a particular cause.

pasteurization   A process of disinfection by which liquids are heated to reduce the number of pathogens.

pathogen-associated molecular patterns (PAMPs)   Molecules not found in or on human cells, but are associated with groups of organisms and are recognized by the innate immune system. The presence of these molecules indicates an infection. Examples include peptidoglycan, LPS, mannose sugar, etc.

pathogens   Microbes capable of producing disease.

patient zero   In an epidemiological investigation, the person with the first documented case of an infectious disease.

pattern recognition receptors (PRRs)   Molecules on the surface of cells that function as part of innate immunity. These receptors recognize pathogen-associated molecular patterns (PAMPs) and, when engaged, activate innate immune cells to act on the infection.

PCR (polymerase chain reaction)   A molecular biology technique that uses a heat stable DNA polymerase produced by *Thermus aquaticus* to generate DNA *in vitro*.

pelvic inflammatory disease (PID)   A condition that occurs in about 50% of untreated female gonorrhea patients; characterized by abdominal pain and sometimes sterility.

penicillin   Group of antibiotics derived from *Penicillium* fungi. It contains a beta lactam ring and is involved in the inhibition of bacterial cell wall synthesis. Was first discovered by Alexander Fleming in 1928. It was then developed into a commercial antibiotic for wide release in the 1940s. Instrumental in reducing the number of deaths from infection.

People's Climate Mobilization Day   Protests against the environmental policies of President Donald Trump and his administration on April 29, 2017, in Washington, D.C., and numerous cities around the globe.

PEP (postexposure prophylaxis)   The use of antiretroviral treatment following potential exposure to HIV (e.g., accidental needlestick injury when caring for a HIV/AIDS patient) to reduce chance of infection. Treatment must be started within 72 hours of possible exposure.

peptic ulcers   Open sores on the inside of the lining of the stomach. Infection by the bacterium *Heliobacter pylori* is the cause of some peptic ulcers.

peptide bonds   Bonds formed between amino acids during protein synthesis.

peptidoglycan   A compound found in Gram (–) and Gram (–) bacterial cell walls that confers rigidity and tensile strength.

perforin   A membrane-penetrating protein released by cytotoxic T cells that leads to lysis of host cells infected by viruses.

period of decline   Phase of course of microbial disease where the number of pathogenic microbes begin to decrease and disease symptoms start to resolve.

peripheral lymphoid tissues   The lymph nodes and the spleen. Immature lymphocytes from primary lymphoid tissues travel to these sites where they then mature and are released when a pathogen is detected.

peritonitis   A potentially fatal condition caused by leakage of intestinal fluids into the abdomen.

personal protective equipment (PPE)   Barriers worn to minimize exposure to pathogens and other hazards. Includes face shields or masks, gloves, and other garments such as Tyvek suits.

personalized medicine   The idea that disease prevention or treatment should be based on the patient's specific characteristics, such as genetic profile, age, environment, etc.

pertussis   An infection of the respiratory tract caused by the bacterium *Bordetella pertussis*. Can cause serious illness, particularly in infants and children, due to violent coughing that makes it difficult to breathe. Also called *whooping cough*. The vaccine is about 70% effective in preventing infection and reduces the symptoms or severity of vaccinated individuals who do get infected.

petechial rash   Pinpoint spots on the skin caused by bleeding from capillaries. Indicative of trauma or infection.

petri dish   A round, platelike container in which bacteria are grown on agar in the laboratory.

Peyer's patches   Small masses of lymphoid tissue in the small intestine that monitor intestinal bacterial populations and act to prevent the growth of pathogenic bacteria. Polioviruses multiply in these masses before they spread in the body.

phage therapy   The use of bacteriophages to treat bacterial infections. Today it is being developed in Western medicine to treat multidrug-resistant bacterial infections but it has been used in Russia for over a century to treat bacterial infections.

phages   Viruses that infect and kill bacteria. Also called *bacteriophages*.

phagocytes   Cells that engulf microbes and bring about their destruction by enzymatic activity; these cells play a vital role in the nonspecific immune system.

phagocytosis   A nonspecific body defense mechanism by which bacteria are ingested and killed by phagocytic cells.

phenol   Surgical antiseptic used to prevent contamination of surgical wounds by pathogens. Use was pioneered by Joseph Lister in the late 1800s. Also called *carbolic acid*.

[^{32}P]   Radioactive isotope of phosphorus used to label DNA in experiments of A. D. Hershey and Martha Chase that proved that DNA is the hereditary material in a cell.

photophobia   Extreme sensitivity of the eyes to light exposure. A symptom of rabies.

photosynthetic autotrophs   Organisms that utilize the energy of the sun and that use carbon dioxide as a carbon source.

PID (pelvic inflammatory disease)   A condition that occurs in about 50% of untreated female gonorrhea patients; characterized by abdominal pain and sometimes sterility.

pigmentation   The secretion or production of colorful pigments by certain microbes, especially bacteria. For example, *Serratia marcescens* colonies secrete a red pigment. Occurs through quorum sensing and plays a role in protection.

pili   Bacterial appendages in Gram (–) bacteria that act as adhesins; some serve as a bridge allowing genetic exchange.

pilin   The protein that composes pili.

pinworm disease   A parasitic infection caused by *Enterobius vermicularis*, a type of roundworm. Common in children; symptoms include itching in the anal area. Also called *enterobiasis*.

plague   A bacterial disease caused by *Yersinia pestis*, a highly virulent bacterium; also known as the *Black Death*.

plantar warts   Deep, painful warts found on the soles of the feet; caused by human low risk types of papillomaviruses.

plaque   A biofilm of bacteria attached to the surface of teeth that may give rise to dental caries.

plasma cells   Antibody-producing cells derived from B cells.

plasmids   Small circular molecules of nonchromosomal dsDNA found in some bacteria.

platelets   Blood cells that initiate the clotting of blood.

pleomorphic   Variation in the shape or size of a microbe or virus (e.g., *Corynebacterium* spp. consist of rod-shaped bacteria of various shapes).

pleurisy   Inflammation of the pleural lining of the lungs.

pneumonia   Inflammation of the lungs; can be caused by a variety of microbes and viruses.

pneumonic plague   A form of bubonic plague that develops into pneumonia.

pneumotropic pathogens   Pathogens that have an affinity for the respiratory tract.

point mutation   A genetic process in which one nucleotide is replaced with another nucleotide; simplest mutation; can occur naturally or through genetic engineering manipulations.

Polio Global Eradication Initiative   Initiative established in 1988 with multiple governments and nongovernmental organizations to eliminate polio worldwide through surveillance and vaccination efforts.

poliomyelitis   An infectious disease caused by poliovirus that is spread by an oral–fecal mode of transmission. In severe cases cases, symptoms of poliomyelitis can involve the weakening or paralysis of the muscles and lungs and extremities of its victim; poliomyelitis is rare today because of vaccination efforts.

polymerase chain reaction (PCR)   A molecular biology technique that uses a heat stable DNA polymerase produced by *Thermus aquaticus* to generate DNA *in vitro*.

polymicrobial   Difficult-to-treat infections caused by combinations of multiple viruses, bacteria, fungi, and parasites.

Pontiac fever   Pneumonialike illness caused by *Legionella pneumophila*.

portal of entry   The body site at which microbes or viruses enter a host.

portal of exit   The site from which microbes or viruses are shed or leave a host and may infect another host.

post–Ebola syndrome   Lingering complications following Ebola virus disease infections. Symptoms include eye problems, neurological problems, and joint and muscle pain.

postexposure prophylaxis (PEP)   The use of antiretroviral treatment following potential exposure to HIV (e.g., accidental needlestick injury when caring for a HIV/AIDS patient) to reduce chance of infection. Treatment must be started within 72 hours of possible exposure.

postherpetic neuralgia   A complication of shingles where the chickenpox virus affects the nerve fibers, resulting in an unpleasant burning sensation.

PPE (personal protective equipment)   Barriers worn to minimize exposure to pathogens and other hazards. Includes face shields or masks, gloves, and other garments such as Tyvek suits.

praziquantel   Medication used to treat parasitic worm infections, including those caused by tapeworms.

prebiotics   Foods that promote growth of probiotic bacteria ingested to promote balance of human gut microbiota.

preclinical testing   In drug development, it involves animal testing, the stage before a drug can be tested on humans where the focus is on determining the safe dose of a medication for human use.

predators   Organisms that capture and feed on other organisms.

prednisone   A type of steroid medication used to reduce inflammation (e.g., to reduce inflamed airways of a patient suffering from a severe respiratory tract infection). It is also used to treat certain autoimmune disorders.

premastication   The practice of prechewing food for infants by which HIV can be transmitted.

prevalence   Epidemiological concept that refers to the total number of cases of a disease in a population and a particular point in time.

primary producers   Photosynthetic organisms that utilize the energy of the sun and produce organic compounds and oxygen.

primary syphilis   An early stage of syphilis characterized by the formation of chancres on the genitals.

primer   A short single strand of DNA that serves as the starting point for DNA synthesis.

prions   Infectious, highly stable, misfolded proteins that can cause neurological disease; they do not contain DNA or RNA.

probiotics   Live microorganisms that are ingested to improve or restore the balance of human gut microbiota (e.g., often consumed during or after antibiotic therapy, which can destroy many commensal bacteria and result in the overgrowth of bacteria that cause diarrhea).

procaryotes   Single-celled bacteria that do not contain membrane-bound organelles in their cytoplasm. Includes the domains *Bacteria* and *Archaea*.

prodromal period   An early stage in a microbial disease characterized by headache, tiredness, and muscle aches. Also the first stage of human immunodeficiency virus infection characterized by fever, diarrhea, rash, aches, fatigue, and lymphadenopathy.

professional antigen-presenting cells   Cells that process and present antigens to T helper cells, including dendritic cells, macrophages, and B cells, all of which express Class II MHC for antigen presentation.

proglottids   Compartment-like segments of tapeworms containing both testes and ovaries.

promastigotes   The invertebrate extracellular form of *Leishmania* spp. protozoans (that contains a flagellum) that enter the skin of a human through the bite of an infected sand fly. Once inside of the body (vertebrate host), it becomes a nonmotile amastigote form (without a flagellum) through binary fission.

ProMED-mail   Event surveillance system that promotes communication among professionals in the international infectious disease community.

promoter site   A site on a gene strand that marks the beginning of mRNA transcription.

proofreading   An error-correcting process by DNA polymerases that occurs during DNA replication.

propagated epidemics   Epidemics resulting from direct person-to-person transmission.

prophage   A fragment of phage DNA integrated into a bacterial chromosome.

prophylactic antibiotic therapy   The use of antibiotics to prevent a bacterial infection (e.g., AIDS patients may receive antibiotics prophylactically to prevent opportunistic infections caused by bacteria).

prophylaxis   Actions taken to prevent an infectious disease ranging from antibiotic or antiviral therapy to reduce infections in AIDS patients or elderly to the use of condoms to prevent the spread of sexually transmitted diseases.

protease inhibitors   A class of drugs used in HIV/AIDS therapy that prevent viral maturation into mature virus particles.

proteases   Enzymes that cleave or degrade proteins.

protein   Biological macromolecule composed of amino acids. Play critical structural and nonstructural roles in all living organisms as building blocks and workhorses of cells. Also plays structural and nonstructural roles in the replication and assembly of viruses.

protozoans   A category of unicellular eucaryotic microbes.

pruno   An alcoholic beverage made by prison inmates from a variety of ingredients, such as fruits, candies, sugar, and yeast. Has been implicated as the source of *Clostridium botulinum* causing small outbreaks of botulism in U.S. prisons.

PRRs (pattern recognition receptors)   Molecules on the surface of cells that function as part of innate immunity. These receptors recognize pathogen-associated molecular patterns (PAMPs) and, when engaged, activate innate immune cells to act on the infection.

psychrophiles   Cold-loving bacteria that grow best at about 15°C (59°F).

public health   The science of protecting and improving the health of people and their communities through education, policy-making and practices (e.g., vaccination) through organized efforts.

puerperal fever   A postpartum streptococci infection characterized by fever and lower abdominal pain that begins the second or third day after delivery. Can result in death. Incidence greatly decreased after Ignaz Semmelweis promoted the practice of handwashing to reduce the transmission of bacteria by obstetricians during the late 1800s in Europe. Also called childbed fever.

PulseNet   A national laboratory network that compares the DNA fingerprints of bacteria from people with foodborne diseases to find clusters of disease that might represent unrecognized disease outbreaks.

pure culture   A bacterial culture composed of only one type of bacterium. Used to perform diagnostic testing in the medical technology laboratory.

purines   One of two types of nitrogenous bases that comprise DNA and RNA. They include adenine and guanine, which then pair with their complementary pyrimidines, thymine and cytosine in DNA and thymine and uracil in RNA.

pyrimidines   One of two types of nitrogenous bases that comprise DNA and RNA. They include cytosine and thymine in DNA and cytosine and uracil in RNA, which then pair with their complementary pyrimidines, adenine and guanine.

quarantine signs   Public health strategy used in the 1920s to the 1950s to identify homes in which an individual was suffering from a contagious disease. Signs were placed on the entrance of homes so that visitors would not enter the homes of quarantined individuals.

quarantine   Practice of isolating a person suffering from an infectious disease in order to prevent the spread of the disease to others.

quorum sensing (QS)   A form of bacterial communication that allows bacteria to perform group behaviors or metabolic activities such as pigment production, biofilm formation, or bioluminescence.

quorum-sensing inhibitors (QSIs)   Drugs currently in development that will halt quorum sensing by bacteria in hopes of halting biofilm formation and thus provide a treatment for various difficult-to-treat and/or antibiotic-resistant bacterial infections.

R plasmid   Conjugative resistance transfer plasmids that contain genes coding for antibiotic resistance or the production of sex pili for conjugation or shuffling of plasmids from one bacterium to another.

R strain of *S. pneumoniae*   "Rough" strain of *S. pneumoniae* that is not encapsulated and does not cause disease.

rabies   A viral disease transmitted by the bite of a rabid animal, resulting in damage to the nervous system and eventual death if not treated promptly by vaccination; formerly called hydrophobia.

rabies vaccine   A vaccine used to prevent rabies. First developed by Louis Pasteur and Emile Roux in 1885 and successfully used to save the life of a 9-year-old boy mauled by a rabid dog.

rapid diagnostic tests   Tests for presence of viral infection that can be completed quickly (within minutes). Often used for diagnosis of influenza A or B and strep throat.

Rapivab   Brand name of peramivir, an antiviral drug used to treat influenza. Blocks activity of influenza A and B neuraminidases.

reactivation TB   When a person experiences latent or secondary tuberculosis in which old lesions caused by *Mycobacterium tuberculosis* are reactivated to cause chronic symptoms of disease.

real-time data collection   The collection of surveillance data (e.g., number of newly identified cases of a particular infectious disease) within milliseconds of its recording in order to identify potential public health threats.

real-time surveillance   The collection of surveillance data (e.g., number of newly identified cases of a particular infectious disease) within milliseconds of its recording in order to identify potential public health threats.

receptors   Perform normal cellular functions. They are comprised of proteins or other chemical moieties present on the surface of host cells. Viruses have evolved to attach to cellular receptors, thus facilitating viral entry into a living host cell in order to replicate (e.g., influenza A viruses bind to sialic acid receptors present on the surface of epithelial cells that line the respiratory tract or HIV binds to the CD4 receptor present on T cells).

recombinant DNA molecules   The recombination of DNA molecules from multiple sources using genetic engineering (e.g., the use of restriction enzymes and/or polymerase chain reaction).

recombinant DNA technology   Means of manipulating genes to bring about a desired function for an intended application.

recombinants   Formation of novel viruses or cells through genetic recombination.

recombination   Production of offspring with different traits from the parents due to exchange of genetic material. In eucaryotes, meiosis and sexual reproduction are key mechanisms of the exchange of genetic material. In bacteria, transformation, transduction, and conjugation are responsible for the exchange of genetic material.

Red Queen hypothesis   The evolutionary hypothesis that organisms must constantly adapt and evolve not only to reproduce in order to survive in the face of an ever-changing environment.

redundancy   Characteristic of the genetic code whereby most amino acids are coded for by more than one codon. For example, serine is coded for by four different codons.

reemerging infectious disease   Infectious diseases that were common in the past but then declined in incidence due to control measures such as the advent of antibiotics or use of vaccination but that now are once again increasing in incidence (e.g., tuberculosis, leprosy, syphilis).

reemerging pathogen   Pathogens responsible for reemerging infectious diseases—diseases that were common in the past but then declined in incidence due to control measures such as the advent of antibiotics or use of vaccination but that now are once again increasing in incidence (e.g., *Mycobacterium tuberculosis*, the cause of tuberculosis; *Mycobacterium leprae*, the cause of leprosy; *Treponema pallidum*; and measles virus, the cause of measles/rubeola).

refractometer   Device that measures the index of refraction (degree to which light changes direction) in a sample. Can be used to determine the salinity of brackish water in order to determine if the salt concentration of coastal waters where seawater mixes with freshwater is conducive to the growth of *Vibrio* spp., which can cause severe infections.

release　The last stage in the viral replication cycle during which mature viruses are released from host cells.

Relenza　An antiflu drug that is effective against both A and B influenza viruses. It reduces the severity of the flu if taken within 48 hours of the appearance of symptoms.

reporter genes　Genes that are attached to a gene of interest that can be monitored for expression because the reporter gene activity (e.g., fluorescence) can be measured or quantitated. Can be used to track patient gene therapy or the production of gene products for functional studies in bacteria or even virally infected cells.

repressor　A protein that turns the expression (transcription) particular gene off.

reservoir　A site in nature where microbes survive and multiply and from which they may be transmitted.

restriction enzymes　An enzyme that cleaves DNA molecules at a particular palindromic site known as a restriction enzyme site. Used in genetic engineering to create recombinant DNA molecules.

retrovirus　A virus that inserts a copy of its genome into the DNA of the host cell in order to replicate. It involves reverse transcribing its RNA genome into DNA before it can be integrated into the host's chromosome as a provirus. HIV is an example.

rheumatic fever　A condition involving the heart and joints resulting from repeated bouts of streptococcal infection.

ribonucleic acid (RNA)　Single stranded nucleic acid composed of adenine, cytosine, guanine, and uracil. It is transcribed from DNA. There are three forms in cells: messenger RNA (mRNA); ribosomal RNA (rRNA), and transfer RNA (tRNA). The mRNA is translated into proteins.

ribose　A pentose sugar that is a component of RNA.

ribosome　Sphere-shaped structures that are composed of two subunits of rRNA and associated proteins that translate mRNA into proteins within the cytoplasm of cells.

rinderpest　An infectious viral disease of cattle and other ruminants. It is one of the most important diseases of livestock due to its high mortality rates.

ring vaccination　Inhibition of the spread of a viral disease by vaccinating all susceptible individuals in the region surrounding a disease outbreak. It was a strategy used to eradicate smallpox and was used to control an Ebola outbreak in the Democratic Republic of Congo in 2018 (using an experimental vaccine).

ringworm　Common fungal infection of the top layer of skin that can affect the groin, scalp, or feet. Causes a red, scaly rash; rash is sometimes raised and circular. Also called *tinia* or *dermatophytosis*. It can spread from animals to humans and person-to-person through the sharing of combs, brushes, or other personal care items.

river blindness　A common type of blindness resulting from the migration of larval forms of the worm *Onchocerca volvulus* into the eye; also called onchocerciasis.

RNA (ribonucleic acid)　Single stranded nucleic acid composed of adenine, cytosine, guanine, and uracil. It is transcribed from DNA. There are three forms in cells: messenger RNA (mRNA); ribosomal RNA (rRNA), and transfer RNA (tRNA). The mRNA is translated into proteins.

RNA polymerase　Enzyme responsible for synthesis of RNA molecules from the DNA template during the process of transcription.

R-nought (R_0)　The mean number of secondary cases occurring in a nonimmune (susceptible) population in the wake of a particular infection. R_0 determines how fast an epidemic will spread. For an epidemic to spread, the R_0 value must be greater than 1: if it is less than 1, the epidemic will die out.

route of entry　The body site whereby a microorganism or virus enters (infects) a host organism.

rumor registers　Epidemiological technique for tracking diseases where unconfirmed cases noted in the news or informal sources are recorded in order to monitor the possible emergence or spread of an outbreak. Were used during the smallpox eradication program to track new cases of the disease and to use ring vaccination as a strategy to prevent its spread.

S strain of *S. pneumoniae*　"Smooth" strain of *S. pneumoniae* that is an encapsulated pathogen.

Saber shins　A condition sometimes seen in syphilis patients in which the shinbone develops abnormally.

Sabin vaccine　A polio vaccine developed by Albert Sabin consisting of attenuated polioviruses.

safranin　A dye used as a counterstain in the Gram stain procedure to differentiate Gram (+) from Gram (–) bacterial cells. Gram (–) cells are stained red by safranin.

Salk vaccine　The first vaccine against polio developed by Jonas Salk containing inactivated polioviruses.

salmonellosis　A condition caused by ingestion of *Salmonella* bacteria resulting in gastroenteritis manifested by nausea, vomiting, abdominal cramps, and diarrhea.

salvarsan　One of the first drugs developed to treat syphilis caused by the bacterium *Treponema pallidum*.

sand flies　Insects that transmit *Leishmania* species.

sanguivores　Mammals that feed on the fresh blood of other mammals (e.g., vampire bats).

sanitarian　Public health figures who emphasized the need for adequate ventilation and cleanliness to promote healing of patients (e.g., nurse Florence Nightingale).

sanitation　Provision of clean water and waste disposal in order prevent the spread of infectious diseases.

sashimi　Thinly sliced strips of raw fish or meat eaten fresh with soy sauce and wasabi paste. Considered a delicacy. However, the raw food may contain parasites such as tapeworms.

scalded skin syndrome　A condition caused by a staphylococcal toxin that results in the skin's becoming blistery with a tendency to peel.

scanning electron microscopy (SEM)　Type of microscopy where a beam of electrons is used to scan surfaces in order to provide a three-dimensional image. It is often performed on biological specimens that are small in size, such as insects, microbes, or viruses in order to obtain information about its topography and composition.

scarlet fever　A disease caused by a strain of streptococcus that produces an erythrogenic toxin leading to the development of a red rash and a strawberry-colored tongue.

scavengers　A synonym for decomposers.

schistosomiasis　A parasitic disease caused by blood flukes that enter the body through the skin.

SCID (severe combined immunodeficiency)　A disease in which individuals lack both functional T and B cells and cannot mount either an antibody or a cell-mediated immune response.

scientific method　Procedure of knowledge acquisition based on the formulation and testing of a hypothesis through experimentation.

scolex　The head of a tapeworm; it attaches to the host intestinal wall by suckerlike projections.

Scotch tape test　Technique used to collect and identify pinworms and pinworm eggs by applying tape to the skin around the anus.

scrapie　A form of transmissible spongiform encephalopathy (TSE) caused by prions that affects sheep. Characterized by weight loss leading to death.

secondary bacterial infections　Infections that result from bacteria in individuals suffering from the flu or other conditions that lower immune resistance.

secondary syphilis   A stage of syphilis characterized by a rash on the palms and soles; during this stage, Trepanema pallidum multiplies and spreads throughout the body.

selective toxicity   Antimicrobials that can target and inhibit the pathogen causing an infection without harming the host.

selectively permeable   Able to be permeated by some but not all types of molecules; a property of cell membranes.

self-infection   Reinfection with pathogenic microbes such as pinworms due to contact with a contaminated body part.

SEM (scanning electron microscopy)   Type of microscopy where a beam of electrons is used to scan surfaces in order to provide a three-dimensional image. It is often performed on biological specimens that are small in size, such as insects, microbes, or viruses in order to obtain information about its topography and composition.

semiconservative   Characteristic of DNA replication where one strand of each new double-stranded DNA molecule is derived from a template strand within the original DNA.

sentinel rhesus monkeys   Rhesus monkeys used in infectious disease research as a host for viruses isolated and identified by investigators as new or emerging viral threats to humans (e.g., the use of such monkeys in the Zika Forest led to the discovery of Zika virus).

sepsis   A complication of an infection that occurs when bacterial toxins are released into the bloodstream, triggering life-threatening inflammatory responses, blood clots and organ failure.

septic   Contamination of a wound by microbes or viruses that then spread to the bloodstream.

septic shock   Dangerously low blood pressure that may result in organ failure due to widespread infection throughout the body.

septicemic plague   A form of the plague resulting from the spread of infection from the lungs to other parts of the body.

septum   During binary fission by bacteria, the structure formed that divides the bacterium into two new bacteria.

serology   Laboratory diagnostic tests performed on blood samples drawn from patients.

serum hepatitis   A serious liver infection caused by hepatitis B virus that can cause cirrhosis, liver cancer, liver failure, and death. Infection can be prevented through vaccination.

serum sickness   A condition characterized by the formation of antigen–antibody complexes that are deposited in the skin, kidney, and other sites, as might occur in treatment with immunoglobulin.

70-70-60 plan   Plan developed to contain the spread of Ebola virus that ought to ensure that 70% of burials were conducted safely and 70% of those infected were isolated, both within 60 days.

severe combined immunodeficiency (SCID)   A disease in which individuals lack both functional T and B cells and cannot mount either an antibody or a cell-mediated immune response.

sexually transmitted infections (STIs)   Diseases caused by transmission of microbes from the warm, moist mucous membranes of one individual to the mucous membranes of another individual during sexual contact; formerly called venereal diseases.

Shiga toxin   Potent bacterial toxin produced by Shigella spp. and some varieties of Escherichia coli.

shigellosis   A bacterial gastrointestinal illness caused by the ingestion of Shigella-contaminated foods and water resulting in symptoms of diarrhea, abdominal cramping, and, in some cases, dysentery.

shingles   A disease caused by the reactivation of varicella zoster virus that occurs in some individuals with a history of chickenpox; the virus infects nerve fibers, creating intense pain.

Shingrix   Trade name of vaccine used to prevent shingles.

sialic acid   Receptor sites present on the surface of epithelial cells that line the respiratory tract to which influenza A viruses can attach and facilitate their entry into host cells.

Siberian ulcer   Skin lesion characteristic of the cutaneous form of anthrax caused by Bacillus anthrasis.

sick building syndrome   When occupants of a building experience headaches and respiratory problems or other symptoms due to air quality issues in the building, such as high levels of the black fungus Stachybotrus. Occurs after flood damage to buildings, which results in conditions conducive to the growth of mold.

Silvadene   Silver sulfadiazine, a topical antibiotic used to prevent bacterial infections on areas of skin burns.

Sin Nombre strain   A strain of New World hantaviruses spread to humans by the inhalation of aerosoled deer mice urine or droppings. It causes a severe pulmonary syndrome/pneumonia. First cases occurred in the Four Corners region (New Mexico, Arizona, Utah, and Colorado) of the United States during the early 1990s.

6-second medical examination   Medical examination on Ellis Island. As an immigrant passed, the doctor examined the person's hair, face, neck, and hands. If the doctor noticed that some area needed to be checked more thoroughly, he wrote a letter on the immigrant's clothes.

16s rRNA gene   Ribosomal ribonucleic acid gene unique to each bacterial species.

sludge   The product of the conversion of waste material that may be used as fertilizer; also called biosolid.

smallpox   Highly contagious, deadly disease caused by variola virus characterized by a disfiguring (pockmarks or scars from scabs that formed on the skin), centrifugal rash, and flulike symptoms. A global vaccination effort resulted in its eradication in 1980.

smallpox recognition cards   Pictures distributed as part of smallpox eradication efforts to help local people recognize the rash associated with the disease. They were used in conjunction with ring vaccination.

specialized transduction   Process whereby a fragment of bacterial DNA is recombined with phage DNA and packaged into its head. The recombinant phage DNA is introduced into another susceptible bacterium and may become incorporated into the bacterial genome as a provirus.

species barrier   Natural mechanisms that prevent the spread of pathogens from one species to another species (e.g., variola virus, the cause of smallpox, only infects humans; variola virus has never crossed the species barrier into any other living species, such as monkeys).

spillover   Epidemiological event where a reservoir population with a high prevalence of a pathogen comes into contact with a new possible host population, resulting of transmission of the pathogen to the new population. Ebola viruses were involved in such an event whereby Ebola viruses crossed the species barrier from bats into humans.

spirochetes   Term used to describe flexible, corkscrew-shaped, motile bacteria (e.g., Treponema pallidum is a spirochete-shaped bacterium).

spleen   An organ that is part of the secondary immune system; it contains phagocytic cells and both mature B and T cells.

splenectomy   Surgical removal of all or part of the spleen.

spontaneous generation   A false but once popular theory that nonlife could give rise to living organisms (e.g., dirty old rags gave rise to rats). Louis Pasteur played a major role in disproving this theory.

sporadic infectious disease   Infectious diseases that appear infrequently or irregularly (e.g., Ebola virus disease).

sporozoites   Forms of Cryptosporidium parvum that penetrate the intestinal cells; also, infective forms of malaria parasites.

sputum   A mixture of mucus and saliva. Generally produced due to infection or disease of the respiratory tract. Examination can aid in diagnosis of a respiratory infectious disease such as tuberculosis.

surgical site infection (SSI)   An infection that develops at a surgical site.

ST-246   An antiviral that is effective against smallpox and monkeypox. The U.S. government has stockpiled millions of doses in the event of a bioterrorism event.

staphylococci   General term used to describe bacteria of the genus *Staphylococcus*.

Staphylococcus aureus   A Gram (+) bacteria that can cause serious infections of the skin, digestive tract, or respiratory tract. Antibiotic-resistant strains such as MRSA can cause life-threatening infections.

starter culture   A preparation used to initiate the fermentation process in the production of a food or fermented drink (e.g., sourdough bread or yogurt).

stationary phase   In the bacterial growth curve, the period during which multiplication or binary fission rate and death rate are equal, often due to a growth-limiting factor such as the depletion of an essential nutrient.

sterilization   A process of disinfection used to destroy all live microbes, endspores, and viruses.

steroids   Medication that can be used to reduce inflammation (e.g., prednisone is used to treat individuals with severe respiratory infections in which airways are inflamed, causing breathing to be difficult).

STIs (sexually transmitted infections)   Diseases caused by transmission of microbes from the warm, moist mucous membranes of one individual to the mucous membranes of another individual during sexual contact; formerly called venereal diseases.

strawberry tongue   A swollen and bumpy tongue; often a symptom of strep throat or other infection.

strep throat   An airborne infection caused by *Streptococcus pyogenes* with characteristic symptoms of red or sore throat, fever, and headache. If untreated, the infection can result in complications such as rheumatic fever and glomerulonephritis.

streptococci   Term used to describe bacteria of the genus *Streptococcus*.

streptokinase   An enzyme produced by *Streptococcus* spp. that dissolves blood clots.

streptomycin   A common antibiotic that inhibits protein synthesis by binding to ribosomes of bacteria. It is used to treat a number of different bacterial infections (e.g., *Klebsiella pneumoniae* and *Haemophilus influenzae* infections).

strict anaerobes   Bacteria that are unable to survive in the presence of oxygen because they lack enzymes such as catalase to detoxify oxygen radicals that will accumulate and kill cells. *Clostridium* spp., which cause a number of different types of infections, and *Bacterioides* spp., a common bacterium present in the human intestines, are strict anaerobes.

strongyloidiasis   Illness caused by the nematode *Strongyloides stercoralis*; symptoms are nausea, vomiting, anemia, weight loss, and chronic bloody diarrhea.

subcellular   Simpler than a cell. A virus is a subcellular pathogen.

subclinical   Term used to describe and infection in which an infected individual is asymptomatic and therefore not diagnosed.

Sulabh International Museum of Toilets   Museum in New Delhi, India, that features a collection of artifacts, pictures, and objects illustrating the historical development of toilets since the year 2500 B.C. In 2018, it was chosen as a winner of TripExpert Experts' Choice Award recognizing it as one of the top tourist attractions around the world. Only 2% of attractions in the world receive this award.

sulfonamide (sulfa) drugs   The first antibiotics used to treat bacterial infections. They contain a sulfonamide functional group and have a bacteriostatic effect. The first drug of this class, prontosil, was discovered in 1935.

[35S]   Radioactive isotope of sulfur used to label proteins in experiments of Alfred Hershey and Martha Chase that proved that DNA is the hereditary material.

superbug   A bacterial pathogen that has developed resistance to multiple antibiotics.

surgical site infection (SSI)   An infection that develops at a surgical site.

surveillance   The monitoring of behavior, activities, or other information (such as the number of people sick) for the purpose of gathering information to protect people. Is used to monitor disease outbreaks in progress or for changes in the environment or human practices that could contribute to a potential disease outbreak.

surveillance systems   Formal system in place to collect data to monitor the presence of infectious disease outbreaks and identify pathogens causing disease outbreaks. The three surveillance strategies are disease specific surveillance, syndrome/symptom-based surveillance, and event surveillance.

sushi   Japanese dish made of rice and a combination of other ingredients such as vegetables (cucumber, carrots, ginger, or eggplant in Japanese sushi and avocado in American-made sushi) and raw fish (e.g., tuna or salmon) or seafood (e.g., cooked shrimp).

swamp fever   A disease caused by the bacterial spirochete *Leptospira* spp. (most commonly by *Leptospira interrogans*). Transmission occurs through the accidental exposure of contaminated water or soil with urine from an infected rodent. Symptoms range from mild to severe; also called *mud fever*, *autumn fever*, *rice-field fever*, *Cane cutter's fever* and *leptospirosis*.

symbiosis   Term used to describe a relationship between two or more organisms that live together; includes mutualism, commensalism, and parasitism.

symbiotic microbes   Microbes that live in a relationship with another microbe where both microbes benefit from the relationship.

syncytia   Giant multinucleated cells formed from the fusion of multiple cells during viral infection caused viruses such as HIV and measles virus; allows viruses to spread from cell to cell more rapidly.

syndrome/symptom-based surveillance   Surveillance systems are based on the real-time (or near real-time) tracking of data on the occurrence of syndromes or symptoms to identify potential public health threats.

synthetic biology   An emerging area of research that uses genetic engineering based on natural biological molecules and systems to redesign and produce new unnatural biological molecules (e.g., enzymes, other proteins or gene circuits) cells, tissues, organs, and whole organisms that did not exist in nature.

synthetic gene circuits   An application in synthetic biology in which bacteria are engineered to perform tasks in a designed, logical manner (acting similar to an electric circuit). Probiotic bacteria can be engineered to produce therapeutic proteins that block the progression of disease.

syphilis   A sexually transmitted disease caused by *Treponema pallidum*; the symptoms of primary syphilis begin as sores on the penis or cervix, rash, and if untreated can lead to degeneration of the spinal cord, organs and tissues and cause dementia. The disease has three stages: primary, secondary, and tertiary (which takes years to develop). Syphilis is treatable with antibiotics.

systemic infections   Bloodborne infections that spread throughout the body.

T cells   Important lymphocytes of adaptive immunity. They mature in the thymus, enter the blood and then settle into secondary lymph tissue. They express antigen-specific T cell receptors that recognize peptide antigens presented by MHC on the surface of antigen-presenting cells. Functional subsets include CD4+ T helper cells and CD8+ cytotoxic T cells. Also called *T lymphocytes*.

T-cell subset   A collective term for the categories of T cells differentiated by clusters of differentiation.

T helper lymphocytes   Type of white blood cell required for adaptive immune response. They activate B cells to produce antibodies and cytotoxic T cells to kill virally infected cells.

T lymphocytes   Important lymphocytes of adaptive immunity. They mature in the thymus, enter the blood and then settle into secondary lymph tissue. They express antigen-specific T cell receptors that recognize peptide antigens presented by MHC on the surface of antigen-presenting cells. Functional subsets include CD4+ T helper cells and CD8+ cytotoxic T cells. Also called *T cells*.

tabes   The slow and progressive degeneration of the spinal cord that occurs in the late or tertiary stage of syphilis; also called *neurosyphilis*.

tachypnea   An elevated respiratory rate; that is, breathing that is faster than normal. Often caused by severe lung infections.

Tamiflu®   Trade name for antiviral drug used to treat the influenza A and B virus infections in people 2 weeks of age and older who have had flu symptoms for no more than 2 days. It inhibits the viral neuraminidase. Can also be used prophylactically to prevent infections (e.g., preventing an influenza outbreak among elderly in nursing homes).

Taq polymerase   Heat stable DNA polymerase produced by *Thermus aquaticus* that is essential to the PCR technique used to generate DNA *in vitro*.

TB disease   A case of tuberculosis where *Mycobacterium tuberculosis* are actively multiplying and the immune system is unable to fight the disease on its own, requiring antibiotic therapy. Symptoms include fever, night sweats, weight loss, fatigue, and coughing up of blood-tinged sputum.

TB infection   A case of tuberculosis where the person has active bacteria but no symptoms of disease and does not spread the infection to other people.

TB sanatoriums   Hospitals dedicated to treatment of tuberculosis that were common the late 19th and early 20th centuries before the discovery of antibiotics.

Td (tetanus and diphtheria) vaccine   A booster vaccine for tetanus and diphtheria.

Tdap (tetanus, diphtheria, and pertussis) vaccine   A combination vaccine the provides protection against tetanus, diphtheria, and pertussis.

tecovirimat   Generic name of an antiviral (also known as ST-246) that is effective against variola virus (cause of smallpox) and monkeypox virus (cause of monkeypox). The FDA approved its use to treat smallpox victims in the event of a bioterrorism event.

TEM (transmission electron microscopy)   Microscopy technique whereby electrons are beams through a specimen in order to create an image. It is used to visualize microbes and viruses (magnifying biological specimens that are nanometers in size by as much as 100,000 magnification or more of their normal size).

temperate zones   Regions that have wider temperature ranges throughout the year with four distinct seasons (winter, fall, spring, summer). Occur between the subtropics and the polar circles.

template DNA   In DNA synthesis (including PCR) or RNA transcription, the strand to which the DNA or RNA polymerase, respectively, attaches each subsequent complementary base, thus growing the new nucleic acid.

terbinafine   A topical antifungal used to treat fungal infections of the scalp such as tinea and ringworm and fungal infections of the toenails and fingernails.

terminator sequence   A site on DNA marking the end point of transcription.

tersorium   In the public latrine in ancient Rome a "butt brush" that was supplied for wiping.

tertiary syphilis   The third or late stage of syphilis during which the spinal cord, organs and tissues undergo degenerative changes causing symptoms such as problems controlling muscles movements, numbness, vision problems, and dementia.

tetanospasmin   A neurotoxin produced by *Clostridium tetani* that results in rigid paralysis.

tetanus   A bacterial disease caused by *Clostridium tetani* or its endospores. Symptoms include stiffness in the jaw and contraction of muscles in the limbs, stomach, and neck. Also called *lockjaw*.

tetrads   Groupings of cocci-shaped bacteria in clusters of four (e.g., characteristic of bacteria such as *Micrococcus luteus*).

Thermus aquaticus   A hyperthermophile isolated from hot springs. Its heat stable DNA polymerase was developed for use in PCR technology.

thrombocytopenia syndrome   A tickborne disease symptom characterized by a drop in platelets and white blood cells, as well as fever, nausea, and vomiting.

thrush   A yeast infection that forms in the mouth or throat (thrush) caused by *Candida albicans*; also called *candidiasis*.

thymus   The organ in which T-cell maturation is completed; it is located behind the sternum and just above the heart.

TKM-Ebola   Experimental drug used to treat Ebola virus disease (EVD) during the 2014–2016 outbreak in West Africa.

toilet phobia   The fear of using public toilets in ancient Rome because of the insects, rodents, and the accumulation of bacteria that produced potentially explosive end products that resided in the sewers and toilets.

tonsillectomy   Surgical removal of the tonsils.

tonsils   Structures of the secondary immune system located at the back of the throat that aid in protection from microbes entering through the nose and throat.

toxic shock syndrome   A condition caused by toxin-producing staphylococci; primarily associated with the use of highly absorbent tampons.

toxigenicity   The ability of microbes to produce toxins.

toxins   Proteins that are virulence factors produced by bacterial (or sometimes fungal) pathogens that cause toxic effects or are harmful to the body; tetanus, botulinum, and erythrogenic toxins are examples.

toxoid   A toxin produced by a pathogen that has been chemically modified so that it is no longer toxic and can be used as a vaccine to prevent an infectious disease caused by a particular pathogen (e.g., use of the *Corynebacterium diphtheriae* toxoid as a vaccine to prevent diphtheria).

toxoplasmosis   A protozoan disease caused by *Toxoplasma gondii* and acquired by the ingestion of oocysts present in cat feces. Symptoms include sore throat, low-grade fever, and lymph node enlargement; can cause congenital defects in newborns.

TPOXX   Brand name of an antiviral that is effective against smallpox and monkeypox. The U.S. government has stockpiled millions of doses in the event of a bioterrorism event. Also known as *tecovirimat* or *ST-246*.

trachoma   An infection of the eye caused by *Chlamydia trachomatis* that can lead to pain and eventual blindness. Immigrants entering the United States through Ellis Island Federal Immigration Station were screened for trachoma.

transcription The synthesis of RNA by which DNA is "read" or transcribed by RNA polymerase into mRNA.

transcription factors Cellular proteins that direct RNA polymerase to transcribe DNA into RNA (e.g., sigma factors in procaryotes).

transduction DNA recombination process whereby bacterial DNA from one bacterium is introduced into another bacterium by a bacteriophage.

transfer RNA (tRNA) A clover leaf-shaped RNA molecule that transfers a single specific amino acid to the ribosome during the translation of proteins.

transformation DNA recombination process characterized by the uptake of "naked" DNA into competent bacterial cells.

transgenic An genetically modified organism that has been altered through genetic engineering techniques to contain foreign DNA.

translation Cellular process in which mRNA is converted by ribosomes into an amino acid sequence, which is assembled into a protein; also called *protein synthesis*.

transmissible spongiform encephalopathies (TSEs) Degenerative brain diseases that are caused by abnormally folded prions that latch onto normal prions and convert them into an altered, defective form (e.g., scrapie, kuru, vCJD, and BSE are examples of TSEs).

transmission electron microscopy (TEM) Microscopy technique whereby electrons are beams through a specimen in order to create an image. It is used to visualize microbes and viruses (magnifying biological specimens that are nanometers in size by as much as 100,000 magnification or more of their normal size).

transpeptidase Bacterial enzyme that cross-links peptidoglycan to form bacterial cell walls.

transposon A DNA segment that can change its position or "jump" to different locations within a genome.

traveler's diarrhea An illness caused by enterotoxigenic *Escherichia coli* strains.

trichinellosis A roundworm disease caused by *Trichinella spiralis* that is transmitted in undercooked pork. Symptoms include nausea, diarrhea, vomiting, fatigue, fever, headaches, chills, aching joints, and itchy skin.

trichomoniasis A sexually transmitted infection caused by *Trichomonas vaginalis*. Symptoms include intense itching, urinary frequency, pain during urination, and vaginal discharge in females. In males, symptoms include pain during urination, inflammation of the urethra, and a thin, milky discharge.

trichuriasis An infection caused by the helminth *Trichuris trichiura*. The disease is common in the tropics and subtropics. The worm infects the large intestine and most infected adults are asymptomatic. Children may experience symptoms of abdominal pain, diarrhea and rectal prolapse.

triclocarban An organohalide with antimicrobial properties that has a number of commercial uses. The FDA recently banned adding it as an ingredient in liquid and hand soaps.

triclosan An organohalide with antimicrobial properties that has a number of commercial uses. The FDA recently banned adding it as an ingredient in liquid and hand soaps.

triotomid insect Another name for kissing bugs. Responsible for transmission *Trypanosoma cruzi*, the protozoan parasite that causes Chagas disease.

trophozoite The reproductive and feeding stage of parasitic amoebae and other protozoan parasites.

trypanosomes Form of parasitic protozoans that cause African sleeping sickness and Chagas disease in humans and trypanosomiasis in livestock. Are spread by insect bites.

tsetse fly Large biting flies common in Africa that are the vector for the trypanosomes that cause African sleeping sickness and trypanosomiasis in livestock.

tube well A simple device constructed of steel pipes sunk deep into the ground and fitted with a pump handle. The pump is sealed topside to prevent water from leaking back down the pipe. Microbes are filtered out as groundwater trickles through the aquifer, resulting in microbiologically safe water.

tuberculin skin test A method for diagnosis of tuberculosis; purified protein derived from *Mycobacterium tuberculosis* is injected into the arm of a patient, and the presence of an induration between 48 and 72 hours after injection indicates past or present exposure to the tubercle bacillus.

tuberculosis A contagious lower respiratory tract disease caused by *Mycobacterium tuberculosis*. Symptoms include fever, night sweats, weight loss, fatigue, and coughing up of blood-tinged sputum; also called *consumption*, *white plague*, or *white death*.

typhoid A disease caused by *Salmonella typhi* that is transmitted by flies and fomites.

Typhoid Mary A nickname for Mary Mallon (1869–1938), a notorious healthy carrier of *Salmonella enterica* serotype *Typhi*, the cause of typhoid fever.

Uganda Virus Research Institute (UVRI) Medical institute of the Ugandan government that focuses on viral infections affecting humans and animals; originally established to isolate new or emerging viruses such as yellow fever virus and Zika virus transmitted by mosquitoes through the bite of sentinel rhesus monkeys placed in cages on treetops in the Zika forest.

uncoating A step in the replication cycle of viruses following viral attachment and entry into host cells in which its genome is released from the surrounding protein coat in order to undergo replication.

uni The edible parts of sea urchins that are used in sushi.

United Nation Children's Fund (UNICEF) The United Nations agency charged with providing humanitarian assistance to mothers and children in developing countries, including providing health care, immunizations, nutrition, access to safe water and sanitation, and emergency relief.

U.S. Department of Agriculture (USDA) Federal agency charged with the safety, labeling, and packaging of the nation's commercial supply of meat, poultry, and eggs.

unsafe burial practices The traditional practice in countries most affected by Ebola virus disease of kissing, washing, redressing, and burial of the deceased person's body. Resulted in spread of Ebola virus disease.

upper respiratory tract The portion of the respiratory tract that includes the nose, nostrils, nasal cavity, mouth, pharynx, and larynx.

urbanization The gradual increase in people moving from rural areas to more populated cities and urban centers.

urease An enzyme that breaks down urea; its detection in a breath test is used in the diagnosis of *Helicobacter pylori* infections, the cause of some peptic ulcers.

urgent care clinic A walk-in clinic with expanded hours that is able to handle a range of minor illnesses and injuries.

urinary tract infection (UTI) Bacterial infection that occurs in any part of the urinary tract.

U-shaped mortality curve Typical disease mortality pattern where those most likely to die from an infectious disease such as influenza are those who are very young or elderly.

UTI (urinary tract infection) Bacterial infection that occurs in any part of the urinary tract.

uveitis An inflammation of the middle layer of the eye that can lead to vision loss if left untreated. Uveitis was a complication experienced by some Ebola virus disease survivors in the recent outbreak in West Africa.

UVRI (Uganda Virus Research Institute) Medical institute of the Ugandan government that focuses on viral infections affecting humans and animals; originally established to isolate new or emerging viruses such as yellow fever virus and Zika virus transmitted by mosquitoes through the bite of sentinel rhesus monkeys placed in cages on treetops in the Zika forest.

vaccination   Administration of a vaccine that contains whole (which is killed or inactivated) or parts of pathogens that will stimulate an immune response and result in the development of adaptive immunity toward a specific pathogen.

Vaccine Adverse Event Reporting System (VAERS)   Database maintained by the CDC and FDA that tracks adverse reactions to vaccines in order to monitor and ensure the safety of all vaccines licensed for use in the United States.

vaginitis   Vaginal inflammation characterized by itching, discharge, and pain. Can be caused by bacterial infections.

Valcyte®   Trade name of antiviral used to treat cytomegalovirus infections.

Valley fever   Another name for *coccidioidomycosis*, a fungal infection caused by *Coccidioides immitis*. The fungus is endemic to parts of the southwestern United States and northern Mexico. Primarily affects the lungs and is caused by inhalation of airborne fungal spores that are swept into the air by disruption of contaminated soils during dust storms.

vancomycin   An antibiotic that is used to treat bacterial infections that do not respond to antibiotic therapy. Its mechanism is to inhibit the synthesis of bacterial cell walls.

vancomycin-intermediate *Staphylococcus aureus* (VISA)   A *Staphylococcus aureus* bacterial infection that has moderate resistance to the antibiotic vancomycin.

vancomycin-resistant *Staphylococcus aureus* (VRSA)   A *Staphylococcus aureus* bacterial infection that is resistant to the antibiotic vancomycin.

VAP (ventilator-associated pneumonia)   A hospital-acquired bacterial infection caused by contaminated ventilator parts, resulting in pneumonia caused by *Pseudomonas aeruginosa*, *Acinetobacter* spp., or *Stenotrophomonas maltophilia*.

variant Creutzfeldt-Jakob disease (vCJD)   A fatal human transmissible spongiform encephalopathic disease caused by the ingestion of beef contaminated with prions. It is similar to mad cow disease in that it causes degeneration of brain tissue and severe neurological damage. Human cases of vCJD were first reported in the United Kingdom in 1996.

varicella zoster virus   The herpesvirus that causes both chickenpox and shingles.

variola major   Strain of variola virus that causes smallpox. This variant was more common and typically presented with the more severe form of the disease that had a 30% or greater mortality rate and survivors had complications such as blindness.

variola minor   Strain of variola virus that causes smallpox. This variant was less common and typically presented with the less severe form of the disease and had about a 1% fatality rate.

variolation   First method developed in the 16th and 17th centuries to immunize people against smallpox. Dried scabs taken Asians or Africans who suffered from variola minor infections were blown into the nose of susceptible persons in hopes that the person would only suffer from a mild form of the disease and would be protected from subsequent exposure variola virus.

vector   Any organism that is a carrier of a pathogen that is capable of transmitting and infecting a host. Examples include arthropod vectors such as mosquitoes, ticks, or fleas that harbor pathogens.

vegetative cells   Non-endospore-forming bacteria.

vehicleborne transmission   A fomite (i.e., contaminated inanimate object such as a facial tissue or door knob) upon which a pathogen is able to survive and then bee pasds to a susceptible person who touches or otherwise comes into contact with the pathogen present on the fomite. A form of indirect disease transmission.

venereal diseases   A former name for sexually transmitted diseases.

ventilator-associated pneumonia (VAP)   A hospital-acquired bacterial infection caused by contaminated ventilator parts, resulting in pneumonia caused by *Pseudomonas aeruginosa*, *Acinetobacter* spp., or *Stenotrophomonas maltophilia*.

vertical transmission   A method of transmission characterized by passage of pathogens from parent to offspring across the placenta, in breast milk, or in the birth canal.

Vessel Sanitation Program   Centers for Disease Control and Prevention program that works with the cruise ship industry to perform inspections of cruise ships in order to prevent the introduction and spread of gastrointestinal illnesses caused by pathogens such as noroviruses on cruise ships.

vibrios   Term used to describe bacteria of the genus *Vibrio* that are comma-shaped rods; *Vibrio cholerae*, the causative agent of cholera, is an example.

vibriosis   An infection caused by *Vibrio* spp. Infection can occur from drinking contaminated water or food such as eating raw or undercooked shellfish (especially oysters) or open cuts or wounds exposed to warm brackish water containing the bacteria.

viral envelope   Part of the plasma membrane from a viral host is wrapped around the protein capsid of a virus as the assembled virus buds out a host cell. The envelope contains phospholipids and proteins, which are part of the stolen host cell membrane. Enveloped viruses are not very stable in the environment and can be disrupted by hand sanitizer.

viral load   The number of viruses in a given clinical sample. Usually expressed as the number of viral particles per milliliter (mL). Measured in HIV and hepatitis C patients to determine if the patient's antiviral treatment must be adjusted. If the treatment is not working to inhibit the viral infection, the attending physician will prescribe changes in the patient's antiviral regime.

virion   A complete infectious virus particle.

virophage   Viruses that infect giruses (e.g., Sputnik, which infects mimivirus).

virotherapy   The use of oncolytic viruses as therapeutic agents to destroy cancer cells.

virulence factors   Proteins such toxins or capsules that enable a pathogen to cause an infection and disease.

virulent   Term used to describe the ability of a pathogen to severely damage the host organism.

viruses   Subcellular agent dependent upon a host cell for its replication. When infected by a virus, the host cell machinery is manipulated by the virus to produce identical copies of the original virus.

virus-like particles (VLPs)   Empty viral capsids that do not contain genetic material. They cannot cause disease. A component of human papillomavirus vaccines.

VISA (vancomycin-intermediate *Staphylococcus aureus*)   A *Staphylococcus aureus* bacterial infection that has moderate resistance to the antibiotic vancomycin.

visceral leishmaniasis   The most severe form of leishmaniasis; symptoms are fever, weakness, weight loss, anemia, and protrusion of the abdomen. Also known as kala-azar.

Vitek systems   A rapid and accurate automated instrumentation system for identification and susceptibility testing of bacteria (e.g., API20E testing for pathogens in the *Enterobacteriaceae* family).

VLPs (virus-like particles)   Empty viral capsids that do not contain genetic material. They cannot cause disease. A component of human papillomavirus vaccines.

VRSA (vancomycin-resistant *Staphylococcus aureus*)   A *Staphylococcus aureus* bacterial infection that is resistant to the antibiotic vancomycin.

wasting syndrome   Rapid weight and muscle loss. Is often associated with end-stage AIDS, various stage 4 cancers, and chronic wasting disease (CWD) of deer.

West Nile encephalitis   Severe cases of people (typically elderly or individuals with preexisting conditions) who are infected with West Nile virus after being bit by a *Culex* mosquito carrying the virus. They develop swelling of the brain or meninges (encephalitis), resulting in seizures or death. Many people infected with West Nile virus are asymptomatic or experience mild signs or symptoms.

wet phase   Second phase of Ebola virus disease (EVD) characterized by severe vomiting and diarrhea. This is the phase in which the patient is most contagious and has a much higher probability of dying from the disease.

whipworms   Helminth *Trichuris trichiura* that commonly causes disease in the tropics and subtropics. It infects the large intestine and most infected adults are asymptomatic. Children may experience symptoms of abdominal pain, diarrhea and rectal prolapse. Also called *trichuriasis*.

white death   Name applied to individuals suffering from tuberculosis because of their white or pale, sickly appearance. Tuberculosis is a lower respiratory tract disease caused by *Mycobacterium tuberculosis*. Symptoms include fever, night sweats, weight loss, fatigue, and coughing up of blood-tinged sputum. Also called *consumption* and *white plague*.

white plague   Name applied to individuals suffering from tuberculosis because of their white or pale, sickly appearance. Tuberculosis is a lower respiratory tract disease caused by *Mycobacterium tuberculosis*. Symptoms include fever, night sweats, weight loss, fatigue, and coughing up of blood-tinged sputum. Also called *consumption* and *white death*.

white-nose syndrome   Fungal disease of bats characterized by a white fungal growth on the bats' muzzles and wings. Responsible for deaths of millions of bats in the United States and Canada.

Whittaker's five-kingdom system   Biological classification system developed by R. H. Whittaker that places organisms into five kingdoms: Animalia, Plantae, Fungi, Protista, and Monera.

whooping cough   An infection of the respiratory tract caused by the bacterium *Bordetella pertussis*. Can cause serious illness, particularly in infants and children, due to violent coughing that makes it difficult to breathe. The vaccine is about 70% effective in preventing infection and reduces the symptoms or severity of vaccinated individuals who do get infected. Also called *pertussis*.

wildlife   The native plants and animals of a region.

wildtype   The typical phenotype of a trait as it appears in nature.

winter vomiting disease   Another name for gastroenteritis caused by noroviruses, which is characterized by diarrhea, vomiting, and stomach pain. The illness is more common in the winter months in temperate regions.

Woese's three-domain system   Biological classification system developed by Carl Woese that places organisms into three domains: Archaea, Bacteria, and Eucarya.

woolsorter's disease   Another name for cutaneous anthrax caused by *Bacillus anthracis*. The most common form of anthrax is cutaneous anthrax. *B. anthracis* is transmitted through contact with injured skin during occupational exposure to infected animals or animal products such as skin or wool. Cutaneous anthrax was known as an occupational hazard for those people who sorted wool. Also called *ragpicker's disease*, *tanner's disease*, and *Black Bain*.

World Bank   An international financial institution that provides loans to developing countries for capital projects, such as the development of sewage disposal systems, water purification systems, and other infrastructure.

World Health Organization (WHO)   United Nations agency charged with monitoring and promoting public health and responding to global health crises. Headquarters is in Geneva, Switzerland.

World Hepatitis Day   Public health campaign devoted to the prevention and treatment of hepatitis worldwide in efforts to reduce the incidence of disease. Celebrated every July 28th.

World Malaria Day   Public health campaign devoted to worldwide efforts to reduce the incidence of malaria. Celebrated every April 25th.

World Mosquito Day   Public health campaign devoted to worldwide efforts to reduce mosquito populations with the goal of reducing the incidence of mosquitoborne diseases. Celebrated every August 20th.

World Rabies Day   Public health campaign devoted to the prevention or rabies worldwide in efforts to reduce the incidence of disease. Celebrated every September 28th.

World TB Day   Public health campaign devoted to raising awareness worldwide on the prevention and treatment of tuberculosis. Celebrated every March 24th.

World Toilet Day   Public health campaign devoted to the promotion of sanitation worldwide in efforts to reduce the incidence of disease. Celebrated every November 19th.

wort   A sugar-containing liquid that is used during the fermentation of brewing yeast to produce beer or malt liquor.

W-shaped mortality curve   Mortality pattern observed during the Spanish Flu of 1918 whereby those in the prime of their lives (ages 20s to 40s) were more likely to die from influenza A virus infection than infants and young children or elderly (their deaths usually result in a U-shaped curve during seasonal influenza epidemics).

XDR TB   Extremely drug-resistant strains of tuberculosis.

x-ray crystallography   A technique that uses x-rays to determine the structure of DNA, other macromolecules, and viruses. This technique was used by Rosalind Franklin to aid in deciphering the structure of DNA and viruses.

yellow fever   Yellow fever virus is carried by and transmitted to humans through the bite of infected *Aedes aegypti* mosquitoes. Symptoms include fever, bloody nose, headache, nausea, muscle pain, vomiting, and jaundice. Also known as *yellow jack*.

yellow jack   Another term used to refer to yellow fever because of the yellow or jaundice appearance of the skin and conjunctiva of those infected with yellow fever virus. Yellow fever virus is carried by *Aedes aegypti* mosquitoes. Also called *black vomit* and *American plague*.

Zidovudine   The first clinically safe and effective drug used for the treatment of AIDS; it acts as an inhibitor of the reverse transcriptase enzyme. Also known as *azidothymidine (AZT)*.

Ziehl-Neelson staining procedure   Acid-fast differential staining method developed by Franz Ziehl and Friedrich Neelsen to identify non-acid fast bacteria that resist decolorization by acid alcohol due to the presence of mycolic acids in their cell walls, which give the walls a waxlike property. *Mycobacterium* spp. and *Nocardia* spp. are acid-fast bacteria that remain fuchsia or pink after completion of this procedure. Non-acid-fast bacteria are blue.

Zika Forest   Forest in Uganda that has been set aside for conservation and scientific research involving in the identification of new or emerging mosquitoborne viruses. First location where Zika virus was first isolated from sentinel rhesus monkeys in 1947 and later identified in humans in 1952.

Zika virus disease (ZVD)   Mosquitoborne viral infection characterized by fever, conjunctivitis, skin rash, muscle and joint pain, and headache. The virus is carried by *Aedes aegypti* and *Aedes albopictus* mosquitoes. Pregnant women who become infected may pass the virus to the fetus, which can result in devastating congenital abnormalities such as microcephaly.

ZMapp   Experimental therapy that is composed of a cocktail of monoclonal antibodies (or plantibodies because of their production in tobacco plants) shown to bind to and inhibit Ebola viruses *in vitro* and to be effective in the treatment of Ebola virus disease in preclinical monkey studies. Doses were provided as an experimental therapy to less than a dozen individuals suffering from Ebola virus disease during the 2014–2016 epidemic in West Africa.

zones of inhibition   A clearing or zone present around an antibiotic disk where no bacterial growth occurs on Mueller Hinton agar. The diameters of these areas are measured in millimeters to determine the antibiotic susceptibility of bacterial isolates isolated from clinical specimens.

zoonoses   Diseases for which domestic and/or wild animals are the reservoirs and which can be transmitted to humans (e.g., rabies, Ebola virus disease, SARS, MERS, swine and avian influenza, Nipah encephalitis). The majority of new and emerging viruses are zoonotic.

Zostovax   Brand name of vaccine licensed in 2006 to prevent shingles (caused by herpes zoster virus) in older adults. The CDC recommends vaccination for people age 60 or older.

Zovirax   Original trade name of antiviral drug (acyclovir) used to treat herpesvirus infections, including herpes simplex, genital herpes, chickenpox, and shingles. Today acyclovir is manufactured by over 60 companies with more than 144 trade names. New generics and trade names are constantly being updated.

ZVD (Zika virus disease)   Mosquitoborne viral infection characterized by fever, conjunctivitis, skin rash, muscle and joint pain, and headache. Zika virus is carried by *Aedes aegypti* and *Aedes albopictus* mosquitoes. Pregnant women who become infected may pass the virus to the fetus, which can result in devastating congenital abnormalities such as microcephaly.

Note: Italicized page locators indicate figures/photos; tables are noted with t.